FIFTEENTH EDITION

Introduction to Geography

Arthur Getis
San Diego State University

Mark D. Bjelland
Calvin College

Victoria Getis
Northwestern University

Mc
Graw
Hill
Education

INTRODUCTION TO GEOGRAPHY, FIFTEENTH EDITION

Published by McGraw-Hill Education, 2 Penn Plaza, New York, NY 10121. Copyright © 2018 by McGraw-Hill Education. All rights reserved. Printed in the United States of America. Previous editions © 2014, 2011, and 2009. No part of this publication may be reproduced or distributed in any form or by any means, or stored in a database or retrieval system, without the prior written consent of McGraw-Hill Education, including, but not limited to, in any network or other electronic storage or transmission, or broadcast for distance learning.

Some ancillaries, including electronic and print components, may not be available to customers outside the United States.

This book is printed on acid-free paper.

1 2 3 4 5 6 7 8 9 LMN 21 20 19 18 17

ISBN 978-1-259-57000-1
MHID 1-259-57000-2

Chief Product Officer, SVP Products & Markets: *G. Scott Virkler*
Vice President, General Manager, Products & Markets: *Marty Lange*
Vice President, Content Design & Delivery: *Betsy Whalen*
Managing Director: *Thomas Timp*
Brand Manager: *Michael Ivanov, PhD*
Director, Product Development: *Rose Koos*
Product Developer: *Jodi Rhomberg*
Marketing Manager: *Noah Evans*
Market Development Manager: *Shannon O'Donnell*
Digital Product Developer: *Joan Weber*
Director, Content Design & Delivery: *Linda Avenarius*
Program Manager: *Lora Neyens*
Content Project Managers: *Jeni McAtee; Tammy Juran; Sandy Schnee*
Buyer: *Sandy Ludovissy*
Content Licensing Specialists: *Lori Hancock, image; Lori Slattery, text*
Cover Image: *© McGraw-Hill Education/Barry Barker*
ISE Cover Image: *© Rodrigo A Torres/Glow Images RF*
Design Elements Images: *© Medioimages/Photodisc/Getty Images RF; ©Goodshot/Alamy Stock Photo RF; ©IMS Communications Ltd/Capstone Design RF; ©Digital Vision/Thinkstock RF*
Compositor: *SPi Global*
Printer: *LSC Communications*

All credits appearing on page or at the end of the book are considered to be an extension of the copyright page.

Library of Congress Cataloging-in-Publication Data

Names: Getis, Arthur, 1934- co-author.
Title: Introduction to geography / Arthur Getis, San Diego State University,
  Mark D. Bjelland, Calvin College, Victoria Getis, Northwestern University.
Description: Fifteenth Edition. | New York : McGraw-Hill Education, [2017] |
  "Previous editions ? 2014, 2011 and 2009"—T.p. verso. | Audience: Ages: 18+
Identifiers: LCCN 2016042240| ISBN 9781259570001 (acid-free paper) | ISBN
  1259570002 (acid-free paper)
Subjects: LCSH: Geography.
Classification: LCC G128 .G495 2018 | DDC 910—dc23 LC record available at
https://lccn.loc.gov/2016042240

The Internet addresses listed in the text were accurate at the time of publication. The inclusion of a website does not indicate an endorsement by the authors or McGraw-Hill Education, and McGraw-Hill Education does not guarantee the accuracy of the information presented at these sites.

Arthur Getis received his B.S. and M.S. degrees from The Pennsylvania State University and his Ph.D. from the University of Washington. He is coauthor of several geography textbooks as well as several books dealing with the analysis of spatial data. Together with Judith Getis, he was among the original unit authors of the High School Geography Project sponsored by the National Science Foundation and the Association of American Geographers (AAG). He has published widely in the areas of infectious diseases, spatial analysis, and geographic information systems. He is honorary editor of the *Journal of Geographical Systems*, and he serves on the executive committee of the *Geographical Analysis* journal and on the editorial board of the *Annals of Regional Science*. He has had administrative appointments at Rutgers University, the University of Illinois, and San Diego State University (SDSU), where he held the Birch Endowed Chair of Geographical Studies. In 2002, he received the AAG Distinguished Scholarship Award. Professor Getis is a member and an elected fellow of the University Consortium of Geographical Information Sciences, the Western Regional Science Association, and the Regional Science Association International. Currently he is Distinguished Professor of Geography Emeritus at SDSU.

Mark Bjelland earned his B.S. degree from the University of Minnesota and an M.S. degree from the University of Washington. He worked for six years as an environmental consultant on projects dealing with environmental justice, pollution cleanup, urban redevelopment, and water resources. He then earned a Ph.D. in geography from the University of Minnesota, writing his dissertation on abandoned, contaminated industrial lands in the U.S. and Canadian cities. He taught geography for 15 years at Gustavus Adolphus College before taking his current position as professor of geography in the Department of Geology, Geography, and Environmental Studies at Calvin College. He has led geography field courses to the Midwest, western Canada, Pacific Northwest, Hawaii, and Great Britain. He has been granted two Fulbright awards to study urban environmental planning in Germany and Great Britain. His research has been published in the *Research Journal of the Water Pollution Control Federation*, the *Journal of Environmental Engineering, The Geographical Review, The Professional Geographer, Urban Geography*, and a number of book chapters.

Victoria Getis received her A.B. degree from Oberlin College and her M.A. and Ph.D. degrees from the University of Michigan. She has been a contributor to *Introduction to Geography* since its 10th edition. She is also the author of *The Juvenile Court and the Progressives* and the coauthor of *Muddy Boots and Ragged Aprons: Images of Working Class Detroit, 1900–1930*. She is currently coauthoring a textbook on America in the 1960s. Victoria is the Manager of Faculty Support Services at Northwestern University; formerly, she was the Director of the Digital Union at Ohio State University.

BRIEF CONTENTS

CONTENTS

Source: NOAA

Chapter 4

Physical Geography: Weather and Climate   75

© Kate Sidlo

Chapter 5

Population Geography   110

© Porterfield/Chickering/Science Source

Chapter 6
Cultural Geography   141

© CelebrityHomePhotos/Newscom

Chapter 7
Human Interaction   185

© dbimages/Alamy Stock Photo RF

Chapter 8
Political Geography   214

© Corbis RF

Chapter 9
Economic Geography: Agriculture and Primary Activities 247

© Moodboard/Thinkstock RF

Chapter 10
Economic Geography: Manufacturing and Services 275

© Jan Hanus/Alamy Stock Photo RF

Chapter 11
An Urban World 299

© Michael Interisano/Design Pics/Getty
Images RF

Chapter 12
The Geography of Natural Resources   334

© Al Franklin/Corbis Premium RF/
Alamy Stock Photo RF

Chapter 13
Human Impact on the Environment   374

"If you build it, they will come" was the message that inspired the character played by Kevin Costner in the movie *Field of Dreams* to create a baseball field in his Iowa cornfield. A similar hope encouraged us when we first began to think about writing *Introduction to Geography* in 1975. At that time, very few departments of geography in the United States and Canada offered a general introductory course for students—that is, one that sought to acquaint students with the breadth of the entire field. Instead, most departments offered separate courses in physical and human or cultural geography. Recognizing that most students will have only a single college course and textbook in geography, we wanted to develop a book that covers all of the systematic topics that geographers study. Our hope, of course, was that the book would so persuasively identify and satisfy a disciplinary instructional need that more departments would begin to offer a general introductory course to the discipline, a dream that has been realized.

Approach

Our purpose is to convey concisely and clearly the nature of the field of geography, its intellectual challenges, and the logical interconnections of its parts. Even if students take no further work in geography, we are satisfied that they will have come into contact with the richness and breadth of our discipline and have at their command new insights and understandings for their present and future roles as informed adults. Other students may have the opportunity and interest to pursue further work in geography. For them, we believe, this text will make apparent the content and scope of the subfields of geography, emphasize its unifying themes, and provide the foundation for further work in their areas of interest.

A useful textbook must be flexible enough in its organization to permit an instructor to adapt it to the time and subject matter constraints of a particular course. Although designed with a one-quarter or one-semester course in mind, this text may be used in a full-year introduction to geography when employed as a point of departure for special topics and amplifications introduced by the instructor or when supplemented by additional readings and class projects.

Moreover, the chapters are reasonably self-contained and need not be assigned in the sequence presented here. The chapters may be rearranged to suit the emphases and sequences preferred by the instructor or found to be of greatest interest to the students. The format of the course should properly reflect the joint contribution of instructor and book, rather than be dictated by the book alone.

New to this Edition

Although we have retained the framework of presentation introduced in the previous edition of this book, we have revised, added, and deleted material for a variety of reasons.

- Current events always mandate an updating of facts and analyses and may suggest discussion of additional topics. Examples include a new chapter opening vignettes on the 2015 earthquake in Nepal and toxic algal blooms in the Great Lakes.

- In every new edition, both changes in spatially variable patterns of demographic parameters and changes in the populations of countries and major urban areas require updating. Maps and tables depicting demographic variables and the populations of the world's largest countries and metropolitan regions were updated based on the most recent data available from the United Nations and Population Reference Bureau in 2016.

- Every table and figure in the book has been reviewed for accuracy and currency and has been replaced, updated, or otherwise revised where necessary.

- As always, we rely on reviewers of the previous edition to offer suggestions and to call our attention to new emphases or research findings in the different topical areas of geography. Our effort to incorporate their ideas is reflected not only in the brief text modifications or additions that occur in nearly every chapter but also in more significant alterations.

- The economic geography chapters (Chapter 9 and Chapter 10) give less attention to centrally planned economies due to their decreased importance today.

- The presentation of urban structure models in Chapter 11 has been revised so that models are presented in the order in which they were developed. This allows the student to see how cities changed over the past century and how, in turn, scholars have developed new models to understand them.

New Figures and Tables

To reflect the most recent data, many figures have been revised or newly drawn for the 14th edition of *Introduction to Geography*. They include:

- Maps representing population distribution in California in 2015 using different cartographic techniques (Chapter 2)
- New photos illustrating place name changes that reflect indigenous toponyms (Chapter 6)
- New photos illustrating urban abandonment, gentrification, Chicago's role as a railroad hub, homelessness, European cities, Canadian cities, slums, and cities of the developing world (Chapter 11)
- All maps, graphs, charts, and tables related to population that required updating (Chapter 5)
- Graph of the distribution of the world's population by latitude using 2010 data (Chapter 5)
- New map of Internet users around the world (Chapter 6)
- Updated map of gender-related development index rankings (Chapter 6)
- Substantially updated map of refugee movements, reflecting the effects of the Syrian civil war (Chapter 7)

- Figures depicting membership in the European Union and NATO were revised due to recent withdrawals and new membership applications (Chapter 8)
- Updated and improved map of women in legislatures around the world (Chapter 8)
- Improved map depicting legislative reapportionment in the United States (Chapter 8)
- New map illustrating the uneven geography of research and innovation in the United States as measured by patents (Chapter 10)
- Improved map of population changes in U.S. metropolitan regions (Chapter 11)
- Figures and tables now reflecting 2015 population data from the the U.S. Census Bureau and new UN Population Division data and projections (Chapter 5 and Chapter 11)
- New map of global soil degradation (Chapter 12)
- All maps, graphs, charts, and tables related to natural resource reserves, production, or use have been updated. Maps depicting trade in crude oil and natural gas have been updated and improved (Chapter 12)
- New satellite images showing toxic algal blooms (Chapter 13)
- New Appendix 3 using 2015 population and economic data

New/Revised Boxes

The boxed elements in the book have been updated if necessary or replaced with new discussion texts.

- New Geography & Public Policy box "Changing Place Names" covering trends in recent geographic name changes (Chapter 6)
- Moved the environmental justice box from the political geography chapter (Chapter 8) to the environmental impacts chapter (Chapter 13)
- Updated data in the Legislative Women box (Chapter 8)
- Revised Geography & Public Policy Box, "Incentives or Bribery?" describing the recent competition between states for a new Boeing airplane assembly plant (Chapter 10)
- Revised Geography & Public Policy Box, "The Homeless" (Chapter 11), presenting new methods of addressing homelessness
- Revised Geography & Public Policy box, "Fuel Economy and CAFE Standards" (Chapter 12)
- New box, "Eating Fossil Fuels," highlighting the connections between fossil fuels and agriculture (Chapter 12)

New/Revised Topical Discussions

- New opening vignette about the 2015 earthquake that struck Nepal
- New opening vignette about harmful algal blooms including unsafe drinking water supplies in Toledo, Ohio in 2014
- New discussion of the refugee crisis and population dislocations caused by the Syrian civil war (Chapter 7)
- Updated discussion of heavily indebted countries and debt-relief programs (Chapter 9)
- Updated discussion of the use of genetically modified crops and related scientific concerns
- Expanded discussion of the interrelated challenges of food production, fertilizer use, energy supplies, water supplies, water pollution, and global climate change
- New discussion summarizing progress in meeting the United Nation's Millennium Development Goals during the 1990–2015 period
- New discussion of changes to the nuclear energy industry after the 2011 Fukushima accident in Japan
- New discussion of harmful algal blooms and expanded discussion of eutrophication due to agriculture and fertilizer use
- The discussion on climate change updated to be in line with current scientific thinking on the subject

Acknowledgments

A number of reviewers have greatly improved the content of this and earlier editions of *Introduction to Geography* by their critical comments and suggestions. Although we could not act upon every helpful suggestion, or adopt every useful observation, all were carefully and gratefully considered. In addition to those acknowledgments of assistance detailed in previous editions, we note the thoughtful advice provided by the following individuals.

Steve Nisbet
Baker College

Michael Caudill
Hocking College

Monica Milburn
Lone Star College – Kingwood

Jeff Bradley
Northwest Missouri State University

Velvet Nelson
Sam Houston State University

Daniel Morgan
Technical College of the Lowcountry

Adil Wadia
The University of Akron Wayne College

Mary Passe-Smith
University of Central Arkansas

Gerald Reynolds
University of Central Arkansas

Brad Watkins
University of Central Oklahoma

Paul C. Vincent
Valdosta State University

We would like to thank the following individuals who wrote and/or reviewed learning goal-oriented content for **LearnSmart**.

Sylvester Allred
Northern Arizona University

Lisa Hammersley
California State University—Sacramento

Arthur C. Lee
Roane State Community College

We gratefully express appreciation to these and unnamed others for their help and contributions and specifically absolve them of responsibility for decisions on content and for any errors of fact or interpretation that users may detect. Finally, we note with deep appreciation and admiration the efforts of the publisher's "book team," separately named on the copyright page, who collectively shepherded this revision to completion. We are grateful for their highly professional interest, guidance, and support.

Arthur Getis
Mark D. Bjelland
Victoria Getis

Pedigogocal content in Introduction to Geography has been created to gain and retain student attention, the essential first step in the learning process.

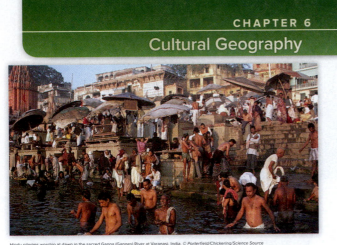

CHAPTER 6
Cultural Geography

Hindu pilgrims worship at dawn in the sacred Ganga (Ganges) River at Varanasi, India. © Porterfield/Chickering/Science Source

CHAPTER OUTLINE

6.1 **Components of Culture**
6.2 **Subsystems of Culture**
   The Technological Subsystem
   The Sociological Subsystem
   The Ideological Subsystem
6.3 **Interaction of People and Environment**
   Environments as Controls
   Human Impacts
6.4 **Culture Change**
   Innovation
   Diffusion
   Acculturation
6.5 **Cultural Diversity**
6.6 **Language**
   Language Spread and Change
   Standard and Variant Languages
   Language and Culture

6.7 **Religion**
   Classification and Distribution of Religions
   The Principal Religions
     Judaism
     Christianity
     Islam
     Hinduism
     Buddhism
     East Asian Ethnic Religions
6.8 **Ethnicity**
6.9 **Gender and Culture**
6.10 **Other Aspects of Diversity**
   SUMMARY OF KEY CONCEPTS
   KEY WORDS
   THINKING GEOGRAPHICALLY

141

Numbered **Chapter Outlines** are included on the opening page of each chapter to clarify the organization of the chapter and to make it easy to locate specific topics of discussion.

Each chapter opens with **Learning Objectives** students can use to guide their study and help them focus on critical concepts. These objectives specify what students are expected to know, understand, and be able to do after studying the chapter.

Vignettes are used to begin each chapter with a brief real-life story intended to capture student interest and prepare them for the subject matter to follow.

Physical Geography: Landforms   47

LEARNING OBJECTIVES

After studying this chapter you should be able to:

3.1 Characterize the three classes of rock.
3.2 Define folding, joint, and faulting.
3.3 Illustrate how plate tectonics relate to earthquakes.
3.4 Explain how a tsunami originates.
3.5 Compare the effect of mechanical and chemical weathering on landforms.
3.6 Compare the effect of groundwater erosion with that of surface water erosion.
3.7 Relate how glaciers form and how their erosion creates landscapes.
3.8 Define landform features such as deltas, alluvial fans, natural levees, and moraines.
3.9 Understand the landform changes due to waves, currents, and wind.

Although too early for sunbathers and snorkelers, the Hawaiian Islands will have a new island to add to their collection, which contains such scenic beauties as Oahu, Maui, and Kauai. It is Loihi, 0.8 kilometer (0.5 mi) below sea level, just 27 kilometers (17 mi) from the big island of Hawaii. Because the speed of its ascent must be measured in geologic time, it probably will not appear above the water surface for another million or so years. It is a good example, however, of the ceaseless changes that take place on the Earth's surface. As the westernmost of the islands erode and sink below sea level, new islands arise at the eastern end. In Loihi's most recent explosion in 1996, scientists feared that a giant wave would be set off at the surface that could devastate the islands, including the city of Honolulu and popular Waikiki Beach. Fortunately, this was not the case.

Humans on their trip through life continuously are in touch with the ever-changing, active, moving physical environment. Most of the time, we are able to live comfortably with the changes, but when a freeway is torn apart by an earthquake, or floodwaters force us to abandon our homes, we suddenly realize that we spend a good portion of our lives trying to adapt to the challenges the physical environment has for us.

For the geographer, things just will not stand still—not only little things, such as icebergs or new islands rising out of the sea, or big ones, such as exploding volcanoes changing their shape and form, but also giant things, such as continents that wander about like nomads and ocean basins that expand, contract, and split in the middle like worn-out coats.

Geologic time is long, but the forces that give shape to the land are timeless and constant. Processes of creation and destruction are continually at work to fashion the seemingly eternal structure upon which humans live and work. Two types of forces interact to produce those infinite local variations in the surface of the Earth called *landforms:* (1) forces that push, move, and raise the Earth's surface and (2) forces that scour, wash, and wear down the surface.

Mountains rise and are then worn away. The eroded materials—soil, sand, pebbles, rocks—are transported to new locations and help create new landforms. How long these processes have worked, how they work, and their effects are the subject of this chapter.

Much of the research needed to create the story of land-forms results from the work of geomorphologists. A branch of the fields of geology and physical geography, *geomorphology* is the study of the origin, characteristics, and development of landforms. It emphasizes the study of the various processes that create landscapes. Geomorphologists examine the erosion, transportation, and deposition of materials and the interrelationships among climate, soils, plant and animal life, and landforms.

In a single chapter, we can only begin to explore the many and varied contributions of geomorphologists. After discussing the contexts within which landform change takes place, we consider the forces that are building up the Earth's surface and then review the forces wearing it down.

3.1 Earth Materials

The rocks of the Earth's crust vary according to mineral composition. Rocks are composed of particles that contain various combinations of such common elements as oxygen, silicon, aluminum, iron, and calcium, together with less-abundant elements. A particular chemical combination that has a hardness, density, and definite crystal structure of its own is called a **mineral.** Some well-known minerals are quartz, feldspar, and micas. Depending on the nature of the minerals that form them, rocks are hard or soft, more or less dense, one color or another, or chemically stable or not. While some rocks resist decomposition, others are very easily broken down. Among the more common varieties of rock are granites, basalts, limestones, sandstones, and slates.

Although one can classify rocks according to their physical properties, the more common approach is to classify them by the way they formed. The three main groups of rock are igneous, sedimentary, and metamorphic.

Igneous Rocks

Igneous rocks are formed by the cooling and solidification of molten rock. Openings in the crust give molten rock an opportunity to find its way into or onto the crust. When the molten rock cools, it solidifies and becomes igneous rock. The name for underground molten rock is *magma;* aboveground, it is *lava. Intrusive* igneous rocks are formed below ground level by the solidification of magma, whereas *extrusive* igneous rocks are created above ground level by the solidification of lava (**Figure 3.1**).

The composition of magma and lava and, to a limited extent, the rate of cooling determine the minerals that form. The rate of cooling is mainly responsible for the size of the crystals. Large crystals of quartz—a hard mineral—form slowly beneath the surface of the Earth. When combined with other minerals, quartz forms the intrusive igneous rock called *granite.*

The lava that oozes out onto the Earth's surface and makes up a large part of the ocean basins becomes the extrusive igneous rock called *basalt,* the most common rock on the Earth's surface. If, instead of oozing, the lava erupts from a volcano crater, it may cool

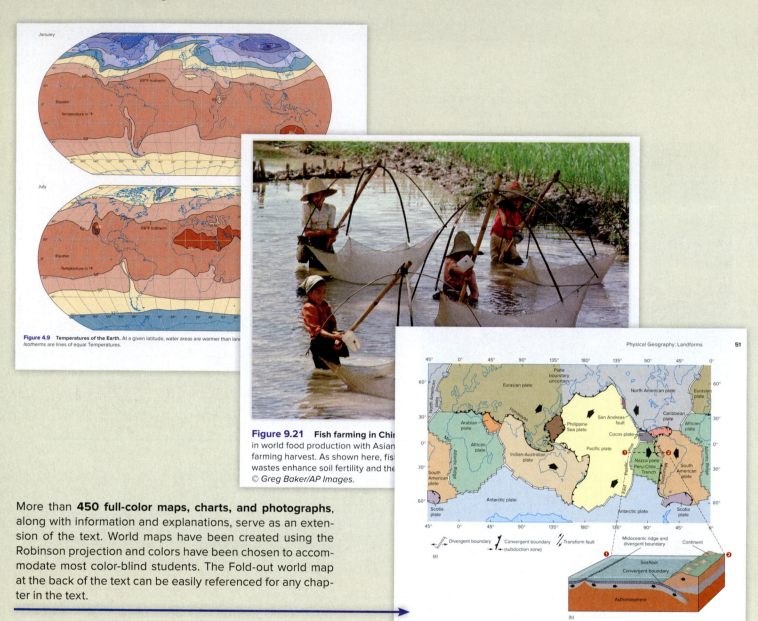

Figure 4.9   **Temperatures of the Earth.** At a given latitude, water areas are warmer than land... Isotherms are lines of equal Temperatures.

Figure 9.21   Fish farming in Chi... in world food production with Asian... farming harvest. As shown here, fis... wastes enhance soil fertility and the... © *Greg Baker/AP Images.*

More than **450 full-color maps, charts, and photographs**, along with information and explanations, serve as an extension of the text. World maps have been created using the Robinson projection and colors have been chosen to accommodate most color-blind students. The Fold-out world map at the back of the text can be easily referenced for any chapter in the text.

Physical Geography: Landforms   **51**

Figure 3.6   **(a) Principal lithospheric plates of the world.** Arrows indicate the direction of plate motion. **(b) Plate motion away from a divergent boundary toward a convergent boundary.**

Physical Geography: Weather and Climate   **89**

EL NIÑO-SOUTHERN OSCILLATION

El Niño is a term coined years ago by fishermen who noticed that the normally cool waters off the coasts of Ecuador and Peru were considerably warmer every 3 or 4 years around Christmas-time, hence the name El Niño, Spanish for "the child," referring to the infant Jesus. The fish catch was significantly reduced during these periods. If fishermen had been able to identify the scientific associations that present-day oceanographers and climatologists make, they would have recognized a host of other effects that follow from El Niño. The **El Niño-Southern Oscillation** is a fluctuation in tropical ocean temperatures off the Pacific coast of South America. The warm phase is called El Niño, the cool phase La Niña.

During the winter of 1997–1998, an unusually severe El Niño caused enormous damage and hundreds of deaths. The West Coast of the United States, especially California, was inundated with rainfall amounts double, triple, and even quadruple the normal. For the November to March winter period, San Francisco received 102.24 centimeters (40.25 in.) of rain—the normal is 41.63 centimeters (16.39 in.). The 38 centimeters (15 in.) in February 1998 was the most for that month in the 150 years of record keeping in San Francisco. The resort city of Acapulco, Mexico, was badly battered by torrential rains and high, wind-blown tides. Parts of South America, especially Ecuador, Peru, and Chile, were ravaged by floods and mud slides, while droughts and fires scorched eastern South America, Australia, and parts of Asia, especially Indonesia. A stronger than usual southern branch of the jet stream generated by El Niño spawned dozens of tornadoes, which killed more than 100 people in Alabama, Georgia, and Florida.

The top diagram shows normal circumstances in the southern Pacific Ocean. Normally trade winds blow warm surface water westward toward Indonesia and allow cold water to rise to the surface along the South American coast. The bottom diagram shows that, during El Niño, trade winds and ocean currents weaken and the warm water shifts eastward, shifting wind belts, and bringing storms to North and South America.

Boxed inserts are written to further develop ideas and enhance student interest in the course material. Chapters generally include three to five boxes and most chapters include a box on gender-related issues.

Geography & Public Policy boxes highlight important or controversial issues, encouraging students to think about the relevance of geography to real-world concerns. Critical thinking questions at the end of each box prompt students to reflect upon and form an opinion about specific issues and serve as a catalyst for class discussion.

GEOGRAPHY & PUBLIC POLICY

International Population Policies

After a sometimes rancorous 9-day meeting in Cairo in September 1994, the United Nations International Conference on Population and Development endorsed a strategy for stabilizing the world's population at 7.27 billion by no later than 2015. The 20-year program of action accepted by 179 signatory countries sought to avoid the environmental consequences of excessive population growth. Its proposals were therefore linked to discussions and decisions of the UN Conference on Environment and Development held in Rio de Janeiro in June 1992.

The Cairo plan abandoned several decades of top-down government programs that promoted *population control* (a phrase avoided by the conference) based on targets and quotas and, instead, for the first time embraced policies giving women greater control over their lives, greater economic equality and opportunity, and a greater voice in reproduction decisions. It recognized that limiting population growth depends on programs that lead women to want fewer children and make them partners in economic development. In that recognition, the conference accepted the documented link between increased educational access and economic opportunity for women and falling birth rates and smaller families. Earlier population conferences—1974 in Bucharest and 1984 in Mexico City—did not fully address these issues of equality, opportunity, education, and political rights; their adopted goals had failed to achieve hoped-for changes in birth rates, in large part because women in many traditional societies had no power to enforce contraception and feared their other alternative, sterilization.

The earlier conferences had carefully avoided or specifically excluded abortion as an acceptable family planning method. It was the more open discussion of abortion in Cairo that elicited much of the spirited debate that registered religious objections by the Vatican and many Muslim and Latin American states to the inclusion of legal abortion as part of health care, and to language suggesting approval of sexual relations outside marriage. Although the final text of the conference declaration did not promote any universal right to abortion and excluded it as a means of family planning, some delegations still registered reservations to its wording on both sex and abortion. At the conference's close, however, the Vatican endorsed the declaration's underlying principles, including the family as "the basic unit of society" and the need to "stimulate economic growth and to promote gender equality, equity, and the empowerment of women."

A special United Nations "Cairo+5" session in 1999 recommended some adjustments to the earlier agreements. It urged an emphasis on measures ensuring safe and accessible abortion in countries where it is legal, called for schoolchildren at all levels to be instructed in sexual and reproductive health issues, and told governments to provide special family planning and health services for sexually active adolescents, with particular stress on reducing their vulnerability to AIDS.

In 2004, the UN reported progress toward reaching Cairo and Cairo+5 goals. The consensus was that much remained to be done to broaden programs for the poorest population groups, to invest in rural development and urban planning, to strengthen laws ending discrimination against women, and to encourage donor countries to fully meet their agreed-upon contributions to the program. Nevertheless, positive Cairo plan results were also seen in declining fertility rates in many of the world's most-populous developing countries. Some demographers and many women's health organizations pointedly claim that those declines had little to do with government planning policies. Rather, they assert, current lower and falling fertility rates were the expected result of women's assuming greater control over their economic and reproductive lives. The director of the UN Population Division noted: "A woman in a village making a decision to have one or two or at most three children is a small decision in itself. But . . . compounded by millions and millions . . . of women in India and Brazil and Egypt, it has global consequences."

That women are making those decisions, population specialists have observed, reflects important cultural factors emerging since Cairo. Satellite television takes contraceptive information even to remote villages and shows programs of small, apparently happy families that viewers think of emulating. Increasing urbanization reduces some traditional family controls on women and makes contraceptives easier to find, and declining infant mortality makes mothers more confident their babies will survive. Perhaps most important, population experts assert, is the dramatic increase in most developing states in female school attendance, with corresponding reductions in the illiteracy rates of girls and young women, who will themselves soon be making fertility decisions.

Considering the Issues

1. Do you think it is appropriate or useful for international bodies to promote policies affecting such purely personal or national concerns as reproduction and family planning? Why or why not?
2. Do you think that current international concerns over population growth, development, and the environment are sufficiently valid and pressing to risk the loss of long-enduring cultural norms and religious practices in many of the world's traditional societies? Why or why not?
3. The Cairo plan called for sizable monetary pledges from developed countries to support enhanced population planning in the developing world. For the most part, those pledges have not been honored. Do you think the financial obligations assigned to donor countries are justified in light of the many other international needs and domestic concerns faced by their governments? Why or why not?
4. Many environmentalists see the world as a finite system unable to support ever-increasing populations; to exceed its limits would cause frightful environmental damage and global misery. Many economists counter that free markets will keep supplies of needed commodities in line with growing demand and that science will, as necessary, supply technological fixes in the form of substitutes or expansion of production. In light of such diametrically opposed views of population growth consequences, is it appropriate or wise to base international programs solely on one of them? Why or why not?

129

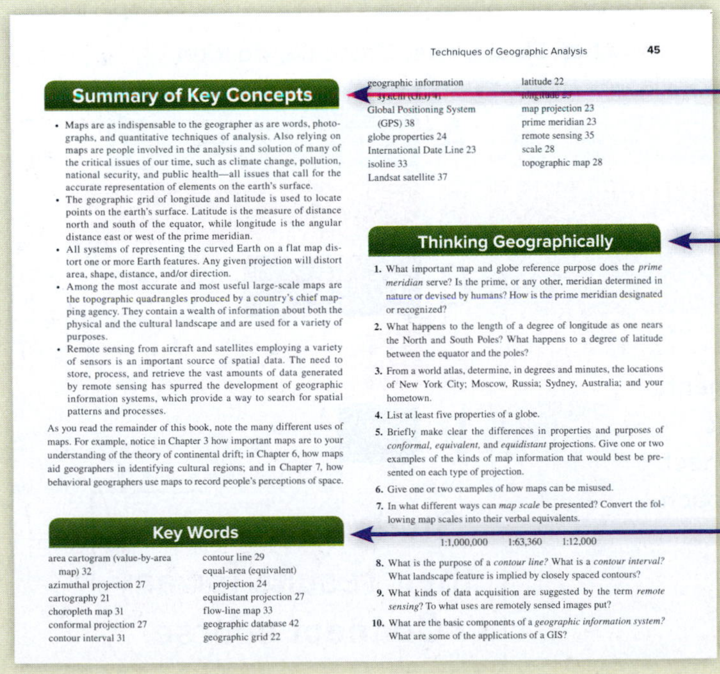

Techniques of Geographic Analysis 45

Summary of Key Concepts

- Maps are as indispensable to the geographer as are words, photographs, and quantitative techniques of analysis. Also relying on maps are people involved in the analysis and solution of many of the critical issues of our time, such as climate change, pollution, national security, and public health—all issues that call for the accurate representation of elements on the earth's surface.
- The geographic grid of longitude and latitude is used to locate points on the earth's surface. Latitude is the measure of distance north and south of the equator, while longitude is the angular distance east or west of the prime meridian.
- All systems of representing the curved Earth on a flat map distort one or more Earth features. Any given projection will distort area, shape, distance, and/or direction.
- Among the most accurate and most useful large-scale maps are the topographic quadrangles produced by a country's chief mapping agency. They contain a wealth of information about both the physical and the cultural landscape and are used for a variety of purposes.
- Remote sensing from aircraft and satellites employing a variety of sensors is an important source of spatial data. The need to store, process, and retrieve the vast amounts of data generated by remote sensing has spurred the development of geographic information systems, which provide a way to search for spatial patterns and processes.

As you read the remainder of this book, note the many different uses of maps. For example, notice in Chapter 3 how important maps are to your understanding of the theory of continental drift; in Chapter 6, how maps aid geographers in identifying cultural regions; and in Chapter 7, how behavioral geographers use maps to record people's perceptions of space.

Key Words

area cartogram (value-by-area map) 32
azimuthal projection 27
cartography 21
choropleth map 31
conformal projection 27
contour interval 31
contour line 29
equal-area (equivalent) projection 24
equidistant projection 27
flow-line map 33
geographic database 42
geographic grid 22
geographic information system (GIS) 41
Global Positioning System (GPS) 38
globe properties 24
International Date Line 23
isoline 33
Landsat satellite 37
latitude 22
longitude [illegible]
map projection 23
prime meridian 23
remote sensing 35
scale 28
topographic map 28

Thinking Geographically

1. What important map and globe reference purpose does the *prime meridian* serve? Is the prime, or any other, meridian determined in nature or devised by humans? How is the prime meridian designated or recognized?
2. What happens to the length of a degree of longitude as one nears the North and South Poles? What happens to a degree of latitude between the equator and the poles?
3. From a world atlas, determine, in degrees and minutes, the locations of New York City; Moscow, Russia; Sydney, Australia; and your hometown.
4. List at least five properties of a globe.
5. Briefly make clear the differences in properties and purposes of *conformal, equivalent,* and *equidistant* projections. Give one or two examples of the kinds of map information that would best be presented on each type of projection.
6. Give one or two examples of how maps can be misused.
7. In what different ways can *map scale* be presented? Convert the following map scales into their verbal equivalents.

   1:1,000,000 1:63,360 1:12,000
8. What is the purpose of a *contour line?* What is a *contour interval?* What landscape feature is implied by closely spaced contours?
9. What kinds of data acquisition are suggested by the term *remote sensing*? To what uses are remotely sensed images put?
10. What are the basic components of a *geographic information system?* What are some of the applications of a GIS?

Chapters **Summaries of Key Concepts** appear at the end of each chapter as a way to reinforce the major ideas of the chapter and guide student understanding of key concepts.

Thinking Geographically questions are easily assignable and provide students an opportunity to check their grasp of chapter material.

A **Key Words** list with page references makes it easy for students to verify their understanding of the most important terms in the chapter.

Appendix 1: Map Projections include a discussion of methods of projection, globe properties and map distortion, and classes of projection.

Appendix 2: Climates, Soils, and Vegetation supplements Chapter 4 *Physical Geography: Weather and Climate* by providing information about soil formation, soil profiles and horizons, soil taxonomy, and natural vegetation regions.

Appendix 3: 2012 World Population Data Sheet for the Population Reference Bureau (a modified version) includes basic demographic data and projections for countries, regions, and continents, as well as selected economic and social statistics helpful in national and regional comparisons. The comparative information in the appendix data is useful for student projects, regional and topical analyses, and the study of world patterns.

McGraw-Hill Connect® Learn Without Limits

Connect is a teaching and learning platform that is proven to deliver better results for students and instructors.

Connect empowers students by continually adapting to deliver precisely what they need, when they need it, and how they need it, so your class time is more engaging and effective.

73% of instructors who use Connect require it; instructor satisfaction increases by 28% when Connect is required.

Connect's Impact on Retention Rates, Pass Rates, and Average Exam Scores

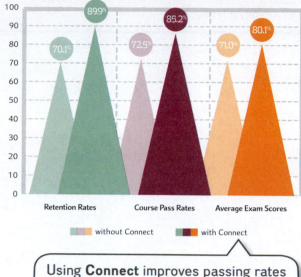

Using **Connect** improves passing rates by **12.7%**, and retention by **19.8%**.

Analytics

Connect Insight®

Connect Insight is Connect's new one-of-a-kind visual analytics dashboard—now available for both instructors and students—that provides at-a-glance information regarding student performance, which is immediately actionable. By presenting assignment, assessment, and topical performance results together with a time metric that is easily visible for aggregate or individual results, Connect Insight gives the user the ability to take a just-in-time approach to teaching and learning, which was never before available. Connect Insight presents data that empowers students and helps instructors improve class performance in a way that is efficient and effective.

Impact on Final Course Grade Distribution

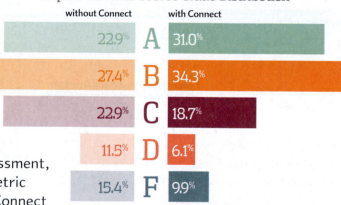

without Connect		with Connect
22.9%	A	31.0%
27.4%	B	34.3%
22.9%	C	18.7%
11.5%	D	6.1%
15.4%	F	9.9%

Students can view their results for any Connect course.

Mobile

Connect's new, intuitive mobile interface gives students and instructors flexible and convenient, anytime–anywhere access to all components of the Connect platform.

Adaptive

More students earn **A's** and **B's** when they use McGraw-Hill Education **Adaptive** products.

SmartBook®

Proven to help students improve grades and study more efficiently, SmartBook contains the same content within the print book, but actively tailors that content to the needs of the individual. SmartBook's adaptive technology provides precise, personalized instruction on what the student should do next, guiding the student to master and remember key concepts, targeting gaps in knowledge and offering customized feedback, and driving the student toward comprehension and retention of the subject matter. Available on tablets, SmartBook puts learning at the student's fingertips—anywhere, anytime.

Over **8 billion questions** have been answered, making McGraw-Hill Education products more intelligent, reliable, and precise.

www.mheducation.com

THE ADAPTIVE READING EXPERIENCE DESIGNED TO TRANSFORM THE WAY STUDENTS READ

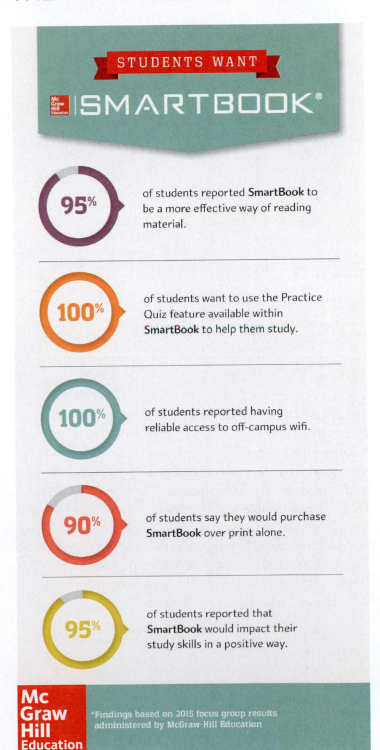

STUDENTS WANT

SMARTBOOK®

95% of students reported **SmartBook** to be a more effective way of reading material.

100% of students want to use the Practice Quiz feature available within **SmartBook** to help them study.

100% of students reported having reliable access to off-campus wifi.

90% of students say they would purchase **SmartBook** over print alone.

95% of students reported that **SmartBook** would impact their study skills in a positive way.

*Findings based on 2015 focus group results administered by McGraw-Hill Education

CHAPTER 1

Introduction

Two Sherpas carry heavy packs in the Himalayas near Mount Everest. Sherpas are an ethnic group who live in the high mountains of Nepal. Many Sherpas are expert mountaineers who work in risky positions as guides and porters for the tourist mountain climbing industry. © *Sean White/Design Pics RF.*

LEARNING OBJECTIVES

After studying this chapter you should be able to:

1.1   Understand what geographers mean when they say that "location matters."

1.2   Describe what is meant by physical and cultural landscapes.

1.3   Discuss how geography aids in understanding national and international problems.

1.4   Explain how the word *spatial* is used in the discipline of geography.

1.5   Appreciate which concepts are used to understand the processes of human interaction.

1.6   Summarize the kinds of understanding encompassed in the National Standards.

Nepal is home to some of the most dramatic landscapes in the world. Rising from tropical lowlands on the border with India to the world's highest peaks in the Himalayas, this small country has been nicknamed "the mountain kingdom." The Himalayas were built and continue to grow due to the collision of two of the Earth's crustal plates. The Indian subcontinent is moving northeast at a rate of about 4 centimeters per year, about the same rate at which human fingernails grow. The collision of the Indian crustal plate with Eurasia pushes up mountains and builds stresses that are periodically relieved by earthquakes. On April 28, 2015, a magnitude 7.9 earthquake struck Nepal, just 77 kilometers from Kathmandu, the country's capital and largest city. Thousands of buildings crumpled and landslides buried villages and roadways, killing over 9,000 people (see **Figure 1.1**). Adding to the tragedy, many cultural treasures such as monumental towers, Buddhist stupas, and Hindu temples were damaged or destroyed. The Kathmandu Valley has seven cultural landscape sites designated by the United Nations as being of global significance. These cultural sites include historic royal palace squares, sacred sites, and celebrated Hindu and Buddhist monuments. The quake also triggered an avalanche on Mount Everest, the world's highest mountain, killing 19 climbers in the base camp and stranding hundreds of climbers and their Sherpa guides on the mountainside. The climbing season for 2015 was shut down. Thus, the earthquake also devastated both the cultural tourism and ecotourism sectors that are vital to Nepal's economy. Long after the world's media attention has moved on, Nepal's residents continue to work resiliently to rebuild their country's homes, businesses, roads, utilities, historic landmarks, and religious structures. As they rebuild their lives, institutions, and economy, they hope that the tourists will soon return.

The study of geography, by combining knowledge of earth systems and human societies, sheds important light on events such as these. Studying geography can also help societies become more resilient and sustainable. In Chapter 3, for example, we discuss the processes that cause earthquakes and build mountains and other landforms. While news reports often refer to catastrophes such as these as "natural disasters," there was more than nature involved in the 2015 Nepal earthquake. For example, natural disasters, such as hurricanes, floods, landslides, and earthquakes that hit poor low-income countries have death tolls many times higher than those that hit high-income countries. Partly to blame for the many deaths in the 2015 Nepal earthquake were construction materials that could not withstand the forces unleashed by this earthquake. Brick, stone, or concrete buildings without steel reinforcing bars cannot handle the ground shaking of an earthquake and often collapse. As discussed in Chapter 7, Nepal is classified by the United Nations as among the least developed countries in the world. Per capita incomes in Nepal average 1/25th of those in the United States of America. Nepali scientists actively monitor the earthquake hazards in their country and in 1994 building regulations were passed to increase earthquake safety. But in many cases, the building code was ignored because the builders simply could not afford the steel reinforcing materials to make homes, schools, and commercial buildings safe.

Relief and recovery efforts were complicated by poor road and utility systems and the lack of emergency services. Extreme poverty meant there was no insurance policy or bank account

Figure 1.1   Earthquake destruction in Kathmandu, Nepal. The April 25, 2015 earthquake toppled many culturally significant buildings and caused 9,000 fatalities and US$4 billion in damage. Natural disasters are tragic reminders of the close relationships between human societies and the natural environment. © *Michelle Fyneweaver.*

to fall back upon for many Nepali residents who lost homes, jobs, and family members. One bright spot was the use of new geographic techniques to assess damage and map the changing locations of people displaced by the earthquake. Satellite imagery provided insights into the condition of buildings and the location of landslides. Anonymous cell phone data from millions of Nepali users were collected by a nonprofit organization and used to track the location of people displaced by the quake so that relief could be directed to where it was needed. The tragic Nepal earthquake of 2015 is a reminder that human actions take place in the context of a particular geographic environment. It is also a reminder that, just as maps were essential tools in the rescue, relief, and recovery processes, many of the world's pressing problems require a geographic understanding that brings together earth systems, human cultures, the locations of human activities, and the relationships between human societies and their environment—all important themes in the study of geography.

1.1 What Is Geography?

Many people associate the word *geography* simply with knowing *where* things are: whether they be countries, such as Myanmar and Uruguay; cities, such as Timbuktu or Almaty; or deposits of natural resources, such as petroleum or iron ore. Some people pride themselves on knowing which rivers are the longest, which mountains are the tallest, and which deserts are the largest. Such factual knowledge about the world has value, permitting us to place current events in their proper spatial setting. When we hear of an earthquake in Turkey or an assault in Chechnya, we at least can visualize where they occurred. Knowing *why* they occurred in those places, however, is considerably more important.

Geography is much more than place names and locations. It is the study of spatial variation, of how and why things differ from place to place on the surface of the Earth. It is, further, the study of how observable spatial patterns evolved through time. Just as knowing the names and locations of organs in the human body does not equip one to perform open-heart surgery, knowing where things are located is only the first step toward understanding why things are where they are, and what events and processes determine or change their distribution. Why are earthquakes common in Turkey but not in Russia, and why is Chechnya but not Tasmania wracked by insurgency? Why are the mountains in the eastern United States rounded and those in the western states taller and more rugged? Why do you find a concentration of French speakers in Quebec but not in other parts of Canada?

In answering questions such as these, geographers focus on the interaction of people and social groups with their environment—planet Earth—and with one another; they seek to understand how and why physical and cultural spatial patterns evolved through time and continue to change. Because geographers study both the physical environment and human use of that environment, they are sensitive to the variety of forces affecting a place and the interactions among them. To explain why Brazilians burn a significant portion of the tropical rain forest each year, for example, they draw on their knowledge of the climate and soils of the Amazon Basin; population pressures, landlessness, and the need for greater agricultural area in rural Brazil; the country's foreign debt status; midlatitude markets for lumber, beef, and soybeans; and Brazil's economic development objectives. Understanding the environmental consequences of the burning requires knowledge of, among other things, the oxygen and carbon balance of the Earth; the contribution of the fires to the greenhouse effect, acid rain, and depletion of the ozone layer; and the relationships among deforestation, soil erosion, and floods.

Geography, therefore, is about Earth space and the content of that space. We think of and respond to places from the standpoint of not only where they are but, what is more important, what they contain or what we think they contain. Reference to a place or an area usually calls up images about its physical nature or what people do there, and this often suggests to us, without our consciously thinking about it, how those physical things and activities are related. Examples include "Bangladesh," "farming," and "flooding" as well as "Colorado," "mountains," and "skiing." That is, the content of an area has both physical and cultural aspects, and geography is always concerned with understanding both (**Figure 1.2**).

1.2 Evolution of the Discipline

Geography's combination of interests was apparent even in the work of the early Greek geographers who first gave structure to the discipline. Geography's name was reputedly coined by the Greek scientist Eratosthenes over 2200 years ago from the words *geo,* "the Earth," and *graphein,* "to write." From the beginning, that writing focused both on the physical structure of the Earth and on the nature and activities of the people who inhabited the various lands of the known world. To Strabo (c. 64 B.C.–A.D. 20), the task of geography was to "describe the several parts of the inhabited world, . . . to write the assessment of the countries of the world [and] to treat the differences between countries." Even earlier, Herodotus (c. 484–425 B.C.) had found it necessary to devote much of his writing to the lands, peoples, economies, and customs of the various parts of the Persian Empire as necessary background to an understanding of the causes and course of the Persian wars.

Greek (and, later, Roman) geographers measured the Earth, devised the global grid of parallels and meridians (marking latitudes and longitudes; see p. 7), and drew upon that grid surprisingly sophisticated maps of their known world (**Figure 1.3**). They explored the apparent latitudinal variations in climate and described in numerous works the familiar Mediterranean Basin and the more remote, partly rumored lands of northern Europe, Asia, and equatorial Africa. Employing nearly modern concepts, they described river systems, explored cycles of erosion and patterns of deposition, cited the dangers of deforestation, described variations in the natural landscape, and noted the consequences of environmental abuse. Against that physical backdrop, they focused their attention on what humans did in home and distant areas—how they lived; what their distinctive similarities and differences were in language, religion, and custom; and how they used, altered, and perhaps destroyed the lands they inhabited.

Figure 1.2 **Aspen, Colorado, demonstrates changing interactions between physical environment and human activity.** Mineral resources, mountainous terrain, and abundant snowfall have made different, specialized human uses attractive and possible. The brick buildings in the foreground are the legacy of its original settlement as a silver mining town, peaking with over 5000 residents in 1890 but declining to about 700 by 1930. The groomed ski slopes in the background represent the town's current identity as a premier ski resort, year-round tourist destination and home to celebrities. © *Punchstock RF.*

Strabo, indeed, cautioned against the assumption that the nature and actions of humans were determined by the physical environment they inhabited. He observed that humans were the active elements in a human-environmental partnership.

The interests guiding the early Greek and Roman geographers were and are enduring and universal. The ancient Chinese, for example, were as involved in geography as an explanatory viewpoint as were westerners, though there was no exchange between them. Further, as Christian Europe entered its Middle Ages between A.D. 800 and 1400 and lost its knowledge of Greek and Roman geographic work, Muslim scholars—who retained that knowledge—undertook to describe and analyze their known world in its physical, cultural, and regional variation.

In the 15th and 16th centuries, European voyages of exploration and discovery put geography at the forefront of the scientific revival. Modern geography had its origins in the surge of scholarly inquiry that, beginning in the 17th century, gave rise to many of the traditional academic disciplines we know today. In its European rebirth, geography from the outset was recognized—as it always had been—as a broadly based integrative study. Patterns and processes of the physical landscape were early interests, as was concern with humans as part of the Earth's variation from place to place. The rapid development of geology, botany, zoology, climatology, and other natural sciences by the end of the 18th century strengthened regional geographic investigation and increased scholarly and popular awareness of the intricate interconnections of things in space and between places. By that time, accurate determination of latitude and longitude and scientific mapping of the Earth had made assignment of place information more reliable and comprehensive. A key figure during this period of geographic research was Alexander von Humboldt. Humboldt, for whom Humboldt University in Berlin, Germany, is named, led ambitious scientific expeditions to distant places and synthesized vast amounts of geographic data in his famous writings.

Subfields of Geography

During the 19th century, national censuses, trade statistics, and ethnographic studies gave firmer foundation to human geographic investigation. By the end of the 19th century, geography had become a distinctive and respected discipline in universities throughout Europe and in other regions of the world where European academic examples were followed. The proliferation of professional geographers and geography programs resulted in the development of a whole series of increasingly specialized disciplinary subdivisions, many represented by separate chapters of this book. Political geography, urban geography, and economic geography are examples of some of these subdivisions.

Geography's specialized subfields are not isolated from one another; rather, they are closely interrelated. Geography in all its subdivisions is characterized by three dominating interests. The first is in the spatial variation of physical and human phenomena on the surface of the Earth; geography examines relationships between human societies and the natural environments that they occupy and modify. The second is a focus on the systems that link physical phenomena and human activities in one area of the Earth with other areas. Together, these interests lead to a third enduring theme, that of regional analysis: geography studies human-environmental (or "ecological") relationships and spatial systems in specific locational settings. This areal orientation pursued by some geographers is called *regional geography.*

Other geographers choose to identify particular classes of things, rather than segments of the Earth's surface, for specialized study. These *systematic geographers* may focus their attention on one or a few related aspects of the physical environment or of human populations and societies. In each case, the topic selected for study is examined in its interrelationships with other spatial systems and areal patterns. *Physical geography* directs its attention to the natural environmental side of the human-environmental structure. Its concerns are with landforms and their distribution,

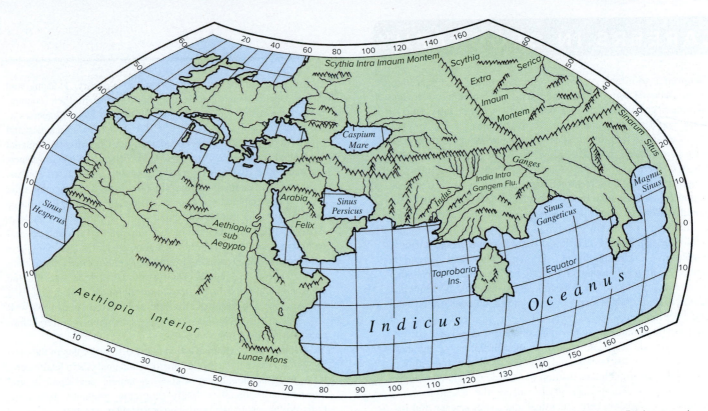

Figure 1.3   World map of the 2nd century A.D. Greco-Egyptian geographer-astronomer Ptolemy. Ptolemy (Claudius Ptolemaeus) adopted a previously developed map grid of latitude and longitude based on the division of the circle into 360°, permitting a precise mathematical location for every recorded place. Unfortunately, errors of assumption and measurement rendered both the map and its accompanying six-volume gazetteer inaccurate. Ptolemy's map, accepted in Europe as authoritative for nearly 1500 years, was published in many variants in the 15th and 16th centuries. The version shown here summarizes the extent and content of the original. Its underestimation of the Earth's size convinced Columbus a short westward voyage would carry him to Asia.

with atmospheric conditions and climatic patterns, with soils or vegetation associations, and the like. The other systematic branch of geography is *human geography*. Its emphasis is on people: where they are, what they are like, how they interact over space, and what kinds of landscapes of human use they erect on the natural landscapes they occupy.

Why Geography Matters

There are three good reasons why people study geography. First, it is the only discipline concerned with understanding why and how both physical and cultural phenomena differ from place to place on the surface of the Earth. Each chapter in this book is designed to give you a basic knowledge of the many processes that shape our world. Chapter 3, for example, introduces you to the tectonic forces that warp, fold, and fault landforms; create volcanoes; and cause earthquakes and tsunami. The discussion of cultural geography in Chapter 6 will give you a framework for understanding the technological, sociological, and ideological components of culture and an awareness of the forces that bring about changes in a culture over time.

Second, a grasp of the broad concerns and topics of geography is vital to an understanding of the national and international problems that dominate daily news reports. Global climate change, the diffusion of HIV-AIDS, Ebola, Zika, and other viruses,

international trade imbalances, inadequate food supply and population growth in developing countries, turmoil in Africa and the Middle East—all of these problems have geographic dimensions, and geography helps explain them. To be geographically illiterate is to deny oneself not only the ability to comprehend local and world problems but also the opportunity to contribute meaningfully to the development of policies for dealing with them.

Third, because geography is such a broad field of study, a great diversity of job opportunities await those who pursue college training in the discipline. Geographic training opens the way to careers in a wide array of fields (see "Careers in Geography"). Geographical techniques of analysis are used for interpreting remotely sensed images, determining the optimum location for new businesses, monitoring the spread of infectious diseases, delineating voting districts, and a host of other tasks. A good book to read is *Why Geography Matters* (Oxford University Press, 2005) by Harm J. de Blij.

1.3 Some Core Geographic Concepts

The topics included within the broad field of geography are diverse. That very diversity, however, emphasizes the reality that all geographers—whatever their particular topical or regional

CAREERS IN GEOGRAPHY

Geography admirably serves the objectives of a liberal education. It can make us better-informed citizens, more able to understand the important issues facing our communities, our country, and our world and better prepared to contribute solutions.

Can it, as well, be a pathway to employment for those who wish to specialize in the discipline? The answer is yes, in a number of different types of jobs. One broad cluster is concerned with supporting the field itself through teaching and research. Teaching opportunities exist at all levels, from elementary to university postgraduate. Teachers with some training in geography are in increasing demand in elementary and high schools in the United States, reflecting geography's inclusion as a core subject in the federally adopted *No Child Left Behind Act* and the national determination to create a geographically literate society (see "The National Standards," p. 16). At the college level, specialized teaching and research in all branches of geography have long been established, and geographically trained scholars are prominently associated with urban, community, and environmental studies; regional science; locational economics; and other interdisciplinary programs.

Because of the breadth and diversity of the field, training in geography involves the acquisition of techniques and approaches applicable to a wide variety of jobs outside the academic world. Modern geography is both a physical and social science and fosters a wealth of technical skills. The employment possibilities it presents are as many and varied as are the public and private agencies and enterprises dealing with the natural environment, with human economic and social activities, and with the acquisition and analysis of spatial data.

Many professional geographers work in government at the federal, state, and local levels and in a variety of international organizations. Indeed, geographers have made careers in essentially all of the many bureaus and offices of the executive departments of the U.S. national government—Agriculture, Commerce, Education,

Health and Human Services, Homeland Security, Housing and Urban Development Interior, and others—and in their counterparts at the state level. Such major independent federal agencies as the Central Intelligence Agency (CIA), National Aeronautics and Space Administration (NASA), Federal Trade Commission, National Geospatial-Intelligence Agency (NGA), Federal Aviation Agency, and many others have steady need for geographically trained workers.

Although many positions do not carry a geography title, physical geographers serve as water and other natural resource analysts, weather and climate experts, soil scientists, and the like. Areas of recent high demand include environmental managers and technicians and geographic information specialists. Geographers who have specialized in environmental studies find jobs in both public and private agencies. Their work may include assessing the environmental impact of proposed development projects on such things as air and water quality and endangered species, as well as preparing the environmental impact statements required before construction can begin.

Human geographers work in many different roles in the public sector. Jobs include data acquisition and analysis in health care, transportation, population studies, economic development, and international economics. Many geography graduates find positions as planners in local and state government agencies concerned with housing and community development, park and recreation planning, and urban and regional planning. They map and analyze land use plans and transportation systems, monitor urban land development, make informed recommendations about the location of public facilities, and engage in basic social science research.

Most of the same specializations are found in the private sector. Geographic training is ideal for such tasks as business planning and market analysis; factory, store, and shopping center site selection; and community and economic development programs for banks, public utilities, and railroads. Publishers of maps,

---

interests—are united by the similar questions they ask and the common set of basic concepts they employ to consider their answers. Of either a physical or cultural phenomenon, they will inquire: What is it? Where is it? How did it come to be what and where it is? Where is it in relation to other physical or cultural realities that affect it or are affected by it? How is it part of a functioning whole? How does its location affect people's lives and the content of the area in which it is found?

These and similar questions are rooted in geography's concern with Earth space and are derived from enduring central themes in geography. In answering them, geographers draw upon a common store of concepts, terms, and methods of study that together form the basic structure and vocabulary of geography. Geographers believe that recognizing spatial patterns is the essential starting point for understanding how people live on and shape the Earth's surface.

Geographers use the word *spatial* as an essential modifier in framing their questions and forming their concepts. Geography, they say, is a *spatial science*. It is concerned with the *spatial*

distribution of phenomena, with the *spatial extent* of regions, the *spatial behavior* of people, the *spatial relationships* between places on the Earth's surface, and the *spatial processes* that underlie those behaviors and relationships. Geographers use *spatial data* to identify *spatial patterns* and to analyze *spatial systems, spatial interaction, spatial diffusion,* and *spatial variation* from place to place.

The word *spatial* comes, of course, from *space,* and to geographers it always carries the idea of the way things are distributed, the way movements occur, and the way processes operate over the whole or a part of the surface of the Earth. The geographer's space, then, is Earth space, the surface area occupied or available to be occupied by humans. Spatial phenomena have locations on that surface, and spatial interactions occur among places, things, and people within the Earth area available to them. The need to understand those relationships, interactions, and processes helps frame the questions that geographers ask.

Those questions have their starting point in basic observations about the location and nature of places and about how places are similar to or different from one another. Such observations, though

atlases, news and travel magazines, and the like employ geographers as writers, editors, and mapmakers.

The combination of a traditional, broad-based liberal arts perspective with the technical skills required in geographic research and analysis gives geography graduates a competitive edge in the labor market. These field-based skills include familiarity with geographic information systems (GIS, explained in Chapter 2), cartography and computer mapping, remote sensing and photogrammetry, and competence in data analysis and problem solving. In particular, students with expertise in GIS, who are knowledgeable about data sources, hardware, and software, are finding they have ready access to employment opportunities. The following table, based on the booklet "Careers in Geography,"[a] summarizes some of the professional opportunities open to students who have specialized in one (or more) of the various subfields of geography. Also, be sure to read the discussion of geography careers accessed on the homepage of the Association of American Geographers at **www.aag.org**. Additional links on the topic of geography careers can be found in the Online Learning Center for this text. The link can be found in the Preface.

Geographic Field of Concentration	Employment Opportunities
Cartography and geographic information systems	Cartographer for federal government (agencies such as Defense Mapping Agency, U.S. Geological Survey, or Environmental Protection Agency) or private sector (e.g., Environmental Systems Research Institute, ERDAS, Intergraph, or Bentley); map librarian; GIS specialist for planners, land developers, real estate agencies, utility companies, local government; remote-sensing analyst; surveyor
Physical geography	Weather forecaster; outdoor guide; coastal zone manager; hydrologist; soil conservation/agricultural extension agent
Environmental studies	Environmental manager; forestry technician; park ranger; hazardous waste planner
Cultural geography	Community developer; Peace Corps volunteer; health care analyst
Economic geography	Site selection analyst for business and industry; market researcher; traffic/route delivery manager; real estate agent/broker/appraiser; economic development researcher
Urban and regional planning	Urban and community planner; transportation planner; housing, park, and recreation planner; health services planner
Regional geography	Area specialist for federal government; international business representative; travel agent; travel writer
Geographic education	Elementary/secondary school teacher; general geography college professor; overseas teacher

[a]"Careers in Geography," by Richard G. Boehm. Washington, D.C.: National Geographic Society, 1996. Previously published by Peterson's Guides, Inc.

simply stated, are profoundly important to our comprehension of the world we occupy.

- Places have location, direction, and distance with respect to other places.
- A place has size; it is large, medium, or small. Scale is important.
- A place has both physical structure and cultural content.
- The attributes, or characteristics, of places develop and change over time.
- The content of places is structured and explainable.
- The elements of places interrelate with other places.
- Places may be generalized into regions of similarities and differences.

These basic notions are the means by which geographers express fundamental observations about the Earth space they examine and put those observations into a common framework of reference. Each of the concepts is worth further discussion, for they are not quite as simple as they seem.

Location, Direction, and Distance

Location, direction, and *distance* are everyday ways of assessing the space around us and identifying our position in relation to other things and places of interest. They are also essential in understanding the processes of spatial interaction that figure so importantly in the study of both physical and human geography.

Location

The location of places and things is the starting point of all geographic study as well as of our personal movements and spatial actions in everyday life. We think of and refer to location in at least two different senses, *absolute* and *relative*.

Absolute location is the identification of place by a precise and accepted system of coordinates; therefore, sometimes it is called *mathematical location*. We have several such accepted systems of pinpointing positions. One of them is the global grid of parallels and meridians—that is, latitude and longitude (discussed in Chapter 2, pp. 21–23). With it, the absolute location of any point on the Earth can be accurately described by reference to its degrees, minutes, and seconds of *latitude* and *longitude*.

Other coordinate systems are also in use. Survey systems such as the township, range, and section description of property in much of the United States give mathematical locations on a regional level, and street address precisely defines a building according to the reference system of an individual town. Absolute location is unique to each described place, is independent of any other characteristic or observation about that place, and has obvious value in the legal description of places, in measuring the distance separating places, or in finding directions between places on the Earth's surface.

When geographers—or real estate agents—remark that "location matters," however, their reference is usually not to absolute but to **relative location**—the position of a place or thing in relation to that of other places or things (**Figure 1.4**). Relative location expresses spatial interconnection and interdependence and may carry social (neighborhood character) and economic (assessed valuations of vacant land) implications. On an immediate and personal level, we think of the location of the school library not in terms of its street address or room number but where it is relative to our classrooms, the cafeteria, or another reference point. On the larger scene, relative location tells us that people, things, and places exist not in a spatial vacuum but in a world of physical and cultural characteristics that differ from place to place.

New York City, for example, may be described in absolute terms as located at (approximately) latitude 40°43′N (read as 40 degrees, 43 minutes north) and longitude 73°58′W. We have a better understanding of the meaning of its location, however, when reference is made to its spatial relationships: to the continental interior through the Hudson-Mohawk lowland corridor or to its position on the eastern seaboard of the United States. Within the city, we gain understanding of the locational significance of Central Park or the Lower East Side not solely by reference to the street addresses or city blocks they occupy but also by their spatial and functional relationships to the total land use, activity, and population patterns of New York City.

In view of these different ways of looking at location, geographers make a distinction between the *site* and the *situation* of a place (**Figure 1.5**). **Site,** an absolute location concept, refers to the physical and cultural characteristics and attributes of the place itself. It is more than mathematical location, for it tells us something about the specific features of that place. **Situation,** on the other hand, refers to the relations between a place and other places. It is an expression of relative location with particular reference to items of significance to the place in question. Site and situation in the city context are further examined in Chapter 11.

Direction

Direction is the second universal spatial concept. Like location, it has more than one meaning and can be expressed in absolute or relative terms. **Absolute direction** is based on the cardinal points of north, south, east, and west. These appear in all cultures, derived from the obvious "givens" of nature: the rising and setting of the sun for east and west, the sky location of the noontime sun and of certain fixed stars for north and south.

We also commonly use **relative,** or *relational,* **directions.** In the United States, we go "out West," "back East," or "down South"; we worry about conflict in the "Near East" or economic competition from the "Far Eastern countries." Despite their reference to cardinal compass points, these directional references are culturally based and locationally variable. The Near East and the Far East locate parts of Asia from the European perspective; they are retained in the Americas by custom and usage, even though one would normally travel westward across the Pacific, for example, to reach the "Far East" from California, British Columbia, or Chile. For many Americans, "back East" and "out West" are reflections of the migration paths of earlier generations for whom home was in the eastern part of the country, to which they might look back. "Up North" and "down South" reflect our accepted custom of putting north at the top and south at the bottom of our maps.

Distance

Distance joins *location* and *direction* as a commonly understood term that has dual meanings for geographers. Like its two companion spatial concepts, distance may be viewed in both an absolute and a relative sense.

Absolute distance refers to the spatial separation between two points on the Earth's surface, measured by an accepted standard unit—such as miles or kilometers for widely separated locales, feet or meters for more closely spaced points. **Relative distance** transforms those linear measurements into other units more meaningful to human experience or decision making.

Figure 1.4   **The reality of *relative location*** on the globe may be strikingly different from the impressions we form from flat maps. The position of Russia with respect to North America when observed from a polar perspective emphasizes that relative location properly viewed is important to our understanding of spatial relationships and interactions between the two world areas.

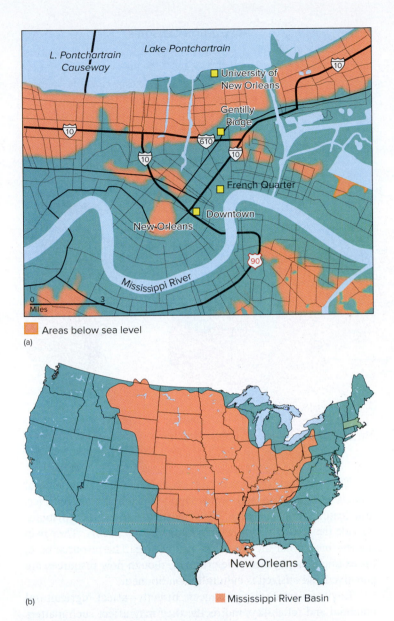

(a)

Areas below sea level

(b)

Mississippi River Basin

Figure 1.5   Site and Situation (a) The *site* of New Orleans is hardly ideal for building a city. The French occupied the most suitable high ground they could find near the mouth of the Mississippi River. The site extends from the "high ground" on the natural levee next to the Mississippi River to former wetlands near Lake Pontchartrain. Much of the city and its suburbs are below sea level on sinking soils composed of soft sediments deposited by past river floods. (b) The *situation* of New Orleans is ideal for building a city. New Orleans is connected to 9000 miles of navigable waterways through the Mississippi River, which drains a basin that stretches from the Rocky Mountains to the Appalachian Mountains.

To know that two competing malls are about equidistant in miles from your residence is perhaps less important in planning your shopping trip than is knowing that, because of street conditions or traffic congestion, one is 5 minutes and the other 15 minutes away (**Figure 1.6**). Most people, in fact, think of time distance rather than linear distance in their daily activities; downtown is 20 minutes by bus, the library is a 5-minute walk. In some instances, money rather than time is the distance transformation.

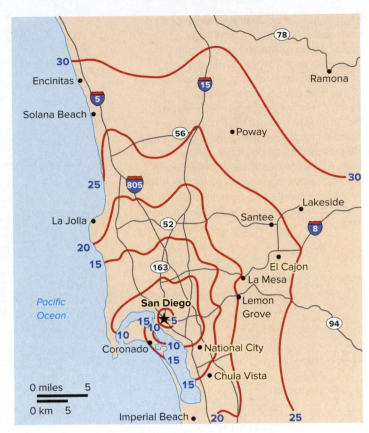

Figure 1.6   Travel times from downtown San Diego, 2002, in minutes. Lines of equal travel time (*isochrones:* from Greek *isos,* equal, and *chronos,* time) mark off the different linear distances accessible within given spans of time from a starting point. The fingerlike outlines of isochrone boundaries reflect variations in road conditions, terrain, traffic congestion, and other aids or impediments to movement. Note the effect of freeways on travel time.

An urban destination might be estimated to be a $10 cab ride away, information that may affect either the decision to make the trip at all or the choice of travel mode to get there. As a college student, you already know that rooms and apartments are less expensive at a greater distance from campus.

A *psychological* transformation of linear distance is also frequent. A solitary late-night walk back to the car through an unfamiliar or dangerous neighborhood seems far longer than a daytime stroll of the same distance through familiar and friendly territory. A first-time trip to a new destination frequently seems much longer than the return trip over the same path. Nonlinear distance and spatial interaction are further considered in Chapter 7.

Size and Scale

When we say that a place may be large, middle size, or small, we speak both of the nature of the place itself and of the generalizations that can be made about it. Geographers are concerned with **scale,** though we may use that term in different ways. We can, for example, study a problem such as population or landforms at the local scale or on a global scale. Here, the reference is purely to the size of unit studied. More technically, scale tells us the relationship between the size of an area on a map and the actual size of the

Figure 1.7   Population density and map scale. "Truth" depends on one's scale of inquiry. Map **(a)** reveals that the maximum year 2010 population density of midwestern states was no more than 123 people per square kilometer (319 per sq mi). From map **(b)**, however, we see that population densities in three Illinois counties exceeded 494 people per square kilometer (1280 per sq mi) in 2010. If we were to reduce our scale of inquiry even further, examining individual city blocks in Chicago, we would find densities reaching 2500 or more people per square kilometer (10,000 per sq mi). Scale matters!

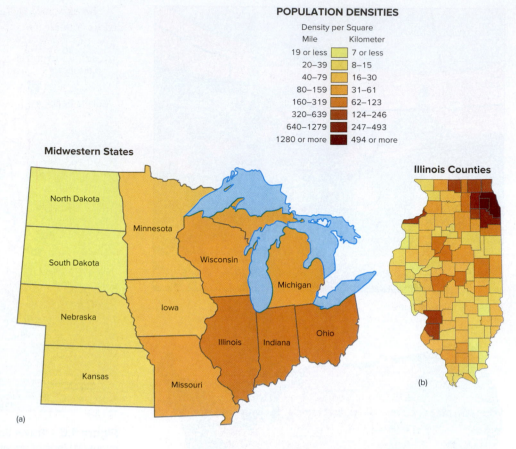

mapped area on the surface of the Earth. In this sense, as Chapter 2 makes clear, scale is a feature of every map and is essential to recognizing what is shown on that map.

In both senses of the word, *scale* implies the degree of generalization represented (**Figure 1.7**). Geographic inquiry may be broad or narrow; it occurs at many different size scales. Climate may be an object of study, but research and generalization focused on climates of the world will differ in degree and kind from study of the microclimates of a city. Awareness of scale is very important. In geographic work, concepts, relationships, and understandings that have meaning at one scale may not be applicable at another.

For example, the study of world agricultural patterns may refer to global climate patterns, cultural food preferences, levels of economic development, and patterns of world trade. These large-scale relationships are of little concern in the study of crop patterns within single counties of the United States, where topography, soil and drainage conditions, farm size, ownership, and capitalization, or even personal management preferences, may be of greater explanatory significance.

Physical and Cultural Attributes

All places have individual physical and cultural attributes distinguishing them from other places and giving them character, potential, and meaning. Geographers are concerned with identifying and analyzing the details of those attributes and, particularly, with recognizing the interrelationship between the physical and cultural components of area: the human-environmental interface.

The physical characteristics of a place are such natural aspects as its climate, soil, water supplies, mineral resources, terrain features, and the like. These **natural landscape** attributes provide the setting within which human action occurs. They help shape—but do not dictate—how people live. The resource base, for example, is physically determined, though how resources are perceived and utilized is culturally conditioned.

Environmental circumstances directly affect agricultural potential and reliability; indirectly, they may affect such matters as employment patterns, trade flows, population distributions, national diets, and so on. The physical environment simultaneously presents advantages and disadvantages with which humans must deal. Thus, most places offer trade-offs in terms of climate favorability, natural hazards, farming and fishing productivity, natural resources, and natural scenery. For example, a scenic volcano may someday erupt, or a mild, coastal location may be vulnerable to hurricanes, and so forth. Physical environmental patterns and processes are explored in Chapters 3 and 4 of this book.

At the same time, by occupying a given place, people modify its physical attributes. The visible imprint of that human activity is called the **cultural landscape.** It, too, exists at different scales and at different levels of visibility. Contrasts in agricultural practices and land use between Mexico and southern California are evident in **Figure 1.8**, whereas the signs, structures, and people of Los Angeles's Chinatown leave a smaller, more confined imprint within the larger cultural landscape of the metropolitan area itself.

The physical and human characteristics of places are the keys to understanding both the simple and the complex interactions and

Figure 1.8 **This Landsat satellite image reveals contrasting cultural landscapes along the Mexico-California border.** Move your eyes from the Salton Sea (the dark patch at the top of the image) southward to the agricultural land extending to the edge of the picture. Notice how the regularity of the fields and the bright colors (representing growing vegetation) give way to a marked break, where irregularly shaped fields and less prosperous agriculture are evident. Above the break is the Imperial Valley of California; below the border is Mexico. *NASA*.

interconnections between people and the environments they occupy and modify. Those interconnections and modifications are not static or permanent but are subject to continual change.

The existence of the U.S. Environmental Protection Agency (and its counterparts elsewhere) is a reminder that humans are the active and frequently harmful agents in the continuing interplay between the cultural and physical worlds (**Figure 1.9**). Virtually every human activity leaves its imprint on the Earth's soil, water, vegetation, animal life, and other resources, as well as on the atmosphere common to all Earth space, as Chapters 12 and 13 make clear.

Attributes of Place Are Always Changing

The physical environment surrounding us seems eternal and unchanging but, of course, it is not. In the framework of geologic time, change is both continuous and pronounced. Islands form and disappear; mountains rise and are worn low to swampy plains; vast continental glaciers form, move, and melt away, and sea levels fall and rise in response. Geologic time is long, but the forces that give shape to the land are timeless and relentless.

Even within the short period of time since the most recent retreat of continental glaciers—12,000 or 13,000 years ago—the environments occupied by humans have been subject to change. Glacial retreat itself marked a period of climatic alteration, extending the area habitable by humans to include vast reaches of northern Eurasia and North America formerly covered by thousands of feet of ice. With moderating climatic conditions came changes in vegetation and fauna. On the global scale, these were natural environmental changes; humans were as yet too few in number and too limited in technology to alter materially the course of physical events. On the regional scale, however, even early human societies exerted an impact on the environments they occupied. Fire was used to clear forest undergrowth, to maintain or extend grassland for grazing animals and to drive them in the hunt, and later to clear openings for rudimentary agriculture.

With the dawn of civilizations and the invention and spread of agricultural technologies, humans accelerated their management and alteration of the now no longer "natural" environment.

Figure 1.9 **Sites such as this Anacortes, Washington, oil refinery are major emitters of potentially toxic chemicals to the atmosphere, land, and water.** Pollution control technologies have significantly reduced, but not eliminated, their negative impacts on the environment. However unsightly or smelly they may be, oil refineries provide the gasoline, diesel, heating oil, jet fuel, and asphalt products that are necessary to economic activity and everyday life in industrialized countries. © *Walter Siegmund*.

Even the classical Greeks noted how the landscape they occupied differed—for the worse—from its former condition. With growing numbers of people, and particularly with industrialization and the spread of European exploitative technologies throughout the world, the pace of change in the content of area accelerated. The built landscape—the product of human effort—increasingly replaced the natural landscape. Each new settlement or city; each agricultural assault on forests; and each new mine, dam, or factory changed the content of regions and altered the temporarily established spatial interconnections between humans and the environment.

Characteristics of places today are the result of constantly changing past conditions. They are the forerunners of differing human-environmental balances yet to be struck. Geographers are concerned with places at given moments of time. But to understand fully the nature and development of places, to appreciate the significance of their relative locations, and to understand the interplay of their physical and cultural characteristics, geographers must view places as the present result of past operation of physical and cultural processes (**Figure 1.10**).

You will recall that one of the questions geographers ask about a place or thing is "How did it come to be what and where it is?" This is an inquiry about process and about becoming. The forces and events shaping the physical and explaining the cultural environment of places today are an important focus of geography and are the topics of most of the chapters of this book. To understand them is to appreciate the changing nature of the spatial order of our contemporary world.

Interrelations between Places

The concepts of relative location and distance that were introduced earlier lead directly to another fundamental spatial reality: places are interrelated with other places in structured and comprehensible ways. In describing the processes and patterns of that **human interaction,** geographers add *accessibility* and *connectivity* to the ideas of location and distance.

Tobler's First Law of Geography tells us that, in a spatial sense, everything is related to everything else but relationships are stronger when things are near one another. Our observation, therefore, is that interaction between places diminishes in intensity and frequency as distance between them increases—a statement of the idea of "distance decay," which we explore in Chapter 7. Are you more likely to go to a fast-food outlet next door or to a nearly identical restaurant across town? Our decision making sometimes is unpredictable, but in this case you can see that most people would probably choose the nearer place more often.

Consideration of distance implies assessment of **accessibility.** How easy or difficult is it to overcome the "friction of distance"? That is, how easy or difficult is it to overcome the barrier of the time and space separation of places? Distance isolated North America from Europe until the development of ships (and aircraft) that reduced the effective distance between the continents. All parts of ancient and medieval cities were accessible by walking; they were "pedestrian cities," a status lost as cities expanded in area and population with industrialization. Accessibility between city districts could be maintained only by the development of

(a)

(b)

Figure 1.10   **The process of change in a cultural landscape.** **(a)** Dubai, a city-state in the United Arab Emirates, was a modest-sized fishing town in the 1960s. **(b)** Discovery of oil and growth as a tax-free haven and cosmopolitan business center led to fantastic growth. Former places are hardly recognizable today. The city is home to the world's tallest building and the world's largest shopping mall, which features an indoor ski resort. *(a) © Chris Ware/Getty Images (b) © Chris Schmid/Getty Images.*

public transit systems whose fixed lines of travel increased the ease of movement between connected points and reduced it between areas not on the transit lines themselves.

Accessibility, therefore, suggests the idea of **connectivity,** a broader concept implying all the tangible and intangible ways in which places are connected: by physical telephone lines, street and road systems, and pipelines and sewers; by unrestrained walking across open countryside; by radio and TV broadcasts; by cell phone service areas; and in nature even by movements of wind systems and ocean currents. Where routes are fixed and flow is channelized, *networks*—the patterns of routes connecting sets of places—determine the efficiency of movement and the connectedness of points. Demand for universal instantaneous connectivity is common and unquestioned in today's advanced societies. Technologies and devices to achieve it proliferate, as our own lifestyles show. Cell phones, e-mail, broadband wireless Internet, instant messaging, and more have erased time and distance barriers formerly separating and isolating individuals and groups and have reduced our dependence on physical movement and on networks fixed in the landscape.

There is, inevitably, interchange between connected places. **Spatial diffusion** is the process of dispersion of an idea or a thing (a new consumer product or a new song, for example) from a center of origin to more distant points. The rate and extent of that diffusion are affected, again, by the distance separating the origin of the new idea or technology and other places where it is eventually adopted. Diffusion rates are also affected by such factors as population densities, means of communication, advantages of the innovation, and importance or prestige of the originating node. Further discussion of spatial diffusion is found in Chapter 7.

Geographers study the dynamics of spatial relationships. Movement, connection, and interaction are part of the social and economic processes that give character to places and regions (**Figure 1.11**). The increasingly global reach of those spatial interactions is expressed in the term *globalization*. **Globalization** implies the increasing interconnection of more and more peoples and parts of the world as the full range of social, cultural, political, economic, and environmental processes becomes international in scale and effect. Promoted by continuing advances in worldwide accessibility and connectivity, globalization encompasses other core geographic concepts of spatial interaction, accessibility, connectivity, and diffusion. More detailed implications of globalization will be touched on in Chapters 7 and 10.

Place Similarity and Regions

The distinctive characteristics of places—physical, cultural, locational—immediately suggest two geographically important ideas. The first is that no two places on the surface of the Earth can be *exactly* the same. Not only do they have different absolute locations, but—as in the features of the human face—the precise mix of physical and cultural characteristics of place is never exactly duplicated. The inevitable uniqueness of place would seem to impose impossible problems of generalizing spatial information.

That this is not the case results from the second important idea, that the natural and cultural characteristics of places show patterns of similarity in some areas. For example, a geographer doing fieldwork in France may find that all farmers in one area use a similar, specialized technique to build fences around their fields. Often, such similarities are striking enough for us to conclude that spatial regularities exist. They permit us to recognize and define **regions,** Earth areas that display significant elements of internal uniformity and external differences from surrounding territories. Places are, therefore, both unlike and like other places, creating patterns of areal differences and coherent spatial similarity.

The problems of the historian and the geographer are similar. Each must generalize about items of study that are essentially unique. The historian creates arbitrary but meaningful and useful historical periods for reference and study. The "Roaring Twenties" and the "Victorian Era" are shorthand summary names for specific

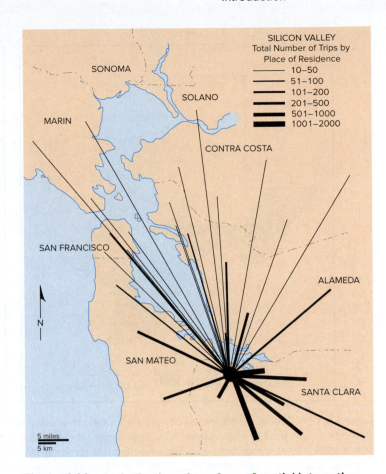

Figure 1.11   An indication of one form of spatial interaction and connectivity is suggested by this "desire line" map recording the volume of daily work trips within the San Francisco Bay area to the Silicon Valley employment node. The ends of the desire lines define the outer reaches of a physical interaction region defined by the network of connecting roads and routes. The region changed in size and shape over time as the network was enlarged and improved, the Silicon Valley employment base expanded, and the commuting range of workers increased. The map, of course, gives no indication of the global reach of the Silicon Valley's *accessibility* and interaction through other means of communication and interchange. *Redrawn with permission from Robert Cervero, Suburban Gridlock. © 1986 Center for Urban Policy Research, Rutgers, the State University of New Jersey.*

time spans, internally quite complex and varied but significantly distinct from what went before or followed after. The region is the geographer's equivalent of the historian's era: a device to classify the complex reality of the Earth's surface into manageable pieces. Just as historians focus on key events to characterize certain historical periods, geographers focus on key unifying elements or similarities to determine the boundaries of regions. By identifying and naming regions, a complex set of interrelated environmental or cultural attributes can easily be conveyed through a simpler construct.

Spatial Distributions

Regions are not "given" in nature any more than "eras" are given in the course of human events. Regions are devised; they are spatial summaries designed to bring order to the infinite diversity of

Data for 1971

Status in 1972

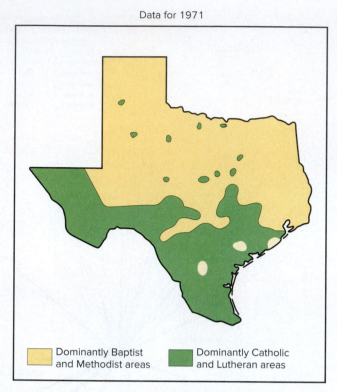
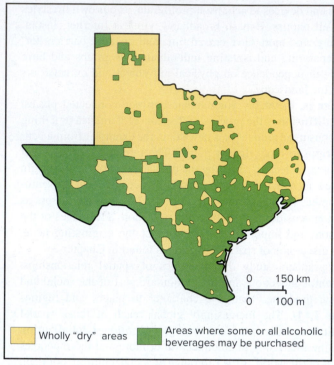

Dominantly Baptist and Methodist areas

Dominantly Catholic and Lutheran areas

Wholly "dry" areas

Areas where some or all alcoholic beverages may be purchased

Figure 1.12   **Spatial distributions of religion and alcohol sales in Texas.** Catholic and Lutheran areas tend to be "wet," and Baptist and Methodist areas tend to retain prohibition. Both Baptist and Methodist churches have traditionally taken a stand against the consumption of alcohol. The two maps suggest a spatial association between religion and alcohol prohibition laws. *Sources: 38th Annual Report of the Texas Alcoholic Beverage Commission, Austin, 1972, p. 49; and Churches and Church Membership in the United States: 1971, National Council of Churches of Christ in the U.S.A., 1974.*

the Earth's surface. At their root, they are based on the recognition and mapping of *spatial distributions*—the spatial arrangement of environmental, human, or organizational features selected for study. For example, the location of Welsh speakers in Great Britain is a distribution that can be identified and mapped. As many spatial distributions exist as there are imaginable physical, cultural, or connectivity elements of area to examine. Those that are selected for study, however, are those that contribute to the understanding of a specific topic or problem.

Let us assume that we are interested in studying burglary rates in the United States. Statistics indicate that some states have significantly higher rates than others. A resident of Arizona is roughly three times as likely to be a victim of a burglary as a resident of North Dakota. We would ask whether the distribution of rates appears random. Mapping the spatial distribution is a first step and shows that the states with the highest rates appear to be clustered along the southern border and the West Coast. Next, we must try to explain the spatial distribution. We would ask what factors account for the observed pattern. We would ask whether the pattern is similar to that for other types of crimes. Since it is commonly assumed that big cities, poverty, and youthful populations are associated with crime, we would want to see whether any of those distributions were correlated with burglary rates.

When two spatial distributions are closely related, they are said to have a *spatial association*. In **Figure 1.12**, we observe that communities in Texas where consuming alcoholic beverages is legal tend to have a majority of Catholic or Lutheran residents, while so-called dry communities are more likely to be predominantly Baptist or Methodist. Geographers attempt to identify spatial associations that are stronger than expected by chance alone.

Types of Regions

Regions may be *administrative, formal, functional,* or *perceptual.* **Administrative regions** are created by laws, treaties, or regulations. Examples include countries, states, counties, cities, and school districts. The political map of the world shows the boundaries of one set of administrative regions. Even the end zone on a football field is an administrative region. Boundaries of administrative regions are as precisely defined as measurement allows, and generally laws and rules are applied uniformly to all places within those boundaries.

A formal (or **uniform**) **region** is an area of essential uniformity for a single physical or cultural feature or a limited combination of physical or cultural features. The name *Corn Belt* suggests a region based on its farm economy and predominant crop. **Figure 1.13** depicts formal regional patterns. The formal region is based on objective, often statistically derived identifiers. Whatever the basis of its definition, the formal region is a sizable area over which a valid generalization of uniformity may be made with respect to some attribute or attributes—that is, the attribute holds true for the entire region.

Figure 1.13   **This generalized land use map of Australia** is composed of *formal regions* whose internal economic characteristics show essential uniformities, setting them off from adjacent territories of different condition or use.

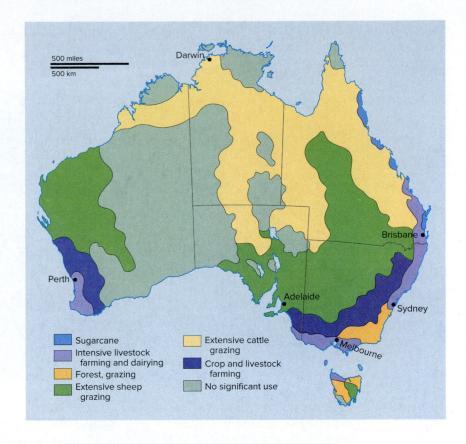

500 miles
500 km

Darwin

Perth

Adelaide

Brisbane

Sydney

Melbourne

Sugarcane

Intensive livestock farming and dairying

Forest, grazing

Extensive sheep grazing

Extensive cattle grazing

Crop and livestock farming

No significant use

A **functional** (or **nodal**) **region,** in contrast, may be visualized as a spatial system. Its parts are interdependent, and throughout its extent the functional region operates as a dynamic, organizational unit. Like the formal region, a functional region is objectively defined; but a functional region has unity in the manner of its operational connectivity, not in the sense of static content. The defining characteristics of interaction and connection of a functional region are most clearly recognized at its node or core and lessen in dominance toward its margins. As the degree and extent of control and interaction in an area change, the boundaries of the functional region change in response; that is, a nodal region's boundaries remain constant only as long as the interchanges that establish it remain unaltered. Examples are the trade areas of towns, the circulation area of a newspaper, the area that receives a television station's signal, and the territories subordinate to the financial, administrative, health care, retail, and service functions of regional capitals, such as Chicago, Atlanta, or Minneapolis (**Figure 1.14**).

Perceptual (or **vernacular** or **popular**) **regions** are less rigorously structured than the formal and functional regions geographers devise. They are regions that exist and have reality in the perceptions of their inhabitants and the general society. As composites of the mental maps of ordinary folk, they reflect feelings and images rather than objective data. Because of that, perceptual regions may be more meaningful in individuals' daily lives than the more objective regions of geographers.

Ordinary people have a clear idea of spatial variation and employ the regional concept to distinguish between territorial entities. People individually and collectively agree on where they live.

The vernacular regions they recognize have reality in their minds and are reflected in regionally based names employed in businesses, by sports teams, or in advertising slogans. The frequency of references to "New England" in the northeastern United States represents that kind of regional consensus and awareness, as does "Midwest" in popular understanding and literary references (**Figure 1.15**). The boundaries of vernacular regions, of course, vary on the mental maps of different groups both within and outside the recognized area. Still, the regions are important for they reflect the way people view space, assign their loyalties, and interpret their world. At a different scale, such urban ethnic enclaves as "Little Italy" and "Chinatown" have comparable regional identity in the minds of their inhabitants. Less clearly perceived by outsiders but unmistakable to their inhabitants are the "turfs" of urban clubs or gangs. Their boundaries are sharp, and the perceived distinctions between them are paramount in the daily lives and activities of their occupants.

As you read the chapters of this book, notice how many examples of regions and regionalism are presented in map form and discussed in the text. Note, too, how those depictions and discussions vary between the three different regional types as the subjects and purposes of the examples change.

1.4 Geography's Themes and Standards

The core geographic concepts discussed so far in this chapter reflect both the "fundamental themes in geography" and the

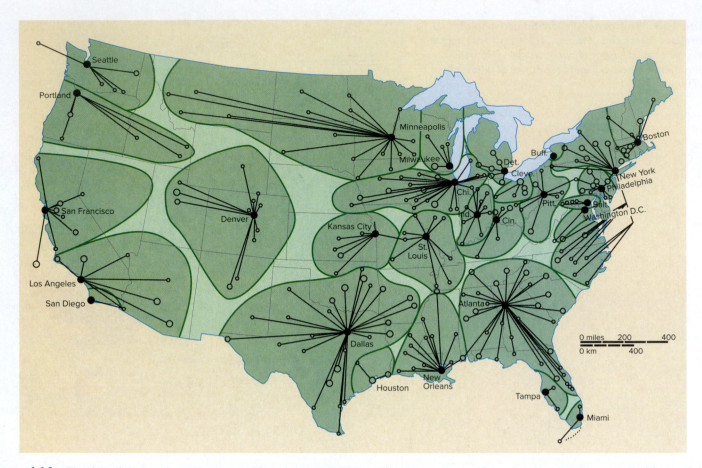

Figure 1.14   **The functional regions shown on this map were based on linkages between large banks of major central cities and the "correspondent" banks they formerly served in smaller towns.** Although the rise of nationwide banks has reduced their role, the regions once defined an important form of *connectivity* between principal cities and locales beyond their own immediate metropolitan area. *Source: Redrawn by permission from* Annals of the Association of American Geographers, *John R. Borchert, Vol. 62, p. 358, Association of American Geographers, 1972.*

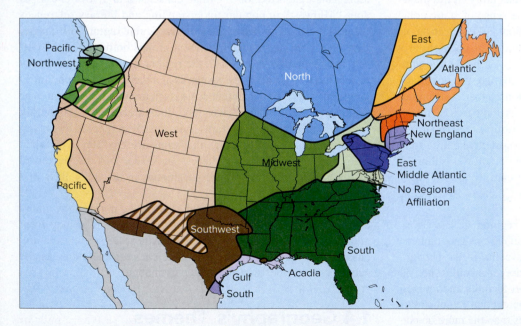

Figure 1.15   **Perceptual (vernacular or popular) regions of North America.** *Source: Wilbur Zelinsky, "North America's Vernacular Regions" in* Annals of the Association of American Geographers, *Vol. 70, Figure I, p. 14, 1980. Redrawn with permission.*

"National Geography Standards." Together, the "themes" and "standards" have helped organize and structure the study of geography over the past several years at all grade and college levels. Both focus on the development of geographic literacy. The former

represent an instructional approach keyed to identification and instruction in the knowledge, skills, and perspectives students should gain from a structured program in geographic education. The latter—"standards"—codify the essential subject matter,

THE NATIONAL STANDARDS

The inclusion of geography as a subject area to be assessed reflects the conviction that a grasp of the skills and understandings of geography is essential in an American educational system "tailored to the needs of productive and responsible citizenship in the global economy." Along with the "basic observations" reviewed in the text, the National Geography Standards 1994 help frame the kinds of understanding we will seek in the following pages and suggest the purpose and benefit of further study of geography.

The 18 geography standards tell us the following: *The geographically informed person knows and understands:*

Geographic Techniques and Skills

1. How to use maps and other geographic tools and technologies to acquire, process, and report information from a spatial perspective.
2. How to use mental maps to organize information about people, places, and environments in a spatial context.
3. How to analyze the spatial organization of people, places, and environments on Earth's surface.

Physical Systems

4. The physical processes that shape the patterns of Earth's surface.
5. The characteristics and spatial distribution of ecosystems on Earth's surface.

Human Systems

6. The characteristics, distribution, and migration of human populations on Earth's surface.

7. The characteristics, distribution, and complexity of Earth's cultural mosaics.
8. The patterns and networks of economic interdependence on Earth's surface.
9. The processes, patterns, and functions of human settlement.
10. How the forces of cooperation and conflict among people influence the division and control of Earth's surface.

Environment and Society

11. How human actions modify the physical environment.
12. How physical systems affect human systems.
13. The changes that occur in the meaning, use, distribution, and importance of resources.

The Uses of Geography

14. How to apply geography to interpret the past.
15. How to apply geography to interpret the present and plan for the future.

Places and Regions

16. The physical and human characteristics of places.
17. That people create regions to interpret Earth's complexity.
18. How culture and experience influence people's perceptions of places and regions.

Source: *Geography for Life: National Geography Standards 1994.* Washington, D.C.: National Geographic Research and Exploration, 1994.

---

skills, and perspectives of geography essential to the mental equipment of all educated adults.

The *five fundamental themes,* as summarized by a joint committee of the National Council for Geographic Education and the Association of American Geographers, are those basic concepts and topics that recur in all geographic inquiry and at all levels of instruction:

- location: the meaning of relative and absolute position on the Earth's surface;
- place: the distinctive and distinguishing physical and human characteristics of locales;
- relationships within places: the development and consequences of human-environmental interactions;
- movement: patterns and change in human spatial interaction on the Earth;
- regions: how they form and change.

The National Geography Standards were established as part of the nationally adopted *Goals 2000: Educate America Act* (see "The National Standards"). Designed specifically as guidelines to the essential geographic literacy to be acquired by students who have gone through the U.S. public school system, the standards address the same conviction underlying this edition of *Introduction to Geography*—that being literate in geography is a necessary part of the mental framework of all informed persons.

1.5 Organization of This Book

Despite its outward appearance of diversity, geography has a broad consistency of purpose achieved by having a limited number of distinct but closely related areas of study. The unifying categories within which geographers work and the plan of this book can be seen in **Figure 1.16**. Not included in this book, however, is an emphasis on specific regions, such as Europe, the Mississippi Valley, the developing world, etc. These are discussed in individual courses where students can gain a deep understanding and knowledge of a particular area.

In this book, we begin with a discussion of fundamental concepts and **techniques of geographic analysis** (Chapters 1 and 2). The emphasis is on the nature of maps and the skills necessary to create them. Basic cartography (mapmaking) is discussed as well as some of the exciting new technologies, such as geographic information systems (GIS) and global positioning systems (GPS), which allow for great detail, flexibility in map use, and creativity in "seeing" the Earth as it has never been seen before. In our introduction to spatial analysis, we consider ways in which maps can be used to understand patterns of change, such as population growth, the spread of a disease, or the diffusion of a commercial enterprise, and so on.

To understand the **physical and human characteristics of the** Earth, we begin, in Chapter 3, Physical Geography: Landforms,

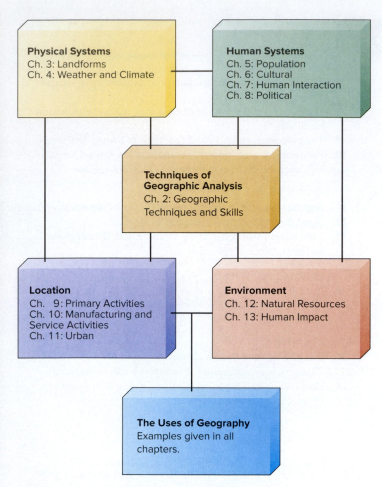

Figure 1.16   The areas of study of geography do not stand alone. Rather, each is interconnected with the others, and all together depend on unifying research techniques and tools. As the diagram indicates, it is possible to group chapters by broad areas of interest: Physical Systems, Human Systems, Location, and Environment, all held together by geographic techniques and skills.

to explore **physical systems.** The approach of geography is to attempt to recognize the processes responsible for surface physical features such as mountains and streams. In Chapter 4, weather and climate are investigated. This requires that we identify the principles of the science of the lower atmosphere where Earth-sun relations are the backbone of any understanding of changes in weather and climate. Also included is a listing of climates based on a geographic classification system that simplifies understanding of the great variation of Earth's weather and climate.

Human systems on Earth are many and complex. In its development, the discipline of geography has found efficient ways to compartmentalize these complexities. Each of the discussions beginning with Chapter 5 and continuing to Chapter 7 represents a singling out of a particular human system. Chapter 5 is a study of population geography. Although there is much in common with the field of demography, the emphasis in population geography is on the location, distribution, and spatial trends in the numbers of people over time. In Chapter 6, we discuss the fundamentals of cultural geography, including the technological, sociological, and

ideological systems of human behavior. Differences in language, religion, and ethnicity play major roles in the human use of the Earth. A section on gender and culture helps to foster a deep appreciation for geographic variations in roles humans play. Chapter 7, on human interaction, emphasizes a conceptual view of human activities to help understand the geographic controls on communication and commerce. A major section concerns the role that migration plays in affecting human systems. Chapter 8 contains a comprehensive look at the geographic characteristics of political systems. Nations, states, cooperative behavior, and local and regional political organizations are understood in terms of culture and past history. In geography, emphasis is on the size of political entities, their locations, and the cohesiveness of their boundaries.

Chapters representing the study of **location** follow in Chapters 9 through 11. These chapters emphasize the location of human activities, the systems that make them function, and the settlements that evolve from their differing levels of human activity. Primary activities relate to the use of the physical environment, such as mining and agriculture. These are discussed in Chapter 9. A major point made is that none of these activities occurred by chance and that the nature of the land in places and the demand for the output of that land have much to do with the kinds of activities in which humans engage.

The output of primary activities gives rise to manufacturing (secondary) and commercial (tertiary) activities. These are explored in Chapter 10. The point of view of geography is to attempt to understand the location of these activities. Principles of location are investigated, and the resulting interaction and commercial trade are discussed. The outputs and demand for primary, secondary, and tertiary activity give rise to settlements of varying size and complexity. These complexities represent the field of urban geography which is studied in Chapter 11. The city and its hinterland are examined; the internal spatial arrangement of cities in different cultures and societies in which they appear is dissected. Principles of city systems and the increasingly intertwined world city system are discussed in the chapter.

The final two chapters explore the interaction of the physical and human **environment** of the Earth and how they are used and abused by humans. Chapter 12 on the geography of natural resources explains the meaning of renewable and nonrenewable resources. Emphasis is on the critical resources of modern times—energy and land. Our hope is that if humans can understand the issues related to the exploitation of resources, they can better cope with their dwindling supplies. Thus, relatively new energy resources such as solar, wind, and natural gas power are discussed as alternatives to coal and oil resources. In Chapter 13, the point is made that humans play a crucial role in altering the environment. Many of the important issues of the day are discussed with regard to negative human impacts. In turn, the human impact on water, air, climate, landforms, plants, and animals are discussed, and possible solutions are introduced.

Throughout the book, current issues are introduced and public policy alternatives explored. An informed citizenry can come to intelligent, rational decisions about the future use of the Earth only if it is well-informed and understands available options.

Key Words

absolute direction 8
absolute distance 8
absolute location 7
accessibility 12
administrative region 14
connectivity 12
cultural landscape 10
environment 18
formal (uniform) region 14
functional (nodal) region 15
globalization 13
human interaction 12
human systems 18
location 18

natural landscape 10
perceptual (vernacular, popular)
    region 15
physical systems 18
region 13
relative direction 8
relative distance 8
relative location 8
scale 9
site 8
situation 8
spatial diffusion 13
techniques of geographic
    analysis 17

Thinking Geographically

1. In what two meanings and for what different purposes do geographers refer to *location?* When geographers say "location matters," what aspect of location commands their interest?

2. What does the term *cultural landscape* imply? Is the nature of the cultural landscape dictated by the physical environment?

3. What kinds of distance transformations are suggested by the term *relative distance?* How is the concept of *psychological distance* related to relative distance?

4. How are the ideas of *distance, accessibility,* and *connectivity* related to processes of *human interaction?*

5. Why do geographers concern themselves with *regions?* How are *formal* and *functional* regions different in concept and definition?

6. What are the National Standards for geography? How does the geographic knowledge encompassed by the National Standards help to understand a specific local, national, or international problem?

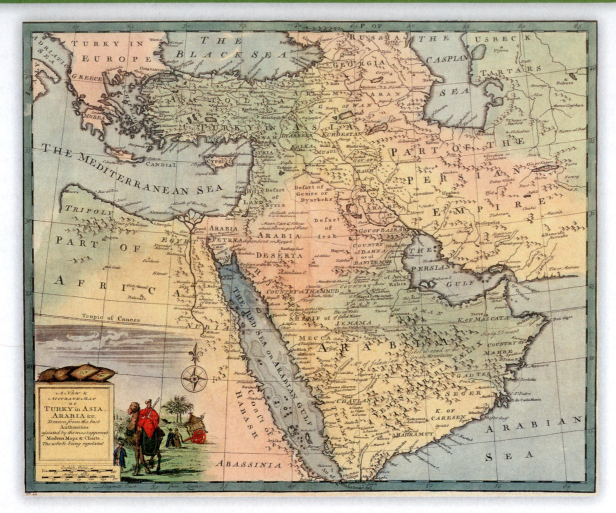

A map of Asia Minor from the mid-1700s. © *Corbis RF.*

CHAPTER OUTLINE

On January 8, 2005, the nuclear attack submarine *San Francisco* sped along at top speed some 150 meters (500 ft) beneath the surface of the South Pacific on its way from Guam to Brisbane, Australia. Many of the 136 crew members were eating lunch when they heard a horrible screeching followed by a thunderous blast. Within seconds, sailors were tossed about like mannequins. The *San Francisco* had crashed head-on into an undersea mountain that is part of a range of undersea volcanoes and reefs. One crewman was killed and 98 were injured, many of them severely. Although the mountain rises to within 30 meters (100 ft) of the ocean surface, it was not on the submarine's navigational charts, which did not show any potential obstacles within 4.7 kilometers (3 mi) of the crash.

Three years earlier, Americans had been riveted by the plight of nine coal miners trapped 73 meters (240 ft) below ground in the flooded Quecreek Mine in southwestern Pennsylvania. On July 24, 2002, a machine operator in the mine broke through to an adjacent, flooded mine that had been abandoned 38 years earlier. As millions of gallons of water rushed in, the nine men looked frantically for a way out, but all the exits were in areas already filled with water.

According to the report issued by investigators from the Department of Labor, at one point the miners

estimated they had about an hour left based on the rate the water was rising. The miners took some time to reflect on their situation and prepared for the worst. Some of the miners tied themselves together so they could be found together in the event they were drowned. They wrote notes to their families and placed them in a plastic bucket. The bucket was closed with a lid, sealed with electrical tape and secured near the roof bolting machine to prevent it from floating away.

Above ground, rescuers drilled holes to pump out the water and sank a shaft 0.8 meter (30 in.) in diameter to reach the miners. After 78 hours underground, all nine were lifted to safety.

This was an accident that should not have happened. The 1957 map the mining companies were using showed the old mine as 138 meters (150 yds) away from Quecreek. But another 421,000 tons of coal had been removed before the mine was closed in 1964, and that additional mining had put one shaft immediately adjacent to Quecreek.

As these examples indicate, accurate maps can literally mean the difference between life and death. Governmental agencies rely on maps of flood-prone areas, of volcanic eruptions, of earthquake hazard zones, and of areas subject to landslides to develop their long-range plans. Epidemiologists map the occurrence of a disease over time and space, which helps them identify the source of the outbreak and create a plan to halt the spread of the disease. Law enforcement agencies use maps to identify patterns of specific types of crime and to help them predict where those crimes are likely to occur in the future. The value of examining information in a spatial context cannot be overstated.

2.1 Maps as the Tools of Geography

"The role of geography is a platform for understanding the world. Geographic Information Systems (GIS) is making geography come alive. It condenses our data, information, and science in a language that we can easily understand: maps." So said Jack Dangermond, President of ESRI, the largest company in the world to provide GIS software (*ArcNews*, Fall 2012). Maps have a special significance for geographers. They are geographers' primary tools of spatial analysis. For a variety of reasons, the spatial distributions, patterns, and relations of interest to geographers usually cannot easily be observed or interpreted in the landscape itself.

- Many phenomena, such as landform or agricultural regions or major cities, are so extensive spatially that they cannot be seen or studied in their totality from one or a few vantage points.
- Many distributions, such as those of language usage or religious belief, are spatial phenomena but are not tangible or visible.
- Many interactions, flows, and exchanges imparting the dynamic quality to spatial interaction may not be directly observable at all.

Even if all things of geographic interest could be seen and measured through field examination, the infinite variety of tangible and intangible content of an area would make it nearly impossible to isolate for study and interpretation the few topics selected for special investigation.

Therefore, the map has become the essential and distinctive tool of geographers. Only through the map can spatial distributions and interactions of whatever nature be reduced to an observable scale, isolated for individual study, and combined or recombined to reveal relationships not directly measurable in the landscape itself.

The art, science, and technology of making maps are called **cartography.** Modern scientific mapping has its roots in the 17th century, although the Earth scientists of ancient Greece are justly famous for their contributions. They recognized the spherical form of the Earth and developed map projections and the grid system. Unfortunately, much of the cartographic tradition of Greece was lost to Europe during the Middle Ages and essentially had to be rediscovered. Several developments during the Renaissance gave an impetus to accurate cartography. Among these were the development of printing, the rediscovery of the work of Ptolemy and other Greeks, and the great voyages of discovery.

In addition, the rise of nationalism in many European countries made it imperative to determine and accurately portray boundaries and coastlines, as well as to depict the kinds of landforms contained within the borders of a country. During the

17th century, important national surveys were undertaken in France and England. Many conventions in the way data are presented on maps had their origin in these surveys.

Knowledge of the way information is recorded on maps enables us to read and interpret them correctly. To be on guard against drawing inaccurate conclusions or to avoid being swayed by distorted or biased presentations, we must be able to understand and assess the ways in which facts are represented. Of course, all maps are necessarily distorted because of the need to portray the round Earth on a flat surface, to use symbols to represent objects, to generalize, and to record features at a different size than they actually are. This distortion of reality is necessary because the map is smaller than the things it depicts and because its effective communication depends upon selective emphasis of only a portion of reality. As long as map readers know the limitations of the commonly used types of maps and understand what relationships are distorted, they can interpret maps correctly.

2.2 Locating Points on a Sphere

As we noted in Chapter 1, the starting point of all geographic study is the location of places and things, and absolute location is the identification of place by a precise and accepted system of coordinates.

The Geographic Grid

In order to visualize the basic system for locating points on the Earth, think of the world as a sphere with no markings whatsoever on it. There would, of course, be no way of describing the exact location of a particular point on the sphere without establishing a system of reference. We use the **geographic grid,** a set of imaginary lines that intersect at right angles to form a system of reference for locating points on the surface of the Earth. The key reference points in that system are the North and South Poles, the equator, and the prime meridian.

The North and South Poles are the end points of the axis about which the Earth spins. The line that encircles the globe halfway between the poles, perpendicular to the axis, is the *equator*. We can describe the location of a point in terms of its distance north or south of the equator, measured as an angle at the earth's center. Because a circle contains 360 degrees, the distance between the two poles is 180 degrees and between the equator and each pole, 90 degrees. **Latitude** is the angular distance north or south of the equator, measured in degrees ranging from 0° (the equator) to 90° (the North and South Poles). As is evident in **Figure 2.1a**, the parallels of latitude are parallel to one another and to the equator and run east-west.

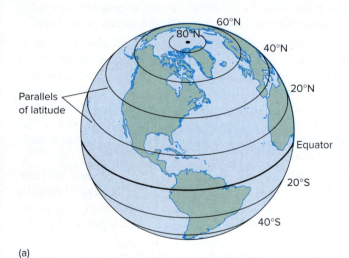

(a)

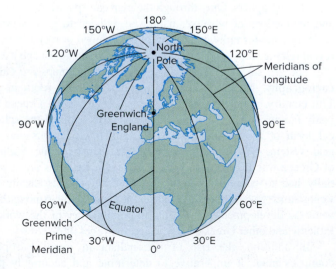

(b)

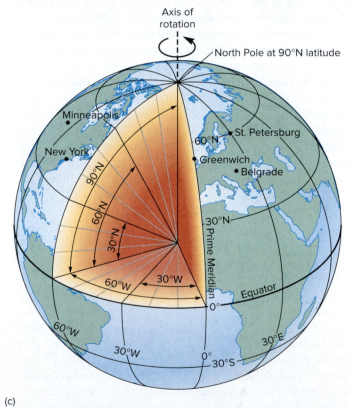

(c)

Figure 2.1 **(a) The grid system: parallels of latitude.** Note that the parallels become increasingly shorter closer to the poles. On the globe, the 60th parallel is only one-half as long as the equator. **(b) The grid system: meridians of longitude.** East-west measurements range from 0° to 180°—that is, from the prime meridian to the 180th meridian in each direction. Because the meridians converge at the poles, the distance between degrees of longitude becomes shorter as one moves away from the equator. **(c) The Earth grid, or graticule, consisting of parallels of latitude and meridians of longitude.**

The polar circumference of the Earth is 40,071 kilometers; thus, the distance between degrees of latitude equals 40,071 4 360, or about 111 kilometers (69 mi). If the Earth were a perfect sphere, all degrees of latitude would be equally long. Due to the slight flattening of the Earth in polar regions, however, degrees of latitude are slightly longer near the poles (111.70 km; 69.41 mi) than near the equator (110.56 km; 68.70 mi).

To record the latitude of a place in a more precise way, degrees are divided into 60 *minutes* ('), and each minute into 60 *seconds* ("), exactly as in an hour of time. One minute of latitude is about 1.85 kilometers (1.15 mi), and one second of latitude is about 31 meters (101 ft). The latitude of the center of Chicago is written 41°52'50"N.

Because the distance north or south of the equator is not by itself enough to locate a point in space, we need to specify a second coordinate to indicate distance east or west from an agreed-upon reference line. As a starting point for east-west measurement, cartographers in most countries use as the **prime meridian,** which is an imaginary line passing through the Royal Observatory at Greenwich, England. This prime meridian was selected as the zero-degree longitude by an international conference in 1884. Like all *meridians,* it is a true north-south line connecting the poles of the Earth (**Figure 2.1b**). ("True" north and south vary from magnetic north and south, the direction of the earth's magnetic poles, to which a compass needle points.) Meridians are farthest apart at the equator, come closer and closer together as latitude increases, and

converge at the North and South Poles. Unlike *parallels* of latitude, all meridians are the same length.

Longitude is the angular distance east or west of the prime (zero) meridian measured in degrees ranging from 0° to 180°. Directly opposite the prime meridian is the 180th meridian, located in the Pacific Ocean. Like parallels of latitude, degrees of longitude can be subdivided into minutes and seconds. However, the distance between the adjacent degrees of longitude decreases away from the equator because the meridians converge at the poles. With the exception of a few Alaskan islands, all places in North and South America are in the area of west longitude; with the exception of a portion of the Chukchi Peninsula of Siberia, all places in Asia and Australia have east longitude.

Time zones and longitude are related. The Earth, which makes a complete 360-degree rotation once every 24 hours, is divided into 24 time zones roughly centered on meridians at 15-degree intervals. *Greenwich mean time (GMT)* is the time at the prime meridian. The **International Date Line,** where each new day begins, generally follows the 180th meridian. As **Figure 2.2** indicates, however, the date line deviates from the meridian in some places in order to avoid having two different dates within a country or an island group. Thus, the International Date Line zigzags so that Siberia has the same date as the rest of Russia and the Aleutian Island and Fiji Island groups are not split. New days begin at the date line and proceed westward, so that west of the line is always 1 day later than east of the line.

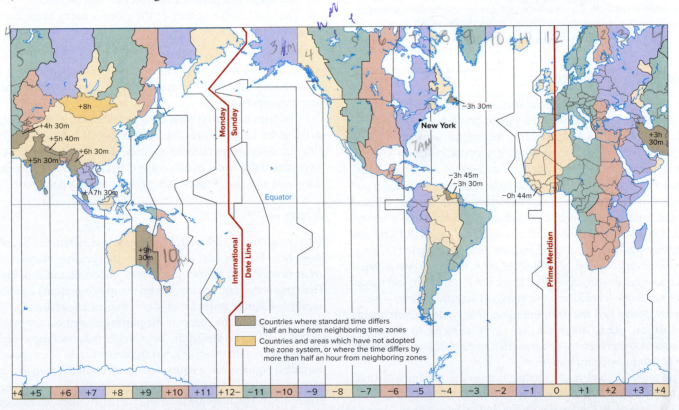

Figure 2.2 **World time zones.** Each time zone is about 15 degrees wide, but variations occur to accommodate political boundaries. The figures at the bottom of the map represent the time difference in hours when it is 12 noon in the time zone centered on Greenwich, England. New York is in column −5, so the time there is 7 A.M. when it is noon at Greenwich. Modifications to the universal system of time zones are numerous. Thus, Iceland operates on the same time as Britain, although it is a time zone away. Spain, entirely within the boundaries of the GMT zone, sets its clocks at +1 hour, whereas Portugal conforms to GMT. China straddles five time zones, but the whole country operates on Beijing time (+8 hours). In South America, Chile (in the −5 hour zone) uses the 24 hour designation, whereas Argentina uses the −3 hour zone instead of the −4 hour zone to which it is better suited.

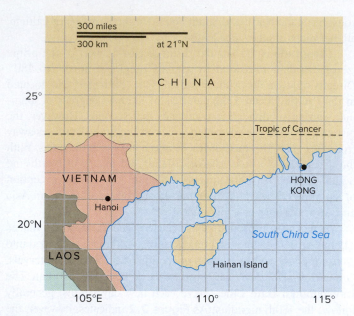

Figure 2.3   The latitude and longitude of Hong Kong are 22°17′N, 114°10′E. What are the coordinates of Hanoi?

By citing the degrees, minutes, and, if necessary, seconds of longitude and latitude, we can describe the location of any place on the earth's surface. To conclude our earlier example, the center of Chicago is located at 41°52′50″N, 87°38′28″W. Hong Kong is at 22°17′40″N, 114°10′26″E (**Figure 2.3**).

Land Survey Systems

When independence from Great Britain was achieved, the federal government decided that the public domain should be surveyed and subdivided before being opened for settlement. The Land Ordinance of 1785 established a systematic survey known as the *township and range* system. It was based on survey lines oriented in the cardinal directions: *base lines* that run east-west and *meridians* that run north-south (**Figure 2.4**). A grid of lines spaced at 6-mile (9.7-km) intervals divided the land into a series of squares. A *township* consisted of a square 6 miles (9.7 km) on a side; this was further divided into 36 *sections* 1 mile (1.6 km) on a side. Every section of 640 acres (259 hectares) was subdivided into quarter-sections of 160 acres (64.8 hectares), and these quarter-sections—considered the standard size for a farm—were originally designated the minimum area that could be purchased for settlement. That minimum was later reduced to 80 acres (32.4 hectares) and then to 40 acres (16.2 hectares). Each parcel of land had a unique identification.

The township and range rectangular survey system was first used in eastern Ohio and later extended across most of the United States, as far west as the Pacific Ocean and as far north as Alaska. The Canada Land Survey System is similar to that developed in the United States, employing base lines and meridians and dividing land into townships, ranges, sections, and subdivisions of sections. The rectangular survey system significantly affected the landscape of the central and western United States and Canada, creating the basic checkerboard pattern of minor civil divisions, the regular pattern of section-line and quarter-line country roads, the block patterns of fields and farms, and the gridiron street systems of towns and cities.

2.3 Map Projections

The Earth can be represented with reasonable accuracy only on a globe, but globes are not as convenient as flat maps to store or use, and they cannot depict much detail. For example, if we had a large globe with a diameter of 1 meter, we would have to fit the details of over 100,000 square kilometers of Earth's surface in an area a few centimeters on a side. Obviously, a globe of reasonable size cannot show the transportation system of a city or the location of very small towns and villages.

In transforming a globe into a map, we cannot flatten the curved surface and keep intact all the properties of the original. The **globe properties** are as follows:

1. All meridians are of equal length; each is one-half the length of the equator.
2. All meridians meet at the North and South Poles and are true north-south lines.
3. All parallels of latitude are parallel to the equator and to one another.
4. Parallels decrease in length with distance from the equator.
5. Meridians and parallels intersect at right angles.
6. The scale on the surface of the globe is everywhere the same in all directions.

Only the globe grid itself retains all of these characteristics. To project it onto a surface that can be laid flat is to distort some or all of these properties and consequently to distort the reality the map attempts to portray.

The term **map projection** designates the way the curved surface of the globe is represented on a flat map. All flat maps distort, in different ways and to different degrees, some or all of the four main properties of the actual Earth's surface relationships: area, shape, distance, and direction. **Figure 2.5** shows how distortion occurs.

Area

Some projections, such as the Mollweide and cylindrical equal-area projection (Figures 2.5a, b), enable the cartographer to represent the *areas* of regions in correct or constant proportion to Earth's reality. That means that any square inch on the map represents an identical number of square miles (or of similar units) anywhere else on the map. As a result, the shape of the portrayed area is inevitably distorted. A square on the Earth, for example, may become a rectangle on the map, but that rectangle has the correct area. Such projections are called **equal-area** or **equivalent projections.** *A map that shows correct areal relationships always distorts the shapes of regions.* Equal-area projections are used when a map is intended to show the actual areal extent of a phenomenon on the earth's surface.

Shape

Although no projection can provide correct shapes for large areas, some accurately portray the shapes of small areas by preserving correct angular relationships (Figure 2.5c). These true-shape projections

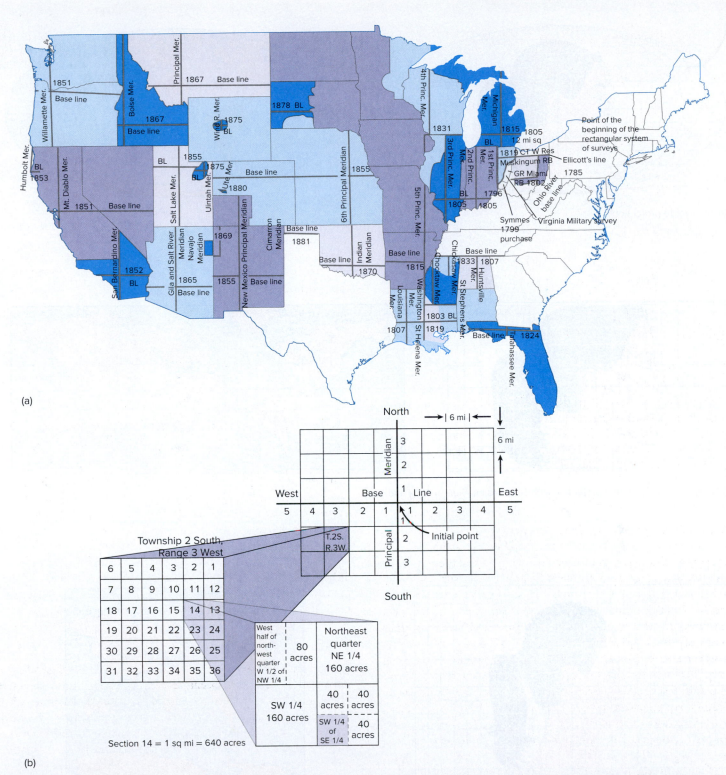

Figure 2.4 **(a) Principal base lines and meridians governing the U.S. Public Land Survey (USPLS). (b) Township, section, and further divisions of the USPLS.** The township and range survey system gives each parcel of land a unique identification. *Townships* are numbered by rows (called tiers) and columns (called ranges). In the example shown here, the township in the second tier south of the base line and in the third range west of the principal meridian is labeled T.2S, R.3W. Every township is divided into sections 1 mile (1.6 km) on a side and numbered from 1 to 36, beginning at the northeast corner of the township. *Sections* can be divided into quarters, eighths ("half-quarters"), and sixteenths ("quarter-quarters"). The Land Office code for the shaded area in the lower right diagram would be SW 1/4 of the SE 1/4 of Sec. 14, T.2S, R.3W. *(a) Source: From U.S. Department of the Interior, Bureau of Land Management, Surveying Our Public Lands. Washington, D.C.: U.S. Government Printing Office, 1980.*

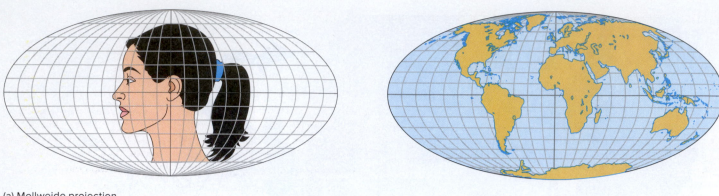

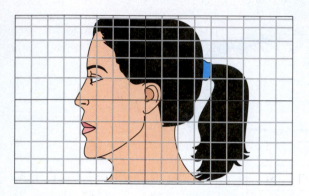
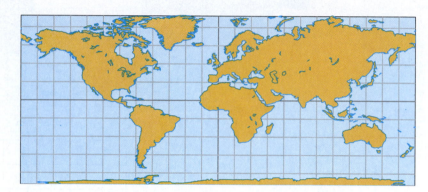

(a) Mollweide projection

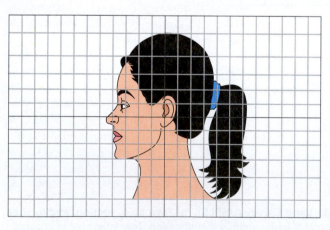
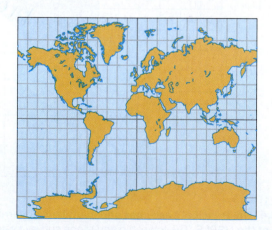

(b) A cylindrical equal-area projection with standard parallels at 30°N and S

(c) Mercator projection

Figure 2.5   **This figure illustrates the distortion inherent in three different map projections.** A head drawn on one projection (Mollweide) has been transferred to two other projections, keeping the latitude and longitude the same as they are found on the first. This does *not* mean the first projection is the best of the three. The head could have been drawn on any one of them and then plotted on the others. *Source: Arthur Robinson et al.,* Elements of Cartography, *5th ed., Fig. 5.6, p. 85. New York, Wiley, © 1984.*

are called **conformal projections,** and the importance of *conformality* is that regions and features "look right" and have the correct directional relationships. They achieve these properties for small areas by ensuring that parallels of latitude and meridians of longitude cross one another at right angles and that the scale is the same in all directions at any given location. Both these conditions exist on the globe but can be retained for only relatively small areas on maps. Because that is so, the shapes of large regions—continents, for example—are always different from their true Earth shapes, even on conformal maps. *A map cannot be both equivalent and conformal.*

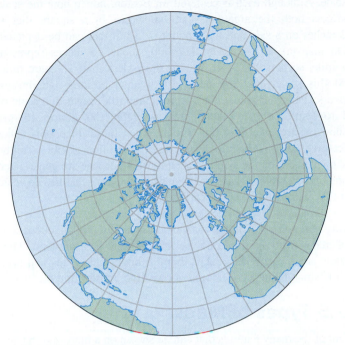

(a) Azimuthal equidistant projection, polar case

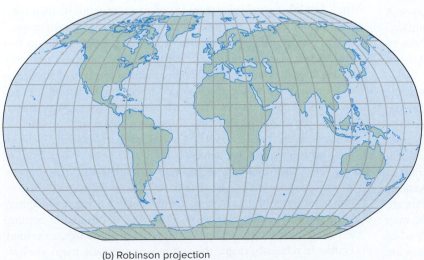

(b) Robinson projection

Figure 2.6   **(a) On this equidistant projection,** distances and directions to all places are true only from the center (North Pole). No flat map can be both equidistant and equal-area. **(b) The Robinson projection,** a compromise between an equal-area and a conformal projection, gives a fairly realistic view of the world. The most pronounced shape distortions are in the less-populated areas of the higher latitudes, such as northern Canada, Greenland, and Russia. On the map, Canada is 21% larger than in reality, while the 48 contiguous states of the United States are 3% smaller than they really are.

Distance

Distance relationships are nearly always distorted on a map, but some projections maintain true distances in one direction or along certain selected lines. Others, called **equidistant projections,** show true distance in all directions, but only from one or two central points (**Figure 2.6a**). Distances between all other locations are incorrect and, quite likely, greatly distorted. An equidistant map centered on Detroit, for example, shows the correct distance between Detroit and the cities of Boston, Los Angeles, and any other point on the map. But it does *not* show the correct distance between Los Angeles and Boston. *A map cannot be both equidistant and equal-area.*

Direction

As is true of distances, directions between all points cannot be shown without distortion. On **azimuthal projections,** however, true directions are shown from one central point to all other points. (An *azimuth* is the angle formed at the beginning point of a straight line, in relation to a meridian.) Directions or azimuths from points other than the central point to other points are not accurate. The azimuthal property of a projection is not exclusive—that is, an azimuthal projection may also be equivalent, conformal, or equidistant. The equidistant map shown in **Figure 2.6a** is a true-direction map from the same North Pole origin.

Not all maps are equal-area, conformal, or equidistant; most are compromises. One example of such a compromise is the *Robinson projection,* which was designed to show the whole world in a visually satisfactory manner and which is used for most of the world maps in this textbook (**Figure 2.6b**). It does not show true distances or directions and is neither equal-area nor conformal. Instead, it permits some exaggeration of size in the high latitudes in order to improve the shapes of land-masses. Size and shape are most accurate in the temperate and tropical zones, where most people live.

Mapmakers must be conscious of the properties of the projections they use, selecting the one that best suits their purposes. If a map shows only a small area, the choice of a projection is not critical—virtually any can be used. The choice is more important when the area to be shown extends over a considerable longitude and latitude; then the selection of a projection depends on the purpose of the map. Some projections are useful for navigation. If numerical data are being mapped, the relative sizes of the areas involved should be correct, so that one of the many equal-area projections is likely to be used. Display maps usually employ conformal projections. Most atlases indicate which projection has been used for each map, thus informing the map reader of the properties of the maps and their distortions. More information about map projections can be found in Appendix 1.

Selection of the map grid, determined by the projection, is the first task of the mapmaker. A second decision involves the scale at which the map is to be drawn.

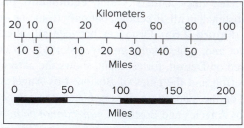

"1 inch to 1 mile"
"1 centimeter to 5 kilometers"

(a) **Verbal scale**

(b) **Graphic scale**

$$\frac{1}{62,500}$$ 1:62,500

(c) **Representative fraction scale**

Figure 2.7   Map scales relate a map distance to a distance on the earth's surface. **(a) A verbal scale** is given in words. **(b) A graphic scale** divides a line into units, each unit representing the distance between two points on the earth's surface. The graphic scale is the only kind of scale to remain correct if the map is reproduced as a different size, *provided that* the scale is enlarged or reduced by the same percentage. **(c) A representative fraction** *(RF) scale* is a simple fraction or ratio. The units of distance on both sides of the scale must be the same; they need not be stated.

2.4 Scale

The **scale** of a map is the ratio between the measurement of something on the map and the corresponding measurement on the Earth. Scale is typically represented in one of the three ways: verbally, graphically, or numerically as a representative fraction (**Figure 2.7**). As the name implies, a *verbal* scale is given in words, such as "1 inch to 1 mile" or "10 centimeters to 1 kilometer." A *graphic* scale, sometimes called a *bar* scale, is a line or bar placed on the map that has been subdivided to show the map lengths of units of the Earth's distance.

A *representative fraction (RF)* scale gives two numbers, the first representing the map distance and the second indicating the ground distance. The fraction may be written in a number of ways. There are 5280 feet in 1 mile and 12 inches in 1 foot; 5280 times 12 equals 63,360, the number of inches in 1 mile. The fractional scale of a map at 1 inch to 1 mile can be written as 1:63,360 or 1/63,360. On the simpler metric scale, 1 centimeter to 1 kilometer is 1:100,000. The units used in each part of the fractional scale are the same; thus, 1:63,360 could also mean that 1 foot on the map represents 63,360 feet on the ground, or 12 miles—which is, of course, the same as 1 inch represents 1 mile. Numerical scales are the most accurate of all scale statements and can be understood in any language.

The map scale, or ratio between the map dimensions and those of reality, can range from very large to very small. A *large-scale map,* such as a plan of a city, shows an area in considerable detail. That is, the ratio of map to ground distance is relatively large—for example, 1:600 (1 in. on the map represents 600 in., or 50 ft, on the ground) or 1:24,000. At this scale, features

such as buildings and highways can be drawn to scale. Figure 2.9 on page 29 is an example of a large-scale map. *Small-scale maps,* such as those of countries or continents, have a much smaller ratio. Buildings, roads, and other small features cannot be drawn to scale and must be magnified and represented by symbols to be seen. Figures 2.2 and 2.3 are small-scale maps. Although no rigid numerical limits differentiate large-scale from small-scale maps, most cartographers would consider large-scale maps to have a ratio of 1:50,000 or less, and maps with ratios of 1:500,000 or more to be small-scale.

Each of the four maps in **Figure 2.8** is drawn at a different scale. Although each is centered on Boston, notice how the scale affects both the area that can be shown in a square that is 2 inches on a side and the amount of detail that can be depicted. On map (a), at a scale of 1:25,000, about 2.6 inches represent 1 mile, so that the 2-inch square shows less than 1 square mile. At this scale, one can identify individual buildings, highways, rivers, and other landscape features. Map (d), drawn to a scale of 1 to 1 million (1:1,000,000, or 1 in. represents almost 16 mi), shows an area of almost 1000 square miles. In this map, only major features, such as main highways and the location of cities, can be shown, and even the symbols used for that purpose are generalized and occupy more space on the map than would the features depicted if they were drawn true to scale.

Small-scale maps such as (c) and (d) in Figure 2.8 are said to be very *generalized.* They give a general idea of the relative locations of major features but do not permit accurate measurement. They show significantly less detail than do large-scale maps and typically smooth out such features as coastlines, rivers, and highways.

2.5 Types of Maps

Out of the many features that can be shown on a map, geographers must first select those that are relevant to the problem at hand and then decide how to display them in order to communicate their message. In that effort, they can choose from different types of maps.

General-purpose, reference, or *location* maps make up one major class of maps familiar to everyone. Their purpose is simply to display one or more natural and/or cultural features of an area or of the world as a whole. Common examples of the natural features shown on maps are water features (coastlines, rivers, lakes, and so on) and the shape and elevation of terrain. Cultural features include transportation routes, populated areas, property-ownership lines, political boundaries, and names.

The other major type of map is called the *thematic,* or *special-purpose,* map, one that shows a specific spatial distribution or category of data. Again, the phenomena being mapped may be physical (climate, vegetation, soils, and so on) and/or cultural (e.g., the distribution of population, religions, diseases, or crime). Unlike in reference maps, the features on thematic maps are limited to just those that communicate the specific spatial distribution.

Topographic Maps and Terrain Representation

As we noted, some general-purpose maps depict the shape and elevation of the terrain. These are called **topographic maps.** They usually portray the surface features of relatively small areas, often

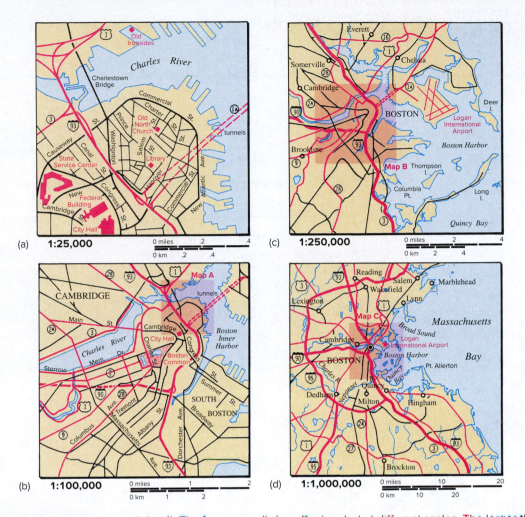

Figure 2.8 **The effect of scale on area and detail.** The four maps all show Boston, but at different scales. The larger the scale, the greater the number and kinds of features that can be included. Among other things, at a scale of 1:25,000, map **(a)** shows streets, street names, and some buildings. Map **(d)**, at the smallest scale, shows only major cities, highways, and water bodies. The area shown in map **(a)** is indicated by the pink square on map **(b)**, the area covered in **(b)** by the pink square on map **(c)**, and the area of map **(c)** by the square on map **(d)**.

with great accuracy (**Figure 2.9**). They not only show landforms, streams, and other natural features but also may display things that people have added to the natural landscape. These include transportation routes, buildings, and such land uses as orchards, vineyards, and cemeteries. Many types of boundaries, from state borders to field or airport limits, are also depicted on topographic maps.

The U.S. Geological Survey (USGS), the chief federal agency for topographic mapping in this country, produces several topographic map series, each on a standard scale. Complete topographic coverage of the United States is available at scales of 1:250,000 and 1:100,000. Maps are also available at various other scales. Scales used for state maps depend on the size of the state and range from 1:125,000 (Connecticut) to 1:500,000 (Alaska).

A single map in one of these series is called a *quadrangle*. Topographic quadrangles at the scale of 1:24,000 exist for the entire area of the 48 contiguous states, Hawaii, and territories, a feat that requires about 57,000 maps. Each map covers a rectangular area that is 7.5 minutes of latitude by 7.5 minutes of longitude. As is evident from Figure 2.9, these 7.5-minute quadrangle maps provide detailed information about the natural and cultural features

of an area. Because of Alaska's large size and sparse population, the primary scale for mapping that state is 1:63,360 (1 in. represents 1 mi). The Alaska quadrangle series consists of more than 2900 maps.

As noted earlier, topographic maps depict the surface of the Earth. Cartographers use a variety of techniques to represent the three-dimensional surface of the Earth on a two-dimensional map. The easiest way to show relief, or variation in elevation, is to use numbers called *spot heights* to indicate the elevation of selected points. A *bench mark* is a particular type of spot height that is used as a reference in calculating elevations of nearby locations (see "Geodetic Control Data" p. 31).

The principal symbol used to show elevation on topographic maps, however, is the **contour line,** along which all points are of equal elevation above a datum plane, usually the mean sea level. Contours are imaginary lines, perhaps best thought of as the outlines that would occur if a series of parallel, equally spaced horizontal slices were made through a vertical feature. **Figure 2.10** shows the relationship of contour lines to elevation for an imaginary island.

The **contour interval** is the vertical spacing between contour lines, and it is normally stated on the map. The more irregular the

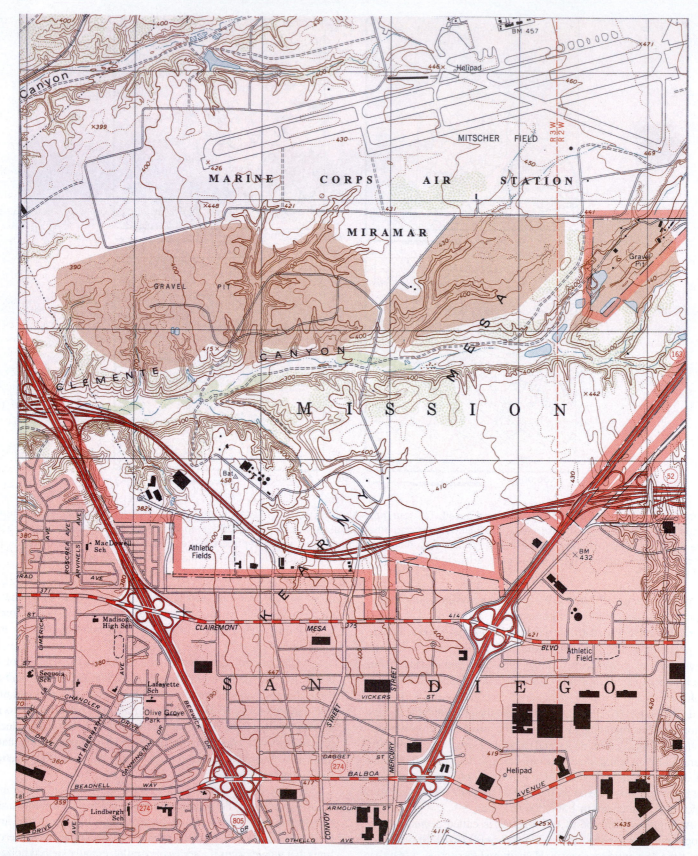

Figure 2.9   **A portion of the La Jolla (San Diego), California, 7.5-minute series of U.S. Geological Survey topographic maps.** The fractional scale is 1:24,000 (1 in. equals about 1/3 mi), allowing considerable detail to be shown. The pink tint denotes built-up areas, in which only schools, churches, cemeteries, parks, and other public facilities are shown. *Source: U.S. Geological Survey.*

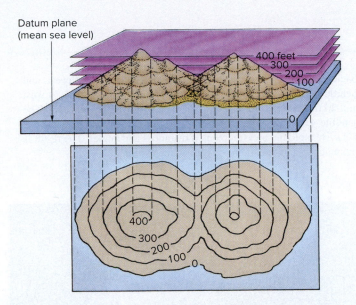

Datum plane
(mean sea level)

400 feet
300
200
100

0

400
300
200
100
0

Figure 2.10   Contours drawn for an imaginary island. The intersection of the landform by a plane held parallel to sea level is a contour representing the height of the plane above sea level.

surface is, usually, the greater will be the number of contour lines that needs to be drawn; the steeper the slope is, the closer will be the contour lines rendering that slope. Contour intervals of 10 and 20 feet are often used, though in relatively flat areas the interval may be only 5 feet. In mountainous areas, the spacing between contours is greater: 40 feet, 100 feet, or more.

Although contour lines represent terrain, giving the map reader information about the elevation of any place on the map and the size, shape, and slope of all relief features, most map readers find it difficult to visualize the landscape from contour lines. To heighten the graphic effect of a topographic map, contours are sometimes supplemented by the use of *shaded relief.* An imaginary light source, usually in the northwest, can be thought of as illuminating a model of the area, simulating the appearance of sunlight and shadows and creating the illusion of three-dimensional topography. Additionally, bands of color for elevation ranges can be used to "color between" the contour lines. These are called elevation, or *hypsometric,* tints.

The tremendous amount of information on topographic maps makes them useful to engineers, regional planners, land use analysts, and developers, as well as to hikers and casual users. Given such a wealth of information, the experienced map reader can make deductions about both the physical character of the area and the cultural use of the land.

Thematic Maps and Data Representation

The study of the spatial patterns and interrelationships of things, whether people, crops, or traffic flows, is the essence of geography. Various kinds of symbols are used to record the location or numbers of these phenomena on thematic maps. The symbols and maps may be either *qualitative* or *quantitative.*

The principal purpose of the qualitative map is to show the distribution of a particular class of information. The world location of producing oil fields, the distribution of national parks, and the pattern of areas of agricultural specialization within a country are

examples. The interest is in where these things are, without reporting about, for example, the barrels of oil extracted, number of park visitors, or value of crops produced.

In contrast, quantitative thematic maps show the spatial characteristics of numerical data. Usually, a single variable, such as population, income, or land value, is chosen, and the map displays the variation of that feature from place to place. Multivariate maps show two or more variables at once.

Point Symbols

Features that occur at a particular point in space are represented on maps by *point symbols.* Thousands of types of such features exist on the Earth: churches, schools, cemeteries, and historical sites, to name a few. Symbols used to represent them include dots, crosses, triangles, and other shapes. On a qualitative thematic map, each such symbol records merely the location of that feature at a particular point on Earth.

Sometimes, however, our interest is in showing the variation in the number of things that exist at several points—for example, the population of selected cities, the tonnage handled at certain terminals, or the number of passengers at given airports.

There are two chief means of symbolizing such distributions, as **Figures 2.11a** and **2.11b** indicate. One method is to choose a symbol, usually a dot, to represent a given quantity of the mapped item (such as 50 people) and to repeat that symbol as many times as necessary.

Such a map is easily understood because the dots give the map reader a visual impression of the pattern. Sometimes pictorial symbols—for example, human figures or oil barrels—are used instead, to mimic the phenomenon being mapped.

If the range of the data is great, the geographer may find it inconvenient to use a repeated symbol. For example, if one country has 500 times the population of another, or one port handles 50 or 100 times as much tonnage as another, that many more dots would have to be placed on the map and could begin to coalesce. To circumvent this problem, the cartographer can choose a second method and use graduated symbols. The size of the symbol is varied according to the quantities represented. Thus, if squares or circles are used, the *area* of the symbol ordinarily is proportional to the quantity shown (**Figure 2.11b**).

There are occasions, however, when the range of the data is so great that even circles or squares would take up too much room on the map. In such cases, symbols such as spheres or cubes are used, and their *volume* is proportional to the data. Unfortunately, many map readers fail to perceive the added dimension implicit in volume, and most cartographers do not recommend the use of such symbols.

Area Symbols

One way to show how the *amount* of a phenomenon varies from area to area is by using **choropleth maps.** The term is derived from the Greek words *choros* ("place") and *pleth* ("magnitude" or "value"). The quantities shown may be absolute numbers (e.g., the population of counties) or derived values, such as percentages, ratios, rates, and densities (e.g., population density by county). The data are grouped into a limited number of classes, each represented

GEODETIC CONTROL DATA

The horizontal position of a place, specified in terms of latitude and longitude, constitutes only two-thirds of the information needed to locate it in three-dimensional space. Also needed is a vertical control point defining elevation, usually specified in terms of altitude above sea level. Together, the horizontal and vertical positions constitute *geodetic control data.* A network of more than 1 million points whose latitude, longitude, and elevation have been precisely determined, recorded, and marked covers the entire United States.

Each point is indicated by a bronze marker fixed in the ground. You may have seen some of the vertical markers, called *bench marks,* on mountaintops, hilltops, or even city sidewalks. The marker indicates which agency put it in place, its elevation, and sometimes the date it was put in place. Every U.S. Geological Survey (USGS) map shows the markers in the area covered by the map, and the USGS maintains Geodetic Control Lists containing the description, location, and elevation of each marker. A bench mark is indicated on the map by the letters *BM,* a small *x,* and the elevation.

These lists were revised in 1987 when, after 12 years' effort, federal scientists completed the recalculation of the precise location of some 250,000 bench marks across the country. In using a satellite locating system for the first national resurvey of control points since 1927, the National Oceanic and Atmospheric Administration (NOAA) found, for example, that New York's Empire State Building is 36.7 meters (120.4 ft) northeast of where it formerly officially stood. The Washington Monument has been moved northeast by 28.8 meters (94.5 ft), the dome of California's state capitol in

Sacramento has been relocated 91.7 meters (300.9 ft) southwest, and Seattle's Space Needle now has a position 93 meters (305.1 ft) west and 20 meters (65.6 ft) south of where maps formerly showed it. The satellite survey provides much more accurate locations than did the old system of land measurement of distances and angles. The result is more accurate maps and more precise navigation.

© Roger Scott.

---

by a distinctive color, shade, or pattern. **Figure 2.11c** is an example of a choropleth map. In this case, the areal units are states. Other commonly used subdivisions are counties, townships, cities, and census divisions.

Features found within defined areas of the earth's surface are represented on maps by *area symbols.* As with point symbols, such maps fall into two general categories: those showing differences in kind and those showing differences in quantity. Atlases contain numerous examples of the first category, such as patterns of religions, languages, political entities, vegetation, or types of rock. Normally, different colors or patterns are used for different areas, as shown in **Figure 2.12**.

As Figures 2.11c and 2.12 reveal, three main problems characterize maps (whether qualitative or quantitative) that show the distribution of a phenomenon in an area:

1. They give the impression of uniformity to areas that may actually contain significant variations.
2. Boundaries attain unrealistic precision and significance, implying abrupt changes between areas, when, in reality, the changes may be gradual.
3. Unless colors are chosen wisely, some areas may look more important than others.

A special type of area map is the **area cartogram,** or **value-by-area map,** in which the areas of units are drawn proportional to the data they represent (**Figure 2.13**). Population, income,

cost, or another variable becomes the standard of measurement. Depending on the idea that the mapmaker wishes to convey, the sizes and shapes of areas may be altered, distances and directions may be distorted, and contiguity may or may not be preserved (see "Red States, Blue States," p. 35).

Line Symbols

As the term suggests, *line symbols* represent features that have length but insignificant width. Some lines on maps do not have numerical significance. The lines representing rivers, political boundaries, roads, and railroads, for example, are not quantitative. They are indicated on maps by such standardized symbols as those that follow.

Often, however, lines on maps do denote specific numerical values. Contour lines that connect points of equal elevation above

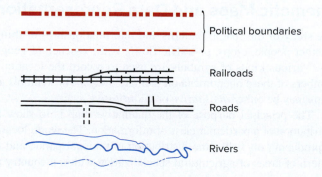

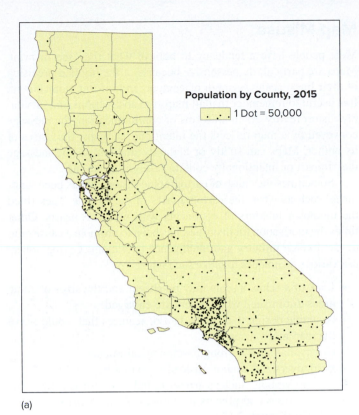

(a)

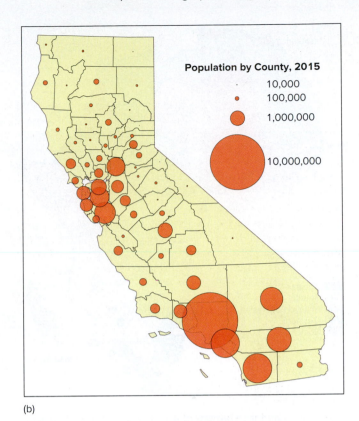

(b)

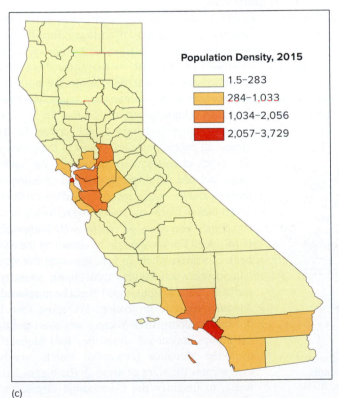

(c)

Figure 2.11   Although population is the theme of each, these different California maps present their information in strikingly different ways. **(a)** In a **dot-distribution map** where large numbers of items are involved, the value or each dot is identical and stated in the map legend. The placement of dots on the map does not indicate precise locations of people within the county, but simply their total number. **(b)** In the **graduated circle map,** the area of the circle is approximately proportional to the absolute number of people within each county. **(c)** This is a **choropleth map** that shows population density by county. Quantitative variation by area is more easily visualized in a map than in a table.

mean sea level are a kind of **isoline,** or line of constant value. Other examples of isolines are *isohyets* (equal rainfall), *isotherms* (equal temperature), and *isobars* (equal barometric pressure).

   Flow-line maps are used to portray linear movement between places. They may be qualitative or quantitative. Examples of *qualitative* flow maps are those showing ocean currents or airline routes.

The lines are of uniform thickness and generally have arrowheads to denote direction of movement. On *quantitative* flow maps, on the other hand, the flow lines are scaled so that their widths are proportional to the amounts they represent. Migration, traffic, and commodity flows are usually portrayed in this way. The location of the route taken, the direction of movement, and the volume of flow

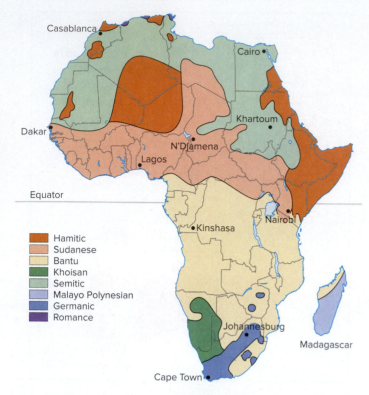

Figure 2.12   **Language regions of Africa.** Maps such as this one may give the false impression of uniformity within a given area—for example, that only Bantu is spoken over much of southern Africa. Such maps are intended to represent only the predominant language in an area.

can all be depicted. The amount shown may be either an absolute or a derived value—for example, the actual traffic flows or a per-mile figure. In **Figure 2.14**, the width of the flow lines is proportional to the number of interstate migrants in the United States.

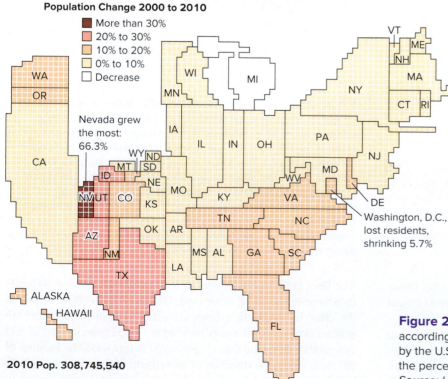

2010 Pop. 308,745,540

Map Misuse

Most people have a tendency to believe what they see in print. Maps are particularly persuasive because of the implied precision of their lines, scales, symbol placement, and information content. It is useful to remember that all maps are abstractions and inevitably distort reality. As in all forms of communication, the message conveyed by a map reflects the intent and, perhaps, the biases of its author. Maps can subtly or blatantly manipulate the message they impart or intentionally contain false information.

Sometimes the cause of cartographic distortion has been ignorance, such as when the cartographers of the Middle Ages filled the unknown interiors of continents with mythical beasts. Other times the motivation for distortion has been to promote a cause or to thwart foreign military and intelligence operations. Cartographers use various techniques to make such maps.

- Lack of a scale: a scale may be absent and the sizes of some areas diminished while others are enlarged.
- A simple design that omits data or features that would make the map more accurate.
- Colors that have a strong psychological impact.
- Bold, oversized, and/or misleading symbols.
- Action symbols, such as arrows to indicate military invasions or repulsions and pincers to show areas threatened by encirclement (**Figure 2.15**).
- Selective omission of data: many governments, for example, do not indicate the location of military installations on their maps; the hub maps in airline magazines typically show lines radiating from the hub to the cities the airline serves, giving the impression that the flights are nonstop.
- Inaccuracies or "disinformation" for military opponents. The chief cartographer of the USSR acknowledged in 1988 that for 50 years the policy of the Soviet Union had been to deliberately falsify almost all publicly available maps of the country. The types of cartographic distortions on Soviet maps included the displacement and omission of features and the use of incorrect grid coordinates. The routes of highways, rivers, and railroads were sometimes altered by as much as 10 kilometers (6 mi). A city or town might be shown on the east bank of a river, when, in fact, it was on the west bank. Even when features were shown correctly, the latitude and longitude grid might be misplaced.
- An inappropriate projection. For more than a decade, the John Birch Society and other political groups concerned about the "Red Menace" used the Mercator projection, which grossly exaggerates the sizes of areas in the higher latitudes, to magnify the Communist threat, and China and Russia were colored red. The Peters projection was developed to promote social justice (see "The Peters Projection" p. 36).

Figure 2.13   **A cartogram** in which each state is sized according to its number of residents in the year 2010 as reported by the U.S. Bureau of the Census. The cartogram also shows the percentage change in population between 2000 and 2010. *Source: U.S. Bureau of the Census.*

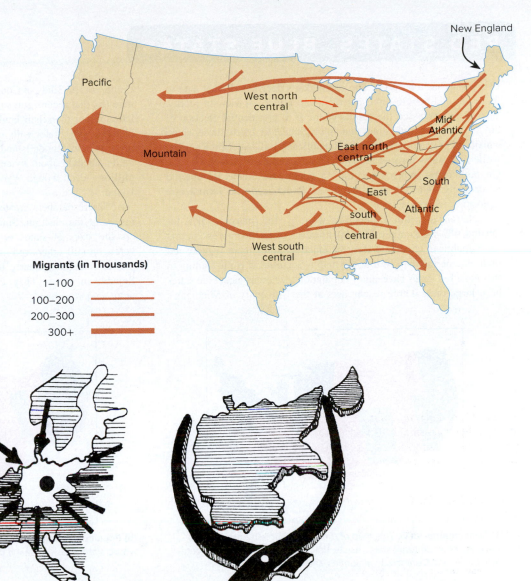

Figure 2.14   A quantitative flow-line map of migration patterns in the United States in the 1950s.

Figure 2.15   The Nazis, who ruled Germany from 1933 to 1945, used maps as tools of propaganda. The maps shown here were designed to increase sympathy for Germany by showing it threatened by encirclement. **(a)** Arrows represent pressure on Germany from all sides. **(b)** Pincers signify pressure against Germany from France and Poland. *Source: Karl Springenschmid,* Die Staaten als Lebewesen: Geopolitisches Skizzenbuch *(Leipzig: Verlag Ernst Wunderlich, 1934).*

In summary, maps can distort and lie as readily as they can convey verifiable spatial data or the results of scientifically valid analyses. The more that map users are aware of those possibilities and the greater understanding they possess of map projections, symbolization, and common forms of thematic and reference mapping standards, the more likely they are to reasonably question and clearly understand the messages maps communicate.

2.6 Contemporary Spatial Technologies

The latter half of the 20th century saw a revolution in the ways geographic data are collected, stored, and analyzed; in the ways maps are produced; in the number and kinds of maps that can be made; and in the applications to which maps are now put. Two of the important new technologies involve remote sensing and global positioning satellites.

Remote Sensing

When topographic maps were first developed, it was necessary to obtain the data for them through fieldwork, a slow and tedious process that involved relating a given point on the earth's surface to other points by measuring its distance, direction, and altitude. Much fieldwork has now been replaced by **remote sensing,** detecting the nature of an object and the content of an area without direct contact with the ground. In the early 20th century, fixed-wing aircraft provided a platform for the camera and the photographer, and by the 1930s aerial photographs from planned positions and routes permitted reliable data gathering for mapping purposes.

RED STATES, BLUE STATES

Every map has a purpose. The mapmaker's decision of what data to represent and how to represent them can affect our view of reality. The media discussed the results of the U.S. presidential election of 2012 in terms of "red" states and "blue" states. Red were those where a majority of voters voted for the Republican candidate, Mitt Romney, while blue states favored the Democratic candidate, Barack Obama. On both maps shown here, the 48 contiguous states are colored accordingly: red and blue to indicate Republican and Democratic majorities, respectively.

Nationwide, the Democrats won by the margin—51% for Obama, 48% for Romney. Map (a), however, gives the impression that the Republican candidate won the election, since there is more red than blue on the map. While the map is accurate, it is misleading in that most red states have small populations, whereas most blue states have large ones. Three researchers at the University of Michigan's Center for the Study of Complex Systems devised an area cartogram, map (b), that reconfigures the states based on the size of their population rather than their land area. On this map, there is clearly more blue than red. States with more people appear larger than states with fewer people. Rhode Island, for example, with its 1.1 million inhabitants, appears about twice the size of Wyoming, which has half a million—even though Wyoming has 60 times the acreage of Rhode Island.

The researchers created a number of other maps depicting the results of the election. Some show election results by county, others the sizes of states proportional to their number of electoral votes; on still others red and blue are joined by a third color, purple, to indicate a nearly balanced percentage of Democratic and Republican votes. They can be viewed at **www.personal.umich .edu/~mejn/election/**.

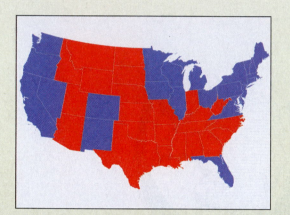

(a) Conventional view. This map of the November 2012 presidential election shows state-by-state results, colored red and blue to indicate Republican and Democratic majorities, respectively.

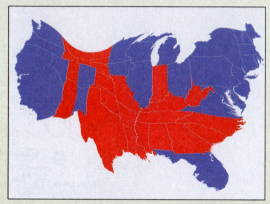

(b) Population cartogram. Election results are shown on a cartogram where state sizes are based on population rather than land area.

Although there are now a variety of sensing devices, aerial photography employing cameras with returned film remains a widely used remote-sensing technique. Mapping from the air has certain obvious advantages over surveying from the ground, the most evident being the bird's-eye view that the cartographer obtains. Using stereoscopic devices, the cartographer can determine the exact slope and size of features such as mountains, rivers, and coastlines. Areas that are otherwise hard to survey, such as mountains and deserts, can be mapped easily from the air. Furthermore, millions of square miles can be surveyed in a very short time.

Standard photographic film detects reflected energy within the visible portion of the electromagnetic spectrum (**Figure 2.16a**). It can be supplemented by special sensitized infrared film that has proved particularly useful for the recording of vegetation and hydrographic features. Color-infrared photography yields what are called *false-color images*—"false" because the film does not produce an image that appears natural (**Figure 2.16b**). For example, leaves of healthy vegetation have a high infrared reflectance and are recorded as red on color-infrared film, while unhealthy or dormant vegetation appears as blue, green, or gray. Clear water appears as black, but sediment-laden water may appear light blue.

For wavelengths longer than 1.2 micrometers (1 micrometer is 1 one-millionth of a meter) on the electromagnetic spectrum, sensing devices other than photographic film must be used. Nonphotographic imaging sensors include thermal scanners, radar, and lidar.

- *Thermal scanners* record the longwave radiation emitted by water bodies, clouds, and vegetation, as well as by buildings and other structures and are used to produce images of thermal radiation (**Figure 2.17**). Unlike conventional photography, thermal sensing can be employed during nighttime as well as daytime, giving it military applications. It is widely used for studying various aspects of water resources, such as ocean currents, water pollution, surface energy budgets, and irrigation scheduling.

- Operating in a different band of the electromagnetic spectrum, *radar* (short for *ra[dio] d[etection] a[nd] r[anging]*)

THE PETERS PROJECTION

Developed and promoted by Arno Peters, a German journalist-historian, the Peters projection purports to reflect concern for the problems of the Third World by providing a less European-centered representation of the world. Because it is an equal-area map, Peters claimed that it shows the densely populated parts of the Earth and the countries of the Third World in proper proportion to one another. He persuaded a number of socially concerned agencies with special interest in the Third World, including the World Council of Churches, the Lutheran Church of America, and UNESCO and several other United Nations organizations, to adopt the map.

Presented in 1973 as a "new invention," the projection aroused a storm of controversy. Critics pointed out that, by saying his projection was fairer and more accurate than the Mercator, Peters used the latter as a meaningless foil, a "straw man" to knock down. If Peters wanted to demonstrate that the less-developed countries deserve a larger share of our attention and resources, he might better have used an area cartogram (see Figure 2.13) in which each country is scaled according to its number of inhabitants, which would do more to call attention to enormous populations of Third World countries, such as India, China, and Indonesia. Detractors also noted that the Peters projection badly distorts shapes in the tropics and at high latitudes and that many equal-area projections yield a world map with less distortion of shapes. In addition, distances and directions cannot be measured except under very limited conditions. Finally, the projection was not new but, in fact, a very slight modification of an equal-area projection developed by James Gall in 1855.

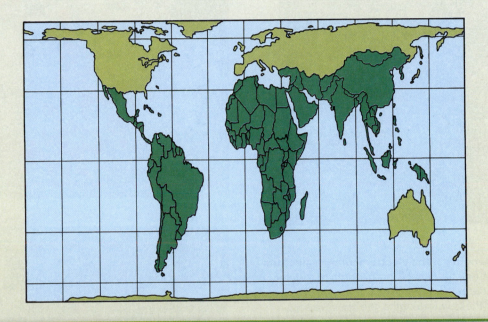

systems also can be used day or night. Because radar can penetrate clouds and vegetation as well as darkness, it is particularly useful for monitoring the locations of airplanes, ships, and storm systems and for mapping parts of the world that are perpetually hazy or cloud-covered (**Figure 2.18**).

- *Lidar* (short for *li[ght] d[etection] a[nd] r[anging]*) is a relatively new remote-sensing technology that utilizes an airborne laser to transmit light out to an object. Some of the light is reflected back to the instrument, where it is analyzed to yield information about the target. Lidar is ideal for any kind of mapping that requires a precise depiction of the ground surface (see Figure 2.21).

Since the 1970s, both staffed and unstaffed spacecraft have supplemented the airplane as the vehicle for imaging Earth features. Many images are now taken either from continuously orbiting satellites or from staffed spacecraft flights, such as those of the *Apollo* and *Gemini* missions. Among the advantages of satellites are the speed of coverage and the fact that views of large regions can be obtained.

In addition, satellites are equipped to record and report back to the Earth digitized information from multiple parts of the electromagnetic spectrum that are outside the range of human eyesight. Satellites enable us to map the invisible, including atmospheric and weather conditions, in addition to providing images with applications in agriculture and forest inventory, identification of geologic structures and mineral deposits, and monitoring of a variety of environmental phenomena, including water pollution and the effects of acid rain. Military applications of remotely sensed images include better aircraft navigation, improved weapons targeting, and enhanced battlefield management and tactical planning, which raises the question of who should have access to the information (see "Civilian Spy Satellites" p. 39).

Perhaps the best known remote-sensing spacecraft are the **Landsat satellites,** the first of which was launched in 1972. The different sensors of the Landsat satellites are capable of resolving objects between 15 and 60 meters (50 and 200 ft) in size. Even sharper images are yielded by the French SPOT satellite; its sensors can show objects that are larger than 10 meters (33 ft).

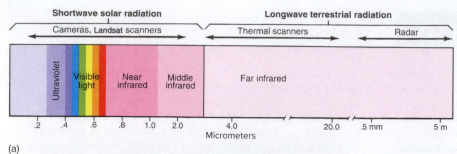

(a)

(b)

Figure 2.16   (a) Wavelengths of the electromagnetic spectrum in micrometers. One micrometer equals one-millionth of a meter. Sunlight is made up of different wavelengths. The human eye is sensitive to only some of these wavelengths, the ones we see in the colors of the rainbow. Although invisible, near infrared wavelengths can be recorded on special sensitized film and by scanners on satellites. The scanners measure reflected light in both the visible and near infrared portions of the spectrum. Wavelengths longer than 4.0 micrometers characterize terrestrial radiation. **(b) A color-infrared aerial photograph of Washington, D.C.** *(b) Source: U.S. Geological Survey.*

Satellite imagery is relayed by electronic signals to receiving stations, where computers convert the signals into photograph like images for use in both long-term scientific research and current-condition mapping programs.

Landsat images have a variety of research applications, including the following:

- tracing ocean currents
- assessing water quality in lakes
- mapping snow cover, glaciers, and polar ice sheets
- analyzing soil and vegetation conditions
- monitoring global deforestation
- monitoring strip-mining reclamation
- identifying geologic structures and associated mineral deposits
- mapping population changes in metropolitan areas

Some Landsat data are utilized not for long-term scientific research but for immediately monitoring, mapping, and responding to natural and human-caused disasters, such as storms, floods, earthquakes, volcanoes, wildfires, and oil spills (**Figure 2.19**).

The Global Positioning System

In recent years, the **Global Positioning System (GPS)** has made the determination of location significantly easier than it used to be. This navigation and positioning system was conceived

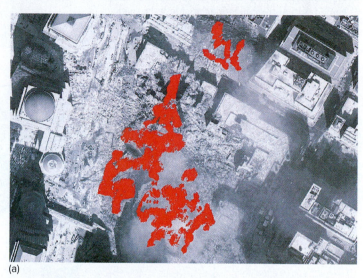

(a)

(b)

Figure 2.17   Thermal radiation images of the World Trade Center site in New York City, 2001. Following the collapse of the twin towers on September 11, 2001, fire-fighting and rescue teams relied on daily thermal images of the site to detect fire patterns in the rubble and underground. Based on the patterns revealed, the teams decided where to work that day. **(a) A few days after the attack,** a nearly constant field of heat (shown in red) covered most of the 16-acre site. **(b) A month later,** the underground blaze was confined largely to where the two towers once stood.
© New York State Office of Cyber Security & Critical Infrastructure Coordination (CSCIC) © 2001

Figure 2.18 **A SLAR mosaic of Los Angeles, California.** Side-looking airborne radar (SLAR) instruments on aircraft or satellites transmit microwave energy to the ground. The portion of the signal that returns to the sensor is recorded as digital values that can be represented on photographic film. The "side-looking" aspect produces shadows of varying lengths, enhancing subtle features of the terrain. This mosaic was compiled from many radar image strips. *Source: U.S. Geological Survey.*

in the 1970s and is maintained by the U.S. Department of Defense (DoD). The technology uses a network of DoD satellites that orbit some 20,000 kilometers (12,500 mi) above the Earth, passing over the same spot every 24 hours. As they orbit, the satellites continuously transmit their positions, time signals, and other data. A GPS receiver records the positions of a number of the satellites simultaneously, then determines its latitude, longitude, and altitude and the time.

Global positioning systems (GPS) rely upon a system of 24 orbiting satellites, Earth-bound tracking stations that control the satellites, and portable receivers that determine exact geographic locations based on the time delay in signals received from three or more satellites (technically, they determine location from an inference based on the time required for several signals to travel from the satellite to Earth and back).

GPS technology was originally designed for military applications, particularly naval and aerial navigation. The technology has facilitated the development of precision-guided weapons, the so-called smart bombs that hone in on a target. Other government applications include the use of GPS receivers for monitoring geologic fault lines and ocean currents, the sensing of global warming in the atmosphere, firefighting, and mapping disaster scenes.

Seeking clues as to why the space shuttle *Columbia* broke apart upon its reentry into the atmosphere on February 1, 2003, for example, federal government investigators used GPS technology to define the debris field, which covered portions of Texas, Louisiana, and several other states. As hundreds of volunteers and law enforcement officers found and collected thousands of pieces of debris, their precise locations were fed into a computer equipped with special mapping software. Just a few days after the tragedy, the Federal Emergency Management Agency's Disaster Field Office in Lufkin, Texas, was printing more than 1000 maps a day. The maps helped emergency workers focus on the areas they needed to search in order to retrieve more of the shuttle debris.

Figure 2.19   **Three major clusters of fires are evident in this satellite image of** southern California taken on October 22, 2007: near Los Angeles, in the San Bernardino Mountains, and in San Diego. Data from remotely sensed images can be used to keep track of the extent and intensity of fires and to update maps of active fires several times a day. They assist fire managers on the ground in determining where best to position firefighters to contain a blaze, to assess damage after a fire is contained, and to plan recovery efforts. *Source: Image courtesy of MODIS Rapid Response Project at NASA/GSFC.*

As GPS receivers have become smaller, lighter, and less expensive, civilian applications have multiplied (see "Geocaching" p. 40). Automobile manufacturers make in-car navigation systems an option in their new cars. The systems tap GPS signals to monitor the car's exact location, comparing it with a computerized atlas stored on a compact disc. The car's location, constantly updated, appears on a computer screen mounted on the dashboard. The navigation systems enable motorists to find out where they are and how to reach their destination. For example, the driver can give a street address or the name of a movie theater, a hospital, or another building, and then the system displays it on the screen, indicates how far away it is, and tells how long it should take to drive there. The system also gives turn-by-turn directions on the screen map or directions "spoken" by an electronic voice.

In the past few years, systems have been developed for building miniaturized GPS receivers into all kinds of things, such as watches, bracelets, cell phones, and even dog collars, in order to ascertain their locations. A number of states use GPS monitoring

Civilian Spy Satellites

Although remote-sensing satellites have been orbiting Earth for about three decades, only since 1999 have detailed satellite images such as that shown here become available to the general public. Until recently, images from commercial satellites were considerably fuzzier and less detailed than those from military satellites. That is no longer the case. In 1994, the federal government lifted technical restrictions on private companies, permitting them to build a new generation of civilian "spy" satellites and to sell the images to any buyer. Space Imaging Corporation, based in Denver, Colorado, launched the first high-resolution commercial satellite, *Ikonos 1* (after the Greek word for "image") in 1999. Two other U.S. firms launched similar satellites the following year. The new imaging technology is so powerful that it can detect and record objects on the ground as small as 1 meter wide: cars, houses, even hot tubs.

Military planners are finding high-resolution images from commercial satellites invaluable, capable of providing up-to-date information and a higher level of detail and precision than conventional maps. Adding multiple layers of data, such as information from a topographic map, current weather conditions, and reports from soldiers in the field, to the images facilitates the planning of missions. Army commanders in Iraq, for example, have used the images to pinpoint tall structures that might be used by snipers, to determine whether bridges have been destroyed, and to identify alleys that could be ambush routes.

The new imagery also has been welcomed by clients such as geologists, city planners, and disaster-relief officials. Nongovernmental public interest groups use the images for such activities as pressuring governments to live up to environmental laws and treaties, tracking refugee movements, detecting illegal waste dumps, and monitoring arms control agreements. At the same time, the sharp visual detail and widespread availability of the images provided by the satellites have raised fears about national security. "Any time a powerful new technology is introduced, there's a battle over the uses to which it's put," one intelligence official noted. "Here, the potential for beneficial uses is very high, on balance. But the potential for abuse certainly exists and we'll no doubt see some of that."

One of the chief concerns is that the images, which rival military reconnaissance in sharpness and accuracy, can be purchased by those who would threaten the national welfare. In wartime, for example, foes could use them to track the location of military troops and equipment. Terrorists could use the imagery to plan surprise attacks. General Richard B. Myers, head of the U.S. Space Command, warns that the government must decide what to do in times of armed conflict. "There's going to be risk in operations like this when you're selling what can be used against you."

A satellite image of San Francisco International Airport. Only recently have detailed, accurate, high-quality images such as this one been taken from commercial satellites and made available to the general public. © 2000 Space Imaging. All Rights Reserved.

American companies, however, operate under a number of security restrictions. They are forbidden to sell images to customers in several countries, including Cuba and North Korea. And in the interests of national security, the government can put any area off limits. Soon after September 11, 2001, for example, the Pentagon purchased exclusive rights to all satellite images of Afghanistan and Pakistan from Space Imaging.

Considering the Issues

1. Do you think that the availability of the new, detailed satellite images is a potential threat to national security? Might access to images of their enemies make belligerent countries more dangerous than they already are? Why or why not?
2. In what ways might access to satellite imagery stem the tides of environmental and social destabilization?
3. Should the federal government, which licenses the satellites, be allowed to exercise "shutter control," cutting off image sales during wartime? Defend your answer.

---

devices as a surveillance system to track the movements of people on parole and probation. The law enforcement use of GPS technology to track suspects is raising legal and ethical questions about privacy rights and public safety (see "An Invasion of Privacy?" p. 41).

Virtual and Interactive Maps

Since its establishment in 1993, the World Wide Web has been instrumental both in disseminating geographic information and in fostering the integration of geospatial data. Maps are easily, freely, and widely available on the Internet, and online mapping systems enable users to view images of nearly any place on Earth. Google is one of the best known of the companies that produce maps on the Web; others are Microsoft and Yahoo.

Google Earth (earth.google.com) combines aerial photographs, satellite images, and maps with street, terrain, and other data. (Google Earth is free; professionals can purchase more sophisticated versions of the software, such as Plus and Pro, from the company.) The program gives the user an aerial view of a place.

GEOCACHING

Geocaching has been variously described as high-tech hide-and-seek, a new twist on an old-fashioned scavenger hunt, and the hottest new leisure activity. The rules are simple. An individual goes outside and in a publicly accessible place hides a waterproof container containing a few items—pens, key chains, small toys, or other inexpensive trinkets—as well as a logbook. This is the *cache.* Having hidden it, the individual goes online to geocaching.com or another website to post the latitude and longitude of the hidden treasure, and perhaps some clues to help searchers find it.

Now it's your turn. You go to one of the websites where the coordinates of caches are posted, pick a site in your area, program the coordinates into your portable GPS device, and use the device to find the cache. If successful, you enter your name, the date, and the time in the logbook; take an item out of the cache; leave something in return; and then replace the cache exactly where you found it.

Geocaching got its start in 2000, when the U.S. government allowed civilian GPS devices to have the same degree of accuracy as military devices, and almost immediately the hobby caught on around the globe. Coordinates of hundreds of thousands of caches have been posted on the Web. Enthusiasts say the hobby gives them exercise, an incentive to explore new areas, and experience in orienteering.

© Arthur Getis.

This can be accessed by entering the name of the place or its longitude and latitude or simply by scrolling across the virtual globe and zooming in on a particular location. The accuracy and sharpness of the images vary. Most of the aerial photographs and satellite images have been taken within the last 3 years, and Google periodically updates them. For some major cities, the resolution is high enough to provide clear three-dimensional images of individual buildings and the color of cars on the street. Special features enable the user to zoom in and out of an image as well as to tilt or rotate it.

The digital maps produced by Google, Microsoft, and some other Internet companies can be merged with data from other sources to create what are called *mashups,* Web applications that combine data from more than one source into an integrated experience—an example of *interactive mapping.* Now anybody with modest programming skills, not just professional cartographers, can make maps, and people have created millions of them. Mashups can be simple or complex. Some people simply overlay the locations of things—crime data, for example—onto an online street map of their area. The user can search for crimes by type, location, or the date they were committed. Other types of information that have been combined with maps are bicycle trails, gas stations with low prices, and school rankings; the list is virtually endless. Some people have annotated digital maps and/or photo images with text (e.g., recent news events, reviews), photographs, sound, and even videos. To get an idea of the number, kinds, and variety of interactive websites, search for "interactive maps" on the Web. It is now possible to ask questions about such things as restaurants and movies. Soon it is expected that phones will signal us in words or sounds when objects of particular interest to us are located nearby.

2.7 Integrating Technology: Geographic Information Systems

The technologies just described would not be possible without computers. They have become an integral part of almost every stage of the cartographic process, from the collection and recording of data to the production and revision of maps. Although the initial cost of equipment is high, the investment is repaid in the more efficient and more accurate production and revision of maps.

Computers are at the heart of what is known as a **geographic information system (GIS),** a computer-based set of procedures for assembling, storing, manipulating, analyzing, and displaying geographically referenced information. Any data that can be located spatially can be entered into a GIS. The five major components of a GIS are as follows:

1. a data input component that converts maps and other data from their existing form into digital or computer-readable form
2. a data management component used to store and retrieve data
3. data manipulation functions that allow data from disparate sources to be used simultaneously
4. analysis functions that enable the extraction of useful information from the data
5. a data output component that makes it possible to visualize maps and tables on the computer monitor or as hard copy (such as paper)

While the use of computers and printers in map production permits increases in the speed, flexibility, and accuracy of many steps in the mapmaking process, that use in no way reduces the

GEOGRAPHY & PUBLIC POLICY

An Invasion of Privacy?

Police investigating a series of burglaries in Latham, New York, in 2005 attached a GPS receiver under the bumper of a suspect's car and left it there for 65 days. Based on the tracking information, police charged the suspect with burglarizing two stores. In another case, someone was attacking women in northern Virginia, grabbing them from behind and molesting them before running away. After logging 11 cases in 6 months, police identified a suspect. They had put a GPS device on the suspect's van, allowing them to track his movements, and soon caught him dragging a woman into a wooded area. In neither case did the police obtain a search warrant.

Law enforcement agents have also used GPS technology embedded in cell phones to link suspects to crime scenes by analyzing their phone records. Such records show a phone's approximate location at the beginning and end of a call. In a murder case in New York City, they were used to convict a nightclub bouncer in the death of a female graduate student whom he had met at the bar. The phone records showed that he had made several calls as he drove from his house in Queens to a deserted street in Brooklyn to dump the body.

In hundreds and perhaps thousands of cases, police have used GPS to catch killers, car thieves, drug dealers, sexual predators, burglars, and robbers, often without obtaining a warrant or court order. Does the covert use of GPS technology to track suspects violate the Fourth Amendment rights of protection against unreasonable searches and seizures? Should the government be required to obtain a search warrant by showing probable cause for connecting a cell phone user to criminal activity?

Existing laws do not provide clear or uniform guidelines. Some jurisdictions allow law enforcement agencies to track the location of cell phone users without search warrants; others do not. Phone companies wish that Congress would clarify the laws, so that they are clear about their legal responsibilities. The U.S. Supreme Court has not addressed the issue, and as of 2009 a few states had reached varied conclusions. In New York, Washington, and Oregon, police may not use GPS receivers to track an individual's movements without a warrant, but the Wisconsin Court of Appeals ruled that GPS tracking does not violate Fourth Amendment rights—that police can attach receivers to cars without obtaining search warrants, even if an individual is not a crime suspect.

Considering the Issues

1. Advocates of the warrantless use of GPS technology argue that GPS tracking is essentially the same as having an officer trail someone; it's just cheaper and more accurate. As one attorney said, "It helps cut down on the number of police officers who would have to be out tracking someone." On the other hand, another attorney contended that "While it may be true that police can conduct surveillance of people on a public street without violating their rights, tracking people everywhere they go and keeping a computer record of it for days or even months without them knowing is a completely different type of intrusion." With which side, if either, do you agree?

2. A district attorney in Queens said, "People who obey the law have nothing to fear from cell phone tracking. Law enforcement has a responsibility to keep pace with the latest advances in technology in order to improve its efficiency in combating crime." Privacy advocates, however, contend that people have a reasonable expectation of privacy regarding their physical location. As one judge wrote, "Most Americans don't know that their cell phones create a record of their movements and would be appalled to learn that the government can access it without showing probable cause." If you have a cell phone, do you care if your movements can be tracked as long as the phone is turned on? Do you know what information your phone is sending about your movements and where that information is going? If locational information is to be collected, should people be given advance notice and the opportunity to opt out?

3. GPS devices give the government and corporations access to much extraneous information about individuals, including the clubs and restaurants they go to, whose houses they visit, and what political meetings they attend. Should the cost of carrying a cell phone include the loss of one's personal privacy, or do you believe citizens do not expect their every move to be continuously and indefinitely monitored by a technical device without their knowledge, except where a warrant has been issued based on probable cause?

---

obligation of the mapmaker to employ sound judgment in the design of the map or the communication of its content.

The Geographic Database

The first step in developing a GIS is to create a **geographic database,** a digital record of geographic information from such sources as maps, field surveys, aerial photographs, and satellite imagery. As long as every item in the database is tied to a precise geographic location, a GIS can use information from many different sources and in many different forms. The purpose of the study will determine the data to be entered into the database. For a physical geographer studying the sensitivity of wetlands to damage in a particular area, the source data might include maps of rainfall amounts at different points, soil types, vegetative cover, originating points of water pollution, contours, and direction of stream flow. An urban geographer or regional planner, on the other hand,

might use a GIS data set that contains the great amount of place-specific information collected and published by the U.S. Census Bureau, including political boundaries, census tracts, population distribution, a building inventory, race, ethnicity, income, housing, employment, and so on.

Once geographic information is in the computer in digital form, the data can be manipulated, analyzed, and displayed with a speed and precision not otherwise possible. Because computers can process millions of facts in seconds, they are particularly useful for researchers who need to analyze many variables simultaneously. The development of geographic information systems has deemphasized the use of maps to store information and has enabled researchers to concentrate on using maps for analyzing and communicating spatial information. With the appropriate software, a computer operator can display any combination of data, showing the relationships among variables almost instantly (**Figure 2.20**). In this sense, a GIS allows an operator to generate maps or perform spatial analyses that were virtually impossible to create or perform only a few decades ago.

GIS operations can produce several types of output: displays on a computer monitor, listings of data, or hard copy. When a map is to be produced, the specialist can quickly call up the desired data. Geographic information systems are particularly useful for revising existing maps because outdated data—for example, population sizes—can be modified or replaced easily. Additionally, a GIS facilitates exploratory analysis by enabling an operator to quickly change variables and/or model parameters and to use multiple spatial scales.

Applications of GIS

Who uses geographic information systems? Tens of thousands of people in a variety of fields use them for a variety of purposes. Every issue of *ArcNews,* a quarterly publication of ESRI, Inc., Redlands, California, describes numerous examples of "GIS in Action" from many fields. In human geography, the vast and growing array of spatial data has encouraged the use of GIS to explore models of regional economic and social structure, to examine transportation systems and urban growth patterns, to study patterns of voting behavior, and so on. For physical geographers, the analytic and modeling capabilities of GIS are fundamental to the understanding of processes and interrelations in the natural environment.

In addition to geographers, researchers in fields ranging from archaeology to zoology use geographic information systems, as a few examples indicate.

- Biologists and ecologists use GIS to study numerous environmental problems, including air and water pollution, landscape conservation, wildlife management, and the protection of endangered species.
- Epidemiologists need accurately mapped information to study both the diffusion of diseases, such as malaria, SARS, AIDS, and dengue fever, and entomological risk factors.
- GIS software has made it possible for political scientists to evaluate existing legislative districts, using criteria such as compactness and contiguity, and to suggest ways the boundaries of the districts might be redrawn.
- Sociologists have used GIS to identify clusters of segregation and to examine the changing structures of segregation over time.

Systems, Maps, and Models

Many companies in the private sector also use computerized map-making systems. Among others, oil and gas companies, restaurant chains, soft-drink bottlers, and car rental companies rely on GIS to perform such diverse tasks as identifying drilling sites, picking locations for new franchises, analyzing sales territories, and calculating optimal driving routes.

Many bureaus and agencies at the local, regional, state, and national levels of government employ geographic information

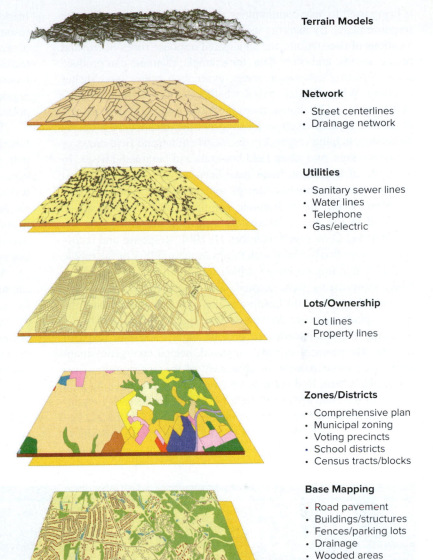

Terrain Models

Network
- Street centerlines
- Drainage network

Utilities
- Sanitary sewer lines
- Water lines
- Telephone
- Gas/electric

Lots/Ownership
- Lot lines
- Property lines

Zones/Districts
- Comprehensive plan
- Municipal zoning
- Voting precincts
- School districts
- Census tracts/blocks

Base Mapping
- Road pavement
- Buildings/structures
- Fences/parking lots
- Drainage
- Wooded areas
- Spot elevation
- Contour lines
- Recreational facilities

Figure 2.20   A model of a geographic information system. Information layering is the essence of a GIS. Map information that has been converted to digital data is stored in the computer in different data "layers." A GIS enables the user to combine just the layers that are desired to produce a composite map and to analyze how those variables relate to one another. In this example, the different layers of information are to be used in different combinations for city planning purposes. What sorts of layers might you need if you were asked to locate a new Starbucks in an urban area? *Source: Reprinted by permission of Shaoli Huang.*

systems. These include such departments as highway and traffic control, public utilities, and planning. Law enforcement agencies use spatially based software packages to analyze patterns of crime, identify "hot spots" of criminal activity, and redeploy police resources accordingly.

Government officials also use GIS to help plan emergency response to natural and human-induced disasters, such as tornadoes, hurricanes, earthquakes, floods, and wildfires. Because disasters usually occur suddenly, threatening people and structures, they create chaos and panic. Increasingly, GIS technology is being used to help communities prepare for disasters and create response plans. By merging information on the types of roads, the locations of fire stations, the anticipated response times of fire and rescue squads, and other data, for example, planners can produce maps depicting evacuation zones, evacuation routes, and shelter locations. When disaster strikes, whether it is a wildfire in Arizona or a tornado in Oklahoma, maps produced using GIS have proved invaluable in tasks such as locating houses, identifying property ownership, helping responders decide where to send field crews or rescue workers, and siting field hospitals and paramedic bases. In the wake of a disaster, maps have been used to locate damaged structures, assess property damage, and prepare for debris removal.

One of the most dramatic examples of GIS applications occurred in the days immediately following the attacks on the World Trade Center on September 11, 2001. Response and recovery teams needed to know such things as the stability of the rubble and the remaining buildings, which subway lines were damaged, where watermains were located, and where there were utility outages. With fires still burning at the site of the attack, and the mayor's planning office destroyed, GIS experts immediately began gathering the data needed to provide highly accurate maps of the site. The maps, continually updated, helped emergency managers track the expansion or abatement of the underground fires and enabled them to determine how rescue equipment could gain access to the site, where to safely position large recovery equipment, and by what route debris could be removed (**Figure 2.21**).

The career opportunities for people skilled in GIS techniques are excellent. A variety of private industries, including insurance, marketing, real estate, epidemiology, health care, environmental resource management, transportation, homeland security, and disaster preparedness and response, use geospatial information technologies. GIS professionals are also widely employed by federal, state, and local government departments that deal with regional and community planning and such services as water, police, fire, sewer, transportation, health, education, and welfare.

The systems of geographic concern are those in which the functionally important variables are spatial: location, distance, direction, density, connectivity, and the other basic concepts. Systems have components, and the analysis of the role of components helps reveal the operation of the system as a whole. To conduct that analysis, individual system elements must be isolated for separate identification and, perhaps, manipulated to see their function within the structure of the system or subsystem. Maps and models are devices that geographers use to achieve that isolation and separate study.

Maps, as we have seen, are effective to the degree that they can segregate at an appropriate scale those system elements selected for examination. By compressing, simplifying, and abstracting reality, maps record in manageable dimension the real-world conditions of interest. A model is a simplified abstraction of reality, designed to clarify relationships between its elements. Maps are a type of model, representing reality in an idealized form to make certain aspects clearer.

The complexities of spatial systems analysis—and the opportunities for quantitative analysis of systems made possible by computers and sophisticated statistical techniques—have led geographers to use other kinds of models in their work. Model building is the technique scientists use to simplify complex situations, to eliminate (as does the map) unimportant details, and to isolate for special study and analysis the role of one or more interacting elements in a total system. Models also allow geographers to conduct experiments on a simulation of a portion of reality instead of the reality itself.

An interaction model discussed in Chapter 7, for instance, suggests that the amount of exchange expected between two places depends on the distance separating them and on their population sizes. The model indicates that the larger the places, and the closer their distance, the greater is the amount of interaction. Such a model helps us to isolate the important components of the spatial system, to manipulate them separately, and to reach conclusions concerning their relative importance. When a model satisfactorily predicts the volume of intercity interaction in the majority of cases, the lack of agreement in a particular case leads to an examination of the circumstances contributing to the disparity. The quality of connection roads, political barriers, or other variables may affect the specific places examined, and these causative elements may be isolated for further study.

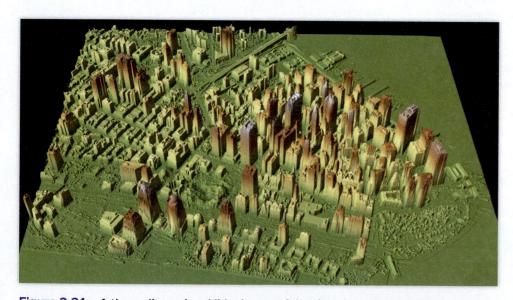

Figure 2.21   A three-dimensional lidar image of the site of the World Trade Center. Accurately mapping the wreckage of the World Trade Center using a combination of remote sensing, GPS, and GIS was invaluable in recovery and cleanup efforts after the attacks of 9/11. The GPS system was used to position both ground and airborne mapping sensors. Aircraft outfitted with three kinds of sensors collected high-resolution aerial photography, thermal imagery, and lidar (light detection and ranging) data. Within hours after the data were collected, GIS professionals from government, industry, and academia merged the digital data to produce large, high-resolution images of the building structures and the surrounding area. The three-dimensional models created by the lidar system enabled engineers to calculate the volume of the rubble piles, track their movement and change, and determine the reach needed by cranes to remove them. *Source: NOAA/U.S. Army JPSD.*

Summary of Key Concepts

- Maps are as indispensable to the geographer as are words, photographs, and quantitative techniques of analysis. Also relying on maps are people involved in the analysis and solution of many of the critical issues of our time, such as climate change, pollution, national security, and public health—all issues that call for the accurate representation of elements on the earth's surface.
- The geographic grid of longitude and latitude is used to locate points on the earth's surface. Latitude is the measure of distance north and south of the equator, while longitude is the angular distance east or west of the prime meridian.
- All systems of representing the curved Earth on a flat map distort one or more Earth features. Any given projection will distort area, shape, distance, and/or direction.
- Among the most accurate and most useful large-scale maps are the topographic quadrangles produced by a country's chief mapping agency. They contain a wealth of information about both the physical and the cultural landscape and are used for a variety of purposes.
- Remote sensing from aircraft and satellites employing a variety of sensors is an important source of spatial data. The need to store, process, and retrieve the vast amounts of data generated by remote sensing has spurred the development of geographic information systems, which provide a way to search for spatial patterns and processes.

As you read the remainder of this book, note the many different uses of maps. For example, notice in Chapter 3 how important maps are to your understanding of the theory of continental drift; in Chapter 6, how maps aid geographers in identifying cultural regions; and in Chapter 7, how behavioral geographers use maps to record people's perceptions of space.

Key Words

area cartogram (value-by-area map) 32
azimuthal projection 27
cartography 21
choropleth map 31
conformal projection 27
contour interval 31

contour line 29
equal-area (equivalent) projection 24
equidistant projection 27
flow-line map 33
geographic database 42
geographic grid 22

geographic information system (GIS) 41
Global Positioning System (GPS) 38
globe properties 24
International Date Line 23
isoline 33
Landsat satellite 37

latitude 22
longitude 23
map projection 24
prime meridian 23
remote sensing 35
scale 28
topographic map 28

Thinking Geographically

1. What important map and globe reference purpose does the *prime meridian* serve? Is the prime, or any other, meridian determined in nature or devised by humans? How is the prime meridian designated or recognized?

2. What happens to the length of a degree of longitude as one nears the North and South Poles? What happens to a degree of latitude between the equator and the poles?

3. From a world atlas, determine, in degrees and minutes, the locations of New York City; Moscow, Russia; Sydney, Australia; and your hometown.

4. List at least five properties of a globe.

5. Briefly make clear the differences in properties and purposes of *conformal, equivalent,* and *equidistant* projections. Give one or two examples of the kinds of map information that would best be presented on each type of projection.

6. Give one or two examples of how maps can be misused.

7. In what different ways can *map scale* be presented? Convert the following map scales into their verbal equivalents.

1:1,000,000   1:63,360   1:12,000

8. What is the purpose of a *contour line?* What is a *contour interval?* What landscape feature is implied by closely spaced contours?

9. What kinds of data acquisition are suggested by the term *remote sensing?* To what uses are remotely sensed images put?

10. What are the basic components of a *geographic information system?* What are some of the applications of a GIS?

CHAPTER 3
Physical Geography: Landforms

Rock spires in Monument Valley, Navajo Tribal Park, UT/AZ. © Thomas Roche/Getty Images RF.

CHAPTER OUTLINE

After studying this chapter you should be able to:

3.1   Characterize the three classes of rock.

3.2   Define folding, joint, and faulting.

3.3   Illustrate how plate tectonics relate to earthquakes.

3.4   Explain how a tsunami originates.

3.5   Compare the effect of mechanical and chemical weathering on landforms.

3.6   Compare the effect of groundwater erosion with that of surface water erosion.

3.7   Relate how glaciers form and how their erosion creates landscapes.

3.8   Define landform features such as deltas, alluvial fans, natural levees, and moraines.

3.9   Understand the landform changes due to waves, currents, and wind.

Although too early for sunbathers and snorkelers, the Hawaiian Islands will have a new island to add to their collection, which contains such scenic beauties as Oahu, Maui, and Kauai. It is Loihi, 0.8 kilometer (0.5 mi) below sea level, just 27 kilometers (17 mi) from the big island of Hawaii. Because the speed of its ascent must be measured in geologic time, it probably will not appear above the water surface for another million or so years. It is a good example, however, of the ceaseless changes that take place on the Earth's surface. As the westernmost of the islands erode and sink below sea level, new islands arise at the eastern end. In Loihi's most recent explosion in 1996, scientists feared that a giant wave would be set off at the surface that could devastate the islands, including the city of Honolulu and popular Waikiki Beach. Fortunately, this was not the case.

Humans on their trip through life continuously are in touch with the ever-changing, active, moving physical environment. Most of the time, we are able to live comfortably with the changes, but when a freeway is torn apart by an earthquake, or floodwaters force us to abandon our homes, we suddenly realize that we spend a good portion of our lives trying to adapt to the challenges the physical environment has for us.

For the geographer, things just will not stand still—not only little things, such as icebergs or new islands rising out of the sea, or big ones, such as exploding volcanoes changing their shape and form, but also giant things, such as continents that wander about like nomads and ocean basins that expand, contract, and split in the middle like worn-out coats.

Geologic time is long, but the forces that give shape to the land are timeless and constant. Processes of creation and destruction are continually at work to fashion the seemingly eternal structure upon which humans live and work. Two types of forces interact to produce those infinite local variations in the surface of the Earth called *landforms:* (1) forces that push, move, and raise the Earth's surface and (2) forces that scour, wash, and wear down the surface.

Mountains rise and are then worn away. The eroded materials—soil, sand, pebbles, rocks—are transported to new locations and help create new landforms. How long these processes have worked, how they work, and their effects are the subject of this chapter.

Much of the research needed to create the story of landforms results from the work of geomorphologists. A branch of the fields of geology and physical geography, *geomorphology* is the study of the origin, characteristics, and development of landforms. It emphasizes the study of the various processes that create landscapes. Geomorphologists examine the erosion, transportation, and deposition of materials and the interrelationships among climate, soils, plant and animal life, and landforms.

In a single chapter, we can only begin to explore the many and varied contributions of geomorphologists. After discussing the contexts within which landform change takes place, we consider the forces that are building up the Earth's surface and then review the forces wearing it down.

3.1 Earth Materials

The rocks of the Earth's crust vary according to mineral composition. Rocks are composed of particles that contain various combinations of such common elements as oxygen, silicon, aluminum, iron, and calcium, together with less-abundant elements. A particular chemical combination that has a hardness, density, and definite crystal structure of its own is called a **mineral.** Some well-known minerals are quartz, feldspar, and mica. Depending on the nature of the minerals that form them, rocks are hard or soft, more or less dense, one color or another, or chemically stable or not. While some rocks resist decomposition, others are very easily broken down. Among the more common varieties of rock are granites, basalts, limestones, sandstones, and slates.

Although one can classify rocks according to their physical properties, the more common approach is to classify them by the way they formed. The three main groups of rock are igneous, sedimentary, and metamorphic.

Igneous Rocks

Igneous rocks are formed by the cooling and solidification of molten rock. Openings in the crust give molten rock an opportunity to find its way into or onto the crust. When the molten rock cools, it solidifies and becomes igneous rock. The name for underground molten rock is *magma;* aboveground, it is *lava. Intrusive* igneous rocks are formed below ground level by the solidification of magma, whereas *extrusive* igneous rocks are created above ground level by the solidification of lava (**Figure 3.1**).

The composition of magma and lava and, to a limited extent, the rate of cooling determine the minerals that form. The rate of cooling is mainly responsible for the size of the crystals. Large crystals of quartz—a hard mineral—form slowly beneath the surface of the Earth. When combined with other minerals, quartz forms the intrusive igneous rock called *granite.*

The lava that oozes out onto the Earth's surface and makes up a large part of the ocean basins becomes the extrusive igneous rock called *basalt,* the most common rock on the Earth's surface. If, instead of oozing, the lava erupts from a volcano crater, it may cool

(a) Basalt (igneous)

(b) Limestone (sedimentary)

(c) Shale (sedimentary)

(d) Gneiss (metamorphic)

Figure 3.1 **Various rock types.**
(a) © McGraw-Hill Education/Bob Coyle; (b) Source: Photo by I.J. Witkind/U.S. Geological Society; (c) © McGraw-Hill Education/ Bob Coyle; (d) © McGraw-Hill Education/Jacques Cornell.

very rapidly. Some of the igneous rocks formed in this manner contain cavities and are light, such as *pumice.* Some may be glassy, as is *obsidian.* The glassiness occurs when lava meets standing water and cools suddenly.

Sedimentary Rocks

Some **sedimentary rocks** are composed of particles of gravel, sand, silt, and clay that were eroded from already existing rocks. Surface waters carry the sediment to oceans, marshes, lakes, or tidal basins. Compression of these materials by the weight of additional deposits on top of them and a cementing process brought on by the chemical action of water and certain minerals cause sedimentary rock to form.

Sedimentary rocks evolve under water in horizontal beds called *strata* (**Figure 3.2**). Usually one type of sediment collects in a given area. If the particles are large and rounded—for instance, the size and shape of gravel—a gravelly rock called *conglomerate* forms. Sand particles are the ingredient for *sandstone,* whereas silt and clay form *shale* or *siltstone.*

Sedimentary rocks also derive from organic material, such as coral, shells, and marine skeletons. These materials settle into beds in shallow seas, forming *limestone.* If the organic material forms mainly from decomposing vegetation, it can develop into a sedimentary rock called *bituminous coal. Petroleum* is also a biological product, formed during the millions of years of burial by chemical reactions that transform some of the organic material into liquid and gaseous compounds. The oil and gas are light; therefore, they rise through the pores of the surrounding rock to places where low-permeability rocks such as shale block their upward movement. Sedimentary rocks vary considerably in color (from coal black to chalk white), hardness, density, and resistance to chemical decomposition.

Large parts of the continents contain sedimentary rocks. For example, nearly the entire eastern half of the United States is overlain with these rocks. Such formations indicate that, in the geologic past, seas covered larger proportions of the Earth than they do today.

Metamorphic Rocks

Metamorphic rocks are formed from igneous and sedimentary rocks by Earth's forces that generate heat, pressure, or chemical reaction. The word *metamorphic* means "changed shape." The internal Earth forces may be so great that heat and pressure change the mineral structure of a rock, forming new rocks. For example, under great pressure, shale, a sedimentary rock, becomes *slate,* a rock with different properties. Limestone, under certain conditions, may become *marble,* and granite may become *gneiss* (pronounced nice). Materials metamorphosed at great depths and exposed only after overlying surfaces have been slowly eroded away are among the oldest rocks known on Earth. Like igneous and sedimentary rocks, however, their formation is a continuing process.

Rocks are the constituent ingredients of most landforms. The strength or weakness, permeability, and mineral content control the way rocks respond to the forces that shape and reshape them. Two principal processes alter rocks: (1) the forces that tend to build landforms up and (2) the gradational processes that wear landforms

Figure 3.2 **The sedimentary rocks of the Grand Canyon** in Arizona are evident in this photograph. © *John Wang/Photodisc / Getty Images RF.*

a) Cementation and compaction (lithification)
b) Heat and pressure
c) Weathering, transportation, and deposition
d) Cooling and solidification

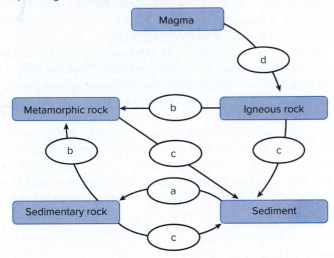

Figure 3.3 **The rock cycle.** *Source: Adapted from McConnell et al., The Good Earth, p. 209, Checkpoint 7.22, © McGraw-Hill, 2008.*

down. All rocks are part of the *rock cycle* through which old rocks are continually transformed into new ones by these processes. No rocks have been preserved unaltered throughout the Earth's history. **Figure 3.3** shows the rock cycle and the processes that shaped it.

3.2 Geologic Time

The Earth was formed about 4.5 billion years ago. When we think of a person who lives to be 100 years old as having had a long life, it becomes clear that the Earth is incredibly old indeed. Because our usual concept of time is dwarfed when we speak of billions of years, it is useful to compare the age of the Earth with something more familiar.

Imagine that the height of the Willis Tower in Chicago represents the age of the Earth. The tower is 110 stories, or 412 meters (1447 ft), tall. In relative terms, even the thickness of a piece of paper laid on the rooftop would be too great to represent an average person's lifetime. Of the total building height, only 4.8 stories represent the 200 million years that have elapsed since the present ocean floors began to form.

At this moment, the landforms on which we live are ever so slightly being created and destroyed. The processes involved have been in operation for so long that any given location most likely was the site of ocean and land at a number of different times in its past. Many of the landscape features on Earth today can be traced back to millions of years. The processes responsible for building up and tearing down those features are occurring simultaneously, but usually at different rates.

Since the 1960s, scientists have developed a useful framework within which one can best study our constantly changing physical environment. Their work is based on the early 20th-century geologic studies of Alfred Wegener, who proposed the theory of **continental drift.** He believed that all landmasses were once united in one supercontinent, which Wegener named Pangaea ("all Earth"), and that over many millions of years the continents broke away from one another, slowly drifting to their current positions. Although Wegener's theory was initially rejected outright, new evidence and

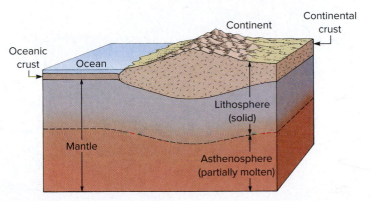

Figure 3.4 **The outer zones of the Earth (not to scale).** The *lithosphere* includes the crust. The *asthenosphere* lies below the lithosphere.

new ways of rethinking old knowledge have led to wide acceptance in recent years by Earth scientists of the idea of moving continents. Wegener's ideas were a forerunner of the broader **plate tectonics** theory, which is explained in the next section, "Movements of the Continents."

3.3 Movements of the Continents

The landforms mapped by cartographers are only the surface features of a thin cover of rock, the Earth's *crust* (**Figure 3.4**). Above the Earth's interior is a partially molten layer called the **asthenosphere.** It supports a thin but strong solid shell of rocks called the **lithosphere,** of which the outer, lighter portion is the Earth's crust. The crust consists of one set of rocks found below the oceans and another set that makes up the continents.

The lithosphere is broken into about 12 large and many small, rigid plates, each of which, according to the theory of plate tectonics, slides or drifts very slowly over the heavy, semimolten asthenosphere. A single plate often contains both oceanic and continental crust. **Figure 3.6a** shows that the North American plate, for example, contains the northwest Atlantic Ocean and most, but not all, of North America. The peninsula of Mexico (Baja California) and part of California are on the Pacific plate.

Scientists are not certain why lithospheric plates move. One reasonable theory suggests that heat and heated material from the Earth's interior rise by convection into particular crustal zones of weakness. These zones are sources for the divergence of the plates. The cooled materials then sink downward in subduction zones. In this way, the plates are thought to be set in motion. Strong evidence indicates that, about 225 million years ago, the entire continental crust was connected in one supercontinent, which was broken into plates as the seafloor began to spread. The divergence came from the widening of what is now the Atlantic Ocean. **Figure 3.5** shows four stages of the drifting of the continents.

Materials from the asthenosphere have been rising along the mid-Atlantic Ocean fracture and, as a result, the seafloor has continued to spread. The Atlantic Ocean is now 6920 kilometers (4300 mi) wide at the equator. If it diverges by a bit less than 2.5 centimeters (1 in.) per year, as scientists have estimated, one could calculate that the separation of the continents did, in fact, begin about 225 million years ago. Notice on **Figures 3.6a** and **3.7** how the ridge line that makes up the axis of the ocean runs parallel to the eastern coast of North and South America and the western coast of Europe and Africa.

Boundaries where plates move away from one another are called *divergent plate boundaries*. *Transform boundaries* occur where one plate slides horizontally past another plate, whereas at *convergent boundaries* two plates move toward each other (**Figure 3.6b**). Collisions sometimes occur as the lithospheric plates move. The pressure exerted at the intersections of plates can cause earthquakes, which over periods of many years change the

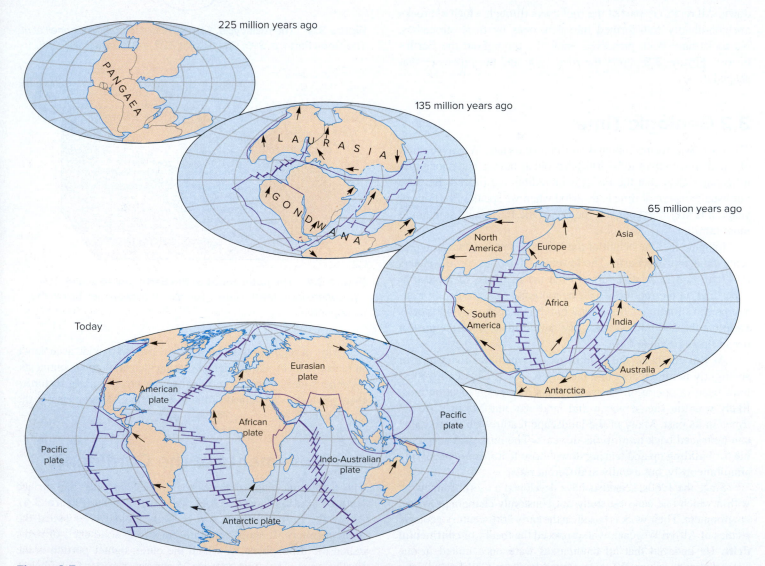

Figure 3.5   Reconstruction of plate movements during the past 225 million years. The northern and southern portions of Pangaea are called Laurasia and Gondwana, respectively. Some 225 million years ago, the continents were connected as one large landmass. After they split apart, the continents gradually moved to their present positions. Notice how India broke away from Antarctica and collided with the Eurasian landmass. The Himalayas were formed at the zone of contact. *Source: American Petroleum Institute.*

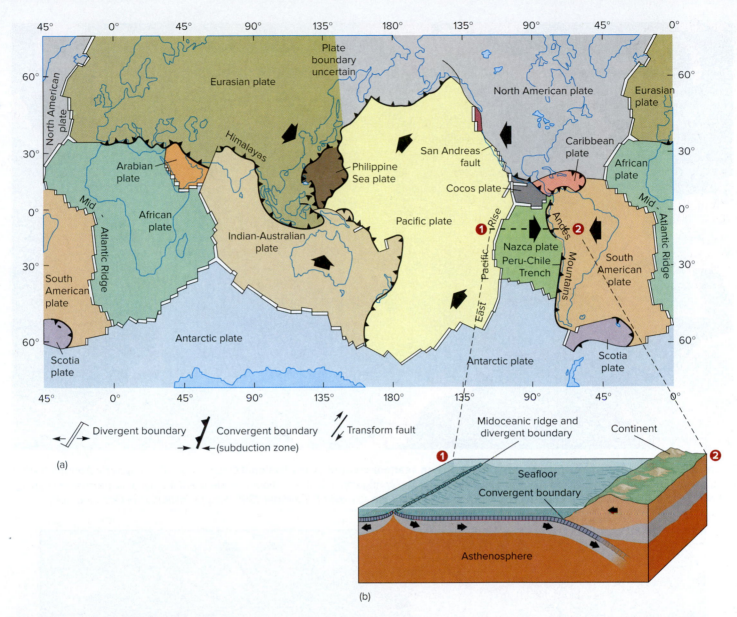

Figure 3.6 **(a) Principal lithospheric plates of the world.** Arrows indicate the direction of plate motion. **(b) Plate motion away from a divergent boundary toward a convergent boundary.**

shape and features of landforms. **Figure 3.8** shows the location of near-surface earthquakes for a recent time period. Comparison with Figure 3.6a illustrates that the areas of greatest earthquake activity are at plate boundaries. An example of this is the Haitian earthquake of January 12, 2010 (see p. 2 in Chapter 1). The capital city of Haiti, Port-au-Prince, lies close to the boundary between the Caribbean and North American plates.

The famous San Andreas fault in California is part of a long fracture separating two lithospheric plates, the North American and the Pacific. Earthquakes occur along **faults** (fractures in rock along which there has been movement) when the tension or compression at the junction becomes so great that only an Earth movement can release the pressure.

Despite the availability of scientific knowledge about earthquake zones, the general disregard for this danger is a difficult cultural phenomenon with which to deal (see the section on diastrophism). Every year there are hundreds and sometimes thousands of casualties resulting from inadequate preparation for earthquakes. In some highly populated areas, the chances that damaging earthquakes will occur are very great. The distribution of earthquakes shown in Figure 3.8 reveals the potential dangers to densely settled areas of Japan, the Philippines, parts of Southeast Asia, and the western rim of the Americas.

Convergent movement of the lithospheric plates results in the formation of deep-sea trenches and continental-scale mountain ranges, as well as in the occurrence of earthquakes. The continental crust is made up of lighter rocks than is the oceanic crust. Where plates with different types of crust at their edges converge, there is a tendency for the denser but thinner oceanic crust to be forced down into the asthenosphere. Deep trenches form below the ocean

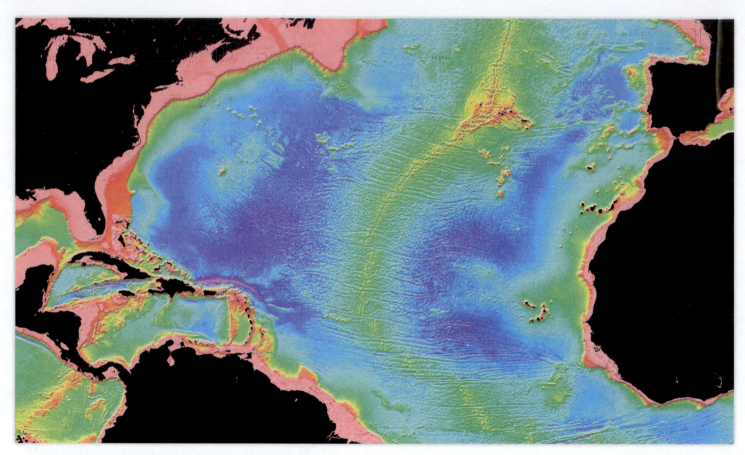

Figure 3.7   An accurate map of the North Atlantic Ocean seafloor created by the National Oceanic and Atmospheric Administration using gravity measurements taken from satellite readings. The configuration of the seafloor is evidence of the dynamic processes shaping continents and ocean basins. Darker ocean colors indicate greater depth. © *David T. Sandwell, 1995. Scripps Institution of Oceanography.*

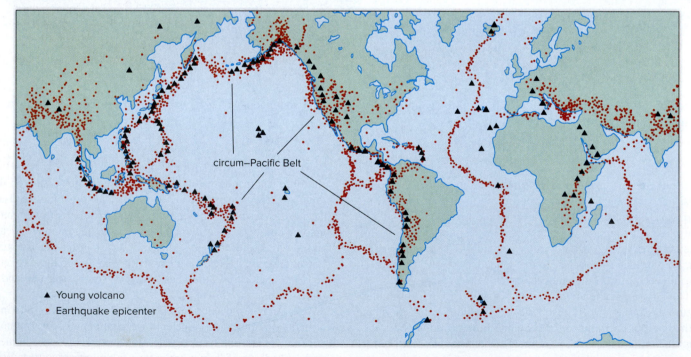

Figure 3.8   Locations of young volcanoes and the epicenters of earthquakes. Notice that they are concentrated at the margins of the lithospheric plates, as comparison with Figure 3.5 reveals. The most important concentration of earthquakes is in the circum-Pacific belt, which encircles the rim of the Pacific Ocean and is popularly known as the "ring of fire." Volcanoes can also grow in the middle of a plate. The volcanoes of Hawaii, for example, are located in the middle of the Pacific plate. *Map plotted by the Environmental Data and Information Service of NOAA; Earthquakes from U.S. Coast and Geodetic Survey.*

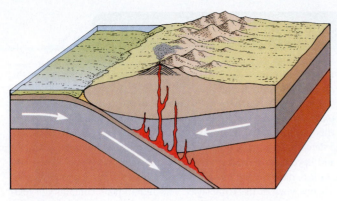

Figure 3.9   **The process of subduction.** When lithospheric plates collide, the denser oceanic crust is usually forced beneath the lighter continental material. See Figure 3.6a for the subduction zones of the world.

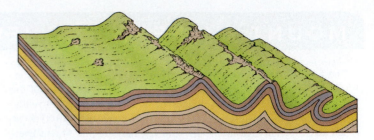

Figure 3.10   **Stylized forms of folding.** Degrees of folding vary from slight undulations of strata with little departure from the horizontal to highly compressed or overturned beds.

at these convergent boundaries. This type of collision is termed **subduction** (**Figure 3.9**). The subduction zones of the world are shown in Figure 3.6a.

Most of the Pacific Ocean is underlain by a plate that, like the others, is constantly pushing and being pushed. The continental crust on adjacent plates is being forced to rise and fracture, making an active volcano zone of the Pacific Ocean rim (sometimes called the "ring of fire"). The tremendous explosion that rocked Mount St. Helens in the state of Washington in 1980 is an example of continuing volcanic activity along the Pacific rim. A number of damaging earthquakes have occurred along the San Andreas fault in recent years. Their epicenters (the points on the Earth's surface directly above the foci of earthquakes) were on the fault. The most recent strong earthquake, in 2004, did considerable damage to the city of Paso Robles, California.

Plate intersections are not the only locations susceptible to readjustments in the lithosphere. As lithospheric plates have moved, the Earth's crust has been cracked or broken in virtually thousands of places. Some breaks are weakened to the point that they become *hot spots,* areas of volcanic eruption due to a rising plume of molten material. The molten material may explode out of a volcano or ooze out of cracks. Later in this chapter, when we discuss the Earth-building forces, we will return to the discussion of volcanic activity.

3.4 Tectonic Forces

The Earth's crust is altered by the constant forces resulting from plate movement. *Tectonic* (generated from within the Earth) forces shaping and reshaping the Earth's crust are of two types, either diastrophic or volcanic. **Diastrophism** is the great pressure acting on the plates that deforms them by folding, twisting, warping, breaking, or compressing rock. **Volcanism** is the force that transports heated material to or toward the surface of the Earth. When particular places on the continents are subject to diastrophism or volcanism, the changes that take place can be as simple as the bending and cracking of rock or as dramatic as lava exploding from the crater and sides of Mount St. Helens.

Diastrophism

In the process of plate tectonics, pressures build in various parts of the Earth's crust, and slowly, usually over thousands of years, the crust is transformed (see "Mount Everest: The Jewel in the Crown," p. 54). By studying rock formations, geologists are able to trace the history of the development of a region. Over geologic time, most continental areas have been subjected to both tectonic and gradational activity—building up and tearing down. They usually have a complex history of broad warping, folding, faulting, and leveling. Some flat plains in existence today may hide a history of great mountain development in the past.

Broad Warping

Great tectonic forces resulting from the movement of continents may bow an entire continent. Also, the changing weight of a large region may result in the **warping** of the surface. For example, the down-warping of the eastern United States is evident in the many irregularly shaped stream estuaries. As the coastal area warped downward, the sea advanced, forming estuaries and underwater canyons.

Folding

When the compressional pressure caused by plate movements is great, layers of rock are forced to buckle. The result may be a warping or bending effect, and a ridge or series of parallel **folds** may develop. **Figure 3.10** shows a variety of structures resulting from folding. The folds can be thrust upward many thousands of feet and laterally for many miles. The Ridge and Valley region of the eastern United States is, at present, low parallel mountains—300 to 900 meters (1000 to 3000 ft) above sea level—but the rock evidence suggests that the tops of the present mountains were once the valleys between 9100-meter (30,000-ft) crests (**Figure 3.11**).

Faulting

A fault is a break or fracture in rock along which movement has taken place. The stress causing a fault results in displacement of the Earth's crust along the fracture zone. **Figure 3.12** diagrammatically shows examples of fault types. There may be uplift on

MOUNT EVEREST: THE JEWEL IN THE CROWN

The fastest growing mountain range on Earth happens also to contain the world's tallest set of mountains. Nearly in the center of the Himalayas stands the world's highest peak, Mount Everest. Currently, Mount Everest is measured as standing 8844 meters (29,010 ft) above sea level. Recent measurements, however, indicate that Mount Everest and many of the other peaks, such as K-2, are growing at about 1 centimeter (1/2 in.) a year.

As the mountains build ever higher, their great weight softens underlying materials, causing the mountains to settle. In other words, there are two forces at work. The force building the mountains is that of the Indian plate moving northward, crashing into the Eurasian plate. The Himalayas, at the edge of the Eurasian plate, react to the great force by pushing higher and higher. But apparently a second force that works against the first keeps mountains on Earth from rising to heights of 15,000 to 18,000 meters (roughly 50,000 to 60,000 ft). One may think of these great mountains as being in a kind of equilibrium—the bigger they get, the heavier they get, and the more likely they are to sag.

The battle between the two plates began about 45 million years ago. Usually, one plate is forced under the more stationary plate (subduction). In this case, however, the rocks of the two plates are similar in weight and density. Thus, subduction has not taken place, and what normally would have been a modest crinkle on the Earth's surface turned into the tallest, most rugged mountain range on Earth. Particularly interesting is one view of the Himalayas. The Indian plate contains the relatively low-elevation Indian subcontinent and, as a result, the view of the Himalayas from the plains of northern India is one of the great sights on Earth. No wonder Mount Everest is the jewel in the mountain climber's crown. Edmund Hillary and Tenzing Norgay were the first to reach the summit, in 1953. Since then, more than 4000 climbers have reached the top, most of them since 1990; 280 have died attempting to scale Mount Everest.

© Dinodia Photo/Corbis/Getty Images RF

one side of the fault or downthrust on the other. In some cases, a steep slope known as a fault *escarpment,* which may be several hundred feet high and several hundred miles long, is formed. The stress can push one side up and over the other side, or a separation away from the fault may cause the sinking of land, creating a *rift valley* (**Figure 3.13**).

Many fractures are merely cracks (called *joints*) with little noticeable movement along them. In other cases, however, mountains such as the Sierra Nevadas of California have risen as the result of faulting. Sometimes, the movement has been horizontal along the surface rather than upward or downward. The San Andreas transform fault, shown in **Figures 3.14** and **3.15**, is such a case.

Whenever movement occurs along a fault, or at another point of weakness, an earthquake results. The greater the movement, the greater the magnitude of the earthquake (see "Scaling Earthquakes," p. 58). Stress builds in rock as tectonic forces are applied and, when a critical point is finally reached, an earthquake occurs and tension is reduced.

The earthquake that occurred in Alaska on Good Friday in 1964 was one of the strongest measured, with a magnitude of 9.2 on the Richter scale. Although the stress point of that earthquake was

Figure 3.11 **(a) The Ridge and Valley region of Pennsylvania,** now eroded to hill lands, is the relic of 9100-meter (30,000-ft) folds that were reduced to form *synclinal* (downarched) hills and *anticlinal* (uparched) valleys. The rock in the original troughs, having been compressed, was less susceptible to erosion. **(b) A syncline in Maryland on the border of Pennsylvania,** exposed by a road cut. *(b) © Mark C. Burnett/Science Source.*

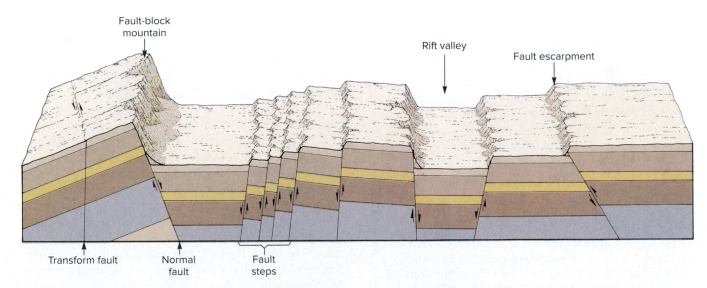

Figure 3.12 **Faults,** in their great variation, are common features of mountain belts where deformation is great. The different forms of faulting are categorized by the direction of movement along the plane of fracture. The features shown here would not occur in a single setting.

below ground 121 kilometers (75 mi) away from Anchorage, vibrations called *seismic waves* caused Earth movement in the weak clays underneath the city. Sections of Anchorage slid downhill, and part of the business district dropped 3 meters (10 ft).

If an earthquake, a volcanic eruption, or an underwater landslide occurs below an ocean, jolting the waters above, the movement can generate sea waves called **tsunami** (from the Japanese *tsu,* for "harbor," and *nami,* for "wave") (see "Tsunami," p. 60). Traveling at great speed in the open ocean, the waves may be hardly noticeable, resembling a fast-moving tide—the reason they often are mistakenly referred to as "tidal waves" in spite of the fact that they are unrelated to tides. As the waves near shore and enter shallower water, however, friction with the ocean floor causes the waves to slow down, producing a buildup of water that

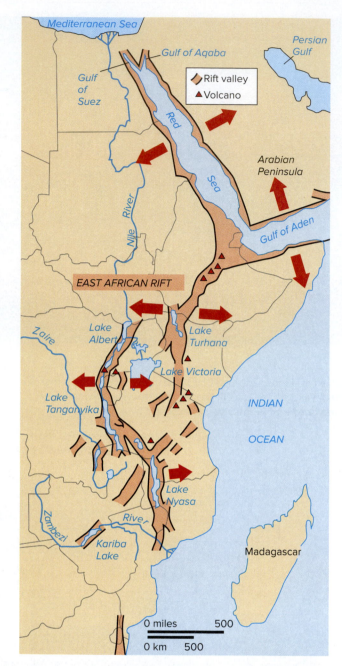

Figure 3.13　The rift valleys of East Africa. Great fractures in the Earth's crust resulted in the creation, through subsidence, of an extensive rift valley system (see Figure 3.12) in East Africa. The parallel faults, some reaching more than 610 meters (2000 ft) below sea level, are bordered by steep walls of the adjacent plateau, which rises to 1500 meters (5000 ft) above sea level and from which the structure dropped.

can reach 15 meters (50 ft) or more above sea level. The water sweeps inland with massive force once it hits the shore, particularly when the narrowed topography of harbors and inlets focuses the waves into smaller spaces.

Earthquakes occur daily in hundreds of places throughout the world. Most are slight and only noticeable on *seismographs,* instruments that record seismic waves. But from time to time, large-scale earthquakes occur, such as those in Iran

in 2003 (35,000 deaths) and China in 1976 (242,000 deaths). Most earthquakes take place on the Pacific rim (see Figure 3.8), where stress from the converging lithospheric plates is greatest. The Aleutian Islands of Alaska, Japan, Central America, and Indonesia experience a number of moderately severe earthquakes each year. The Haitian earthquake did massive amounts of damage (see p. 2 in Chapter 1). Major earthquakes in the San Francisco Bay area caused considerable damage (**Figure 3.16**). In recent years, major earthquakes and volcanic activity have also occurred in non-Pacific areas, such as Turkey, Iran, and Algeria.

The huge earthquake that struck Kashmir on October 8, 2005, was a reminder that the Himalayas are along the boundary of three tectonic plates, making countries all the way from Afghanistan and Pakistan in the west to Myanmar in the east subject to quakes. For further information on earthquakes and other earth processes go to **usgs.gov.**

Volcanism

The second tectonic force is volcanism. The most likely places through which molten materials can move toward the surface are at or near the intersections of plates. However, other zones, such as hot spots, are also subject to volcanic activity (**Figure 3.17**). The volcanoes of Hawaii, for example, formed above a relatively stationary hot spot in the Earth's interior.

If sufficient internal pressure forces the magma upward, weaknesses in the crust, or faults, enable molten materials to reach the surface. The material ejected onto the Earth's surface may arrive as a series of explosions, forming a steep-sided cone composed of alternate layers of solidified lava and ash and cinders, termed a *strato* or *composite volcano* (**Figure 3.18a**). The eruption may also be without explosions, forming a gently sloping *shield volcano* (**Figure 3.18b**).

The major volcanic belt of the world coincides with the major earthquake and fault zones. This belt occurs at the convergence of plates. A second zone of volcanic activity is at diverging plate boundaries, such as in the center of the Atlantic Ocean.

Molten material can either flow smoothly out of a crater or be shot into the air with explosive force. Some relatively quiet volcanoes have long, gentle slopes indicative of smooth flow, whereas explosive volcanoes have steep sides. Steam and gases are constantly escaping from the nearly 300 active volcanoes in the world today. The city of Pompeii, Italy, was partially buried under about 5 meters (16 ft) of ash and pumice as a result of the eruption of Mt. Vesuvius in A.D. 79.

When pressure builds, a crater can become a boiling cauldron with steam, gas, lava, and ash billowing out (**Figure 3.19**). In the case of Mount St. Helens in 1980, a large bulge had formed on the north slope of the mountain. An earthquake occurred and an explosion followed, shooting debris into the air, completely devastating an area of about 400 square kilometers (150 sq mi), causing about 1 centimeter (0.4 in.) of ash to rain down on eastern Washington and parts of Idaho and Montana and reducing the elevation of the mountain by more than 300 meters (1000 ft).

In many cases, the pressure beneath the crust is not intense enough to allow magma to reach the surface. In these instances,

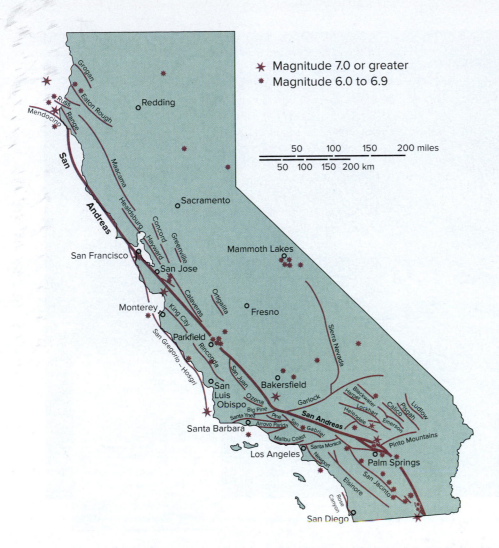

Figure 3.14 **The San Andreas fault system in California, with the epicenters of magnitude 6.0 and greater earthquakes since 1900.** *Source: Map updated from "The San Andreas Fault System, California," ed. by Robert E. Wallace, U.S. Geological Survey Professional Paper 1515, 1990.*

magma solidifies into a variety of underground formations of igneous rock that on occasion affect surface landform features. However, gradational forces may erode overlying rock, so that igneous rock, which is usually hard and resists erosion, becomes a surface feature. The Palisades, a rocky ridge facing New York City from the west, and Stone Mountain, near Atlanta, Georgia, are this type of landform.

On other occasions, a rock formation below the Earth's surface may allow the growth of a mass of magma but denies exit to the surface because of dense overlying rock. Through the pressure it exerts, however, the magmatic intrusion may still buckle, bubble, or break the surface rocks. In addition, domes of considerable size may develop, such as the Black Hills of South Dakota. A by-product of volcanic eruptions is the large amount of ash that spews into the atmosphere, sometimes affecting weather and climate patterns thousands of miles away. The famous volcano Krakatoa, near Java, exploded in 1883, affecting world climate patterns for over a year after the eruption.

Evidence from the past shows that lava has sometimes flowed through fissures or fractures without forming volcanoes. These oozing lava flows have covered ocean floors. On continents, the Deccan Plateau of India and the Columbia Plateau of the Pacific Northwest in the United States are examples of this type of process (**Figure 3.20**).

3.5 Gradational Processes

Gradational processes are responsible for the reduction of land surface. If a land surface where a mountain once stood is now a low, flat plain, gradational processes have been at work. The worn, scraped, or blown away material is deposited in new places and, as a result, new landforms are created. In terms of geologic time, the Rocky Mountains are a recent phenomenon; gradational processes are active there, just as on all land surfaces, but they have not yet had time to reduce these huge mountains.

Three kinds of gradational processes occur: *weathering, mass movement,* and *erosion.* Both mechanical and chemical weathering processes play a role in preparing bits of rock for the creation of soils and for movement to new sites by means of gravity or erosion. Mass movement transfers downslope by gravity any loosened, higher-lying material, including rock debris and soil; and the agents of running water, moving ice, wind, waves, and currents erode and carry these loose materials to other areas, where landforms are created or changed.

Figure 3.15   **View along part of the San Andreas fault, looking northward toward San Francisco.** Here, Upper Crystal Springs Reservoir occupies part of the fault zone. A transform fault (see Figure 3.12), the San Andreas marks a part of the slipping boundary between the Pacific and the North American plates. The inset map shows the relative southward movement of the North American plate; dislocation averages about 1 centimeter (0.4 in.) per year. © *Georg Gerster/Science Source.*

Figure 3.16   During the 6.7 (Richter scale) Loma Prieta earthquake of 1989 in the San Francisco Bay area, the double-deck Cypress Freeway collapsed. © *Lloyd Cluff/Corbis Documentary/Getty Images.*

Weathering

The breakdown and decomposition of rocks and minerals at or near the Earth's surface in response to atmospheric factors (water, air, and temperature) is called **weathering.** It occurs as a result of both mechanical and chemical processes.

Mechanical Weathering

Mechanical weathering is the physical disintegration of earth materials at or near the surface; that is, larger rocks are broken into smaller pieces. A number of processes cause mechanical weathering. The three most important are frost action, the development of salt crystals, and root action.

If water that soaks into a rock (between particles or along joints) freezes, ice crystals grow and exert pressure on the rock. When the process is repeated—freezing, thawing, freezing, thawing, and so on—the rock begins to disintegrate. Potholes on city streets are often created in this way. Salt crystals act similarly in

SCALING EARTHQUAKES

In 1935, C. F. Richter devised a scale of earthquake *magnitude,* a measure of the energy released during an earthquake. An earthquake is really a form of energy expressed as wave motion passing through the surface layer of the Earth. Radiating in all directions from the earthquake focus, seismic waves gradually dissipate their energy at increasing distances from the *epicenter.* On the Richter scale, the amount of energy released during an earthquake is estimated by measurement of the ground motion that occurs. Seismographs record earthquake waves, and by comparison of wave heights, the relative strength of quakes can be determined. Although Richter scale numbers run from 0 to 9, there is no absolute upper limit to earthquake severity. Presumably, nature could outdo the magnitude of the most intense earthquakes recorded so far, which reached 9.5.

Because magnitude, as opposed to intensity (a measure of an earthquake's size by its effect on people and buildings), can be measured accurately, the Richter scale has been widely adopted. Nevertheless, it is still only an approximation of the amount of energy released in an earthquake. In addition, the height of the seismic waves can be affected by the rock materials under the seismographic station, and some seismologists believe that the Richter scale underestimates the magnitude of major tremors.

Richter Scale[a]	Urban Effects of Earthquakes Occurring near the Earth's Surface
1, 2	Not felt
3	Felt by some
4	Windows rattle
5	Felt widely, slight damage near epicenter
6	Poorly constructed buildings destroyed; others damaged within 10 km (6 mi)
7	Widespread damage up to 100 km (62 mi)
8	Great damage over several hundred kilometers
9 and above	Rare great earthquakes: Chile (1960), 9.5; Alaska (1964), 9.2; Sumatra (2004), 9.2

[a]The Richter scale is logarithmic; each increment of a whole number signifies a 10-fold increase in magnitude. Thus, a magnitude 4 earthquake has an effect 10 times greater than a magnitude 3 earthquake. The actual impact of earthquakes on humans varies not only with the severity of the quake and with such secondary effects such as tsunami or landslides, but also with the density of population and the quality of the buildings in the affected areas.

dry climates, where groundwater is drawn to the surface by *capillary action* (water rising because of surface tension). This action is similar to the process in plants whereby liquid plant nutrients move upward through the stem and leaf system. Evaporation leaves behind salt crystals that form, expand, and disintegrate rocks. Roots of trees and other plants may also find their way into rock joints; as they grow, they break and disintegrate the rock. These are all mechanical processes because they are physical in nature and do not alter the chemical composition of the material upon which they act.

Chemical Weathering

A number of **chemical weathering** processes cause rock to decompose rather than disintegrate. In other words, the minerals composing rocks separate into component parts by chemical reaction rather than fragmentation. The three most important processes are oxidation, hydrolysis, and carbonation. Because each of these depends on the availability of water, less chemical weathering occurs in dry and cold areas than in moist and warm ones.

Oxidation occurs when oxygen combines with mineral components, such as iron, to form oxides. As a result, some rock areas in contact with oxygen begin to decompose. Decomposition also results when water comes into contact with certain rock minerals, such as aluminosilicates. The chemical change that occurs is called *hydrolysis.* When carbon dioxide gas from the atmosphere dissolves in water, a weak carbonic acid forms. The action of the acid, called *carbonation,* is particularly evident on limestone because the calcium bicarbonate salt created in the process readily dissolves and is removed by groundwater and surface water.

Weathering, either mechanical or chemical, does not itself create distinctive landforms. Nevertheless, it prepares rock particles for erosion and for the creation of soil. After the weathering process decomposes rock, the force of gravity and the erosional agents of running water, wind, and moving ice are able to carry the weathered material to new locations.

Mechanical and chemical weathering creates *soil,* the thin layer of fine material containing organic matter, air, water, and weathered rock materials that rests on the solid rock below it. The type of soil formed is a function of the climate of the region in which it occurs and of the kind of bedrock below it. Temperature and rainfall act on minerals, in conjunction with the decaying of overriding vegetation, to form soils. The topic of soils is discussed in greater detail in Chapter 4.

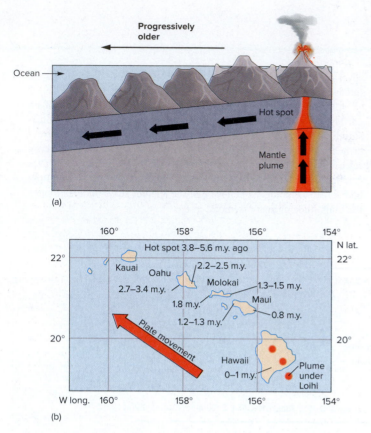

(a)

(b)

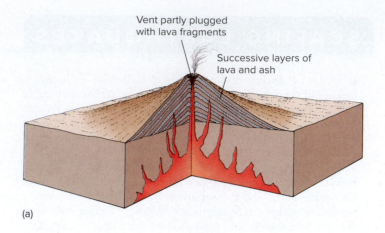

(a)

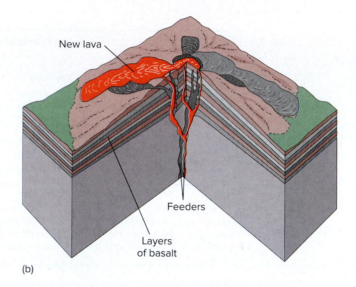

(b)

Figure 3.17 **(a) Plumes,** narrow columns of hot mantle rock, can form "hot spots" of volcanism on the Earth's surface. Some plumes rise beneath the centers of oceanic plates rather than at their intersections. A plume under Hawaii rises in the middle of the Pacific plate. As the plate moved over the plume, a chain of volcanoes formed. Each was carried away from the hot spot as the Pacific plate moved northwestward. **(b) Ages of volcanic rock in the Hawaiian islands.** Note that the volcanoes become progressively older to the northwest. M.Y. means million years. The island of Hawaii contains two active volcanoes, shown by red dots, Mauna Loa and Mauna Kea.

Figure 3.18 **(a) Composite volcanoes,** such as the one diagrammed, are composed of alternate layers of solidified lava and of ash and cinders. Sudden decompression of gases contained within lavas results in explosions of rock material to form ashes and cinders. **(b) Cutaway view of a shield volcano.** Composed of solidified lava flows, shield volcanoes are broad and gently sloping.

Mass Movement

The force of gravity—that is, the attraction of the Earth for bodies at or near its surface—is constantly pulling on all materials. The downslope movement of material due to gravity is called *mass wasting* or **mass movement.** Because it is more descriptive, this book uses the latter term. Small particles or huge boulders, if not held back by solid rock or other stable material, will fall down slopes. Spectacular acts of mass movement include avalanches and landslides. More widespread, but less noticeable, are mass movements such as soil-creep and the flow of mud down hillsides (**Figure 3.21**). In general, all nonrigid formations, such as water, ice, and wind, move downslope under the influence of gravity.

Especially in dry areas, a common but very dramatic landform created by the accumulation of rock particles at the base of hills and mountains is *talus,* pictured in **Figure 3.21a**. As pebbles, particles of rock, or even larger stones break away from exposed bedrock on a mountainside because of weathering, the masses fall and

accumulate, producing large, conelike landforms. The larger rocks travel farther than the fine-grained sand particles, which remain near the top of the slope.

Erosional Agents and Deposition

Erosional agents, such as wind, water, and glaciers, carve already existing landforms into new shapes. Fast-moving agents carry debris, and slow-moving agents drop it. The material that has been worn, scraped, or blown away is deposited in new places, and new landforms are created. Each erosional agent is associated with a distinctive set of landforms.

Running Water

Running water is a powerful erosional agent. Water, whether flowing across land surfaces or in stream channels, plays an enormous role in wearing down and building up landforms.

TSUNAMI

The Japanese earthquake of March 11, 2011, registered magnitude 9.0 on the Richter scale (greater than 8 on the moment magnitude scale) and was responsible for tsunami as high as a 10-story building (40.5 m or 133 ft). Over 15,000 people died, and about 1 million buildings were either damaged or destroyed. The tsunami caused the Fukushima Daiichi nuclear disaster. In like manner but resulting in an even greater loss of life was the Indian Ocean tsunami of December 26, 2004, in which 283,000 people were killed. The effects of the Indian Ocean tsunami were felt mainly along the coasts of Indonesia—the island of Sumatra, in particular—Thailand, Sri Lanka, India, and South Africa. Images of the city of Banda Aceh, at the northern tip of Sumatra, show the area several months before the tsunami struck and two days afterward.

Tsunami are generated by an underwater seismic disturbance such as a submarine earthquake, eruption, or landslide. In the Indian Ocean case, water rushed in to fill the depression caused by the falling away of the ocean bottom, a 9.2 magnitude earthquake. The quake occurred along a 1300-kilometer (800-mi) boundary between the Indian-Australian tectonic plate and the Eurasian plate, pushing the former some 15 meters (50 ft) beneath the latter. At first, water was drawn out from the coasts and then rushed back in as tsunami. In the Japan case, subduction at the boundary between the Pacific plate and the Eurasian plate caused by the earthquake forced the seafloor to rise, pushing water in all directions. The resulting tsunami was devastating. Many towns in Japan were destroyed. Places as far away as British Columbia, Washington, and Oregon were hit by waves as high as 2.4 meters (8 ft). Damaged materials washed ashore on North America's west coast. In both cases, the initial waves traveled very fast, about the speed of a jet plane (640 km or 440 mi per hour), but were not very high. As the waves approached land and the water became shallower, their speed decreased to about 48 kilometers (30 mi) per hour, but their height increased. The water came ashore with terrific force. One cubic meter of water weighs nearly a ton. Trees, automobiles, pieces of roadway, or buildings became lethal projectiles as the rushing water pushed inland, inundating low-lying areas along the shore.

In an attempt to minimize loss of life in Pacific coastal communities, the U.S. Coast and Geodetic Survey established a tsunami early warning system in 1948. Any major seismic disturbance in the Pacific region that is capable of generating tsunami is reported to the Tsunami Warning Center in Honolulu. If tsunami are detected, the center relays information about their source, speed, and estimated time of arrival to low-lying coastal communities in danger. No doubt this system saved many lives in Japan. It was the Indian Ocean tsunami that prompted the creation of an Indian Ocean tsunami warning system.

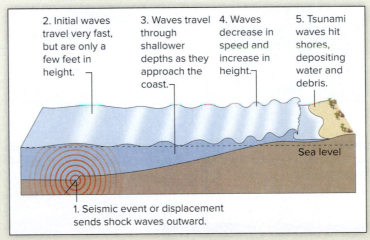

2. Initial waves travel very fast, but are only a few feet in height.

3. Waves travel through shallower depths as they approach the coast.

4. Waves decrease in speed and increase in height.

5. Tsunami waves hit shores, depositing water and debris.

Sea level

1. Seismic event or displacement sends shock waves outward.

© Digital Globe

(a)

(b)

Satellite images of Banda Aceh on Sumatra **(a)** on June 23, 2004 and **(b)** two days after the tsunami of December 25. *Source: USGS Center for Coastal Geology.*

(a)

(b)

(c)

Figure 3.19 **(a) Mount St. Helens, Washington, before it erupted on May 18, 1980. (b) The eruption** sent a cloud of steam and hot ash out of the cone 15 kilometers (9 mi) into the atmosphere, and a landslide carried ash and coarser debris down the slopes of the mountain. **(c) The explosion** blew away the top and much of one side of the volcano, lowering its summit by nearly 400 meters (1300 ft). *(a–c) Source: Courtesy U.S. Geological Survey.*

Figure 3.20 **Fluid lavas created the Columbia Plateau, covering an area of 130,000 square kilometers (50,000 sq mi).** Some individual flows were more than 100 meters (300 ft) thick and spread up to 60 kilometers (40 mi) from their original fissures. *Source: © McGraw-Hill Education.*

(a)

(b)

Figure 3.21 **(a) Rockfall from this butte has created talus,** an accumulation of broken rock at the base of a cliff. **(b) Creeping soil** has caused trees to tilt. *(a) © Robert N. Wallen; (b) © Victoria Getis.*

Running water's ability to erode depends upon several factors: (1) the amount of precipitation; (2) the length and steepness of the slope; and (3) the kind of rock and vegetative cover. Steeper slopes and faster-flowing water result, of course, in more rapid erosion. Vegetative cover sometimes slows the flow of water. When this vegetation is reduced, perhaps because of farming or livestock grazing, erosion can be severe, as shown in **Figure 3.22**.

Even the impact of precipitation—heavy rain or hail—can cause erosion. After hard rain dislodges soil, the force of the rain causes the surface to become more compact; therefore, further precipitation fails to penetrate the soil. The result is that more water, prevented from seeping into the ground, becomes available for surface erosion. Soil and rock particles in the water are carried to streams, leaving behind gullies and small stream channels.

Both the force of water and the particles contained in the stream are agents of erosion. Abrasion, or wearing away, takes place when particles strike against stream channel walls and along the streambed. Because of the force of the current, large particles, such as gravel, slide along the streambed, grinding rock on the way.

Floods and rapidly moving water are responsible for dramatic changes in channel size and configuration, sometimes forming new channels. In cities where paved surfaces cover soil that would otherwise have absorbed or held water, runoff is accentuated so that nearby rivers and streams rapidly increase in size and velocity after heavy rains. Oftentimes, flash floods and severe erosion result.

Small particles, such as clay and silt, are suspended in water and constitute—together with material dissolved in the water or dragged along the bottom—the *load* of a stream. Rapidly moving floodwaters carry huge loads. As high water or floodwater recedes, and stream velocity decreases, sediment contained within the stream no longer remains suspended, and particles begin to settle. Heavy, coarse materials drop the quickest; finer particles are carried longer and transported farther. The decline in velocity and the resulting deposition are especially pronounced and abrupt when streams meet slowly moving water in bays, oceans, and lakes. Silt and sand accumulate at the intersections, creating *deltas,* as pictured in **Figure 3.23**.

A great river, such as the Chang Jiang (Yangtze) in China, has a large, growing delta, but less prominent deltas exist at the mouths of many streams. Until the recent completion of the Aswan Dam, the huge delta of the Nile River had been growing. Now much of the silt is being dropped in Lake Nasser behind the dam, which is far upstream from the delta.

Figure 3.22 **Gullying** can result from heavy rain and poor farming techniques, including overgrazing by livestock or many years of continuous row crops. Surface runoff removes topsoil easily when vegetation is too thin to protect it. *Source: USDA-Natural Resources Conservation Service.*

Figure 3.23   **The Ganges and Brahmaputra Rivers are fed by the streams from the Himalayas.** Note the brown sediment entering the Bay of Bengal. *Source: NASA.*

In plains adjacent to streams, land is sometimes built up by the deposition of *stream load*. If the deposited material is rich, it may be a welcome and necessary part of farming activities, such as that historically known in Egypt along the Nile. Should the deposition be composed of sterile sands and boulders, however, formerly fertile bottomland may be destroyed. By drowning crops or inundating inhabited areas, the floods themselves, of course, may cause great human and financial loss. More than 900,000 lives were lost in the floods of the Huang He (Yellow River) of China in 1887.

Stream Landscapes

A landscape is in a particular state of balance between the uplift of land and its erosion. Rapid uplift is not followed by nicely ordered stages of erosion. Recall that uplift and erosion take place simultaneously. At a given location, one force may be greater than the other at a given time, but as yet there is no way to predict accurately the "next" stage of landscape evolution.

Perhaps the most important factor differentiating the effect of streams on landforms is whether the recent climate (for example, the past several million years) has tended to be humid or arid.

Stream Landscapes in Humid Areas   Perhaps weak surface material or a depression in rock allows the development of a stream channel. In its downhill run in mountainous regions, a stream may flow over precipices, forming *falls* in the process. The steep downhill gradient allows streams to flow rapidly, cutting narrow, V-shaped channels in the rock (**Figure 3.24a**). Under these conditions, the erosional process is greatly accelerated. Over time, the stream may have worn away sufficient rock for the falls to become rapids, and the stream channel becomes incised below the height of the surrounding landforms. This is evident in the upper reaches of the Delaware, Connecticut, and Tennessee rivers.

In humid areas, the effect of stream erosion is to round landforms. Streams flowing down moderate gradients tend to carve valleys that are wider than those in mountainous areas. Surrounding hills become rounded, and valleys eventually become **floodplains** as they broaden and flatten. Streams work to widen the floodplain. Their courses meander, constantly carving out new river channels. The channels left behind as new ones are cut become *oxbow* lakes, hundreds of which are found in the Mississippi River floodplain. An oxbow lake is crescent-shaped and occupies the abandoned channel of a stream meander (**Figure 3.24b**).

In nearly flat floodplains, the highest elevations may be the banks of rivers, where *natural levees* are formed by the deposition of silt at river edges during floods. Floods that breach levees are particularly disastrous because floodwaters fill the floodplain as they equalize its elevation with that of the swollen river. The U.S. Army Corps of Engineers has augmented the natural levees in particularly susceptible areas, such as the banks of the lower Mississippi River, but some of these barriers failed during Hurricane Katrina in August 2005 (**Figure 3.25**).

Stream Landscapes in Arid Areas   A distinction must be made between the results of stream erosion in humid regions and those in arid areas. The lack of vegetation in arid regions greatly increases the erosional force of running water. Water originating in mountainous areas sometimes never reaches the sea if the channel runs through a desert. In fact, stream channels may be empty except during rainy periods, when water rushes down hillsides to collect and form temporary lakes called *playas*. In the process, **alluvium** (sand and mud) builds up in the lakes and at lower elevations, and *alluvial fans* are formed along hillsides (**Figure 3.26**). The fan is produced by the deposition of silt, sand, and gravel outward as the stream reaches lowlands at the base of the slope it traverses. If the process has been particularly long-standing, alluvial deposits may bury the eroded mountain masses. In desert regions in Nevada, Arizona, and California, it is not unusual to observe partially buried mountains poking through eroded material.

(a)

(b)

Figure 3.24 **(a) V-shaped valley** of a rapidly downcutting stream, the Yellowstone River in Wyoming. **(b) An oxbow-shaped lake** adjoining a meandering stream in Wyoming. *(a) © Robert N. Wallen; (b) Source: U.S. Geological Survey.*

Figure 3.25 **Aerial view of damage caused by Hurricane Katrina** the day after the hurricane hit, August 30, 2005. *Source: FEMA-Federal Emergency Management Agency.*

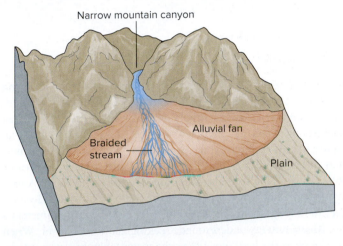

Figure 3.26 **Alluvial fans** are built where the velocity of streams is reduced as they flow out upon the more level land at the base of the mountain slope. The abrupt change in slope and velocity greatly reduces the stream's capacity to carry its load of coarse material. Deposition occurs, choking the stream channel and diverting the flow of water. With the canyon mouth fixing the head of the alluvial fan, the stream sweeps back and forth, building and extending a broad area of deposition. *Redrawn from Charles C. Plummer and David McGeary,* Physical Geology, *8th ed.*

Because streams in arid areas have only a temporary existence, their erosional power is less consistent than that of the freely flowing streams of humid areas. In some instances, they barely mark the landscape; in other cases, swiftly moving water may carve deep, straight-sided *arroyos*. Water may rush onto an alluvial plain in a complicated pattern resembling a multistrand braid, leaving in its wake an alluvial fan. The channels resulting from this rush of water are called *washes*. The erosional power of unrestricted running water in arid regions is dramatically illustrated by the steep-walled configuration of *buttes* and *mesas* (large buttes), such as those in Utah shown in **Figure 3.27**.

Figure 3.27 **Canyonlands National Park, Utah.** The resistant caprock of the mesa protects softer, underlying strata from downward erosion. Where the caprock is removed, lateral erosion lowers the surface, leaving the mesa as an extensive and pronounced relic of the former higher-lying landscape. © *SuperStock/Corbis RF.*

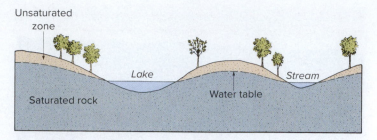

Figure 3.28   **The groundwater table** generally follows surface contours but in subdued fashion. Water flows slowly through the saturated rock, emerging at Earth depressions that are lower than the level of the water table. During a drought, the table is lowered and the stream channel becomes dry.

Groundwater

Some of the water supplied by rain and snow sinks underground into pores and cracks in rocks and soil, not in the form of an underground pond or lake but simply as subsurface material. When underground water accumulates, a zone of saturation called an *aquifer* forms, through which water can move readily. As indicated in **Figure 3.28**, the upper level of this zone is the **water table;** below it, the soils and rocks are saturated with water. A well may be drilled into an aquifer to ensure a supply of water. Groundwater moves constantly but very slowly (usually only centimeters a day). Most remains underground, seeking the lowest level. When the surface of the land dips below the water table, however, ponds, lakes, and marshes form. Some water finds its way to the surface by capillary action in the ground or in vegetation. When the ground surface extends below the level of the water table, the most common feature to develop is a stream.

Groundwater, particularly when combined with carbon dioxide, dissolves soluble materials by a chemical process called *solution.* Although groundwater tends to decompose many types of rocks, its effect on limestone is most spectacular. Many of the great caves of the world have been created by the underground movement of water through limestone regions. Water sinking through the overlying rock leaves carbonate deposits as it drips into empty spaces. The deposits hang from cave roofs (*stalactites*) and build upward from cave floors (*stalagmites*). In some areas, the uneven effect of groundwater erosion on limestone leaves a landscape pockmarked by a series of *sinkholes,* surface depressions in an area of collapsing caverns.

Karst topography refers to a large limestone region marked by sinkholes, caverns, and underground streams, as shown in **Figure 3.29**. East central Florida, a karst area, has suffered considerable damage from the creation and widening of sinkholes.

This type of topography gets its name from a region on the Adriatic Sea at the Italy-Slovenia border. The Mammoth Cave region of Kentucky, another karst area, has many kilometers of interconnected limestone caves.

Glaciers

Another agent causing erosion and deposition to occur is glaciers. Although they are much less extensive today, glaciers covered a large part of the Earth's land area as recently as 10,000 to 15,000 years ago. Many landforms were created by the erosional or depositional effects of glaciers.

Glaciers form only in very cold places with short or nonexistent summers, where annual snowfall exceeds annual snowmelt and evaporation. The weight of the snow causes it to compact at the base and form ice. When the snowfall reaches a thickness of about 100 meters (328 ft), ice at the bottom becomes like thick toothpaste and begins to move slowly. A **glacier,** then, is a large body of ice moving slowly down a slope or spreading outward on a land surface (**Figure 3.30**). Some glaciers appear to be stationary simply because the melting and evaporation at the glacier's edge equal the speed of the ice advance. Glaciers can, however, move as much as a meter per day.

Most theories of glacial formation concern Earth's climatic cooling. Perhaps a combination of the following theories explains the evolution of glaciers. The first theory attributes cooling to periods when there may have been excessive amounts of volcanic dust in the atmosphere. The argument is that the dust, by reducing the amount of solar energy reaching the Earth, effectively lowered temperatures at the surface. A second theory attributes the ice ages to known changes in the shape, tilt, and seasonal positions of the Earth's orbit around the sun over the last half-million years. Such changes alter the amount of solar radiation received by the Earth and its distribution over the Earth. A recent theory suggests that, when large continental plates drift over polar regions, temperatures on Earth become more extreme and, as a result, induce the development of glaciers. This theory, of course, cannot explain the most recent ice ages.

Today, continental-size glaciers exist on Antarctica, Greenland, and Baffin Island in Canada, but mountain glaciers

Figure 3.29 **Limestone erodes easily** in the presence of water. **(a) Karst topography,** such as that shown here, occurs in humid areas where limestone in flat beds is at the surface. **(b)** This satellite photo of east central Florida shows the many **round lakes formed in the sinkholes** of a karst landscape. *(b) Source: NASA.*

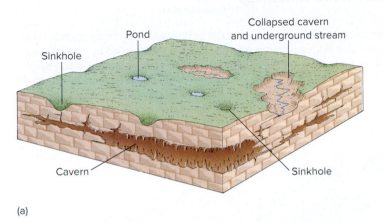

(a)

(b)

are found in many parts of the world. About 10% of the Earth's land area is under ice. During the most recent advance of ice, the continental ice of Greenland was part of an enormous glacier that covered nearly all of Canada (**Figure 3.31**) and the northernmost portions of the United States and Eurasia. The giant glacier reached thicknesses of 3000 meters (10,000 ft) (the depth in Greenland today), enveloping entire mountain systems. Another feature of the last ice age was the development of **permafrost,** a permanently frozen layer of ground that can be as much as 1500 meters (4900 ft) deep. Because the permafrost layer prevents the downward percolation of moisture, the surface soil may become saturated with water during the brief summer season, when only a thin surface thaws. In recent years, glaciers have been melting at a faster rate than their development due to a decided warming of the far northern and southern latitudes. Climate change is discussed in Chapter 4.

The weight of glaciers breaks up underlying rock and prepares it for transportation by the moving mass of ice. Consequently, glaciers change landforms by erosion. Glaciers scour the land as they move, leaving surface scratches, or striations, on rocks that remain. Much of eastern Canada has been scoured by glaciers that left little soil but many ice-gouged lakes and streams. The erosional forms created by glacial scourings have a variety of names. A *glacial trough* is a deep, U-shaped valley visible only

after the glacier has receded. If the valley is below sea level today, as in Norway or British Columbia, *fiords,* or arms of the sea, are formed. Some of the landforms created by scouring are shown in Figure 3.30. **Figure 3.32** shows horns, *cirques,* and arêtes, sharp ridges that separate adjacent glacially carved valleys. Cirques are formed by ice erosion at the head of a glacial valley.

Glaciers create landforms when they deposit the debris they have transported. These deposits, called *till,* consist of rocks, pebbles, and silt. As the great tongues of ice move forward, debris accumulates in parts of the glacier. The ice that scours valley walls and the ice at the tip of the advancing tongue are particularly filled with debris. As a glacier melts, it leaves behind hills of till of different sizes and shapes, such as moraines, eskers, and drumlins (**Figure 3.33**).

Many other landforms have been formed by glaciers. The most important is the *outwash plain,* a gently sloping area in front of a melting glacier. The melting along a broad front sends thousands of small streams running out from the glacier in braided fashion, streams that deposit neatly stratified drift made up of sand and gravel. Outwash plains, which are essentially great alluvial fans, cover a wide area and provide new parent-material for soil formation. Most of the midwestern part of the United States owes some of its soil fertility to the effects of wind on glacial deposition (see Figure 3.38).

Figure 3.30 **Alpine glacial landforms.** Frost shattering and ice movement carve *cirques,* the irregular bottoms of which may contain lakes (*tarns*) after glacial melt. Where cirque walls adjoin from opposite sides, knifelike ridges called *arêtes* are formed, interrupted by overeroded passes, or *cols.* The intersection of three or more creates a pointed peak, or *horn.* Rock debris falling from cirque walls is carried along by the moving ice. *Lateral moraines* form between the ice and the valley walls; *medial moraines* mark the union of such debris where two valley glaciers join. *Recessional moraines* form where the end of the ice remains stationary long enough for an accumulation of sediment to form, while a *terminal moraine* marks the glacier's farthest advance. Small, conical hills of sediment are called *kames.*

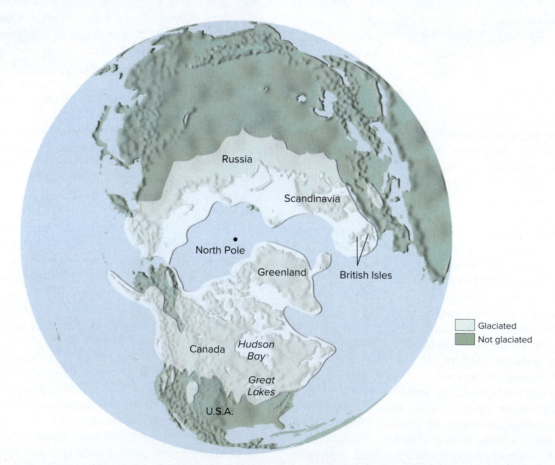

Figure 3.31 **Maximum extent of continental glaciation in the Northern Hemisphere (about 15,000 years ago).** Because sea level was lower than at present, due to the large volume of moisture trapped as ice on the land, glaciation extended beyond present continental shorelines. Separate centers of snow accumulation and ice formation developed. Large lakes were created between the western mountains of North America and the advancing ice front. To the south, huge rivers carried away glacial meltwaters.

Figure 3.32 **Horns, cirques, and arêtes created by glaciation,** San Juan Mountains, Colorado. © *McGraw-Hill Education.*

Waves, Currents, and Coastal Landforms

Whereas glacial action is intermittent in Earth's history, the breaking of ocean waves on continental coasts and islands is unceasing and causes considerable change in coastal landforms. As waves reach shallow water close to shore, they are forced by friction along the seafloor to heighten until a breaker is formed, as shown in **Figure 3.34**. The uprush of water not only carries sand for deposition but also erodes the landforms at the coast, while the backwash carries the eroded material away. This type of action results in different kinds of landforms, depending on conditions.

If land at the coast is well above sea level, the wave action causes cliffs to form. Cliffs then erode at a rate dependent on the rock's resistance to the constant assault of salt water. During storms, a great deal of power is released by the forward thrust of waves, and much erosion takes place. Landslides are a hazard during coastal storms, and they occur particularly in areas where weak sedimentary rock or till exists.

Beaches are formed by the deposition of sand grains contained in the water. The sand originates from the vast amount of coastal erosion and streams (**Figure 3.35**). *Longshore currents,* which move roughly parallel to the shore, transport the sand, forming beaches and *spits.* A more sheltered area increases the chances of a beach being built.

The backwash of waves, however, takes sand away from beaches if no longshore current exists. As a result, *sandbars* can develop a short distance away from the shoreline. If sandbars become extensive, they can eventually close off the shore, creating a new coastal configuration that encloses *lagoons* or *inlets. Salt marshes* very often develop in and around these areas. For example, the Outer Banks of North Carolina, composed of long ribbons of sand, are constantly shifting as a result of currents and storms.

Before the end of the most recent ice age, at least three previous major advances occurred during the 1.5 million years of the Pleistocene period. Firm evidence is not available on whether we have emerged from the cycle of ice advance and retreat. Factors concerning the Earth's changing temperature, which are discussed in Chapter 4, must be considered before assessing the likelihood of a new ice advance. For the first half of the 20th century, as has been the case in recent years, the world's glaciers were melting faster than they were building up.

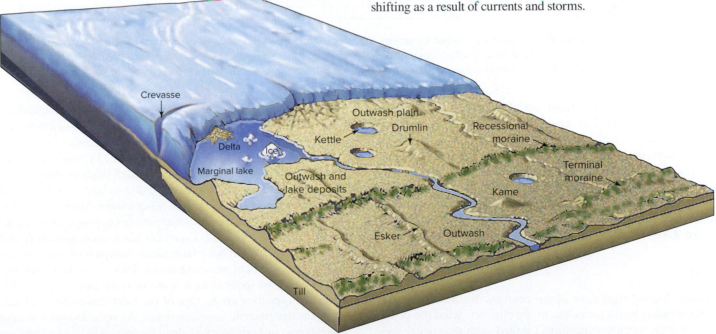

Figure 3.33 **Depositional features formed by ice sheets.** The debris carried by continental glaciers and left behind as they retreat creates various landforms. Moraines of *till,* unsorted glacial drift, form at the retreating edge. Streams carry silt to form *outwash plains. Kames* are small, conical hills formed of outwash deposits; *drumlins* are elongated hills made of till and oriented in the direction of ice movement; and *eskers* are long ridges of sediment deposited by glacial meltwater. Enclosed depressions caused by the melting of a stagnant block of ice that was surrounded and buried by sediment are called *kettles.*

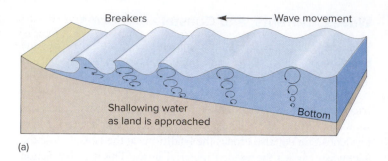

Figure 3.34 **Formation of waves and breakers. (a)** As the offshore swell approaches the gently sloping beach bottom, **sharp-crested waves** form, build up to a steep wall of water, and break forward in plunging surf. **(b) Evenly spaced breakers form** as successive waves touch bottom along a regularly sloping shore. *(b) © Carla Montgomery.*

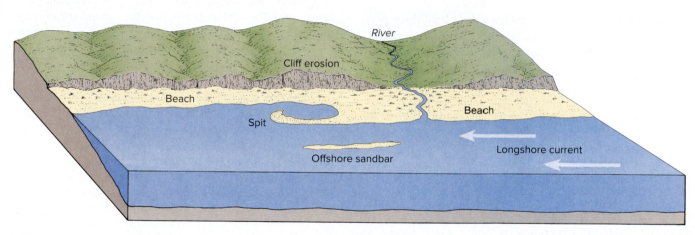

Figure 3.35 **The cliffs behind the shore are eroded by waves during storms and high water.** Sediment from the cliff and river forms the beach deposit; the longshore current moves some sediment downcurrent to form a *spit*. *Offshore sandbars* are created from material removed from the beach and deposited by retreating waves. Generally, not all the features shown here would occur in a single setting.

Coral reefs, made not from sand but from coral organisms growing in shallow tropical water, are formed by the secretion of calcium carbonate in the presence of warm water and sunlight. Reefs, consisting of millions of colorful coral skeletons, develop short distances offshore. Off the coast of northeastern Australia lies the most famous coral reef, the Great Barrier Reef. *Atolls,* found in the South Pacific, are reefs formed in shallow water around a volcano that has since been covered or nearly covered by water (**Figure 3.36**).

Wind

In humid areas, vegetation cover confines the effect of wind mainly to sandy beach areas, but in dry climates, wind is a powerful agent of erosion and deposition. Limited vegetation in dry areas leaves exposed particles of sand, clay, and silt subject to movement by wind. Thus, many of the sculptured features found in dry areas result from mechanical weathering, that is, from the abrasive action of sand and dust particles as they are blown against rock surfaces. Sand and dust storms occurring in a drought-stricken farm area

may make it unusable for agriculture. Inhabitants of Oklahoma, Texas, and Colorado suffered greatly in the 1930s when their farmlands became the "Dust Bowl" of the United States.

Several types of landforms are produced by wind-driven sand. **Figure 3.37** depicts one of these. Although sandy deserts are much less common than gravelly deserts, also called *desert pavement,* their characteristic landforms are better known. Most of the Sahara, the Gobi, and the western U.S. deserts are covered with rocks, pebbles, and gravel, not sand. Each also has a small portion (and the Saudi Arabian Desert has a large area) covered with sand blown by wind into a series of waves, or *dunes.* Unless vegetation stabilizes them, the dunes move as sand is blown from their windward faces onto and over their crests. One of the most distinctive sand desert dunes is the crescent-shaped *barchan.* Along seacoasts and inland lakeshores, in both wet and dry climates, wind can create sand ridges that reach a height of 90 meters (300 ft). Sometimes, coastal communities and farmlands are threatened or destroyed by moving sand (see "Beaches on the Brink" p. 71).

Another kind of wind-deposited material, silty in texture, is called **loess.** Encountered usually in midlatitude westerly wind

Figure 3.36   **Coral reef,** part of the Great Barrier Reef off the east coast of Australia. © *Fuse/Getty Images RF.*

major deposits are assumed to have resulted from wind erosion of nonvegetated sediment deposited by meltwater from retreating glaciers. Because rich soils usually form from loess deposits, if climatic circumstances are appropriate, these areas are among the most productive agricultural lands in the world.

3.6 Landform Regions

Every piece of land not covered by buildings and other structures contains clues as to how it has changed over time. Geomorphologists interpret these clues, studying such things as earth materials and soils, the availability of water, drainage patterns, evidences of erosion, and glacial history. The scale of analysis may be as small as a stream or as large as a *landform region,* a large section of the Earth's surface where a great deal of homogeneity occurs among the types of landforms that characterize it.

A foldout map inside the back cover of this book shows, in a general way, the kinds of landform regions found in different parts of the world. Note how the *mountain* belts generally coincide with convergent plate boundaries not found beneath the sea (see Figure 3.6) and with the earthquake-prone areas (see Figure 3.8). Vast *plains* exist in North and South America, Europe, Asia, and Australia. Many of these regions were created under former seas and appeared as land when seas contracted. These, and the smaller plains areas, are the drainage basins for some of the great rivers of the world,

belts, it covers extensive areas in the United States (**Figure 3.38**), central Europe, central Asia, and Argentina. It has its greatest development in northern China, where loess covers hundreds of thousands of square miles, often to depths of more than 30 meters (100 ft). The windborne origin of loess is confirmed by its typical occurrence downwind from extensive desert areas, though

Figure 3.37   **The prevailing wind from the left has given these transverse dunes** a characteristic gentle windward slope and a steep, irregular leeward slope. © *James A. Bier.*

Beaches on the Brink

Headlines such as "Beaches on the Brink," "Storm-Lashed Cape Is a Fragile Environment," "Fighting the Development Tide," and "State Looks for Money to Restore the Shore" signal a growing concern with the condition of coastlines. In addition, they raise the central question of how we can utilize coastlines without, at the same time, destroying them. Hurricane Sandy played havoc with beachfront property in densely settled New York and New Jersey.

Because many of the world's people live or vacation on coasts, and because such areas are often densely populated, coastal processes have a considerable impact on humans. Nature's forces are continually shaping and reshaping the coasts; they are dynamic environments, always in a state of flux. Some processes are dramatic and induce rapid change: tropical cyclones (hurricanes and typhoons), tsunami, and floods can wreak havoc, take thousands of lives, and cause millions of dollars in damage in a matter of hours. A less hazardous process is beach erosion, although it tends to magnify the effects of storms.

Some beach erosion is caused by natural processes, both marine and land. Waves carry huge loads of suspended sand, and longshore currents constantly move sand along the shoreline. The weathering and erosion of sea cliffs produces sediment, and rivers carry silt from mountains to beaches.

Human activities affect both erosion and deposition, however. Dams, for example, decrease the flow of sand to the water's edge by trapping sediment upstream. We fill marshes, construct dikes, and bulldoze dunes or remove their natural stabilizing vegetation. We hasten erosion when we build roads, houses, and other structures on cliffs and dunes or when we plant trees and lawns on top of them.

Especially vulnerable to erosion are barrier islands, narrow strips of sand parallel to the mainland. Under natural conditions, they are not stable places; their ends typically migrate, and during storms, waves can wash right over them. Some barrier islands are highly developed and densely populated, including Atlantic City and Miami Beach.

Once hotels and condominiums, railroads, and highways have been built along the waterfront, people attempt to protect their investments by preventing beaches from eroding. Breakwaters, offshore structures built to absorb the force of large, breaking waves and to provide quiet water near shore, are designed to trap sand and thus retard erosion, but they are not always successful. Although some are locally beneficial by forming new areas of deposition, they almost invariably accelerate erosion in adjacent areas. The efforts of one community are often negated by those of nearby towns and cities, underscoring the need for agencies to coordinate efforts and develop comprehensive land use plans for an extensive length of shoreline.

An alternative to erecting artificial structures is beach replenishment: adding sand to beaches to replace that lost by erosion. This provides recreational opportunities; at the same time, it buffers property from damage by storms. Sand may be dredged from harbors or pumped from offshore sandbars. The disadvantages of this technique are that dredging disrupts marine organisms, sand of the right texture may be hard to obtain, and replenished beaches can be short-lived. For example, more than $5 million was spent replenishing the beach at Ocean City, New Jersey, in 1982. The beach disappeared within 3 months, after a series of northeasters hit the area. Similarly, a massive dredging project that cost more than $17 million replenished 8 kilometers (5 mi) of beaches along the San Diego coast in 2000. Within months, after the largest waves of the year pounded the coast, more than one-half of the sand was gone.

The expense of maintaining coastal zones raises two basic questions: Who benefits? Who should pay? Some people contend that the interests of those who own coastal property are not compatible with the public interest, and it is unwise to expend large amounts of public funds to protect the property of only a few. People argue that shorefront businesses and homeowners are the chief beneficiaries of shore protection measures, that they often deny the public access to the beaches in front of their property, and therefore they should pay for the majority of the cost of maintaining the shoreline. At present, however, this is rarely the case; costs are typically shared by communities, states, and the federal government. Indeed, 51 federal programs subsidize coastal development and redevelopment. The largest is the National Flood Insurance Program, which offers low-cost insurance to homeowners in flood-prone areas. People have used the protection of that insurance to build anywhere, even in high-risk coastal areas.

Considering the Issues

1. From 1994 to the present, the Army Corps of Engineers has spent millions of dollars annually to pump sand from the ocean bottom onto eroded beaches in New Jersey. This ongoing project is based on the assumption that additional replenishment will be

(a)

(b)

Storm surge damage on North Carolina's Outer Banks **(a)** before and **(b)** after Hurricane Fran in 1996. *Source: USGS Center for Coastal Geology.*

(continued)

(continued)

required every 5 or 6 years as beaches erode. Currently, the federal government pays 65% of the cost, the state government 25%, and local governments 10%. New Jersey Senator Frank Lautenberg contends that beach replenishment is critical for the region's future. "People's lives and property are at stake. Jersey's beaches bring crucial tourist dollars to the state." But James Tripp of the Environmental Defense Fund argues that beach rebuilding is simply throwing taxpayers' money into the ocean. "Pumping all the sand in the world is not going to save the day." Do you think the beach replenishment project is a wise use of taxpayers' money? Does the federal government have an obligation to protect or rebuild storm-damaged beaches? Why or why not?

2. Coastal erosion is not a problem for beaches, only for people who want to use them. Do you believe that we should learn to live with erosion, not building in the coastal zone unless we are prepared to consider our structures temporary and expendable? Should communities adopt zoning plans that prohibit building on undeveloped lands within, say, 50 meters (164 ft) of the shore?

3. Should the federal government curtail programs that provide inexpensive storm insurance for oceanfront houses and businesses, as well as speedy grants and loans for storm repairs not covered by insurance? Should people be allowed to rebuild storm-damaged buildings even if they are vulnerable to future damage? Why or why not?

4. Coastal erosion will become more serious if the current rise in sea level—about 2.5 centimeters (1 in.) every 12 years—continues or even increases as global warming causes seawater to expand or the polar ice caps to melt. Many of the world's major cities would be threatened by this rise in sea level. What are some of those cities? How might they protect themselves?

such as the Mississippi-Missouri, Amazon, Volga, Nile, Ganges, and Tigris-Euphrates. The valleys carved by these rivers and the silt deposited by them are among the most agriculturally productive areas in the world. The *plateau* regions are many and varied. The African plateau region is the largest. Much of the African landscape is characterized by low mountains and hills whose base is about 700 meters (2300 ft) above sea level. Generally quiet from the standpoint of tectonic activity, Africa is largely made up of geologically ancient continental blocks that have been in an advanced stage of erosion for millions of years.

Humans affect and are affected by the landscape, landforms, moving continents, and earthquakes; however, except at times of natural disaster, these elements of the physical world are, for most of us, quiet, accepted background. More immediate in affecting our lives and fortunes are the great patterns of climate. Climate helps define the limits of the economically possible with present levels of technology, the daily changes of weather that affect the success of picnics and crop yield alike, and the patterns of vegetation and soils. We turn our attention to these elements of the natural environment in Chapter 4.

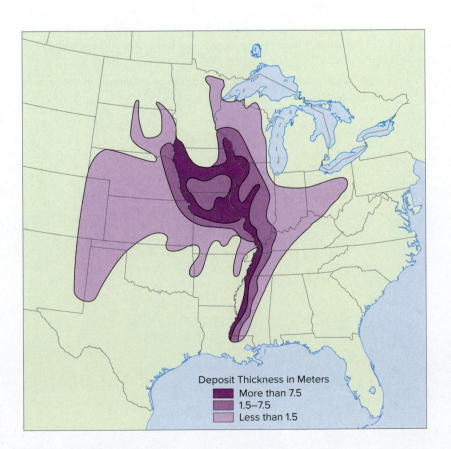

Deposit Thickness in Meters
- More than 7.5
- 1.5–7.5
- Less than 1.5

Figure 3.38 **Location of windblown silt deposits, including loess, in the United States.** The thicker layers, found in the upper Mississippi Valley area, are associated with the wind movement of glacial debris. Farther west, in the Great Plains, wind-deposited materials are sandy in texture, not loessial. *Source: Adapted from* Geology of Soils *by Charles B. Hunt, copyright 1972 W. H. Freeman and Company.*

Summary of Key Concepts

- Rocks, the materials that constitute the Earth's surface, are classified as igneous, sedimentary, and metamorphic.

- In the most recent 200 million of the Earth's 4.5 billion years, continental plates drifted on the asthenosphere to their present positions.

- At or near plate intersections, tectonic activity is particularly in evidence in two forms. Diastrophism, such as faulting, results in earthquakes and, on occasion, tsunami. Volcanism moves molten materials toward the Earth's surface.

- The building up of the Earth's surface is balanced by three gradational processes—weathering, mass movement, and erosion. Weathering, both mechanical and chemical, prepares materials for transport by disintegrating rocks and is instrumental in the development of soils. Talus slopes and soil-creep are examples of the effect of mass movement. The erosional agents of running water, groundwater, glaciers, waves and currents, and wind move materials to new locations.

- Examples of landforms created by the collection of eroded materials are alluvial fans, deltas, natural levees, moraines, and sand dunes.

Key Words

alluvium 64	gradational processes 57
asthenosphere 49	igneous rock 47
chemical weathering 59	karst topography 66
continental drift 49	lithosphere 49
diastrophism 53	loess 70
erosional agent 60	mass movement 60
fault 51	mechanical weathering 58
floodplain 64	metamorphic rock 48
fold 53	mineral 47
glacier 66	permafrost 67

plate tectonics 49	volcanism 53
sedimentary rock 48	warping 53
subduction 53	water table 66
tsunami 55	weathering 58

Thinking Geographically

1. How can rocks be classified? List three classes of rocks according to their origin. In what ways can they be distinguished from one another?

2. What evidence makes the theory of plate tectonics plausible?

3. What is *subduction?* What are its effects?

4. Explain what is meant by *gradation* and *volcanism.*

5. What is meant by *folding, joint,* and *faulting?*

6. Draw a diagram indicating the varieties of ways *faults* occur.

7. With what earth movements are earthquakes associated? What are *tsunami* and how do they develop?

8. What is the distinction between *mechanical* and *chemical weathering?* Is weathering responsible for landform creation? In what ways do glaciers engage in mechanical weathering?

9. Explain the origin of the various landforms one usually finds in desert environments.

10. How do *glaciers* form? What landscape characteristics are associated with glacial erosion? With glacial deposition?

11. How are alluvial fans, deltas, natural levees, and moraines formed?

12. How is groundwater erosion differentiated from surface water erosion?

13. How are the processes that bring about change due to waves and currents related to the processes that bring about change by the force of wind?

14. What processes account for the landform features of the area in which you live?

CHAPTER 4

Physical Geography: Weather and Climate

Flooding in New Orleans following the passage of Hurricane Katrina in August 2005. *Source: NOAA*.

CHAPTER OUTLINE

LEARNING OBJECTIVES

After studying this chapter you should be able to:

4.1   Understand the difference between weather and climate.

4.2   Define atmospheric terminology such as insolation, lapse rate, and temperature inversion.

4.3   Contrast the way in which land and water areas respond differently to equal insolation.

4.4   Explain how the planetary wind and pressure system works.

4.5   Understand the origin of the different types of large-scale precipitation.

4.6   Explain in as great detail as possible a current weather map that includes fronts, temperatures, and precipitation. Using the map, attempt to predict weather conditions in the next 48 hours.

4.7   Summarize the distinguishing characteristics of each type of climate.

4.8   Discuss the factors that may be responsible for climate change.

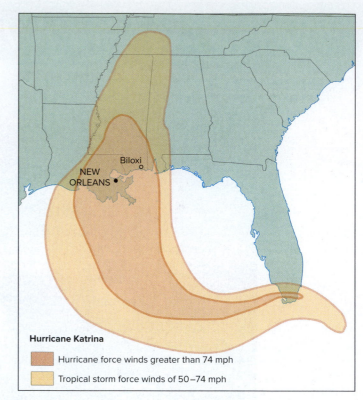

Figure 4.1   **Path of hurricane Katrina in August 2005.**

On August 29, 2005, the "Big Easy," New Orleans, a city built mostly at and below sea level near the delta of the Mississippi River, lay in the path of Hurricane Katrina. Because most residents had obeyed prehurricane evacuation orders, the city was largely deserted. It was mostly the poor and immobile who were left behind as the storm hit. Winds of about 200 kilometers per hour (125 mph) pounded the city, downing trees and power lines, and wind-driven high water poured ashore.

At first it looked as though New Orleans had escaped with no more damage than might have been expected, but the worst was yet to come. In several places, the sea walls and levees that held back the Mississippi River and Lake Pontchartrain collapsed. The breaches proved catastrophic; for 2 days after the storm had passed, waters continued to rise and pour into the city. About 80% of New Orleans was flooded with anywhere from 1 to 3 meters (3 to 10 ft) of water (**Figure 4.1**). Approximately, 1800 people are believed to have died in New Orleans and in other towns of Louisiana, Mississippi, and Alabama before the storm's fury dissipated over Tennessee.

When it was over, New Orleans was devastated. Katrina had proved to be one of the worst natural disasters in the history of the United States. All of the systems that make a city function had failed, including the water and sewer systems, the electric grid, the transportation network, and the telephone system. To add insult to injury, 3 weeks later Hurricane Rita attacked the Gulf Coast. Although Rita struck farther to the west, New Orleans suffered from renewed flooding. More than a month after the two hurricanes, most residents had not yet been allowed to return, because power was still out in large sections of the city and water for drinking had not yet been restored.

It would take billions of dollars to rebuild New Orleans. There were many tasks that had to be done before rebuilding could even begin: removal of debris, draining of floodwater, repair of levees and sea walls, cleanup of toxic materials, and restoration of electricity and the sewage system.

Whether storms such as Hurricanes Katrina and Sandy occur in Asia or North America, they do great damage, affecting the lives of everyone in their paths. Tropical storms are one extreme type of weather phenomenon. Most people are "weather watchers"—they watch television forecasts with great interest and plan their lives around weather events. In this chapter, we review the subsection of physical geography concerned with weather and climate. It deals with normal, patterned phenomena from which such extreme weather events as Hurricane Katrina occasionally emerge.

A weather forecaster describes current conditions for a limited region, such as a metropolitan area, and predicts future weather conditions. If the elements that make up the **weather,** such as temperature, wind, and precipitation, are recorded at specified moments in time, such as every hour, an inventory of weather conditions can be developed. By finding trends in data that have been gathered over an extended period of time, we can speak about typical conditions. These characteristic circumstances describe the **climate** of a region. Weather is a moment's view of the lower atmosphere, whereas climate is a description of typical weather conditions in an area or at a place over a period of time. Geographers analyze the differences in weather and climate from place to place in order to understand how climatic elements affect human occupance of the Earth.

In geography, we are particularly interested in the physical environment that surrounds us. That is why the **troposphere,** the lowest layer of the Earth's atmosphere, attracts our attention. This layer, extending about 10 kilometers (6 mi) above the ground, contains virtually all of the air, clouds, and precipitation of the Earth (**Figure 4.2**).

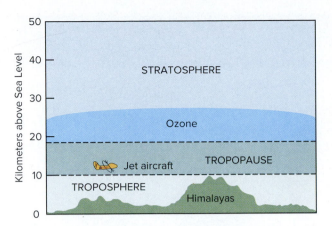

Figure 4.2   The troposphere. Virtually all of the air, clouds, and precipitation of the Earth are contained in the troposphere.

In this chapter, we try to answer the questions usually raised regarding characteristics of the lower atmosphere. By discussing these answers from the viewpoint of averages or average variations, we attempt to give a view of the Earth's climatic differences, a view held to be very important for understanding the way people use the land. Climate is a key to understanding, in a broad way, the distribution of world population. People have great difficulty living in areas that are, on average, very cold, very hot, very dry, or very wet. They are also negatively affected by huge storms or flooding. In this chapter, we first discuss the elements that constitute weather conditions and then describe the various climates of the Earth.

4.1 Air Temperature

Perhaps the most fundamental question about weather is, "Why do temperatures vary from place to place?" The answer to this question requires the discussion of a number of concepts to help focus on the way heat accumulates on the Earth's surface.

Energy from the sun, called *solar energy,* is transformed into heat, primarily at the Earth's surface and secondarily in the atmosphere. Not every part of the Earth or its overlying atmosphere receives the same amount of solar energy. At any given place, the amount of incoming solar radiation, or **insolation,** available depends on the intensity and duration of radiation from the

sun. These are determined by both the *angle at which the sun's rays strike the Earth* and the *number of daylight hours.* These two fundamental factors, plus the following five modifying variables, determine the temperature at any given location:

1. the amount of water vapor in the air
2. the degree of cloud cover
3. the nature of the surface of the Earth
4. the elevation above sea level
5. the degree and direction of air movement

Let us look at these factors briefly.

Earth Inclination

The axis of the Earth—that is, the imaginary line connecting the North Pole to the South Pole—always remains in the same position. It is tilted about 23.5° away from the perpendicular (**Figure 4.3**). Every 24 hours, the Earth rotates once on that axis, as shown in **Figure 4.4**. While rotating, the Earth is slowly revolving around the sun in a nearly circular annual orbit (**Figure 4.5**).

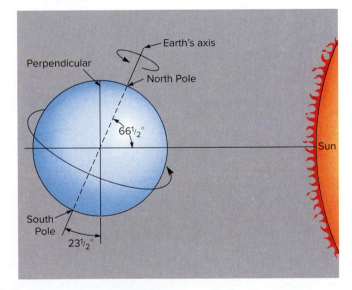

Figure 4.3   The Earth's position relative to the sun. The Earth is in its summer position in the Northern Hemisphere (winter in the Southern Hemisphere).

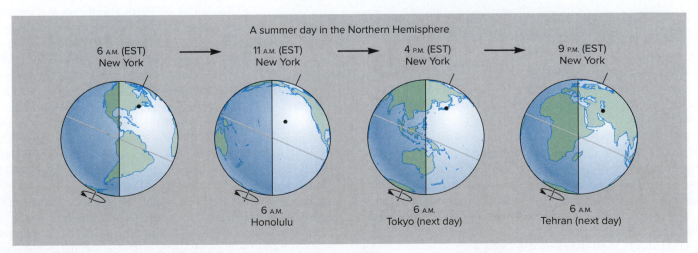

Figure 4.4   The process of the 24-hour rotation of the Earth on its axis.

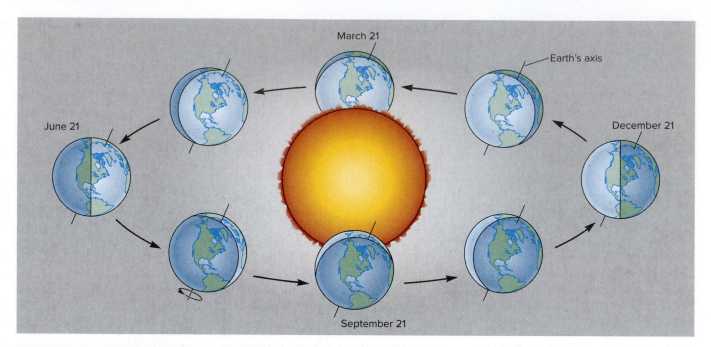

Figure 4.5   The process of the yearly revolution of the Earth around the sun. The sun, which is about 93 million miles from Earth, is not drawn to scale; it is much larger relative to the size of Earth.

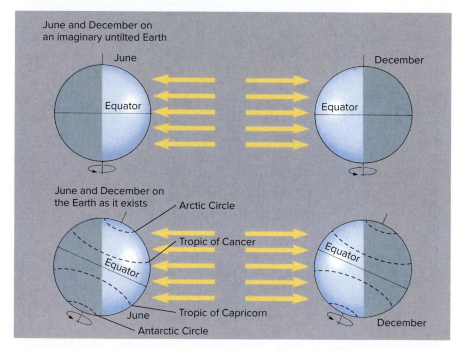

Figure 4.6   Notice on the bottom two diagrams that, as the Earth revolves, the north polar area in June is bathed in sunshine for 24 hours, while the south polar area is dark. The most intense of the sun's rays are felt north of the equator in June and south of the equator in December. None of this is true in the untilted examples shown on the upper two diagrams.

If the Earth were not tilted from the perpendicular, the solar energy received *at a given latitude* would not vary during the course of the year. The rays of the sun would directly strike the equator, and as distance away from the equator became greater, rays would strike the Earth at ever-decreasing angles, therefore diminishing the intensity of the energy and giving climates a latitudinal standardization (**Figures 4.6** and **4.7**).

Because of the inclination, however, the location of highest incidence of incoming solar energy varies during the course of the year. When the Northern Hemisphere is tilted directly toward the sun, the vertical rays of the sun are felt as far north as 23.5°N latitude (Tropic of Cancer). This position of the Earth occurs about June 21, the summer *solstice* for the Northern Hemisphere and the winter solstice for the Southern Hemisphere. About December 21, when the vertical rays of the sun strike near 23.5°S latitude (Tropic of Capricorn), it is the beginning of summer in the Southern Hemisphere and the onset of winter in the Northern Hemisphere. During the rest of the year, the position of the Earth relative to the sun results in direct rays migrating from about 23.5°N to 23.5°S and back again. On about March 21 and September 21 (the spring and autumn *equinoxes*), the vertical rays of the sun strike the equator.

The tilt of the Earth also means that the length of days and nights varies during the year. One-half of the Earth is always illuminated by the sun, but only at the equator is it light for 12 hours each day of the year. As distance away from the equator becomes greater, the hours of daylight or darkness increase, depending on whether the direct rays of the sun are north or south of the equator. In the summer, daylight increases to the maximum of 24 hours from the Arctic Circle (66.5°N) to the North Pole, and during the same period, nighttime finally reaches 24 hours in length from the Antarctic Circle (66.5°S) to the South Pole.

Because of the 24-hour daylight, it would seem that much solar energy should be available in the summer polar region. The angle of the sun is so narrow (the sun is low in the sky), however, that solar energy is spread over a wide surface. By contrast, the combination of

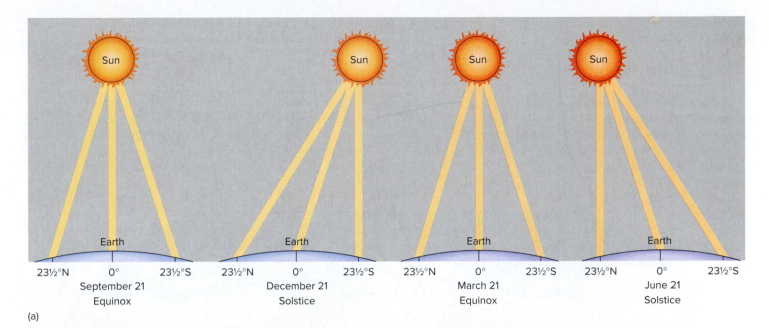

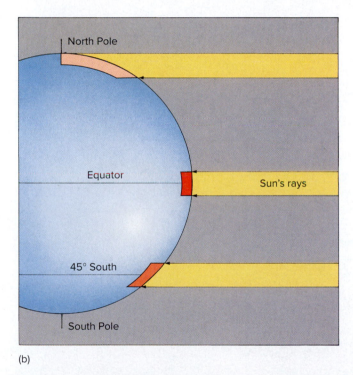

(b)

Figure 4.7   **Imaginary rays from the sun. (a)** Imaginary rays from the sun at spring and autumn equinoxes and summer and winter solstices. **(b)** Three equal, imaginary rays from the sun are shown striking the Earth at different latitudes at the time of the equinox. As distance away from the equator increases, the rays become more diffused, showing how the sun's intensity is diluted in the high latitudes.

relatively long days and sun angles close to 90° makes an enormous amount of energy available to areas in the neighborhood of 15° to 30° north and south latitude during each hemisphere's summer.

Reflection and Reradiation

Much of the potentially receivable solar radiation is, in fact, sent back to outer space or diffused in the troposphere in a process known as **reflection.** Clouds, which are dense concentrations of suspended, tiny water or ice particles, reflect a great deal of energy. Light-colored surfaces, especially snow cover, also reflect large amounts of solar energy.

Energy is lost through reradiation as well as reflection. In the **reradiation** process, the Earth's surface acts as a communicator of energy. As indicated in **Figure 4.8**, the energy that

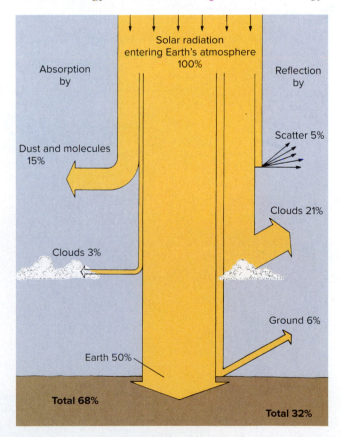

Figure 4.8   **Consider the incoming solar radiation** as 100%. The portion that is absorbed into the Earth (50%) is eventually released to the atmosphere and then reradiated into space.

is absorbed into the land and water is returned to the atmosphere in the form of terrestrial radiation. On a clear night, when no clouds can block or diffuse movement, temperatures continually decrease, as the Earth reradiates as heat the energy it has received and stored during the course of the day.

Some kinds of Earth surface material, especially water, store solar energy more effectively than others. Because water is transparent, solar rays can penetrate a great distance below its surface. If water currents are present, heat is distributed even more effectively. On the other hand, land surfaces are opaque, so all of the energy received from the sun is concentrated at the surface. Land, having more heat available at the surface, reradiates its energy faster than water. Air is heated by the process of reradiation from the Earth and not directly by energy from the sun passing through it. Thus, because land heats and cools more rapidly than water, hot and cold temperature extremes recorded on Earth occur on land and not the sea.

Temperatures are moderated by the presence of large bodies of water near land areas. Note in **Figure 4.9** that coastal areas have lower summer temperatures and higher winter temperatures than those places at the same distance from the equator,

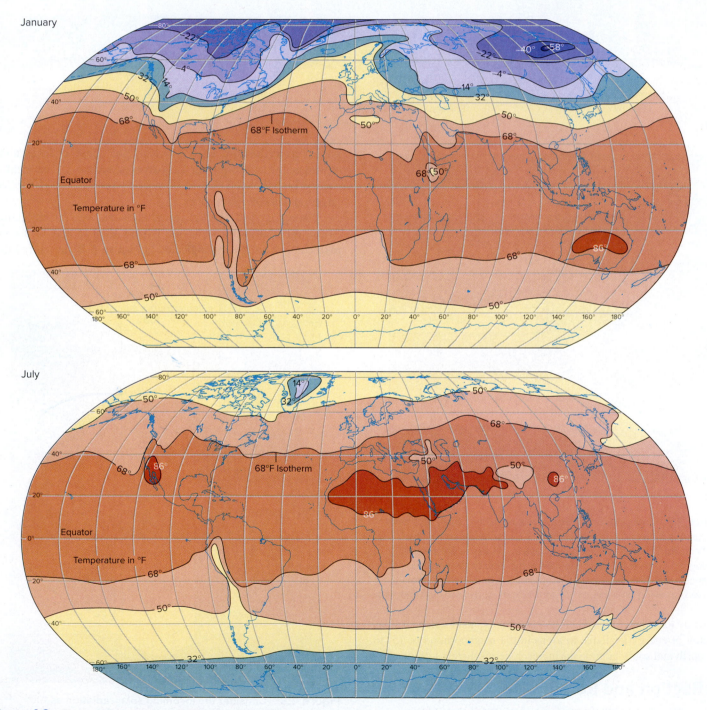

Figure 4.9   Temperatures of the Earth. At a given latitude, water areas are warmer than land areas in winter and cooler in summer. *Isotherms* are lines of equal temperatures.

excluding seacoasts. Land areas affected by the moderating influences of water are considered *marine* environments; those areas not affected by nearby water are *continental* environments.

Temperatures vary in a cyclical way each day. In the course of a day, as incoming solar energy exceeds energy lost through reflection and reradiation, temperatures begin to rise. The ground stores some heat, and temperatures continue to rise until the angle of the sun becomes so narrow that energy received no longer exceeds that lost by the reflection and reradiation processes. Not all of the heat loss occurs during the night, but long nights appreciably deplete stored energy.

Lapse Rate

We may think that, as we move vertically away from the Earth toward the sun, temperatures would increase. However, this is not true within the troposphere. The Earth absorbs and reradiates heat; therefore, temperatures are usually warmest at the Earth's surface and lower as elevation increases. Note in **Figure 4.10** that this temperature **lapse rate** (the rate of temperature change with altitude in the troposphere) averages about 6.4°C per 1000 meters (3.5°F per 1000 ft). For example, the difference in elevation between Denver and Pikes Peak is about 2700 meters (9000 ft), which normally results in a 17°C (32°F) difference in temperature. Jet planes flying at an altitude of 9100 meters (30,000 ft) are moving through air that is about 56°C (100°F) colder than ground temperatures.

The normal lapse rate does not always hold, however. Rapid reradiation sometimes causes temperatures to be higher above the Earth's surface than at the surface itself. This particular condition, in which air at lower altitudes is cooler than air aloft, is called a **temperature inversion.** An inversion is important because of its effect on air movement. Warm air at the surface, which normally rises, may be blocked by the even warmer air of a temperature inversion (**Figure 4.11**). Thus, surface air is trapped; if it is filled with automobile exhaust emissions or smoke, a serious smog condition may develop (see "The Donora Tragedy" p. 82). Because of the configuration of nearby mountains, Los Angeles often experiences temperature inversions, causing sunlight to be reduced to a dull haze (**Figure 4.12**).

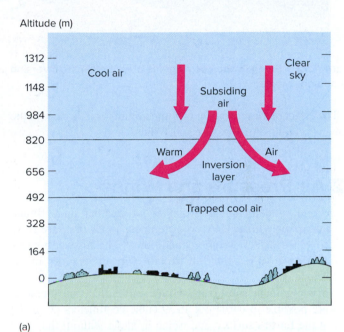

(a)

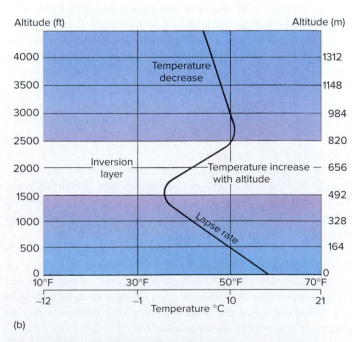

(b)

Figure 4.11   **Temperature inversion. (a)** A layer of warm, subsiding air acts as a cap, temporarily trapping cooler air close to the ground. **(b)** Note that air temperature decreases with distance from the ground until the warm inversion layer is reached, at which point the temperature increases.

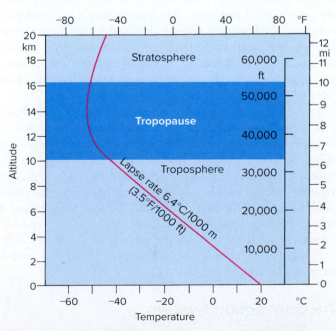

Figure 4.10   **The temperature lapse rate under typical conditions.** The *tropopause* is a transition zone between the troposphere and the stratosphere. It marks the level at which temperature ceases to fall with altitude.

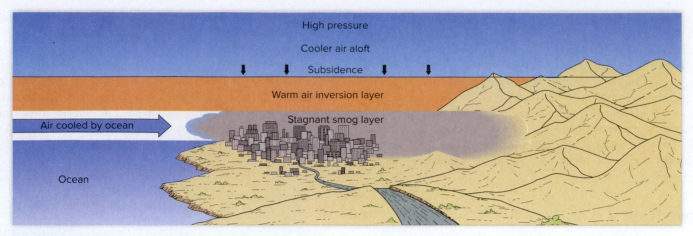

Figure 4.12   Smog in the Los Angeles area. Below the inversion layer, stagnant air holds increasing amounts of pollutants, caused mainly by automobile exhaust.

The effect of air movement on temperature is made clear in the following section, "Air Pressure and Winds."

4.2 Air Pressure and Winds

The second fundamental question about weather and climate concerns **air pressure:** How do differences in air pressure from place to place affect weather conditions? The answer to this question first requires that we explain why differences in air pressure occur.

Air is a gaseous substance whose weight affects air pressure. If it were possible to carve out 16.39 cubic centimeters (1 cu in.) of air at the Earth's surface and weigh it, along with all the other cubic centimeters of air above it, under normal conditions the total weight would equal approximately 6.67 kilograms (14.7 lbs) of air as measured at sea level. Actually, this is not very heavy when you consider the dimensions of the column of air: 2.54 centimeters by 2.54 centimeters (1 in. by 1 in.) by about 9.7 kilometers (6 mi), or about 6.2 cubic meters (220 cu ft). The weight of air 4.8 kilometers (3 mi) above the Earth's surface, however, is considerably less than 6.67 kilograms (14.7 lbs) because there is correspondingly less air above it. Thus, it is clear that air is heavier and air pressure is higher closer to the Earth's surface.

It is a physical law that, for equal amounts of cold and hot air, the cold air is denser. This law exemplifies why hot-air balloons, filled with lighter air, can rise into the atmosphere. A cold morning is characterized by relatively heavy air, but as afternoon temperatures rise, air becomes lighter.

Barometers of various types are used to record changes in air pressure. Barometric readings in inches of mercury or millibars are a normal part, along with recorded temperatures, of every weather report. Air pressure at a given location changes as surfaces heat or cool. Barometers record a drop in atmospheric pressure when air heats and a rise in pressure when air cools.

In order to visualize the effect of air movements on weather, it is useful to think of air as a liquid made up of two fluids with different densities (representing light air and heavy air)—for example, water and gasoline. If the fluids are put into a tank at the same time, the lighter liquid will move to the top and the heavier liquid will move to the bottom, representing the vertical motion of air. The heavier liquid spreads out horizontally along the bottom of the tank, becoming the same thickness everywhere. This flow represents the horizontal movement of the air or wind on the Earth's surface. Air attempts to achieve an equilibrium by evening out pressure imbalances that result from the heating and cooling processes. Air races from heavy (cold) air locations to light (warm) air locations. Thus, the greater the differences in air pressure between places, the greater the wind.

Pressure Gradient Force

Because of differences in the nature of the Earth's surface—water, snow cover, dark green forests, cities, and so on—and the other factors that affect energy receipt and retention, zones of high and low air pressure develop. Sometimes, these high- and low-pressure zones cover entire continents, but usually they are considerably smaller—several hundred miles wide—and within these regions, small differences are noted over short distances. When pressure differences exist between areas, a **pressure gradient force** causes air to blow from an area of high pressure toward an area of low pressure.

In order to balance pressure differences that have developed, air from the heavier high-pressure zones flows to low-pressure zones. Heavy air stays close to the Earth's surface as it moves, producing winds, and forces the upward movement of warm air. The velocity, or speed, of the wind is in direct proportion to pressure differences. Winds are caused by pressure differences that induce airflow from zones of high pressure to zones of low pressure. If distances between high- and low-pressure zones are short, pressure gradients are steep and wind velocities are great. Gentler air movements occur when zones of different pressure are far apart and the degree of difference is not great.

The Convection System

A room's temperature is lower near the floor than at the ceiling because warm air rises and cool air descends. The circulatory motion of descending cool air and ascending warm air is known as **convection** (**Figure 4.13**). A convectional wind system results

THE DONORA TRAGEDY

A heavy fog settled over the valley town of Donora, Pennsylvania, in late October 1948. Stagnant, moisture-filled air was trapped in the valley by surrounding hills and by a temperature inversion that held cooler air, gradually filling with smoke and fumes from the town's zinc works, against the ground under a lid of lighter, warmer upper air. For 5 days, the smog increased in concentration; the sulfur dioxide emitted from the zinc works continually converted to deadly sulfur trioxide by contact with the air.

Both old and young, with and without histories of respiratory problems, reported to doctors and hospitals a difficulty in breathing and unbearable chest pains. Before the rains washed the air clean nearly a week after the smog buildup, 20 died and hundreds of others were hospitalized. A normally harmless, water-saturated inversion had been converted to deadly poison by a tragic union of natural weather processes and human activity.

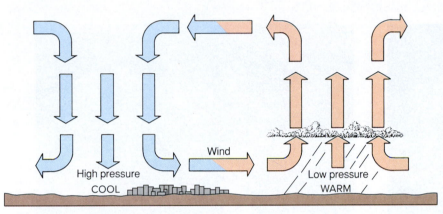

Figure 4.13 **A convection system.** Descending cool air flows toward low pressure. Precipitation very often occurs in low-pressure zones. As warm air rises, it cools and can become supersaturated, resulting in precipitation.

when surface-heated warm air rises and is replaced by cool air from above.

Land and Sea Breezes

A good example of a convectional system is **land** and **sea breezes** (**Figures 4.14a, b**). Close to a large body of water, the differential daytime heating between land and water is great. As a result, warmer air over the land rises vertically, only to be replaced by cooler air from over the sea. At night, just the opposite occurs; the water is warmer than the land, which has reradiated much of its heat, and the result is a land breeze toward the sea. These two winds make seashore locations in warm climates particularly comfortable.

Mountain and Valley Breezes

Gravitational force causes the heavy cool air that accumulates over snow in mountainous areas to descend into lower valley locations, as suggested in Figure 4.14c. Consequently, valleys can become much colder than the slopes, and a temperature inversion occurs. Slopes are the preferred sites for agriculture in mountainous regions because cold air from **mountain breezes** can cause freezing conditions in the valleys. In densely settled narrow valleys where industry is concentrated, air pollution can become

particularly dangerous. Mountain breezes usually occur during the night; **valley breezes**—caused by warm air moving up slopes in mountainous regions—are usually a daytime phenomenon (Figure 4.14d). The canyons of southern California are the scenes of strong mountain and valley breezes. In addition, during the dry season, they become dangerous areas for the spread of brush and forest fires.

The Coriolis Effect

In the process of moving from high to low pressure, wind veers toward the right of the direction of travel in the Northern Hemisphere and toward the left in the Southern Hemisphere. This deflection is called the **Coriolis effect.** Were it not for this effect, winds would move in exactly the direction specified by the pressure gradient.

To illustrate the impact of the Coriolis effect upon winds, a familiar example may be helpful. Imagine a line of ice skaters holding hands while skating in a circle, with one skater nearest the center of the circle. This skater turns slowly, while the outermost skater must skate very rapidly in order to keep the line straight. In a similar way, because the Earth rotates on its axis, the equatorial regions are rotating at a much faster rate than the areas around the poles.

Next, suppose that the skater at the center threw a ball directly toward the skater at the end of the line. By the time the ball arrived, it would pass behind the outside skater. If the skaters are going in a counterclockwise direction—as the Earth appears to be moving viewed from the position of the North Pole—the ball appears to the person at the North Pole to pass to the right of the outside skater. If the skaters are going in a clockwise direction—as the Earth appears to be moving viewed from the South Pole—the deflection is to the left. Because air (like the ball) is not firmly attached to the Earth, it, too, will appear to be deflected. The air maintains its direction of movement, but the Earth's surface moves out from under it. Since the position of the air is measured relative to the Earth's surface, the air appears to have diverged from its straight path.

The Coriolis effect and the pressure gradient force produce spirals rather than simple, straight-line patterns of wind, as indicated

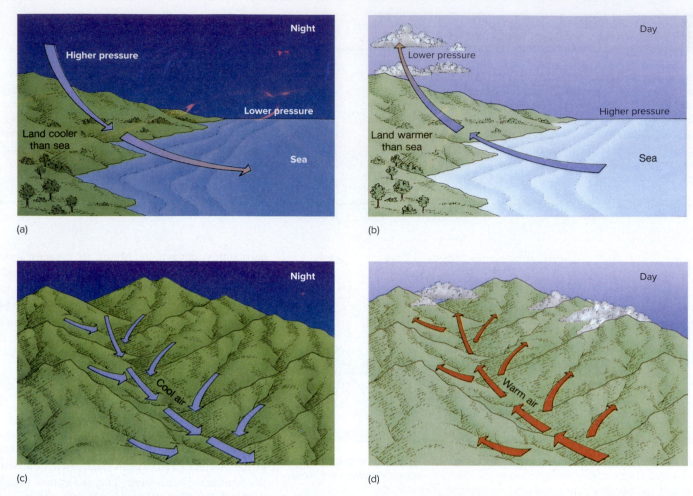

Figure 4.14   Convectional wind effects due to differential heating and cooling. (a) Land breeze. **(b)** Sea breeze. **(c)** Mountain breeze. **(d)** Valley breeze.

in **Figure 4.15**. The spiral of wind is the basic form of the many storms that are so important to the Earth's air-circulation system. These storm patterns are discussed later in this chapter.

The Frictional Effect

Wind movement is slowed by the frictional drag of the Earth's surface. The effect is strongest at the surface and declines until it becomes ineffective at about 1500 meters (about 1 mi) above the surface. Not only is wind speed decreased at the surface, but wind direction is changed as well. Instead of following a course exactly dictated by the pressure gradient force or by the Coriolis effect, the **frictional effect** causes wind to follow an intermediate path.

The Global Air-Circulation Pattern

Equatorial areas of the Earth are zones of low pressure. Intense solar heating in these areas is responsible for a convectional effect. Note in **Figure 4.16** how the warm air rises and tends to move away from the *equatorial low* pressure in both northerly and southerly directions. As equatorial air rises, it cools and eventually becomes dense. The lighter air near the surface cannot support the cool, heavy air. The heavy air falls, forming surface

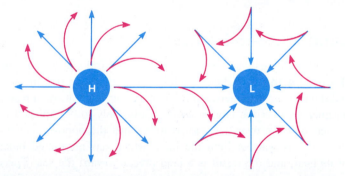

Figure 4.15   The Coriolis effect on flowing air in the Northern Hemisphere. The straight arrows indicate the paths that winds would follow flowing out of an area of high (H) pressure or into one of low (L) pressure, were they to follow the paths dictated by pressure differentials. The curved arrows represent the apparent deflection of the Coriolis effect. Wind direction—indicated on the diagram by curved arrows—is always given by the direction *from* which the wind is coming.

zones of high pressure. These areas of *subtropical high* pressure are located at about 30°N and 30°S of the equator.

When this cooled air reaches the Earth's surface, it, too, moves in both northerly and southerly directions. The Coriolis effect,

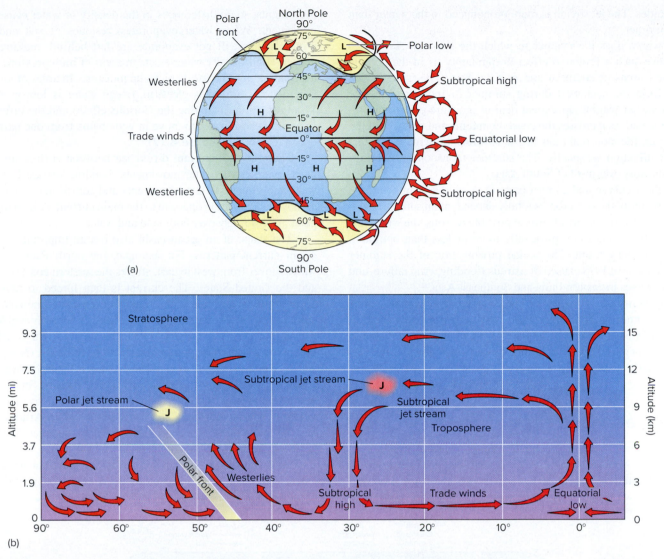

Figure 4.16   **The planetary wind and pressure belts as they would develop on an Earth of homogeneous surface. (a) The high- and low-pressure belts** represent surface pressure conditions; the wind belts are prevailing surface wind movements responding to pressure gradients and the Coriolis effect. Contrasts between land and water areas on the Earth's surface, particularly evident in the Northern Hemisphere, create complex distortions of this simplified pattern. **(b) The general pattern of winds at increasing altitudes above the Earth's surface.** When air descends, high pressure results; when air ascends, as at the equator, low pressure results.

however, modifies wind direction and creates, in the Northern Hemisphere, belts of winds called the *northeast trades* in the tropics and the *westerlies* (really the southwesterlies) in the midlatitudes. The names refer to the direction from which the winds come. Most of the United States lies within the belt of westerlies; that is, the air usually moves across the country from southwest to northeast. A series of ascending air cells also exists over the oceans to the north of the westerlies, called the *subpolar low*. These areas tend to be cool and rainy. The *polar easterlies* connect the subpolar low areas to the *polar high*. The general global air-circulation pattern is modified by local wind conditions.

It should be clear that these belts move in unison as the vertical rays of the sun change position. For example, equatorial low conditions are evident in the area just north of the equator during the Northern Hemisphere summer and just south of the equator during

the Southern Hemisphere summer. Air circulation will be discussed in greater detail in the section "Types of Precipitation."

The strongest flows of upper air winds, 9 to 12 kilometers (30,000 to 40,000 ft), are the **jet streams.** These air streams, moving at 160 to 320 kilometers per hour (100 to 200 mph), from west to east in both the Northern and Southern Hemispheres, circle the Earth in an undulating pattern, first north then south, as they move westward. There are three to six undulations at any one time in the Northern Hemisphere, but the waves are not always continuous. These undulations, or waves, control the flow of air masses on the Earth's surface. More stable undulations are likely to create similar day-to-day weather conditions. These waves tend to separate cold polar air from warm tropical air. When a wave dips far to the south in the Northern Hemisphere, cold air is taken equatorward and warm air moves poleward, bringing severe weather changes to the midlatitudes. The jet stream is more pronounced in the winter than in the summer.

Nowhere does the manner in which the seasonal shift takes place have such a profound effect on humanity as in the densely populated areas of southern and eastern Asia. The wind, which comes from the southwest during summer in India, reaches the landmass after picking up a great deal of moisture over the warm Indian Ocean. As it crosses the coast mountains and foothills of the Himalayas, the monsoon rains begin. A **monsoon** wind is one that changes direction seasonally. The summer monsoon wind brings heavy showers over most of South Asia.

In the southern and eastern parts of Asia, the farm economy, and particularly the rice crop, is totally dependent on summer monsoon rainwater. If, for any of several possible reasons, the wind shift is late or the rainfall is significantly more or less than optimum, crop failure may result. The undue prolongation of the summer monsoon rains in 1978 caused disastrous flooding, crop failure, and the loss of lives in eastern India and Southeast Asia.

The transition to dry northerly monsoon winter winds occurs gradually across the region, first becoming noticeable in the north in September. By January, most of the subcontinent is dry. Then, beginning in March in southern areas, the yearly cycle repeats itself.

4.3 Ocean Currents

Surface ocean currents correspond roughly to global wind direction patterns because the winds of the world set ocean currents in motion. In addition, just as differences in air pressure cause wind movements, so do differences in the density of water cause water movement. When water evaporates, residues of salt and other minerals that will not evaporate are left behind, making water denser. High-density water exists in areas of high pressure, where descending dry air readily picks up moisture. In areas of low pressure, where rainfall is plentiful, ocean water is low in density. Wind direction (including the Coriolis effect) and the differences in density cause water to move in wide paths from one part of the ocean to another (**Figure 4.17**).

There is an important difference between surface air movements and surface water movements. Landmasses are barriers to water movement, deflecting currents and sometimes forcing them to move in a direction opposite to the main current. Air, on the other hand, moves freely over both land and water.

The shape of an ocean basin also has an important effect on ocean current patterns. For example, the north Pacific current, which moves from west to east, strikes the western coast of Canada and the United States. The current is then forced to move both north and south, although the major movement is the cold ocean current that moves south along the California coast. In the Atlantic Ocean, however, as Figure 4.17 indicates, the current is deflected in a northeasterly direction by the shape of the coast (Nova Scotia and Newfoundland jut far into the Atlantic). It then moves freely across the ocean, past the British Isles and Norway, finally reaching the extreme northwest coast of Russia. This massive movement of warm water to northerly lands, called the **North Atlantic drift,** has enormous significance to inhabitants of those areas. Without it, northern Europe would be much colder.

Ocean currents affect not only the temperature but also the precipitation on land areas adjacent to the ocean. A cold ocean current

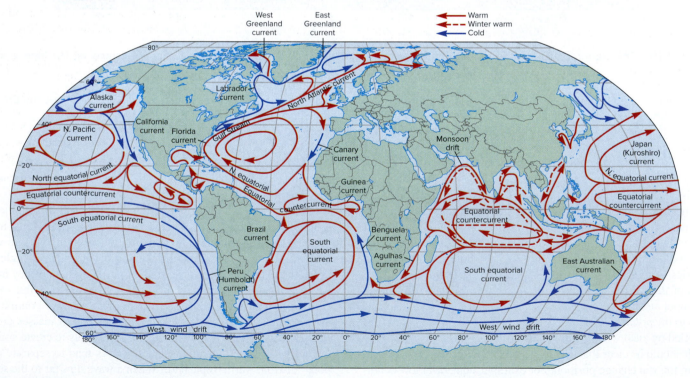

Figure 4.17   The principal surface ocean currents of the world. Notice how the warm waters of the Gulf of Mexico, the Caribbean, and the tropical Atlantic Ocean drift to northern Europe.

near land causes the air just above the water to be cold while the air above is warm. There is very little opportunity for convection, thus denying moisture to nearby land. Coastal deserts of the world usually border cold ocean currents. On the other hand, warm ocean currents—such as those off the coast of India—take moisture to the adjacent land area, especially when prevailing winds are landward (see "El Niño-Southern Oscillation," p. 89).

The earlier question about ways in which differences in air pressure affect weather conditions is now answered, in terms of warm and cool air movement over various surfaces at different times of the year and different times of the day. A more complete answer regarding the causes of different types of weather conditions, however, requires an explanation of the susceptibility of places to receive precipitation, because rainfall and wind patterns are highly related.

4.4 Moisture in the Atmosphere

Air contains water vapor (what we feel as humidity), which is the source of all precipitation. **Precipitation** is any form of water particles—rain, sleet, snow, or hail—that falls from the atmosphere and reaches the Earth's surface. Ascending air can expand easily because less pressure is on it. When heat from the lower air spreads through a larger volume in the troposphere, the mass of air becomes cooler. Cool air is less able to hold water vapor than warm air (**Figure 4.18**).

Air is said to be *supersaturated* when it contains so much water vapor that the vapor condenses (changes from a gas to a liquid) and forms droplets if fine particles, called *condensation nuclei*, are present. These particles, mostly dust, pollen, smoke, and salt crystals, nearly always exist. At first, the tiny water droplets are usually too light to fall. As many droplets coalesce into larger drops, and become too heavy to remain suspended in air, they fall as rain. When temperatures below the freezing point cause water vapor to form ice crystals instead of water droplets, snow is created (**Figure 4.19**).

Large numbers of rain droplets or ice crystals form clouds, which are supported by slight upward movements of air. The form and altitude of clouds depend on the amount of water vapor in the air, the temperature, and wind movement. Descending air in high-pressure zones usually yields cloudless skies. Whenever warm,

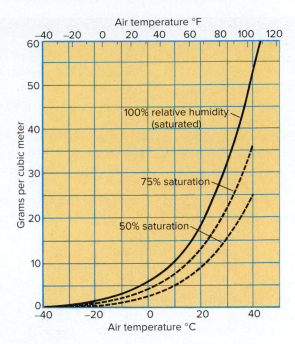

Figure 4.18   The water-carrying capacity of air and relative humidity. The actual water in the air (water vapor) divided by the water-carrying capacity (×100) equals the relative humidity. The solid line represents the maximum water-carrying capacity of air at different temperatures.

moist air rises, clouds form. The most dramatic cloud formation is probably the *cumulonimbus,* pictured in **Figure 4.20**. This is the anvil-head cloud that often accompanies heavy rain. Low, gray *stratus* clouds appear more often in cooler seasons than in warmer months. The very high, wispy *cirrus* clouds that appear in all seasons are made entirely of ice crystals. *Cumulus* clouds are often called fair-weather clouds.

Relative humidity is a percentage measure of the moisture content of the air, expressed as the amount of water vapor present relative to the maximum that can exist at the current temperature. As air gets warmer, the amount of water vapor it can contain increases. If the relative humidity is 100%, the air is completely saturated with water vapor. A relative humidity value of 60% on a hot day means the air is extremely humid and very uncomfortable. A

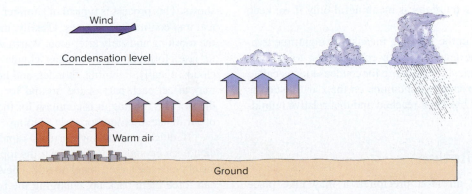

Figure 4.19   Precipitation. As warm air rises, it cools. As it cools, its water vapor condenses and clouds form. If the air becomes supersaturated, precipitation occurs.

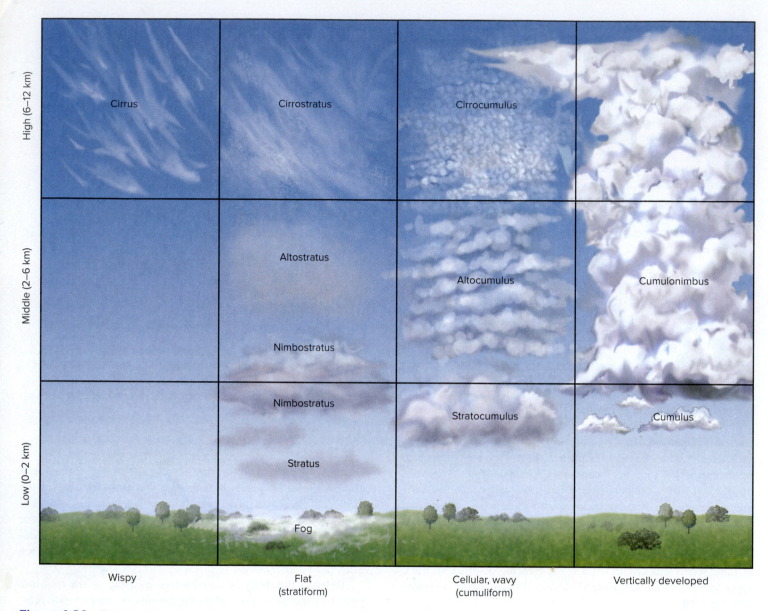

Figure 4.20   Types of clouds. Clouds are classified on the basis of their altitude and appearance. Most clouds occur at a specific range of altitudes, but some, such as cumulonimbus clouds associated with thunderstorms, may span several levels.

60% reading on a cold day, however, indicates that, although the air contains relatively large amounts of water vapor, it holds, in absolute terms, much less vapor than on a hot, muggy day. This example demonstrates that relative humidity is meaningful only if we keep air temperature in mind.

Dew on the ground in the morning means that nighttime temperatures dropped to the level at which condensation took place (see Figure 4.18). The critical temperature for condensation is called the **dew point.** Foggy or cloudy conditions on the Earth's surface imply that the dew point has been reached and that relative humidity is valued at 100%.

Types of Precipitation

When large masses of air rise, precipitation may take place in one of three types: (1) convectional, (2) orographic, or (3) cyclonic, or frontal.

The first type, **convectional precipitation,** results from rising, heated, moisture-laden air. As air rises, it cools. When its dew point is reached, condensation and precipitation occur, as **Figure 4.21** shows. This process is typical of summer storms or showers in tropical and continental climates. Usually, the ground is heated during the morning and early afternoon. Warm air that accumulates begins to rise, forming first cumulus clouds and then cumulonimbus clouds. Finally, lightning, thunder, and heavy rainfall occur, which may affect each part of the ground for only a brief period when the storm is moving. It is common for these convectional storms to occur in late afternoon or early evening.

If quickly rising air currents rapidly circulate air within a cloud, ice crystals may form near the top of the cloud. When these ice crystals are large enough to fall, a new updraft containing water can force them back up, enlarging the pieces of ice. This process may occur repeatedly until the updrafts can no longer sustain the ice pieces as they fall to the ground in the form of hail.

EL NIÑO-SOUTHERN OSCILLATION

El Niño is a term coined years ago by fishermen who noticed that the normally cool waters off the coasts of Ecuador and Peru were considerably warmer every 3 or 4 years around Christmas time, hence the name El Niño, Spanish for "the child," referring to the infant Jesus. The fish catch was significantly reduced during these periods. If fishermen had been able to identify the scientific associations that present-day oceanographers and climatologists make, they would have recognized a host of other effects that follow from El Niño. The **El Niño-Southern Oscillation** is a fluctuation in tropical ocean temperatures off the Pacific coast of South America. The warm phase is called El Niño, the cool phase La Niña.

During the winter of 1997–1998, an unusually severe El Niño caused enormous damage and hundreds of deaths. The West Coast of the United States, especially California, was inundated with rainfall amounts double, triple, and even quadruple the normal. For the November to March winter period, San Francisco received 102.24 centimeters (40.25 in.) of rain—the normal is 41.63 centimeters (16.39 in.). The 38 centimeters (15 in.) in February 1998 was the most for that month in the 150 years of record keeping in San Francisco. The resort city of Acapulco, Mexico, was badly battered by torrential rains and high, wind-blown tides. Parts of South America, especially Ecuador, Peru, and Chile, were ravaged by floods and mud slides, while droughts and fires scorched eastern South America, Australia, and parts of Asia, especially Indonesia. A stronger than usual southern branch of the jet stream generated by El Niño spawned dozens of tornadoes, which killed more than 100 people in Alabama, Georgia, and Florida.

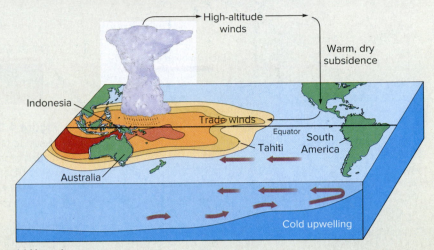

(a) Normal

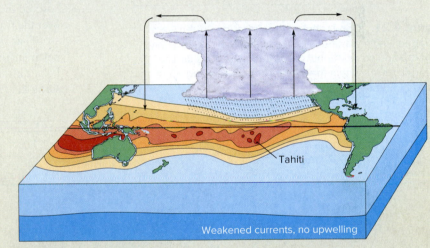

(b) El Niño

The top diagram shows normal circumstances in the southern Pacific Ocean. Normally trade winds blow warm surface water westward toward Indonesia and allow cold water to rise to the surface along the South American coast. The bottom diagram shows that, during El Niño, trade winds and ocean currents weaken and the warm water shifts eastward, shifting wind belts, and bringing storms to North and South America.

Orographic precipitation, the second type and depicted in **Figure 4.22**, occurs as warm air is forced to rise because hills or mountains block moisture-laden winds. This type of precipitation is typical in areas where mountains and hills are downwind from oceans or large lakes. Saturated air from over the water blows onshore, rising as the land rises. Again, the processes of cooling, condensation, and precipitation take place. The *windward* side—the side exposed to the prevailing wind—of the hills and mountains receives a great deal of precipitation. The opposite side, called the *leeward* side or rain shadow, and the adjoining regions downwind are very often dry. The air that passes over the mountains or hills descends and warms. As we have seen, descending air does not produce precipitation, and warming air absorbs moisture from surfaces it passes over. A graphic depiction of the great

differences in rainfall over very short distances is shown on the map of the state of Washington in **Figure 4.23**.

Cyclonic, or **frontal, precipitation,** the third type, is common to the midlatitudes, where cool and warm air masses meet. Although it is less frequent there, this type of precipitation also occurs in the tropics as the originator of hurricanes and typhoons. In order to understand cyclonic, or frontal, precipitation, first visualize the nature of air masses and the way cyclones develop.

Air masses are large bodies of air with similar temperature and humidity characteristics throughout; they form over a **source region.** Source regions include large areas of uniform surface and relatively consistent temperatures, such as the cold land areas of northern Canada, the north central part of Russia, and the warm tropical water areas in oceans close to the equator. Source regions

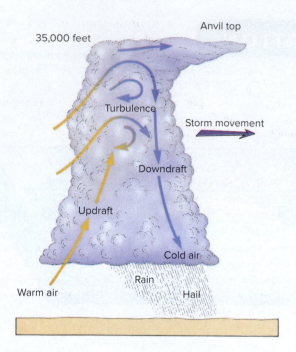

Figure 4.21 **Convectional precipitation.** When warm air laden with moisture rises, a cumulonimbus cloud may develop and convectional precipitation may occur. The falling particles within the system create a downdraft of cold upper-altitude air.

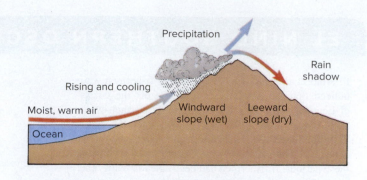

Figure 4.22 **Orographic precipitation.** Surface winds may be raised to higher elevations by hills or mountains lying in their paths. If such orographically lifted air is sufficiently cooled, precipitation occurs. Descending air on the leeward side of the upland barrier becomes warmer, its capacity to retain moisture is increased, and water absorption rather than release takes place.

for North America are shown in **Figure 4.24**. During a period of a few days or a week, an air mass may form in a source region. For example, in the fall in northern Canada, when snow has already covered the vast subarctic landscape, cold, dense, dry air develops over the frozen land surface.

Four major types of source regions are recognized: continental polar, maritime polar, continental tropical, and maritime tropical. Polar air masses are continental when they develop over land or ice surfaces in high latitudes; these are cold and dry. They are maritime

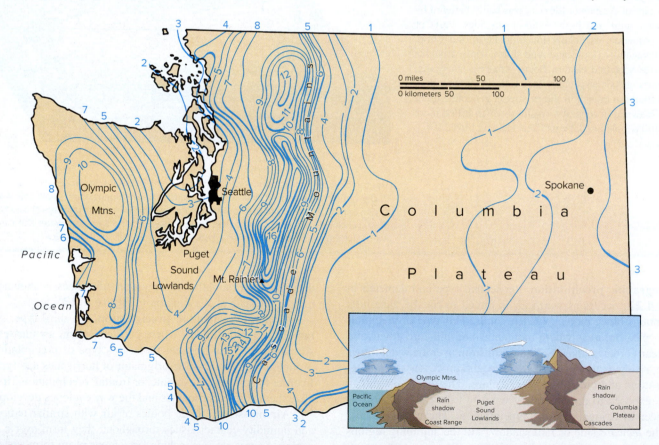

Figure 4.23 **Precipitation in inches for a typical November for the state of Washington.** The moisture-laden Pacific air is first forced up over the 1500- to 2100-meter (5000- to 7000-ft) Olympic Mountains; then it descends into the Puget Sound lowlands; next it goes up over the 2700- to 4300-meter (9000- to 14,000-ft) Cascade Mountains and finally down into the Columbia Plateau of eastern Washington. *From Robert N. Wallen,* Introduction to Physical Geography. *Copyright © 1993. McGraw-Hill Company, Inc., Dubuque, Iowa. All Rights Reserved. Reprinted by permission.*

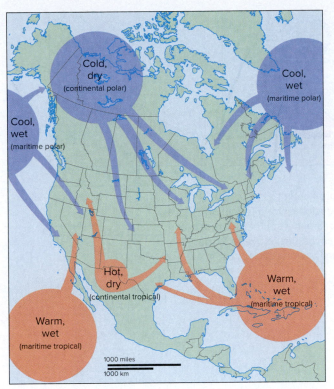

Figure 4.24   Source regions for air masses in North America. The United States and Canada, lying between major contrasting air-mass source regions, are subject to numerous storms and changes in weather. *From T. McKnight,* Physical Geography: A Landscape Appreciation, *4th ed. Copyright © 1993. Adapted by permission of Prentice Hall, Englewood Cliffs, New Jersey.*

when they form over the oceans in high latitudes. An air mass from these sources is cold and moist. Similarly, tropical air is continental when it originates along the Tropics of Cancer and Capricorn over northern Africa and northern Australia and is therefore warm and dry. It is maritime when it forms along the Tropics over the oceans, where it develops as a mass of warm, moist air. A single air mass usually covers thousands of square miles of the Earth's surface when fully formed.

When a continental polar air mass begins to move toward the lighter, warmer air to the south, the leading edge of a tongue of air is called a **front.** The front, in this case, separates the cold, dry air from whatever other air is in its path. If a warm, moist air mass is in front of a polar air mass, then denser, cold air hugs the ground and forces lighter air above it upward. The rising moist air condenses, and frontal precipitation occurs. On the other hand, the movement of rising warm air over cold air pushes the cold air back, again causing precipitation. In the first case, when cold air moves toward warm air, cumulonimbus clouds form and precipitation is brief and heavy. As the front passes, temperatures drop appreciably, the sky clears, and air becomes noticeably drier. In the second case, when warm air moves over cold air, then steel-gray nimbostratus (*nimbo* means "rain") clouds form and precipitation is steady and long-lasting. As the front passes, warm, muggy air becomes characteristic of the area. **Figure 4.25** summarizes the movement of fronts.

Storms

Two air masses coming into contact (a front) creates the possibility of storms developing. If the contrasts in temperature and humidity

are sufficiently great, or if wind directions of the two touching masses are opposite, a wave might develop in the front, as shown in **Figure 4.26.** Once established, the waves may enlarge. On one side of the front, cooler air moves along the surface, while on the other side, warm air moves up and over the cold air. The rising warm air creates a low-pressure center. Considerable precipitation is accompanied, in the Northern Hemisphere, by counterclockwise winds around the low-pressure area. A large system of air circulation centered on a region of low atmospheric pressure is called a midlatitude **cyclone,** which can develop into a storm.

An intense tropical cyclone, or **hurricane,** begins in a low-pressure zone over warm waters, usually in the Northern Hemisphere. In the developing hurricane, the warm, moist air at the surface rises, which helps suck up air from the surface. As a result, tall cumulonimbus clouds form. The energy released by these towering cloud formations warms the center of the growing storm. A feature of a hurricane is the calm, clear central core, called the *eye* (**Figure 4.27a**). The name given to a hurricane in the western Pacific is **typhoon.**

Figure 4.27b shows the usual paths of hurricanes in the world. The winds of these storms move in a counterclockwise

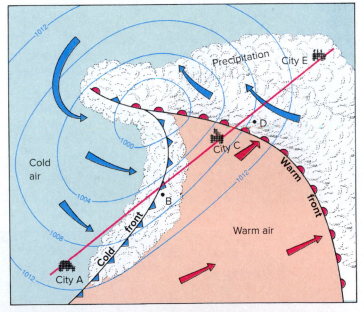

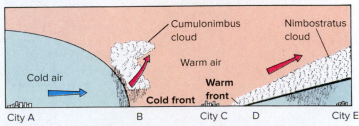

Figure 4.25   Weather fronts. In this diagram, a cold front in the Northern Hemisphere has recently passed over city A and is heading in the direction of B. The meeting lines of unlike air masses are called *fronts*. The warm front is moving away from B and toward city C. The wind direction is shown by arrows and the air pressure by isobars, lines of equal atmospheric pressure. The isobars indicate that the lowest pressure is found at the intersection of the warm and cold fronts.

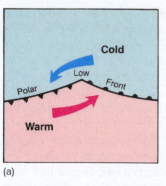
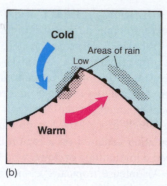
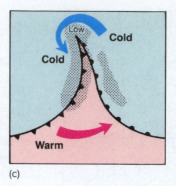
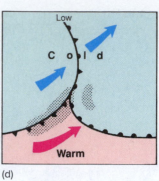

(a)   (b)   (c)   (d)

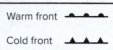

| Warm front | •––•––•––• |
| Cold front | ▲––▲––▲––▲ |

Figure 4.26   Cyclonic storm formation. When warm and cold air come into contact along a low-pressure trough in the Northern Hemisphere midlatitudes, the possibility of cyclonic storm formation occurs. **(a)** A wave begins to form along the polar front. **(b)** Cold air begins to turn in a southerly direction, while warm air moves north. **(c)** Cold air, generally moving faster than warm air, begins to overtake the warm air, forcing it to rise and, in the process, the storm deepens. **(d)** Eventually, two sections of cold air join. The cyclonic storm dissipates as the cold front is reestablished.

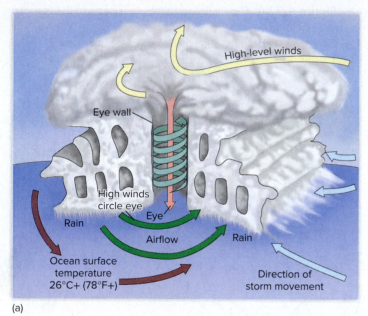

(a)

Figure 4.27   (a) Characteristics of a mature hurricane. Spiral bands of cumulonimbus clouds spawn heavy rainfall. Air ascends in the clouds near the center of the storm. The descending, warming air at the center creates an *eye,* a small area of calm on the surface. Intense convectional circulation creates strong winds away from the eye. **(b) General hurricane paths.** *(a) and (b) Michael Bradshaw and Ruth Weaver,* Physical Geography: An Introduction to Earth Environment, *pp. 177, 179. Reprinted by permission of The McGraw-Hill Companies, Inc.*

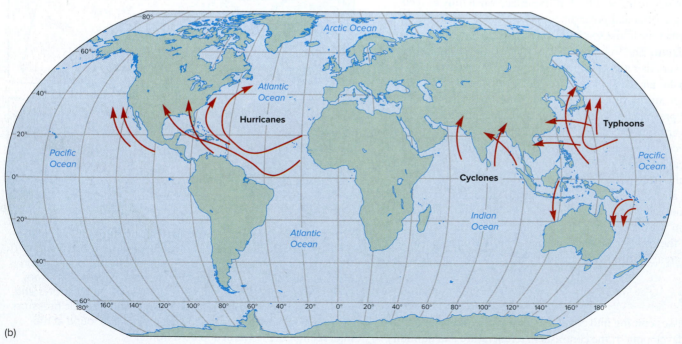

(b)

direction, converging near the center and rising in several concentric belts. Great damage is caused by the high winds (greater than 119 kilometers per hour [74 mph]) and the surge of ocean water into coastal lowlands. At the hurricane's center, the eye, air descends and results in gentle breezes and relatively clear skies. Over land, these storms lose their warm-water energy source and subside quickly. If they move farther into colder northern waters, they are pushed or blocked by other air masses and lose their energy source and abate. **Table 4.1** describes increasingly devastating hurricanes.

A **blizzard** is the occurrence of snow, often heavy, and high winds. On the evening of February 8, 2013, two storm systems collided along the New England coast. New Englanders call such

a situation a nor'easter. This disturbance occurred when moisture-laden storm winds from the south met with a fast-developing polar storm over New England. By evening, the blizzard was well developed. Intense convective snows fell, hurricane force winds blew, and destructive tides occurred, so that by the morning of February 9 about 0.8 meter (30 in.) of snow was on the ground and large areas lost power.

The most violent of all storms is the **tornado.** It is also the smallest storm (**Figure 4.28b**), typically measuring less than 30 meters (100 ft) in diameter. Tornadoes are spawned in the huge cumulonimbus clouds that sometimes travel in advance of a cold front along a squall line. During the spring or fall, when adjacent air masses differ the most, the central United States is

Table 4.1	The Saffir–Simpson Hurricane Scale		
Category	**Barometric Pressure (Inches)**	**Wind (mph)**	**Damage**
1	Over 28.94	74–95	Damage mainly to trees, shrubbery, unanchored mobile homes; storm surge damage for all categories. Storm surge generally 1.2 to 1.5 meters (4 to 5 ft) above normal.
2	28.50–28.94	96–110	Some trees blown down; major damage to exposed mobile homes; some damage to roofs. Storm surge: 1.8 to 2.4 meters (6 to 8 ft).
3	27.91–28.49	111–130	Trees stripped of foliage, large trees blown down; mobile homes destroyed; some structural damage to small buildings. Storm surge: 2.7 to 3.7 meters (9 to 12 ft).
4	27.17–27.90	131–155	All signs blown down; extensive damage to windows, doors, and roofs; flooding inland as far as 10 kilometers (6 mi); major damage to lower floors of structures near shore. Storm surge: 4.0 to 5.5 meters (13 to 18 ft).
5	Less than 27.17	Over 155	Severe damage to windows, doors, and roofs; small buildings overturned and blown away; major damage to structures less than 4.6 meters (15 ft) above sea level within 458 meters (500 yds) of shore. Storm surge: greater than 5.5 meters (18 ft) above normal.

(a)

(b)   (c)

Figure 4.28   Storms. **(a)** Blizzards bring the transportation systems of cities to a halt. **(b)** In the United States, tornadoes occur most frequently in the central and south central parts of the country (especially in Oklahoma, Kansas, and the Texas Panhandle), where cold polar air very often meets warm, moist Gulf air. **(c)** The Oklahoma City tornado of May 3, 1999, was an EF5; it leveled neighborhoods such as this one. *(a) © Brand X/Veer RF; (b) © Corbis RF; (c) © Jeff Mitchell/Reuters.*

prone to many of these funnel-shaped killer clouds. Although winds can reach 500 kilometers per hour (about 300 mph), these storms are small and usually travel on the ground for less than a mile, so only limited areas are affected, though they may be devastated (**Figure 4.28c**). A tornado over water is called a *waterspout*.

The *Enhanced Fujita (EF) scale* of tornado intensity links reported damage to wind speed. It ranges from EF0, a "weak" tornado with wind speeds up to 137 kilometers per hour (85 mph), to EF5, a "violent" one, with winds as great as 322 kilometers per hour (200 mph). Most (74%) tornadoes are either EF0 or EF1, whereas 25% are classified as EF2 or EF3, "strong" tornadoes capable of causing major structural damage. Only 1% are in the ultra-violent (EF4 and EF5) categories.

4.5 Climate Regions

We have traced some of the causes of weather changes that occur as air from high-pressure zones flows toward low-pressure areas, fronts pass and waves develop, dew points are reached, and sea breezes arise. Some parts of the world experience these changes more rapidly and more often than do other parts.

Day-to-day weather conditions can be explained by the principles discussed in this chapter. However, the effect of weather elements—temperature, precipitation, and air pressure and winds—cannot be understood unless a person is conscious of the Earth's surface features. Weather forecasters in each location on Earth must deal with weather elements in the context of their local physical and built environments.

The complexities of daily weather conditions may be summarized by statements about climate. The climate of an area is a generalization based on daily and seasonal weather conditions.

Are summers warm, on the average? Is heavy snow in the winter likely? Are winds normally from the southeast? Are climatic averages typical of daily weather conditions, or are the day-to-day or week-to-week variations so great that one should speak of average variations rather than just averages? These are the questions we must ask in order to form an intelligent description of the differences in conditions from place to place.

The two most important elements that differentiate weather conditions are temperature and precipitation. Although air pressure is also an important weather element, differences in air pressure are hardly noticeable without the use of a barometer. Thus, we may regard warm, moderate, cold, or very cold temperatures as characteristic of a place or region. In addition, high, moderate, and low precipitation levels are good indicators of the degree of humidity or aridity in a place or region.

Figure 4.30 (see pp. 96–97) depicts the various climates of the world and is based on the type of information presented in **Figure 4.31** (p. 98). Called the Köppen system, it is the best known of a number of similar climate classification schemes. Developed in 1918, it is based on natural vegetation in addition to temperature and precipitation criteria.

Table 4.2 on page 95 shows the multilevel system developed by Köppen. There are six broad categories, designated as A, B, C, D, E, and H. The A climates are tropical, B are dry, C are mild climates in the midlatitudes, D climates of the midlatitudes have severely cold winters, E are polar, and H are highland climates.

The column on the right in Table 4.2 lists the kinds of soils and vegetation typical of different climates. Soil types and various forms of vegetation go hand in hand with temperature levels and the seasonal distribution of precipitation. For a discussion of the relationship of climate to soils and vegetation, see Appendix 2.

The letters of the Köppen system in the section headings that follow refer to those in Figure 4.30 and Table 4.2.

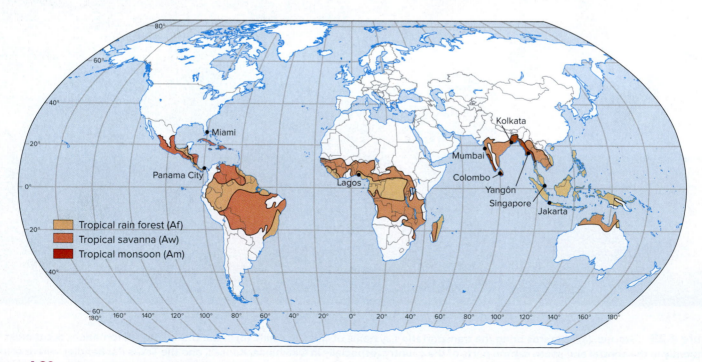

Figure 4.29   **The location of tropical climates.**

Table 4.2	Climate Characteristics		
Climate Type	**Köppen Classification**	**Temperature and Precipitation**	**Soil and Vegetation**
Tropical	**A**		
Tropical rain forest	**Af**	Constant high temperatures Rainfall heavy all year; convectional High amount of cloud cover High humidity	Dense, many species of trees Jungle where light penetrates Most soil nutrients absent
Savanna	**Aw**	High temperatures Rainfall heavy in summer high-sun period; convectional Dry in winter low-sun period	Forests to grassland, depending on rainfall amount Soils respond to fertilization
Monsoon	**Am**	Highest temperature just before rainy season	
Semidesert and Dryland	**B**		
Hot deserts	**BWh**	Extremely high temperatures in summer; warm winters Very little rainfall Low humidity	Shrubs in gravelly or sandy environments Reptiles Soils respond to irrigation
Steppe and desert	**BS** **BWk**	Warm to hot summers Cold winters Some convectional rainfall in summer Some frontal snowfall in winter	Grass and desert shrubs Naturally fertile soils
Humid Midlatitude	**C**		
Mediterranean	**Cs**	Warm to hot summers Mild to cool winters Dry summers Frontal precipitation in winter Generally low humidity	Chapparal vegetation (scrub oak trees and bushes) Soils respond to irrigation
Humid subtropical	**Cfa**	Hot summers Mild winters Convectional showers in summer Frontal precipitation in winter	Deciduous forests, soils rich Coniferous forests, especially in sandy soils
Marine west coast	**Cfb**	Westerly winds year-round Mild summers Cool to cold winters Low rainfall in summer Frontal rainfall in winter	Vast coniferous forests in orographic regions Deciduous forest in plains Acidic soils require fertilization for agriculture
Humid Continental	**D**		
Subarctic	**Dfa and Dfb** **Dfc and Dfd**	Hot to mild summers Cool to very cold winters Convectional showers in summer Frontal snowfall in winter	Coniferous forests Generally infertile soils
Arctic	**E**		
Tundra	**ET**	Cold summers, extremely cold winters	Mosses and lichens
Ice cap	**EF**	Extremely cold with light precipitation	
Highlands	**H**	Great variety of conditions based on elevation, prevailing winds, sun- or nonsun-facing slopes, latitude, valley or nonvalley, ruggedness	

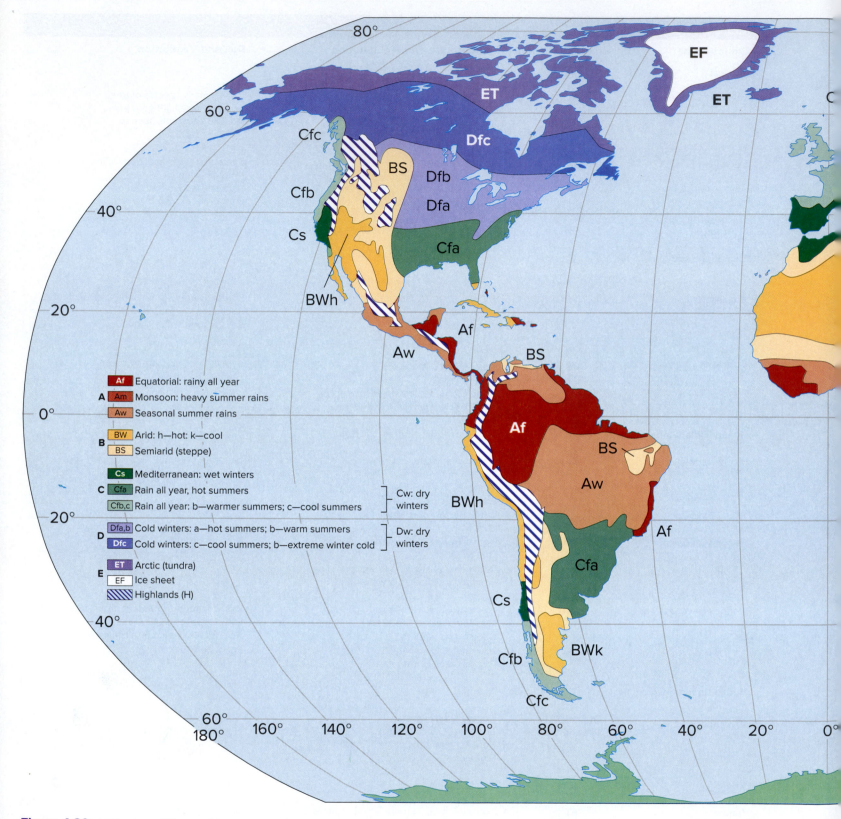

Figure 4.30   **Climates of the world.**

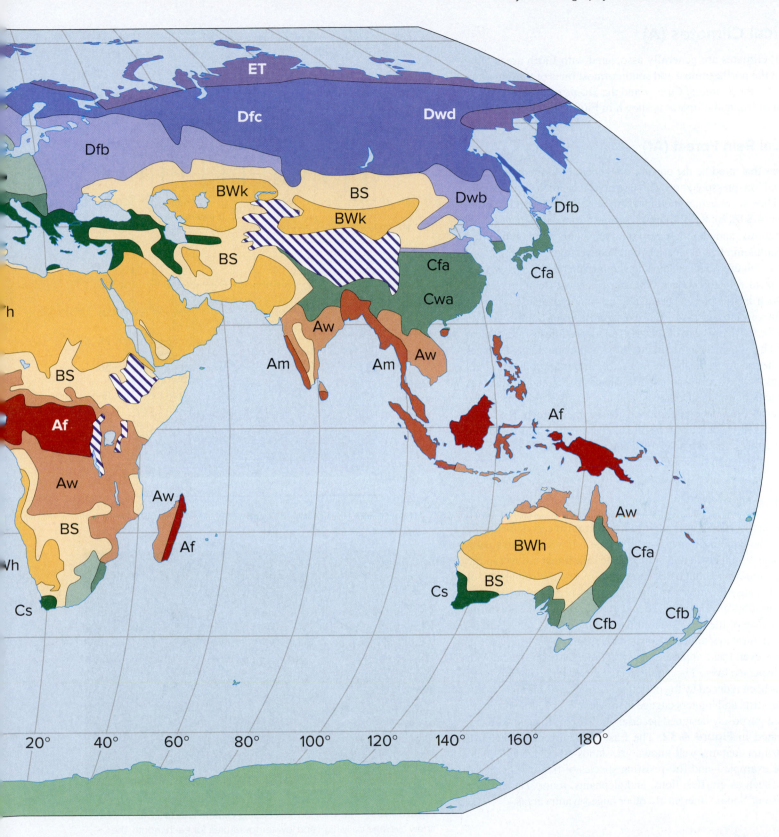

Tropical Climates (A)

Tropical climates are generally associated with Earth areas lying between the northernmost and southernmost lines of the sun's vertical rays—the *Tropic of Cancer* and the *Tropic of Capricorn*. The location of tropical climates is shown in **Figure 4.29** (see p. 94).

Tropical Rain Forest (Af)

The areas that straddle the equator are generally located within the equatorial low-pressure zone. These regions are called **tropical rain forest.** They are warm, wet climates in both the winter and summer (**Figures 4.31a, b**). Rainfall usually comes from daily convectional thunderstorms, and although most days are sunny and hot, by afternoon, cumulonimbus clouds form and convectional rain falls. The caption for Figure 4.31a explains how to interpret a *climagraph*.

Tropical rain forests are typically filled with natural vegetation, which is still present but declining rapidly because of intentional fires in large areas of the Amazon Basin of South America and the Zaire River Basin of Africa. Tall, dense forests of broadleaf trees and heavy vines predominate. Among the hundreds of species of trees found in tropical rain forests, both dark woods and light woods, as well as spongy softwoods, such as balsa, and hardwoods, such as teak and mahogany, exist (Figure 4.31b). Rain forests also extend away from the equator along coasts where prevailing winds supply a constant source of moisture to coastal uplands. In addition, the orographic effect provides sufficient precipitation for heavy vegetation to develop in these forests.

Savanna (Aw)

As the sun's vertical rays extend farther from the equator in the summer, the equatorial low-pressure zone follows the sun's path. Thus, areas to the north and south of the rain forest are wet in the summer months, although still hot, but are dry the remainder of the year because the moist equatorial low has been replaced by the dry air of subtropical highs. These areas are known as **savanna** lands because of the kind of natural vegetation that grows here.

The natural vegetation of savannas resembles a form of scrub forest; however, these areas are now recognized as a grassland with widely dispersed trees. The natural tendency toward a more forested cover has been reduced by the periodic clearing by burning that local agriculturalists and hunters engage in. Savannas sometimes seem to have been purposely designed because of their parklike appearance, as indicated in **Figure 4.32**. The East African region of Kenya and Tanzania contains well-known grasslands—Serengeti National Park, for example—and fire-resisting species of trees, where large animals, such as giraffes, lions, and elephants, roam. The *campos* and *llanos* of South America are other huge savanna areas.

Monsoon (Am)

A special case in Asia needs mentioning. When summer monsoon winds carry water-laden air to the mainland, a significant increase in rainfall occurs on the hills, mountains, and adjacent plains. Notice the pattern of precipitation in **Figure 4.33**. As a result, vegetation is dense, even though the winters are dry. Jungle growth

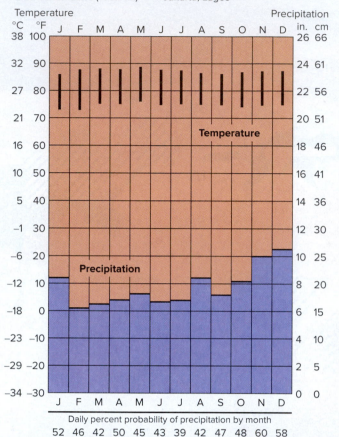

City: Singapore
Latitude: 1°20′N
Altitude: 11 meters (33 ft)
Yearly precipitation:
  256 centimeters (100.7 in.)

Climate designation: Af
Climate name: Tropical rain forest
Other cities with similar climates:
  Colombo, Panama City,
  Jakarta, Lagos

Daily percent probability of precipitation by month
52 46 42 50 45 43 39 42 47 48 60 58

(a)

(b)

Figure 4.31   **(a)** This and succeeding **climate charts** (*climagraphs*) show average daily high and low temperatures for each month, the average precipitation for each month, and the probability of precipitation on any particular day in a designated month. For Singapore, the average daily high temperature in August is 30.5°C (87°F); the low is 24°C (75°F). The rainfall for the month, on average, is 21 centimeters (8.4 in.), and on a given day in August, there is a 42% chance of rainfall. **(b) Tropical rain forest.** The vegetation is characterized by tall, broadleaf, hardwood trees and vines. © *GeoStock/Getty Images RF.*

Figure 4.32   The parklike landscape of grasses characteristic of the drier savanna. © *McGraw-Hill Education/Jill Wilson.*

City: Yangôn, Myanmar
Latitude: 16°46′N
Altitude: 5.5 meters (18 ft)
Yearly precipitation:
  252 centimeters (99.2 in.)

Climate designation: Am
Climate name: Monsoon
Other cities with similar climates:
  Mumbai, Calcutta

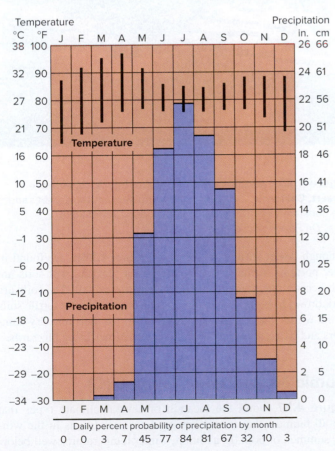

Figure 4.33   **Climagraph for Yangôn, Myanmar.**

and large forests are the natural vegetation. Much of this vegetation, however, has ceased to exist because people have been using the land mainly for rice and tea production for many generations.

Dryland Climates (B)

The location of these climates is shown in **Figure 4.34**. In the interior of continents where mountains block west winds, or in lands far from the reaches of moist tropical air, extensive regions of desert and *semidesert* conditions appear.

Hot Deserts (BWh)

On the poleward side of the savannas, grasses begin to shorten, and desert shrubs become evident. This is where we approach the belt of subtropical high pressure that brings considerable sunshine, hot summer weather, and very little precipitation. Note the minute amount of rainfall shown in **Figure 4.35a**. The precipitation that does fall is convectional but sporadic. As conditions become drier, fewer and fewer drought-resistant shrubs appear and, in some areas, only gravelly and sandy deserts exist, as suggested in **Figure 4.35b**.

The great, hot deserts of the world, such as the Sahara, the Arabian, the Australian, and the Kalahari, are all the products of high-pressure zones. Often, the driest parts of these deserts are along the western coasts, where cold ocean currents are found. Earlier, mention was made of the relationship between cold ocean currents and deserts.

Midlatitude Deserts and Semideserts (BWk, BS)

Figure 4.36a illustrates typical temperature and precipitation patterns in these midlatitude drylands. Occasionally, a summer convectional storm or a frontal system with some moisture occurs. The extreme dry areas are known as cold deserts. The moderately dry lands are called **steppes**. The natural vegetation is grass,

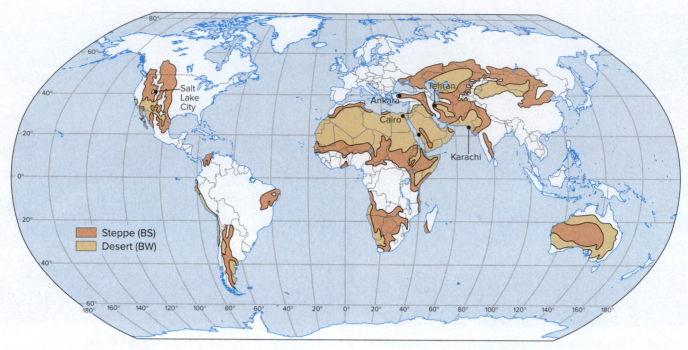

Figure 4.34 The location of steppe and desert climates.

City: Cairo, Egypt
Latitude: 29°52′N
Altitude: 116 meters (381 ft)
Yearly precipitation: 2 centimeters (0.7 in.)

Climate designation: BWh
Climate name: Hot desert
Other cities with similar climates: Mecca, Karachi

Temperature
°C °F

Precipitation
in. cm

Daily percent probability of precipitation by month

J	F	M	A	M	J	J	A	S	O	N	D
3	3	3	0	0	0	0	0	0	0	3	3

(a)

(b)

Figure 4.35 **(a) Climagraph for Cairo, Egypt. (b) Mohave Desert, California.** Devoid of stabilizing vegetation, desert sands are constantly rearranged in complex dune formations. *(b) © Dr. Parvinder Sethi RF.*

although desert shrubs, pictured in **Figure 4.36b**, are found in drier portions of the steppes. Rain is not plentiful, but soils are rich because the grasses return nutrients to the soil. The soils are dark brown to black and are among the most naturally fertile soils in the world. The steppes are also known for their hot, dry summers and biting winter winds, which sometimes bring blizzards.

Humid Midlatitude Climates (C)

Figure 4.37 shows the location of several climate types that are all humid—that is, not having desert conditions in the winter, summer, or both. In addition, winter temperatures well below

City: Tehran, Iran
Latitude: 35°41′N
Altitude: 1220 meters (4002 ft)
Yearly precipitation:
  26 centimeters (10.1 in.)

Climate designation: BS
Climate name: Midlatitude dryland
Other cities with similar climates:
  Salt Lake City, Ankara

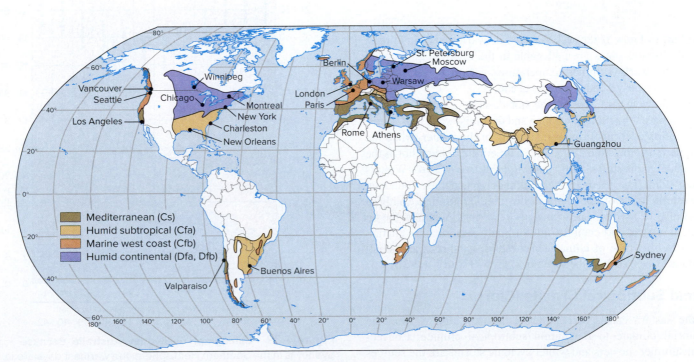

(a)

(b)

Figure 4.36 **(a) Climagraph for Tehran, Iran. (b) Desert shrubs in the midlatitude drylands of northern Mexico.** © *Steven P. Lynch RF.*

those of the tropical climates are characteristic of the humid mid-latitudes. These climate types would be neatly defined, paralleling the lines of latitude, were it not for mountain ranges, warm or cold ocean currents, and, particularly, land-water configurations. These factors cause the greatest variations in the middle latitudes.

Mediterranean Climate (Cs)

Midlatitude winds generally blow from the west in both the Northern and Southern Hemispheres, and a significant amount of the precipitation is produced from frontal systems. Thus, it is

Figure 4.37   **The location of humid midlatitude and continental climates.**

City: Rome, Italy
Latitude: 41°48′N
Altitude: 115 meters (377 ft)
Yearly precipitation:
  85 centimeters (33.3 in.)

Climate designation: Cs
Climate name: Mediterranean
Other cities with similar climates:
  Athens, Los Angeles,
  Valparaiso

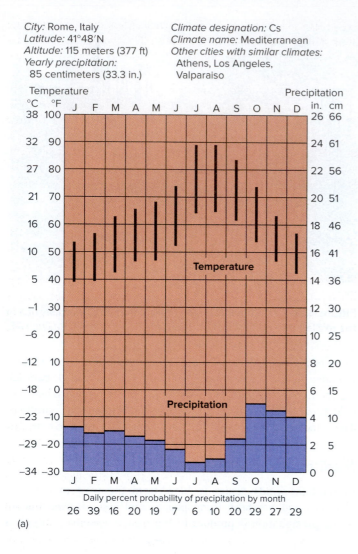

Daily percent probability of precipitation by month
26 39 16 20 19 7 6 10 20 29 27 29

(a)

(b)

Figure 4.38   **(a) Climagraph for Rome, Italy. (b) Vegetation typical of an area with a Mediterranean climate** (coastal California). Trees and brush are short and scattered. © *Digital Vision RF.*

City: Sydney, Australia
Latitude: 33°58′S
Altitude: 9 meters (29 ft)
Yearly precipitation:
  116 centimeters (46.5 in.)

Climate designation: Cfa
Climate name: Humid subtropical
Other cities with similar climates:
  Guangzhou, Charleston,
  New Orleans

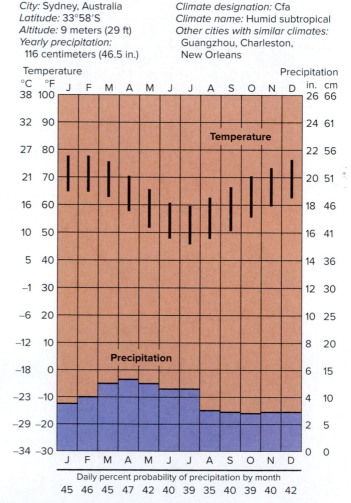

Daily percent probability of precipitation by month
45 46 45 47 42 40 39 35 40 39 40 42

Figure 4.39   **Climagraph for Sydney, Australia.** Because Sydney is in the Southern Hemisphere, the warmest days are in January and the coldest are in July.

important to know if the water is cold or warm near land areas. Several climatic zones are noticeable in the middle latitudes, all marked by warm summer temperatures except those in areas cooled by westerly winds from the ocean.

To the poleward side of the hot deserts, a transition zone occurs between the subtropical high and the moist westerlies zones. Here, cyclonic storms bring rainfall only in the winter, when the westerlies shift toward the equator. Summers are dry and hot as the subtropical highs shift slightly poleward (**Figure 4.38a**). Winters are not cold. The Mediterranean climate is generally found on the western coasts of continents in the middle latitudes. Southern California, the Mediterranean area itself, western Australia, the tip of South Africa, and central Chile in South America are characterized by this type of climate. In these areas, shrubs and small deciduous trees, such as the scrub oak, grow (**Figure 4.38b**).

Humid Subtropical Climate (Cfa)

On the eastern coasts of continents, the transition is from the equatorial climate to the humid subtropical climate. Convectional summer showers and winter cyclonic storms are the sources of precipitation. As illustrated in **Figure 4.39**, this climate is characterized by hot, moist summers and moderate, moist winters. In the fall, on occasion, hurricanes that develop in tropical waters strike the coastal areas.

The generally even distribution of rainfall allows for the presence of deciduous forests containing hardwood trees, such as oak and maple, whose leaves turn orange and red before falling in autumn. In addition, conifers become mixed with deciduous trees as a second-growth forest. Southern Brazil, the southeastern United States, and southern China all have a humid subtropical climate.

Marine West Coast Climate (Cfb)

Closer to the poles, but still within the westerly wind belt, are areas of **marine west coast climate.** Here, cyclonic storms and orographic precipitation play a relatively large role. In the winter, more rainfall and cooler temperatures prevail than in the Mediterranean zones. Compare the patterns in **Figures 4.38** and **4.40**. In the transitional zone just poleward of the Mediterranean climate, little rainfall occurs during the summer. Closer to the poles, however, rainfall increases appreciably in the summer and

even more so in the winter. Marine winds from the west moderate both summer and winter temperatures. Thus, summers are pleasantly cool, and winters, though cold, do not normally produce freezing temperatures.

This climate affects relatively small land areas in all but one region. Because northern Europe contains no great mountain belt to thwart the west-to-east flow of moist air, the marine west coast climate stretches well across the continent to Poland. In Poland, cyclonic storms originating in the Arctic regions are noticeable. Northern Europe's moderate climate also owes its existence to a relatively warm ocean current whose influence is felt for nearly 1600 kilometers (1000 mi), from Ireland to central Europe.

The orographic effect from mountains in areas such as the northwestern United States, western Canada, and southern Chile produces enormous amounts of precipitation, often in the form of snow on the windward side (see Figure 4.23). Vast coniferous forests—needle-leaf trees, such as pines, spruces, and firs—cover the mountains' lower elevations. Because the mountains prevent moist air from continuing to the leeward side, midlatitude deserts are found to the east of these marine west coast areas.

Humid Continental Climates (D)

The poleward transition to the continental climates is accompanied by increasingly colder winters and shorter summers. In this direction as well, cyclonic storms become more responsible for rainfall than are convectional showers. The region can no longer be characterized as humid subtropical; rather, it is described as a **humid continental climate** (Dfa, Dfb).

Air masses that originate close to the poles and drift toward the equator and other air masses that drift toward the poles from the tropics produce frontal precipitation. Whenever warmer air or marine air blocks cold continental air masses, or vice versa, frontal storms develop. **Figure 4.41** shows the range and the dominance of winter temperatures within this climate type.

The continental climate may be contrasted to marine west coast climates in that the former has prevailing winds from the land, the latter from the sea. Coniferous forests become more plentiful in the direction of the poles, until temperatures become so low that trees are denied an adequate growing season (**Figure 4.42**).

Three huge areas of the world are characterized by a humid continental climate: (1) the northern and central United States and southern Canada, (2) most of the European portion of Russia, and (3) northern China. Because there are no land areas at a comparable latitude in the Southern Hemisphere, this climate is not represented there. In fact, the only nonmountain cold climate in the Southern Hemisphere is the polar climate of Antarctica.

Subarctic Climates (Dfc, Dfd, Dwb)

Toward northern areas and into the interior parts of the North American and Eurasian landmasses, increasingly colder temperatures prevail (**Figures 4.43** and **4.44a**). Trees become stunted, and eventually only mosses and other cool-weather plants of the type shown in **Figure 4.44b** will grow.

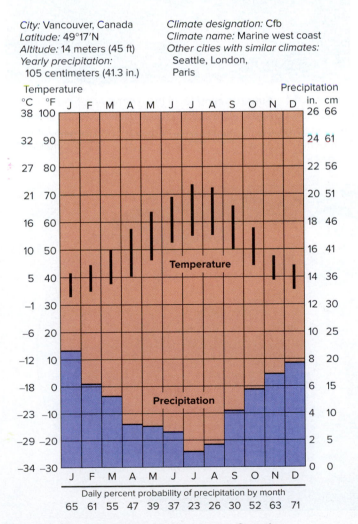

Figure 4.40   Climagraph for Vancouver, Canada.

City: Vancouver, Canada
Latitude: 49°17'N
Altitude: 14 meters (45 ft)
Yearly precipitation:
   105 centimeters (41.3 in.)

Climate designation: Cfb
Climate name: Marine west coast
Other cities with similar climates:
   Seattle, London,
   Paris

Daily percent probability of precipitation by month
65 61 55 47 39 37 23 26 30 52 63 71

City: Chicago, Illinois
Latitude: 41°52′N
Altitude: 181 meters (595 ft)
Yearly precipitation:
  85 centimeters (33.3 in.)

Climate designation: Dfa
Climate name: Humid continental
  (warm summer)
Other cities with similar climates:
  New York, Berlin,
  Warsaw

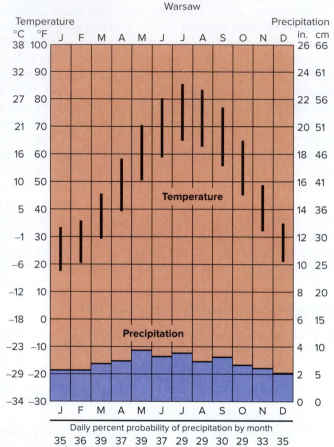

Daily percent probability of precipitation by month
35 36 39 37 39 37 29 29 30 29 33 35

Figure 4.41 **Climagraph for Chicago, Illinois.**

Arctic Climates (E)

The word **tundra** (ET) is often used to describe the northern boundary zone beyond these treed subarctic regions. Because long, cold winters predominate, the ground is frozen most of the year. A few cool summer months, with an abundant supply of mosquitoes and flies, break up the monotony of extreme cold. Although very cold temperatures characterize the tundra, snowfall is not very abundant. Strong easterly winds blow snow, which, combined with ice fogs and little winter sunlight, contributes to a very bleak climate. Alaska, northern Canada, and northern Russia are covered with either the stunted trees of the subarctic climate or the

Figure 4.42 **Subartic climate.** In the extensive region of east central Canada and the area around Moscow, Russia, the summers are long and warm enough to support a dense coniferous forest. Farther north, growth is less luxuriant. © *Bruce Heinemann/Getty Images RF.*

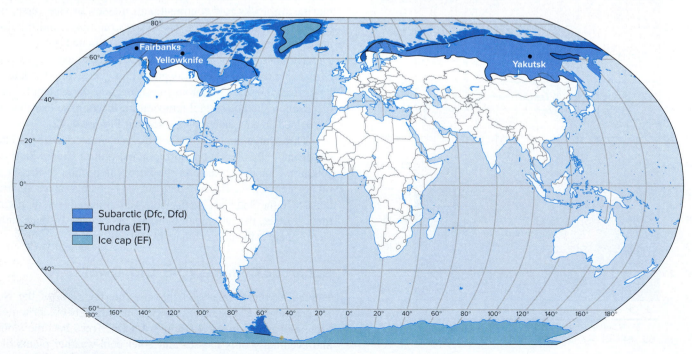

Figure 4.43 **The location of subarctic and Arctic climates.**

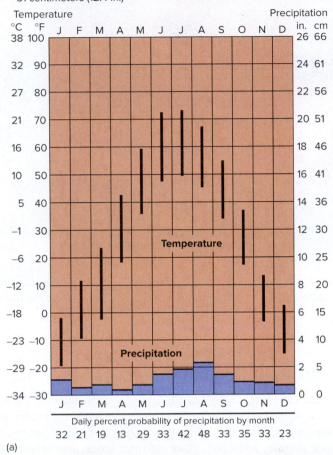

City: Fairbanks, Alaska
Latitude: 64°51'N
Altitude: 134 meters (440 ft)
Yearly precipitation:
   31 centimeters (12.4 in.)

Climate designation: Dfd
Climate name: Subarctic
Other cities with similar climates:
   Yellowknife, Yakutsk

Daily percent probability of precipitation by month
32 21 19 13 29 33 42 48 33 35 33 23

(a)

(b)

Figure 4.44 **(a) Climagraph for Fairbanks, Alaska. (b) Tundra vegetation in Canada.** (b) © George F. Mobley/Getty Images.

bleak, treeless expanse of the tundra. Antarctica and Greenland, however, are icy deserts (EF).

Highland Climates (H)

We mentioned earlier (p. 80) that, under the normal lapse rate, temperatures decrease as altitude increases. As a result, highlands have lower temperatures than do lowlands at the same latitude. Highland climates are complex, however, because elevation and latitude are only two of the factors that determine their nature and the plant and animal life they can support. Some mountain slopes face the prevailing wind, while others are lee slopes. Some face the sun; others are shadowed and cool. Some are sunlit in the cool of the morning; others receive the hot afternoon sun. Mountain valleys have a different climate than do rugged peaks. Every mountain range, then, contains a mosaic of climates far too detailed to show on a map such as Figure 4.29.

These thumbnail sketches of climatic conditions throughout the world give us the basic patterns of large regions. On any given day, conditions may be quite different from those discussed or mapped in this chapter. However, the physical climatological processes, in general, are what concern us. We can deepen our understanding of climates by applying our knowledge of the elements of weather.

4.6 Climate Change

We have stressed that climates are only averages of, perhaps, greatly varying day-to-day conditions. **Figure 4.45** (p. 106) illustrates the global variation in yearly precipitation. Temperatures are less changeable than precipitation on a year-to-year basis, but they, too, vary. How can we account for these variations? Scientists in research stations all around the world are investigating this question. The data they use range from daily temperature and precipitation records to calculations concerning the position of the Earth in relation to the sun. Because day-to-day records for most places date back only 50 to 100 years, scientists look for additional information about past climates in rock formations, the chemical composition of earth materials, ice cores, lake-floor sediments, tree rings, and other sources.

Long-Term Climate Change

Significant variations in climate have occurred over geologic time. For example, approximately 65 million years ago, there was a sudden cooling of the Earth's climate. This cooldown is thought to have caused the extinction of some 75% of all existing plant and animal species, including most dinosaurs. To take another example, cycles of ice sheet formation and breakup occurred at least five times during the last ice age, which lasted 100,000 years and ended only 11,000 years ago.

Climatologists have identified two major climatic periods just in the past 1000 years: a medieval warm period and a "little ice age." Between about A.D. 800 and 1200, during the medieval warm period, temperatures were as warm as or warmer than they are now.

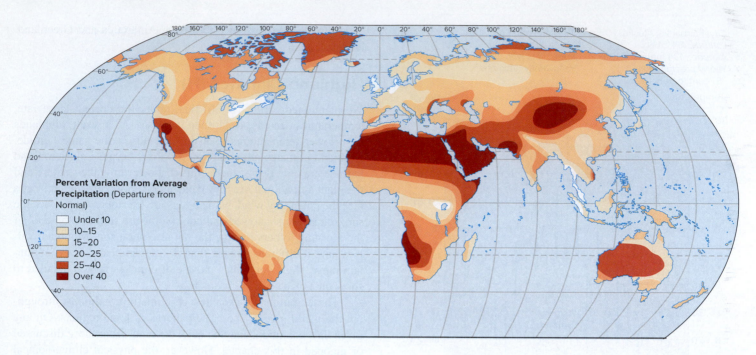

Figure 4.45   **The world pattern of precipitation variability.** Regions of low total precipitation tend to have high variability. In general, the drier the climate, the greater is the probability that there will be considerable differences in rainfall and/or snowfall from one year to the next.

Settlement and farming expanded northward and to higher altitudes, the Vikings colonized Iceland and Greenland, and vineyards flourished in Britain. During the little ice age, which lasted from about A.D. 1300 to 1850, Arctic ice expanded, glaciers advanced, drier areas of the Earth were desiccated, and crop failures and starvation were common.

Scientists have suggested several explanations for such long-term periodic changes in climate. Some of the climatic variations are thought to be due to three aspects of the Earth's motion, all of which affect the amount of solar radiation reaching the planetary surface. First is the shape of the Earth's orbit around the sun, which varies from nearly circular to more elliptical over a period of about 100,000 years. When the orbit is nearly circular, the Earth experiences relatively cold temperatures. When it is elliptical, as it is now, the Earth is closer to the sun for several months, is exposed to greater total solar radiation, and thus has higher temperatures.

Another cycle corresponds to the tilt of the Earth's axis relative to the orbital plane. The tilt varies from 21.5° to 24.5° every 41,000 years. The amount of solar radiation striking polar regions changes as the angle of tilt changes. A low tilt position—that is, a more perpendicular position of the Earth—is accompanied by periods of colder climate. Cooler climates are thought to be critical in the formation of ice sheets.

Finally, like an unbalanced spinning top, the Earth wobbles slightly as it rotates, changing the Earth's orientation to the sun. The gyration of the rotation axis repeats every 23,000 years. When the tilt of the axis is greatest, the polar regions receive less solar radiation than they do at other times and become colder.

Short-Term Climate Change

Climate can change more quickly and irregularly than the Earth's cycles suggest. Great volcanic eruptions can alter climates for several years. They spew enormous amounts of ash, water vapor, sulfur dioxide, and other gases into the upper atmosphere. As these solid and liquid particles spread over much of the planet, they block some of the incoming solar radiation that normally would reach the Earth's surface, producing a cooling effect. The famous "year without a summer"—1816—in New England, when snow fell in June and frost came in July, probably was caused by the eruption a year earlier of the Indonesian volcano Tambora. The explosion ejected an estimated 200 million tons of gaseous aerosols and 50 cubic kilometers (30 cu mi) of dust and ash into the atmosphere. The reflective cooling effect lasted for years. A similar decline in temperatures occurred after the 1883 volcanic eruption of Krakatoa, also in Indonesia. A less extreme drop in temperatures in the early 1990s was attributed to the July 1991 eruption of Mount Pinatubo in the Philippines, which lowered average global temperatures by about 0.5°C (about 1°F).

Two other factors responsible for short-term climate changes are alterations in patterns of oceanic circulation and sunspot activity. As described on page 89, for example, during an El Niño event, warm surface waters from the western Pacific Ocean move eastward, changing the climate along the western coasts of South and North America. Sunspots, relatively cool regions on the surface of the sun, vary in number and intensity over periods of years. They affect both the output of solar energy and the concentrations of ozone in the Earth's upper atmosphere.

The Greenhouse Effect and Global Warming

All of the cycles and factors we have discussed are natural processes. In contrast, more recently, one of the most important questions has been whether human beings are contributing to climate change through what is popularly termed the **greenhouse effect.** Put simply, scientists have found that certain gases in the atmosphere function as an insulating barrier, trapping infrared radiation that would otherwise be radiated back into the upper atmosphere and reradiating it earthward. In other words, like glass in a greenhouse, the gases admit incoming solar radiation but retard its reradiation back into space. You have experienced such a greenhouse effect if you have gotten into a car that has been in the sun; the car's interior is warmer than the outside air.

The Earth has a *natural* greenhouse effect, provided mainly by water vapor that has evaporated from the ocean or evapotranspired from land. The water vapor remains, but during the past 150 years or so, human activities have increased the amount of other greenhouse gases in the atmosphere, augmenting its heat-trapping ability. Scientists fear that an *enhanced* greenhouse effect is responsible for a gradual increase in the Earth's average surface temperature, with significant impacts on the Earth's ecosystems, a process called **global warming.** *That* greenhouse effect is far less benign and nurturing than the name implies.

The Intergovernmental Panel on Climate Change (IPCC), made up of thousands of scientists from about 120 countries, reported in May 2007 that

> *Global atmospheric concentrations of carbon dioxide (CO_2), methane, and nitrous oxide have increased markedly as a result of human activities since 1750 and now far exceed preindustrial values determined from ice cores spanning many thousands of years. The global increases in CO_2 concentration are due primarily to fossil fuel use and land use change, while those of methane and nitrous oxide are primarily due to agriculture.*

Carbon dioxide (CO_2) is the primary greenhouse gas whose amount has been increased by human activities. Although it occurs naturally, excessive quantities of it are released by burning fossil fuels. Beginning with the Industrial Revolution in the mid-1700s, large amounts of coal, petroleum, and natural gas have been burned to power industry, to heat and cool cities, and to drive vehicles. Their combustion has turned fuels into carbon dioxide and water vapor. At the same time, much of the world's forests have been destroyed by logging and to clear land for agriculture. Deforestation adds to the greenhouse effect in two ways: it means there are fewer trees to capture carbon dioxide and produce oxygen, and burning the wood sends CO_2 back into the atmosphere at an accelerated rate. The relative contribution of carbon dioxide to the potential for global warming is about 55%.

Other important greenhouse gases influenced by human activity are

1. methane, from natural gas and coal mining, agriculture and livestock, swamps, and landfills

2. nitrous oxides, from motor vehicles, industry, and nitrogen-containing fertilizers

3. chlorofluorocarbons, hydrofluorocarbons, and perfluorocarbons, widely used industrial chemicals

Although these gases may be present in small amounts, some of them trap heat much more effectively than does CO_2. Nitrous oxide, for example, has 360 times the capacity of CO_2 to trap heat, and even methane is 24 times more potent than CO_2 in absorbing heat close to the Earth.

As the Industrial Revolution gained momentum in Europe and North America during the 19th century, the concentration of CO_2 in the atmosphere rose from its preindustrial level of about 274 parts per million (ppm) to 315 ppm in 1958; it rose since then to 379 ppm in 2005. The methane concentration in the lower atmosphere has *already* more than doubled from its preindustrial level and is currently increasing by just over 1% per year.

Scientists fear that the accelerated warming trend of the past 50 years has exceeded typical climate shifts and cite evidence such as the following:

- The 20th century was the warmest century of the past 600 years, and most of its warmest years were concentrated near its end. The world's average surface temperature rose about 0.6°C (a bit over 1°F) in the 20th century, and the 1990s were the hottest decade of the century. This pattern of warming has continued into the current century. The year 2015 was the warmest on record (**Figure 4.46**).

- Winter temperatures in the Arctic have risen about 4°C (7°F) since the 1950s. The Arctic as a whole is losing its ice cap. Between 1978 and 2000, the coverage of Arctic sea ice in winter decreased by 6%, and the average thickness of Arctic ice declined by 42%, from 3.1 to 1.8 meters (10.2 to 5.9 ft). Similarly, the sea ice west of the Antarctic Peninsula has diminished more than 20% since 1970.

- On every continent, glaciers are thinning and retreating. For example, glaciers atop Mount Kilimanjaro and Mount Kenya

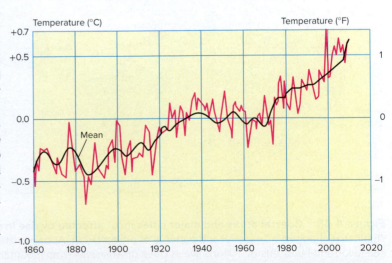

Figure 4.46 **Global temperature difference above and below the long-term average (set to 0), 1860–2010.** *Source: NOAA.*

in Africa shrank by 70% or more during the 20th century, and glaciers in the Swiss Alps are estimated to have lost half their volume since 1850. These patterns of glacial thinning and retreat are repeated in Alaska, Peru, Russia, India, China, Irian Jaya, New Zealand, and elsewhere. In some places (e.g., portions of Montana's Glacier National Park and the eastern Himalayas), glaciers melted and disappeared altogether in the 20th century. Although glaciers have grown and retreated for thousands of years, the rate of melt has accelerated in the past few decades and now exceeds anything in recent centuries.

Whatever the attributable causes of global warming, most climatologists agree on certain of its general consequences, should it continue (**Figure 4.47**). Increases in sea temperatures would cause

ocean waters to expand slightly (thermal expansion) and the polar ice caps to melt significantly. More serious consequences would result from the melting of the Greenland ice sheet and the rapid retreat or total melting of glaciers throughout the world. Although melting sea ice has no effect on sea levels, water melted from continental sources adds to ocean volumes. Inevitably, sea levels would rise, perhaps 1 meter (3.3 ft) in 100 years, with devastating impacts, especially in the tropics and warm temperate regions, where many coastlines are heavily settled. As **Figure 4.48** shows, the most vulnerable areas are along the north and west coasts of Africa, South and Southeast Asia, and low-lying coral atolls in the Pacific and Indian Oceans. In addition, higher temperatures bring on more extreme heat events, causing increased deaths among the elderly, infants, and the infirm. The effects of Hurricane Sandy may have been a product of global warming.

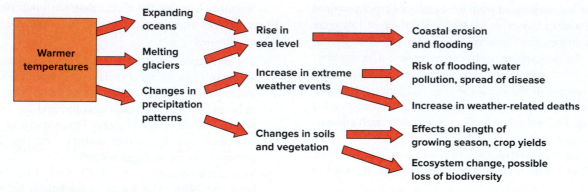

Figure 4.47   Potential effects of global warming. Global warming is projected to cause an increase in the frequency and severity of extreme weather events and weather-related disasters: heavy downpours, floods, hurricanes, heat waves, drought, and wildfires. Secondary effects of short-term extremes of weather would be shifts in plant and animal ranges; an increase in water pollution; and the spread of infectious diseases as warmer, wetter weather conditions widen the range of disease-carrying insects.

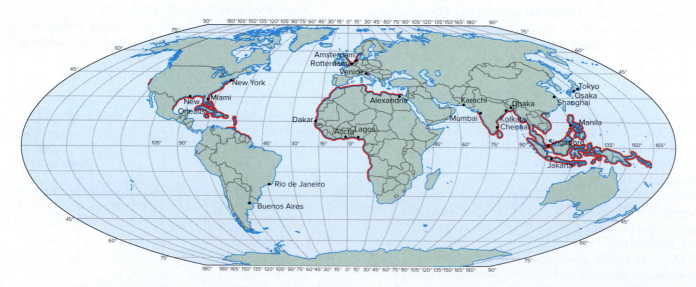

Figure 4.48   Coastal areas and major cities most affected by rise in sea level. The acceleration of global warming could produce significant changes in sea level and patterns of precipitation. Approximately one-sixth of the world's people live at or just a few meters above sea level, and hundreds of millions live in cities downstream from mountains where accelerated melting of glaciers or snow could contribute to severe flooding, especially if the river basins have been heavily deforested. Even a 1-meter (3-ft) rise in sea level would be enough to cover the Maldives and other low-lying island countries. The homes of between 50 and 100 million people would be inundated, one-fifth of Egypt's arable land in the Nile delta would be flooded, and the impact on the people of Bangladesh who live on thousands of alluvial islands known as "chars" would be catastrophic.

Other problems would result from changes in precipitation patterns. The warming of lakes and oceans would speed evaporation, causing more active convection currents in the atmosphere. It is important to note that changes in precipitation would be regional, with some areas receiving more and others less precipitation than they do now. Polar and equatorial regions might get heavier rainfall, whereas the continental interiors of the midlatitudes could become drier and suffer at least periodic drought. More northerly agricultural regions, such as parts of Canada, Scandinavia, and Russia, might benefit from general temperature rises; the longer growing season would make them more productive. Changes in

temperature and precipitation would affect soils and vegetation. The composition of forests would change, as some areas would become less favorable for certain species of plants but more hospitable to others.

Many climatologists point out that climate prediction is not an exact science. Temperature differences are the engine driving the global circulation of winds and ocean currents and help create conditions inducing or inhibiting winter and summer precipitation and daily weather conditions. Exactly how those vital climate details would express themselves locally and regionally is not well understood.

Summary of Key Concepts

- Solar energy is the great generator of the main weather elements of temperature, moisture, and atmospheric pressure. Spatial variation in these elements is caused by two factors: the Earth's broad physical characteristics, such as greater solar radiation at the equator than at the poles, and local physical characteristics, such as the effect of water bodies or mountains on local weather conditions.
- Climate regions help us simplify the complexities that arise from such special conditions as monsoon winds in Asia or cold ocean currents off the western coast of South America.
- Knowledge of climate tells us about the conditions within which one carries out life's daily tasks.
- Climate change results from both long-term and short-term natural processes.
- Human use of the Earth affects the climate.

Although the following chapters focus mainly on the characteristics of human cultural landscapes, one should keep in mind that the physical landscape significantly affects human behavior.

Key Words

air mass 89
air pressure 82
blizzard 93
climate 76
convection 82
convectional precipitation 88
Coriolis effect 83
cyclone 91
cyclonic (frontal) precipitation 89
dew point 88
El Niño-Southern Oscillation 89
frictional effect 84
front 91
global warming 107
greenhouse effect 107
humid continental climate 103
hurricane 91

insolation 77
jet stream 85
land breeze 83
lapse rate 81
marine west coast climate 103
monsoon 86
mountain breeze 83
North Atlantic drift 86
orographic precipitation 89
precipitation 87
pressure gradient force 82
reflection 79
relative humidity 87
reradiation 79
savanna 98
sea breeze 83
source region 89

steppe 99
temperature inversion 81
tornado 93
tropical rain forest 98
troposphere 76

tundra 104
typhoon 91
valley breeze 83
weather 76

Thinking Geographically

1. What is the difference between *weather* and *climate?*

2. What determines the amount of *insolation* received at a given point? Does all potentially receivable solar energy actually reach the Earth? If not, why?

3. How is the atmosphere heated? What is the *lapse rate* and what does it indicate about the atmospheric heat source? Describe a *temperature inversion*.

4. What is the relationship between atmospheric pressure and surface temperature? What is a *pressure gradient* and of what concern is it in weather forecasting?

5. In what ways do land and water areas respond differently to equal insolation? How are these responses related to atmospheric temperatures and pressures?

6. Draw and label a diagram of the planetary wind and pressure system. Account for the occurrence and character of each wind and pressure belt. Why are the belts latitudinally ordered?

7. What is *relative humidity?* How is it affected by changes in air temperatures? What is the *dew point?*

8. What are the three types of large-scale *precipitation?* How does each occur?

9. What are *air masses?* What is a *front?* Describe the development of a cyclonic storm, showing how it relates to air masses and fronts.

10. What factors were chiefly responsible for today's weather?

11. Summarize the distinguishing temperature, moisture, vegetation, and soil characteristics of each type of climate.

12. What is the climate at Tokyo, London, São Paulo, and Bangkok?

13. What causes the greenhouse effect? What impact does it have on the environment?

CHAPTER 5

Population Geography

Cairo (Al Qahirah), Egypt, the largest city in Africa, has been nicknamed the "city of a thousand minarets." The United Nations estimated its metropolitan regional population at 18.4 million in 2014. The slender tower in the background of this bustling shopping street is a distinctive feature of Islamic architecture used in the call to prayer. © Kate Sidlo.

LEARNING OBJECTIVES

After studying this chapter you should be able to:

5.1 Summarize how the world's population has grown over time and how it is forecast to grow during the 21st century.

5.2 Explain the difference between birth rates, fertility rates, and natural increase rates, and illustrate how are they used to predict population change.

5.3 Sketch a sample population pyramid for populations experiencing rapid growth, stability, and decline, and explain how the pyramid expresses the composition and growth rate of the population.

5.4 Summarize the demographic transition model, and explain how birth rates, death rates, and natural increase rates change through each stage of the transition.

5.5 On a world map, identify the world's four great population concentrations.

5.6 Summarize and critique Malthus's view of population and resources.

5.7 Summarize the population policies that have been implemented in different countries.

5.8 Compare and contrast the major population challenges facing the developing and developed regions of the world.

"Zero, possibly even negative [population] growth" was the 1972 slogan proposed by the prime minister of Singapore, an island country in Southeast Asia. His nation's population, which stood at 1 million at the end of World War II (1945), had doubled by the mid-1960s. To avoid the overpopulation he foresaw, the government decreed, "Boy or girl, two is enough" and refused maternity leaves and access to health insurance for third or subsequent births. Abortion and sterilization were legalized, and children born fourth or later in a family were to be discriminated against in school admissions policy. In response, by the mid-1970s, birth rates had fallen below the level necessary to replace the population, and abortions were terminating more than one-third of all pregnancies.

"At least two. Better three. Four if you can afford it" was the national slogan proposed by the same prime minister in 1986, reflecting fears that the stringencies of the earlier campaign had gone too far. From concern that overpopulation would doom the country to perpetual Third World poverty, Prime Minister Lee Kuan Yew was moved to worry that population limitation would deprive it of the growth potential and national strength implicit in a youthful, educated workforce adequate to replace and support the present aging population. His 1990 national budget provided for sizable, long-term tax rebates for second children born to mothers under 28. Concerned that university-educated residents were having fewer children than those with less education, the government set up programs to encourage love matches among university graduates. When financial inducements proved insufficient to increase the population, the Singapore government allowed in more immigrants to offset the lack of births.

The policy reversal in Singapore reflects an inflexible population reality: the structure of the present controls the content of the future. The size, characteristics, growth trends, and migrations of today's populations help shape the well-being of peoples yet unborn but whose numbers and distributions are now being determined. The numbers, age, and sex distribution of people; the patterns and trends in their fertility and mortality; and their density of settlement and rate of growth all affect and are affected by the social, political, and economic organization of a society. Through population data we begin to understand how the people in a given area live, how they may interact with one another, how they use the land, what pressure on resources exists, and what the future may bring. These are important concerns, as attested to by the frequent popular reference to a "population explosion," public debate about legal and undocumented immigration, and speculation about population growth and resource availability. More fundamentally, of course, numbers and locations of people are the essential background to all other aspects of human geography.

Population geography provides the concepts and theories to understand and forecast the size, composition, and geographic distribution of the human population. It differs from **demography,** the statistical study of human population, in its concern with *spatial* analysis—location, density, and patterns of human population. Regional natural resources, stage of economic development, standard of living, food supply, and conditions of health and well-being are basic to geography's population concerns. They are, as well, fundamental expressions of the human–environment relationships that are central to human geography.

5.1 Population Growth

Sometime in early 2012, a human birth raised the Earth's population to 7 billion people. In 1999, the count was 6 billion and in 2025, it is forecast to hit 8 billion. While the increases are substantial, they have been declining over the years. During the late 1980s, the United Nations Population Division reported yearly growth exceeding 90 million compared to 78 million per year in 2016. Even with the slower pace of increase, in 2015 the United Nations projected that the world would likely contain about 9.7 billion inhabitants by 2050. UN projections of the world population for the year 2100 ranged from a low of 9.5 billion to a high of 13.3 billion, with a median value of 11.2 billion.

All demographic forecasts agree, however, that essentially all or any future growth will occur in countries now considered "less developed" (**Figure 5.1**), with especially rapid growth in the 48 least-developed states. The majority of the world's most populous countries are developing countries, and that trend will become more pronounced by 2050 (**Table 5.1**). By 2050, Africa will have three countries on the list of the ten most populous countries in the world and Russia and Japan will have dropped off the list. We will return to these projections later in this chapter and to the difficulties and controversies inherent in making them.

Just what is implied by numbers in the millions and billions? With what can we compare the 2015 population of Estonia in

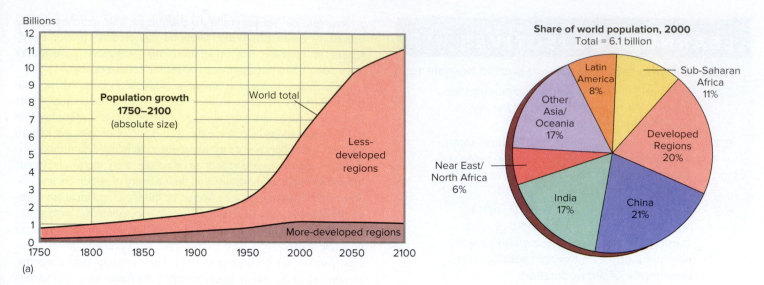

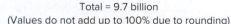

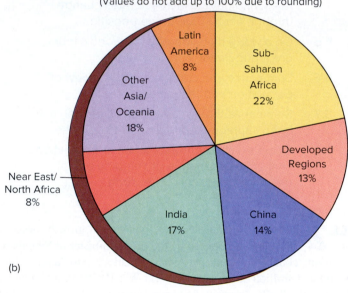

Figure 5.1   World population numbers and projections.
(a) After two centuries of slow growth, the world population began explosive expansion after World War II (1939–1945). United Nations demographers project a global population of 9.7 billion in 2050. Numbers in more-developed regions at the middle of the century will be the same as or lower than at its start, due to population stability or decline below 2000 levels. However, higher fertility rates and immigration are projected to increase the U.S. population by about one-third between 2010 and 2050, and large-volume immigration into Europe could alter its projected population decline. **(b)** Between today and 2100, nearly all of the world's population growth will take place in a group of "less-developed" high-fertility countries, many of which are in Africa. China and Europe are forecast to decline slightly in total population and see a decline in their share of world population. *Sources: (a) Estimates from Population Reference Bureau and United Nations Population Division; (b) based on United Nations and U.S. Bureau of the Census data and projections.*

Europe (about 1.3 million), Mexico (about 130 million), or India (about 1.3 billion)? It is difficult to appreciate a number as vast as 1 million or 1 billion and the great distinction between them. This example offered by the Population Reference Bureau may help in visualizing their immensity and implications:

A 2.5-centimeter (1-in.) stack of U.S. paper currency contains 233 bills. If you had a *million* dollars in thousand-dollar bills, the stack would be 11 centimeters (4.3 in.) high. If you had a *billion* dollars in thousand-dollar bills, your pile of money would reach 109 meters (358 ft)—about the length of a football field.

The implications of the present numbers and the potential increases in population are of vital current social, political, and ecological concern. Population numbers were much smaller some 12,000 years ago, when continental glaciers began their retreat, people spread to formerly unoccupied portions of the globe, and human experimentation with food sources initiated the Agricultural Revolution. The 5 or 10 million people who then constituted all of humanity obviously had considerable potential to expand their

numbers. In retrospect, we see that the natural resource base of the Earth had a population-supporting capacity far in excess of the pressures exerted on it by early hunting and gathering groups.

Some observers maintain that, despite present numbers or even those we can reasonably anticipate for the future, the adaptive and exploitative ingenuity of humans is in no danger of being taxed. Others, however, frightened by the resource demands of a growing world population that had already expanded almost four-fold—from 1.6 billion to 6.1 billion—in the century from 1900 to 2000, compare Earth to a self-contained spaceship and declare with chilling conviction that a finite vessel cannot bear an ever-increasing number of passengers. They point to recurring problems of malnutrition and starvation (though these are realistically more a matter of failures of distribution than of inability to produce enough foodstuffs worldwide). They cite global climate change, air and water pollution, the loss of forest and farmland, rising prices of many minerals and fossil fuels, and other evidence of strains on world resources as signs that the world population has reached the Earth's physical capacity.

Table 5.1	World's Most Populous Countries in 2016 and 2050		
2 0 1 6		**2 0 5 0**	
Country	**Population (Millions)**	**Country**	**Population (Millions)**
China	1372	India	1660
India	1314	China	1366
United States	321	United States	398
Indonesia	256	Nigeria	397
Brazil	205	Indonesia	366
Pakistan	199	Pakistan	344
Bangladesh	160	Brazil	226
Nigeria	182	Bangladesh	202
Russia	144	Congo, Dem. Rep.	194
Japan	127	Ethiopia	165

Source: Population Reference Bureau.

On a worldwide basis, populations grow only one way: the number of births in a given period exceeds the number of deaths. Current estimates of slowing world population growth and eventual stability or decline clearly indicate that humans, by their individual and collective decisions, may effectively limit growth and control global population numbers. The implications of these observations will become clearer after we define some terms important in the study of world population and explore their significance.

5.2 Population Definitions

Demographers use a wide range of measures of population composition and trends, though all their calculations start with a count of events: of individuals in the population, births, deaths, marriages, and so on. To those basic counts, demographers add refinements that make the figures more meaningful and useful in population analysis. Among them are *rates* and *cohort* measures.

Rates simply record the frequency of the occurrence of an event during a given time frame for a designated population—for example, the marriage rate as the number of marriages performed per 1000 population in the United States last year. **Cohort** measures refer data to a population group unified by a specified common characteristic—the age cohort of 1–5 years, perhaps, or the college class of 2014 (**Figure 5.2**). Basic numbers and rates useful in the analysis of world population and population trends have been reprinted in this book with the permission of the Population Reference Bureau as Appendix 3. Examination of them will help illustrate the discussion that follows.

Birth Rates

The **crude birth rate (CBR),** often referred to simply as the *birth rate,* is the annual number of live births per 1000 population. It is "crude" because it relates births to total population without regard

Figure 5.2 Whatever their differences may be by race, sex, or ethnicity, these children will forever be a single *birth cohort.* © *Blend Images/Ariel Skelley/Getty Images RF.*

to the age or sex composition of that population. A country with a population of 2 million and with 40,000 births a year has a crude birth rate of 20 per 1000:

$$\frac{40{,}000}{2{,}000{,}000} = 20 \text{ per } 1000.$$

The birth rate of a country is strongly influenced by the age and sex structure of its population, by the customs and family size expectations of its inhabitants, and by its population policies. Because these conditions vary widely, recorded national birth rates varied in 2015 from 40 or more in some sub-Saharan African states to values below 10 per 1000 in China, Japan, South Korea, Taiwan, and numerous European countries including Germany and Italy. Birth rates of 30 or above per 1000 are considered *high* and are found in sub-Saharan Africa and other countries where poverty is widespread and a high proportion of the female population is young

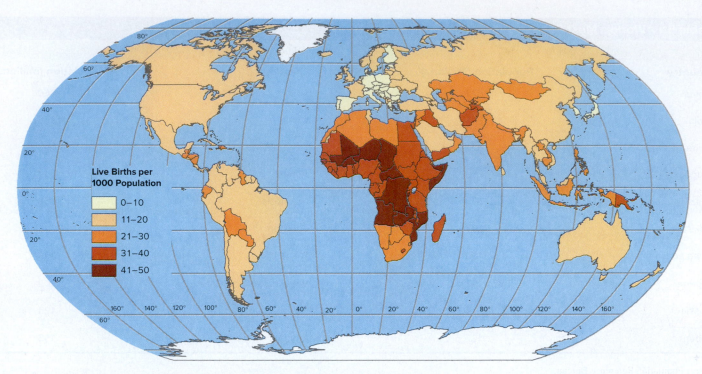

Figure 5.3 **Crude birth rates, 2015.** The map suggests a degree of precision that is misleading in the absence of reliable, universal registration of births. The pattern shown serves, however, as a generally useful summary of comparative reproduction patterns.
Source: Data from the Population Reference Bureau

(**Figure 5.3**). In many of them, birth rates may be significantly higher than official records indicate due to unregistered births.

Birth rates of less than 18 per 1000 are reckoned *low* and are characteristic of industrialized, urbanized countries. All European countries including Russia, the United States, Canada, Australia, New Zealand, and the more industrialized countries of Asia have low crude birth rates. Some countries such as China have adopted effective family planning programs (see "China's Way—and Others," p. 115). In others, changed cultural norms have reduced desired family size. *Transitional* birth rates (between 18 and 30 per 1000) characterize some, mainly smaller developing and newly industrializing countries, although giant India entered that group in 1994.

As the recent population histories of Singapore and China indicate, birth rates are subject to change. The decline to current low birth rates of European countries and of some of the areas that they colonized is usually ascribed to industrialization, urbanization, and in recent years maturing populations. While restrictive family planning policies in China rapidly reduced the birth rate from over 33 per 1000 in 1970 to 18 per 1000 in 1986, industrializing Japan experienced a comparable 15-point decline in the decade 1948–1958 with little government intervention. Indeed, the stage of economic development appears closely related to variations in birth rates among countries, although rigorous testing of this relationship proves it to be imperfect (Figure 5.3). As a group, the more-developed states of the world showed a crude birth rate of 11 per 1000 in 2015; less-developed countries (excluding China) registered 24 per 1000 (down from 35 in 1990).

Religious and political beliefs can also affect birth rates. The convictions of many Roman Catholics and Muslims that their religion forbids the use of artificial birth control techniques often lead to high birth rates among believers. However, predominantly Catholic Italy has one of the world's lowest birth rates. Islam itself does not prohibit contraception and several Muslim countries have low birth rates. Similarly, Singapore and some European governments—concerned about birth rates too low to sustain present population levels—subsidize births in an attempt to raise those rates. Regional variations in projected percentage contributions to world population growth are summarized in **Figure 5.4**.

Fertility Rates

Crude birth rates may display such regional variability because of differences in age and sex composition or disparities in births among the reproductive-age population. The rate is "crude" because its denominator contains persons who have no chance at all of giving birth—males, young girls, and elderly women. The **total fertility rate (TFR),** which is the average number of children a woman will have over the course of her childbearing years, is a more-refined and thus more satisfactory statement than the crude birth rate.

The TFR (**Figure 5.5**) is calculated by assuming that, over her childbearing years, a woman bore children at the current year's rate for women that age. Thus, a TFR of 3 means that the average woman in a population would be expected to have three offspring in her lifetime. The fertility rate minimizes the effects of fluctuation in the population structure and summarizes the demonstrated and expected reproductive behavior of women. Thus, for regional comparative and predictive purposes, it is a more useful and more reliable figure than the crude birth rate.

CHINA'S WAY—AND OTHERS

An ever-larger population is "a good thing," Chairman Mao announced in 1965, when China's birth rate was 37 per 1000 and its population totaled 540 million. At Mao's death in 1976, numbers reached 852 million, although the birth rate then had dropped to 25. During the 1970s, when it became evident that population growth was consuming more than one-half the annual increase in the country's gross domestic product, China introduced a well-publicized campaign advocating the "two-child family" and providing services, including abortions, supporting that program. In response, China's birth rate dropped to 19.5 per 1000 by the late 1970s.

"One couple, one child" became the slogan of a new and more vigorous population control drive launched in 1979, backed by both incentives and penalties to ensure its success in China's tightly controlled society. Late marriages were encouraged; free contraceptives, cash awards, abortions, and sterilizations were provided to families limited to a single child. Penalties, including steep fines, were levied for second births. At the campaign's height in 1983, the government ordered the sterilization of either husband or wife for couples with more than one child. Infanticide—particularly the abandonment or murder of female babies—was a reported means of both conforming to a one-child limit and increasing the chances that the one child would be male.

By 1986, China's officially reported crude birth rate had fallen to 18 per 1000, far below the 37 per 1000 then registered among the rest of the world's less-developed countries. The one-child policy was relaxed in 1984 to permit two-child limits in rural areas, and eliminated in 2016. With the end of the one-child policy, all couples are permitted two children. The government still requires birth permits and those without permits can face government sanctions including forced abortion.

China's population controls were so successful that, by 2000, China's population was 300 million less than it otherwise would have been. Indeed, the one-child policy was ended because demographers and government officials expressed serious concern over population decrease and an aging society. Even in the absence of population controls, newly prosperous urbanites have voluntarily reduced their fertility to well below replacement levels and childless couples are increasingly common. Projections suggest that, after 2026, because of lowered fertility rates, China's population numbers will actually start falling. The country is already beginning to face a pressing social problem: a declining proportion of working-age persons and an absence of an adequate welfare network to care for a rapidly growing number of senior citizens.

Concerned with their own increasing numbers, many developing countries have introduced their own less-extreme programs of family planning, stressing access to contraception and sterilization. International agencies have encouraged these programs, buoyed by such presumed success as the 21% fall in fertility rates in Bangladesh from 1970 to 1990 as the proportion of married women of reproductive age using contraceptives rose from 3% to 40% under intensive family planning encouragement and frequent adviser visits.

Research suggests that fertility falls because women decide they want smaller families, not because they have unmet needs for contraceptive advice and devices. Nineteenth-century northern Europeans without the aid of science, it is observed, had lower fertility rates than their counterparts today in middle-income countries. With some convincing evidence, improved women's education has been proposed as a surer way to reduce births than either encouraged contraception or China's coercive efforts. Studies from individual countries indicate that 1 year of female schooling can reduce the fertility rate by between 5% and 10%. However, the fertility rate of uneducated Thai women is only two-thirds that of Ugandan women with secondary education. Obviously, other cultural factors besides educational levels are at work.

© McGraw-Hill Education/Alasdair Drysdale RF.

A useful concept here is *replacement level fertility*—the level of fertility at which each successive generation of women produces exactly enough children to ensure that the same number of women survive in that generation to have offspring themselves. Although a TFR of 2.0 would seem sufficient exactly to replace present population (one baby to replace each parent), in reality, replacement fertility levels must be slightly higher to compensate for the higher percentage of boys that are born and for mortalities among women before they complete their childbearing years. In developed countries, replacement levels of fertility are assumed to be 2.1. However, the higher the level of mortality in a population, the higher the replacement level of fertility will be. For Mozambique early in the 21st century, for example, the replacement level fertility was 3.4 children per woman.

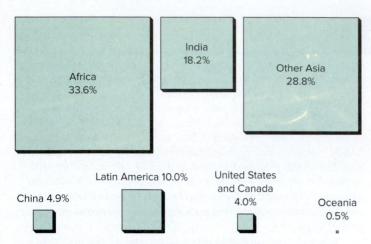

Figure 5.4   Projected percentage contributions to world population growth by region, 2002 and 2050. The annual increase in the world population is projected to slow in the first half of the 21st century. Today, all world regions are growing except Eastern Europe and the newly independent states of the former Soviet Union. By 2050, sub-Saharan Africa is projected to be the major contributor to world population growth. By then, virtually all population growth will be taking place in the world's less-developed countries. *Source: Projections based on World Bank and United Nations figures.*

On a worldwide basis, the TFR in 2015 was 2.5; 20 years earlier, it was 3.6. The more-developed countries recorded a 1.7 TFR in 2015 and have been below replacement levels since the 1980s. That decrease has been dwarfed by the rapid changes in reproductive behavior in much of the developing world. Since 1960, the average TFR in the less-developed world fell by one-half from the traditional 6 or more children to 2.6 today. That dramatic decline reflects the fact that women and men in developing countries are marrying later and having fewer children, following the pattern set earlier in the developed world. There has been, as well, a great increase in family planning and contraceptive use. According to a UN world fertility report, nearly all national governments supported family planning and distributed contraceptives, either directly or indirectly.

The recent fertility declines in developing states have been more rapid and widespread than anyone expected. The TFRs for so many of them have dropped so dramatically since the early 1960s (**Figure 5.6**) that many scholars are skeptical of the United Nations' world population projections anticipating 11.2 billion or more at the end of this century. Indeed, worldwide in 2010, the United Nations estimated that nearly 50% of the world's population lived in countries with fertility rates of 2.1 or less. China's decrease from a TFR of 5.9 births per woman in the period 1960–1965 to 1.7 and comparable drops in Bangladesh, Brazil, Mozambique, and other states demonstrate that fertility reflects cultural values, not biological imperatives. If those values now favor fewer children than formerly, population projections based on earlier, higher TFR rates must be adjusted.

In fact, United Nations population forecasts assume that fertility rates will continue to drop in high-fertility countries and will rise in sub-replacement fertility countries and, in the long run, lead to stable population numbers. However, nothing in logic or history requires population stability at any level. Rather than assume, as in the past, convergence on the replacement fertility value of 2.1,

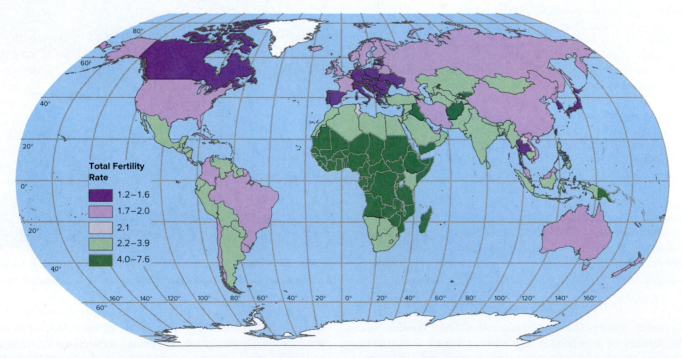

Figure 5.5   Total fertility rate (TFR) 2015 indicates the average number of children that would be born to each woman if, during her childbearing years, she bore children at the same rate as women of those ages actually did in a given year. Because the TFR is age-adjusted, two countries with identical birth rates may have quite different fertility rates and therefore different prospects for growth. Depending on mortality conditions, a TFR of 2.1 to 2.3 or more children per woman is considered the *replacement level*, at which a population will eventually stop growing. *Source: 2015 data from Population Reference Bureau.*

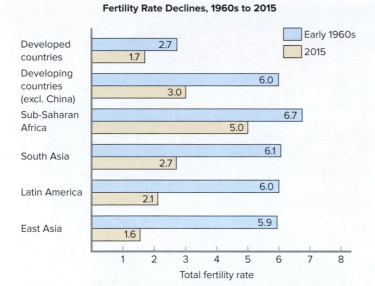

Fertility Rate Declines, 1960s to 2015

Legend: Early 1960s, 2015

Region	Early 1960s	2015
Developed countries	2.7	1.7
Developing countries (excl. China)	6.0	3.0
Sub-Saharan Africa	6.7	5.0
South Asia	6.1	2.7
Latin America	6.0	2.1
East Asia	5.9	1.6

Total fertility rate

Figure 5.6   Differential fertility rate declines, 1960s to 2015. Fertility declined most rapidly in Latin America and East Asia and much more slowly in sub-Saharan Africa. Europe is far below replacement, with a 2015 TFR of 1.4; the United States, however, with a TFR of 1.9, is just below the replacement point. *Sources: Population Reference Bureau, 2015 and United Nations Population Fund.*

demographers disagree about future fertility trends. Fertility rates may remain above replacement levels indefinitely in some world regions while in other regions, sub-replacement fertility may persist, leading to inevitable population decline (see "A Population Implosion?" p. 119). Of course, should cultural values change to again favor children, growth would eventually resume.

Individual country projections based on current fertility rates may be wrong due to migration. Massive international population movements are occurring in response to political instabilities and, particularly, to differentials in perceived economic opportunities. For example, the European Union in recent years has had a negative rate of natural increase, yet since 2000 it has experienced essentially a constant population solely because of immigrant influx from Eastern Europe, Asia, and Africa.

World regional and national fertility rates reported in Appendix 3 and other sources are summaries that conceal significant variations within countries. India's published TFR of 2.3 is a pooled number that hides the variation between a low value of 1.6 in the state of West Bengal and 3.4 in Bihar. In the United States in 2010, fertility rates varied from 2.4 for Hispanics to 2.0 for African Americans, 1.8 for non-Hispanic whites, and just 1.7 for Asian Americans.

Death Rates

The **crude death rate (CDR),** also called the **mortality rate,** is calculated in the same way as the crude birth rate: the annual number of events per 1000 population. In the past, a valid generalization was that the death rate, like the birth rate, varied with national levels of development. Characteristically, the highest rates (over 20 per 1000) were found in the less-developed countries of Africa, Asia, and Latin America; the lowest rates (less

than 10) were associated with the developed states of Europe and the United States and Canada. That correlation became weaker as dramatic reductions in death rates occurred in developing countries in the years following World War II. Infant mortality rates and life expectancies improved as antibiotics, vaccinations, and pesticides to treat diseases and control disease carriers were made available in almost all parts of the world, and as increased attention was paid to funding improvements in urban and rural sanitary facilities and safe water supplies.

Distinctions between more-developed and less-developed countries in mortality (**Figure 5.7**) have been so reduced that, since the 1990s, death rates for less-developed countries as a group actually are lower than those for more-developed states. Notably, that reduction did not extend to infant or maternal mortality rates (see "The Risks of Motherhood" p. 120). Like crude birth rates, death rates are meaningful for comparative purposes only when we study identically structured populations. Countries with a high proportion of elderly people, such as Denmark and Sweden, would be expected to have higher death rates than those with a high proportion of young people, such as Iceland, assuming equality in other national conditions affecting health and longevity. The pronounced youthfulness of populations in developing countries, as much as improvements in sanitary and health conditions, is an important factor in the recently reduced mortality rates of those areas.

To overcome that lack of comparability, death rates can be calculated for specific age groups. The *infant mortality rate,* for example, is the ratio of deaths of infants age 1 year or under per 1000 live births:

$$\frac{\text{deaths age 1 year of less}}{1000 \text{ live births}}.$$

Infant mortality rates are significant because it is at these ages that the greatest declines in mortality have occurred, largely as a result of better health services. The drop in infant mortality accounts for a large part of the decline in the general death rate in the last few decades, because mortality during the first year of life is usually greater than in any other year.

Two centuries ago, it was not uncommon for 200 to 300 infants per 1000 to die in their first year. Today, rates are in the single digits for developed countries and 37 for the world as a whole (**Figure 5.8**). Still, striking world regional and national variations remain. For all of Africa, infant mortality rates are 59 per 1000, and individual African states plagued by recent civil wars or other violence (for example, Central African Republic, the Democratic Republic of Congo, Sierra Leone, Somalia, and South Sudan) show rates above 100. Nor are rates uniform within single countries. Poorer areas within countries often register higher infant mortality rates than national averages.

Modern medicine and sanitation have increased life expectancy and altered age-old relationships between birth and death rates. In the early 1950s, only five countries, all in northern Europe, had life expectancies at birth of more than 70 years. By 2015, the global average life expectancy had risen to 71 years, although Africa still lagged considerably. The availability and employment of modern methods of health and sanitation have varied regionally, and the least-developed countries have least

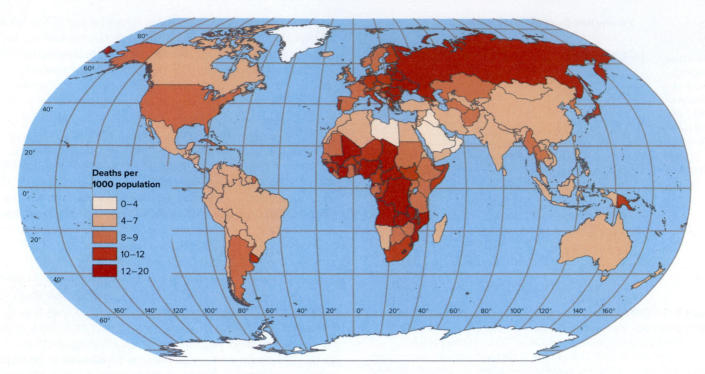

Figure 5.7 **Crude death rates** in 2015 showed less worldwide variability than the birth rates displayed in Figure 5.3. The widespread availability of at least minimal health protection measures and a generally youthful population in the developing countries yield death rates frequently lower than those recorded in "old age" Europe. *Source: 2015 data from Population Reference Bureau.*

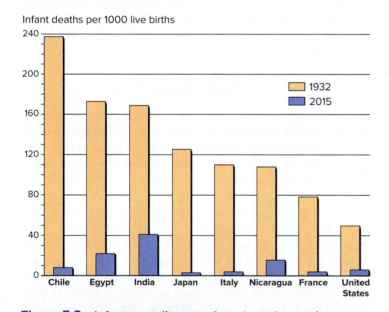

Figure 5.8 **Infant mortality rates for selected countries.** Dramatic declines in the rate have occurred in all countries, a result of international programs of health care delivery aimed at infants and children in developing states. Nevertheless, the decreases have been proportionately greatest in the urbanized, industrialized countries, where sanitation, safe water, and quality health care are widely available. *Source: Data from U.S. Bureau of the Census and Population Reference Bureau.*

benefited from them. In such underdeveloped and impoverished areas as much of sub-Saharan Africa, the chief causes of death other than HIV/AIDS are those no longer of immediate concern in

more-developed lands: diseases such as malaria; intestinal infections; typhoid; cholera; and, especially among infants and children, malnutrition and dehydration from diarrhea.

HIV/AIDS is the tragic and, among developing regions particularly, widespread exception to observed global improvements in life expectancies and reductions in adult death rates and infant and childhood mortalities. AIDS has become the fourth most common cause of death worldwide and is forecast to surpass the Black Death of the 14th century—which caused an estimated 25 million deaths in Europe and 13 million in China—as history's worst-ever epidemic. While the rates of new infections and deaths are slowing, according to a report by UNAIDS, AIDS-related illnesses kill about 1.1 million people per year; most of those deaths occur in sub-Saharan Africa. The United Nations estimated approximately 37 million people were HIV positive in 2015. Some 90% of those infected lived in developing countries and two-thirds in sub-Saharan Africa, where women account for 60% of all cases. In that hardest-hit region, as much as one-fourth of the adult population in some countries is HIV positive, and average life expectancy has been cut drastically. In South Africa, the life expectancy of a baby born in the early 21st century should have been 66 years; AIDS cut that down to 47. In Botswana, it is 34 years instead of 70; in Zimbabwe, the decline has been from 69 years to 37. Overall, AIDS has killed some 17 million Africans since the 1950s, when HIV (originally a disease of monkeys) appears to have established itself in Africa as a virulent human epidemic strain. Along with the deep cuts in sub-Saharan life expectancies, total population by 2015 is now projected to be 60 million less than it would have been in the absence of the disease.

Economically, AIDS cut national incomes in sub-Saharan countries. Southern Africa's economies are based on farming, and

A POPULATION IMPLOSION?

For much of the last half of the 20th century, demographers and economists focused on a "population explosion" and its implied threat of a world with too many people and too few resources to sustain them. For some observers, by the end of that century, those fears were replaced by a new prediction of a world with too few rather than too many people.

That possibility was suggested by two related trends. The first had become apparent by 1970, when it was noted that the total fertility rates (TFRs) of 19 countries, almost all of them in Europe, had fallen below the **replacement level**—the level of fertility at which populations replace themselves—of 2.1. Simultaneously, Europe's population pyramid began to become noticeably distorted, with a smaller proportion of young and a growing share of middle-age and retirement-age inhabitants. By 1970, the decrease in native working-age cohorts had already encouraged the influx of non-European "guest workers" whose labor was needed to maintain economic growth and to sustain the generous security provisions guaranteed to what was becoming the oldest population of any continent.

Many countries of Western and Eastern Europe sought to reverse their birth rate declines by adopting pronatalist policies. The communist states of the East rewarded pregnancies and births with generous family allowances, free medical and hospital care, extended maternity leaves, and child care. France, Italy, the Scandinavian countries, and others gave similar bonuses or awards for first, second, and later births. Despite those inducements, however, reproduction rates continued to fall. Every one of the 43 European countries and territories has fertility rates below replacement levels due to a host of cultural and personal lifestyle decisions. Those decisions were influenced by the increased educational levels of women, with longer years of schooling and deferred marriage ages; opportunities for women to work in professional careers; and the increasing cost of rearing children. Cultural expectations about ideal family size have shifted toward smaller families, and an increasing number of adults are choosing personal pursuits over family obligations. The effect on national growth prospects has been striking. For example, the populations of Spain and Italy are projected to shrink by one-quarter between 2000 and 2050 (and that of Ukraine by 43% between 2005 and 2050). Europe as a whole is forecast to shrink by 70 million people by mid-century. "In demographic terms," France's prime minister remarked, "Europe is vanishing."

Europe's experience soon was echoed in other societies of advanced economic development on all continents. By the 1990s, the United States, Canada, Australia, New Zealand, Japan, Taiwan, South Korea, Singapore, and other older and newly industrializing countries (NICs) had registered fertility rates below the replacement level. As they have for Europe, simple projections foretold their aging and declining population. Japan's numbers, for example, began to decline in 2006, when 21% of its population was age 65 or older. Taiwan forecasts negative growth by 2035. Even China's population will be declining by the late 2020s or early 2030s, and the UN projects that by 2050 populations will be declining in 50 countries.

The second indication that world population numbers should stabilize and even decline during the lifetimes of today's college cohort is a simple extension of the first: TFRs are being reduced to or below the replacement levels in countries at all stages of economic development in all parts of the world. While only 18% of the total world population in 1975 lived in countries with a fertility rate below replacement level, 48% did by 2010. Exceptions to the trend are found in Africa, especially sub-Saharan Africa, and in some areas of South, Central, and West Asia; however, even in those regions, fertility rates have been decreasing in recent years. "Powerful globalizing forces [are] at work pushing toward fertility reduction everywhere" was a 1997 observation from the French National Institute of Demographic Studies.

That conclusion is plausibly supported by assumptions of the United Nations 2008 forecast of a decline of long-term fertility rates in less-developed regions to 2.05. The same UN assessment envisions that those countries will reach those below-replacement fertilities by or before 2050. Should those assumptions prove valid, global depopulation could commence before mid-century. Between 2040 and 2050, one projection indicates, world population will fall by about 85 million (roughly the amount of its annual growth during much of the 1990s) and will shrink further by about 25% with each successive generation.

If that scenario is realized in whole or in part, a much different worldwide demographic and economic future is promised than that prophesied recently by "population explosion" forecasts. Declining rather than increasing pressure on world food and mineral resources will be in our future, along with shrinking rather than expanding world, regional, and national economies. Achievement of **zero population growth (ZPG)** has social and economic consequences not always perceived by its advocates. These inevitably include an increasing proportion of older citizens, fewer young people, a rise in the median age of the population, and a growing old-age dependency ratio with ever-increasing pension and social services costs borne by a shrinking labor force.

women do much of the farming as well as run households. Because AIDS kills more women than men, sub-Saharan food insecurity is rising and food shortages result because many young adults are too feeble to farm. Thus, malnutrition, starvation, and susceptibility to other diseases are AIDS costs added to national income reductions. Nonetheless, because of their high fertility rates, populations in all sub-Saharan countries except South Africa are still expected to grow significantly between 2000 and 2050, adding nearly 1 billion to the continent's total.

Population Pyramids

A **population pyramid** is a powerful tool for visualizing and comparing a population's age and sex composition. The term *pyramid* describes the diagram's shape for many countries in the 1800s, when the display was created: a broad base of younger age groups and a progressive narrowing toward the apex as older populations were thinned by death. Now, many different shapes are formed, each reflecting a different population history

THE RISKS OF MOTHERHOOD

The worldwide leveling of crude death rates does not apply to pregnancy-related deaths. In fact, the maternal mortality ratio—maternal deaths per 100,000 live births—is the single greatest health disparity between developed and developing countries. According to the World Health Organization, approximately 830 women die each day from causes related to pregnancy or childbirth. Each year more than a million children are left motherless by maternal deaths. As shown in the chart, the geography of maternal mortality is highly uneven; 99% of maternal deaths take place in less-developed states where the risks of motherhood are many times greater than in the more-developed countries. One third of all maternal deaths worldwide occur in just two countries: India and Nigeria. According to 2015 data, the risk is 1 death in 74 live births in Sierra Leone, versus 1 in 33,000 in Finland.

Complications of pregnancy, childbirth, and abortions are the leading slayers of women of reproductive age throughout the developing world, although the incidence of maternal mortality is by no means uniform, as the chart indicates. Countries with extraordinarily high maternal death ratios are generally found in war-torn or politically unstable regions such as Burundi, Chad, the Central African Republic, Liberia, Sierra Leone, Somalia, and South Sudan.

The vast majority of maternal deaths in the developing world are preventable. Most result from causes rooted in the social, cultural, and economic barriers confronting females in their home environment throughout their lifetimes: malnutrition, anemia, lack of access to timely basic maternal health care, physical immaturity due to stunted growth, and unavailability of adequate prenatal care or trained medical assistance at birth. Part of the problem is that women are considered expendable in societies where their status is low, although the correlation between women's status and maternal mortality is not exact. In those cultures, little attention is given to women's health or their nutrition, and pregnancy, although a major cause of death, is simply considered a normal condition warranting no special consideration or management.

Recognizing the immense human suffering associated with preventable maternal deaths, the United Nations included maternal health among its eight Millennium Development Goals (MDGs). Specifically, the goal called for reducing the maternal mortality rate by three-quarters between 1990 and 2015. Substantial progress was made in all world regions. Globally, the maternal mortality rate dropped from 385 in 1990 to 216 in 2015. Even war-torn Afghanistan recorded a 70% decline in maternal deaths during that period. The United Nation's Sustainable Development Goal for 2030 is to achieve a global rate below 70.

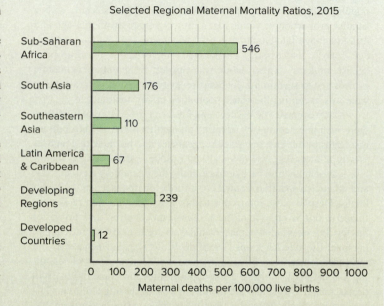

Selected Regional Maternal Mortality Ratios, 2015

Sub-Saharan Africa — 546
South Asia — 176
Southeastern Asia — 110
Latin America & Caribbean — 67
Developing Regions — 239
Developed Countries — 12

Maternal deaths per 100,000 live births

Source: Graph data from Trends in Maternal Mortality: From 1990 to 2015, *Geneva, Switzerland: WHO, 2015.*

(**Figure 5.9**), and *population profile* may be a more appropriate label. By showing the size of different age cohorts, the pyramids highlight the impact of "baby booms," population-reducing wars, birth rate reductions, and external migrations.

A rapidly growing country, such as Uganda, has most people in the lowest age cohorts; the percentage in older age groups declines successively, yielding a pyramid with markedly sloping sides. Typically, female life expectancy is reduced in older cohorts of less-developed countries, so that for Uganda the proportion of females in older age groups is lower than in, for example, Sweden. Female life expectancy and mortality rates may also be affected by cultural rather than economic development causes (see "Millions of Women Are Missing" p. 123). In New Zealand, a wealthy country with a very slow rate of growth, the population is nearly equally divided among the age groups, giving a "pyramid" with almost vertical sides. Among older cohorts, as Germany shows, there may be an imbalance between men and women because of the greater life expectancy of the latter. The impacts of war, as Russia's 1992 pyramid vividly demonstrated, were evident in that country's depleted age cohorts and male–female disparities. The sharp contrasts

between the pyramids for Uganda and Germany summarize the differing population concerns of the developing and developed regions of the world. Even within a single country, migrations, urbanization, and ethnic differences may result in substantial variation in the population profile. The unique demographics of college towns, retirement communities, Indian reservations, and mostly Hispanic border towns are clearly revealed in their population pyramids (**Figure 5.10**).

The population profile provides a quickly visualized demographic picture of immediate practical and predictive value. For example, the percentage of a country's population in each age group strongly influences demand for goods and services within that national economy. A country with a high proportion of young people has a high demand for educational facilities and certain types of health delivery services. In addition, of course, a large portion of the population is too young to be employed (**Figure 5.11**). On the other hand, a population with a high percentage of elderly people also requires medical goods and services specific to that age group, and these people must be supported by a smaller proportion of workers. As the profile of a national

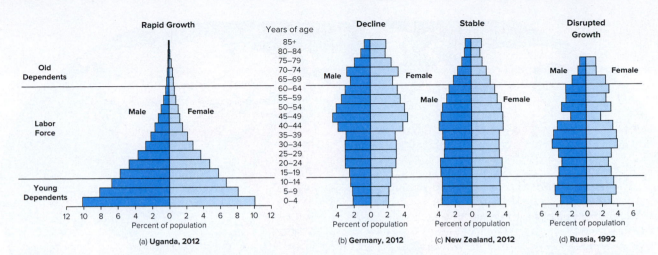

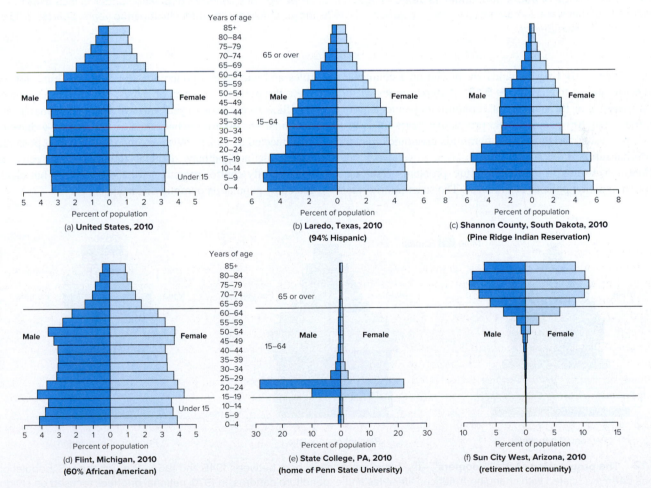

Figure 5.9 **Four patterns of population structure.** These diagrams show that population "pyramids" assume many shapes. The age distribution of national populations reflects the past, records the present, and foretells the future. In countries such as Uganda, social costs related to the young are important, and economic expansion is vital to provide employment for new entrants into the labor force. Germany's negative growth means a future with fewer workers to support a rising demand for social services for the elderly. New Zealand exemplifies a stable population structure. The 1992 pyramid for Russia reported the sharp decline in births during World War II as a "pinching" of the 45–49 cohort and showed in the large deficits of men above age 65 the heavy male mortality of World Wars I and II and late-Soviet period sharp reductions in Russian male longevity. *Sources: U.S. Bureau of the Census, International Data Base; and for Russia: Carl Haub, "Population Change in the Former Soviet Republics," Population Bulletin 49, no. 4 (1994).*

Figure 5.10 **Population pyramids for different communities.** These 2010 pyramids show the dramatic differences in the population structure across the United States. Laredo, Texas, on the U.S.-Mexico border, and Shannon County, South Dakota, on the Pine Ridge Indian Reservation, both display youthful populations with high birth rates. Flint, Michigan, displays the effects of out-migration of working-age adults due to poor economic conditions. College towns and retirement communities create age-segregated places with distinctive population profiles. *Source: U.S. Bureau of the Census, 2010.*

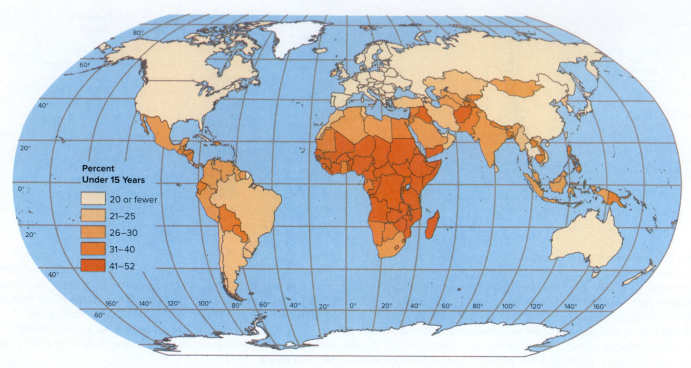

Figure 5.11 **Percentage of population under 15 years of age.** A high proportion of a country's population under 15 increases the dependency ratio of that state and promises future population growth as the youthful cohorts enter childbearing years. *Source: 2015 data from Population Reference Bureau.*

population changes, differing demands are placed on a country's social and economic systems (**Figure 5.12**). A **dependency ratio** is a simple measure of the number of economic dependents, old or young, that each 100 people in the productive years (usually, ages 15–64) must support. Population pyramids give quick visual evidence of that ratio.

Population pyramids also foretell future problems resulting from present population policies or practices. The strict family-size rules and widespread preferences for sons in China, for example, skew the pyramid in favor of males. On current evidence, about 1 million excess males a year enter an imbalanced marriage market in China. The 40 million bachelors China is likely to have in 2020, unconnected to society by wives and children, may pose threats to social order and, perhaps, national stability not foreseen or planned when family control programs were put in place but clearly suggested by population pyramid distortions.

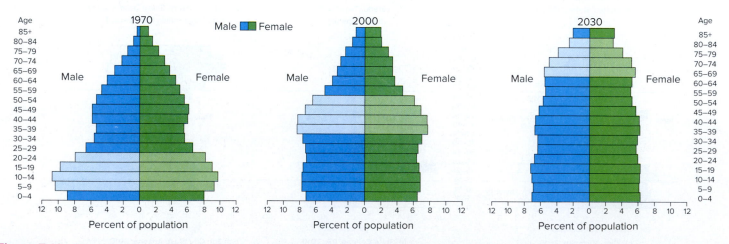

Figure 5.12 **The progression of the "boomers"**—the baby boom cohort born between 1946 and 1964—through the U.S. population pyramid has been associated with changing American lifestyles and expenditure patterns. In 1970, national priorities focused on childhood and young adult interests and the needs, education, and support of younger age groups. At the start of the 21st century, boomers formed the largest share of the working-age adult population, and their wants and spending patterns shaped the national culture and economy. By 2030, the pyramid foretells, their desires and support needs—now for retirement facilities and old-age care—will again be central concerns. *Source: Redrawn from Christine L. Himes, "Elderly Americans." Population Bulletin 56, no. 4 (Dec. 2001), Fig. 1.*

MILLIONS OF WOMEN ARE MISSING

Worldwide, according to one UN estimate, between 113 million and 200 million women are demographically missing, victims of nothing more than their sex. Their absence is the result of sex-selective abortions, female infanticide, abandonment, mistreatment, and violence in countries where boys are favored. If the UN estimate is correct, between 1.5 million and 3 million women and girls are lost to gender-based discrimination and violence each year in a regionally variable pattern. The majority of that loss appears to be due to birth disparities.

In China, India, Pakistan, New Guinea, and many other developing countries, a traditional preference for boys has meant neglect and death for girls, millions of whom are killed at birth, deprived of adequate food, or denied the medical attention provided to sons favored as old-age and wealth-gathering insurance for parents. In both China and India, ultrasound and amniocentesis tests are employed, often against government directives, to determine the sex of a fetus, so that it can be aborted if it is a female.

The evidence for the missing women starts with one fact: about 106 males are conceived and born for every 100 females. Normally, girls are hardier and more resistant to disease than boys, and in populations where the sexes are treated equally in matters of nutrition and health care, the number of males and females tends to equalize as the cohort ages. However, in many Asian countries, the ratio of males to females has been rising since the 1980s. East Asia has the most dramatic imbalances with an overall male-to-female ratio

of 119 baby boys for every 100 baby girls. China's male-to-female ratio at birth has increased from a normal ratio in the early 1980s to 121 boys for every 100 girls during 2000 to 2005. Even higher year 2000 differentials were registered in Hainan and Guangdong provinces in southeastern China, with newborn ratios of between 130 and 140 males to 100 females.

Ratio deviations are most striking for second and subsequent births. In China, South Korea, Taiwan, and Hong Kong, for example, the most recent figures for first-child sex ratios are near normal but rise to 121 boys per 100 girls for a second Chinese child and to 185 per 100 for a third Korean child. On that evidence, the problem of missing females is getting worse. Conservative calculations suggest there are more than 60 million females missing in China alone, almost 5% of the national population and more than are unaccounted for in any other country. India and Pakistan have the greatest percentage of "missing" women at about 8% of their total female population.

But not all poor countries show the same disparities. In sub-Saharan Africa, where poverty and disease are perhaps more prevalent than on any other continent, but there is no tradition of deadly violence against women, there are 102 females for every 100 males; and in Latin America and the Caribbean, there are equal numbers of males and females. Cultural norms and practices, not poverty or underdevelopment, seem to determine the fate and swell the numbers of the world's millions of missing women.

Natural Increase and Doubling Times

Knowledge of a country's sex and age distributions also enables demographers to forecast its future population levels, though the reliability of projections decreases with increasing length of forecast (**Figure 5.13**). Thus, a country with a high proportion of young people will experience a high rate of natural increase unless there is a very high mortality rate among infants and juveniles or fertility and birth rates change materially. The **rate of natural increase** of a population is derived by subtracting the crude death rate from the crude birth rate. *Natural* means that increases or decreases due to migration are not included. If a country had a birth rate of 22 per 1000 and a death rate of 12 per 1000 for a given year, the rate of natural increase would be 10 per 1000. This rate is usually expressed as a percentage—that is, as a rate per 100 rather than per 1000. In the example given, the annual increase would be 1%.

The rate of natural increase can be related to the time it takes for a population to double if the current growth rate remains constant—that is, the **doubling time**. **Table 5.2** shows that it would take 70 years for a population with a rate of increase of 1% (approximately the rate of natural growth of Argentina at the start of the 21st century) to double. A 2% rate of increase—recorded in 2015 by Guatemala—means that the population would double in only 35 years. (Population doubling time can be roughly determined by applying the Rule of 70, which simply involves dividing

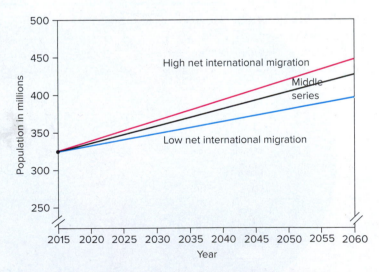

Figure 5.13   Possible population futures for the United States. As these population projections to 2060 illustrate, expected future numbers vary greatly. These three projections take into account the changing age-specific fertility rates for different ethnic and racial groups as well as estimates of future mortality and immigration. Longer term projections often diverge widely because of the effect of slightly different assumptions about fertility, death rates, or immigration. *Source: U.S. Bureau of the Census, 2012.*

Table 5.2	Doubling Time in Years at Different Rates of Increase
Annual Percentage Increase	**Doubling Time (years)**
0.5	140
1.0	70
2.0	35
3.0	24
4.0	17
5.0	14
10.0	7

Table 5.3	Population Growth and Approximate Doubling Times since A.D. 1	
Year	**Estimated Population**	**Doubling Time (years)**
1	250 million	
1650	500 million	1650
1804	1 billion	154
1927	2 billion	123
1974	4 billion	47
World population may reach:		
2025	8 billion	NA[a]

Source: United Nations.
[a]The final estimate of doubling reflects assumptions of decreasing and stabilizing fertility rates. No current projections contemplate a further doubling to 16 billion people.

70 by the growth rate.) How could adding only 20 people per 1000 cause a population to grow so quickly? The principle is the same as that used to compound interest in a bank. Until recently, for the world as a whole, the rates of increase have risen over the span of human history. Therefore, the doubling time has steadily decreased (**Table 5.3**). Growth rates vary regionally, and in countries with high rates of increase (**Figure 5.14**), the doubling time is less than the 54 years projected for the world as a whole at current growth rates. Should world fertility rates decline (as they have in recent years), the population doubling time would increase, as it has since 1990.

Here, then, lies the answer to the earlier question. Even small annual additions accumulate to large total increments because we are dealing with geometric, or exponential (1, 2, 4, 8), rather than arithmetic (1, 2, 3, 4), growth. The ever-increasing world base population has reached such a size that each additional doubling would, if actually achieved, result in an astronomical increase in the total.

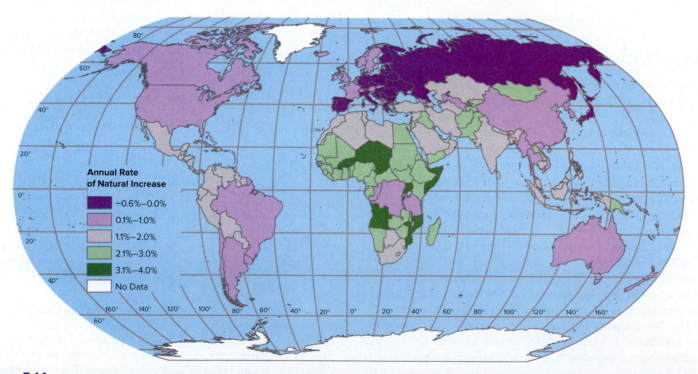

Figure 5.14   Annual rates of natural increase. The world's 2015 rate of natural increase (1.3%) would mean a doubling of population in 54 years. Because demographers now anticipate the world population will stabilize at around 9.5 (in about A.D. 2100) and perhaps actually decline after that, the "doubling" implication and time frame of current rates of natural increase reflect mathematical, not realistic, projections. Many individual continents and countries, of course, deviate widely from the global average rate of growth and have vastly different potential doubling times. Africa as a whole has the highest rates of increase. Europe as a whole (including Russia) had a natural increase rate of 0.0. *Source: 2015 data from Population Reference Bureau.*

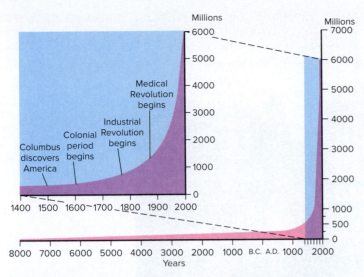

Figure 5.15   **World population growth, 8000 B.C. to A.D. 2000.** Notice that the bend in the J-curve begins in about the mid-1700s when industrialization started to provide new means to support the population growth made possible by revolutionary changes in agriculture and food supply. Improvements in medical science, sanitation, and nutrition reduced death rates near the opening of the 20th century in the industrializing countries.

A simple mental exercise suggests the inevitable consequences of such doubling, or **J-curve,** growth. Take a very large sheet of the thinnest paper you can find and fold it in half. Fold it in half again. After 7 or 8 folds, the sheet will have become as thick as a book—too thick for further folding by hand. If you could make 20 folds, the stack would be nearly as high as a football field is long. From then on, the results of further doubling would be astounding. At 40 folds, the stack would be well on the way to the moon, and at 70 it would reach twice as far as the distance to the nearest star. After 1950, rounding the bend on the J-curve, which **Figure 5.15** suggests world population did around 1900, fostered dire predictions of inevitable, unsupportable pressures on the planet's population support capabilities.

Today, however, rates of natural increase in developed countries, particularly Europe, are approaching zero or even negative values. But individual country growth is also dependent on patterns of immigration and emigration and of changes in life expectancy. That is, a country's "natural" population growth based solely on births and deaths may yield significantly lower population projections and longer doubling times than does the same country's "overall" growth, taking migration into account. The contrast may be striking. For example, Canada has a 0.4% rate of natural increase and a doubling time of 175 years; but with its high rates of immigration, it had an overall growth rate of just over 1% with a doubling time of 64 years.

With replacement or below-replacement fertility rates throughout the developed world and declining fertility rates in most of the developing world, doubt is cast on the applicability of long-term doubling-time projections. Although doubling times are easier to understand than growth rates, they can be misleading because they are based on the dubious assumption that present growth rates will continue indefinitely.

5.3 The Demographic Transition

Exponential population growth cannot continue indefinitely on a finite planet. Some form of braking mechanism must operate to control population growth. If voluntary population limitation is not undertaken, involuntary controls such as famine, disease, or resource wars may take place.

One attempt to summarize the observed voluntary population control that has accompanied economic development and modernization is the **demographic transition** model. The model traces the changing levels of human fertility and mortality associated with industrialization, health care improvements, urbanization, and changing cultural attitudes regarding childbearing. As societies move through the model's stages, high birth and death rates are replaced by low rates (**Figure 5.16**). During the intermediate stages of the model, populations grow rapidly before stabilizing in the final stage.

The first stage of the demographic transition model is characterized by high birth and high but fluctuating death rates. Birth rates and death rates are similar and population grows slowly. Demographers think that it took from approximately A.D. 1 to 1500 for the population to increase from 250 million to 500 million, a doubling time of a millennium and a half. Growth was not steady, of course. There were periods of regional expansion that were usually offset by sometimes catastrophic decline. Wars, epidemics, poor harvests, and natural disasters took heavy tolls. For example, the bubonic plague (the Black Death), which swept across Europe in the 14th century, is estimated to have killed between one-third and one-half of the population of that continent; and epidemic diseases brought by Europeans to the Western Hemisphere are believed to have reduced New World native populations by 95% within a century or two of first contact. The high birth rates, high death rates, and slow population growth of the first stage describe all of human history

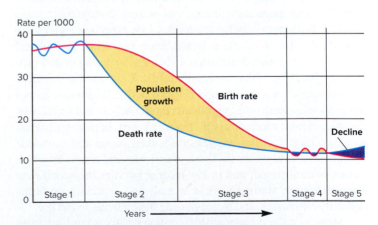

Figure 5.16   **Stages in the demographic transition.** During stage 1, birth and death rates are both high, and population grows slowly. When the death rate drops and the birth rate remains high, there is a rapid increase in numbers. During stage 3, birth rates decline and population growth is less rapid. Stage 4 is marked by low birth and death rates and, consequently, by a low rate of natural increase or even by decrease if death rates exceed those of births. The negative growth rates of many European countries and the falling birth rates in other regions suggest that a stage 5, one of population decline, is regionally—and ultimately worldwide—a logical extension of the transition model.

Figure 5.17 **London, England's Piccadily Circus in the 19th century.** A modernizing Europe experienced improved living conditions and declining death rates during that century of progress. *Source: Library of Congress Prints and Photographs Division [LC-DIG-ppmsc-08577].*

until about A.D. 1750 when Western Europe entered stage 2 of the transition. Over the past 150 years, the decline in death rates that signals the transition to stage 2 diffused around the globe. Today, death rates in even the poorest countries are below 20 per 1000, and no country remains in stage 1 of the demographic transition.

Stage 2 of the transition model is usually associated with the consequences of Europe's industrial revolution and modernization. Rapidly rising population during the second demographic stage occurs as birth rates outstrip death rates. Life expectancies increase dramatically due to advances in medicine, sanitation, agricultural productivity, and food distribution. Urbanization during stage 2 provides the stage on which sanitation, medical, and food distribution improvements are concentrated (**Figure 5.17**). Birth rates do not fall as soon as death rates; ingrained cultural beliefs and social relationships change more slowly than technologies. In most traditional societies, large families are valued for their social and economic advantages. Children are the focus of activities and rituals by which culture is transmitted, and in low-income families they contribute economically by starting work at an early age, especially on farms or family businesses, and by supporting their parents in old age.

Stage 3 of the demographic transition occurs when birth rates decline as people begin to control family size. In urbanized, industrialized cultures, raising children is expensive. In fact, such cultures may view children as economic liabilities rather than assets. In addition, declining childhood mortality rates give parents confidence to have fewer children, knowing that most are likely to survive. When the birth rate falls and the death rate remains low, the rate of population increase slows.

The classic demographic transition model ends with stage 4 characterized by very low, nearly equal birth and death rates. This stage approaches a condition of zero population growth as natural increase

rates drop to zero. With longer life expectancies and fewer births, a significant aging of the population accompanies stage 4 of the model.

In some countries that have completed the demographic transition, death rates equal or exceed birth rates and populations are actually declining. This extension of stage 4 into stage 5 of population decrease so far has been largely confined to the rich, industrialized world—notably Europe and Japan—but increasingly promises to affect much of the rest of the world as well. The dramatic decline in fertility recorded in almost all countries since the 1980s means that a majority of the world's population resides in areas where the only significant population growth is from demographic momentum.

The Western Experience

The demographic transition model was originally devised to describe the experience of Western European countries as they transitioned from rural-agrarian life to urban-industrial societies. Thus, the model may not accurately predict the course of events for all parts of the world. In Europe, church and municipal records, some dating from the 16th century, show that prior to the demographic transition people tended to marry late or not at all. In England before the Industrial Revolution, as many as one-half of all women in the 15–50 age cohort were unmarried. Infant mortality was high, and life expectancies were low. Around 1800, 25% of Swedish infants died before their first birthday. With the coming of industrialization in the 18th and 19th centuries, immediate factory wages instead of long apprenticeship programs permitted earlier marriage and more children.

Beginning about 1860, first death rates and then birth rates began their significant, though gradual, decline. This "mortality revolution" came first, as an epidemiological transition: formerly fatal epidemic

Figure 5.18 Pure piped water replacing individual or neighborhood wells, and sewers and waste treatment plants instead of privies, became increasingly common in urban Europe and North America during the 19th century. Their modern successors, such as the Las Vegas, Nevada, treatment plant shown here, helped complete the *epidemiologic transition* in developed countries. *Source: USDA-Natural Resources Conservation Service.*

diseases became endemic, that is, essentially continual within a population. As people developed partial immunities, mortality rates declined. Improvements in livestock raising, crop rotation, fertilizer use, and new crops (the potato was an early example) from overseas colonies raised the level of health of the European population in general.

At the same time, sewage systems and sanitary water supplies became common in larger cities, reducing the frequency of waterborne illnesses such as cholera and typhoid (**Figure 5.18**). Deaths due to infectious, parasitic, and respiratory diseases and to malnutrition declined, while those related to chronic illnesses associated with an aging population increased. Western Europe passed from a first stage "age of pestilence and famine" to the "age of degenerative and human-origin diseases." However, recent increases in drug- and antibiotic-resistant diseases, pesticide resistance of disease-carrying insects, and such new scourges of both the less-developed and more-developed countries as AIDS (acquired immune deficiency syndrome) cast doubt on the stability of that ultimate stage (see "Our Delicate State of Health" p. 128). Nevertheless, even the resurgence of old and the emergence of new scourges such as malaria, tuberculosis, and AIDS are unlikely to have decisive demographic consequences on the global scale.

In Europe, the striking reduction in death rates was echoed by similar declines in birth rates as societies began to alter their traditional concepts of ideal family size. In cities, child labor laws and mandatory schooling meant that children no longer were important contributors to family economies. As "poor-relief" legislation and other forms of social welfare substituted for family support structures, the value of children as a social safety net declined. Family consumption patterns altered as the Industrial Revolution made widely available the consumer goods once considered luxuries. For some, children came to be seen as a hindrance rather than an aid to social mobility, lifstyle improvement, and self-expression. Perhaps most important were changes in the education levels, career opportunities, and social status of women that spread the conviction that control over childbearing was within their power and to their benefit.

A Divided World, A Converging World

The dramatic decline in mortality that had emerged only gradually throughout the European world occurred with startling speed in developing countries after 1950. The introduction of Western technologies of medicine and public health, including antibiotics, insecticides, sanitation, immunization, infant and child health care, and eradication of smallpox, quickly and dramatically increased life expectancies in developing countries. Such imported technologies and treatments accomplished in a few years what it took Europe 50 to 100 years to experience. Sri Lanka, for example, sprayed extensively with DDT to combat malaria; life expectancy jumped from 44 years in 1946 to 60 only 8 years later. With similar public health programs, India also experienced a steady reduction in its death rate after 1947. Simultaneously, with international sponsorship, food aid cut the death tolls due to droughts and other disasters. Thus, stage 2 of the demographic transition—declining death rates accompanied by continuing high birth rates—diffused worldwide.

Some countries in Africa, Latin America, and southern Asia display the characteristics of stage 2 in the demographic transition model. Ghana, with a birth rate of 31 and a death rate of 8, and Guatemala, with rates of 30 and 6, are typical. The annual rates of increase for such countries are 2.3% to 2.4%, and their populations will double in about 30 years. Such rates, of course, do not mean

OUR DELICATE STATE OF HEALTH

Death rates have plummeted, and the benefits of modern medicines and sanitary practices have enhanced both the quality and the expectancy of life in the developed and much of the developing world. Far from being won, however, the struggle against infectious and parasitic diseases is growing in intensity and is, perhaps, unwinnable. More than a half-century after the discovery of antibiotics, the diseases they were to eradicate are on the rise, and both old and new disease-causing microorganisms are emerging and spreading all over the world. Infectious and parasitic diseases kill between 17 and 20 million people each year; they officially account for one-quarter to one-third of global mortality and, because of poor diagnosis, certainly are responsible for far more. And their global incidence is rising.

The six leading infectious killers are, in order, acute respiratory infections (such as pneumonia), diarrheal diseases, AIDS, tuberculosis, malaria, and measles. The incidence of infection, of course, is much greater than the occurrence of deaths. Nearly 30% of the world's people, for example, are infected with the bacterium that causes tuberculosis, but only 2 to 3 million are killed by the disease each year. More than 500 million people are infected with such tropical diseases as malaria, sleeping sickness, schistosomiasis, and river blindness, with perhaps 3 million annual deaths. Newer pathogens are constantly appearing, such as those causing the Zika virus, Lassa fever, Rift Valley fever, Ebola fever, hantavirus pulmonary syndrome, West Nile encephalitis, hepatitis C, and severe acute respiratory syndrome (SARS), incapacitating and endangering far more than they kill. In fact, at least 30 previously unknown infectious diseases have appeared since the mid-1970s.

The spread and virulence of infectious diseases are linked to the dramatic changes occurring in the Earth's physical and social environments. Climate change permits temperature-restricted pathogens to expand their geographic range. Deforestation, water contamination, wetland drainage, and other human-induced alterations to the physical environment disturb ecosystems and simultaneously disrupt the natural system of controls that keeps infectious diseases in check. Rapid population growth and explosive urbanization, increasing global tourism, population-dislocating wars and migrations, and expanding world trade all increase interpersonal disease-transmitting contacts and the mobility and range of disease-causing microbes, including those brought from previously isolated areas by newly opened road systems and air routes. Add in natural disasters, poorly planned public health programs, inadequate investment in sanitary infrastructures, and inadequate medical personnel and facilities, and the human role in many of the current disease epidemics is clearly visible.

The most effective weapons in the battle against epidemics are already known. They include improved health education; disease prevention and surveillance; research on disease vectors and incidence areas (using geographic information systems and spatial analysis); careful monitoring of drug therapy; mosquito control programs; provision of clean water supplies; and distribution of such simple and cheap remedies and preventives as childhood immunizations, oral rehydration therapy, and vitamin A supplementation. All, however, require expanded investment and attention to those spreading infectious diseases.

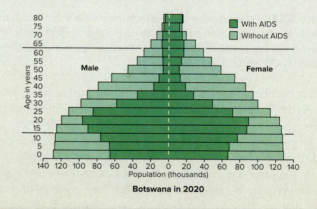

Botswana in 2020

that the full impact of modernization has been felt worldwide; they do mean that the underdeveloped societies have been beneficiaries of the life preservation techniques it generated.

Birth rate declines, of course, depend in part on birth control technology, but even more important is social acceptance of the idea of fewer children and smaller families. That acceptance has diffused unevenly around the world. The steep declines in birth rates in developing countries as a group indicate that most of them have moved into stage 3 or 4 of the demographic transition (**Figure 5.19**). In 1984, only 18% of the world population lived in countries with fertility rates at or below replacement levels (that is, countries that had achieved stage 4 in the demographic transition). By 2015, however, 46% lived in such countries, and it is increasingly difficult to distinguish between developed and developing societies on the basis of their fertility rates. The world's lowest-fertility countries are evenly split between Europe and Asia. Significant decreases to near the replacement level have also occurred in the space of a single generation in many other Asian and Latin American states with high recent rates of economic growth. Increasingly, it appears, low fertility is becoming a feature of both rich and poor, developed and developing states.

The demographic transition model assumes an inevitable course of events from the high birth and death rates of premodern (nonindustrialized) societies to the low and stable rates of advanced (industrialized) countries. The model failed to anticipate, however, that the European experience would not be matched by all developing countries. Some developing societies remain in stage 2 or early stage 3 of the model, unable to realize the economic gains and social changes necessary to complete the demographic transition. Thus, despite a substantial convergence of fertility rates, many observers point to a continuing and growing demographic divide. On one side of the divide are the low-fertility countries that are home to 46% of the world's population. The low fertility rates of these mainly wealthy countries guarantee future population decline and rapid aging. On the other side of the divide are a group of high-fertility countries that account for just 8% of the world population. Nearly all of the high-fertility countries are included on the United Nation's list of least-developed countries. Most of them are in sub-Saharan Africa, and all suffer from low per capita income, illiteracy, low standards of living, and inadequate health facilities and care. In the middle, containing 46% of the world's population, is a group of countries that have

International Population Policies

After a sometimes rancorous 9-day meeting in Cairo in September 1994, the United Nations International Conference on Population and Development endorsed a strategy for stabilizing the world's population at 7.27 billion by no later than 2015. The 20-year program of action accepted by 179 signatory countries sought to avoid the environmental consequences of excessive population growth. Its proposals were therefore linked to discussions and decisions of the UN Conference on Environment and Development held in Rio de Janeiro in June 1992.

The Cairo plan abandoned several decades of top-down government programs that promoted *population control* (a phrase avoided by the conference) based on targets and quotas and, instead, for the first time embraced policies giving women greater control over their lives, greater economic equality and opportunity, and a greater voice in reproduction decisions. It recognized that limiting population growth depends on programs that lead women to want fewer children and make them partners in economic development. In that recognition, the conference accepted the documented link between increased educational access and economic opportunity for women and falling birth rates and smaller families. Earlier population conferences—1974 in Bucharest and 1984 in Mexico City—did not fully address these issues of equality, opportunity, education, and political rights; their adopted goals had failed to achieve hoped-for changes in birth rates, in large part because women in many traditional societies had no power to enforce contraception and feared their other alternative, sterilization.

The earlier conferences had carefully avoided or specifically excluded abortion as an acceptable family planning method. It was the more open discussion of abortion in Cairo that elicited much of the spirited debate that registered religious objections by the Vatican and many Muslim and Latin American states to the inclusion of legal abortion as part of health care, and to language suggesting approval of sexual relations outside marriage. Although the final text of the conference declaration did not promote any universal right to abortion and excluded it as a means of family planning, some delegations still registered reservations to its wording on both sex and abortion. At the conference's close, however, the Vatican endorsed the declaration's underlying principles, including the family as "the basic unit of society" and the need to "stimulate economic growth and to promote gender equality, equity, and the empowerment of women."

A special United Nations "Cairo+5" session in 1999 recommended some adjustments to the earlier agreements. It urged an emphasis on measures ensuring safe and accessible abortion in countries where it is legal, called for schoolchildren at all levels to be instructed in sexual and reproductive health issues, and told governments to provide special family planning and health services for sexually active adolescents, with particular stress on reducing their vulnerability to AIDS.

In 2004, the UN reported progress toward reaching Cairo and Cairo+5 goals. The consensus was that much remained to be done to broaden programs for the poorest population groups, to invest in rural development and urban planning, to strengthen laws ending discrimination against women, and to encourage donor countries to fully meet their agreed-upon contributions to the program. Nevertheless, positive Cairo plan results were also seen in declining fertility rates in many of the world's most-populous developing countries. Some demographers and many women's health organizations pointedly claim that those declines had little to do with government planning policies. Rather, they assert, current lower and falling fertility rates were the expected result of women's assuming greater control over their economic and reproductive lives. The director of the UN Population Division noted: "A woman in a village making a decision to have one or two or at most three children is a small decision in itself. But . . . compounded by millions and millions . . . of women in India and Brazil and Egypt, it has global consequences."

That women are making those decisions, population specialists have observed, reflects important cultural factors emerging since Cairo. Satellite television takes contraceptive information even to remote villages and shows programs of small, apparently happy families that viewers think of emulating. Increasing urbanization reduces some traditional family controls on women and makes contraceptives easier to find, and declining infant mortality makes mothers more confident their babies will survive. Perhaps most important, population experts assert, is the dramatic increase in most developing states in female school attendance, with corresponding reductions in the illiteracy rates of girls and young women, who will themselves soon be making fertility decisions.

Considering the Issues

1. Do you think it is appropriate or useful for international bodies to promote policies affecting such purely personal or national concerns as reproduction and family planning? Why or why not?

2. Do you think that current international concerns over population growth, development, and the environment are sufficiently valid and pressing to risk the loss of long-enduring cultural norms and religious practices in many of the world's traditional societies? Why or why not?

3. The Cairo plan called for sizable monetary pledges from developed countries to support enhanced population planning in the developing world. For the most part, those pledges have not been honored. Do you think the financial obligations assigned to donor countries are justified in light of the many other international needs and domestic concerns faced by their governments? Why or why not?

4. Many environmentalists see the world as a finite system unable to support ever-increasing populations; to exceed its limits would cause frightful environmental damage and global misery. Many economists counter that free markets will keep supplies of needed commodities in line with growing demand and that science will, as necessary, supply technological fixes in the form of substitutes or expansion of production. In light of such diametrically opposed views of population growth consequences, is it appropriate or wise to base international programs solely on one of them? Why or why not?

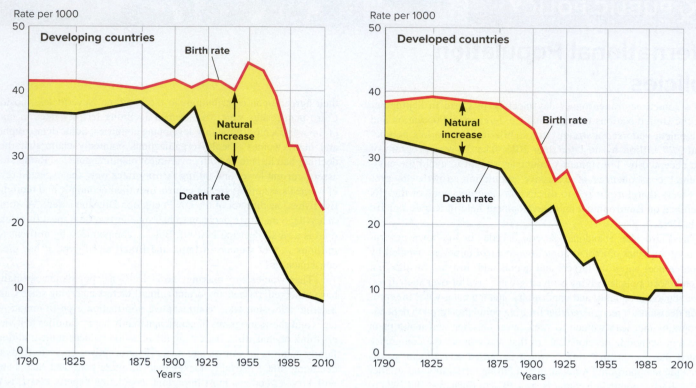

Figure 5.19   World birth and death rates. The "population explosion" after World War II (1939–1945) reflected the effects of drastically reduced death rates in developing countries without simultaneous and compensating reductions in births. Today, however, three interrelated trends appear in many developing world countries: (1) fertility has dropped further and faster than had been predicted 25 years earlier, (2) contraceptive acceptance and use has increased markedly, and (3) age at marriage is rising. In consequence, the demographic transition had been compressed from a century to a generation in some developing states. In others, fertility decline began to slacken in the mid-1970s but continued to reflect the average number of children—four or more—still desired in many societies. *Source: Revised and redrawn from Elaine M. Murphy,* World Population: Toward the Next Century, *revised ed. (Washington, D.C.: Population Reference Bureau, 1989) and World Population Data Sheet (Washington, D.C.: annual).*

made substantial reductions in fertility rates and are approaching replacement fertility.

The established patterns of both high- and low-fertility regions tend to be self-reinforcing. Low growth permits the expansion of personal income and the accumulation of wealth to enhance the quality of life and make large families less attractive or essential. In contrast, in high-fertility regions, population growth requires spending on social services that otherwise might have been invested to promote economic expansion. Growing populations place ever-greater demands on limited natural resources. As the environmental base deteriorates, productivity declines, undermining the economic progress on which the demographic transition depends (see "International Population Policies" p. 129). The vastly different future prospects for personal and national prosperity between the high-fertility countries and the rest of the world make the demographic divide a matter of continuing international concern.

5.4 The Demographic Equation

Changes in a region's population stem from three basic life events: people are born, they migrate, and they die. Migration involves the long-distance movement of people from one residence to another. When that relocation occurs across political boundaries, it affects the population structure of both the origin and destination

jurisdictions. The **demographic equation** summarizes the contribution made to regional population change over time by the combination of *natural change* (births minus deaths) and *net migration* (difference between in-migration and out-migration).[1] On a global scale, of course, all population change is accounted for by natural change. The impact of migration on the demographic equation increases as the population size of the areal unit decreases.

Population Relocation

Emigration proved an important device for relieving the pressures of rapid population growth during the demographic transition for at least some European countries (**Figure 5.20**). For example, in one 90-year span, 45% of the natural increase in the population of the British Isles emigrated, and between 1846 and 1935, some 60 million Europeans of all nationalities left that continent. Despite recent massive movements of economic and political refugees across Asian, African, and Latin American boundaries, emigration today provides no comparable relief valve for developing countries. Total population numbers are too great to be affected much by migrations of even millions of people. In only a few countries—Afghanistan, Cuba, El Salvador, and Haiti, for example—have as many as 10% of the population emigrated in

[1] See the Glossary definition for the calculation of the equation.

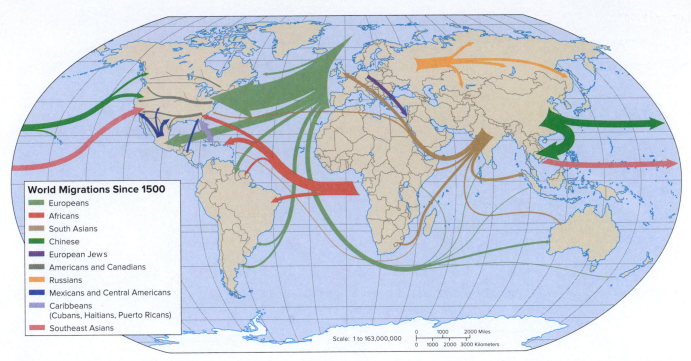

World Migrations Since 1500

- Europeans
- Africans
- South Asians
- Chinese
- European Jews
- Americans and Canadians
- Russians
- Mexicans and Central Americans
- Caribbeans (Cubans, Haitians, Puerto Ricans)
- Southeast Asians

Scale: 1 to 163,000,000

Figure 5.20   Principal migrations since 1500. The arrows suggest the direction and relative size of the major voluntary and forced international population movements since about 1500. Not shown are the more-recent migrations into the United Kingdom, France, Belgium, and the Netherlands from their former colonies and the less-prosperous countries of Europe. *Source: John Allen and Christopher Sutton,* Student Atlas of World Geography, *7th edition, p. 36.*

recent decades. A more detailed treatment of the processes and patterns of international and intranational migrations as expressions of human interaction is presented in Chapter 7.

Immigration Impacts

Where cross-border movements are massive enough, migration may have a pronounced impact on the demographic equation and result in significant changes in the population structures of both the origin and destination regions. Past European and African migrations, for example, not only altered but substantially created the population structures of new, sparsely inhabited lands of colonization in the Western Hemisphere and Australasia. In some decades of the late 18th and early 19th centuries, 30% to more than 40% of the population increase in the United States was accounted for by immigration. Similarly, eastward-moving Slavs colonized underpopulated Siberia and overwhelmed native peoples.

Migrants are rarely a representative cross section of the population group they leave, and they add an unbalanced age and sex component to the group they join. A recurrent research observation is that emigrant groups are heavily skewed in favor of young singles. Whether males or females dominate the outflow varies with circumstances. Although males traditionally have far exceeded females in international flows, in recent years females have accounted for between 40% and 60% of all transborder migrants.

At the least, then, the receiving country will have its population structure altered by an outside increase in its younger age and, probably, unmarried cohorts. The results are both immediate in a modified population pyramid and potential in their future impact on birth rates and natural increase rates. The origin area will have lost a portion of

its young, active members of childbearing years. Perhaps it will have suffered distortion in its young adult sex ratios, and it certainly will have recorded a statistical aging of its population. The destination society will likely experience increases in births associated with the youthful newcomers and, in general, have its average age reduced.

5.5 World Population Distribution

The world's population is not uniformly distributed over the Earth. The most striking feature of the world population distribution map (**Figure 5.21**) is the very unevenness of the pattern. Some land areas are nearly uninhabited, others are sparsely settled, and still others contain dense agglomerations of people. Until recently, rural people—unevenly concentrated—always outnumbered urban people. After 2007, however, urbanites remain more numerous, with cities capturing most of the world's population growth.

Earth regions of apparently very similar physical makeup show quite different population numbers and densities, perhaps the result of differently timed settlement or of settlement by different cultural groups. Northern and Western Europe, for example, inhabited thousands of years before North America, contain as many people as the United States on 70% less land; the present heterogeneous population of the Western Hemisphere is vastly denser overall than was that of earlier Native Americans.

We can draw certain generalizations from the uneven but far from irrational distribution of population shown in Figure 5.21. First, almost 90% of all people live north of the equator, and two-thirds of the total dwell in the midlatitudes between

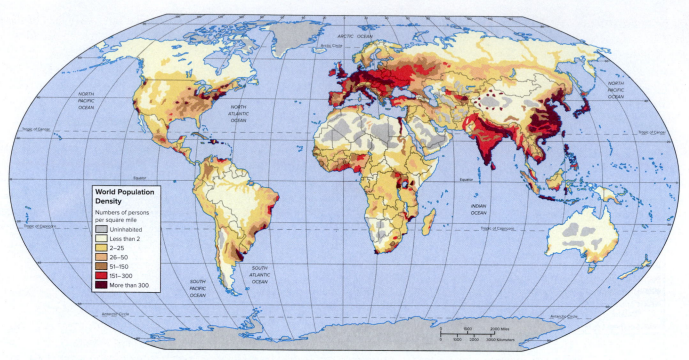

Figure 5.21 **World population density.** *Source: Christopher Sutton,* Student Atlas of World Geography, *8th edition, p. 28.*

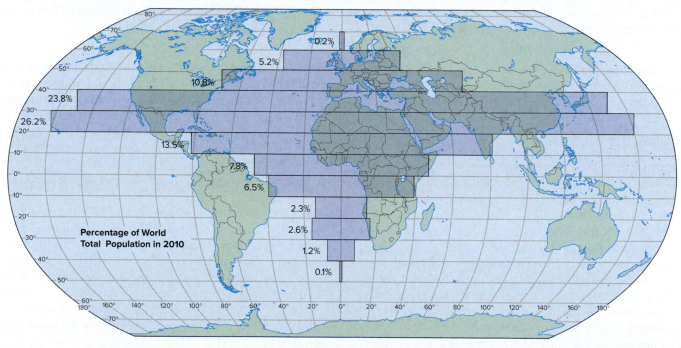

Figure 5.22 **The population dominance of the Northern Hemisphere** is strikingly evident from this bar chart. Only 12% of the world's population live south of the Equator—not because the Southern Hemisphere is underpopulated but because it is mainly water. *Data Source: Center for International Earth Science Information Network (CIESIN), Columbia University, 2005. Gridded Population of the World: Future Estimates (2011).*

20° and 60° North (**Figure 5.22**). Second, a large majority of the world's inhabitants occupy only a small part of its land surface. More than one-half the people live on about 5% of the land, two-thirds on 10%, and almost nine-tenths on less than 20%. Third, people congregate in lowland areas; their numbers decrease sharply with increases in elevation. Temperature, length of growing season, slope and erosion problems, even oxygen reductions at very high altitudes all appear to limit the

habitability of higher elevations. One estimate is that 34% of all people live below 100 meters (330 ft), a zone containing about 17% of the total land area.

Fourth, although low-lying areas are preferred settlement locations, not all such areas are equally favored. Continental margins have attracted the densest settlement. On average, density in coastal areas is about twice the world's average population density. Latitude, aridity, and elevation, however, limit the attractiveness of many seafront locations. The low temperatures and infertile soils of the extensive Arctic coastal lowlands of the Northern Hemisphere have restricted settlement there. Mountainous or desert coasts are sparsely occupied at any latitude, and some tropical lowlands and river valleys that are marshy, forested, and disease-infested are also unevenly settled.

Within the sections of the world generally conducive to settlement, four areas contain great clusters of population: South Asia, East Asia, Europe, and the northeastern United States/southeastern Canada. The *South Asia* cluster is composed primarily of countries associated with the Indian subcontinent—Bangladesh, India, Nepal, Pakistan, and the island state of Sri Lanka. The South Asia cluster is the world's most populous, containing one-fourth of the world's inhabitants. The *East Asia* zone, which includes Japan, China, Taiwan, and South Korea, is the second-largest cluster numerically. The four countries forming it contain nearly 25% of all people on Earth; China alone accounts for one in five of the world's inhabitants. Thus, the South and the East Asian concentrations are home to nearly one-half of the world's people.

Europe—southern, western, and eastern through Ukraine and much of European Russia—is the third extensive population concentration, with another 11% of the world's inhabitants. Much smaller in extent and total numbers is the cluster in the *northeastern United States/southeastern Canada*. Other smaller but pronounced concentrations are found around the globe: on the island of Java in Indonesia and along the Nile River in Egypt.

The **ecumene** comprises the permanently inhabited areas of the Earth's surface. The ancient Greeks used the word, derived from their verb for "to inhabit," to describe their known world between what they believed to be the unpopulated, searing southern equatorial lands and the permanently frozen northern polar reaches of the Earth. Clearly, natural conditions are less restrictive than Greek geographers believed. Both ancient and modern technologies have rendered habitable the areas that natural conditions make forbidding. Irrigation, terracing, diking, and draining are among the methods devised to extend the ecumene locally (**Figure 5.23**).

At the world scale, the ancient observation of habitability appears remarkably astute. The Earth's **nonecumene,** the uninhabited or very sparsely occupied zone, includes the permanent ice caps of the Far North and Antarctica and large segments of the tundra and coniferous forest of northern Asia and North America. But the nonecumene is not continuous, as the ancients supposed. It is discontinuously encountered in all portions of the globe and includes parts of the tropical rain forests of equatorial zones, midlatitude deserts of both the Northern and Southern Hemispheres, and high mountain areas.

Even parts of these unoccupied or sparsely occupied districts have localized, dense settlement nodes, or zones, based on irrigation agriculture, mining and industrial activities, and the like. Perhaps the most striking case of settlement in an environment elsewhere considered part of the nonecumene world is that of the dense population in the Andes Mountains of South America and the plateau of Mexico. There, Native Americans found temperate conditions away from the dry coastal regions and the hot, wet Amazon Basin. The fertile high basins have served a large population for more than a thousand years.

Even with these locally important exceptions, the nonecumene portion of the Earth is extensive; 35% to 40% of all the world's land surface is inhospitable and without significant settlement. Admittedly, this is a smaller proportion of the Earth than would have qualified as uninhabitable in ancient times or even during the 19th century. Since the end of the ice age some 12,000 years ago, humans have steadily expanded their areas of settlement.

Population Density

The boundaries of the ecumene could only be extended as humans learned to support themselves from the resources of new settlement areas. The size of the population that could be supported depended upon the resource potentials of those areas and the cultural levels and technologies possessed by the occupying populations. A **population density** is the relationship between the number of inhabitants and the area they occupy.

Density figures are useful, if sometimes misleading, representations of regional variations of human distribution. The **crude density,** or **arithmetic density,** of population is the most common and least satisfying expression of that variation. It is the calculation of the number of people per unit area of land, usually within the boundaries of a political entity. It is an easily reckoned figure; all that is required is information on total population and total land area, both commonly available for national or other political units. The figure can, however, be misleading and may obscure more of reality than it reveals. The calculation is an average that blankets a country's largely undevelopable or sparsely populated regions along with its intensively settled and developed districts. In general, the larger the political unit for which crude density is calculated, the less useful the figure.

Various modifications may be made to refine density as a useful description of distribution. Its precision is improved if the area in question can be subdivided into comparable regions or units. Thus, it is more revealing to know that, in 2010, New Jersey had a density of 467 and Wyoming of 2 persons per square kilometer (1196 and 5.8 per sq mi) of land area than to know only that the figure for the United States was 32 per square kilometer (83.8 per sq mi). The calculation may also be modified to provide density distinctions between classes of population—rural versus urban, for example. Rural densities in the United States rarely exceed 115 per square

Figure 5.23　Terracing of hillsides is one device to extend a naturally limited productive area. The technique is used effectively on the island of Bali, Indonesia. © *Glen Allison/Getty Images RF.*

Table 5.4   Comparative Densities for Selected Countries

Country	CRUDE DENSITY		PHYSIOLOGICAL DENSITY [a]		AGRICULTURAL DENSITY [b]	
	sq mi	km²	sq mi	km²	sq mi	km²
Argentina	41	16	280	108	26	10
Australia	8	3	132	51	13	5
Bangladesh	2870	1110	5410	2089	3720	1450
Canada	8	3	205	79	36	14
China	370	143	3350	1293	1390	540
Egypt	231	89	8280	3196	3890	1520
India	1040	400	2180	842	1290	502
Japan	870	336	7770	3000	2550	995
Nigeria	510	197	1350	520	543	212
United Kingdom	694	268	2720	1050	261	102
United States	88	34	536	207	84	33

Sources: World Bank, *Little Green Data Book 2010;* and Population Reference Bureau, *World Population Data Sheet 2015.*
[a]Total population divided by area of arable land.
[b]Rural population divided by area of arable land.
Rounding may produce apparent conversion discrepancies.

kilometer (300 per sq mi), while portions of major cities can have tens of thousands of people in an equivalent space.

Another useful refinement of crude density relates population not simply to total national territory but to the area of a country that is or may be cultivated—that is, to *arable* land. When total population is divided by arable land area alone, the resulting figure is the **physiological density,** which is, in a sense, an expression of population pressure exerted on agricultural land. Countries differ in physiological density and the contrasts between the crude and physiological densities of countries point up actual settlement pressures that are not revealed by crude densities alone (**Table 5.4**). The calculation of physiological density, however, depends on uncertain definitions of arable and cultivated land, assumes that all arable land is equally productive.

Agricultural density is another useful variant. It simply excludes city populations from the physiological density calculation and reports the number of rural residents per unit of agriculturally productive land. It is, therefore, an estimate of the pressure of people on the rural areas of a country. See Table 5.4.

Overpopulation?

After one compares population density values, it is common to draw conclusions about overpopulation or overcrowding. It is wise to remember that **overpopulation** is a value judgment reflecting an observation or a conviction that an environment or a territory is unable to support its present population. (The related concept of *underpopulation* is the circumstance of too few people

to sufficiently develop the resources of a country or region to improve the level of living of its inhabitants.)

Overpopulation is a reflection, not of numbers per unit area but of the **carrying capacity** of the land, the prevailing agricultural technology, and the ability to afford imported food. A region devoted to energy-intensive commercial agriculture that makes heavy use of irrigation, fertilizers, and biocides can support more people at a higher standard of living than one engaged in the nomadic herding or shifting cultivation styles of agriculture described in Chapter 9. An industrial society that takes advantage of resources such as coal and iron ore and has access to imported food will not feel population pressure at the same density levels as a country with rudimentary technology. In fact, we should be careful in borrowing the concept of carrying capacity from ecology since compared to other animals, humans are distinguished by their ability to adopt new production technologies, increase their appetite for consumption above biological requirements, and trade resources globally.

In a world of growing global interdependence, high physiological densities alone do not indicate overpopulation. Few countries are agriculturally self-sufficient. Japan has a very high physiological density, as Table 5.4 indicates, and only produces about 40% of the calories its population consumes. Nonetheless Japan ranks high on all indicators of national well-being and prosperity. For countries such as Japan, South Korea, Malaysia, and Taiwan—all of which currently import the majority of the grain they consume—a sudden cessation of the international trade that permits the exchange of industrial products for imported food and raw materials would be disastrous. Domestic food production could not maintain the dietary

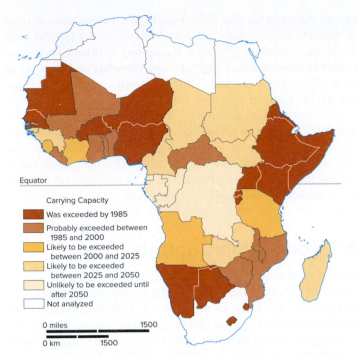

Figure 5.24   Carrying capacity and potentials in sub-Saharan Africa. The map assumes that (1) all cultivated land is used for growing food; (2) food imports are insignificant; and (3) agriculture is conducted by low-technology methods. *Source: World Bank; United Nations Development Programme; Food and Agriculture Organization (FAO); and Bread for the World Institute.*

levels now enjoyed by their populations, and they, more starkly than many less-developed countries, would be "overpopulated."

One measure of overpopulation might be a lack of food security. **Food security** means having access to safe and nutritious food supplies sufficient to meet individual dietary needs in accord with cultural preferences. Unfortunately, dietary insufficiencies—with long-term adverse implications for life expectancy, physical vigor, and mental development—are most likely to be encountered in the developing countries, where much of the population is young and vulnerable (Figure 5.11). Over the past decades, sub-Saharan Africa saw its population grow faster than food production, widening the population-food gap and increasing reliance on imports (**Figure 5.24**). For poor countries, reliance on food imports means vulnerability to price volatility in international markets and rising rates of undernourishment whenever food prices are high. The international Food and Agriculture Organization (FAO) estimates that about 800 million people are undernourished, down from about 1 billion in the early 1990s. While it may be tempting to conclude that a country with a high rate of undernourishment is overpopulated, the underlying causes are found in the interaction of poverty and population pressure.

It is difficult to draw meaningful conclusions about the relationship between population density and levels of development. As a group, the less-developed countries have higher densities than the more-developed countries. Densities in Australia, Canada, New Zealand, Norway, and the United States, where there is a great deal of unused and unsettled land, are considerably lower than those in Bangladesh, where essentially all land is arable and which, with some 1110 people per square kilometer (2870 per sq mi), is the most

densely populated non-island state in the world. However, counterexamples abound. Mongolia, a sizable state between China and Siberian Russia, has 1.8 persons per square kilometer (4.6 per sq mi), while Singapore, a highly urbanized island country in Southeast Asia, has 8040 persons per square kilometer (20,800 per sq mi). Incomes are about 7 times higher in Singapore than in Mongolia. Many African countries have low population densities and low levels of living, whereas Japan combines both high densities and high incomes.

5.6 Population Data and Projections

Population geographers, demographers, planners, government officials, and a host of others rely on detailed population data to make their assessments of present national and world population patterns and to estimate future conditions. Birth rates and death rates, rates of fertility and of natural increase, age and sex composition of the population, and other items are all necessary ingredients for their work.

Population Data

The data that students of population use come primarily from the United Nations Statistical Office, the World Bank, the Population Reference Bureau, and ultimately national censuses and sample surveys. Unfortunately, the data are far from perfect. For much of the developing world, a national census is a massive undertaking. Isolation and poor transportation, insufficiency of funds and trained census personnel, high rates of illiteracy limiting the type of questions that can be asked, and populations suspicious of government data collectors restrict the frequency, coverage, and accuracy of population reports.

However derived, detailed data are published by the major reporting agencies for all national units, even when those figures are poorly based on fact or are essentially fictitious. For example, for years, data on the total population, birth and death rates, and other vital statistics for Somalia were regularly reported and annually revised. The fact was, however, that Somalia had never had a census and had no system for recording births. Seemingly precise data were regularly reported as well for Ethiopia. When that country had its first-ever census in 1985, at least one data source had to drop its estimate of the country's birth rate by 15% and increase its figure for Ethiopia's total population by more than 20%. The 2006 census in Nigeria was surrounded by protests, boycotts, and fraud charges, despite a lack of questions about religious and ethnic identity, which are controversial in a country with over 250 ethnic groups and a population nearly evenly divided between Muslims and Christians.

Fortunately, census coverage on a world basis is improving. Almost every country has now had at least one census of its population, and most have been subjected to periodic sample surveys (**Figure 5.25**). However, only about 10% of the developing world's population live in countries with anything approaching complete systems for registering births and deaths. Estimates are that 40% or fewer of live births in Indonesia, Pakistan, India, and the Philippines are officially recorded; sub-Saharan Africa has the highest percentage of unregistered births (65%), according to UNICEF.

Figure 5.25 Nearly all countries in the world conduct regular censuses, although some are of doubtful completeness or accuracy. This photo shows Indian census-takers collecting data in a rural community. To count the world's second-largest population for the 2011 Indian census, each person over the age of 15 was photographed and fingerprinted. As in many countries, the Indian census collects additional data beyond population counts such as economic activity, literacy, education levels, housing conditions, and the availability of drinking water and electricity. © *Deshakalyan Chowdhury/Getty Images.*

Throughout Asia, apparently deaths are even less completely reported than births. And whatever the deficiencies of Asian states, African statistics are still less complete and reliable. It is on just these basic birth and death data that projections about population growth and composition are founded.

Population Projections

For all their inadequacies and imprecisions, current data reported for country units form the basis of **population projections,** estimates of future population size, age, and sex composition. Projections are not forecasts, and demographers are not the social science equivalent of meteorologists. Weather forecasters work with a myriad of accurate observations applied against a known, tested model of the atmosphere. Demographers, in contrast, work with sparse, imprecise, out-of-date, and missing data applied to human actions that will be unpredictably responsive to stimuli not yet evident.

Population projections, therefore, are based on assumptions for the future applied to current data, which themselves are frequently suspect. Since projections are not predictions, they can never be wrong. They are simply the inevitable result of calculations about fertility, mortality, and migration rates applied to each age cohort of a population now living, as well as the making of birth rate, survival, and migration assumptions about cohorts yet unborn. Of course, the perfectly valid *projections* of future population size and structure resulting from those calculations may be dead wrong as *predictions.*

Because those projections are invariably treated as scientific expectations by a public that ignores their underlying qualifying assumptions, agencies (such as the UN) that estimate the population

of, say, Africa in the year 2025 do so by not one but three or more projections: high, medium, and low, for example. For areas as large as Africa, a medium projection is assumed to benefit from compensating errors and statistically predictable behaviors of very large populations. For individual African countries and smaller populations, the medium projection may be much less satisfying. The usual tendency in projections is to assume that something like current conditions will be applicable in the future. Obviously, the more distant the future, the less likely is that assumption to remain true. The resulting observation should be that, the further into the future the population structure of small areas is projected, the greater is the implicit and inevitable error (see Figure 5.13).

5.7 Population Controls

As the number of humans and the extent of the ecumene have expanded, attention has turned to the possibility of overpopulation and the need for population control. Ancient Chinese thinker Confucius warned against rapid population growth, and ancient Greek philosophers Plato and Aristotle gave careful thought to the ideal population size for a city-state. All population projections include an assumption that at some point in time fertility rates will stabilize at replacement levels and population growth will cease. Otherwise, future numbers become unthinkably large. At present growth rates, in four centuries from now there would be 1 trillion people living on Earth, a sum beyond our imagination.

Population pressures do not come from the amount of physical space humans occupy. It has been calculated, for example, that the entire human race could easily be accommodated within the boundaries of the state of Delaware. The problems stem from the energy, food, water, and other resources necessary to support the population and absorb its waste products. Clearly, at some point population will have to stop increasing. The demographic transition model provides the reassuring message that the decreases in birth rates that accompany modernization will automatically bring about a condition of zero population growth. However, the presence of a significant demographic divide between the nearly stable populations of the developed countries and rapidly growing populations in some developing regions lead some observers to question the model's universality. According to these observers, completion of the demographic transition in the least-developed countries will not take place soon enough, if at all. If the self-induced limitations on growth implicit in the demographic transition model do not take place, environmental limits will appear in more dramatic fashion.

Thomas Robert **Malthus** (1766–1834), an English clergyman and economist, published *An Essay on the Principle of Population* in 1798, which set the framework for ongoing debates on population and resources. Prior to Malthus, most European economic and political thinkers were pronatalist; that is, they supported population growth to increase the number of workers and military might of a state. According to Malthus, the biological potential for population growth outstripped the potential for increasing food supplies to meet human subsistence needs. In essence, Malthus argued that if unchecked, human population would increase at a geometric rate while food supplies expanded at an arithmetic rate. If humans did not restrain their reproductive capacity with "private" means of moral restraint such as late marriage, or celibacy, nature would enact "destructive" checks on overpopulation. He wrote (p. 44):

> *The power of population is so superior to the power in the earth to produce subsistence for man that premature death must in some shape or other visit the human race. The vices of mankind are active and able ministers of depopulation. . . . But should they fail in the war of extermination, sickly seasons, epidemics, pestilence, and plague, advance in terrific array, and sweep off their thousands and ten thousands. Should success be still incomplete, gigantic inevitable famine stalks in the rear, and with one mighty blow, levels the population with the food of the world.*

Malthus was not alone in drawing pessimistic conclusions about population growth and resources; Hung Liangchi of China wrote in the 19th century that "Within a hundred years or so, the population can increase from fivefold to twentyfold, while the means of subsistence . . . can increase only from three to five times."

However, inevitably—following the logic of Malthus, the apparent evidence of history, and observations of animal populations—equilibrium must be achieved between numbers and support resources. When overpopulation of any species occurs, a population dieback is inevitable. The madly ascending leg of the J-curve is bent to the horizontal, and the J-curve is converted to an **S-curve.** It has happened before in human history, as **Figure 5.26** summarizes. The top of the S-curve represents a population size consistent with and supportable by the exploitable resource base. When the population is equivalent to the carrying capacity of the occupied area, it is said to have reached a **homeostatic plateau.**

Malthus did not have the benefit of witnessing the demographic transition that was just getting underway in Europe or the coming advances in contraceptives and food production technology. Yet, that did not stop the revival of his ideas after World War II when many developing countries entered stage 2 of the demographic transition. Neo-Malthusians, most notably Paul Ehrlich, a Stanford University biologist and author of the best-seller *The Population Bomb,*

updated Malthus' arguments for the 20th century. Sounding the alarm in 1968, Ehrlich wrote (p. xi):

> *The battle to save humanity is over. In the 1970s and 1980s hundreds of millions of people will starve to death despite any crash programs embarked upon now. At this late date, nothing can be done to prevent a substantial increase in the world death rate, although many lives could be saved through dramatic programs to "stretch" the carrying capacity of the earth by increasing food production and providing for more equitable distribution of whatever food is available. But these programs will only provide a stay of execution unless they are accompanied by determined and successful efforts at population control.*

Neo-Malthusianism gained popularity among environmentalists and international development specialists who saw rapid population growth damaging the environment and diverting scarce resources away from capital investment. In order to lift living standards, the existing national efforts to lower mortality rates had to be balanced by government programs to reduce birth rates. Thus, neo-Malthusian thinking became the basis for national and international programs of population control. These programs were promoted around the world by the United Nations and nongovernmental organizations, and they were especially influential in Asian countries such as China and India that adopted comprehensive family planning programs (**Figure 5.27**). In some instances, success has been declared complete. Singapore established its Population and Family Planning Board in 1965, when its fertility rate was 4.9 lifetime births per woman. By 1986, that rate had declined to 1.7, well below the 2.1 replacement level for developed countries, and the board was abolished as no longer necessary.

Africa and the Middle East have generally been less responsive to the neo-Malthusian arguments because of ingrained cultural convictions among people, if not in all government circles, that large

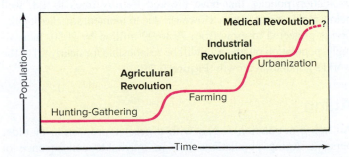

Figure 5.26 The steadily higher *homeostatic plateaus* (states of equilibrium) achieved by humans are evidence of their ability to increase the carrying capacity of the land through technological advance. Each new plateau represents the conversion of the J-curve into an S-curve. "Medical Revolution" implies the range of modern sanitary and public health technologies and disease preventive and curative advances that materially reduced morbidity and mortality rates.

Figure 5.27 A Mumbai, India, sign promoting the government's continuing program to reduce the country's high fertility rate. Female sterilization is the world's most popular form of birth prevention, and in Brazil, China, the Dominican Republic, and India, a reported one-third or more of all married women have been sterilized. © *Carl Purcell..*

families are desirable. Although total fertility rates have begun to decline in most sub-Saharan African states, they still remain above replacement levels nearly everywhere. Islamic fundamentalism opposed to birth restrictions also is a cultural factor in the Near East and North Africa. However, the Muslim theocracy of Iran has endorsed a range of contraceptive procedures and developed one of the world's more aggressive family planning programs. Predominantly Roman Catholic countries in southern Europe and Latin America have witnessed substantial fertility declines despite official pronatalist policies of the church.

Other barriers to fertility control exist. When first proposed by Western states, neo-Malthusian arguments that family planning is necessary for development were rejected by many less-developed countries. Reflecting both nationalistic and Marxist concepts, they maintained that remnant colonial-era social, economic, and class structures rather than population increase hinder development. Some government leaders think there is a correlation between population size and power and pursue pronatalist policies, as did Mao's China during the 1950s and early 1960s.

A group of economists labeled *cornucopians* offered a contrasting perspective. Danish economist Esther Boserup concluded, on the basis of detailed historical and field studies, that past agricultural improvements had occurred as a result of population pressure. In order to feed more people, farmers had developed new ways to use their land and labor more intensively. In other words, the **Boserup thesis** is that population growth acts as a stimulus, not a deterrent, to development. American economist Julian Simon went further, arguing that resources do not exist in nature, but are created by human ingenuity. For cornucopians like Simon, more people means more scientists and innovations. Since the time of Malthus, they observed, world population had grown from 900 million to 7 billion without the predicted dire consequences—proof that Malthus failed to recognize the importance of technology in raising the carrying capacity of the Earth. Still higher population numbers, they suggest, are sustainable, perhaps even with improved standards of living for all.

An intermediate view admits that products of human ingenuity such as the Green Revolution (Chapter 9) allowed food production to keep pace with rapid population growth, but warns that continued gains in food production technology are not guaranteed. Both complacency and inadequate research support have hindered continuing progress in recent years. And even if further advances are made, they observe, not all countries or regions have the social and political will or capacity to take advantage of them. Those that do not, will fail to keep pace with the needs of their populace and will sink into varying degrees of poverty and environmental decay, creating national and regional—though not necessarily global—crises.

5.8 Population Prospects

Regardless of population philosophies, theories, or cultural norms, the fact remains that many or most developing countries are showing significantly declining population growth rates. Global fertility and birth rates appear to be falling to an extent not anticipated by pessimistic neo-Malthusians and at a pace that suggests a peaking of world population numbers sooner—and at lower totals—than previously projected (see "A Population Implosion?" p. 119). In all world regions, steady fertility declines have been recorded over the past years, reducing fertility from global levels of 5 children per woman in the early 1950s to 2.5 per woman today. Most future population growth in developed countries will be due to momentum from past high fertility, and aging will be an inevitable consequence of the recent changes in fertility.

Momentum

A bike will continue to coast long after the rider stops pedaling. Similarly, reducing fertility levels even to the replacement level of about 2.1 births per woman does not mean an immediate end to population growth. Because of the age composition of many societies, numbers of births will continue to grow even as fertility rates decline. The reason is to be found in **population (demographic) momentum,** and the key to that is the age structure of a country's population.

When a high proportion of the population is young, the product of past high fertility rates, larger and larger numbers enter the childbearing age each year; that is the case for major parts of the world today. The populations of developing countries are far younger than those of the established, industrially developed regions (see Figure 5.11), with over 40% in Africa below the age of 15. The consequences of the fertility of these young people have yet to be realized. A population with a greater number of young people tends to grow rapidly, regardless of the number of children per woman. The results will continue to be felt as the large youthful cohorts mature and work their way through the population pyramid.

Inevitably, while this is happening, even the most stringent national policies limiting growth cannot stop it entirely. A country with a large population of young people will experience large numerical increases despite declining birth rates. Indeed, the higher fertility is to begin with and the sharper its drop to low levels, the greater will be the role of momentum even after rates drop below replacement. For example, Iran has recently implemented strong population policies that have lowered fertility rates to 1.8, well below replacement values. However, due to momentum, the population is projected to grow from 79 to 99 million by 2050. Increasingly, population momentum will be responsible for nearly all of the world's population growth (**Figure 5.28**).

Aging

An aging population is one with a greater proportion of older persons. Aging populations are an inevitable consequence of the decreased fertility rates and increased life expectancies that accompany the demographic transition. Population aging is unprecedented in human history and has profound social and economic consequences. The problems of a rapidly aging population that already confront the industrialized economies are now being realized in the developing world as well. Throughout human history, young people outnumbered the elderly; sometime between

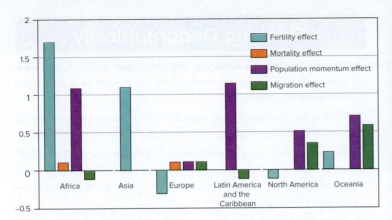

Figure 5.28 **Components of population change, 2010–2020.** Population growth can occur due to high fertility rates, decreasing mortality rates, migration, or the momentum effect. Population momentum stems from large numbers of people of childbearing age as a result of high fertility rates several decades ago. In Asia and Latin America, population growth is due solely to population momentum. In North America and Oceania, population momentum is the largest contributor to growth.

2015 and 2020, persons 65 years and older will outnumber those under 5 years of age. Europe is the "oldest" world region and Africa is the youngest. Japan leads the world in the percentage of its population aged 65 and older at 26%. Italy and Germany are close behind.

The progression toward older populations is considered irreversible; the youthful majorities of the past are unlikely to occur again. By 2050, the UN projects, one out of five persons worldwide will be 60 years old or older.

In the developing world, older persons are projected to make up 20% of the population by 2050, in contrast to the 8% over age 60 there in 2000. Because the demographic transition took place more quickly, the pace of aging is much faster in the developing countries. Thus, they will have less time than the developed world did to adjust to the consequences of that aging. And those consequences will be experienced at lower levels of personal and national income and economic strength.

In both rich and poor countries, the working-age populations will face increasing burdens and obligations. The potential support ratio (PSR)—the number of persons aged 15 to 64 years per one citizen aged 65 or older—has steadily fallen. Between 1950 and 2000, it dropped from 12 to 9 workers for each older person; by mid-century, the PSR is projected to drop to 4. The implications for Social Security programs, health insurance, government finances, and social support obligations are serious. Exacerbating the strain on resources, the sharp increase in the oldest population (80 years or older) will require greater expenditures for health and long-term care. The consequences of population aging appear most intractable for the world's poorest developing states that generally lack health, income, housing, and social service support systems adequate for the needs of their older citizens. Therefore, to the social and economic implications of their present population momentum, developing countries must add the aging consequences of past patterns and rates of growth (**Figure 5.29**).

Figure 5.29 These senior residents of a Moroccan nursing home are part of the rapidly aging population found in many developing countries. Worldwide, the over-60 cohort will number some 22% of total population by 2050 and will be larger than the number of children less than 15 years of age. But by 2020, China will have as large a share of its population over 60—about one in four—as will Europe. Already the numbers of old people in the world's poorer countries are beginning to dwarf those in the rich world. There are nearly twice as many persons over 60 in developing countries as in the advanced ones, but most are without the old-age assistance and welfare programs that the developed countries have put in place. © *Nathan Benn/Corbis Documentary/Getty Images.*

Summary of Key Concepts

- Birth, death, fertility, migration, and growth rates are basic for understanding the numbers, composition, distribution, and spatial trends of population.
- Predictions of ever-increasing population numbers now appear faulty. Birth rates, already below replacement levels in most developed countries, are dropping as well in less-developed regions. Even so, essentially all future numerical increases will occur in areas currently considered underdeveloped where total fertility rates still remain significantly above the replacement level.
- Death rates everywhere have declined steeply as the availability of preventive and curative health care has spread to all parts of the world. Reductions in birth and death rates result in a *demographic transition* from high to low levels of both and a corresponding stabilization or reduction in numbers regionally and globally. Areal numbers are also determined by a *demographic equation,* which includes the effects of population relocations and migrations.
- People are unevenly distributed globally, latitudinally, and areally. The Northern Hemisphere, midlatitudes, lowlands, and coastal locations house two-thirds of the world's people on less than 10% of its land. East Asia, South Asia, Europe, and eastern North America together contain two-thirds of the total. Urban areas everywhere are growing and now hold more than one-half of the total world population.
- Various measures of density have been devised, but differences in population numbers and densities have no necessary correlation

with observations about over- or underpopulation; these are concepts relative to means of support, not to absolute numbers.

- Population projections are based on assumptions about future conditions, which may not be realized. The more distant the future for which projections are made, the less accurate they are likely to be.

- All projections must take into account cultural factors such as religion, education, and the status of women that influence individual and group reproduction decisions. Future population levels will be significantly affected by the current age and sex structure summarized by a region's *population pyramid*. That structure influences the inevitable growth through the *population momentum* implicit in a youthful majority or the numerical decline suggested by regional and global aging of populations.

Human populations cannot be understood solely through statistical analysis. Societies are distinguished not just by the abstract data of their numbers, rates, and trends but also by the experiences, beliefs, understandings, and aspirations that collectively constitute the human spatial and behavioral variable called *culture*. It is to that fundamental human diversity that we next turn our attention.

Key Words

agricultural density 134	Malthus 136
Boserup thesis 138	neo-Malthusianism 137
carrying capacity 134	nonecumene 133
cohort 113	overpopulation 134
crude birth rate (CBR) 113	physiological density 134
crude death rate (CDR)	population density 133
(mortality rate) 117	population geography 111
crude density (arithmetic	population (demographic)
density) 133	momentum 138
demographic equation 130	population projection 136
demographic transition 125	population pyramid 119
demography 111	rate 113
dependency ratio 122	rate of natural increase 123
doubling time 123	replacement level 119
ecumene 133	S-curve 137
food security 135	total fertility rate (TFR) 114
homeostatic plateau 137	zero population growth (ZPG) 119
J-curve 125	

Thinking Geographically

1. How do the *crude birth rate* and the *fertility rate* differ? Which measure is the more accurate statement of the amount of reproduction occurring in a population?

2. How is the *crude death rate* calculated? What factors account for the worldwide decline in death rates since 1945?

3. How is a *population pyramid* constructed? What shape of "pyramid" reflects the structure of a rapidly growing country? Of a population with a slow rate of growth? What can we tell about future population numbers from those shapes?

4. What variations do we discern in the spatial pattern of the *rate of natural increase* and, consequently, of population growth? What rate of natural increase would double population in 35 years?

5. How are population numbers projected from present conditions? Are projections the same as predictions? If not, in what ways do they differ?

6. Describe the stages in the *demographic transition*. Where has the final stage of the transition been achieved? What appears to be the applicability of the demographic transition to other parts of the world?

7. Contrast *crude population density, physiological density,* and *agricultural density.* For what different purposes might each be useful? How is *carrying capacity* related to the concept of density?

8. What was Malthus's underlying assumption concerning the relationship between population growth and food supply? In what ways do the arguments of *neo-Malthusians* differ from the original doctrine? What government policies are implicit in neo-Malthusianism?

9. Why is *population momentum* a matter of interest in population projections? In which world areas are the implications of demographic momentum most serious in calculating population growth, stability, or decline?

Hindu pilgrims worship at dawn in the sacred Ganga (Ganges) River at Varanasi, India. © Porterfield/Chickering/Science Source

CHAPTER OUTLINE

LEARNING OBJECTIVES

After studying this chapter you should be able to:

6.1   Explain how social scientists define culture.

6.2   Explain how the technological, sociological, and ideological subsystems of culture are distinct, and give an example of how are they integrated.

6.3   Compare the three major frameworks that social scientists have used to understand the relationship between culture and environment.

6.4   Explain how ethnicity differs from race.

6.5   Explain how globalization has altered the spatial distribution of languages.

6.6   Identify the places of origin for the world's major religions, and trace their paths of diffusion.

The Crow country. The Great Spirit put it exactly in the right place; while you are in it, you fare well; whenever you get out of it, whichever way you travel, you fare worse. . . . The Crow country is in exactly the right place. It has snowy mountains and sunny plains; all kinds of climates and good things for every season. When the summer heats scorch the prairies, you can draw up under the mountains, where the air is sweet and cool. . . . In the autumn when your horses are fat and strong from the mountain pastures, you can go down on the plains and hunt the buffalo or trap beaver on the streams. And when winter comes on, you can take shelter in the woody bottoms along the rivers. The Crow country is exactly in the right place. Everything good is found there. There is no country like the Crow country.

Such was the opinion of Arapoosh, chief of the Crows, speaking of the Big Horn basin country of Wyoming in the early 19th century. In contrast, in the 1860s, Captain Raynolds reported to the Secretary of War that the basin was "repelling in all its characteristics, surrounded on all sides by mountain ridges [and offering] but few agricultural advantages."

In Chapters 3 and 4, our primary concern was with the physical landscape. The approximately 7 billion people who were the subject of Chapter 5 are of a single human family, but it is a family differentiated into many branches, each characterized by a distinctive *culture*. Arapoosh and Raynolds viewed the same physical landscape, but from the standpoints of their separate cultures and conditioning. The world is a mosaic of culture groups that invite geographic study. Culture is like looking through tinted glass, affecting and distorting our view of the Earth. Culture conditions the way people think about the land, the way they use and alter the land, and the way they interact with one another on the land. Such conditioning is the focus of the culture-environment tradition of geography—a tradition concerned with the landscape, but not in the physical science sense. In this chapter, landscapes take on an added dimension and become cultural rather than purely physical.

To some writers and commentators, *culture* means the arts (literature, painting, music, etc.). To a social scientist, **culture** is the specialized behavioral patterns, understandings, and adaptations that summarize the way of life of a group of people. In this broader sense, culture is as much a part of the regional differentiation of the Earth as are topography, climate, and other aspects of the physical environment. The visible and invisible evidences of culture—buildings and farming patterns, language and political organization—are elements in the spatial diversity that invites and is subject to geographic inquiry. Cultural differences in area result in human landscapes with variations as subtle as the differing "feel" of urban Paris, Moscow, and New York or as obvious as the sharp contrasts of rural Zimbabwe and a cash grain farm in America's Midwest (**Figure 6.1**).

(a)

(b)

Figure 6.1   **Cultural and economic contrasts** are clearly evident between **(a)** a subsistence maize plot in Zimbabwe and **(b)** the extensive fields and mechanized farming of the U.S. Midwest. *(a) © Ian Murphy/The Image Bank/Getty Images; (b) Source: USDA-Natural Resources Conservation Service.*

Because such differences exist, cultural geography exists, and one branch of cultural geography addresses a whole range of "Why?" and "What?" and "How?" questions. Why, since humankind constitutes a single species, are cultures so varied? What are the most pronounced ways in which cultures and culture regions are distinguished? What were the origins of the different culture regions we now observe? Where did culture traits and innovations develop and how did they diffuse over a wider portion of the globe? Why do cultural contrasts persist between recognizably distinct groups even in such presumed "melting pot" societies as that of the United States or in the outwardly homogeneous, long-established countries of Europe? How is knowledge about cultural differences important to us today? These and similar questions are the concerns of cultural geography.

6.1 Components of Culture

Culture is transmitted within a society to succeeding generations by imitation, instruction, and example. It is learned, not biological, and has nothing to do with instinct or with genes. As members of a social group, individuals acquire integrated sets of behavioral patterns, environmental and social perceptions, and knowledge of existing technologies. Of necessity, we learn the culture in which we are born and reared. But we need not—indeed, cannot—learn its totality. Age, sex, status, and occupation dictate the aspects of the cultural whole in which we become fully indoctrinated.

A culture displays a social structure—a framework of roles and interrelationships of individuals and established groups. Despite overall generalized and identifying characteristics and even an outward appearance of uniformity, a culture is not homogeneous. The "American" culture, for example, contains innumerable complex, composite, and often competing subgroups: farmers and city dwellers; females and males; teenagers and retirees; liberals and conservatives; owners and employees; members of different religious, political, social, or other formal organizations; and the like. Each individual learns and is expected to adhere to the rules and conventions not only of the culture as a whole but also specific to the subgroups to which he or she belongs. And those subgroups may have their own recognized social structures.

Culture is a complexly interlocked web of behaviors and attitudes. Its full and diverse content cannot be appreciated, and in fact may be wholly misunderstood, if we concentrate our attention only on limited, obvious traits. Distinctive eating utensils, the use of gestures, or the ritual of religious ceremony may characterize a culture for the casual observer. These are, however, individually insignificant parts of a much more complex structure that can be appreciated only when the whole is experienced.

Out of the richness and intricacy of human life we seek to isolate for special study those more-fundamental cultural variables that give structure and spatial order to societies. We begin with culture traits, the smallest distinctive items of culture. **Culture traits** are units of learned behavior ranging from the language spoken to the tools used or to the games played. A trait may be an object (a fishhook, for example), a technique (weaving and knotting of a fish net), a belief (in the spirits resident in water bodies), or an attitude (a conviction that fish is superior to other animal flesh). Of course, the same trait—the Christian religion, perhaps, or the

Spanish language—may be part of more than one culture. Traits are the most elementary expressions of culture, the building blocks of the complex behavioral patterns of distinctive groups of peoples.

Individual culture traits that are functionally interrelated constitute a **culture complex.** The existence of such complexes is universal. At one time, keeping cattle was a *culture trait* of the Masai of Kenya and Tanzania. Related traits included the measurement of personal wealth by the number of cattle owned; a diet containing the milk and blood of cattle; and disdain for labor unrelated to herding. The assemblage of these and other related traits yielded a *culture complex* descriptive of one aspect of Masai society (**Figure 6.2**). In exactly the same way, religious complexes, business behavior complexes, sports complexes, and others can easily be recognized in American or any other society.

A **culture system** is a broader generalization and refers to a collection of interacting culture traits and culture complexes that are shared by a group within a particular territory. Multiethnic societies, perhaps further subdivided by linguistic differences, varied food preferences, and a host of other internal differentiations, may nonetheless share enough joint characteristics to be recognizably distinctive cultural entities to themselves and to others. Certainly, citizens of the "melting pot" United States would identify themselves as *Americans,* together constituting a unique culture system on the world scene.

Culture traits, complexes, and systems have spatial extent. When they are plotted on maps, the regional character of the components of a culture is revealed. Geographers are interested in the spatial distribution of these individual elements, but their usual concern is with the **culture region,** a portion of the Earth's surface occupied by people sharing recognizable and distinctive cultural

Figure 6.2 **The formerly migratory Masai of eastern Africa are now largely sedentary, partially urbanized, and frequently owners of fenced farms.** Cattle formed the traditional basis of Masai culture and were the evidence of wealth and social status. They also provided the milk and blood important in the Masai diet. © McGraw-Hill Education/Barry Barker RF.

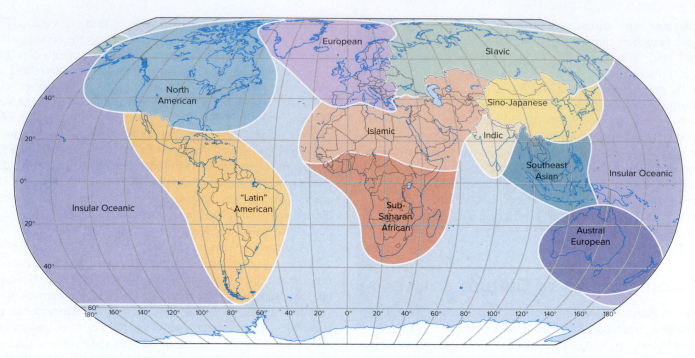

Figure 6.3 **Culture realms of the modern world.** Culture realms, regions, systems, complexes, and traits are all elements in the spatial hierarchy of cultural geography. This map proposes just one of many possible subdivisions of the world into multifactor culture realms.

characteristics. Examples include the political organizations societies devise, their religions, their form of economy, and even their clothing, eating utensils, and housing. There are as many such culture regions as there are separate culture traits and complexes of population groups.

Finally, a set of culture regions showing related culture complexes and landscapes may be grouped to form a **culture realm.** The term recognizes a large segment of the Earth's surface having basic uniformity in its cultural characteristics and significant differences from adjacent realms. Culture realms are, in a sense, culture regions at the broadest scale of generalization. In fact, the scale is so broad and the diversity within the recognized realms so great that the very concept of realm may mislead more than it informs.

Indeed, the current validity of distinctive culture realms such as those suggested in **Figure 6.3** is challenged by the assumed globalization of all aspects of human society and economy. The result of that globalization, some observers maintain, is a homogenization of cultures as economies are integrated and uniform consumer demands are satisfied by standardized commodities produced by international corporations. Others insist, however, that the world is far from homogenized and observe that globalization continues to be countered by powerful forces of regionalism, place identity, and ethnicity. One of the many possible divisions of human cultures into realms is offered in Figure 6.3. Culture realms, regions, systems, complexes, and traits are all elements in the spatial hierarchy of cultural geography.

6.2 Subsystems of Culture

Understanding a culture fully is, perhaps, impossible for one who is not part of it. For analytic purposes, however, the traits and complexes of culture—its building blocks and expressions—can be grouped and examined as subsets of the whole. The anthropologist Leslie White suggested that a culture can be viewed as a three-part structure composed of subsystems that he termed *technological, sociological,* and *ideological.* Together, the subsystems constitute the structure of culture as a whole. The subsystems are integrated; each reacts to the others and is affected by them in turn.

The **technological subsystem** is composed of the material objects, also known as *artifacts*, and the techniques of their use. Such objects include the tools and other instruments that enable us to feed, clothe, house, defend, transport, and amuse ourselves. The **sociological subsystem** of a culture is the sum of those expected and accepted patterns of interpersonal relations that find their outlet in economic, political, military, religious, kinship, and other associations. The **ideological subsystem** consists of the ideas, beliefs, and knowledge of a culture and of the ways in which they are expressed in speech or other forms of communication.

The Technological Subsystem

Examination of variations both in culture and in the manner of human existence from place to place centers on a series of commonplace questions: How do the people in an area make a living? What resources and what tools—what artifacts—do they use to feed, clothe, and house themselves? Is a larger percentage of the population engaged in agriculture than in manufacturing? Do people travel to work in cars, on bicycles, or on foot? Do they shop for food or grow their own?

These questions concern the adaptive strategies used by different cultures in "making a living." In a broad sense, they address the technological subsystems at the disposal of those cultures—the instruments and tools people use in the daily cycle of existence. For most of human history, people lived by hunting and gathering,

CHACO CANYON DESOLATION

It is not certain when they first came, but by A.D. 1000 the Anasazi people were building a flourishing civilization in what are now the states of Arizona and New Mexico. They were corn farmers thriving during the 300 years or so of the medieval warm period beginning about A.D. 900 in the American Southwest. In Chaco Canyon alone, they erected as many as 75 towns, all centered around pueblos, huge stone-and-adobe apartment buildings as tall as five stories and with as many as 800 rooms. These were the largest and tallest buildings of North America prior to the construction of iron-framed "cloud scrapers" in major cities at the end of the 19th century. An elaborate network of roads and irrigation canals connected and supported the pueblos. At about A.D. 1200, the settlements were abruptly abandoned. The Anasazi, advanced in their skills of agriculture and communal dwelling, were—according to some scholars—forced to move on by the ecological disaster their pressures had brought to a fragile environment.

They needed forests for fuel and for the hundreds of thousands of logs used as beams and bulwarks in their dwellings. The pinyon-juniper woodland of the canyon was quickly depleted. For larger timbers needed for construction, they first harvested stands of ponderosa pine found some 40 kilometers (25 mi) away. As early as A.D. 1030, these, too, were exhausted, and the community switched to spruce and Douglas fir from distant mountaintops surrounding the canyon. When they were gone by 1200, the Anasazi's fate was sealed—not only by the loss of forest but also by the irreversible ecological changes deforestation and agriculture had wrought. With forest loss came erosion that destroyed the topsoil. The surface water channels that had been built for irrigation were deepened by accelerated erosion, converting them into enlarging arroyos, useless for agriculture.

The material roots of their culture destroyed, the Anasazi turned on themselves and warfare convulsed the region. Smaller groups sought refuge elsewhere, recreating on a reduced scale their pueblo way of life, but now in nearly inaccessible, highly defensible mesa and cliff locations. The destruction they had wrought destroyed the Anasazi in turn.

© Corbis RF.

---

taking the bounty of nature with only minimal dependence on weaponry, implements, and the controlled use of fire. Their adaptive skills were great, but their technological level was low. They had few specialized tools, could exploit only a limited range of potential resources, and had little or no control of nonhuman sources of energy. Their impact on the environment was small, but at the same time the "carrying capacity" of the land discussed in Chapter 5 was low everywhere, for technologies and artifacts were essentially the same among all groups.

The retreat of the last glaciers about 12,000 to 13,000 years ago marked the start of a period of unprecedented cultural development. It led from primitive hunting and gathering economies at the outset through the evolution of agriculture and animal husbandry to, ultimately, urbanization, industrialization, and the intricate complexity of the modern technological subsystem. Since not all cultures passed through all stages at the same time, or even at all, *cultural divergence* between human groups became evident.

Cultural diversity among ancient societies reflected the proliferation of technologies that followed a more assured food supply and made possible a more intensive and extensive utilization of resources. Different groups in separate environmental circumstances developed specialized tools and behaviors to exploit resources they recognized. Beginning with the Industrial Revolution of the 18th century, however, a reverse trend—toward commonality of technology—began.

Today, advanced societies are nearly indistinguishable in the tools and techniques at their command. They have experienced *cultural convergence*—the sharing of technologies, organizational structures, and even culture traits and artifacts that is so evident among widely separated societies in a modern world united by instantaneous communication and efficient transportation. The differences in technology that still exist between developed and underdeveloped societies reflect, in part, national and personal wealth, stage of economic advancement and complexity, and, importantly, the level and type of energy used (**Figure 6.4**).

In technologically advanced countries, many people are employed in manufacturing or services. Per capita incomes tend

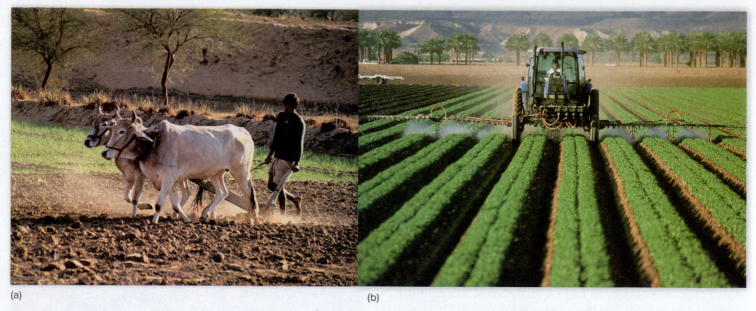

(a)   (b)

Figure 6.4   Contrasting Technological Subsystems. (a) This Rajasthan, India, farmer working with draft animals employs tools typical of the low technological levels of subsistence economies. **(b)** Cultures with advanced technological subsystems increase production by using complex machinery, fossil fuels, and chemicals such as the pesticides being applied in this photograph. *(a) © Santokh Kochar/Getty Images RF; (b) Source: Photo by Jeff Vanuga, USDA Natural Resources Conservation Service RF.*

to be high, as do levels of education and nutrition, life expectancies, and medical services. These countries wield great economic and political power and have access to the latest technologies. These technologically advanced countries are classified by the United Nations as *developed economies.* In contrast, technologically less-advanced countries have a high percentage of people engaged in farming, with much of the agriculture at a subsistence level (**Figure 6.5**). Per capita incomes, life expectancies, literacy rates, technology levels, and even Internet access rates also tend to be low (**Figure 6.6**). The United Nations refers to these less-advanced countries as *developing economies* and also classifies 52 countries as *least developed countries.*

Labels such as *advanced–less-advanced, developed–developing,* and *industrial–nonindustrial* can be misinterpreted to mean "superior–inferior." This belief is totally improper because the terms relate solely to economic and technological circumstances and bear no qualitative relationship to such vital aspects of culture as music, art, religion, and personal relationships.

Properly understood, however, terms and measures of economic development can reveal important national and world regional contrasts in the technological subsystems of different cultures and societies. **Figure 6.7** suggests that technological status is relatively high in nearly all European countries and Japan, the United States, and Canada—the "North." Most of the less-developed countries are in Latin America, Africa, and southern Asia—the "Global South."[1]

It is important, however, to recognize that these national averages conceal internal contrasts. All countries include areas that are

at different levels of development. We must also remember that technological development is a dynamic concept. It is most useful and accurate to think of the countries of the world as arrayed along an ever-changing continuum of technological levels and subsystems.

The Sociological Subsystem

Continuum and change also characterize the religious, political, formal and informal educational, and other institutions that constitute the sociological subsystem of culture. Together, these *sociofacts* define the social organizations of a culture. They regulate how the individual functions relative to the group, whether it be family, religious institution, or state.

There are no "givens" as far as patterns of interaction in any of these associations are concerned, except that most cultures possess a variety of formal and informal ways of structuring behavior. The importance to the society of the differing behavior sets varies among, and constitutes obvious differences between, cultures. Differing patterns of behavior are learned expressions of culture and are transmitted from one generation to the next by formal instruction or by example and expectation (**Figure 6.8**).

Social institutions are closely related to the technological system of a culture group. Thus, hunter-gatherers have one set of institutions; industrial societies have quite different ones. Preagricultural societies tended to be composed of small bands based on kinship ties, with little social differentiation or specialization of function in the band; the San and Hadza (Bushmen) of Africa and isolated

---

[1] The use of the terms *Global North* and *Global South* to describe relative developmental levels was introduced in *North–South: A Programme for Survival* (sometimes called the *Brandt Report*), published in 1980 by the Independent Commission on International Development Issues. The former Soviet Union at that time was included within the North; since its breakup in 1991, Georgia, Uzbekistan, and other former Soviet republics in Asia have also been classified as economies in transition by the United Nations. Likewise, the countries formed after the breakup of Yugoslavia are part of the Global North and are classified by the United Nations as economies in transition.

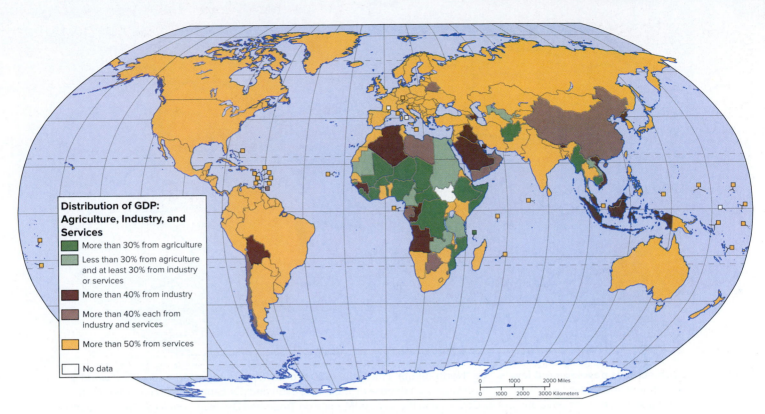

Figure 6.5 **Distribution of gross domestic product among agriculture, industry, and services.** The percentage of a country's economic activity devoted to agriculture, industry, and services is a good measure of its stage of development. Less-developed countries often have most of their workers in the agricultural sector. Highly developed economies usually are focused on manufacturing or services and have very low proportions of their labor forces in the agricultural sector. *Source: Sutton, Atlas of World Geography 8e, MAP 56, p. 76.*

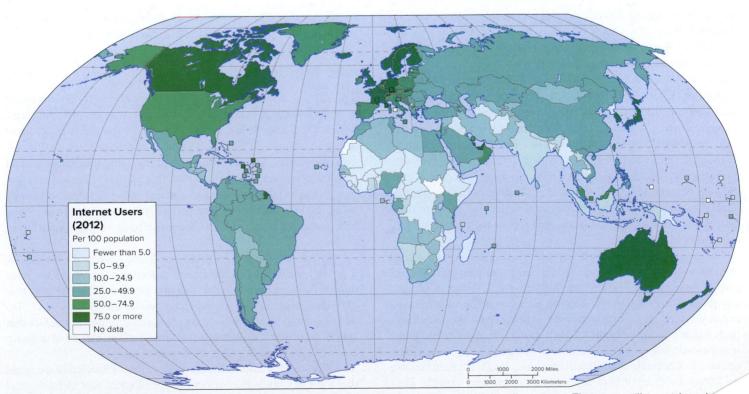

Figure 6.6 **Internet use** has diffused to all corners of the Earth, yet its diffusion is geographically uneven. There are still countries where less than 5% of the population has used the Internet. *Source: Sutton, Student Atlas of World Geography 8e, Map 70, p. 90.*

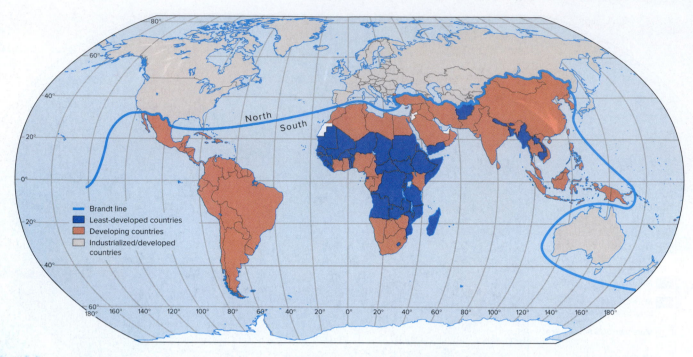

Figure 6.7   Comparative development levels. The "North–South" line of the 1980 *Brandt Report* suggested a simplified world contrast of development and underdevelopment based largely on degree of industrialization and per capita wealth recorded then. As of 2016, the United Nations Economic and Social Council and the UN Conference on Trade and Development (UNCTAD) recognized 48 "least-developed countries." That recognition now reflects low ratings in three criteria: gross domestic income per capita, "quality of life" indicators, and vulnerability to economic disruptions and natural disasters. The broad category of "developing countries" ignores recent significant economic and social gains in several Asian and Latin American states, raising them now to "industrialized/developed" status. Some "least-developed" states are small island countries not shown at this map scale. *Sources: UNCTAD and United Nations Development Programme.*

rain forest groups in Amazonia might serve as modern examples (**Figure 6.9**).

The revolution in food production occasioned by plant and animal domestication beginning around 10,000 years ago touched off a social transformation that included increases in population, urbanization, work specialization, and structural differentiation within the society. Politically, the rules and institutions by which people were governed changed with the formation of sedentary, agricultural societies. Loyalty was transferred from the kinship group to the state; resources became possessions rather than the common property of all. Equally far-reaching changes occurred after the 18th-century Industrial Revolution, leading to the complex of human social organizations that we experience and are controlled by today in "developed" states and that increasingly affect all cultures everywhere.

Culture is a complexly intertwined whole. Each organizational form or institution affects, and is affected by, related culture traits and complexes in intricate and variable ways. Systems of land and property ownership and control, for example, are institutional expressions of the sociological subsystem. They are simultaneously explicitly central to the classification of economies and to the understanding of spatial and structural patterns of economic development, as Chapters 9 and 10 will examine. Again, for each country the adopted system of laws and justice is a cultural variable identified with the sociological subsystem but extending its influence to all aspects of economic and social organization, including the political geographic systems discussed in Chapter 8.

The Ideological Subsystem

The third class of elements defining and identifying a culture is the ideological subsystem. This subsystem consists of ideas, beliefs, and knowledge, as well as the ways we express these things in our speech and other forms of communication. Mythologies, theologies, legend, literature, philosophy, folk wisdom, and commonsense knowledge make up this category. Passed on from generation to generation, these abstract belief systems, or *mentifacts,* tell us what we ought to believe, what we should value, and how we ought to act. Beliefs form the basis of the socialization process.

Often, we know—or think we know—what the beliefs of a group are from written sources. Sometimes, however, we must depend on the actions or objectives of a group to tell us what its true ideas and values are. "Actions speak louder than words" and "Do as I say, not as I do" are commonplace recognitions of the fact that actions and words do not always coincide. The values of a group cannot be deduced from the written record alone.

Nothing in a culture stands totally alone. Changes in the ideas that a society holds may affect the sociological and technological systems, just as, for example, changes in technology force changes

Figure 6.8   **All societies prepare their children for membership in the culture group.** Culture is what is learned, not inherited. In each of these settings, certain values, beliefs, skills, and proper ways of acting are being transmitted to young people. *(a) © Jonathan Kingston/Getty Images; (b) © gulfimages/Alamy Stock Photo; (c) © Sidney Bahrt/Science Source; (d) © Ryan McVay/Getty Images RF.*

in the social system. The abrupt alteration after World War I (1914–1918) of the ideological structure of Russia from a monarchical, agrarian, capitalist system to an industrialized, communist society involved sudden alteration in all facets of that country's culture system. The equally abrupt disintegration of Russian communism in the early 1990s was similarly disruptive of all its established economic, social, and administrative organizations. The interlocking nature of all aspects of a culture is termed **cultural integration.**

The recognition of three distinctive subsystems of culture, helping us appreciate its structure and complexity, can at the same time obscure the many-sided nature of individual elements *cultural integration* means that any cultural object or act may have a number of meanings. Clothing, for example, serves as bodily protection appropriate to climatic conditions, and techniques, or the activity in which the

wearer is engaged. But garments, houses, and other artifacts also may be sociofacts, identifying an individual's role and status within the social structure of the community or culture, and mentifacts, evoking larger community value systems (**Figure 6.10**).

6.3 Interaction of People and Environment

Culture develops in a physical environment that, in its way, contributes to differences among people. In premodern subsistence societies, the acquisition of food, shelter, and clothing—all parts of culture—depends on the utilization of the natural resources at hand. The interrelations of people with the environment of a given area, their perceptions and use of it, and their impact on it are interwoven themes of geography. They are the special concerns

Figure 6.9   Hunter-gatherers practiced the most enduring lifestyle in human history, trading it for the more arduous life of farmers in order to provide larger quantities of food for a growing population. Unlike agricultural or urban societies, there is little social or economic stratification among hunter-gatherers. Instead, status and the division of labor are based on one's age and sex. This Hadza (Bushman) youth from Tanzania is a member of one of the very few groups in Africa or the world still pursuing this ancient lifestyle. The Hadza grew no food, raise no livestock, and live in a way that has changed little in 10,000 years. © *ranplett/Vetta/Getty Images.*

of those geographers exploring **cultural ecology,** the study of the relationship between a culture group and the natural environment it occupies.

Geographers and anthropologists see evidence that subsistence pastoralists, hunter-gatherers, and gardeners adapted their productive activities—and, by extension, their social organizations and relationships—to the specific physical limitations of their different local habitats. Presumably, similar natural environmental conditions influenced the development of similar adaptive responses and cultural outcomes in distant, unconnected locales. That initial influence, of course, does not predetermine the details of the subsequent culture.

Environments as Controls

At first glance, the fact that most of the world's least-developed countries are located in the Tropics while most of the developed countries are in the middle or high latitudes suggests that that the environment played a key role in cultural development. However, geographers have long dismissed as invalid and intellectually limiting the ideas of **environmental determinism.** Environmental determinism, which peaked in popularity in the late 19th and early 20th centuries, taught that the physical environment—in particular, the climate—determined which cultures would become the most advanced and economically developed. However, the cultural variations that occur around the world are not determined by a society's physical surroundings. Levels of technology, systems of social organization, and ideas about what is true and right have no obvious relationship to environmental circumstances. Development differences between the North and the Global South have many causes, including the history of colonialism and the slave trade.

The environment does place certain limitations on the human use of territory. However, such limitations must be seen not as absolute, enduring restrictions but as relative to technologies, cost considerations, national aspirations, and linkages with the larger world. Human choices in the use of landscapes are affected by group perception of the possibility and desirability of their settlement and exploitation. These are not circumstances inherent in the land.

Possibilism is the viewpoint that people, not environments, are the dynamic forces of cultural development. The needs, traditions, and technological level of a culture affect how that culture assesses the possibilities of an area and shapes the choices that it makes regarding them. As shown in the chapter opening story about Crow country, each society uses natural resources in accordance with its culture. Changes in a group's technical abilities or objectives bring about changes in its perceptions of the usefulness of the land. Of course, there are some environmental limitations on use of area. For example, if resources for feeding, clothing, or housing ourselves within an area are lacking, or if we do not recognize them there, there is no inducement for people to occupy that territory. Environments that do contain such recognized resources provide the framework within which a culture operates.

Human Impacts

People are also able to modify their environment, and this is the other half of the human-environmental relationship studied by geographers. Geography, including cultural geography, examines both the reactions of people to the physical environment and their impact on that environment. By using it, we modify our environment—in part, through the material objects we place on the landscape: cities, farms, roads, and so on (**Figure 6.11**). The form these take is the product of the kind of culture group in which we live. The **cultural landscape,** the Earth's surface as modified by human action, is the tangible, physical record of a given culture. House types, transportation networks, parks and

(a)

(c)

(b)

Figure 6.10 (**a**) Houses are important artifacts, providing shelter for their occupants. These traditional houses on Nias Island off the west coast of Sumatra, Indonesia, reflect an ingenious cultural adaptation to a hot, humid tropical climate. The elevated floor keeps occupants dry in heavy downpours and allows cooling breezes to circulate. The steep roof sheds rain and shades the interior from the tropical sun. (**b**) Houses are also sociofacts, reflecting the family, kinship, and communal ideals of a culture. This landscape of detached, single-family houses and backyard swimming pools in Las Vegas, Nevada, reflects social relations in the United States, with an emphasis on individualism, privacy, and the nuclear family. (**c**) Houses are also mentifacts, reflecting a culture group's ideas about appropriate design, orientation, and building materials. The White House in Washington, D.C., reflects cultural ideals of beauty and symmetry drawn from classical Greek architecture and Palladian and Georgian houses in Europe. *(a) © Jane Sweeney/Getty Images; (b) Source: Lynn Betts/USDA Natural Resources Conservation Service; (c) © PhotoDisc/Getty Images RF.*

cemeteries, and the size and distribution of settlements are among the indicators of the use that humans have made of the land.

As a rule, the more technologically advanced and complex the culture, the greater its impact on the environment, although preindustrial societies can and frequently do exert destructive pressures on the lands they occupy (see "Chaco Canyon Desolation," p. 145). In sprawling urban industrial societies, the cultural landscape has come to outweigh the natural physical environment in its impact on people's daily lives. It interposes itself between "nature" and humans. Residents of the cities of such societies—living and working in climate-controlled buildings, driving to enclosed shopping malls, surfing websites from around the world—can go through life having very little contact with or concern about the physical environment.

6.4 Culture Change

The recurring theme of cultural geography is change. No culture is, or has been, characterized by a permanently fixed set of

material objects, systems of organization, or ideologies, although all of these may be long-enduring within a stable, isolated society. Such isolation and stability has always been rare. On the whole, although cultures are essentially conservative, they are always in a state of flux.

Many individual changes, of course, are so slight that they initially are almost unnoticed, though collectively they may substantially alter the affected culture. Think of how the culture of the United States differs today from what it was in 1940—not in essentials, perhaps, but in the innumerable electric, electronic, and transportation technologies that have been introduced and in the recreational, social, and behavioral adjustments they and other technological changes have wrought. Among the latter have been shifts in employment patterns to include greater participation by women in the workforce and associated changes in attitudes toward the role of women in the society at large. Such cumulative changes occur because the culture traits of any group are not independent; they are clustered in a coherent and integrated pattern. Change on a small scale will have wide repercussions as associated traits also change

Figure 6.11   The physical and cultural landscapes in juxtaposition. Advanced societies are capable of so altering the circumstances of nature that the cultural landscapes become controlling environments. The city of Los Angeles, California, is a built environment constructed atop and often obscuring its physical surroundings. Consider how much this landscape has changed with human settlement. © *Robert Landau/Corbis/Getty Images RF.*

to accommodate the adopted adjustment. Change, both major and minor, within cultures is induced by *innovation, spatial diffusion,* and *acculturation.*

Innovation

Innovation implies changes to a culture that result from ideas created within the social group itself. The innovation may be an improvement in material technology, such as the bow and arrow or the jet engine. It may involve the development of non-material forms of social structure and interaction: feudalism or Christianity, for example.

Premodern and traditional societies characteristically are slower to innovate or accept change. In societies at equilibrium with their environment and with no unmet needs, change has no adaptive value and has no reason to occur. Indeed, all societies have an innate resistance to change, since innovation inevitably creates tensions between the new reality and other established socioeconomic conditions. Those tensions can be resolved only by adaptive changes elsewhere in the total system. The gap that may develop between, for example, a newly adopted technology and other, slower-paced social traits has been called *cultural lag.* Complaints about youthful fads and the glorification of times past are familiar examples of reluctance to accept or adjust to change.

Innovation (invention), frequently under stress, has marked the history of humankind. An expanded food base accompanied the pressures of growing populations at the end of the Ice Age. Domestication of plants and animals appears to have occurred

independently in several recognizable areas of "invention" of agriculture, shown in **Figure 6.12**. From them, presumably, came a rapid diffusion of food types, production techniques, and new modes of economic and social organization as the majority of humankind changed from hunting–gathering to sedentary farming at least 2000 years ago. All innovation has a radiating impact on the web of culture; the more basic the innovation, the more pervasive its consequences.

Few innovations in human history have been more basic than the Agricultural Revolution. It affected every aspect of society and created irreconcilable conflict between preagricultural hunter-gatherers and sedentary farming cultures. Where those two groups came into contact, farmers were the victors and hunter-gatherers the losers in competition for territorial control. The contest continued into modern times. During the past 500 years, European expansion totally dominated the hunting and gathering cultures it encountered in large parts of the world, such as North America and Australia (see "Is Geography Destiny?"). With agriculture and its sedentary way of life, culture altered at an accelerating pace, and change itself became a way of life. Humans learned the arts of spinning and weaving plant and animal fibers. They learned to use the potter's wheel, to fire clay, and to make utensils. They developed the techniques of brick making, mortaring, and building construction. They discovered the skills of mining, smelting, and casting metals. Special local advantages in resources or products promoted the development of long-distance trading connections. On the foundation of such technical advancements a more complex, exploitative culture appeared, including a stratified society to replace the rough equality of hunting and gathering economies.

The source regions of such social and technical revolutions were initially spatially confined. The term **culture hearth** is used to describe those areas of innovation from which key culture elements diffused to exert an influence on surrounding regions. The hearth may be viewed as the "cradle" of any culture group whose developed systems of livelihood and life created a distinctive cultural landscape. Most of the thousands of hearths that evolved across the world remained at low levels of social and technical development. Only a few produced the trappings of *civilization,* which are usually assumed to include writing (or other forms of record keeping), metallurgy, long-distance trade connections, astronomy and mathematics, social stratification and labor specialization, formalized government systems, and a structured urban society. Several major culture hearths emerged, some as early as 7000 to 8000 years ago, following the initial revolution in food production. Prominent centers of early creativity were located in Egypt, Mesopotamia, the Indus Valley of the Indian subcontinent, northern China, southeastern Asia,

IS GEOGRAPHY DESTINY?

In his 1997 Pulitzer Prize-winning book *Guns, Germs and Steel: The Fates of Human Societies,* Jared Diamond argues, "History followed different courses for different peoples because of differences among peoples' environments, not because of biological differences among peoples themselves." Diamond, who is sometimes accused of environmental determinism (see p. 150), believes that what counted—and led to world dominance by Eurasians—was the availability in Eurasia of an abundance of plants and animals suitable for domestication and a landmass whose east–west orientation enabled long-distance transfer of animals, food crops, and technologies. No other continent had either of those advantages.

Food production was the key. Although agriculture was independently developed in several world areas after the end of the Ice Age, the inhabitants of the Middle East were fortunate in having an abundance of plants suitable for domestication, including six of the eight most important food grasses, among them ancestral wheat. These plants adapted easily to cultivation, grew rapidly, had high nutritive value, and could support large populations. Eurasia also had an abundance of large animals that could be domesticated, including the cow, goat, pig, sheep, and horse, further spurring population growth. In addition, by living in close proximity to animals, Eurasians contracted the epidemic diseases that were to prove so devastating later to the inhabitants of other continents and simultaneously developed their own immunities to those maladies.

The food-producing technologies developed in such hearth regions as the Middle East were easily diffused along the immense east–west axis of Eurasia, where roughly similar climates suited to the same mix of crops and livestock extended from China to Spain. Eurasia's great size meant, as well, a great number of different peoples, each capable of developing new technologies that in turn could be diffused over long distances. Population growth, agricultural productivity, and inventive minds led to civilizations—central governments, cities, labor specializations, textiles, pottery, writing, mathematics, long-distance trade, metalworking, and eventually, the guns that conquering Eurasians carried to other continents.

No other world region enjoyed Eurasia's environmental and subsequent technological advantages. The few food crops developed in Africa or the Americas could not effectively diffuse across the climatic and ecological barriers in those north–south aligned continents. Because of accidents of nature or massive predation of large animals by early inhabitants, sub-Saharan Africa and Australia yielded no domesticated animals and the Americas had only the localized llama. Without the food base and easy east–west movement of Eurasia, populations elsewhere remained smaller, more isolated, and collectively less inventive. When the voyages of discovery and colonization began in the 15th century A.D., Eurasian advantages proved overwhelming. Decimated by diseases against which they had no resistance, and lacking the horses, armor, firearms, and organization of their conquerors, inhabitants of other continents found themselves quickly subdued and dominated—not, in Jared Diamond's opinion, because of innate inferiority but because of geographical disadvantages that limited or delayed their development prospects.

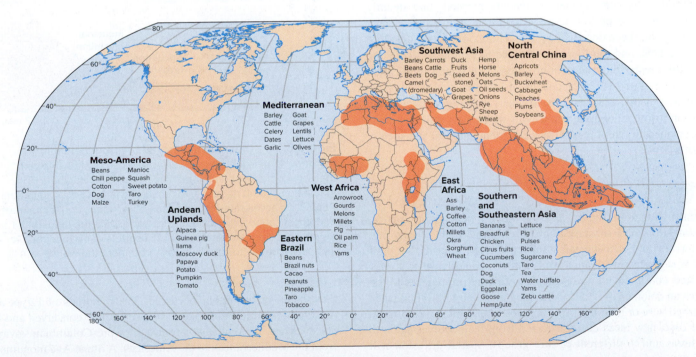

Figure 6.12   Chief centers of plant and animal domestication. The southern and southeastern Asia center was characterized by the domestication of plants, such as taro, that are propagated by the division and replanting of existing plants (vegetative reproduction). Reproduction by the planting of seeds (e.g., maize and wheat) was more characteristic of Meso-America and Southwest Asia. The African and Andean areas developed crops reproduced by both methods. The lists of crops and livestock associated with the separate origin areas are selective, not exhaustive.

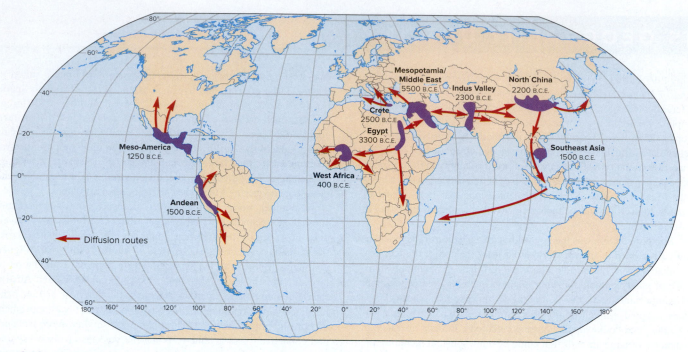

Figure 6.13 **Early culture hearths of the Old World and the Americas.** The B.C.E. (Before the Common Era) dates approximate times when the hearths developed complex social, intellectual, and technological bases and served as cultural diffusion centers.

and several locations in Africa, the Americas, and elsewhere (**Figure 6.13**).

In most modern societies, innovative change has become common, expected, and inevitable, though it may be rejected by some separate culture subgroups. The rate of invention, at least as measured by the number of patents granted, has steadily increased, and the period between idea conception and product availability has been decreasing. A general axiom is that the more ideas available and the more minds able to exploit and combine them, the greater the rate of innovation. The spatial implication is that large cities tend to be centers of innovation because of the efficient exchange of ideas. Indeed, ideas not only stimulate new ideas but also create circumstances in which new solutions must be developed to maintain the forward momentum of the society (**Figure 6.14**).

Diffusion

Spatial diffusion is the process by which a concept, a practice, an innovation, or a substance spreads from its point of origin to new territories. Diffusion can assume a variety of forms, but basically two processes are involved. Either people move to a new area and take their culture with them (as the immigrants to the American colonies did), or information about an innovation (such as barbed wire or hybrid corn) can spread throughout a culture. In either case, new ideas are transferred from their source region to new areas and to different culture groups. Spatial diffusion will be discussed in greater detail in Chapter 7. It is not always possible to determine whether the existence of a culture trait in two different areas is the result of diffusion or of independent (or *parallel*) innovation. Cultural similarities do not necessarily prove

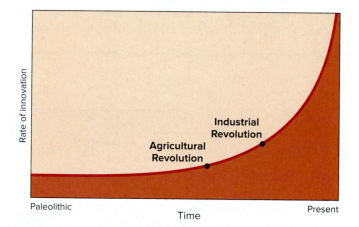

Figure 6.14 **The trend of innovation through human history.** Hunter-gatherers, dwelling in close relationship with their environment and their resource base, had little need for innovation and no necessity for cultural change. The Agricultural Revolution accelerated the diffusion of the ideas and techniques of domestication, urbanization, and trade. With the Industrial Revolution, dramatic increases in all aspects of socioeconomic innovation began to alter cultures throughout the world.

that spatial diffusion has occurred. The pyramids of Egypt and of Central America most likely were separately conceived and are not evidence, as some have proposed, of pre-Columbian voyages from the Mediterranean to the Americas. A Stone Age monument-building culture, after all, has only a limited number of shapes from which to choose.

Historical examples of independent, parallel invention are numerous: logarithms by Napier (1614) and Burgi (1620), calculus by Newton (1672) and Leibnitz (1675), and the telephone by Elisha Gray and Alexander Graham Bell (1876) are commonly cited examples. It appears beyond doubt that agriculture was independently invented not only both in the New World and in the Old but also in more than one culture hearth in each of the hemispheres.

All cultures are mixtures of innumerable innovations spread spatially from their points of origin and integrated into the structure of the receiving societies. It has been estimated that no more than 10% of the cultural items of any society are traceable to innovations created by its members and that the other 90% come to the society through diffusion (see "A Homemade Culture," p. 157).

Barriers to diffusion do exist, of course, as Chapter 7 explains. Generally, the closer and the more similar two cultural areas are to each other, the lower those barriers and the greater the likelihood of adoption of an innovation. Of course, the receiver culture may selectively adopt some goods or ideas from the donor society and reject others. The decision to adopt is governed by the receiving group's own culture. Political restrictions, religious taboos, and other social customs are cultural barriers to diffusion. French Canadians, although geographically close to many centers of diffusion, such as Toronto, New York, and Boston, are selective in their acceptance of Anglo influences. Traditional groups, perhaps controlled by firm religious conviction—such as the Hasidic Jewish community of New York City—may very largely reject culture traits and technologies of the larger society in whose midst they live (see "Folk and Popular Culture," p. 163, and **Figure 6.15**).

Diffused ideas and artifacts commonly undergo some sort of alteration of meaning or form that makes them acceptable to a borrowing group. The process of joining the old and new, called **syncretism,** is a major feature of cultural change. It can be seen in alterations to religious ritual and dogma made by convert societies seeking acceptable conformity between old and new beliefs. For example, slaves brought voodoo from West Africa to the Americas, where it thrived in Haiti and Louisiana. Over the years it absorbed influences from French and Spanish Catholicism, American Indian spiritual practices, and even Masonic tradition. Despite those adaptive mixings, many believers consider themselves to be Catholics and see no contradiction between Christianity and their faith in protective spirits and other tenets of voodoo. On a more familiar level, syncretism is reflected in subtle or blatant alterations of imported ethnic cuisines to make them conform to the demands of America's fast-food franchises.

Acculturation

Acculturation is the process by which one culture group undergoes a major modification by adopting many of the characteristics of another, usually dominant culture group. In practice, acculturation may involve changes in the original cultural patterns of one or both of two groups involved in prolonged firsthand contact. Such contact and subsequent cultural alteration may occur in a conquered or colonized region. Very often, the subordinate or subject population is forced to acculturate or does so voluntarily, overwhelmed by the superiority in numbers or the technical level of the conqueror.

The tribal Europeans in areas of Roman conquest, native populations in the wake of Slavic occupation of Siberia, and Native Americans following European settlement of North America experienced this kind of acculturation. In a different fashion, it is evident in the changes in Japanese political organization and philosophy imposed by occupying Americans after World War II or in the Japanese adoption of some more recreational aspects of American life (**Figure 6.16**). In turn, American life was enriched by awareness of Japanese food, architecture, philosophy, and popular culture, demonstrating the two-way nature of acculturation.

On occasion, the invading group is assimilated into the conquered society, as the older, richer Chinese culture prevailed over that of the conquering tribes of invading Mongols during the 13th and 14th centuries. The relationship of a mother country to its colony may also result in permanent changes in the culture of the colonizer, even though little direct population contact is involved. The European spread of tobacco addiction (see "Documenting Diffusion," in Chapter 7), potatoes, maize, and turkeys reflected colonial contact with the Americas. In Great Britain, drinking tea and eating curry are national traditions, but both are imports, reflecting the colonial experience in India.

In the modern period, the population relocations and immigration impacts discussed in Chapter 5 (pp. 130–131) have resulted in unprecedented cultural mixings throughout the world. The traditional

Figure 6.15   Motivated by religious conviction that the "good life" must be reduced to its simplest forms, the **Amish community of east central Illinois** shuns the modern luxuries of the dominant secular society around them. Children use horse and buggy, not school bus or automobile, on the daily trip to their rural school. © *Jean Fellmann.*

Figure 6.16   Baseball, an import from America, is one of the most popular sports in Japan, attracting millions of spectators annually. Similarly, a generation ago most Americans considered soccer a foreign and exotic game of little or no interest. Today, soccer is one of the most popular youth sports in the United States. © *Anonymous/AP Images.*

"melting pot" view—more formally, **amalgamation theory**—of immigrant integration into, for example, U.S. society suggests that the receiving society and the varied arriving newcomer groups eventually merge into a composite mainstream culture, incorporating the many traits of its collective components. More realistically, to be accepted, newcomer groups must learn the accustomed patterns of behavior and response and the dominating language of the workplace and government of the culture they have entered. Acculturation for them involves the adoption of the values, attitudes, ways of behavior, and speech of the receiving society, which itself undergoes change from absorption of the arriving group. In that process, the immigrant group loses much of its separate cultural identity to the extent that it accepts over time the culture of the larger host community.

Although acculturation usually involves a minority group adopting the patterns of the dominant population, the process can be reciprocal. That is, the dominant group may also adopt at least some patterns and practices identified with new minority groups as a broader, more diverse composite culture is created. Instead of the presumed ideal of the melting pot, a "salad bowl" or "lumpy stew" cultural mixture results.

When the integration process is completed, **assimilation** has occurred. But assimilation does not necessarily mean that consciousness of original cultural identity is reduced or lost. *Competition theory,* in fact, suggests that, as cultural minorities begin to achieve success and enter into mainstream social and economic life, awareness of cultural differences may be heightened, transforming the strengthening immigrant group into a self-assertive minority, pursuing goals and interests that defend and protect its position within the larger society. Carried to extremes, militant minority self-assertion may result in the loss of the larger social and cultural integration the acculturation process seeks to ensure—a circumstance of increasing concern in Western European and North American destination countries.

6.5 Cultural Diversity

We began our discussion of culture with its technological, sociological, and ideological subsystems. We have learned that the distinctive makeup of those subsystems—the combinations and interactions of traits and complexes characteristic of particular cultures—is subjected to, and is the product of, change through innovation, spatial diffusion, adoption, and acculturation. Those processes of cultural development and alteration have not, however, led to a homogenized global culture, even after thousands of years of cultural contact and exchange.

As we earlier observed, in an increasingly integrated world, access to the lifestyles and technologies of modern life is widely available to most peoples and societies. As a result, important cultural commonalities have developed. Nevertheless, the world remains divided, not unified, in culture. In fact, some observers note that as the world becomes homogenized through globalization processes, some groups cling ever more tightly to local, ethnic, religious, or other cultural identities. Our concern as geographers is to identify culture traits that have spatial expression, starting with traits that vary over extensive regions of the world and differentiate societies in a broad, summary fashion.

Language, religion, ethnicity, and gender meet our criteria and are among the most prominent of the differentiating culture traits of societies and regions. Language and religion are basic components of culture, helping identify who and what we are as individuals and clearly placing us within larger communities of persons with similar characteristics. In our earlier terminology, they are components of the ideological subsystem of culture that help shape the belief system of a society which in turn influences the technological and sociological subsystems of culture.

Ethnicity is a cultural summary rather than a single trait. It is a shared identity based on distinguishing characteristics, which may include a combination of language, religion, national origin, unique customs, or other identifiers. Like language and religion, ethnicity displays clear geographic patterns. Like them, too, it may serve as an element of diversity within multicultural societies and states.

Cultural identities based on language, religion, or ethnicity are open to both males and females. However, among the most prominent strands in the fabric of culture are the social structures (sociofacts) that assign particular duties, relationships, and standards, based on whether a person is male or female. *Gender* is the reference term recognizing those socially created distinctions. Gender conditions the way people use space and assess economic and cultural opportunities, and thus the status of women is a cultural variable of fundamental geographical significance. In the following sections we examine each of these components of cultural diversity: language, religion, ethnicity, and gender.

A HOMEMADE CULTURE

Reflecting on an average morning in the life of a "100% American," Ralph Linton noted:

> Our solid American citizen awakens in a bed built on a pattern which originated in the Near East but which was modified in northern Europe before it was transmitted to America. He throws back covers made from cotton, domesticated in India, or linen, domesticated in the Near East, or wool from sheep, also domesticated in the Near East, or silk, the use of which was discovered in China. All of these materials have been spun and woven by processes invented in the Near East. . . . He takes off his pajamas, a garment invented in India, and washes with soap invented by the ancient Gauls. . . .
>
> Returning to the bedroom, . . . he puts on garments whose form originally derived from the skin clothing of the nomads of the Asiatic steppes [and] puts on shoes made from skins tanned by a process invented in ancient Egypt and cut to a pattern derived from the classical civilizations of the Mediterranean. . . . Before going out for breakfast he glances through the window, made of glass invented in Egypt, and if it is raining puts on overshoes made of rubber discovered by the Central American Indians and takes an umbrella invented in southeastern Asia. . . .

> [At breakfast,] a whole new series of borrowed elements confronts him. His plate is made of a form of pottery invented in China. His knife is of steel, an alloy first made in southern India, his fork a medieval Italian invention, and his spoon a derivative of a Roman original. He begins breakfast with an orange, from the eastern Mediterranean, a cantaloupe from Persia, or perhaps a piece of African watermelon. With this he has coffee, an Abyssinian plant. . . . [H]e may have the egg of a species of bird domesticated in Indo-China, or thin strips of flesh of an animal domesticated in Eastern Asia which have been salted and smoked by a process developed in northern Europe.
>
> When our friend has finished eating . . . he reads the news of the day, imprinted in characters invented by the ancient Semites upon a material invented in China by a process invented in Germany. As he absorbs the accounts of foreign troubles he will, if he is a good conservative citizen, thank a Hebrew deity in an Indo-European language that he is 100 percent American.

Source: Ralph Linton, *The Study of Man* © 1936, renewed 1964, pp. 326–327, Prentice Hall, Inc., Englewood Cliffs, NJ.

6.6 Language

Forever changing and evolving, language in spoken or written form makes possible the cooperative efforts, the group understandings, and the shared behavioral patterns that distinguish culture groups. *Language,* defined simply as an organized system of speech by which people communicate with one another with mutual comprehension, is the most important medium by which culture is transmitted. Language enables parents to teach their children about the world and their roles and responsibilities as functioning members of society. Some argue that the language of a society structures the perceptions of its speakers. By the words it contains and the concepts it can formulate, language is said to determine the attitudes, the understandings, and the responses of the society. Language, therefore, can be both a cause of and a symbol of cultural differentiation (**Figure 6.17**).

If that conclusion is true, one aspect of cultural heterogeneity may be easily understood. More than 7 billion people on Earth speak about 7100 different languages. Knowing that Africa as a whole contains nearly one-third of all living languages (though 85% of Africans speak one or more variants of 15 core languages) gives us a clearer appreciation of the political and social division in that continent. Europe alone has some 287 languages and dialects (**Figure 6.18**). Language is a hallmark of cultural diversity, and the present world distribution of major languages (**Figure 6.19**) records not only the migrations and conquests of our linguistic ancestors but also the continuing dynamic pattern of human movements, settlements, and colonizations of more-recent centuries.

Languages differ greatly in their relative importance, if *importance* can be taken to mean the number of people using them. More than one-half of the world's inhabitants are native speakers of just eight of its thousands of tongues, and at least one-half regularly use or have competence in just four of them. **Table 6.1**, p. 162, lists those languages spoken by more than 100 million people. At the other end of the scale are a number of rapidly declining languages whose speakers number in the hundreds or, at most, the few thousands. Indeed, the world today has far less linguistic diversity than it has had historically, and each year sees the loss of additional tongues, displaced through the spread of English and the other major languages. Scholars estimate that about one-half of the currently spoken languages are endangered—meaning that children no longer learn them and their youngest speakers are now middle-aged—and will disappear within the next century.

The diversity of languages is simplified when we recognize among them related families. A **language** (or *linguistic*) **family** is a group of languages thought to have a common origin in a single, earlier tongue. The Indo-European family of languages is among the most prominent of such groupings, embracing most of the languages of Europe and a large part of Asia, as well as the introduced—not the native—languages of the Americas (Figure 6.19). All told,

Figure 6.17 The survival of Welsh (Cymraeg), a Celtic language, within Great Britain illustrates the **importance of language to cultural identity**. After more than a century of decline as English came to dominate education and public life, laws passed in the 1990s gave Welsh equal status with English in Wales. Instruction in Welsh is compulsory up to age 16, and most government publications and road signs in Wales are bilingual. The British government has been supportive of Wales, devolving power to the Welsh Government. © Mark Bjelland.

"Bantu line" in sub-Saharan Africa, for example, are variants of a proto-Bantu carried by an expanding, culturally advanced people who displaced linguistically different preexisting populations. More recently, the languages of European colonists similarly replaced native tongues in their areas of settlement in North and South America, Australasia, and Siberia. That is, languages may spread because their speakers occupy new territory.

Latin, however, replaced earlier Celtic languages in Western Europe not by force of numbers—Roman legionnaires, administrators, and settlers never represented a majority population—but by the gradual abandonment of their former tongues by native populations brought under the influence and control of the Roman Empire. Adoption rather than eviction of language appears the rule followed in the majority of historical and contemporary instances of language spread. That is, languages may spread because they acquire new speakers.

Either form of language spread—the dispersion of speakers or the acquisition of speakers—may, through segregation and isolation, give rise to separate, mutually incomprehensible tongues. Comparable changes occur normally and naturally within a single language in word meaning, pronunciation, vocabulary, and *syntax* (the way words are put together in phrases and sentences). Because they are gradual, such changes tend to go unremarked, yet cumulatively they can result in language change so great that, in the course of centuries, an essentially new language has been created. The English of 17th-century Shakespearean writings or the King James Bible (1611) sounds stilted to our ears, and 8th-century *Beowulf* is practically unintelligible. Language evolution may be gradual and cumulative, with each generation deviating in small degree from the speech patterns and vocabulary of its parents, or it may be massive and abrupt—reflecting conquests, migrations, new trade contacts, and other disruptions of cultural isolation.

English is, in many respects, a multicultural language. It owes its form to the Celts, the original inhabitants of the British Isles, and to successive waves of invaders, including the Latin-speaking Romans and the Germanic Angles, Saxons, and Danes. The French-speaking Norman conquerors of the 11th century added about 10,000 new words to the evolving English tongue. Discovery and colonization of new lands in the 16th and 17th centuries greatly expanded English as new foods, plants, animals, and artifacts were encountered along with their existing aboriginal American, Australian, Indian, or African names. The Indian languages of the Americas alone brought more than 200 relatively common daily words to English, 80 or more from the North American native tongues and the rest from Caribbean, Central, and South American languages. More than 2000 more specialized or localized words were also added. *Moose, raccoon, skunk,*

languages in the Indo-European family are spoken by about one-half of the world's peoples.

By recognizing similar words in most Indo-European languages, linguists deduce that these languages derived from a common ancestor tongue called *proto-Indo-European*, which was spoken by people living somewhere in Eastern Europe about 5000 years ago (though some conclude that central Turkey was the more likely site of origin). By at least 2500 B.C., their society had apparently fragmented; the homeland was left and segments of the parent culture migrated in different directions. Wherever this remarkable people settled, they appear to have dominated local populations and imposed their language on them.

Within a language family, we can distinguish *subfamilies*. The Romance languages (including French, Spanish, and Italian)—offsprings of Latin—and the Germanic languages (such as English, German, and Dutch) are subfamilies, or branches, of Indo-European. The languages in a subfamily often show similarities in sounds, grammatical structure, and vocabulary, even though they are mutually unintelligible. English *daughter*, German *Tochter*, and Swedish *dotter* are Germanic examples.

Language Spread and Change

Language spread as a geographic event represents the increase or relocation through time in the area over which a language is spoken. The more than 300 Bantu languages found south of the

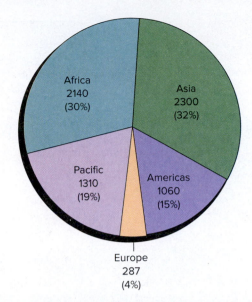

Figure 6.18   World distribution of living languages, 2009.
Of the approximately 7100 languages still spoken today, one-third are found in Asia. Linguists' estimates of the number of languages ever spoken on Earth range from 31,000 to as many as 300,000 or more. Assuming the lower estimate or even one considerably smaller, dead languages far outnumber the living. One or two additional tongues, most spoken in the forests of Papua, New Guinea, or in Indonesia, are lost each week. In contrast, as diverse peoples and cultures have mixed more than 100 new languages have been created over the past four centuries. *Source: Estimates based on Lewis, M. Paul (ed.), 2016. Ethnologue: Languages of the World, Nineteenth edition. Dallas, Tex.: SIL International.*

maize, squash, succotash, igloo, toboggan, hurricane, blizzard, hickory, pecan, and a host of other names were taken directly into English; others were adopted secondhand from Spanish variants of South American native words: *cigar, potato, chocolate, tomato, tobacco,* and *hammock.*

A worldwide diffusion of the language resulted as English colonists carried it to the Western Hemisphere and Australasia and as trade, conquest, and territorial claim took it to Africa and Asia. In that areal spread, English was further enriched by its contacts with other languages. By becoming the accepted language of commerce and science, it contributed in turn to the common vocabularies of other tongues. Within some 400 years, English has developed from a localized language of 7 million islanders off the European coast to a truly international language with some 400 million native speakers, perhaps 600 million who use it as a second, sometimes nationally official lauguage, and additional millions who have reasonable competence in English as a foreign language. English serves as an official language of some 60 countries (**Figure 6.20**), far exceeding in that role French (32), Arabic (25), or Spanish (21), the other leading current international languages. English has become the dominant language facilitating increased global interaction. It is used in many international scientific conferences, air traffic control, and international diplomacy. In the 1990s, English web pages dominated the Internet, although that dominance is declining as other major languages gain ground.

Standard and Variant Languages

People who speak a common language, such as English, are members of a *speech community,* but membership does not necessarily imply linguistic uniformity. A speech community usually possesses both a **standard language** comprising the accepted community norms of syntax, vocabulary, and pronunciation and a number of more or less distinctive **dialects,** the ordinary speech of areal, social, professional, or other subdivisions of the general population.

An official or unofficial standard language is the form carrying government, educational, or societal sanction. In Arab countries, for example, classical Arabic is the language of the mosque, of education, and of the newspapers and is standardized throughout the Arabic-speaking world. Colloquial Arabic is used at home, in the street, and at the market—and in its regional variants, it may be as widely different as are, for example, Portuguese and Italian. On the other hand, the United States, English-speaking Canada, Australia, and the United Kingdom all have only slightly different forms of standard English.

Just as no two individuals talk exactly the same, all but the smallest and most closely knit speech communities display recognizable speech variants called dialects. Vocabulary, pronunciation, rhythm, and the speed at which the language is spoken may clearly set groups of speakers apart from one another. Most dialects exhibit clear spatial patterns. Speech is a geographic variable because each locale is likely to have its own, perhaps slight language differences from neighboring places. Such differences in pronunciation, vocabulary, word meanings, and other language characteristics help define the *linguistic geography*—the study of the character and spatial pattern of *geographic,* or *regional, dialects*—of a generalized speech community. **Figure 6.21** records the variation in usage associated with just one phrase. The regional dialects of the South are most easily recognized by their distinctive accents. In some instances, there is so much variation among geographic dialects that some are almost foreign tongues to other speakers of the same language. Effort is required for Americans to understand Australian English or that spoken in Liverpool, England, or in Glasgow, Scotland (see "World Englishes," p. 166).

In the United States, distinct regional dialects were established along the Atlantic and Gulf coasts and were carried inland by settlers (**Figure 6.22**). However, local dialects and accents do not display predictable patterns of consistency or change. We might expect that the influence of mass media and the rise of domestic migration would tend to level out regional dialects and there is evidence of declining use of local dialect pronunciations in southern cities such as Atlanta and Dallas, which have received major influxes of northerners. On the other hand, there are reports of increasing contrasts between the speech patterns and accents of Chicago, New York, Birmingham, St. Louis, and other cities.

Language is rarely a total barrier to communication between peoples. Bilingualism or multilingualism may permit skilled linguists to communicate in a jointly understood third language, but long-term contact between less-able populations may require the creation of a new language—a pidgin—learned by both parties. A **pidgin** is a mixture of languages, usually a simplified form of one of them, such as English or French, with borrowings from another,

Figure 6.19 **World language families.** Language families are groups of individual tongues that have a common but remote ancestor. By suggesting that the area assigned to a language or language family uses that language exclusively, the map pattern conceals important linguistic details. Many countries and regions have local languages spoken in territories too small to be recorded at this scale. The map also fails to report that the population in many regions is fluent in more than one language, or that a second language serves as the necessary vehicle of commerce, education, or government. Nor is important information given about the number of speakers of different languages; the fact that there are more speakers of English in India or Africa than in Australia is not even hinted at by a map at this scale.

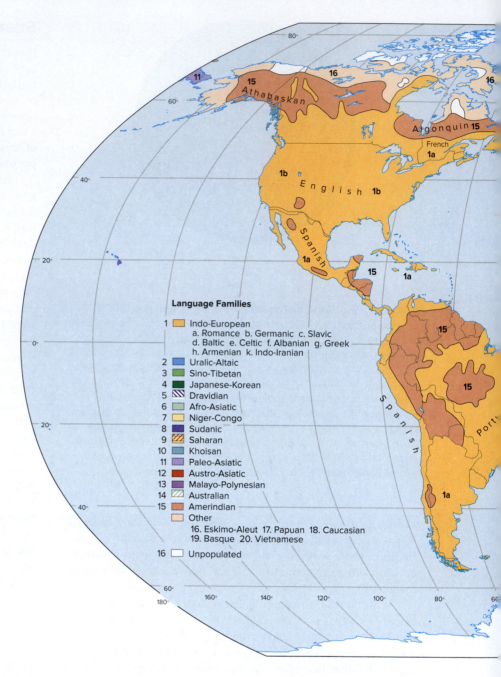

Language Families
1. Indo-European
   a. Romance b. Germanic c. Slavic
   d. Baltic e. Celtic f. Albanian g. Greek
   h. Armenian k. Indo-Iranian
2. Uralic-Altaic
3. Sino-Tibetan
4. Japanese-Korean
5. Dravidian
6. Afro-Asiatic
7. Niger-Congo
8. Sudanic
9. Saharan
10. Khoisan
11. Paleo-Asiatic
12. Austro-Asiatic
13. Malayo-Polynesian
14. Australian
15. Amerindian
Other
   16. Eskimo-Aleut 17. Papuan 18. Caucasian
   19. Basque 20. Vietnamese
16. Unpopulated

perhaps non-European local language. In its original form, a pidgin is not the mother tongue of any of its speakers; it is a second language for everyone who uses it, one generally restricted to such specific functions as commerce, administration, and work supervision.

Pidgins are characterized by a highly simplified grammatical structure and a sharply reduced vocabulary adequate to express basic ideas but not complex concepts. If a pidgin becomes the first language of a group of speakers—who may have lost their former native tongue through disuse—a **creole** has evolved. Creoles invariably acquire a more complex grammatical structure and enhanced vocabulary.

Creole languages have proved to be useful integrative tools in linguistically diverse areas; several have become symbols of nationhood. Swahili, a pidgin formed from a number of Bantu dialects with major vocabulary additions from Arabic, originated in the

coastal areas of East Africa and spread inland first by Arab ivory and slave caravans and later by trade during the period of English and German colonial rules. When Kenya and Tanzania gained independence, they made Swahili the national language of administration and education. Other examples of creolization are Afrikaans (a pidginized form of 17th-century Dutch used in the Republic of South Africa); Haitian Creole (the language of Haiti, derived from the pidginized French used in the slave trade); and Bazaar Malay (a pidginized form of the Malay language, a version of which is the official national language of Indonesia).

A **lingua franca** is an established language used habitually for communication by people whose native tongues are mutually incomprehensible. For them, it is a *second language,* one learned in addition to the native tongue. Lingua franca (literally, "Frankish tongue") was named from the French dialect adopted as a common

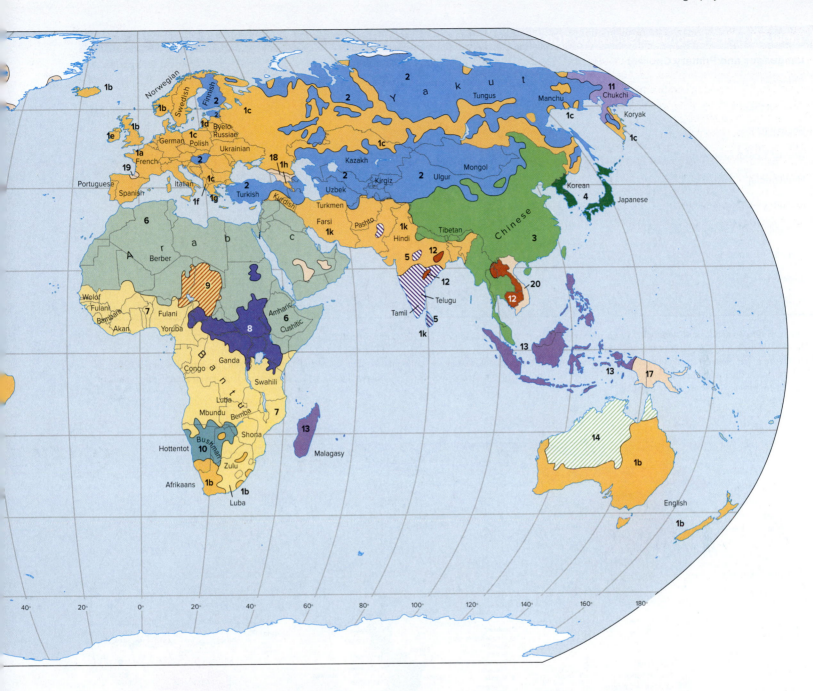

language by the Crusaders at war in the Holy Land. Later, Latin became the lingua franca of the Mediterranean world until, finally, it was displaced by vernacular European tongues. Arabic followed Muslim conquest as the unifying language of that international religion after the 7th century. Mandarin Chinese and Hindi in India have traditionally had a lingua franca role in their linguistically diverse countries. The immense linguistic complexity of Africa has made regional lingua francas there necessary and inevitable—Swahili in East Africa, for example, and Hausa in parts of West Africa.

Language and Culture

Language embodies the culture complex of a people, reflecting both environment and technology. Arabic has 80 words related to camels,

an animal on which a regional culture relied for food, transport, and labor, and Japanese contains more than 20 words for various types of rice. Russian is rich in terms for ice and snow, indicative of the prevailing climate of its linguistic cradle; and the 15,000 tributaries and subtributaries of the Amazon River have obliged the Brazilians to enrich Portuguese with words that go beyond *river*. Among them are *paraná* (a stream that leaves and reenters the same river), *igarapé* (an offshoot that runs until it dries up), and *furo* (a waterway that connects two rivers).

Most—perhaps all—cultures display subtle or pronounced differences in ways males and females use language. Most have to do with vocabulary and with grammatical forms peculiar to individual cultures. For example, among the Caribs of the Caribbean, the Zulu of Africa, and elsewhere, men have words that women through custom or taboo are not permitted to use, and the women have words

Table 6.1 First Languages Spoken by 100 Million or More People, 2016

Languages and Primary Country	Millions of Speakers	Total Countries
Chinese[a] (China)	1302	35
Spanish (Spain)	427	31
English (U.K.)	339	106
Arabic[b] (Saudi Arabia)	267	58
Hindi (India)	260	4
Portuguese (Portugal)	202	12
Bengali (Bangladesh)	189	4
Russian (Russian Federation)	171	17
Japanese (Japan)	128	2
German	117	8

Sources: Based on data from *Ethnologue: Languages of the World,* 19th ed.; *Linguasphere 2000;* and other sources.

[a]The official dialect of Mandarin is spoken by perhaps 900 million.

[b]The figure given includes speakers of the many, often mutually unintelligible versions of colloquial Arabic. Classical or literary Arabic, the language of the Koran, is uniform and standardized but restricted to formal usage as a spoken tongue. Because of its religious association, Arabic is a second language for many inhabitants of Muslim countries with other native tongues.

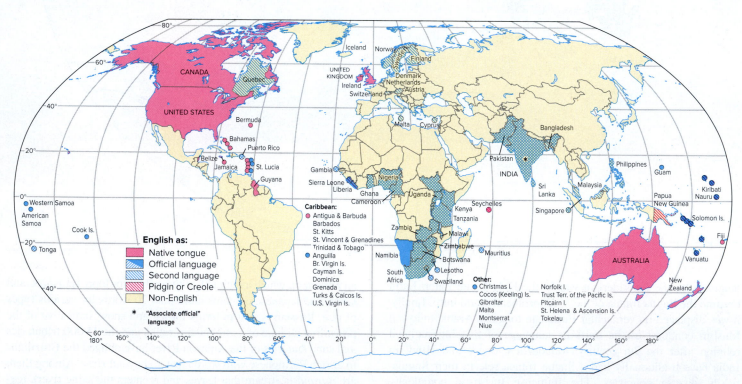

Figure 6.20 **International English.** In worldwide diffusion and acceptance, English has no past or present rivals. Along with French, it is one of the two working languages of the United Nations and the effective common language of the workers and committees of European Union institutions; some two-thirds of all scientific papers are published in it, making it the first language of scientific discourse. In addition to being the accepted language of international air traffic control, English is the sole or joint official language of more nations and territories, some too small to be shown here, than any other tongue. It also serves as the effective unofficial language of administration in other multilingual countries with different formal official languages. "English as a second language" is indicated for countries with near-universal or mandatory English instruction in public schools. Not evident on this map is the full extent of English penetration of Continental Europe, where more than 80% of secondary school students (and 92% of those of European Union states) study it as a second language and more than one-third of European Union residents can easily converse in it.

FOLK AND POPULAR CULTURE

Not all culture groups, even in "developed" societies, readily adopt or adjust to cultural change. In general understanding, *culture* is taken to mean "our way of life"—how we act (and why), what we eat and wear, how we amuse ourselves, what we believe, whom we admire. There are distinctions to be made, however, on the universality of the "way of life" that is accepted.

Folk groups exist in either spatial or self-imposed social isolation from the common culture of the larger societies of which they are presumably a part. **Folk culture** connotes the traditional and nonfaddish way of life characteristic of a homogeneous, cohesive, largely self-sufficient group that is essentially isolated from or resistant to outside influences. Tradition controls folk culture, and resistance to change is strong. The homemade and handmade dominate in tools, food, music, story, and ritual. *Folk life* is a cultural whole composed of both tangible and intangible elements. **Material culture** is the tangible part, made up of physical, visible things: everything from musical instruments to furniture, tools, and buildings. In folk societies, these things are products of the household or community itself, not of commercial mass production. Their intangible **nonmaterial culture** comprises the mentifacts and sociofacts expressed in oral tradition, folk song and folk story, and customary behavior; ways of speech, patterns of worship, outlooks, and philosophies are passed to following generations by teachings and examples.

Within the United States and Canada, true folk groups are few and dwindling. Their earlier, larger numbers were based on the customs and beliefs brought to the New World by immigrant groups distinguished by their language, religious beliefs, and areas of origin. With time, many of their imported ethnic characteristics became transmuted into American "folk" features. For example, the traditional songs of western Virginia can be considered both nonmaterial folk expressions of the Upland South and evidence of an immigrant ethnic heritage derived from rural English forebears.

In that respect, each of us bears the evidence of ethnic origin and folk life. Each of us uses proverbs traditional to our family or culture; each is familiar with childhood nursery rhymes and fables. We rap wood for luck, have heard how to plant a garden by phases of the moon, and know what is the "right" way to celebrate a holiday or prepare a favorite dish. For most, however, such evidences of folk culture are minor elements in our life, and only a few groups—such as the Old Order Amish, with their rejection of electricity, the internal combustion engine, and other "worldly" accouterments in favor of buggy, hand tools, and traditional dress—remain in the United States as reminders of the folk cultural distinctions formerly widely recognizable. Canada, on the other hand, has retained a greater number of clearly recognizable ethnically unique folk and decorative arts traditions.

Popular culture stands in opposition to—and as the replacement for—folk culture. *Popular* implies the general mass of people, rather than the small-group distinctiveness of folk culture. It suggests a process of constantly adopting, conforming to, and quickly abandoning ever-changing fads and common modes of behavior. In that process, locally distinctive lifestyles and material and nonmaterial folk culture traits are largely replaced and lost; uniformity replaces variety, and small-group identity is eroded. For most of us, it is a sought-after conformity. In the 1750s, George Washington wrote to his British agent to request ". . . two pair of Work'd Ruffles . . . ; if work'd Ruffles shou'd be out of fashion send such as are not . . ." and "whatever goods you may send me . . . you will let them be fashionable." His desire, echoed today, was to fit in with the peer group and larger social milieu of which he was a part.

Popular culture may be seen as both a leveling and a liberating force. On the one hand, it obliterates those locally distinctive folk culture lifestyles that emerge when groups remain isolated and self-sufficient. At the same time, however, individuals are exposed to a broader range of available opportunities—in clothing, food, tools, recreation, and lifestyles—than were ever available to isolated folk culture groups. Broad geographical uniformity—in the form of the seemingly endless repetition of chain discount stores, duplicate national or global retailers in identical shopping malls, and the same group of fast-food chains—often displaces distinctive local character. While the diffusion of popular culture opens an immense and ever-changing range of possibilities, it undermines traditional folk practices as well as social and religious values that are not always fully appreciated until they have been lost.

and phrases that the men never use "or they would be laughed to scorn," an informant reports. Evidence from English and many other unrelated tongues indicates that, as a rule, female speakers use forms considered to be "better" or "more correct" than do males of the same social class. The greater and more inflexible the difference in the social roles of men and women in a particular culture, the greater and more rigid are the observed linguistic differences between the sexes.

A common language fosters unity among people. It promotes a feeling for a region; if it is spoken throughout a country, it fosters nationalism. For this reason, languages often gain political significance and serve as a focus of opposition to what is perceived as foreign domination. Although nearly all people in Wales speak English, many also want to preserve Welsh because they consider it an important aspect of their culture. They think that, if the language is forgotten, their entire culture may also be threatened. French Canadians received government recognition of their language and established it as the official language of Quebec Province; Canada itself is officially bilingual. In India, with 18 constitutional languages and 1652 other tongues, serious riots have occurred by people expressing opposition to the imposition of Hindi as the single official national language.

Bilingualism or multilingualism complicates national linguistic structure. Areas are considered bilingual if more than one language is spoken by a significant proportion of the population. In some countries—Belgium and Switzerland, for example—there is more than one official language. In many others, such as the United States, only one language may have implicit or official government sanction, although several others are spoken. Speakers of one of these may be concentrated in restricted areas (e.g., most speakers of French in Canada live in the Province of Quebec). Less often, they are distributed fairly evenly throughout the country. In some countries, the language in which instruction, commercial transactions, and government business take place

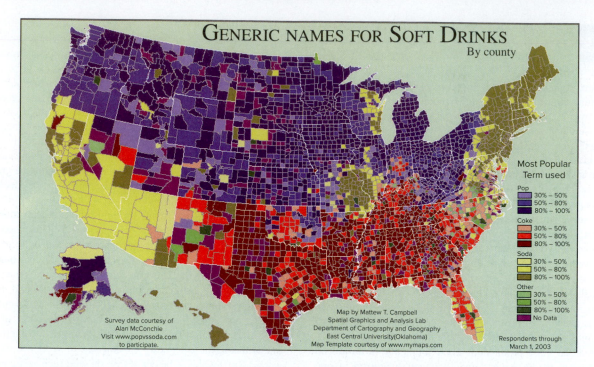

Figure 6.21 **Dialect differences.** Descriptive terms for everyday items help identify dialect differences. The generic term for a soft drink varies regionally across the United States, from *soda* to *pop* to *Coke*. Despite the influence of national mass media in promoting a "standard" American word usage and pronunciation, regional variations persist. *Source: M. Campbell and G. Plumb, Web Atlas of Oklahoma, East Central University.*

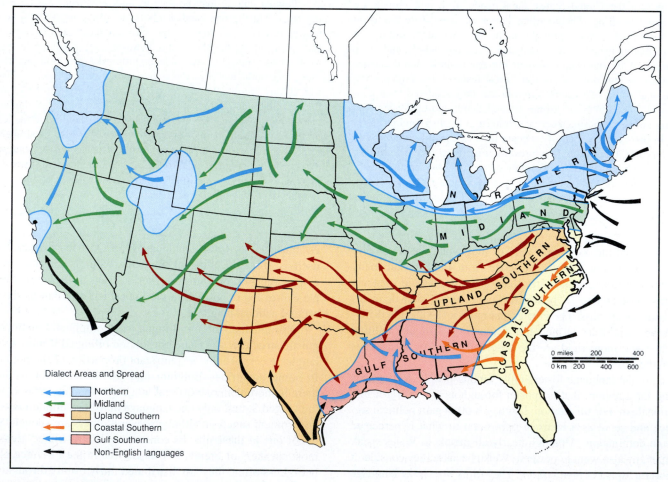

Figure 6.22 **Dialect areas and spread in the United States.** Distinct regional dialects developed along the Atlantic and Gulf coasts and diffused inland.

is not a domestic language at all. In linguistically complex sub-Saharan Africa, nearly all countries have selected a European tongue—usually that of their former colonial governors—as an official language (**Figure 6.23**).

Toponyms—place-names—are language on the land, the record of those persons and groups who gave names to geographic features that endure as reminders of their culture. **Toponymy,** the study of place-names, therefore, is a revealing tool of historical cultural geography, because place-names become a part of the cultural landscape that remains long after the name givers have passed from the scene.

In England, for example, place-names ending in *chester* (as in Winchester and Manchester) evolved from the Latin *castra,* meaning "camp." Common Anglo-Saxon suffixes for tribal and family settlements were *ing* (people or family) and *ham* (hamlet or, perhaps, meadow), as in Birmingham and Gillingham. Norse and Danish settlers contributed place-names ending in *thwaite* (meadow) and others denoting such landscape features as *fell* (an uncultivated hill) and *beck* (a small brook). The Arabs, sweeping out from Arabia across North Africa and into Iberia, left their imprint in place-names to mark their conquest and control. *Cairo* means "victorious," *Sudan* is "the land of the blacks," and *Sahara* is "wasteland" or "wilderness." In Spain, a corrupted version of the Arabic *wadi,* "watercourse," is found in *Guadalajara* and *Guadalquivir.*

In the New World, not one people but many people placed names on landscape features and new settlements. In doing so, they remembered their homes and homelands, honored their monarchs

and heroes, borrowed and mispronounced from rivals, adopted and distorted Amerindian names, followed fads, and recalled the Bible. Homelands were honored in New England, New France, and New Holland; settlers' hometown memories brought Boston, New Bern, and New Rochelle from England, Switzerland, and France. Monarchs were remembered in Virginia for the Virgin Queen Elizabeth, Carolina for one English king, Georgia for another, and Louisiana for a king of France. Washington, D.C.; Jackson, Mississippi and Michigan; Austin, Texas; and Lincoln, Illinois, memorialized heroes and leaders.

Names given by the Dutch in New York were often distorted by the English; Breukelyn, Vlissingen, and Haarlem became Brooklyn, Flushing, and Harlem. French names underwent similar twisting or translation, and Spanish names were adopted, altered, or later put into such bilingual combinations as Hermosa Beach. Amerindian tribal names—the Yenrish, Maha, Kansa—were modified, first by French and later by English speakers, to Erie, Omaha, and Kansas. A faddish classical revival after the American Revolution gave us Troy, Athens, Rome, Sparta, and other ancient town names. Bethlehem, Ephrata, Nazareth, and Salem came from the Bible.

Of course, European colonists and their descendants gave place-names to a physical landscape already named by indigenous peoples. Those names were sometimes adopted but often shortened, altered, or—certainly—mispronounced. The vast territory that local Amerindians called *Mesconsing,* meaning "the long river," was recorded by Lewis and Clark as "Quisconsing," later to be further distorted into "Wisconsin." Milwaukee, Winnipeg, Potomac, Niagara, Adirondack, Chesapeake, Shenandoah, and Yukon; the names of 28 of the 50 United States; and the present identity of thousands of North American places and features, large and small, had their origin in Native American languages. The United States Board on Geographic Names is charged with maintaining uniform names of geographic features. Recently it has reviewed numerous proposals to change toponyms back to their original indigenous names. For example, in 2015 the name of North America's highest mountain was officially changed from Mount McKinley back to Denali.

6.7 Religion

Enduring place-names are one measure of the importance of language as a powerful, unifying cultural thread. But language is not alone in that role. At times religion complements language or even replaces it as a dominant cultural rallying point. For example, traditionally, the cultural identity of the Québécois in Canada was rooted in both the French language and the dominance of Roman Catholicism. However, unlike language, which is an attribute of all people, religion varies in its cultural role—dominating in some societies, unimportant, rejected, or even repressed in others. All societies have value systems—common beliefs, understandings, expectations, and controls—that unite their members and set them off from other, different culture groups. Such a value system is termed a *religion* when it involves systems of formal or informal worship of and faith in the sacred and divine. In a more inclusive sense, religion may be viewed as a unified system of beliefs and

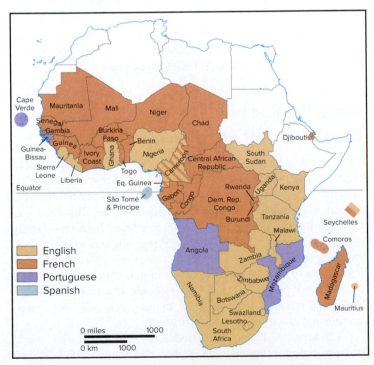

Figure 6.23 **Europe in Africa through official languages.** Both the linguistic complexity of sub-Saharan Africa and the colonial histories of its present political units are implicit in the designation of a European language as the sole or joint "official" language of the different countries.

WORLD ENGLISHES

Nonnative speakers of English far outnumber those for whom English is the first language. Most of the more than 1 billion people who speak and understand at least some English as a second language live in Asia; they are appropriating the language and remaking it in regionally distinctive fashions to suit their own cultures, linguistic backgrounds, and needs.

It is inevitable that widely spoken languages separated by distance, isolation, and cultural differences will fragment into dialects, which in turn evolve into new languages. Latin splintered into French, Spanish, Italian, and other Romance languages. English is similarly experiencing that sort of regional differentiation, shaped by the variant life worlds of its far-flung community of speakers and following the same path to mutual unintelligibility. Although standard English may be one of or the sole official language of their countries of birth, millions of people around the world claiming proficiency in English or English as their national language cannot understand one another. Even teachers of English from India, Malaya, Nigeria, or the Philippines, for example, may not be able to communicate in their supposedly common tongue—and find cockney English of London utterly alien.

The splintering of spoken English is a fact of linguistic life, and its offspring—called "World Englishes" by linguists—defy frequent attempts by various governments to remove localisms and encourage adherence to international standards. Singlish (Singapore English) and Taglish (a mixture of English and Tagalog, the dominant language of the Philippines) are commonly cited examples of the multiplying World Englishes, but equally distinctive regional variants have emerged in India, Malaysia, Hong Kong, Nigeria, the Caribbean, and elsewhere. One linguist suggests that, beyond an "inner circle" of countries where English is the first and native language—for example Canada, Australia, and the United States—lies an "outer circle" where English is a second language (Bangladesh, Ghana, India, Kenya, Pakistan, Zambia, and many others) and where the regionally distinctive World Englishes are most obviously developing. Even farther out is an "expanding circle" of such countries as China, Egypt, Korea, Nepal, and Saudi Arabia where English is a foreign language and distinctive local variants in common usage have not yet developed.

Although the constant stream of print and digital communications between the variant regional Englishes make it likely that the common language will remain universally intelligible, it also seems probable that mutually incomprehensible forms of English will become entrenched as the language is taught, learned, and used in world areas far removed from contact with first-language users. "Our only revenge," said a French official deploring the declining role of French within the European Union, "is that the English language is being killed by all these foreigners speaking it so badly."

practices that join all those who adhere to them into a single moral community.

Religion may intimately affect all facets of a culture. Religious belief is, by definition, an element of the ideological subsystem; formalized and organized religion is an institutional expression of the sociological subsystem. And religious beliefs strongly influence attitudes toward the tools and rewards of the technological subsystem.

Nonreligious value systems—humanism or Marxism, for example—can be just as binding on the societies that espouse them as more-traditional religious beliefs. Even societies that largely reject religion, however, are strongly influenced by traditional values and customs set by predecessor religions—in days of work and rest or in legal principles, for example.

Because religions are formalized views on questions of ultimate significance, each carries a distinct conception of the meaning and value of this life, and most contain strictures about what must be done to achieve salvation (**Figure 6.24**). These beliefs become interwoven with the traditions of a culture. One cannot understand India without a knowledge of Hinduism, or Israel without an appreciation of Judaism.

Economic patterns may be intertwined with past or present religious beliefs. Traditional restrictions on food and drink may affect the kinds of animals that are raised or avoided, the crops that are grown, and the importance of those crops in the daily diet. Occupational assignment in the Hindu caste system is, in part, religiously supported. In many countries, there is a state religion; that is, religious and political structures are intertwined. Buddhism, for example, has been the state religion in Myanmar, Laos, and Thailand. By their official names, the Islamic Republic of Pakistan and the Islamic Republic of Iran proclaim their identity of religion and government. Despite the country's overwhelming Muslim majority, Indonesia sought and formerly found domestic harmony by recognizing five official religions and a state ideology—*pancasila*—whose first tenet is belief in one god.

Classification and Distribution of Religions

Religions may be unique to a single culture group, closely related to the faiths professed in nearby areas, or derived from belief systems in distant locations. Although interconnections and derivations among religions can frequently be discerned—as Christianity and Islam can trace descent from Judaism—family groupings are not as useful in classifying religions as they are in studying languages. A distinction between *monotheism*, belief in a single deity, and *polytheism*, belief in many gods, is frequently made but not particularly spatially relevant. It is more relevant to the spatial interests of geographers to categorize religions as *universalizing, ethnic*, or *tribal (traditional)*.

Christianity, Islam, and Buddhism are the major world **universalizing religions**, faiths that claim applicability to all humans and that seek to transmit their beliefs to all lands through missionary work and conversion. Membership in universalizing religions is open to anyone who chooses to make a symbolic commitment, such as baptism in Christianity. No one is excluded because of nationality, ethnicity, or previous religious belief.

Figure 6.24 **Worshipers gathered during hajj, the pilgrimage to Mecca.** The black structure is the Ka'ba, the symbol of Allah's (God's) oneness and of the unity of God and humans. Many rules concerning daily life are given in the Koran, the holy book of the Muslims. All Muslims are expected to observe the five pillars of the faith: (1) repeated saying of the basic creed; (2) prayers five times daily, facing Mecca; (3) a month of daytime fasting (Ramadan); (4) almsgiving; and (5) if possible, a pilgrimage to Mecca. © Rabi Karim Photography/Getty Images RF.

Ethnic religions have strong territorial and cultural group identification. One usually becomes a member of an ethnic religion by birth or by adoption of a complex lifestyle and cultural identity, not by a simple declaration of faith. These religions do not usually proselytize (attempt to convert nonbelievers), and their members often form distinctive closed communities identified with a particular ethnic group, region, or political unit. An ethnic religion—for example, Judaism, Indian Hinduism, or Japanese Shinto—is an integral element of a specific culture. To be part of the religion is to be immersed in the totality of the culture.

Tribal (or *traditional*) **religions** are special forms of ethnic religions distinguished by their small size, their unique identity with localized culture groups not yet fully absorbed into modern society, and their close ties to nature. *Animism* is the name given to their belief that life exists in all objects, from rocks and trees to lakes and mountains, or that such objects are the abode of the dead, of spirits, and of gods. *Shamanism* is a form of tribal religion that involves community acceptance of a *shaman* who, through special powers, can intercede with and interpret the spirit world.

The nature of the different classes of religions is reflected in their distributions over the world (**Figure 6.25**) and in their number of adherents. Universalizing religions tend to be expansionary, carrying their message to new peoples and areas. Ethnic religions, unless their adherents are dispersed, tend to be regionally confined or to expand only slowly and over long periods. Tribal religions tend to contract spatially as their adherents are incorporated into modern society and/or converted by proselytizing faiths.

As we expect in cultural geography, the map records only the latest stage of a constantly changing reality. While established religious institutions tend to be conservative and resistant to change, religion as a culture trait is dynamic. Personal and collective beliefs may alter in response to developing individual and societal needs and challenges. Religions may be imposed by conquest, adopted by conversion, defended in the face of hostility, suppressed by foes, or weakened by secularism.

Nor does the map present a full picture even of current religious regionalization or affiliation. Few societies are homogeneous, and most modern ones contain a variety of faiths or, at least, variants of the dominant professed religion. Some of those variants in many religions are intolerant or antagonistic toward other faiths or toward the sects and members of their own faith deemed insufficiently committed or orthodox (see "Militant Fundamentalism," p. 172).

Despite its many weaknesses, the world map of principal religions is important to understanding world events. For example, differences between Muslims and Hindus forced the partition of the Indian subcontinent after the departure of the British in 1947. While contemporary conflicts are mostly driven by economic or political strife rather than religious disagreements, people are often mobilized along ethnic or religious lines. Thus, a number of recent conflicts have occurred along religious boundaries such as those between Catholic and Protestant Christian groups in Northern Ireland; Muslim sects in Lebanon, Iran, and Iraq; Muslims and Jews in Palestine; Christians and Muslims in the Balkans, the Philippines, Nigeria, and Lebanon; and Buddhists and Hindus in Sri Lanka.

Frequently, members of a particular religion show areal concentration within a country. Thus, in urban Northern Ireland, Protestants and Catholics reside in separate areas whose boundaries are clearly understood and respected. The "Green Line" in Beirut, Lebanon, marked a guarded border between the Christian east and the Muslim west sides of the city, whereas, within the country as a whole, regional concentrations of adherents of different faiths and sects are clearly recognized (**Figure 6.26**). Religious diversity within countries may reflect the degree of toleration a majority culture affords minority religions. In dominantly (55% to 88%, depending on the definition) Muslim Indonesia, Christian Bataks, Hindu Balinese, and Muslim Javanese lived in peaceful coexistence for many years. By contrast, the fundamentalist Islamic regime in Iran has persecuted and executed those of the Baha'i faith.

Changing Place Names

Place names are powerful. The act of naming a place is to define its identity, at least in part. Toponyms demonstrate the power of the person or country that assigned the name and the legitimacy of the person for whom the place was named. Local governments have discovered that they can raise revenues by selling the naming rights to important, publicly-owned facilities such as stadiums, public buildings, and transit stations. In Dubai, in the United Arab Emirates, the names of several transit stations have been sold. Instead of being named for the streets or neighborhood where they are located, the stations carry corporate names such as First Gulf Bank station. The names of sports stadium names once reflected their location such as Riverfront Stadium in Cincinnati. Other stadiums were named as patriotic memorials such as Soldier Field in Chicago, or were named for the home team such as Yankee Stadium or Dodger Stadium. More recently, stadium names have become market commodities, reflecting corporate control over popular culture and the cultural landscape. Three Rivers Stadium in Pittsburgh was replaced by PNC Park, named for a financial service corporation. Philadelphia's Veterans Stadium was replaced by Citizens Bank Park and Lincoln Financial Field and the Hoosierdome in Indianapolis (Hoosier is an old nickname for Indiana residents) was first renamed RCA Dome and then replaced by Lucas Oil Field.

Toponyms reflect political power. Grosse Frankfurter Strasse, one of the most important streets in Berlin, Germany, was renamed Stalinallee by the communists in 1949, and then renamed Karl Marx Allee in 1961 when Stalin fell out of fashion. Proposals to revert to the original street name have been discussed. Colonial names imposed a new language and erased signs of indigenous culture. Thus, it is not surprising that post-colonial struggles would result in calls to change place names. In India, rising post-colonial pride resulted in decisions to undo anglicized city names: Bombay changed to *Mumbai*, Madras to *Chennai*, and Calcutta to *Kolkata*. In northern Canada, Indian and Inuit (Eskimo) place names are returning. The town of Frobisher Bay has reverted to its Eskimo name Iqaluit ("place of the fish") and Resolute Bay has become Kaujuitok ("place where the sun never rises") in Inuktitut, the language of the Canadian Eskimos.

The U.S. Board on Geographic Names is charged with standardizing foreign toponyms and designating official names of geographic features. Recently, it has received many proposals to eliminate derogatory or racist names and recover indigenous names. In 2015, proposals before the U.S. Board on Geographic Names included changing Squaw Valley Peak in California to Delunga Peak, Squaw Creek in Wisconsin to its Ho-Chunk name *Suukjak Sep* Creek, and Negro Run in Virginia to Freedom Run. The most significant recent action was the official name change of North America's highest mountain from Mount McKinley to *Denali*. The name "Mount McKinley" was given to commemorate President McKinley who was assassinated in 1901. However, President McKinley, a native of Ohio, had never travelled to Alaska and had no meaningful connections to the mountain. The name *Denali* means "the tall one" in the indigenous Athabascan language and had been used for centuries. Thus, in keeping with the wishes of Alaskans who had been petitioning for the change since the 1970s and against the wishes of Ohioans who had annually introduced legislation to stop the change, the name of the mountain was officially changed in 2015.

Minneapolis' Lake Calhoun was named in honor of Secretary of War John C. Calhoun who commissioned the 1817 army survey that was the first to map the lake. Calhoun served for South Carolina in both the U.S. Senate and the House and as Vice President of the United States where he was an outspoken defender of slavery. Almost two centuries after the lake was named in his honor, public pressure to eliminate symbols of racism began to focus on Lake Calhoun, even though most Minneapolis residents had no idea where the name came from. Activists

© Mark Bjelland.

succeeded in convincing the City of Minneapolis to install new signs listing both the official name and the original indigenous Dakota name—*Bde Maka Ska* which means "white earth lake."

Considering the Issues

1. Have any stadiums or other public facilities in your city or university sold their naming rights? What do the new names suggest about who holds cultural and economic power? Where would you draw the line in allowing public facilities and landscape features to have their naming rights sold to the highest bidder?

2. Birch Creek in Alaska recently reverted to its indigenous name K'iidòotinjik River. Some detractors might complain that a non-English name is difficult to spell and pronounce. Is maintaining language diversity and indigenous cultural identity worth the inconvenience to English speakers who must learn new, unfamiliar names?

3. Minneapolis' Lake Calhoun Parkway is home to many apartments, houses, and businesses. Is eliminating a symbol of racism worth the costs of changing road signs, maps, and street addresses.

4. Some defenders of the status quo argue that keeping old toponyms is an important part of remembering the past in order to avoid repeating it, even if that past was racist, sexist, oppressive, or violent. Do you think it is important to keep offensive toponyms in order to remember the past or is eliminating those toponyms an important step in overcoming the evils of the past? Where do we draw the line in renaming places?

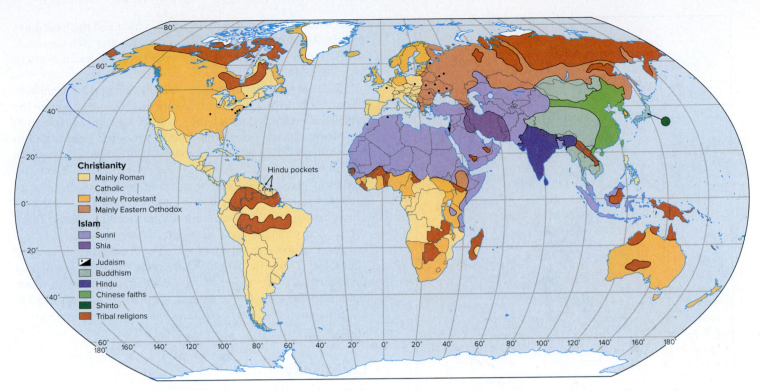

Figure 6.25   Principal world religions. The assignment of individual countries to a single religion category conceals a growing intermixture of faiths in countries that have experienced major immigration flows or religious change. In some instances, those influxes are altering the effective, if not the numerical, religious balance. In nominally Christian, Catholic France, for example, low church-going rates suggest that now more Muslims than practicing Catholics reside there and, considering birth rate differentials, that someday Islam may be the country's predominant religion as measured by the number of practicing adherents. Secularism—rejection of religious belief—is common in many countries but is not locationally indicated on this map. Areas of sub-Saharan Africa labeled Christian are intermixed with tribal religions they have displaced.

One cannot assume that all people within a mapped religious region are adherents of the designated faith, nor can it be assumed that membership in a religious community means active participation in its belief system. *Secularism,* an indifference to or rejection of religion and religious belief, is an increasing part of many modern societies, particularly of the industrialized countries and those now or recently under communist regimes. In England, for example, the state Church of England claims 20% of the British as communicants, but only 2% of the population attends its Sunday services. Even in devoutly Roman Catholic South American states, low church attendance attests to the rise of at least informal secularism. In Colombia, only 18% of the people attend Sunday services; in Chile, the figure is 12%; in Mexico, 11%; and in Bolivia, 5%.

The Principal Religions

Each of the major religions has its own unique mix of cultural values and expressions, each has had its own pattern of innovation and spatial diffusion (**Figure 6.27**), and each has had its own impact on the cultural landscape. Together, they contribute importantly to the worldwide pattern of human diversity.

Judaism

We can begin our review of world faiths with *Judaism,* whose belief in a single God laid the foundation for both Christianity and Islam. Unlike its universalizing offspring, Judaism is closely identified with a single ethnic group and with a complex and restrictive set of beliefs and laws. It emerged some 3000 to 3500 years ago in the Near East, one of the ancient culture hearth regions (see Figure 6.13).

Judaism is a distinctively *ethnic* religion, the determining factors of which are descent from Israel (the patriarch Jacob), the Torah (law and scripture), and the traditions of the culture and the faith. Early military success gave the Jews a sense of territorial and political identity to supplement their religious self-awareness. Later conquests by nonbelievers led to their dispersion (*diaspora*) to much of the Mediterranean world and farther east into Asia by A.D. 500 (**Figure 6.28**).

During the 13th and 14th centuries, many Jews sought refuge in Poland and Russia from persecution in Western and Central Europe; during the later 19th and early 20th centuries, Jews were important elements of the European immigrant stream to the Western Hemisphere. The mass annihilation of Jews in Europe before and during World War II—the Holocaust—drastically reduced their presence on that continent. The establishment of the state of Israel in 1948 was a fulfillment of the goal of *Zionism,* the belief in the need to create an autonomous Jewish state in Palestine. It demonstrated a determination that Jews not lose their identity by absorption into alien cultures and societies. It also spatially united two earlier, separated Jewish communities: the Sephardim, who were expelled from Iberia in the late

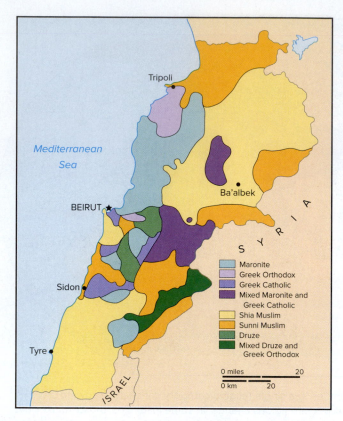

Figure 6.26   Religious regions of Lebanon. Long-standing religious territoriality and rivalry led, in the 1960s and 1970s, to open conflict between Muslims and Christians and among various branches of each major faith in this eastern Mediterranean country.

15th century, fleeing initially to North Africa and the Near East, and the Ashkenazim, who, between the 13th and 16th centuries, sought refuge in Eastern Europe from persecutions in Western and Central Europe.

Judaism's imprint upon the cultural landscape has been subtle and unobtrusive. The Jewish community reserves space for the practice of communal burial; the spread of the cultivated citron in the Mediterranean area during Roman times has been traced to Jewish ritual needs; and the religious use of grape wine ensured the cultivation of the vine in their areas of settlement. The synagogues as places of worship have tended to be less elaborate or architecturally distinctive than those of other major world religions. Synagogues feature an ark (cabinet) containing the Torah scrolls and generally face Jerusalem. However, what is essential for a religious service is merely the presence of at least 10 adults, not a specific structure. Orthodox Jews are a subgroup that adheres to a stricter set of beliefs and practices, one of which forbids driving a car on the Sabbath—the day of worship. To follow this simple rule, Orthodox Jews tend to live close together in cities.

Christianity

Christianity had its origin in the life and teachings of Jesus, a Jewish preacher during the 1st century of the common era, whom his followers believed was the messiah promised by God. The new covenant he preached was not a rejection of traditional Judaism but a promise of salvation to all humankind, rather than to just a chosen people.

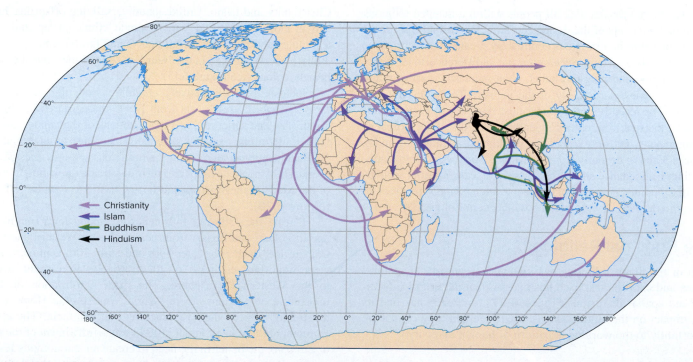

Figure 6.27   Innovation areas and diffusion routes of major world religions. The monotheistic (single-deity) faiths of Judaism, Christianity, and Islam arose in southwestern Asia, the first two in Palestine in the eastern Mediterranean region and the latter in western Arabia near the Red Sea. Hinduism and Buddhism originated within a confined hearth region in the northern part of the Indian subcontinent. Their rates, extent, and directions of spread are suggested here and detailed on later maps.

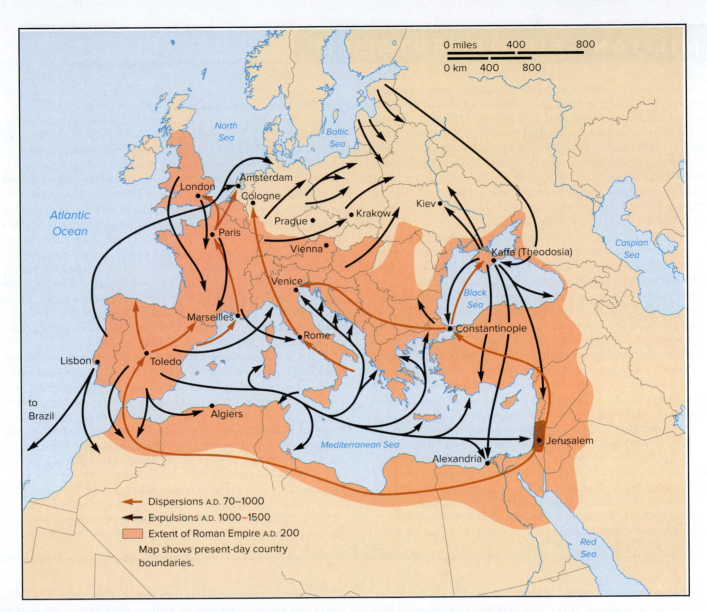

Figure 6.28   Jewish dispersions, A.D. **70–1500.** A revolt against Roman rule in A.D. 66 was followed by the destruction of the Jewish Temple 4 years later and an Imperial decision to Romanize the city of Jerusalem. Judaism spread from the hearth region, carried by its adherents dispersing from their homeland to Europe, Africa, and eventually in great numbers to the Western Hemisphere. Although Jews established themselves and their religion in new lands, they did not lose their sense of cultural identity nor did they seek to attract converts to their faith.

Christianity's mission was conversion, and missionary work was critical in its diffusion. As a universal religion of salvation and hope, it spread quickly among the underclasses of both the eastern and western parts of the Roman Empire, carried to major cities and ports along the excellent system of Roman roads and sea lanes (**Figure 6.29**). In A.D. 313, Emperor Constantine proclaimed Christianity the state religion. Much later, of course, the faith was brought to the New World with European settlement (see Figure 6.27).

The dissolution of the Roman Empire into a western half and eastern half after the fall of Rome also divided Christianity. The Western Church, based in Rome, was one of the very few stabilizing and civilizing forces uniting Western Europe during the Dark Ages. Its bishops became the civil as well as ecclesiastical authorities over vast areas devoid of other effective government. Parish churches were the focus and organizational structure for rural and urban life, and cathedrals replaced Roman monuments and temples as the symbols of the social order.

Secular imperial control endured in the Eastern Empire, whose capital was Constantinople (now known as Istanbul). Thriving under its protection, the Eastern Church expanded into the Balkans, Eastern Europe, Russia, and the Near East. The fall of the Eastern Empire to the Turks in the 15th century opened Eastern Europe temporarily to Islam, though the Eastern Orthodox Church (the direct descendant of the state church of the Eastern Roman Empire) remains, in its various ethnic branches, a major component of Christianity.

The Protestant Reformation of the 15th and 16th centuries split the western church, leaving Roman Catholicism supreme in Southern Europe but installing a variety of Protestant denominations and national churches in Western and Northern Europe.

MILITANT FUNDAMENTALISM

The term *fundamentalism* describes reactionary, ultraconservative religious movements. Originally, it designated a Christian movement in the United States named after a set of book—*The Fundamentals: A Testimony of the Truth*—published between 1910 and 1915. These early fundamentalists reasserted traditional religious orthodoxy in response to challenges posed by science and secularism. More recently, fundamentalism has become a generic description for all religious movements that seek to regain and publicly institutionalize traditional social and cultural values that are usually rooted in the teachings of a sacred text or written dogma.

While most religious believers have been able to adjust their beliefs and practices to accommodate modernization, scientific understandings of the world, and the presence of other faiths, some have reacted strongly against such changes. Fundamentalism is now found in every dominant religion, wherever a modernized society has developed, including Islam, Christianity, Hinduism, Judaism, Sikhism, Buddhism, Confucianism, and Zoroastrianism. As a reaction against the modern world, fundamentalism represents an effort to draw upon the religious traditions of a "golden age" in order to counteract a changing society that is believed to undermine the true faith. The near universality of fundamentalist movements can be seen as an expression of a widespread rebellion against the disruptions and presumed evils fostered by secular globalization.

Fundamentalists place a high priority on doctrinal conformity. Further, they are convinced of the correctness of their beliefs and the necessity of the unquestioned acceptance of those beliefs. Fundamentalists have gained public attention when they have attempted to have their beliefs taught in schools or enforced through government legislation. To some observers, fundamentalism is, by its nature, undemocratic, and states controlled by fundamentalist regimes combining politics and religion of necessity stifle debate and punish dissent. In the modern world,

that rigidity seems most apparent in Islam where, it is claimed, "all Muslims believe in the absolute inerrancy of the Quran . . ." (*The Islamic Herald,* April 1995) and several countries—for example, the Islamic Republic of Iran and the Islamic Republic of Pakistan—proclaim by official name their administrative commitment to religious control.

In most of the modern world, however, such commitment is not overt or official, and fundamentalists often believe that they and their religious convictions are under mortal threat. They view modern secular society—with its assumption of equality of competing voices and values—as trying to eradicate the true faith and religious verities. Initially, therefore, every fundamentalist movement begins as an intrareligious struggle directed against its own coreligionists and countrymen in response to a felt assault by the liberal or secular society they inhabit. At first, group members may blame their own weakness and irresolution for the oppression they feel and the general social decay they perceive. To restore society to its idealized standards, the aroused group may exhort its followers to ardent prayer, ascetic practices, and physical or military training.

If it is unable to impose its beliefs on others peacefully, the fundamentalist group—seeing itself as the savior of society—may justify other, more-extreme actions against perceived oppressors. Initial protests and nonviolent actions may escalate to attacks on corrupt public figures thwarting their vision and to outright terrorism. That escalation is advanced and gains willing supporters when inflexible fundamentalism is combined with the unending poverty and political impotence felt in many, particularly Middle Eastern, societies today. When an external culture or power—commonly, a demonized United States—is seen as the unquestioned source of the corruption and exploitation frustrating their social vision, some fundamentalists have been able to justify any extreme action and personal sacrifice for their cause. In their struggle, it appears an easy progression from domestic dispute to international terrorism.

The split was reflected in the subsequent worldwide dispersion of Christianity. Catholic Spain and Portugal colonized Latin America, taking both their languages and the Roman Church to that area (see Figure 6.27), as they did to colonial outposts in the Philippines, India, and Africa. Catholic France colonized Quebec in North America. Protestants, many of them fleeing Catholic or repressive Protestant state churches, were primary early settlers of the United States, Canada, Australia, New Zealand, Oceania, and South Africa.

Although religious intermingling rather than rigid territorial division is characteristic of the contemporary American scene (**Figure 6.30**), the beliefs and practices of various immigrant groups and the innovations of domestic congregations have created a particularly varied spatial patterning of "religious regions" in the United States (**Figure 6.31**).

The mark of Christianity on the cultural landscape has been conspicuous and enduring. In pre-Reformation Catholic Europe, the parish church formed the center of life for small neighborhoods of every town; the village church was the centerpiece of every rural community; and in larger cities the central cathedral served simultaneously as a glorification of God, a symbol of piety, and the focus of religious and secular life (**Figure 6.32a**).

Protestantism placed less importance on the church as a monument and symbol, although in many communities—colonial New England, for example—the churches of the principal denominations were at the village center (**Figure 6.32b**). Many were adjoined by a cemetery, because Christians—in common with Muslims and Jews—practice burial in areas reserved for the dead. In Christian countries, particularly, the cemetery—whether connected to the church, separate from it, or unrelated to a specific denomination—has traditionally been a significant land use within urban areas.

Islam

Islam springs from the same Judaic roots as Christianity and embodies many of the same beliefs: there is only one God, who can be revealed to humans through prophets; Adam was the first human; Abraham was one of his descendants. Mohammed is revered as the prophet of *Allah* (God), succeeding and completing the work of earlier prophets of Judaism and Christianity, including Moses, David, and Jesus. The Koran, the word of Allah revealed to Mohammed, contains not only rules of worship and details of doctrine but also instructions on the conduct of human affairs. For fundamentalists, it thus becomes the unquestioned guide to

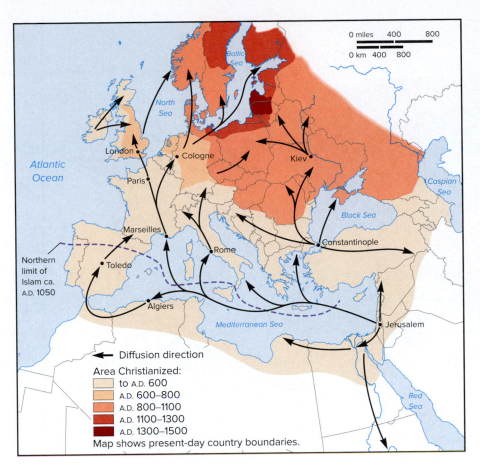

Figure 6.29 Diffusion paths of Christianity, A.D. 100–1500. Routes and dates are for Christianity as a composite faith. No distinction is made between the Western Church and the various subdivisions of the Eastern Orthodox denominations.

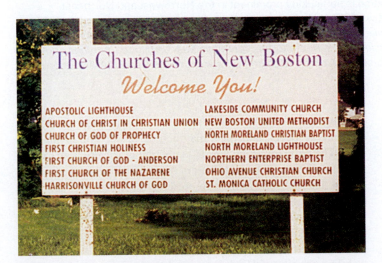

Figure 6.30 Advertised evidence of religious diversity in the United States. The sign details only a few Christian congregations. In reality, the United States has become the most religiously diverse country in the world with essentially all of the world's faiths represented within its borders. Welcoming signs for other, particularly larger, towns might also announce Muslim, Hindu, Buddhist, and many other congregations in their varied religious mix. © Susan Reisenweaver.

matters both religious and secular. Of central importance is observance of the "five pillars" of Islam (see Figure 6.24), two of which are distinctly geographical—prayer facing toward Mecca (based on the shortest route determined using a globe) and the pilgrimage to Mecca. Surrender to the will of Allah and prayer and Koran recitations in Arabic unite the faithful into a community that has no concern with race, color, language, or caste.

It was that law of community that unified an Arab world sorely divided by tribes, social ranks, and multiple local deities. Mohammed was a resident of Mecca but fled in A.D. 622 to Medina, where the Prophet proclaimed a constitution and announced the universal mission of the Islamic community. That flight—*Hegira*—marks the starting point of the Islamic (lunar) calendar. By the time of Mohammed's death in A.H. 11 (*Anno*—in the year of—*Hegira*, or A.D. 632), all of Arabia had joined Islam. Islam diffused rapidly through Islamic political and military expansion. It swept quickly outward from its Arabian hearth region across North Africa, much of Central Asia, and, at the expense of Hinduism, into northern India (**Figure 6.33**). Later, Islam dispersed into Indonesia, southern Africa, and the Western Hemisphere. It continues its spatial spread and numerical growth as the fastest-growing major religion at the present time.

Disagreements over the succession of leadership after the Prophet led to a division between two primary groups, Sunnis and Shi'ites. Sunnis, the majority (80% to 85% of Muslims) recognize the first four *caliphs* (originally, "successor" and later the title of the religious and civil head of the Muslim state) as Mohammed's rightful successors. The Shi'ites reject the legitimacy of the first three and believe that Muslim leadership rightly belonged to the fourth caliph, the Prophet's son-in-law, Ali, and his descendants. Today, Sunnis constitute the majority of Muslims in all countries except Iran, Iraq, Bahrain, and perhaps Yemen.

The mosque—place of worship, community clubhouse, meeting hall, and school—is the focal point of Islamic communal life and the primary imprint of the religion on the cultural landscape. Its principal purpose is to accommodate the Friday communal service, mandatory for all male Muslims. It is the congregation rather than the structure that is important; small or poor communities are as well served by a bare, whitewashed room as are larger cities by architecturally splendid mosques. With its perfectly proportioned, frequently gilded or tiled domes; its graceful, soaring towers and minarets (from which the faithful are called to prayer); and its delicately wrought parapets and cupolas, the carefully tended mosque is frequently the most elaborate and imposing structure of the town (**Figure 6.34**).

Hinduism

Hinduism is the world's oldest major religion. Though it has no datable founding event or initial prophet, some evidence traces its origin

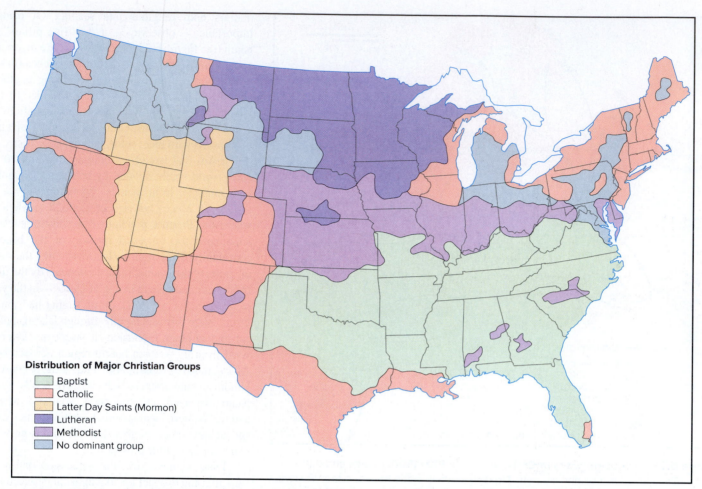

Distribution of Major Christian Groups

- ▢ Baptist
- ▢ Catholic
- ▢ Latter Day Saints (Mormon)
- ▢ Lutheran
- ▢ Methodist
- ▢ No dominant group

Figure 6.31   Religious affiliation in the conterminous United States. The greatly generalized areas of religious dominance shown conceal the reality of immense diversity of church affiliations throughout the United States. "Major" simply means that the indicated category had a higher percentage response than any other affiliation; usually the value was below 50%. A sizable and growing number of Americans claim to have "no religion." Secularism or the absence of religious affiliation is particularly prominent in the Northwest and Northeast. Rates of "no religious affiliation" vary from a high of 83% in Maine to a low of 21% in Utah. *Sources: Based on data or maps from: the 2001 "American Religious Identity Survey" by the Graduate School at City University of New York; religious denomination maps prepared by Ingolf Vogeler of the University of Wisconsin, Eau-Claire, based on data compiled by the Roper Center for Public Research; and Churches and Church Membership in the United States (Atlanta, Georgia: Glenmary Research Center, 1992) and U.S. Membership Report (Association of Religion Data Archives, 2010).*

back 4000 or more years. Hinduism is an ethnic religion, an intricate web of religious, philosophical, social, economic, and artistic elements constituting a distinctive Indian civilization. Its estimated 1 billion adherents are primarily Asian and largely confined to India, where it claims 80% of the population.

From its cradle area in the valley of the Indus River, Hinduism spread eastward down the Ganges River and southward throughout the subcontinent and adjacent regions by amalgamating, absorbing, and eventually supplanting earlier native religions and customs. Its practice eventually spread throughout Southeast Asia and into Indonesia, Malaysia, Cambodia, Thailand, Laos, and Vietnam, as well as into neighboring Myanmar and Sri Lanka. The largest Hindu temple complex is in Cambodia, not India, and Bali remains a Hindu pocket in dominantly Islamic Indonesia.

There is no common creed, single doctrine, or central ecclesiastical organization defining the Hindu. A Hindu is one born into a caste, a member of a complex social and economic—as well

as religious—community. Hinduism accepts and incorporates all forms of belief; adherents may believe in one god or many or none. The *caste* (meaning "birth") structure of society is an expression of the eternal transmigration of souls. For Hindus, the primary aim of this life is to conform to prescribed social and ritual duties and to the rules of conduct for the assigned caste and profession. Those requirements constitute that individual's *dharma*—law and duties. Traditionally, each craft or profession is the property of a particular caste.

The practice of Hinduism is rich with rites and ceremonies, festivals and feasts, pilgrimages to holy rivers and sacred places, processions, and ritual gatherings of millions of celebrants. It involves careful observance of food and marriage rules and the performance of duties within the framework of the hierarchical caste system. Worship in the temples or shrines (**Figure 6.35**) and the leaving of offerings to secure merit from the gods are required. The temples, shrines, daily rituals and worship, numerous specially

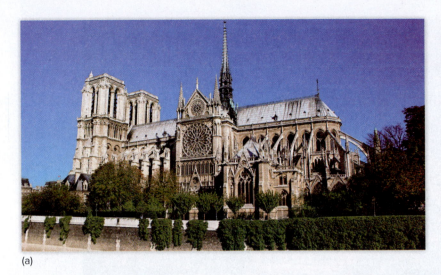
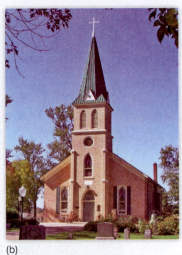

(a)

(b)

Figure 6.32 **In Christian societies, the church assumes a prominent central position in the cultural landscape.** (a) The building of Nôtre Dame Cathedral of Paris, France, begun in 1163, took more than 100 years to complete. Between 1170 and 1270, some 80 cathedrals were constructed in France alone. The cathedrals in all of Catholic Europe were located in the centers of major cities. Their plazas were the sites of markets, public meetings, and religious ceremonies. (b) Individually less imposing than the central cathedral of Catholic areas, the several Protestant churches common in small and large American towns collectively constitute an important land use frequently sited in the center of the community. The Lutheran church and cemetery shown here were built by Swedish immigrants to Minnesota. *(a) © Fuse/Getty Images RF; (b) © Mark Bjelland.*

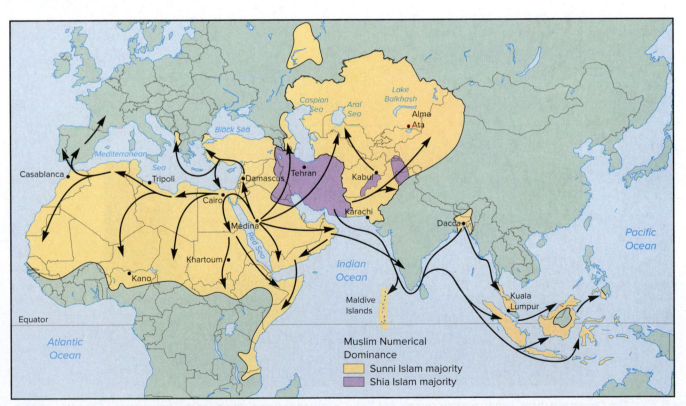

Figure 6.33 **Spread and extent of Islam.** Islam predominates in over 35 countries along a band across northern Africa to central Asia and the northern part of the Indian subcontinent. Still farther east, Indonesia has the largest Muslim population of any country. Islam's greatest development is in Asia, where it is second only to Hinduism, and in Africa, where some observers suggest it may be the leading faith. Current Islamic expansion is particularly rapid in the Southern Hemisphere.

garbed or marked holy men and ascetics, and ever-present sacred animals mark the cultural landscape of Hindu societies, a landscape infused with religious symbols and sights that are part of a total cultural experience.

Buddhism

Numerous reform movements have derived from Hinduism over the centuries, some of which have endured to the present day as

Figure 6.34   The Blue Mosque in Istanbul, Turkey. The common architectural features of the mosque, particularly the slender minaret towers, make it an unmistakable landscape feature of Islam. © *Glen Allison/Photodisc/Getty Images RF.*

Figure 6.35   The Chennakeshava Hindu temple complex at Belur, Karnataka, in southern India. The creation of temples and the images they house has been a principal outlet of Indian artistry for more than 3000 years. At the village level, the structure may be simple, containing only the windowless central cell housing the divine image, a surmounting spire, and the temple porch or stoop to protect the doorway of the cell. The great temples, of immense size, are ornate extensions of the same basic design. © *Allison Bohn.*

major religions on a regional or world scale. For example, *Sikhism* developed in the Punjab area of northwestern India in the late 15th century A.D., combining elements of both Hinduism and Islam and generally understood to be a syncretism of them. Sikhism rejects the formalism of both and proclaims a gospel of universal tolerance. The great majority of some 20 million Sikhs still live in India, mostly in the Punjab, though others have settled in Malaysia, Singapore, East Africa, the United Kingdom, and North America.

The largest and most influential of the dissident movements is *Buddhism,* a universalizing faith founded in the 6th century B.C. in northern India by Siddhartha Gautama, the Buddha ("Enlightened One"). The Buddha's teachings were more like a moral philosophy that offered an explanation for evil and human suffering than a formal religion. He viewed the road to enlightenment and salvation to lie in understanding the "four noble truths": existence involves suffering; suffering is the result of desire; pain ceases when desire is destroyed; the destruction of desire comes through knowledge of correct behavior and correct thoughts. The Buddha instructed his followers to carry his message as missionaries of a doctrine open to all castes, for no distinction among people was recognized. In that message, all could aspire to ultimate enlightenment, a promise of salvation that raised the Buddha in popular imagination from teacher to inspiration and Buddhism from philosophy to universalizing religion.

The belief system spread throughout India, where it was made the state religion in the 3rd century B.C. It was carried elsewhere into Asia by missionaries, monks, and merchants. While expanding abroad, Buddhism began to decline at home as early as the 4th century A.D., slowly but irreversibly reabsorbed into a revived Hinduism. By the 8th century, its dominance in northern India had been broken by conversions to Islam, and by the 15th century, it had essentially disappeared from all of the subcontinent.

Present-day spatial patterns of Buddhist adherence reflect the schools of thought, or *vehicles,* that were dominant during different periods of dispersion of the basic belief system (**Figure 6.36**).

In all of its many variants, Buddhism imprints its presence vividly on the cultural landscape. Buddha images in stylized human form began to appear in the 1st century A.D. and are common in painting and sculpture throughout the Buddhist world. Equally widespread are the three main types of buildings and monuments: the *stupa* (**Figure 6.37**), a commemorative shrine; the temple or pagoda enshrining an image or a relic of the Buddha; and the monastery, some of them the size of small cities.

East Asian Ethnic Religions

When Buddhism reached China from the south some 1500 to 2000 years ago and was carried to Japan from Korea in the 6th century, it encountered and later amalgamated with already well-established ethical belief systems. The Far Eastern ethnic religions are syncretisms. In China, the union was with Confucianism and Taoism, themselves becoming intermingled by the time of Buddhism's arrival, and in Japan, it was with Shinto, a polytheistic animism and shamanism.

Chinese belief systems address not so much the hereafter as the achievement of the best possible way of life in the present existence. They are more ethical or philosophical than religious in the pure sense. Confucius (K'ung Fu-tzu), a compiler of traditional wisdom who lived about the same time as Gautama Buddha, emphasized

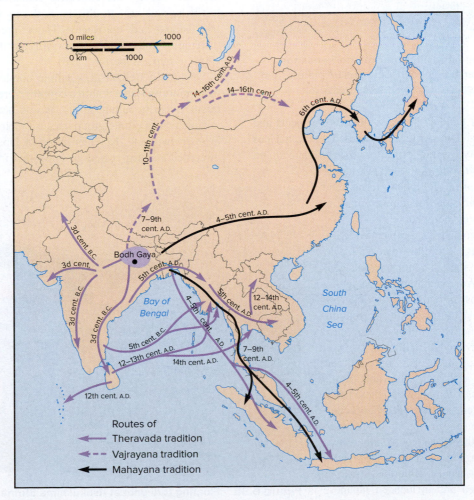

Figure 6.36   Diffusion paths, times, and "vehicles" of Buddhism.

Figure 6.37   **The gold-embellished stupa at the Swedagon pagoda in Yangon, Myanmar** (Rangoon, Burma) is 98 meters (322 ft) tall. © *Medioimages/Photodisc/Getty Images RF.*

the importance of proper conduct between ruler and subjects and between family members. The family was extolled as the nucleus of the state, and filial piety was the loftiest of virtues. There are no temples or clergy in *Confucianism,* though its founder believed in a heaven seen in naturalistic terms, and the Chinese custom of ancestor worship as a mark of gratitude and respect was encouraged.

Confucianism was joined by, or blended with, *Taoism,* an ideology that, according to legend, was first taught by Lao Tsu in the 6th century B.C. Its central theme is *Tao* (the Way), a philosophy teaching that eternal happiness lies in totally identifying with nature and deploring passion, unnecessary invention, unneeded knowledge, and government interference in the simple life of individuals. Buddhism, stripped by Chinese pragmatism of much of its Indian other-worldliness and defining a *nirvana* achievable in this life, was easily accepted as a companion to these traditional Chinese belief systems. Along with Confucianism and Taoism, Buddhism became one of the honored Three Teachings, and to the average person there was no distinction in meaning or importance between a Confucian temple, Taoist shrine, or Buddhist stupa.

Buddhism also joined and influenced Japanese Shinto, the traditional religion of Japan, which

developed out of nature and ancestor worship. *Shinto*—the Way of the Gods—is basically a structure of customs and rituals rather than an ethical or a moral system. It observes a complex set of deities, including deified emperors, family spirits, and the divinities residing in rivers, trees, certain animals, mountains, and particularly the sun and moon. At first resisted, Buddhism was later amalgamated with traditional Shinto. Buddhist deities were seen as Japanese gods in a different form, and Buddhist priests formerly but no longer assumed control of most of the numerous Shinto shrines in which the gods are believed to dwell and which are approached through ceremonial *torii,* or gateway arches (**Figure 6.38**).

6.8 Ethnicity

Any discussion of cultural diversity would be incomplete without the mention of **ethnicity.** Based on the root word *ethnos,* meaning "people" or "nation," the term is usually used to refer to the ancestry of a particular people who have in common distinguishing characteristics associated with their heritage. No single trait denotes ethnicity. Recognition of ethnic communities may be based on language, religion, national origin, unique customs, or, improperly, an ill-defined concept of "race" (see "The Matter of Race," p. 179). Whatever the unifying thread, ethnic groups may strive to preserve their special shared ancestry and cultural heritage through the collective retention of language, religion, festivals, cuisines, traditions, and in-group work relationships, friendships, and marriages. Those preserved associations are fostered by and support *ethnocentrism,* the feeling that one's own ethnic group is superior.

Normally, reference to ethnic communities is recognition of their minority status within a country or region dominated by a different, majority culture group. We do not identify Koreans living in Korea as an ethnic group because theirs is the dominant culture in their own land. Koreans living in Japan, however, constitute a discerned and segregated group in that foreign country. Ethnicity,

Figure 6.38   **Floating torii gate at Itsukushima Shrine on Miyahima Island, Japan.** © *GeoStock/Getty Images RF.*

THE MATTER OF RACE

Human populations may be differentiated from one another on any number of bases: gender, nationality, stage of economic development, and so on. Race and ethnicity are common ways to differentiate people groups and are frequently equated with each other when in fact race and ethnicity are very different concepts. **Race** is an outdated categorization of humans based on visible characteristics such as skin color, hair texture, or eye color and shape.

While humans are all one species that can freely interbreed and produce fertile offspring, there is obvious variation in our physical characteristics. The spread of human beings over the Earth and their occupation of different environments were accompanied by the development of variations in visible characteristics, such as skin pigmentation, hair and eye color, and hair texture, as well as internal differences, such as blood composition or lactose intolerance. Physical differentiation among human groups is old and can reasonably be dated to the Paleolithic (100,000 to about 11,000 years ago) spread and isolation of population groups.

Geographic patterns of distinct combinations of physical traits emerged due to causative forces of evolutionary **natural selection** or **adaptation**, and **genetic drift**. Natural selection favors the transmission of characteristics that enable humans to adapt to a particular environmental feature, such as climate. Studies have suggested some plausible relationship between, for example, solar radiation and skin color and between temperature and body size. Dark skin indicates the presence of melanin, which protects against the penetration of damaging ultraviolet rays from the sun. Conversely, the production of vitamin D in the body, which is necessary to good health, is linked to the penetration of ultraviolet rays. In high latitudes where winter days are short and the sun is low in the sky, light skin confers an adaptive advantage by allowing the production of vitamin D.

Genetic drift is the process by which a heritable trait appears by chance in a group and is accentuated by inbreeding. If two populations are too spatially separated for much interaction to occur (*isolation*), a trait may develop in one but not in the other. Unlike natural selection, genetic drift differentiates populations in nonadaptive ways. Natural selection and genetic drift promote differentiation. Countering them is *gene flow* via interbreeding, which homogenizes neighboring populations. Opportunities for interbreeding, always part of the spread and intermingling of human populations, have accelerated with the growing mobility and migrations of people in the past few centuries.

Racial categorization is a scientifically outdated way of making sense of human variation. Focusing on visible physical characteristics, anthropologists in the 18th and 19th centuries created a variety of racial classification schemes, most of which derived from geographic variations of populations. Some anthropological studies at that time attempted to link physical traits with mental ability in order to construct racial hierarchies that were used to justify slavery, imperialism, immigration restrictions, anti-miscegenation laws, and eugenics. Contemporary biology has rejected racial categorization as a meaningful description of human variation. Skin color does not correspond to genetic closeness between "racial" groups. Further, pure races do not exist, and DNA-based evidence shows that there is greater variation within the so-called racial groups than there is between the groups.

Living in a society where racial categorization has been widespread, we may be tempted to group humans racially and attribute intellectual ability, athletic prowess, or negative characteristics to particular racial groups. This is problematic for many reasons, the most important being that geneticists have rejected race as a scientific concept—there is only one race, the human race. Second, intellectual ability as measured on standardized tests is strongly influenced by socioeconomic status. Third, the athletic abilities displayed by top athletes are the property of particular individuals, not a group trait; and like intellectual ability, they are strongly influenced by social factors.

Nor does race have meaningful application to any human characteristics that are culturally acquired. That is, race is *not* equivalent to ethnicity or nationality and has no bearing on differences in religion or language. There is no "Irish" or "Hispanic" race, for example. Such groupings are based on culture, not genes. Culture summarizes the way of life of a group of people, and members of the group may adopt it irrespective of their individual genetic heritage, or race. Nevertheless, despite the fact that the older view of race as a biological category has been thoroughly discredited, race and ethnicity remain as defining and divisive realities in American society. Both are deeply rooted in individual and group consciousness, and both are strongly ingrained in the country's social and institutional life. While biological notions of race have little meaning, the society itself is extremely "racialized."

If racial categorization were scientifically valid, the categories would be universal. Instead, they vary widely from country to country, reflecting the unique history and geography of particular states. In 2000, the U.S. Bureau of the Census asked respondents to classify themselves into one of five racial categories and answer a separate question about Hispanic status (which is considered an ethnic category). For the 2010 census, people had their choice of 14 racial categories, and continuing an option started in 2000, they could classify themselves as belonging to more than one racial group.

---

therefore, is an evidence of areal cultural diversity and a reminder that culture regions are rarely homogeneous in the characteristics displayed by all of their occupants.

Territorial segregation is a strong and sustaining trait of ethnic identity, one that helps groups retain their distinction. On the world scene, indigenous ethnic groups have developed over time in specific locations and have established themselves in their own and others' eyes as distinctive peoples with defined homeland areas. The boundaries of most countries of the world encompass a number of racial or ethnic minorities (**Figure 6.39**). Their demands for special territorial recognition have sometimes increased with advances in economic development and self-awareness, as Chapter 8 points out. Where clear territorial separation does not exist but ethnic identities are distinct and animosities bitter, tragic conflict within single political units can erupt. Recent histories of deadly warfare between Tutsi and Hutu in Rwanda, or Serb and Croat in Bosnia, make vivid the often continuing reality of ethnic discord and separatism.

Increasingly in a world of immigration and refugee movements, ethnicity is less a matter of indigenous populations and more one of outsiders in an alien culture. Immigrants, legal and illegal,

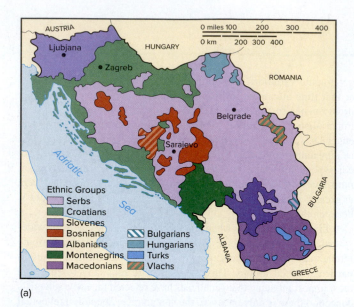

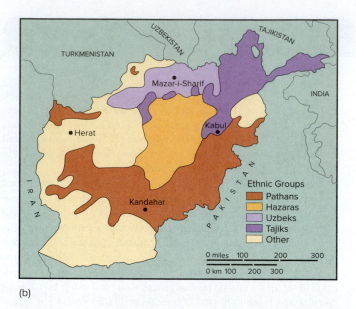

(a)   (b)

Figure 6.39   (a) Ethnicity in former Yugoslavia. Yugoslavia was formed after World War I (1914–1918) from a patchwork of Balkan states and territories, including the former kingdoms of Serbia and Montenegro, Bosnia-Herzegovina, Croatia-Slavonia, and Dalmatia. The authoritarian central government created in 1945 began to disintegrate in 1991 as non-Serb minorities voted for regional independence and national separation. Religious differences between Eastern Orthodox, Roman Catholic, and Muslim adherents compounded resulting conflicts rooted in nationality and rival claims to ethnic homelands. **(b) Afghanistan** houses Pathan, Tajik, Uzbek, and Hazara ethnic groups (among others) speaking Pashto, Dari Persian, Uzbek, and several minor languages and split between majority Sunni and minority Shia Muslim believers.

and refugees from war, famine, or persecution are a growing presence in countries throughout the world. Immigrants to a country typically have two choices. They may hope for *assimilation* by giving up many of their past culture traits, losing their distinguishing characteristics, and merging into the mainstream of the dominant culture, as reviewed earlier on pages 155–156. Or they may try to retain their distinctive cultural heritage. In either case, they usually settle initially in an area where other members of their ethnic group live, as a place of refuge and learning (**Figure 6.40**). With the passage of time, they may leave their protected community and move out among the general population.

The Chinatowns, Little Havanas, and Little Italys of North American cities have provided the support systems essential to new immigrants in an alien culture region. Japanese, Italians, Germans, and other ethnics have formed agricultural colonies in Brazil in much the same spirit. Such ethnic enclaves may provide an entry station, allowing both individuals and the groups to which they belong to undergo cultural and social modifications sufficient to enable them to operate effectively in the new, majority society.

Due to rising affluence among immigrants, increasingly ethnic communities are found in the suburbs of major metropolitan areas. This has given rise to the **ethnoburb,** a suburban community with a significant, though not exclusive, concentration of a single ethnic group. Monterey Park, outside Los Angeles, California, and Richmond, British Columbia, outside Vancouver are examples of Chinese

Figure 6.40   The distinctive architecture, lightposts, and signs in the Chinatown area of Vancouver, Canada, proclaim the existence of a long-established Chinese community. Newer, suburban Chinatowns have also cropped up in the suburbs of Vancouver. Chinatowns, Little Italys, and other ethnic enclave counterparts in cities and their suburbs throughout the United States and Western Europe provide both the spatial refuge and the support systems essential to new arrivals in an alien culture realm. © Mark Bjelland.

ethnoburbs. Ethnoburbs attract relatively prosperous, well-educated, highly mobile immigrants. Many of the immigrants in ethnoburbs display *transnationalism;* that is, they maintain strong ties with more than one country, often in the form of ongoing social and business connections with their homeland.

Sometimes, of course, settlers have no desire to assimilate or are not allowed to assimilate, so that they and their descendants form a more or less permanent subculture in the larger society. The Chinese in Malaysia belong to this category. Ethnicity in the context of nationality is discussed more fully in Chapter 8.

6.9 Gender and Culture

Gender refers to socially created—not biologically based—distinctions between femininity and masculinity. Because gender relationships and role assignments differ between societies, the status of women is a cultural variable of geographic interest and inquiry. Gender distinctions are complex, and the roles and privileges assigned to males and females differ from society to society. In fundamental ways, those assignments are conditioned by geographically uneven levels of economic development. Therefore, we might well assume a close similarity between the economic roles of males and females—and the status of women—in different cultures that are at the same level of technological advancement. Indeed, it has been observed that modern African or Asian subsistence agricultural groups and those of 18th-century frontier American farm families show similarities in gender roles.

It may further be logical to believe that advancement in the technological sense would be reflected in an enhancement of the status and rewards of both men and women in all developing societies. The pattern that we actually observe is not quite that simple or straightforward, however. In addition to a culture's economic stage, religion and custom play important roles in determining gender relationships and female prestige. Further, it appears that, at least in the earlier phases of technological change and development, women generally lose rather than gain in standing and rewards. Only recently and only in the most-developed countries have gender contrasts been reduced within and between societies.

Hunting and gathering cultures observed a general egalitarianism; each sex had a respected, productive, coequal role in the kinship group. Gender is more involved and changeable in agricultural societies (see "Women and the Green Revolution," Chapter 9). The Agricultural Revolution—a major change in the technological subsystem—altered the earlier structure of gender-related responsibilities. In the hoe agriculture that was the first advance over hunting and gathering and is today found in much of sub-Saharan Africa and in South and Southeast Asia, women became responsible for most of the actual fieldwork, while retaining their traditional duties in child rearing, food preparation, and the like; their economic role and status remained equivalent to those of males. Plow agriculture, on the other hand, tended to subordinate the role of women and diminish their level of equality. Women might have hoed, but men plowed, and female participation in farmwork was drastically reduced. This is the case today in Latin America and, increasingly, in sub-Saharan Africa, where women are often more visibly productive in the market than in

Figure 6.41 Women dominate the once-a-week *periodic* markets in nearly all developing countries. Here they sell produce from their gardens or the family farm and frequently offer for sale processed goods they have made. The market shown here is in Mali. More than one-half of the economically active women in the developing world are self-employed, working primarily in the informal sector. © *McGraw-Hill Education/Lissa Harrison RF.*

the field (**Figure 6.41**). As women's agricultural productive role declined, they were afforded less domestic authority, less control over their own lives, and few, if any, property rights independent of male family members.

At the same time, considering all their paid and unpaid labor, women spend more hours per day working than do men. In developing countries, the UN estimates, when unpaid agricultural work and housework are considered along with wage labor, women's work hours exceed men's by 30% and may involve at least as arduous—or more arduous—physical labor. The UN's Food and Agriculture Organization reports that "rural women in the developing world carry 80 tons or more of fuel, water, and farm produce for a distance of 1 km during the course of a year. Men carry much less. . . ." Everywhere, women are paid less than men for comparable employment.

Western industrial—"developed"—society emerged directly from the agricultural tradition of the subordinate female who was not considered an important element in the economically active population, no matter how arduous or essential the domestic tasks assigned. With the growth of cities and industry in 19th-century America, a "cult of true womanhood" developed as a reaction to the competitive pressures of the marketplace and factory floor. It held that women were morally superior to men; their role was in the

LEVELING THE FIELD

Beginning in Mexico City in 1975, the United Nations has sponsored four world conferences on women. The Fourth World Conference on Women, held in Beijing in 1995, produced a strong and broad declaration of women's rights and equality. The Beijing Conference called on all governments to formulate strategies, programs, and laws designed to ensure women of their full human rights to equality and development. Its final declaration detailed recommended policies in the areas of sexuality and childbearing, violence against women, discrimination against girls, female inheritance rights, and family protection. Its particular emphasis focused on efforts to "ensure women's equal access to economic resources including land, credit, science, and technology, vocational training, information, communication and markets as a means to further advancement and empowerment of women and girls."

The conference did not reach its decisions and recommendations without spirited debate clearly rooted in the contrasting ways that different major cultures assign gender roles. Those contrasts, and the relationship between human rights and the national customs they imply, were put in international perspective in the preamble to the conference platform: "While the significance of national and regional particularities and various historical, cultural, and religious background must be borne in mind, it is the duty of states regardless of their political, economic, and cultural systems to promote and protect all human rights and fundamental freedoms."

At the "Beijing Plus Five" Conference held at the United Nations in June 2000, delegates called for the punishment of domestic violence and for the criminalization of the cross-border trafficking of women and girls for sexual and wage slavery purposes.

In 2000, the 189 member states of the United Nations adopted the Millennium Declaration to help the world's poorest countries make progress toward eliminating extreme poverty by 2015. The declaration adopted eight interlinked goals—the Millennium Development Goals—one of which was to "promote gender equality and empower women." The three indicators used to measure progress toward meeting that goal were the ratio of girls to boys in primary, secondary, and higher education; the share of women in nonagricultural employment; and the share of women in national parliaments. By 2015, much progress had been made on all three gender equality indicators. The school enrollment disparities in developing countries were largely eliminated. In southern Asia in 1990, there were 74 girls in primary school for every 100 boys. By 2015, there were more girls in primary school than boys. The proportion of seats in national parliaments held by women increased in all world regions, but parity is a long ways off in some regions.

---

private space of the home, not the public sphere. A woman's job was to rear children, attend church, and above all keep a sober, virtuous, and cultured home, a place that offered the male breadwinner refuge, security, and privacy. In the second half of the 20th century, beginning in the more-developed countries, that subordinate role pattern began to change. Women achieved higher levels of education, became increasingly economically active, and entered the public spaces of work and politics. The feminist movement in modern industrialized societies was the direct response to the barriers that formerly restricted favored economic and legal positions to men.

The present global pattern of gender-related institutional and economic role assignments is influenced not only by a country's level of economic development but also by the persistence of restrictions its culture imposes on women and by the nature of its economic—particularly, agricultural—base (see "Leveling the Field," p. 182). The first control is reflected in contrasts between the developed and developing world; the second and third are evidenced in variations within the developing world itself.

In any case, economic globalization has had a mixed impact on female participation in the paid labor force and on gender economic differentiation. In developed countries, deindustrialization (see Chapter 10) has had a negative effect on men's employment and wages, while the shift to services has favored women. Still, executive and top managerial workers are disproportionately male. On a worldwide basis, it appears that globalization's greater trade openness has increased women's share of paid employment; and in developing countries, firms that are producing for export employ more female workers, often in skilled activities. But growing labor force participation has not necessarily reduced gender discrimination. In developing countries, women also make up the largest share of workers in expanding informal subcontracting and sweatshop piecework—often in the shifting and uncertain garment and shoe industries—with low pay and poor working conditions.

A distinct gender-specific regionalization has emerged. Among the Arab and Arab-influenced Muslim areas of western Asia and North Africa, the proportion of the female population that is economically active is low; religious tradition restricts women's acceptance in economic activities outside the home. The same cultural limitations do not apply under the different rural economic conditions of Muslims in southern and southeastern Asia, where labor force participation by women in Indonesia and Bangladesh, for example, is much higher than it is among the western Muslims. In Latin America, well known for a patriarchal social structure, women have been overcoming cultural restrictions on their growing employment outside the home.

Sub-Saharan Africa, highly diverse culturally and economically, in general is highly dependent on female farm labor and market income. The traditional role of strongly independent, property-owning women formerly encountered under traditional agricultural and village systems, however, has increasingly been replaced by the subordination of women with modernization of agricultural techniques and introduction of formal, male-dominated financial and administrative agricultural institutions. For countries for which sufficient data are available, a set of indicators has been created to establish a "gender-related development index" and ranking (**Figure 6.42**); it clearly displays regional differentials in the position of women in different cultures and world areas.

Regional and cultural contrasts in gender relationships are also encountered in the advanced economies and the industrial components of developing countries. In modernizing eastern Asian states, for example, women have yet to achieve the status they enjoy in

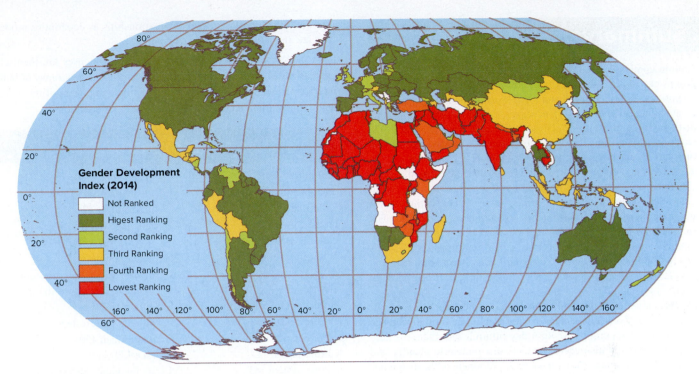

Figure 6.42   The gender-related development index (GDI) is a composite index that attempts to measure inequalities between men and women through differences in their life expectancy, educational attainment, and earned income. The GDI includes both disparities that favor males and those that favor females. Countries are placed by the United Nations into five groups based on their level of gender inequality. The gender-related development index rankings suggest that gender equality does not necessarily depend on the overall income level of a country. *Source: 2015 rankings from United Nations Development Programme, Human Development Report 2015.*

most Western economies. In China, although it ranks in the second quartile in the GDI (see Figure 6.42), women are generally not effectively competitive with men and are largely absent from the highest managerial and administrative levels; in Japan, males nearly exclusively run the huge industrial and political machinery of the country. In contrast, economic and social gender equality is more advanced in the Scandinavian countries than perhaps in any other portion of the industrialized world.

6.10 Other Aspects of Diversity

Culture is the total way of life of a society. It is misleading to isolate, as we have done in this chapter, only a few elements of the technological, sociological, and ideological subsystems and imply that they are identifying characteristics differentiating culture groups. Economic development levels, language, religion, ethnicity, and gender all are important and common distinguishing culture traits, but they tell only a partial story. Other suggestive, though perhaps less-pervasive, basic elements exist.

Architectural styles in public and private buildings are evocative of region of origin, even when they are indiscriminately mixed together in American cities. The Gothic and New England churches, the neoclassical bank, and the skyscraper office building suggest not only the functions they house but also the culturally and regionally variant design solutions that gave them form. The Spanish, Tudor, French provincial, or ranch-style residence may not reveal the ethnic background of its American occupant, but it does constitute a culture statement of the area and the society from which it diffused.

Music, food, games, and other evidence of the joys of life, too, are cultural indicators associated with particular world or national areas. Music is found in all societies. But because it is symbolically expressive of the experiences and emotions of people with particular regional and group identities, it displays wide variations from place to place. Folk music styles and instrumentation differ greatly, but even popular music exhibits geographic variations and distinct hearth regions. Country music lyrics, for example, contain themes that resonate with the experiences of the rural working class. Both hip hop and grunge originated as expressions of the alienation felt by particular segments of the youth population in the Bronx and Pacific Northwest, respectively. Regional musical expressions of culture also include Cajun music of south Louisiana, Tejano music of the Tex-Mex borderlands, and the polka of the Upper Midwest. Where there is sufficient similarity between musical styles and instrumentation, blending (syncretism) and transfer may occur. American jazz represents a blend of musical influences, and even reggae, which is perceived as an authentic Jamaican musical form, is a blend of African rhythms, Caribbean ska and rock steady, and American soul and rhythm and blues music. Foods identified with other culture regions have similarly been transferred to become part of the culinary environment of the American "melting pot."

These are but a few additional minor statements of the variety and intricate interrelationships of that human mosaic called culture. Individually and collectively, however, in their areal expressions and variations they are only part of the subject matter of the cultural geographer. Patterns of human interaction and spatial behavior, national and international political structures, economic activities, and patterns of urbanization—all fundamental aspects of contemporary culture—are the subjects of the following chapters.

Summary of Key Concepts

- Culture comprises the learned behaviors and beliefs of distinctive groups of people. Culture traits combine to shape integrated culture complexes. Together, traits and complexes in their spatial patterns create human—"cultural"—landscapes, define culture regions, and distinguish culture groups. Those landscapes, regions, and group characteristics change through time as human societies interact with their environment, develop new solutions to collective needs, or are altered through innovations adopted from outside the group.

- Understanding of culture is aided by recognition of its component subsystems. The *technological subsystem* is composed of the material objects (artifacts) and techniques of livelihood. The *sociological subsystem* comprises the formal and informal institutions (sociofacts) that control the social organization of a culture group. The *ideological subsystem* consists of the ideas and beliefs (mentifacts) that a culture expresses in speech and through belief systems.

- Domestication of plants and animals led to the emergence of *culture hearths* of development innovation and to a cultural divergence between different groups. Modern-day cultural convergence is lessened by the many distinctive elements that remain to identify and separate social groups. Among the most prominent of the differentiating culture traits are language, religion, ethnicity, and gender.

- Language and religion are both transmitters of culture and identifying traits of separate culture groups. Both have distinctive spatial patterns, reflecting past and present processes of interaction and change.

- Languages can be grouped by origins and historical development, but their world distributions depend as much on the movements of peoples and histories of conquest and colonization as they do on linguistic evolution. *Toponymy,* the study of place-names, helps document that history of movement. Linguistic geography studies spatial variations in languages, variations that may be minimized by the encouragement of standard languages or overcome by pidgins, creoles, and lingua francas.

- Spatial patterns of religion are distinct and reveal histories of migration, conquest, and diffusion. Those patterns are important components in the spatially distinctive cultural landscapes created in response to various religious belief systems. Even in secular societies, religion may influence economic activities, legal systems, holiday observances, and the like.

- Ethnicity, affiliation in a group sharing identifying culture traits, is fostered by territorial separation or isolation and is preserved in ethnically complex societies by a feeling that one's own ethnic group is superior to others. Ethnic diversity is a reality in most countries of the world and is increasing in many of them. Many ethnic minority groups seek absorption into their surrounding majority culture through acculturation and assimilation, but other groups choose to preserve their identifying distinctions through spatial separation or overt rejection of the majority culture traits.

- Gender, the culturally based social distinctions between men and women, reflects religion, custom, and, importantly, the stage of economic development of a society and the productive role assigned to women within that economy. Gender roles change as the economic structure changes, though their modification is often resisted by conservative forces within a culture.

- Culture realms are ever-changing reflections of the migrations of ethnic and cultural groups, the diffusion or adoption of languages and religions, the spread and acceptance of new technologies, and the alteration of gender relationships as economies modernize and cultural traditions respond.

Such movements, diffusions, adoptions, and responses are themselves expressions of broader concepts and patterns of the geography of human interaction and spatial behavior, an essential component of the culture-environment tradition to which we next turn our attention.

Key Words

acculturation 155	genetic drift 179
amalgamation theory 156	ideological subsystem 144
assimilation 156	innovation 152
creole 160	language family 157
cultural ecology 150	lingua franca 160
cultural integration 149	material culture 163
cultural landscape 150	natural selection (adaptation) 179
culture 142	nonmaterial culture 163
culture complex 143	pidgin 159
culture hearth 152	popular culture 163
culture realm 144	possibilism 150
culture region 143	race 179
culture system 143	sociological subsystem 144
culture trait 143	spatial diffusion 154
dialect 159	standard language 159
environmental determinism 150	syncretism 155
ethnicity 178	technological subsystem 144
ethnic religion 167	toponymy 165
ethnoburb 180	tribal religion 167
folk culture 163	universalizing religion 166
gender 181	

Thinking Geographically

1. What is included in the concept of *culture?* How is culture transmitted? What personal characteristics affect the aspects of culture that any individual acquires or fully masters?

2. What is a *culture hearth?* What new traits of culture characterized the early hearths? In the cultural geographic sense, what is meant by *innovation?*

3. Differentiate between *culture traits* and *culture complexes* and between *environmental determinism* and *possibilism.*

4. What are the components or subsystems of the three-part system of culture? What characteristics—aspects of culture—are included in each of the subsystems?

5. Why might one consider language the dominant differentiating element of culture separating societies?

6. In what way may religion affect other culture traits of a society?

7. How does the classification of religions as *universalizing, ethnic,* or *tribal* help us understand their patterns of distribution and spread?

8. How does *acculturation* occur? Is *ethnocentrism* likely to be an obstacle in the acculturation process? How do acculturation and *assimilation* differ?

9. How are the concepts of *ethnicity, race,* and *culture* related and distinct?

CHAPTER 7

Human Interaction

A new office and apartment complex in a suburban area helps create new patterns of human interaction. © *CelebrityHomePhotos/Newscom.*

CHAPTER OUTLINE

Figure 7.1   **A poster advertising the services of the Pony Express, whose riders sped across the West on horses bred to run fast.** In operation from only 1860 to 1861, the Pony Express was rendered obsolete by railroads and the telegraph. In 1861, a telegraph wire strung from New York City to San Francisco reduced the time for communication between the coasts from days to seconds. *Courtesy of Pony Express Museum, St. Joseph, MO.* © *Bettmann/Getty Images.*

❝**F**or a brief 18-month period between April 1860 and October 1861, an undying blend of courage and endurance was created by the Pony Express riders. Racing through Nebraska, Wyoming, Utah, and Nevada, these horsemen carried letters written on lightweight paper (postage ranged from $2 to $10 an ounce) from one relay station to another. They covered the 1966 miles from St. Joseph, Missouri, to Sacramento, California, in 10 days. The nature of the work is implied in the newspaper advertisements seeking riders who weighed less than 135 pounds, did not drink or carouse, and were 'daring young men—preferably orphans.'❞[1] The twice-weekly trip took longer if riders had to avoid Indians or a herd of a million buffalo (**Figure 7.1**).

The $10 charge for an ounce of mail in 1860 is equivalent to $220 in today's money. Only such things as diamonds are as expensive to ship today. It is interesting to note that there are two non-stop and a dozen one-stop flights leaving St. Louis every day for San Francisco (cities near St. Joseph and Sacramento, respectively), and an average one-way fare is about $150 per person (the average weight of a person with baggage is about 180 pounds). It takes approximately 4 hours for a plane to make the trip. The cost for an ounce is 5 cents by plane versus $220 by Pony Express.

This comparison shows unmistakably that the interaction between the Midwest part of the United States and the West has

grown enormously over the past 144 years. The level of interaction is a function of the demand for travel, its speed, and its cost. All of these factors are conditioned by technology. In this case, the technology changed from fresh horses spaced at 16- to 24-kilometer (10- to 15-mi) intervals in the Pony Express days to jet planes. Today, with populations in each area in the millions and the cost of travel low, it is no wonder that so many planes leave daily from each city for the other.

7.1 The Definition of *Human Interaction*

Simply put, **human interaction** is the communication and interdependencies between people. In this book we consider human interaction as a geographic process; that is, we view movements, communications, and interdependencies between places that can be identified on a map. Sometimes the term **spatial interaction** is used to more specifically identify the locations that are the termini of the interaction. For example, a cell phone call is a human interaction, but when we view it as a communication between two

---

[1] *The Story of America*, The Reader's Digest Association, Pleasantville, New York, 1975, p. 199.

places, it is also a spatial interaction. Human interactions include all communications (phone, cell, Facebook, Twitter, mail); all travels of humans and their goods and services; all spreading of communicable diseases, ideas, and knowledge. All of these interactions involve two or more places. When we view these interactions over Earth space (on maps), we can call human interaction spatial interaction. Spatial interaction implies that there is some sort of a flow over a distance separating people.

If there is no one at a site (for example, on an iceberg), there can be no spatial interaction between the site and any other site. On the other hand, if there are a great number of people at one site—for example, Chicago—and a great number of people at another site— say, New York—there will be a great deal of spatial interaction among them. But if the second site is a city as far away as London is from Chicago, there will be fewer interactions between Chicago and London than between Chicago and New York. The attractive force between areas is akin to the force of gravity. Thus, the amount of spatial interaction is a function of the size of the interdependent populations and the distance between them.

7.2 Distance and Human Interaction

People engage in many more short-distance interdependencies than long ones. This is the principle of **distance decay,** the decline of an activity, a function, or an amount of interaction with increasing distance from the point of origin. The tendency is for the frequency of trips to decrease very rapidly beyond an individual's **critical distance,** the distance beyond which cost, effort, means, and perception play an overriding role in our willingness to interact. **Figure 7.2** illustrates this principle with regard to journeys from a homesite. The critical distance is different for each person. The variables of a person's age, mobility, and opportunity, together with an individual's interests and demands, help define how much and how far a person will travel. For example, students on campus might not think twice about visiting the campus library, but off-campus students might wait for a convenient time for a visit, or not visit it at all. The campus library may be beyond the off-campus students' critical distance. Because distance retards human interaction, we can say there is a *friction of distance* in our lives and activities.

In general, the amount of human interaction between two places is a function of the distance separating the places and the populations of the two places. **Figure 7.3** shows the migration from New Orleans just after Hurricane Katrina in 2005. Because they are nearby, places in unflooded parts of New Orleans and in the state of Louisiana received most of the migrants, but many journeyed to homes of friends and family in the metropolitan areas of Houston, Dallas, Chicago, and New York.

A small child will make many trips up and down the block, but he or she will be inhibited by parental admonitions from crossing the street. Different but equally effective constraints control adult behavior. Daily or weekly shopping may be within the critical distance of an individual, and little

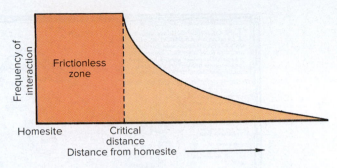

Figure 7.2   The critical distance. This general diagram indicates how most people observe distance. For each activity, there is a distance beyond which the intensity of contact declines. This is called the *critical distance*. The distance up to the critical distance is a frictionless zone, in which time or distance considerations do not effectively figure into the trip decision.

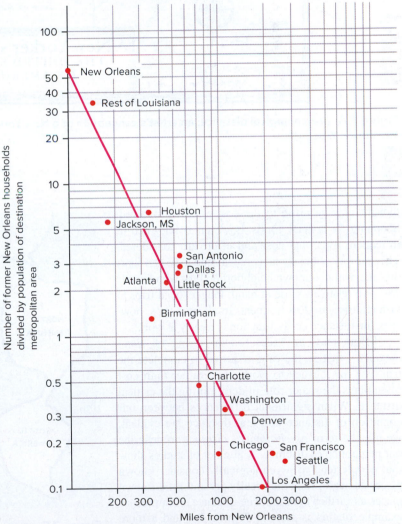

Figure 7.3   Migration of the victims of Hurricane Katrina. This human interaction diagram uses a log-log scale so that distances from New Orleans can be shown with the number of migrants divided by the populations of metropolitan areas. *Source: Data from FEMA, U.S. Census Bureau and Queens College Sociology Department.*

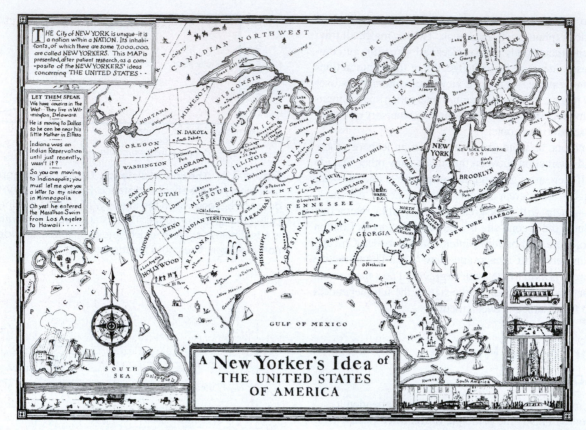

Figure 7.4 **Psychological distance: an artist's conception of a New Yorker's view of the United States.** © *Florence Thierfeldt, Milwaukee, Wisconsin.*

thought may be given to the cost or effort involved. Shopping for special goods, however, is relegated to infrequent trips, and cost and effort are considered. The majority of our social contacts tend to be at a short distance within our own neighborhoods or with friends who live relatively close at hand; longer social trips to visit relatives are less frequent. In all such trips, however, the distance decay function is clearly at work.

Effort may be measured in terms of *time-distance*—that is, the time required to complete the trip. For the journey to work, time rather than cost often plays the major role in determining the critical distance. When significant differences between our cognition of distance and real distance are evident, we use the term *psychological distance* to describe our perception of distance. A number of studies show that people tend to psychologically consider known places as nearer than they really are and little-known places as farther than the true distance. A humorous example of this is seen in **Figure 7.4** and a more serious one in **Figure 7.5**. Also, see "Mental Maps," pp. 190–191.

We gain information about the world from many sources. Although information obtained from radio, television, the Internet, and newspapers is important to

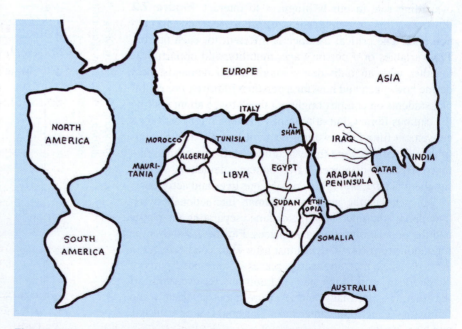

Figure 7.5   **A mental map of the world.** This map was drawn by a Palestinian high school student from Gaza. The map reflects the secondary school education the author is receiving, which conforms to the Egyptian national school curriculum and thus is influenced by the importance of the Nile River and pan-Arabism. Al Sham is the old, but still used, name for the area including Syria, Lebanon, and Palestine. The map might be quite different if it were designed by an Israeli student.

us, face-to-face contact is assumed to be the most effective means of communication. The distance decay principle implies that, as the distance away from the home or workplace increases, the number of possible face-to-face contacts usually decreases. We expect greater human interaction at short distances than at long distances. Where population densities are high, such as in cities (particularly central business districts during business hours), the human interaction between individuals can be at a very high level, which is one reason these centers of commerce are often also centers for the development of new ideas.

7.3 Barriers to Interaction

Recent changes in technology permit us to travel farther than ever before, with greater safety and speed, and to communicate without physical contact more easily and completely than previously possible. This intensification of contact has resulted in an acceleration of innovation and in the rapid spread of goods and ideas. Several millennia ago, innovations such as the smelting of metals took hundreds of years to spread. Today, worldwide diffusion may be almost instantaneous.

The fact that the possible number of interactions is high, however, does not necessarily mean that the effective occurrence of interactions is high. That is, a number of barriers to interaction exist. Such barriers are any conditions that hinder either the flow of information or the movement of people and thus retard or prevent the acceptance of an innovation. Distance itself is a barrier to interaction. Generally, the farther the two areas are from each other, the less likely will be the interaction. The concept of distance decay says that, all else being equal, the amount of interaction decreases as the distance between two areas increases.

Cost is another barrier to interaction. Relatives, friends, and associates living long distances apart may find it difficult to afford to see one another. The frequency and time allocated to telephone and e-mail communication, relatively inexpensive forms of interaction, are very much a function of the location of friends and relatives—which, of course, favors short-distance interaction.

Interregional contact may also be hindered by the physical environment and by the cultural barriers of differing religions, languages, ideologies, genders, and political systems. Mountains and deserts, oceans and rivers can, and have, acted as physical barriers slowing or impeding interaction. Cultural barriers may be equally impenetrable. Those nearby who practice religions differently or speak another language may not be in touch with their neighbors. Governments that interfere with Internet communication, control the flow of foreign literature, and discourage contact between their citizens and foreign nationals impede cultural contact.

In crowded areas, people commonly set psychological barriers around themselves, so that only a limited number of interactions take place. The barriers are raised in defense against information overload and for psychological well-being. We must have a sense of privacy in order to filter out information that does not directly concern us. As a result, we tend to reduce our interests to a narrow range when we find ourselves in crowded situations, allowing our wider interests to be satisfied by our use of the communications media.

7.4 Human Interaction and Innovation

The probability that new ideas will be generated out of old ideas is a function of the number of available old ideas in contact with one another. People who specialize in a particular field of interest seek out others with whom they wish to interact. Crowded central cities are characteristically composed of specialists in very narrow fields of interest. Consequently, under short-distance, high-density circumstances, the old ideas are given a hearing and new ideas are generated by the interaction. New inventions and new social movements usually arise in circumstances of high spatial interaction. An exception, of course, is the case of intensely traditional societies—Japan in the 17th and 18th centuries, for example—where the culture rejects innovation and clings steadfastly to customary ideas and methods.

The culture hearths of an earlier day (see Chapter 6) were the most densely settled, high-interaction centers of the world. At present, the great national and regional capital cities attract people who want or need to interact with others in special-interest fields. The association of population concentrations and the expression of human ingenuity have long been noted. The home addresses recorded for patent applicants by the U.S. Patent Office over the past century indicate that inventors were typically residents of major urban centers, presumably people in close contact and able to exchange ideas with those in shared fields of interest. It still appears that the metropolitan centers of the world attract those who are young and ambitious and that face-to-face or word-of-mouth contact is important in the creation of new ideas and products. The relatively recent revolution in communications, which allows for inexpensive interaction through a variety of phone and Internet services, has suggested to some that the traditional importance of cities as collectors of creative talent may decline in the future.

7.5 Individual Activity Space

We will see in Chapter 8 that groups and countries draw boundaries around themselves to divide space into territories that are defended if necessary. The concept of **territoriality**—the emotional attachment to, and the defense of, home ground—has been seen by some as a root explanation of many human actions and responses. It is true that some collective activity appears to be governed by territorial defense responses: the conflict between street groups in claiming and protecting their "turf" (and the fear for their lives when venturing beyond it) and the sometimes violent rejection by ethnic urban neighborhoods of any different encroaching population group they consider threatening.

But for most of us, our personal sense of territoriality is a tempered one. We regard our homes and property as defensible private domains but open them to innocent visitors, known or unknown, or to those on private or official business. Nor do we confine our activities so exclusively within controlled home territories as street-gang members do within theirs. Rather, we have a more or less extended home range, an **activity space** or area within which we move freely on our rounds of regular activity, sharing that space with others who are also involved in their daily affairs.

Figure 7.6 suggests the activity spaces for a family of five for one day. Note that the activity space for each individual for one day is rather limited, even though two members of the family use

MENTAL MAPS

Human interaction is affected by the way people perceive places. When information about a place is sketchy, blurred pictures develop. These influence the impression we have of places. Our willingness or ability to interact with what might be thought of as strange or unknown places cannot be discounted. We may say that each individual has a **mental map** of the world. No single person, of course, has a true and complete image of the world; therefore, having a completely accurate mental map is impossible. In fact, the best mental map that most individuals have is that of their own residential neighborhood, the place where they spend the most time.

Whenever individuals think about a place or how to get to a place, they produce a mental map. What are believed to be unnecessary details are left out, and only the important elements are incorporated. Those elements usually include awareness that the object or the destination does indeed exist, some conception of the distances separating the starting point and the named object(s), and a feeling for the directional relationships between points. A mental route map may also include reference points to be encountered on the chosen path of connection or on alternate lines of travel. Although mental maps are highly personalized, people with similar experiences tend to give similar answers to questions about the environment and to produce roughly comparable sketch maps.

Awareness of places is usually accompanied by opinions about them, but there is no necessary relationship between the depth of knowledge and the perceptions held. In general, the more familiar we are with a locale, the better will be the factual basis of our mental image of it. But we form firm impressions of places totally unknown by us personally, and these may affect travel, migration decisions, or other forms of human interaction.

One way to ascertain how individuals envisage the environment is to ask them what they think of various places. For instance, they might be asked to rate places according to desirability—perhaps residential desirability—or to make a list of the best and worst places in a region such as the United States. Certain regularities appear in such studies.

Three children, aged 6, 10, and 13, who lived in the same house, were asked to draw maps of their neighborhood. No further instructions were given. Notice how perspectives broaden and neighborhoods expand with age. For the 6-year-old (bottom left), the neighborhood consisted of the houses on each side of her own. The square block on which she lived was the neighborhood for the 10-year-old (top right). The wider activity space of the 13-year-old is also evident (bottom right). The square block that the 10-year-old drew is shaded in the 13-year-old's sketch.

10-year-old

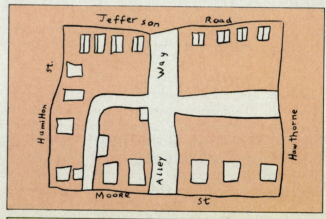

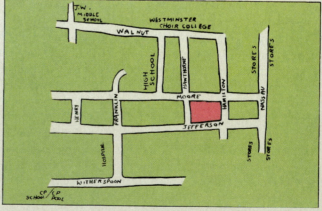

13-year-old

6-year-old

The accompanying figure presents some residential desirability data elicited from college students in three provinces of Canada. These and comparable mental maps suggest that near places are preferred to far places unless much information is available about the far places. Places with similar cultural forms are preferred, as are places with high standards of living. Individuals tend to be indifferent to unfamiliar places and to dislike unfamiliar areas that have competing cultural interests (such as disliked political and military activities) or a physical environment known to be unpleasant.

On the other hand, places perceived to have superior climates or landscape amenities are rated highly in mental map studies and favored in tourism and migration decisions. The southern and southwestern coast of England is attractive to citizens of generally wet and cloudy Britain; and holiday tours to Spain, the south of France, and the Mediterranean islands are heavily booked by the English. A U.S. Bureau of the Census study indicates that "climate" is, after work and family proximity, the most often reported reason for interstate moves by adults of all ages. International studies reveal a similar migration motivation based not only on climate but also on concepts of natural beauty and amenities.

Each of these maps shows the residential preference of a sampled group of Canadians from the provinces of Quebec, Ontario, and British Columbia. Note that each group of respondents prefers its own area but that all like the Canadian and U.S. west coasts. *Source: Herbert A. Whitney, "Preferred Locations in North America: Canadians, Clues, and Conjectures," in Journal of Geography, Vol. 83, no. 5, 1984 p. 222. National Council for Geographic Education, Indiana, PA.*

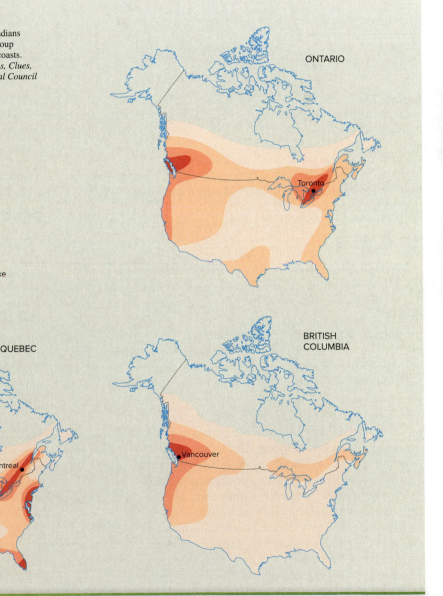

Preference
- ■ Strong like
- ■ Like
- ■ Neutral
- □ Dislike
- □ Strong dislike

ONTARIO

Toronto

QUEBEC

Montreal

BRITISH COLUMBIA

Vancouver

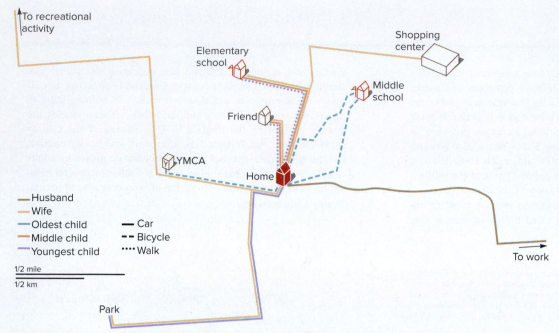

Figure 7.6   Activity space for each member of a family of five for a typical weekday. One parent commutes to work, while the other parent works at home. Routes of regular movement and areas recurrently visited help to foster a sense of territoriality and to affect one's perceptions of space.

automobiles. If one week's activity were shown, more paths would have to be added to the map, and in a year's time, several long trips would probably have to be noted. Long trips are usually taken irregularly.

The kind of activities individuals engage in can be classified according to type of trip: journeys to work, school, shopping center, recreational area, and so on. People in nearly all parts of the world make these same types of journeys, though the spatially variable requirements of culture, economy, and personal circumstances dictate their frequency, duration, and significance in an individual's time budget.

Figure 7.7 suggests the importance of the journey to work in an urban population. The journey to work plays a decisive role in defining the activity space of most adults. Formerly restricted by walking distance or by the routes and schedules of mass transit systems, the critical distances of work trips have steadily increased in European and Anglo American cities as the private automobile has figured more importantly in the movement of workers (**Figure 7.8**). In more recent years, however, it has become evident that, for many, the journey to work is really a multipurpose trip, which may include side trips to day-care centers, cleaners, schools, and shops of various kinds.

The types of trips individuals make, and thus the extent of their activity space, are partly determined by three variables: people's stage in life (age); the means of mobility at their command; and the opportunities implicit in their daily activities.

Stage in Life

The first variable determining the types of trips individuals make, **stage in life,** refers to membership in specific age groups. Stages include preschool

age, school age, young adult, adult, and elderly. Preschoolers stay close to home unless they accompany their parents or caregivers. School-age children usually travel short distances to lower-level schools and longer distances to upper-level schools. After-school activities tend to be limited to walking, bicycle, or automobile trips provided by parents to nearby locations. High school students and other young adults are usually more mobile and take part in more activities than do younger children. They engage in greater spatial interaction. Adults responsible for household duties make shopping trips and trips related to child care, as well as journeys away from home for social, cultural, and recreational purposes. Wage-earning adults usually travel farther from home than other family members. Elderly people may, through infirmity or interests, have less-extensive activity spaces.

Mobility

The second variable that affects the extent of activity space is *mobility,* or the ability to travel. An informal consideration of the cost and effort required to overcome the friction of distance is implicit. Where incomes are high, automobiles and trains are available, and the cost of fuel is a minor item in the family budget, mobility may be great and individual action space can be large. In societies where cars and trains are not a standard means of personal conveyance, the daily activity space may be limited to the shorter range afforded by bicycles or walking. Obviously, both intensity of purpose and the condition of the roadway affect the execution of movement decisions.

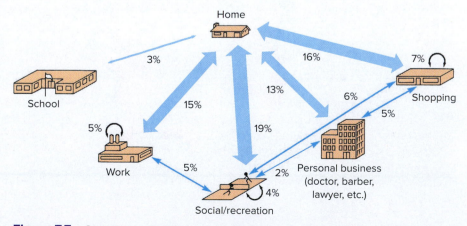

Figure 7.7   Seven-county Minneapolis Metropolitan Area travel patterns. The numbers are the percentages of all urban trips taken on a typical weekday. *Source: Data from* Metropolitan Council: The 2000 Travel Behavior Inventory.

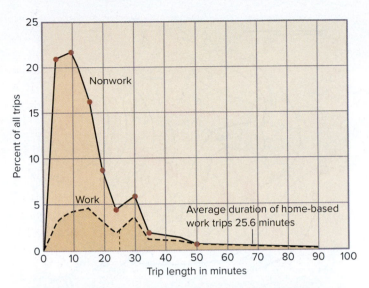

Figure 7.8 **The frequency distribution of work and nonwork trip lengths in minutes in the seven-county Minneapolis Metropolitan Area.** Studies in various metropolitan areas support the conclusions documented by this graph: work trips are usually longer than other recurring journeys. In the United States in 2009, the average work trip covered 19.6 kilometers (12.2 mi), and one-half of all trips to work took under 25 minutes; for suburbanites commuting to the central business district, the journey to work involved between 30 and 45 minutes. *Source: U.S. Department of Energy, 2009 National Household Travel Survey.*

The mobility of individuals in countries or in sections of countries with high incomes is relatively great; people's activity space horizons are broad. These horizons, however, are not limitless. There are a fixed number of hours in a day, most of them consumed in performing work, preparing and eating food, and sleeping. In addition, there are a fixed number of road, rail, and air routes, so even the most mobile individuals are constrained in the amount of activity space they can use. No one can easily claim the world as his or her activity space. One example of this limitation is many women living in suburban communities who must balance family obligations, such as preparing meals and caring for children, with their workforce activities. In this case, women's mobility is restricted; as a result, their occupational opportunities are limited.

Opportunities

The third factor limiting activity space is the individual assessment of the availability of possible activities or *opportunities.* In the teeming cities of Asia, for example, the very poor satisfy their daily needs nearby; the impetus for journeys away from the residence is minimal. In impoverished countries and neighborhoods, low incomes limit the inducements, opportunities, destinations, and necessity of travel. Similarly, if one lives in a remote, sparsely settled area, with few or no roads, schools, factories, or stores, one's expectations and opportunities are limited, and activity space is therefore reduced. Opportunities plus mobility conditioned by life stage bear heavily on the amount of spatial interaction in which individuals engage.

7.6 Diffusion and Innovation

As we noted in Chapter 6, **spatial diffusion** is the process by which a concept, practice, or substance spreads from its point of origin to new territories. A *concept* is an idea or invention, such as a new way of thinking—for example, deciding that shopping on the Internet is worth doing. A *practice* might be the actual process of shopping on the Internet. A *substance* is a tangible thing, such as the goods bought by means of the Internet. Diffusion is at the heart of the geography of spatial interaction.

Ideas generated in a center of activity will remain there unless some process is available for their spread. *Innovations,* the changes to a culture that result from the adoption of new ideas, spread in various ways. Some new inventions are so obviously advantageous, such as the iPad, that they are put to use quickly by those who can afford and profit from them. A new development in petroleum extraction may promise such material reward as to lead to its quick adoption by all major petroleum companies, irrespective of their distance from the point of introduction. The new strains of wheat and rice that significantly increased agricultural yields in much of the world and that were part of the Green Revolution (discussed in Chapter 10) were quickly made known to agronomists in all cereal-producing countries. However, they were more slowly taken up in poor countries, which could benefit most from them, partly because farmers had difficulty paying for them and spatial interaction was limited.

Many innovations are of little consequence by themselves, but sometimes the widespread adoption of seemingly inconsequential innovations brings about large changes when viewed over a period of time. A new musical tune, "adopted" by a few people, may lead many individuals to fancy that tune plus others of a similar sound. This, in turn, may have a bearing on dance routines, which may then bear on clothing selection, which in turn may affect retailers' advertising campaigns and consumers' patterns of expenditures. Eventually, a new cultural form is identified that may have an important impact on the thinking processes of the adopters and on those who come into contact with the adopters. Notice that a broad definition of *innovation* is used, but notice also that what is important is whether or not innovations are adopted.

In spatial terms, we can identify a number of processes for the diffusion of innovations. Each is based on the way innovations spread from person to person and, therefore, from place to place. These processes are discussed in the following sections "Contagious Diffusion" and "Hierarchical Diffusion."

Contagious Diffusion

Let us suppose that a scientist develops a gasoline additive that noticeably improves the performance of her or his car. Assume further that the person shows friends and associates the invention and that they, in turn, tell others. This process is similar to the spread of a contagious disease. The innovation will continue to diffuse until barriers (that is, people not interested in adopting the new idea) are met or until the area is saturated (that is, all available people have adopted the innovation). This **contagious diffusion** process follows the rules of distance decay spatial interaction at each step. Short-distance contacts are more likely than long-distance contacts, but over time the idea may have spread far

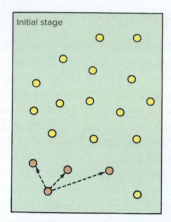
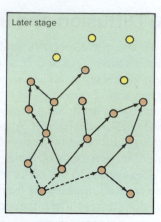

Figure 7.9   Contagious diffusion. A phenomenon spreads from one place to neighboring locations, but in the process it remains and is often intensified in the place of origin. *Source:* College Geography, Spatial Diffusion, *by Peter R. Gould, p. 4. Association of American Geographers, 1969.*

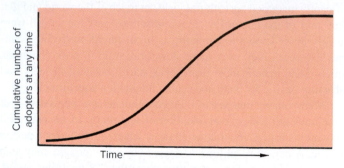

Figure 7.10   The diffusion of innovations over time. The number of adopters of an innovation rises at an increasing rate until the point at which about one-half the total who ultimately decide to adopt the innovation have made the decision. At this point, the number of adopters increases at a decreasing rate.

from the original site. **Figure 7.9** illustrates a theoretical contagious diffusion process.

A number of characteristics of this kind of diffusion are worth noting. If an idea has merit in the eyes of potential adopters, and they become adopters, then the number of contacts of adopters with potential adopters will compound. Consequently, the innovation will spread slowly at first and then more and more rapidly, until saturation occurs or a barrier is reached. The incidence of adoption is represented by the S-shaped curve in **Figure 7.10**. The area in which those affected are located will be small at first, and then the area will enlarge at a faster and faster rate. The spreading process will slow as the available areas and/or people decrease. **Figure 7.11** shows the contagious diffusion of a disease.

If an inventor's idea falls into the hands of a commercial distributor, the diffusion process may follow a somewhat different course than that just discussed. The distributor might "force" the idea into the minds of individuals by using the mass media. If the media were local in impact, such as newspapers, then the pattern of adoptions would be similar to that just described (**Figure 7.12**). If, however, a nationwide television, newspaper, or magazine advertising campaign were undertaken, the innovation would become known in numbers roughly corresponding to the population density.

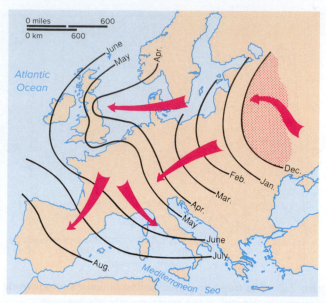

Figure 7.11   The process of contagious diffusion is sensitive to both time and distance, as suggested by the diffusion of the European influenza pandemic of 1781. The flu began in Russia and moved westward, covering Europe in about 8 months. *Source: Based on Gerald F. Pyle and K. David Patterson,* Ecology of Disease 2, *no. 3 (1984): 179.*

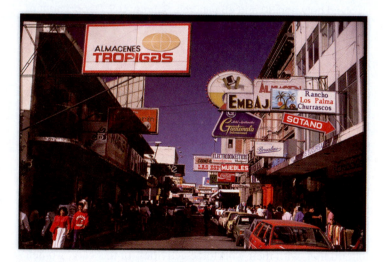

Figure 7.12   A street scene in Guatemala City. In modern society, advertising is a potent force for diffusion. Advertisements over radio and television, in newspapers and magazines, on the Internet, and on billboards and signs communicate information about many different products and innovations. © *Laurence Fordyce/Corbis Documentary/Getty Images.*

Where more people live, there would, of course, be more potential adopters. Economic or other barriers may also affect the diffusion. One immediately sees, however, why large television markets are so valuable and why national advertising is so expensive.

This type of contagious diffusion process may act together with the distance-decay process. Many of those who accept the innovation after learning of it in the mass media will tell others, so that a locally contagious effect will begin to take over soon after the original contact is made. Each type of medium has its own level of effectiveness. Advertisers have found they must repeat messages time and again before the messages are accepted as important information. This fact says something about the effectiveness of the mass media as opposed to, say, face-to-face contact.

Hierarchical Diffusion

A second way innovations are spread combines some aspects of contagious diffusion with the inclusion of a new element: a hierarchy. A *hierarchy* is a classification of objects into categories, so that categories are increasingly complex or have increasingly higher status. Hierarchies are found in many systems of organization, such as government offices (an organization chart), universities (instructors, professors, deans, and the president), and cities (villages, towns, regional centers, and metropolises). **Hierarchical diffusion** is the spread of innovation up or down a hierarchy of places.

As an example, let us suppose that a new way of automobile traffic control is adopted in a major city. Information on the innovation is spread, but only officials in comparably sized cities are in a position to accept the idea at first. It may be that the quality of information diffused to larger cities is better or that larger cities are more financially able to adopt the idea than smaller cities. Eventually, the innovation is adopted in smaller cities, and so on down the hierarchy, as it becomes better known or more financially feasible. A hypothetical scheme showing how a four-level hierarchy may be connected in the flow of information is presented in **Figure 7.13**. Note that the lowest-level centers are connected to higher-level centers but not to one another. Observe, too, that connections may bypass intermediate levels and link only with the highest-level center.

Many times, hierarchical diffusion takes place simultaneously with contagious diffusion. One might expect variations when the density of high-level centers is great and when distances between centers are short. A quick and inexpensive way to spread an idea is to communicate information about it at high-order hierarchical levels. Then the three types of diffusion processes can be used most effectively; even while an idea is diffusing through a high level in the hierarchy, it is also spreading outward from high-level centers. Consequently, low-level centers that are a short distance from high-level centers may be apprised of the innovation before more distant medium-level centers. People living in suburbs and small towns near a large city are privy to much that is new in the large city, as are individuals in other large cities half a continent away. **Figures 7.14 and 7.15** show these patterns for a case taken from Japan.

These forms of diffusion operate in the spread of culture. The consequences are the spatial interaction and innovation already

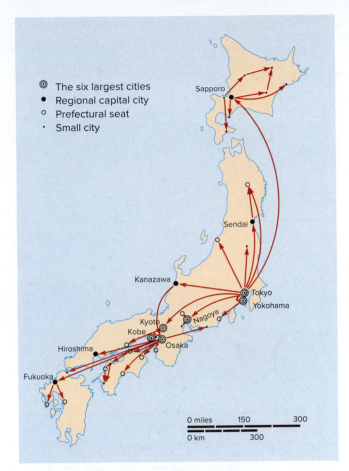

Figure 7.14 **This map shows both a hierarchical and a contagious pattern of diffusion.** Rotary Clubs, members of the international service association, were established in the large cities of Japan during the 1920s. New clubs were established under the sponsorship of the original clubs. *Source: Yoshio Sugiura, "Diffusion of Rotary Clubs in Japan, 1920–1940: A Case of Non-Profit Motivated Innovation Diffusion under a Decentralized Decision-Making Structure," in* Economic Geography, *Vol. 62, no. 2, p. 128. Copyright © 1986 Clark University, Worcester, MA.*

discussed (see p. 189). Also, recall from Chapter 6 that migration, invasions, selective cultural adoptions, and cultural transference aid the diffusion of innovation. These broader movements and exchanges represent interactions of people beyond their usual activity spaces (see "Documenting Diffusion," p. 197).

7.7 Human Interaction and Technology

When opportunities for spatial interaction abound, spatial interaction becomes a major part of people's lives. Opportunities to interact are based not only on the monetary ability to engage in spatial interaction but also on the means of interaction. In the 20th century, the ability of average wage earners in the industrialized countries to own automobiles greatly increased the degree of spatial interaction. At the end of the 20th century and into the 21st century, we have witnessed how cellular phones, low-cost telephone communication, and the Internet have appreciably increased interaction.

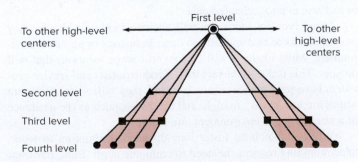

Figure 7.13 **A four-level communication hierarchy.**

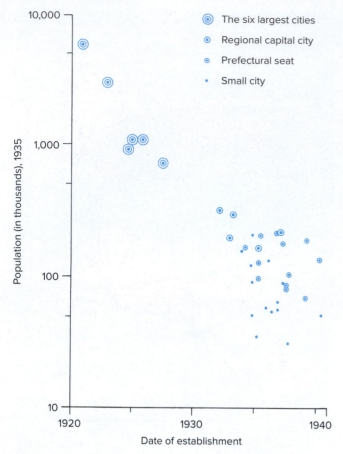

Figure 7.15 **This diagram shows the hierarchical diffusion component** of the spread of Rotary Clubs in Japan. The largest cities were the first centers of Rotary Club activity, followed by cities at lower and lower levels of urban population and city function. *Source of data: Yoshio Sugiura, "Diffusion of Rotary Clubs in Japan, 1920–1940: A Case of Non-Profit Motivated Innovation Diffusion under a Decentralized Decision-Making Structure," in Economic Geography, Vol. 62, no. 2, 1986, p. 128. Clark University, Worcester, MA.*

Automobiles

The automobile provides individuals with fast and flexible transportation on a daily basis. It has increased the ability to overcome spatial separation and has had a profound effect on the location of jobs and services. Much employment has decentralized to suburban locations, creating sprawling cities. Unfortunately, the side effects of this decentralization have been a decrease in opportunities for those without automobiles, those who must depend on public transportation. Societies have accommodated automobiles by building highways and freeways, as opposed to further developing public transportation systems. Those with automobiles are able to commute, shop, see friends and family, and engage in group activities with little inconvenience. As automobiles become more comfortable and high technology-oriented, they further encourage people to seek more opportunities for interaction. Unless local governments control urban growth and development, the result is a sprawling urban environment where people appear to be constantly on the move from one place to

another. This process that increases spatial interaction has been ongoing since the automobile became affordable to large numbers of people.

Telecommunications

For information flows, space has a different meaning than it does for the movement of people or commodities. Communication, for example, does not necessarily imply the time-consuming physical relocation of things (though in the case of letters and print media, it usually does). Indeed, in modern telecommunications, the process of information flow can be instantaneous regardless of distance. The result is space-time convergence to the point of the obliteration of space. A Bell System report tells us that, in 1920, putting through a transcontinental telephone call took 14 minutes and eight operators and cost more than $15.00 for a 3-minute call. By 1940, the call completion time had been reduced to less than 1½ minutes, and the cost had fallen to $4.00. In the 1960s, direct distance dialing allowed a transcontinental connection in less than 30 seconds, and electronic switching has now reduced the completion time to that involved in dialing a number and answering a phone. The price of long-distance conversation essentially disappeared with the advent of voice communication over the Internet in the late 1990s.

The Internet and communication satellites have made worldwide personal and mass communication immediate and data transfers instantaneous. The same technologies that have led to communication space-time convergence have tended toward a space-cost convergence. Domestic mail, which once charged a distance-based postage, is now carried nationwide or across town for the same price.

It is conceivable that the current revolution in telecommunications technology will have a more profound effect on people's lives, and thus social and industrial structure, than the automobile. For those with telecommunications capability, the level of spatial interaction has increased appreciably (**Figure 7.16**). Cellular phones, e-mail, Facebook, Twitter, and low-cost phone services have created lifestyles in which some people spend the better part of their days communicating with others. Because businesses have taken advantage of the technology by offering goods and services online, the number of shopping trips of all sorts has declined. In addition, a number of individuals make a living in a telecommuting environment. That is, they conduct business on the Internet and therefore do not take part in the morning and evening journey to and from work. The implication is that many people will find it less compelling to live in a crowded urban environment, so that sprawl is likely to increase at an accelerated pace.

Many types of industries will become footloose; that is, they won't need to be tied spatially to other industries or an urban environment. Most likely, it will be low-cost wage locations that will prosper. This is due to the fact that, if industrialists and service providers have their choice of any location, they will seek one where wages are low, skills are high, and amenities, such as the existence of a warm tropical environment, are plentiful.

While automobiles foster long-distance commuting to work, telecommuting reduces the need to commute at all. Both, however, encourage urban sprawl. Because the telecommunications revolution

DOCUMENTING DIFFUSION

The places of origin of many ideas, items, and technologies important in contemporary cultures are only dimly known or supposed, and their routes of diffusion are speculative at best. Gunpowder, printing, and spaghetti are presumed to be the products of Chinese inventiveness; the lateen sail has been traced to the Near Eastern culture world. The moldboard plow is ascribed to 6th-century Slavs of northeastern Europe. The sequence and routes of the diffusion of these innovations have not been documented.

In other cases, such documentation exists, and the process of diffusion is open to analysis. Clearly marked is the diffusion path of the custom of smoking tobacco, a practice that originated with Amerindians. Sir Walter Raleigh's Virginia colonists, returning home to England in 1586, introduced smoking in English court circles, and the habit very quickly spread among the general populace. England became the source region of the new custom for northern Europe; English medical students introduced smoking to Holland in 1590. Dutch and English together spread the habit to the Baltic and Scandinavian areas and overland through Germany to Russia. The innovation continued its eastward diffusion, and within 100 years, tobacco spread across Siberia and was, in the 1740s, reintroduced to the American continent at Alaska by Russian fur traders. A second route of diffusion for tobacco smoking can be traced from Spain through the Mediterranean area into Africa, the Near East, and Southeast Asia.

In more recent times, hybrid corn was originally adopted by imaginative farmers of northern Illinois and eastern Iowa in the mid-1930s. By the late 1930s and early 1940s, the new seeds were being planted as far east as Ohio and north to Minnesota, Wisconsin, and northern Michigan. By the late 1940s, all commercial corn-growing

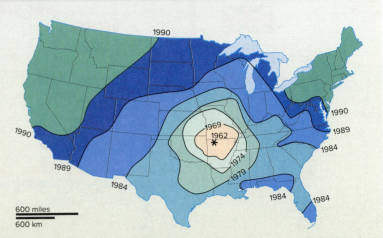

Source: Thomas O. Graff and Dub Ashton, "Spatial Diffusion of Wal-Mart," *The Professional Geographer*, 46, no. 1, 1994, pp. 19–29. Association of American Geographers, 1994.

districts of the United States and southern Canada were cultivating hybrid varieties.

A similar pattern of contagious diffusion marked the expansion of the Wal-Mart stores chain. From its origin in northwest Arkansas in 1962, the discount chain had dispersed throughout the United States by the 1990s to become the country's largest business in sales volume. In its expansion, Wal-Mart displayed a "reverse hierarchical" diffusion, initially spreading by way of small towns before opening its first stores in larger cities and metropolitan areas.

---

began in the mid-1990s and is still in its infancy, we are just beginning to recognize how it affects all aspects of our lifestyles.

7.8 Migration

An important aspect of human history has been the **migration** of peoples—that is, the permanent relocation of both place of residence and activity space. It has contributed to the evolution of separate cultures, to the relocation diffusion of those cultures, and to the complex mix of peoples and cultures found in various parts of the world. The settlement of North America, Australia, and New Zealand involved great long-distance movements of peoples. The flight of refugees from past and recent wars, the settlement of Jews in Israel, the current migration of workers to the United States from Mexico and Central America, and innumerable other examples of mass movement come quickly to mind. In all cases, societies transplanted their cultures to the new areas, their cultures therefore diffused and intermixed, and history was altered.

Massive movements of people within countries, across national borders, and between continents have emerged as a pressing concern of recent decades. They affect national economic structures; determine population density and distribution patterns; alter

traditional ethnic, linguistic, and religious mixtures; and inflame national debates and international tensions. Migration patterns and conflicts touch many aspects of social and economic relations and have become an important part of current geographic realities. In this chapter, our interest is in migration as an unmistakable, recurring, and near-universal expression of human spatial behavior.

Types of Migration

Migration flows can be discussed at different scales, from massive intercontinental torrents to individual decisions to move to a new house or apartment within the same metropolitan area. At each level, although the underlying controls on spatial behavior remain constant, the immediate motivating factors influencing the spatial interaction are different, with differing impacts on population patterns and cultural landscapes.

Naturally, the length of a move and its degree of disruption of people's lives raise distinctions important in the study of migration. A change of residence from the central city to the suburbs certainly changes the activity space of schoolchildren and of adults in many of their nonworking activities, but the workers may still retain the city—indeed, the same place of employment there—as

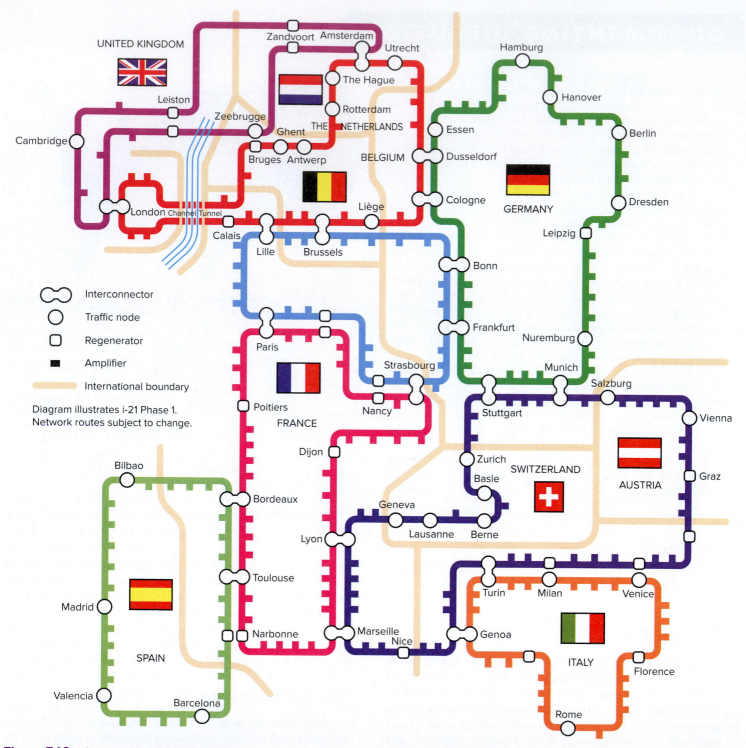

Figure 7.16   **A map of the Internet i-21 network of part of Europe, a fiber-optic network that connects 61 cities in 16 countries.** It has the capacity to carry more than 1 petabit (1 billion megabits) per second of communication traffic. *Source: Martin Dodge and Rob Kitchin,* An Atlas of Cyberspaces, *Copyright Martin Dodge, 2004. This work is licensed under a Creative Commons License. Available at http://www .cybergeography.org/atlas/interoute_large.gif.*

an activity space. On the other hand, migration from Europe to the United States and the massive farm-to-city movements of rural Americans late in the 19th and early 20th centuries meant a total change of all aspects of behavioral patterns.

At the broadest scale, *intercontinental* movements range from the earliest peopling of the habitable world to the most recent flight

of Asian and African refugees to countries of Europe and the Western Hemisphere. The population structure of the United States, Canada, Australia and New Zealand, Argentina, Brazil, and other South American countries is a reflection and result of massive intercontinental flows of immigrants that began as a trickle during the 16th and 17th centuries and reached a flood during the 19th and

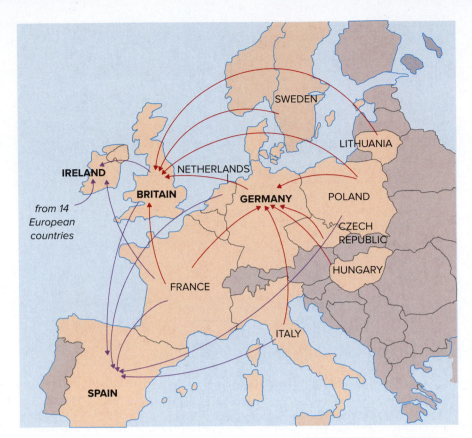

Figure 7.17   **Immigration to four European countries—Britain, Ireland, Germany, and Spain—from other European countries, mid-2000s.** *Source: Data from Eurostat.*

early 20th centuries. Later in the 20th century, World War II (1939–1945) and its immediate aftermath involved more than 25 million permanent population relocations, all of them international but not all intercontinental.

Intracontinental and *interregional* migrations involve movements between countries and within countries, most commonly in response to individual and group assessments of improved economic prospects, but often reflecting flight from difficult or dangerous environmental, military, economic, or political conditions. The millions of refugees leaving their homelands following the dissolution of Eastern European communist states, including the former USSR and Yugoslavia, exemplify that kind of flight. Between 1980 and 2000, Europe received some 20 million newcomers, often refugees, who joined the 15 million labor migrants ("guest workers") already in Western European countries by the early 1990s. The European Union makes it possible for labor to move easily among its many member countries (**Figure 7.17**). About 175 million people—3% of the world population—live in a country other than the country of their birth in the early 2000s, and migration has become a world social, economic, and political issue of first priority.

Migrations may be forced or voluntary, or, in many instances, reluctant relocations imposed on the migrants by circumstances. In *forced,* or *involuntary, migrations,* the relocation decision is made solely by people other than the migrants themselves (**Figure 7.18**). Between 10 and 12 million Africans were forcibly transferred as slaves to the Western Hemisphere from the late 16th to the early 19th centuries. One-half or more were destined

for the Caribbean and most of the remainder for Central and South America, though nearly a million arrived in the United States. Australia owed its earliest European settlement to convicts transported after the 1780s to the British penal colony established in southeastern Australia (New South Wales). More recent involuntary migrants include millions of Soviet citizens forcibly relocated from countryside to cities and from the western areas to labor camps in Siberia and the Russian Far East beginning in the late 1920s.

Less than fully voluntary relocation—*reluctant migration*—of some 8 million Indonesians has taken place under an aggressive government campaign begun in 1969 to move people from densely settled Java to other islands and territories of the country in what has been called the "biggest colonization program in history." International refugees from war and political turmoil or repression numbered some 15 million in 2003, according to the World Refugee Survey—1 out of every 415 people on the planet. In the past, refugees sought asylum mainly in Europe and other developed areas. Between 2003 and 2007, Iran, Syria, and Jordan became host to millions of Iraqis fleeing persecution, terrorism, and war. Sub-Saharan Africa alone houses more than 3 million refugees (**Figure 7.19**).

The civil war in Syria involving government forces (the Assad regime and Russian allies), representing a minority (the Alawites) living in and near the capital, Damascus, are fighting against anti-government rebels, al Queda, Islamic State extremists, and Kurds (now largely autonomous), has taken a country of 22 million people in 2010 and reduced it to about 12 million. Many have been killed, but about 7 million have escaped to neighboring Turkey,

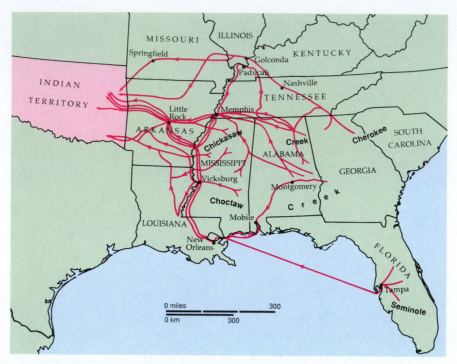

Figure 7.18   Forced migrations: the Five Civilized Tribes. Between 1825 and 1840, some 100,000 southeastern Amerindians were removed from their homelands and transferred by the army across the Mississippi River to "Indian Territory" in present-day Oklahoma. By far, the largest number were members of the Five Civilized Tribes of the South: Cherokees, Choctaws, Chickasaws, Creeks, and Seminoles. Settled, Christianized, literate small farmers, their forced eviction and arduous journey—particularly along what the Cherokees named their "Trail of Tears" in the harsh winter of 1837–1838—resulted in much suffering and death.

Lebanon, and Jordan, and to European countries, mainly Germany and Sweden. The migratory paths of the Syrians in Europe is highlighted by uncooperative treatment in many instances. Worldwide, an additional 22 million people were "internally displaced," effectively internal refugees within their own countries. In a search for security or sustenance, they have left their home areas but not crossed an international boundary.

The great majority of migratory movements, however, are *voluntary*, representing individual response to the factors influencing all spatial interaction decisions. At root, migrations take place because the migrants believe that their opportunities and lives will be better at their destination than they are at their present location.

Incentives to Migrate

The decision to move is a cultural and temporal variable. Nomads fleeing famine and spreading deserts in the Sahel of Africa obviously are motivated by considerations different from those of an executive receiving a job transfer to Chicago, a resident of Appalachia seeking factory employment in the city, or a retired couple searching for sun and sand. In general, people who voluntarily decide to migrate are seeking better economic, political, or cultural conditions or certain amenities. For many, the reasons for migration are frequently a combination of several of these categories.

Negative home conditions that impel the decision to migrate are called **push factors.** They include loss of job, lack of professional opportunity, overcrowding, and a variety of other influences, including poverty, war, and famine. The presumed positive attractions of the migration destination are known as **pull factors.** They include all the attractive attributes perceived to exist at the new location—safety and food, perhaps, or job opportunities, better climate, lower taxes, more room, and so forth. Very often, migration is a result of both perceived push and pull factors (**Figure 7.20**). It is the perception of the opportunities and want satisfaction that is important, regardless of whether the perception is supported by objective reality.

Economic considerations have impelled more migrations than any other single incentive. If migrants face unsatisfactory conditions at home (e.g., unemployment or famine) and believe that the economic opportunities are better elsewhere, they will be attracted to the thought of moving. Poverty is the great motivator. Some 30% of the world's population—nearly 2 billion people—have less than $1.00 per day income. Many additionally are victims of drought, floods, and other natural catastrophes, or of wars and terrorism. Poverty in developing countries is greatest in the countryside; rural areas are home to around 750 million of the world's poorest people. Of these, 20 to 30 million move each year to towns and cities—many as "environmental refugees," abandoning land so eroded or exhausted that it can no longer support them. In the cities, they join the 40% or more of the laborers who are unemployed or underemployed in their home country and seek legal or illegal entry into the more-promising economies of the developed world. All, rural or urban, respond to the same basic forces: the push of poverty and the pull of perceived or hoped-for opportunity.

The desire to escape war and persecution at home and to pursue the promise of freedom in a new location is a *political incentive* for

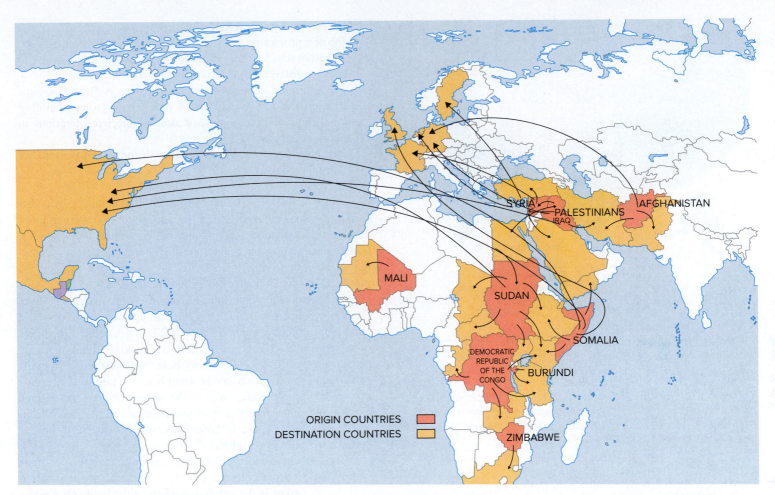

Figure 7.19   Major flows of refugees. The map shows the origin and destination countries of the largest refugee populations as of 2016. In recent years, political upheavals have forced the migration of millions of people from their homes and across international borders. *Source: Data from United Nations High Commissioner for Refugees.*

migration. Americans are familiar with the history of settlers who migrated to North America seeking religious and political freedom (**Figure 7.21**). In more recent times, the United States has received hundreds of thousands of refugees from countries such as Hungary, following the uprising of 1956; Cuba, after its takeover by Fidel Castro; and Vietnam, after the fall of South Vietnam. The massive movements of Hindus and Muslims across the Indian subcontinent in 1947, when Pakistan and India were established as governing entities, and the exodus of Jews fleeing persecution in Nazi Germany in the 1930s are other examples of politically inspired moves. More recently, about 1 million Hutus fled into neighboring African countries after ethnic Tutsis took over Rwanda's government; ethnic Muslims living in Bosnia were pushed out of their ancestral homes by Serbs; and many Haitians, under severe economic privation during a political crisis, have left for the United States.

Migration normally but not always involves a hierarchy of decisions. Once people have decided to move and have selected a general destination (e.g., the United States or the Sunbelt), they must still choose a particular site in which to settle. At this scale, *cultural variables* can be important pull factors. Migrants tend to be attracted to areas where the language, religion, and racial or ethnic background of the inhabitants are similar to their own. This similarity can help migrants feel at home when they arrive at their

destination, and it may make it easier for them to find a job and to become assimilated into the new culture. The Chinatowns and Little Italys of large cities attest to the drawing power of cultural factors, as discussed in Chapter 6.

Another set of inducements is grouped under the heading *amenities,* the particularly attractive or agreeable features that are characteristic of a place. Amenities may be natural (mountains, oceans, climate, and the like) or cultural (e.g., the arts and music opportunities available in large cities). They are particularly important to relatively affluent people seeking "the good life." Amenities help to account for the attractiveness of the so-called Sunbelt states in the United States for retirees; a similar movement to the southern coast has also been observed in countries such as the United Kingdom and France.

The significance of the various incentives varies according to the age, sex, education, and economic status of the migrants (see "Gender and Migration" p. 203). For the modern American, reasons to migrate have been summarized into a limited number of categories that are not mutually exclusive:

1. changes in life course, such as getting married, having children, getting a divorce, or needing less dwelling space when the children leave home

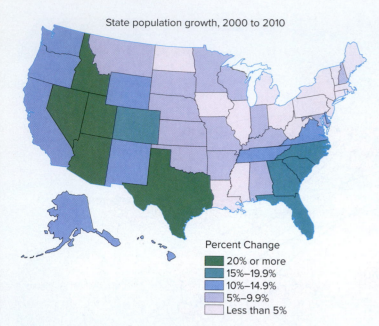

State population growth, 2000 to 2010

Percent Change
- 20% or more
- 15%–19.9%
- 10%–14.9%
- 5%–9.9%
- Less than 5%

Figure 7.20 **Migration in the United States.** Although birth and death rates have a strong bearing on population growth, the growth pattern in the United States from 2000 to 2010 was largely a result of net migration to the West, Southwest, and Southeast. The *pull factors* included job opportunities and mild climates; the *push factors* included loss of job opportunities and harsh winters. *Source: Population Division, U.S. Bureau of the Census.*

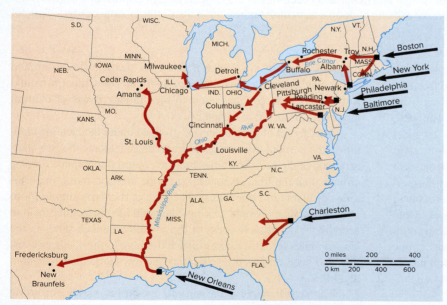

Figure 7.21 **The major paths of the early migration of Germans to America.** Most emigrants left Germany because of religious and political persecution. They chose the United States not only because immigrants were made welcome but also because labor was in demand and farmland was available. The first immigrants landed, and many settled, in Boston, New York, Philadelphia, Baltimore, Charleston, and New Orleans. The migrants carried with them such aspects of their culture as religion, language, and food preferences.

2. changes in career cycle, such as leaving college, getting a first job or a promotion, receiving a career transfer, or retiring
3. forced or reluctant migrations associated with urban development, construction projects, and the like
4. neighborhood changes from which there is flight, perhaps pressures from new and unwelcome ethnic groups, building deterioration, street gangs, or similar rejected alterations in activity space
5. changes of residence associated with individual personality (chronic mobility)

Some people simply tend to move often for no easily discernible reasons, whereas others, *stayers*, settle into a community permanently. Of course, for a country such as China, with its limitations on emigration and severe housing shortages, a totally different set of summary migration factors is present. By *emigration* we mean leaving one's country or region to settle elsewhere.

The factors that contribute to mobility tend to change over time. However, in most societies, one group has always been the most mobile: young adults (**Figure 7.22**). They are the members of society who are launching careers and making initial decisions about occupation and location. They have the fewest responsibilities of all adults; thus, they are not as strongly tied to family and institutions as older people are. Most of the major voluntary migrations have been composed primarily of young people who suffered from a lack of opportunities in the home area and who were easily able to take advantage of opportunities elsewhere.

The concept of **place utility** helps us understand the decision-making process that potential voluntary migrants undergo. Place utility is the value that an individual puts on a given residential site. The decision to migrate is a reflection of the appraisal—the perception—by the prospective migrant of the current homesite as opposed to other sites of which something is known or hoped for. The individual may adjust to conditions at the homesite and decide not to migrate.

In evaluating comparative place utility, the decision maker considers not only the perceived value of the present location but also the expected place utility of each potential destination. The evaluations are matched with the individual's *aspiration level*—that is, the level of accomplishment or ambition that the individual sees for himself or herself. Aspirations tend to be adjusted to what an individual considers attainable. If the person is satisfied with present circumstances, then he or she does not initiate search behavior. If, on the other hand, the person is dissatisfied with the home location, she or he assigns a utility to each possible new site. The utility is based on past or expected future rewards at the various sites. Because the new places are unfamiliar to the individual, the information received about them acts as a substitute for the personal experience of the homesite. The decision maker can do no more than sample information about the new sites, and, of course, there may be errors in both information and interpretation.

GENDER AND MIGRATION

Gender is involved in migration at every level. In a household or family, women and men are likely to play different roles regarding decisions or responsibilities for activities such as child care. These differences, and the inequalities that underlie them, help determine who decides whether the household moves, which household members migrate, and the destination for the move. Outside the household, societal norms about women's mobility and independence often restrict their ability to migrate.

The economies of both sending and receiving areas play a role as well. If jobs are available for women in the receiving area, women have an incentive to migrate, and families are more likely to encourage the migration of women as necessary and beneficial. Thousands of women from East and Southeast Asia have migrated to the oil-rich countries of the Middle East, for example, to take service jobs.

The impact of migration is also likely to be different for women and men. Moving to a new economic or social setting can affect the regular relationships and processes that occur within a household or family. In some cases, women might remain subordinate to the men in their families. A study of Greek-Cypriot immigrant women in London and of Turkish immigrant women in the Netherlands found that although these women were working for wages in their new societies, these new economic roles did not affect their subordinate standing in the family in any fundamental way.

In other situations, however, migration can give women greater power in the family. In Zaire, women in rural areas move to towns to take advantage of job opportunities there, and they gain independence from men in the process.

One of the keys to understanding the role of gender in migration is to disentangle household decision-making processes. Many researchers see migration as a family decision or strategy, but some family members will have more sway than others, and some members will benefit more than others will from those decisions.

For many years, men predominated in the migration streams flowing from Mexico to the United States. Women played an important role in this migration stream, even when they remained in Mexico. Mexican women influenced the migration decisions of other family members; they married migrants to gain the benefits from and opportunity for migration; and they resisted or accepted the new roles in their families that migration created.

In the 1980s, Mexican women began to migrate to the United States in increasing numbers. Economic crises in Mexico and an increase in the number of jobs available for women in the United States, especially in factories, domestic service, and service industries, have changed the backdrop of individual migration decisions. Now, women often initiate family moves or resettlement efforts.

Mexican women have begun to build their own migration networks, which are key to successful migration and resettlement in the United States. Networks provide migrants with information about jobs and places to live and have enabled many Mexican women to make independent decisions about migrating.

In immigrant communities in the United States, women are often the vital links to social institution services and to other immigrants. Thus, women have been instrumental in the way that Mexican immigrants have settled and become integrated into new communities.

Source: Nancy E. Riley, "Gender, Power, and Population Change," *Population Bulletin,* Vol. 52, No. 1, May 1997, pp. 32–33. Copyright © 1997 by the Population Reference Bureau, Inc.

One goal of the potential migrant is to minimize uncertainty. Most decision makers either elect not to migrate or postpone the decision unless uncertainty can be lowered sufficiently. That objective may be achieved either by going through a series of transitional relocation stages or by following the example of known predecessors. **Step migration** involves the place transition from, for example, rural to central city residence through a series of less-extreme location changes—from farm to small town to suburb and, finally, to the major central city itself. The term **chain migration** indicates that the mover is part of an established migrant flow from a common origin to a prepared destination. An advance group of migrants, having established itself in a new home area, is followed by second and subsequent migrations originating in the same home district and frequently united by kinship or friendship. Public and private services for legal migrants and informal service networks for undocumented or illegal migrants become established and contribute to the continuation or expansion of the chain migration flow. Ethnic and foreign-born enclaves in major cities and rural areas in a number of countries are the immediate result.

Sometimes the chain migration is specific to occupational as well as ethnic groups. For example, nearly all newspaper vendors in New Delhi, in the north of India, are reported to come from one small district in Tamil Nadu, in the south of India. Most construction workers in New Delhi come from either Orissa, in the east of India, or Rajasthan, in the northwest, and taxicab drivers originate in the Punjab area. A network of about 250 related families who come from a small town several hundred miles to the north dominates the diamond trade of Mumbai (Bombay), India.

A similar domination of businesses by a single ethnic group occurs in the United States. Familiar examples of these *ethnic niche businesses* include fruit stores owned by Koreans and diners owned by Greeks, but there are many others. Female immigrants from Vietnam dominate the manicure trade, accounting for more than one-half the women in that profession. A large proportion of female immigrants from the Philippines find employment as nurses, particularly in New York City and the West Coast. About 30% of the Filipinos in New York City and its suburbs work as nurses or other health practitioners, the result of their aggressive recruitment by American hospitals and the fact that U.S. immigration authorities have made it easy for nurses to obtain work visas and green cards giving them permanent resident status. Immigrants from India, mostly from the state of Gujarat, now own more than one-third of the hotels in the United States, most of them budget and midpriced franchises such as Holiday Inns, Days Inns, and Ramadas.

Another goal of the potential migrant is to avoid physically dangerous or economically unprofitable outcomes in the final

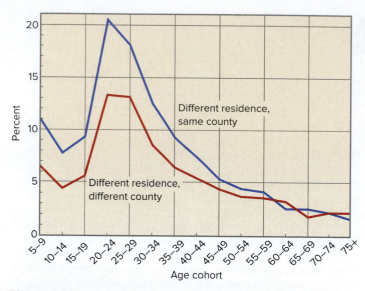

Figure 7.22   Percentage of 2000 population over 5 years of age with a different residence than in 1999. Young adults figure most prominently in both short- and long-distance moves in the United States, an age-related pattern of mobility that has remained constant over time. For the sample year shown, 33% of people in their twenties moved, whereas fewer than 5% of those 65 and older did so. Short-distance moves predominated; 56% of the 43 million U.S. movers between March 1999 and March 2000 relocated within the same county, and another 20% moved to another county in the same state. Some two-thirds of intracounty (mobility) moves were made for housing-related reasons; long-distance moves (migration) are likely to be made for work-related reasons. *Source: U.S. Bureau of the Census.*

migration decision. Place utility evaluation, therefore, requires assessments not only of perceived pull factors of new sites but also of the potentially negative economic and social reception the migrant might experience at those sites. An example of some of these observations can be seen in the case of the large numbers of young people from the Caribbean, Mexico, and Central America who have migrated both legally and illegally to the United States over the past 30 years (**Figure 7.23**).

Faced with poverty at home, these young adults regard the place utility in their own country as minimal. Their space-searching ability is limited, however, by both the lack of money and the lack of alternatives in the land of their birth. With a willingness to work and with aspirations for success—perhaps wealth—in the United States, they learn from friends and relatives of job opportunities north of the border, low paying though they may be. Hundreds of thousands quickly place high utility on perhaps a temporary relocation (maybe 5 or 10 years) to the United States. Many know that dangerous risks are involved if they attempt to enter the country illegally, but even legal immigrants face legal restrictions designed to reduce the pull attractions of the United States (see "Broken Borders," pp. 206–207). The arrival of those who consider the rewards worth the risk indicates their assignment of higher utility to the new site than to the old one.

In the 20th century, nearly all countries experienced a great movement of people from agricultural areas to the cities, continuing the pattern of *rural-to-urban migration* that first became prominent during the 18th- and 19th-century Industrial Revolution in advanced economies. The migration presumably paralleled the number of perceived opportunities within cities and convictions of absence of place utility in the rural districts. Perceptions, of course, do not necessarily accord with reality. Rapid increases in impoverished rural populations of developing countries put increasing and unsustainable pressures on land, fuel, and water in the countryside. Landlessness and hunger, as well as the loss of social cohesion that the growing competition for declining resources induces, help force migration to cities. As a result, while the rate of urban growth is decreasing in the more-developed countries, urbanization in the developing world continues apace, as will be discussed more fully in Chapter 11.

Barriers to Migration

Paralleling the incentives to migration is a set of disincentives, or barriers, to migration. They help account for the fact that many people do not choose to move even when conditions are bad at home and are known to be better elsewhere. Migration depends on knowledge of the opportunities in other areas. People with a limited knowledge of the opportunities elsewhere are less likely to migrate than are those who are better informed. Other barriers include physical features, the costs of moving, ties to individuals and institutions in the original activity space, and government regulations.

Physical barriers to travel include seas, mountains, swamps, deserts, and other natural features. In prehistoric times, physical barriers played an especially significant role in limiting movement. Thus, the spread of the ice sheets across most of Europe in Pleistocene times was a barrier to both migration and human habitation. Physical barriers to movement have probably assumed reduced importance only within the past 400 years. The developments that made possible the great age of exploration, beginning about A.D. 1500, and the technological developments associated with industrialization have enabled people to conquer space more easily. With industrialization came improved forms of transportation, which made travel faster, easier, and cheaper. Even then, as the conditions the Pony Express riders experienced only a century and a half ago show, travel could be arduous, and in some parts of the world, it remains so today.

Economic barriers to migration include the cost of both traveling and establishing a residence elsewhere. Frequently, the additional expense of maintaining contact with those left behind is a pertinent cost factor. Normally, all of these costs increase with the distance traveled and are a more significant barrier to travel for the poor than for the rich. Many immigrants to the United States were married men who came alone; when they had acquired enough money, they sent for their families to join them. This phenomenon is still evident among recent immigrants from the Caribbean area to the United States and among Turks, Yugoslavs, and West Indians who have settled in northern and Western European countries. The *cost factor* limits long-distance movement, but the larger the differential is between present circumstances and perceived opportunities, the more will be the individuals willing to spend on moving. For many, especially older people, the differentials must be extraordinarily high for movement to take place.

Cultural factors also contribute to decisions not to migrate. Family, religious, ethnic, and community relationships defy the principle of differential opportunities. Many people will not migrate under any but the most pressing of circumstances. The fear of

(a)

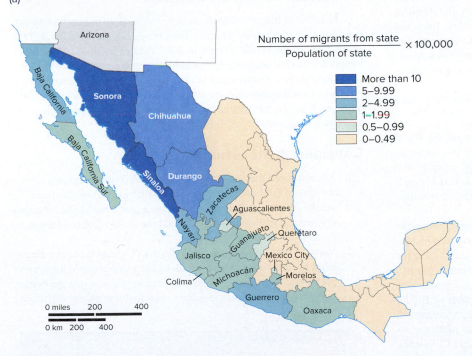

(b)

Figure 7.23 **(a) Illegal Mexican immigrants running from the Border Patrol. (b) Undocumented migration rate to Arizona.** The Arizona region and nearby Mexican states have historical ties that go back to the early 1800s. In many respects, the international border cuts through a culture region. Note that distance plays a large role in the decision to migrate to the United States, with over one-half of the migrants coming from four nearby Mexican states: Sonora, Sinaloa, Durango, and Chihuahua. *(a) © DENIS POROY/AP Images; (b) Source: John P. Harner, "Continuity Amidst Change," The Professional Geographer, 47, no. 4, 1995, Fig. 2, p. 403. Association of American Geographers.*

change and human inertia—the fact that it is easier not to move than to do so—may be so great that people consider, but reject, a move. Ties to one's own country, culture group, neighborhood, or family may be so strong as to compensate for the disadvantages of the home location. Returning migrants may convince potential leavers that the opportunities are not, in fact, better elsewhere—or that, even if they are better, they are not worth entering an alien culture or sacrificing home and family.

Restrictions on immigration and emigration constitute *political barriers* to migration. Many governments frown on

movements into or outside their own borders and restrict outmigration. These restrictions may make it impossible for potential migrants to leave, and they certainly limit the number who can do so. On the other hand, countries suffering from an excess of workers often encourage emigration. The huge migration of people to the Americas in the late 19th and early 20th centuries is a good example of perceived opportunities for economic gain far greater than in the home country. Many European countries were overpopulated, and their political and economic systems stifled domestic economic opportunity at a time when people

Broken Borders

Migrants can enter a country legally—with passport, visa, working permit, or other authorization—or illegally. Some aliens enter a country legally but on a temporary basis (as a student or tourist, for example), but then remain after their official departure date. Others arrive claiming the right of political asylum but actually seeking economic opportunity. The Department of Homeland Security estimates that between one-fourth and one-third of the illegal residents in the United States entered the country legally but then overstayed their visas.

Although it is impossible to determine the precise number of people residing illegally in the United States, a number that changes daily, most authorities estimate there are anywhere from 10 to 12 million. About 55% have come from Mexico, another 20% to 25% from other Latin American countries and the Caribbean, 13% from Asia, and the remainder from Canada, Europe, and Africa. The rising tide of emotion against unauthorized immigrants is directed mainly against those from Latin America, most of whom are unskilled workers.

Once they are in the United States, Latin American immigrants find work in agricultural fields, animal slaughtering and meat packing facilities, construction, hotels, and restaurants. Many work in private residences as maids, nannies, and gardeners. While most undocumented workers initially enter California, Arizona, Texas, or Florida, and many remain in those states, they go where they find jobs. Many blend into the large migrant communities not just of Los Angeles, El Paso, and Houston, for example, but also those of Chicago and New York City. Iowa and North Carolina have some of the fastest-growing populations of illegal immigrants. The demand for their labor is great enough that nearly all of the males have higher labor force participation rates than native-born men.

Concern over the growing number of illegal aliens has been reflected in a number of actions in recent years.

- Security fears since the September 11, 2001, assaults on the World Trade Center and Pentagon have led to more stringent visa applicant background checks, greater restrictions on admitting refugees and asylum seekers, stricter enforcement of Immigration and Naturalization Service (INS) rules on alien residency reports and visa time restrictions, and stricter border controls.
- Greater efforts are being made to deter illegal crossings along the 3380-kilometer (2100-mi) U.S.-Mexico border. To supplement the 120 kilometers (75 mi) of steel fences near a few U.S. cities, the Bush administration in 2007 proposed 1125 kilometers (700 mi) of new fencing, and Congress approved money for a small portion of it. The administration also increased the number of Border Patrol agents, who use automotive vehicles, helicopters, unmanned aircraft known as *drones,* night-vision cameras, and hidden electronic sensors for surveillance. In parts of Arizona and California, self-appointed Minutemen—groups of volunteer militia—patrol the border "to protect our country from a 40-year long invasion across our southern border with Mexico," as one vigilante put it. A candidate for president (2016), Donald J. Trump, proposes building a wall along the entire border.
- Because the burden of coping with illegal immigration falls mostly on state and local governments, four states (Florida, Texas, Arizona, and California) have sued the federal government—so far unsuccessfully—to win reimbursement for the costs of illegal immigration. Similarly, the U.S.-Mexico Border Counties Coalition, composed of representatives from the 24 counties in the United States that abut Mexico, has not yet succeeded in getting the federal

government to reimburse local administrations for money spent on legal and medical services for illegal aliens. These services include the detention, prosecution, and defense of immigrants; emergency medical care; ambulance service; and autopsies and burials for the hundreds who die each year while trying to cross the border.
- Concerned that large numbers of unauthorized immigrants impose a financial burden on taxpayers, congest schools and public health clinics, and result in the reduction of services to legal residents, voters in California approved Proposition 187, which barred illegal immigrants and their families from public schools, social services, and nonemergency health care. The measure also required state and local agencies to report suspected illegal immigrants to the Immigration and Naturalization Service and to certain state officials. The courts struck down Proposition 187, declaring most of the provisions to be unconstitutional, and the measure was never implemented.
- In Arizona, a law targeting employers of undocumented workers took effect on January 1, 2008. Employers are required to check the status of each worker with the federal government. A company found to have knowingly hired an illegal resident has its business license suspended for 10 days for the first offense. A second offense results in the permanent revocation of the business license. Other states are considering similar legislation.

Millions of illegal residents managed to legalize themselves by taking advantage of government amnesties offered between 1984 and 2000. In his State of the Union address in 2013, President Obama called for a temporary-worker program that would legalize the presence of millions of undocumented workers and lead toward citizenship. Congressional efforts to pass a comprehensive immigration reform bill failed in 2007, however. It called for greater border security, a guest worker program, and a path to citizenship for those already here.

Considering the Issues

1. It is estimated that some 800,000 unauthorized immigrants enter the United States each year, and most find gainful employment, yet the country issues only 5000 visas a year for unskilled foreigners seeking year-round work. Should the United States increase the number of visas available?
2. Making illegal crossings more difficult in California has not diminished the number of migrants making the journey north; it has simply pushed *coyotes,* the people who lead migrants across the border, into Arizona. Some people believe resourceful migrants will always

Existing and proposed steel fences near U.S. cities along the Mexico border, 2013.

(continued)

(continued)

© MATT YORK/AP Images.

find a way to get across the border. As one observer noted, "It's like putting rocks in a river—the water just goes around it." Do you think there is any way to seal the entire U.S.-Mexico border, or will immigration continue as long as the income gap between the United States and Latin America remains great? Could the United States reduce immigration pressures by improving the Mexican economy?

3. What reasons can you think of to explain why anti-immigrant sentiment is directed chiefly against unauthorized immigrants from Latin America? Will such feelings eventually fade, as did earlier ones about the immigration of Irish Catholics, Chinese, Eastern Europeans, and other groups?

4. Some argue that, when there is no immigration barrier, *circular migration* occurs, with migrant workers entering and leaving almost at will. It cites Puerto Rico as an example; many who move to the mainland stay for just a few years, and out-migration from the island is very low. A temporary-worker program, on the other hand, encourages migrants to move north with their entire families, and those who are already in the United States stay for good, because border crossings become more expensive and dangerous. If you were a member of Congress, would you be in favor of creating a guest-worker program? Why or why not?

5. It is often said that illegal aliens perform jobs Americans won't do. Why do you think the immigrants are willing to work for low wages, often under poor working conditions? Would Americans take the jobs if they were paid, say, $20 per hour and offered health care and other benefits?

6. What would happen if all states followed Arizona's lead in passing and enforcing laws targeting employers of undocumented workers? One in 10 or 11 workers in Arizona is undocumented, and opponents of the legislation contend that the state has put its economy in jeopardy. The workers tend to be reliable, they fill necessary jobs, they spend their wages in the communities in which they live and work, and most pay taxes because employers in the construction, hotel, restaurant, and other industries withhold taxes from paychecks.

7. Should illegal residents already here be given the opportunity to get work permits and the possibility of eventual citizenship if they have no criminal record, pay a fine, and demonstrate that they are gainfully employed and have paid taxes?

8. Do you believe the federal government has an obligation to fully or partially reimburse state and local governments for the costs of education, medical care, incarceration, and other legal services for unauthorized immigrants? Why or why not?

9. Should the United States require citizens to have a national identification card and to present it to officials upon demand? If so, would all people who look like they may have been born abroad feel it necessary to carry proof of citizenship at all times?

were needed by American entrepreneurs hoping to increase their wealth in untapped resource-rich areas.

The most-developed countries where per capita incomes are high, or perceived as high, are generally the most desired international destinations. To protect themselves against overwhelming migration streams, such countries as the United States, Australia, France, and Germany have restrictive policies on immigration. In addition to absolute quotas on the number of immigrants (usually classified by country of origin), a country may impose other requirements, such as the possession of a labor permit or sponsorship by a recognized association.

Patterns of Migration

Several geographic concepts deal with patterns of migration. The first of these is the **migration field,** an area or areas that dominate a locale's in- and out-migration patterns. For any single place, the origin of its in-migrants and the destination of its out-migrants remain fairly stable spatially over time. As would be expected, areas near the point of origin make up the largest part of the migration field (**Figure 7.24**). However, places far away, especially large cities, may also be prominent. These characteristics of migration fields are functions of the hierarchical movement to larger places and the fact that so many people live in large metropolitan areas that one may expect some migration into and out of them from most areas within a country.

Migration fields do not conform exactly to the diffusion concepts mentioned earlier in Section 7.6, "Diffusion and Innovation." As **Figure 7.25** shows, some migration fields reveal a distinctly *channelized* pattern of flow. The channels link areas that are socially and economically tied to one another by past migration patterns, economic trade considerations, or some other affinity.

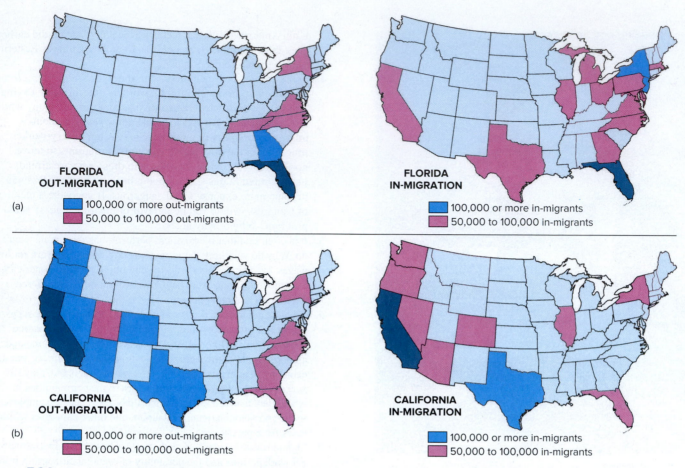

Figure 7.24   **The migration fields of Florida and California in 2005–2009. (a)** For Florida, nearby Georgia receives most out-migrants, but in-migrants originate in large numbers from the northeastern United States. **(b)** For California, the nearby western states receive large numbers of out-migrants, and there are fewer in-migrants from those states. *Source: American Community Survey.*

As a result, flows of migration along these channels are greater than would otherwise be the case. The former movements of blacks from the southern United States to the North; of Scandinavians to Minnesota and Wisconsin; of Mexicans to such border states as California, Texas, and New Mexico; and of retirees to Florida and Arizona are all examples of **channelized migration** flows.

Of course, not all immigrants stay permanently at their first destination. Of the approximately 80 million newcomers to the United States between 1900 and 1980, some 10 million returned to their homelands or moved to another country. Estimates for Canada indicate that perhaps 40 of each 100 immigrants eventually leave, and about 25% of newcomers to Australia also depart permanently. A corollary of all out-migration flows is, therefore, **return migration,** or **countermigration,** the return of migrants to the regions from which they had earlier emigrated (**Figure 7.26**).

Within the United States, return migration—moving back to one's state of birth—makes up about 20% of all domestic moves. That figure varies dramatically among states, however. More than one-third of recent in-migrants to West Virginia, for example, were returnees, as were over 25% of those moving to Pennsylvania, Alabama, Iowa, and a few other states. Such widely different states as New Hampshire, Maryland, California, Florida, Wyoming, and Alaska were among those that found returnees were fewer than 10% of their in-migrants.

Interviews suggest that states deemed attractive draw new migrants in large numbers, whereas those with high proportions of returnees in the migrant stream are not perceived as desirable destinations by other than former residents.

If freedom of movement is not restricted, it is not unusual for as many as 25% of all migrants to return to their place of origin. Unsuccessful migration is sometimes due to an inability to adjust to the new environment. More often, it is the result of false expectations based on distorted mental images of the destination at the time of the move. Myths, secondhand and false information, and people's own exaggerations contribute to what turn out to be a mistaken decision to move. Although return migration often represents the unsuccessful adjustment of individuals to a new environment, it does not necessarily mean that negative information about a place returns with the migrant. It usually means a reinforcement of the channel, as communication lines between the unsuccessful migrant and would-be migrants take on added meaning and understanding.

In addition to channelization, the influence of large cities causes migration fields to deviate from the distance-decay pattern. The concept of **hierarchical migration** assists in understanding the nature of migration fields. Earlier we noted that sometimes information diffuses according to a hierarchical rule—that is, from city to city at the highest level in the hierarchy and then to lower levels.

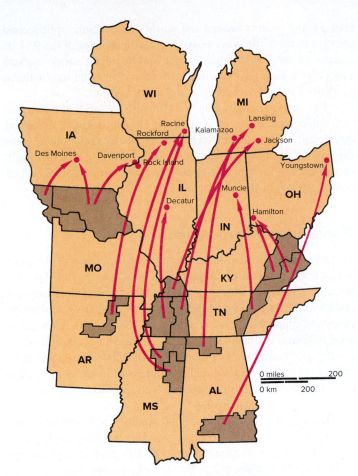

Figure 7.25 **Channelized migration flows from the rural South to midwestern cities of medium size.** Distance is not necessarily the main determinant of flow direction. Perhaps through family and friendship links, the rural southern areas are tied to particular midwestern destinations. *Source:C. C. Roseman, Proceedings of the Association of American Geographers, Vol. 3, p. 142.*

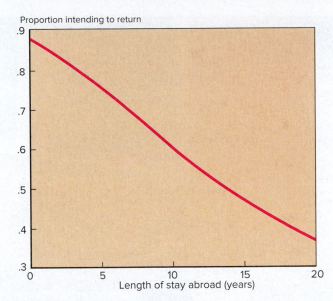

Figure 7.26 **Migrants originally from the former Yugoslavia intending to return home from Germany.** As the length of stay in Germany increases, the proportion intending to return decreases, but even after 10 years abroad, more than one-half intend to return. *Source: B. Waldorf, "Determinants of International Return Migration Intentions," The Professional Geographer, 47, no. 2, Fig. 2, p. 132. Association of American Geographers, 1995.*

Hierarchical migration, in a sense, is a response to that flow. The tendency is for individuals in domestic relocations to move up the level in the hierarchy, from small places to larger ones. Very often, levels are skipped on the way up; only in periods of general economic decline is there considerable movement down the hierarchy. The suburbs of large cities are considered part of the metropolitan area, so the movement from a town to a suburb is considered a move up the hierarchy. From this hierarchical pattern, we can envisage information flowing down to small places from cities and metropolitan areas, and migration flowing up from rural to urban regions.

7.9 Globalization, Integration, and Interaction

We have seen how the cost of communication affects the degree of spatial interaction between places. In the past 20 years, we have witnessed the development of the Internet and have benefited from relatively low transportation costs. Increased computerization of transactions has made it easy to buy goods from abroad

and to travel abroad. During this period, there also has been a strong movement throughout the world to reduce the barriers to trade and foreign investment and ownership. The European Union is a good case in point. Its currency, the *euro,* now makes it possible to conduct financial transactions in a single currency, much as is the case in the United States. Computer technology enables investors to buy stock on foreign stock exchanges or to buy mutual funds that represent firms whose headquarters are in many parts of the world. The new technologies have helped bring about a globalizing world where people are more interdependent than ever before. **Globalization**—the increasing interconnection of all parts of the world—affects economic, political, and cultural patterns and processes.

Economic Integration

One might view the world of the 1950s through the 1980s as a period of division, when there was a wide breach between the peoples of the Western world and those of the communist world. Each side had opposing views about how economic and political systems should be organized. It was a world of division, not integration.

Integration and interdependence characterize globalization. The fact that Eastern and Western Europe are coming together as a single economic entity, and that the East and South Asian countries' economies are being integrated into those of Europe and North America, is as much a function of the revolution in communication and computer technologies as it is the will of the world's political and financial leaders. Low-cost, high-speed computers; communications satellites; fiber-optic networks; and the Internet are the main technologies of the revolution, but other technologies, such as

brokers and traders around the world. As a result, split-second changes in all markets are possible. Within minutes of the 9/11/01 attacks on the World Trade Center and Pentagon, stock markets everywhere went down as investors sensed that the international marketplace was in jeopardy of losing its stability.

The internationalization of finance is also reflected in the immense amount of money in foreign investments. Many Americans, for example, own foreign stocks and bonds, either directly or through mutual funds and pension plans. Similarly, people outside the United States have significant holdings in U.S. companies and in U.S. Treasury bonds.

Transnational Corporations

The past 20 years have seen a tremendous increase in the number of **transnational corporations (TNCs),** companies that have headquarters in one country and subsidiary companies, factories, and other facilities (laboratories, offices, warehouses, and so on) in several countries. As many as 65,000 TNCs—with several hundred thousand affiliates worldwide—engage in economic activities that are international in scope (**Figure 7.28**). They account for trillions of dollars in sales and, by some estimates, control about one-third of the world's productive assets (see also "Transnational Corporations" in Chapter 10).

The way that TNCs produce and provide goods and services is also a part of the globalization process. By exploiting the large differential in wage rates around the world, they keep their production costs down, which has led to the decentralization of manufacturing. More and more, American, Japanese, and Western European companies produce their manufactured goods in lower-labor-cost countries, such as China, Thailand, and Mexico, integrating these developing countries into the global economy.

In the United States, the service sector of employment has risen dramatically as the manufacturing sector has declined. Instead of producing heavy manufactured goods, the U.S. economy has moved toward the production of high-technology goods and services. The service sector has blossomed as the number of financial transactions has increased. With the increased wealth resulting from the effects of new technologies on production processes, Americans are able to travel more and to stay at hotels and eat out more often. This has a great bearing on the types of jobs people hold, not only in the United States but also throughout the business and tourist world. These developments are discussed in greater detail in Chapter 10.

Figure 7.27   The old and the new: a gondolier in Venice, Italy, conducting business on a cellular phone. © *Arthur Getis.*

robotics, microelectronics, electronic mail, fax machines, and cellular phones, also play an important role (**Figure 7.27**). The fact that a consumer in Athens, Greece, can order a book from Amazon.com or clothes from Lands' End, obtain news from CNN, or engage in a transaction on the London Stock Exchange while talking on a cellular phone to a colleague in Tokyo is revolutionary. The forces at work on Americans, the Japanese, and the British have the same effect on Greek consumers as they go about their daily activities. Thus, globalization brings about greater integration and a great deal more spatial interaction.

International Banking

International banking has become a complex system of integrated and interdependent financial markets. It is no wonder that the various stock exchanges of the world tend to rise and fall together. There are exceptions, of course, but when a pharmaceutical company such as Pfizer buys Warner-Lambert Pharmaceuticals, it affects pharmaceutical stock trading throughout the world. The financial balances between companies are monitored closely by

Global Marketing

The greater wealth of those who have benefited from globalization has created a huge new market for goods and services. TNCs market their products around the world, whether they be goods such as Swiss watches, Italian shoes, or Coca-Cola or the services provided by, for example, worldwide hotel chains or cellular phone companies. Japanese cars are found in all parts of the world, just as are fast-food chains. Some of the ramifications of this are discussed in the section "Cultural Integration."

(a) Nokia—Finland (Pakistan)

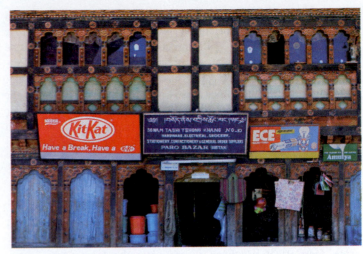

(b) KitKat—U.K. (Bhutan)

(c) McDonald's—United States (Japan)

(d) Ford—United States (China)

(e) IBM—United States (Sudan)

Figure 7.28   The number of the world's transnational corporations increased from about 7000 in 1970 to over 77,000 in 2005. Ninety of the top 100 TNCs are headquartered in the European Union, the United States, and Japan. Their impact, however, is global, as suggested by these billboards advertising just a sample of leading TNCs in distant settings. Corporate names and headquarters countries are followed by billboard locations in parentheses. *(a) © Robert Nickelsberg/Getty Images News/ Getty Images; (b) © Tim Graham/Getty Images; (c) © McGraw-Hill Education/Christopher Kerrigan RF; (d) © Keith Bedford/Bloomberg/ Getty Images; (e) © Barry Iverson/The LIFE Images Collection/Getty Images.*

It is important to remember, however, that globalization is a fairly recent development, and at present its benefits accrue to a minority of the world's 7 billion people. Only one-tenth of those people, about 700 million, are affluent enough to live comfortably and to purchase the goods and services alluded to earlier. According to the United Nations Development Fund, one-quarter of humanity lives on less than $1 per day. An illiterate farmer in a remote village in Tibet more than likely lacks access to a telephone, much less the new technologies of the revolution in communications.

Political Integration

The flow of money (capital), goods, ideas, and information around the globe links people in ways that transcend national boundaries. One effect has been to stimulate the formation of new, supranational alliances or the restructuring of older ones. To enhance commerce, countries are signing free-trade agreements and joining economic organizations such as the North American Free-Trade Agreement (1994) and the World Trade Organization (1995). As will be discussed in Chapter 8, many other international alliances—military, political, and cultural—have been created since 1980, and some that are older, such as NATO, are enlarging their membership.

Another effect of globalization has been the enormous increase in the number of international *nongovernment organizations* (NGOs). Their number more than quadrupled during the 1990s, from about 6000 to more than 26,000. As the names of such well-known organizations as Amnesty International, Doctors without Borders, and Greenpeace indicate, the concerns of international NGOs range widely, from human rights and acid rain to famine relief and resource depletion. What they have in common is that they bring together people in far-flung parts of the world in pursuit of mutual goals.

The transmission of news has never been wider or faster. *Time Magazine* can be bought as easily in New Delhi, India, as in New York City, and CNN broadcasts around the world, informing people about current events and, sometimes, helping bring about government intervention in places where it might not have occurred. Coverage of the wars in Bosnia and Kosovo, for example, stimulated the United Nations and NATO to send in peacekeeping troops to stop the carnage. More recently, in 2002, satellite television's graphic imagery of the havoc wrought by Palestinian suicide bombers and Israeli retaliation fueled support for the Palestinian cause throughout the Arab world.

Finally, the Internet has given people a powerful tool to affect government policies. To give just a few examples, in 1989, pro-democracy students in Beijing, China, used the Internet to publicize what was happening in that city and to rally support for their cause. The woman who won the Nobel Peace Prize in 1997 for her contribution to the international campaign to ban land mines, Jody Williams, used e-mail to organize 1000 human rights and arms control groups on six continents. Similarly, NGOs use the Internet to coordinate massive protests against meetings of the World Trade Organization (**Figure 7.29**).

Figure 7.29   **Some of the thousands of people who demonstrated in Geneva, Switzerland, in 2009 against the World Trade Organization (WTO) and its policies.** Among the protesters were those representing hundreds of NGOs. © *FABRICE COFFRINI/Getty Images.*

Cultural Integration

Imagine this scene: wearing a Yankees baseball cap, a GAP shirt, Levis, and Reebok shoes, a teenager in Lima, Peru, goes with her friends to see the latest thriller. After the movie, they plan to eat at a nearby McDonald's. Meanwhile, her brother sits at home, listening to his iPod while playing a video game. Both children are evidence of the globalization of culture, particularly pop culture. That culture is Western in origin and chiefly American. U.S. movies, television shows, software, music, food, and fashion are marketed worldwide. They influence the beliefs, tastes, and aspirations of people in virtually every country, although their effect is most pronounced on young people. They, rather than their elders, are the ones who want to emulate the stars they see in movies and on YouTube and to adopt what they think are Western lifestyles, manners, and modes of dress.

Another indication of cultural integration is the worldwide spread of the English language. It has become *the* medium of communication in economics, technology, and science.

Both the dominance of English and the globalization of popular culture are resented by many people and rejected by some. Iran, Singapore, and China attempt to restrict the programming that reaches their people, although their citizens' access to satellite dishes and the Internet means those governments are not succeeding in stopping the spread of Western culture. French ministers of culture try to keep the French language pure, unadulterated by English. Other people decry what they see as the homogenization of culture, and still others find the spread of what they define as Western cultural values abhorrent. Whether or not movies, music, and other communications media accurately reflect Western culture, critics argue that they reflect values such as materialism, innovation, self-indulgence, sexuality, spontaneity, and defiance of authority and tradition. Many movies and television shows appear to glorify violence and rebellion and to promote the new over the old, leisure over work, and wealth over self-fulfillment.

Summary of Key Concepts

- The term *human interaction* refers to the movement of people, goods, information, and ideas between one place and another. Many factors influence how individuals view and use places. The concept of psychological distance is helpful in considering how individuals view their environment. Other important variables are the nature of the information available, people's age, their past experiences, and their values.

- The concept of activity space helps us understand differences in the extent of space people use. The age of an individual, his or her degree of mobility, and the availability of opportunities all play a role in defining the limits of individual activity space.

- *Distance decay* refers to the decline of human interaction with increasing distance between places. The concept of critical distance identifies the distance beyond which the decrease in familiarity with places begins to be significant. Unfamiliarity with distant points outside one's normal activity space affects space evaluations, interaction flows, and travel patterns.

- How space is used is a function of all of these factors, but certain factors having to do with the diffusion of innovations indicate what opportunities exist for individuals living in various places. Contagious diffusion and hierarchical diffusion influence the geographic direction that cultural change will take. Effort and cost are among the barriers to diffusion.

- A special type of human interaction is migration. When strong enough, various push and pull forces motivate a long-distance, permanent move. Migration fosters the spread of culture by means of a relocation diffusion process. The ability of many to make well-reasoned, meaningful decisions is a function of the utility they assign to places and the opportunities at those places.

- Economic, political, and cultural patterns and processes are becoming more integrated and people more interdependent as a result of globalization, the increasing interconnection of all parts of the world.

Thinking Geographically

1. What is the role of distance in helping us understand human interaction?

2. Think of the various conflicts that have taken place in the world in the past century. Does the concept of distance decay bear on the location of adversaries? Why?

3. On a blank piece of paper, and without any maps to guide you, draw a map of the United States, putting in state boundaries wherever possible; this is your mental map of the country. Compare it with a standard atlas map. What conclusions can you reach?

4. What is meant by *activity space?* What factors affect the spatial extent of the activity space of an individual? What is your activity space?

5. Recall the places you have visited in the past week. In your movements, were the distance-decay and critical-distance rules operative? What variables affect an individual's critical distance?

6. Briefly distinguish between *contagious diffusion* and *hierarchical diffusion.* In what ways, if any, were these forms of diffusion in operation in the culture hearths discussed in Chapter 6?

7. What considerations affect a decision to migrate? What is *place utility* and how does its perception induce or inhibit migration?

8. What common barriers to migration exist? Why do most people migrate within their own country?

9. Define the term *migration field.* Some migration fields show a *channelized* flow of people. Select a particular channelized migration flow (such as the movement of Scandinavians to the United States, people from the Great Plains to California, or southern blacks to the North) and explain why a channelized flow developed.

10. Identify the effects of globalization on your lifestyle and the patterns of trade in your urban area.

Key Words

activity space 189
chain migration 203
channelized migration 208
contagious diffusion 193
critical distance 187
distance decay 187
globalization 209
hierarchical diffusion 195
hierarchical migration 208
human interaction 186
mental map 190
migration 197
migration field 207
place utility 202
pull factor 200
push factor 200
return migration (countermigration) 208
spatial diffusion 193
spatial interaction 186
stage in life 192
step migration 203
territoriality 189
transnational corporation (TNC) 210

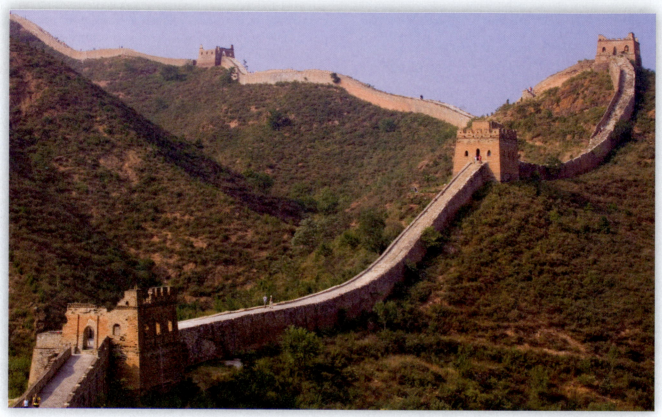

The Great Wall of China remains one of the world's largest construction projects. The wall protected China's northern border and marked the boundary between an agricultural civilization and a nomadic civilization. The wall contributed to the development of Chinese national identity and has come to symbolize China. © dbimages/Alamy Stock Photo RF.

CHAPTER OUTLINE

The process of the political organization of space is as old as human history. From clans to kingdoms, human groups have laid claim to territory and have organized themselves and administered their affairs within it. Indeed, the political organizations of society are as fundamental an expression of culture and cultural differences as are forms of economy or religious beliefs. Geographers are interested in that structuring because it is an expression of the human organization of space and is closely related to other spatial evidences of culture, such as religion, language, and ethnicity.

Political geography is the study of the organization and spatial distribution of political phenomena. Nationality is a basic element in cultural variation among people, and political geography traditionally has had a primary interest in country units, or *states* (**Figure 8.2**). Of central concern have been spatial patterns that reflect the exercise of central government control, such as questions of boundary delimitation and effect. Increasingly, however, attention has shifted both upward and downward on the political scale. On the world scene, international alliances, regional compacts, and producer cartels have increased in prominence since 1945, representing new forms of spatial interaction. At the local level, voting patterns, constituency boundaries and districting rules, and political fragmentation have directed public attention to the significance of area in the domestic political process.

In this chapter, we discuss some of the characteristics of political entities, examine the problems involved in defining jurisdictions, seek the elements that lend cohesion to a political entity, explore the implications of partial surrender of sovereignty, and consider the significance of the fragmentation of political power. We begin with states and end with local political systems.

The emphasis in this chapter on political entities should not make us lose sight of the reality that states are rooted in the operations of the economy and society they represent, that social and economic disputes are as significant as border confrontations, and that in some regards transnational corporations and other nongovernmental agencies may exert greater influence in international affairs than do the separate states in which they are housed or operate. Some of those expanded political considerations are alluded to in the discussions that follow; others are developed more fully in Chapter 10.

LEARNING OUTCOMES

After studying this chapter you should be able to:

8.1 Understand the differences between state and nation

8.2 Describe the evolution of the modern state

8.3 Summarize the importance of size, shape, and location to a state

8.4 Outline different types of conflicts that arise between states and explain their origins

8.5 Identify centripetal forces that promote state cohesion

8.6 Identify centrifugal forces that challenge state authority

8.7 Describe efforts toward transnational cooperation

It was the autumn of 1918. War had been raging for 4 years, and Germany and its allies, the Ottoman Empire and Austria-Hungary, were suffering huge losses. In the Middle East, the British and their Arab allies had pinned down and defeated the Ottoman Turks. T. E. Lawrence, a British officer, hurried straight from the battlefields of Damascus to the British War Cabinet meeting in London, intent on arguing the cause of Arab sovereignty.

The Middle East, Lawrence knew, had been under the domination of the Ottoman Empire for centuries. He felt, however, that with British and Arab military victories the time had come to rewrite the map of the Middle East. Accordingly, Lawrence presented his map, in which he proposed new national boundaries for the Middle East (**Figure 8.1**).

Lawrence argued that the Kurds should have a separate state and that the people of what are now Syria, Jordan, and Saudi Arabia belonged together in one country. He felt that Sunnis and Shi'ites should not be separated in the area that is now Iraq, which is one of the issues that divides the region today. He also placed "Palestine," or a Jewish homeland, on the map.

His reasoning fell on deaf ears. The great powers represented during the peace process carved the region up to reward war allies with resource-rich areas. The peace treaties divided the region into Greek, British, French, and Italian areas of influence and occupation. Only the areas south of what is now Iraq and Jordan were demarcated as independent Arab states.

Would Lawrence's map have changed history? There is no way to know for certain, of course, but some scholars concede that Lawrence's division of the area might have provided a more stable Middle East than the one based on the extent of influence of European powers. Between the end of the war and the current day, wars, revolutions, and alliances have redrawn the borders in the area many times. The region is home to different ethnic groups, religions, and political systems and has been a part of a continuing process of the political organization of space.

8.1 National Political Systems

One of the most significant elements in cultural geography is the nearly complete division of the Earth's land surface into separate national units, as shown on the Countries of the World map foldout at the end of the book. Even Antarctica is subject to the rival territorial claims of seven countries, although these claims have not been pressed because of the Antarctic Treaty of 1959 (**Figure 8.3**). Another element is that this division into country units is relatively recent. Although countries and empires have existed since the days of ancient Egypt and Mesopotamia, only in the past century has the world been almost completely divided into independent governing entities. Now, people everywhere accept the idea of the state, and its claim to sovereignty within its borders, as normal.

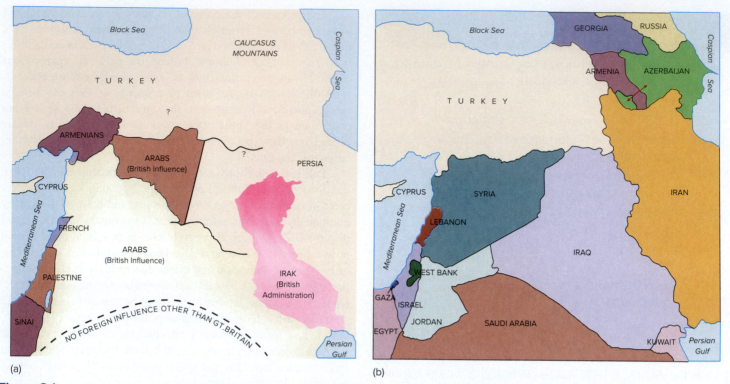

Figure 8.1 **(a) T. E. Lawrence's Peace Map and (b) a current map of the Middle East.** T. E. Lawrence's map of the Middle East was misfiled in the British national archives for decades and only recently rediscovered. Compare it to a current map of the region. Of note:

- Lawrence envisioned a Jewish state called Palestine; the state of Israel was not established until 1948, after 6 million Jews had died in the Holocaust.
- Lawrence included a separate state for Armenians; the territory of historic Armenia had been divided by the Ottomans and the Russians. Now, that area is part of Turkey.
- Lawrence included an area of French administration along the Mediterranean Sea; the British and French agreed during World War I that the French would control the area that is now made up of Lebanon and Syria. Lebanon gained independence in 1943, Syria in 1944, while France was occupied by Nazi Germany.
- Lawrence envisioned at least three Arab states under British influence; settlements at the end of World War I gave Britain a mandate over what is now Iraq, Jordan, Israel, the West Bank, and the Gaza Strip. British companies had freedom to explore and develop Iraqi oil fields.

States, Nations, and Nation-States

Before we begin our consideration of political systems, we need to clarify some terminology. Geographers use the words *state* and *nation* somewhat differently than the way they are used in everyday speech; the confusion arises because each word has more than one meaning. A state can be defined as either (1) any of the political units forming a federal government (e.g., one state of the United States) or (2) an independent political unit holding sovereignty over a territory (e.g., the United States). In the latter sense, *state* is synonymous with *country* or *nation*. That is, a nation can also be defined as (1) an independent political unit holding sovereignty over a territory (e.g., a member of the United Nations). But it can also be used to describe (2) a community of people with a common culture and territory (e.g., the Kurdish nation). The second definition is *not* synonymous with *state* or *country*.

To avoid confusion, we will define a **state** on the international level as an independent political unit occupying a defined, permanently populated territory and having full sovereign control over its internal and foreign affairs. We will use *country* as a synonym for the territorial and political concept of "state." Not all recognized territorial entities are states. Antarctica, for example, has neither an established government nor a permanent population; it is, therefore, not a state. Nor are *colonies* or *protectorates* recognized as states. Although they have defined extent, permanent inhabitants, and some degree of separate government structure, they lack full control over all of their internal and external affairs.

We use *nation* in its second sense, as a reference to people, not to political structure. A **nation** is a group of people with a common culture occupying a particular territory, bound together by a strong sense of unity arising from shared beliefs and customs. Language and religion may be unifying elements, but even more important are an emotional conviction of cultural distinctiveness and a sense of ethnocentrism. For example, the Cree nation exists because of its cultural uniqueness, not by virtue of territorial sovereignty.

The composite term **nation-state** properly refers to a state whose territorial extent coincides with that occupied by a distinct

Figure 8.2   These flags, symbols of separate member states, grace the front of the **United Nations building** in New York City. Although central to political geographic interest, states are only one level of the political organization of space. © *Arthur Getis*.

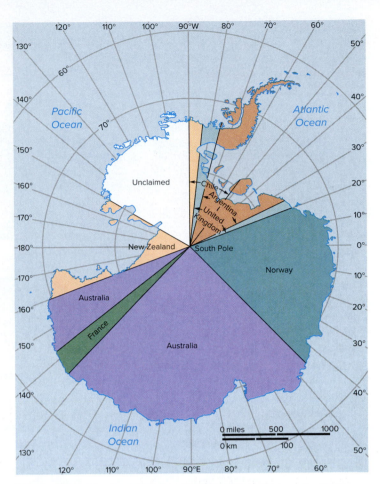

Figure 8.3   **Territorial claims in Antarctica.** Seven countries claim sovereignty over portions of Antarctica, and those of Argentina, Chile, and the United Kingdom overlap. The Antarctic Treaty of 1959 banned further land claims and made scientific research the primary use of the continent. Antarctica is neither a sovereign state—it has no permanent inhabitants or local government—nor a part of one.

nation or people or, at least, whose population shares a general sense of cohesion and adherence to a set of common values (**Figure 8.4a**). That is, a nation-state is an entity whose members feel a natural connection with one another by virtue of sharing language, religion, or some other cultural characteristic strong enough both to bind them together and to give them a sense of distinction from all others outside the community. In reality, very few countries can claim to be true nation-states, since few are or have ever been wholly uniform ethnically or culturally. Iceland, Slovenia, Poland, and the two Koreas are often cited as acceptable examples.

A *binational* or *multinational state* is one that contains more than one nation (**Figure 8.4b**). Often, no single ethnic group dominates the population. In the constitutional structure of the Soviet Union before 1988, one division of the legislative branch of the government was termed the Soviet of Nationalities. It was composed of representatives from civil divisions of the Soviet Union populated by groups of officially recognized "nations": Ukrainians, Kazakhs, Estonians, and others. In this instance, the concept of nationality was territorially less than the extent of the state.

Alternatively, a single nation may be dispersed across and be predominant in two or more states. This is the case with a *part-nation state* (**Figure 8.4c**). Here, a people's sense of nationality exceeds the areal limits of a single state. An example is the Arab nation, which dominates 17 states.

Finally, there is the special case of the *stateless nation*, a people without a state. The Kurds, for example, are a nation of approximately 20 million people divided among six states and dominant in none (**Figure 8.4d**). Kurdish nationalism has survived over the centuries, and many Kurds nurture a vision of an independent Kurdistan. Other stateless nations are the Roma (gypsies), Basques, and Palestinians.

Evolution of the Modern State

The concept and practice of the political organization of space and people arose independently in many parts of the world. Our Western orientations and biases may incline us to trace ideas of spatial political organization through their Near Eastern, Mediterranean, and Western European expressions. Mesopotamian and classical Greek city-states, the Roman Empire, and European colonizing kingdoms and warring principalities were, however, not unique. Southern, southeastern, and eastern Asia had their counterparts, as did sub-Saharan Africa and the Western Hemisphere. Although Western European models and colonization strongly influenced the forms and structures of modern states around the world, the cultural roots of statehood run deeper and reach further back in many parts of the world than the European example alone suggests.

The now universal idea of the modern state was developed by European political philosophers in the 18th century. Their views advanced the concept that people owe allegiance to a state and the people it represents, rather than to its leader, such as a king or a feudal lord. The new concept coincided in France with the French

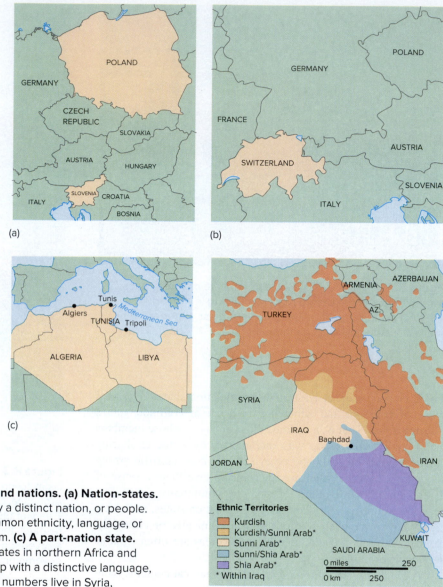

Figure 8.4 Types of relationships between states and nations. (a) Nation-states.
Poland and Slovenia are examples of states occupied by a distinct nation, or people.
(b) A multinational state. Switzerland shows that a common ethnicity, language, or
religion is not necessary for a strong sense of nationalism. **(c) A part-nation state.**
The Arab nation extends across and dominates many states in northern Africa and
the Middle East. **(d) A stateless nation.** An ancient group with a distinctive language,
Kurds are concentrated in Turkey, Iran, and Iraq. Smaller numbers live in Syria,
Armenia, and Azerbaijan.

Revolution and spread over Western Europe, to England, Spain, and Germany.

Many states are the result of European expansion during the 17th, 18th, and 19th centuries, when much of Africa, Asia, and the Americas was divided into colonies. Usually these colonial claims were given fixed and described boundaries where none had earlier been formally defined. Of course, precolonial native populations had relatively fixed home areas of control within which there was recognized dominance and border defense and from which there were, perhaps, raids of plunder or conquest of neighboring "foreign" territories. Beyond understood tribal territories, great empires arose, again with recognized outer limits of influence or control: Mogul and Chinese; Benin and Zulu; Incan and Aztec. Upon them where they still existed, and upon the less formally organized spatial patterns of effective tribal control, European colonizers imposed their arbitrary new administrative divisions of the land. In fact, groups that had little in common were often joined in the

same colony (**Figure 8.5**). The new divisions, therefore, were not usually based on meaningful cultural or physical lines. Instead, the boundaries simply represented the limits of the colonizing empire's power.

As these former colonies have gained political independence, they have retained the idea of the state. They have generally accepted—in the case of Africa, by a conscious decision to avoid precolonial territorial or ethnic claims that could lead to war—the borders established by their former European rulers. The problem that many of the new countries face is "nation-building"—developing feelings of loyalty to the state among their arbitrarily associated citizens. Julius Nyerere, president of Tanzania, noted in 1971, "These new countries are artificial units, geographical expressions carved on the map by European imperialists. These are the units we have tried to turn into nations."

The idea of separate statehood grew slowly at first and more recently has accelerated rapidly. At the time of the Declaration of

Figure 8.5 **The discrepancies between ethnic groups and national boundaries in Africa.** Cultural boundaries were ignored by European colonial powers. The result was significant ethnic diversity in nearly all African countries and conflicts between countries over borders. *Source: from* World Regional Geography: A Question of Place *by Paul Ward English, with James Andrew Miller.*

Independence of the United States in 1776, there were only about 35 empires, kingdoms, and countries in the entire world. By the beginning of World War II in 1939, their number had only doubled to about 70. Following that war, the end of the colonial era brought a rapid increase in the number of sovereign states. With the disintegration of the USSR, Czechoslovakia, and Yugoslavia, more than 20 countries were created. In 2013, there are nearly 200 independent states, and scholars predict that the number will continue to rise as small territories achieve independence, large states break up, and suppressed peoples claim autonomy (**Figure 8.6**).

Challenges to the State

The state and nation-state have long been the focus of political geography, and we will keep that focus in much of the following discussion. We should also realize, however, that the validity of that state-centric view of the world is increasingly under assault

from multiple new agents of economic and social power. These agents include the following:

- The globalization of economies and the emergence of transnational corporations whose economic and production decisions are unrelated to the interests of any single state, including their home office state. Those decisions—outsourcing of production and services, for example—may be detrimental to the employment structure, tax base, and national security of any single state and may limit the applicability of national economic planning and control.

- The proliferation of international and supranational institutions initially concerned with financial or security matters but all representing the voluntary surrender of some traditional state autonomy. The World Trade Organization, the European Union, and regional trade blocs such as the North American Free-Trade Agreement and a host of other international agreements all limit the independence of action of each of

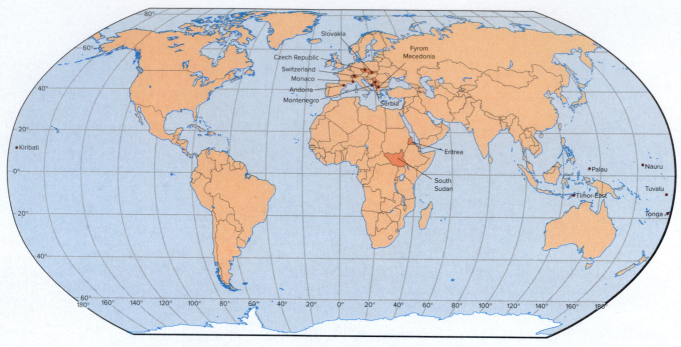

Figure 8.6 **New nations.** Since 1993, sixteen new countries have been admitted to the United Nations. Some of these are new countries entirely (South Sudan and Timor-Leste, for example); others are older countries that have only recently sought UN recognition (Andorra and Switzerland, for example).

their members and thus diminish absolute state primacy in economic and social matters.

- The emergence and multiplication of nongovernmental organizations (NGOs) whose specific interests and collective actions cut across national boundaries and unite people sharing common concerns about issues such as globalization, AIDS efforts, and economic and social injustice. The well-publicized protests and pressures exerted by NGOs channel social pressures to influence or limit government actions.

- The massive international migration flows that tend to undermine the state as a cultural community with assured and expected common values and loyalties. The Internet, inexpensive communication methods, and easy international travel permit immigrants' retention of primary ties with their home culture and state, discouraging their full assimilation into their new social environment or the transfer of their loyalties to their adopted country.

- The increase in nationalist and separatist movements in culturally composite states, weakening through demands for independence or regional autonomy the former unquestioned primacy of the established state.

Some of these agents and developments have been touched on in earlier chapters; others will be reviewed in this chapter (see especially "Centrifugal Forces," p. 232). All represent recent and strengthening forces that, in some assessments, weaken the validity of a worldview in which national governments and institutionalized politics are all-powerful.

Geographic Characteristics of States

Every state has certain geographic characteristics by which it can be described and that set it apart from all other states. A look at the world political map inside the cover of this book confirms that

every state is unique. The size, shape, and location of any one state combine to distinguish it from all others. These characteristics also affect the power and stability of states.

Size

The area that a state occupies may be large, as is true of China, or small, as is Liechtenstein. The world's largest country, Russia, occupies more than 17 million square kilometers (6.5 million sq mi), or some 11% of the land surface of the world. It is more than a million times as large as Nauru, one of the ministates found in all parts of the world (see "The Ministates").

One might assume that the larger a state's area, the greater is the chance that it will have useful resources, such as fertile soil and minerals. In general, that assumption is valid, but much depends upon accidents of location. Mineral resources are unevenly distributed, and a state's size alone does not guarantee their presence. Australia, Canada, and Russia, though large, have relatively small areas capable of supporting productive agriculture. Great size, in fact, may be a disadvantage. A very large country may have vast areas that are inaccessible, sparsely populated, and hard to integrate into the mainstream of economy and society. Small states are more apt than large ones to have a culturally homogeneous population. They find it easier to develop transportation and communication systems to link the sections of the country, and, of course, they have shorter boundaries to defend against invasion. Size alone, then, is not critical in determining a country's stability and strength, but it is a contributing factor.

Shape

Like size, a country's shape can foster or hinder a state's organizational effectiveness. Assuming no major topographical barriers,

THE MINISTATES

Totally or partially autonomous political units that are small in area and population pose some intriguing questions. Should size be a criterion for statehood? What is the potential of ministates to cause friction among the major powers? Under what conditions are they entitled to representation in international assemblies such as the United Nations?

Of the world's growing number of small countries, more than 40 have less than 1 million people, the population size adopted by the United Nations as the upper limit defining "small states," though not too small to be members of that organization. Nauru has about 9000 inhabitants on its 21 square kilometers (8.2 sq mi). Other areally small states, such as Singapore, covering 697 square kilometers (269 sq mi), have populations (5.4 million) well above the UN criterion. Many are island territories located in the Caribbean and the Pacific Ocean (such as Grenada and Tonga Islands), but Europe (Vatican City and Andorra), Asia (Bahrain and Brunei), and Africa (Djibouti and Equatorial Guinea) have their share.

Many ministates are vestiges of colonial systems that no longer exist. Some of the small countries of West Africa and the Arabian peninsula fall into this category. Others, such as Mauritius, served primarily as refueling stops on transoceanic voyages. However, some occupy strategic locations (such as Bahrain, Malta, and Singapore), and others contain valuable minerals (Kuwait and Trinidad). The possibility of claiming 370-kilometer-wide (200-nautical-mile-wide) zones of adjacent seas (see p. 238) adds to the attraction of yet others.

Their strategic or economic value can expose small islands and territories to unwanted attention from larger neighbors. The 1982 war between Britain and Argentina over the Falkland Islands (called the Islas Malvinas by Argentina) and the Iraqi invasion of Kuwait in 1990 demonstrate the ability of such areas to bring major powers into conflict and to receive world attention that is out of proportion to their size and population.

The proliferation of tiny countries raises the question of their representation and their voting weight in international assemblies. Should there be a minimum size necessary for participation in such bodies? Should countries receive a vote proportional to their population? Within the United Nations, the Alliance of Small Island States (AOSIS) has emerged as a significant power bloc, controlling more than one-fifth of UN General Assembly votes. AOSIS is particularly concerned with issues of climate change and promotes initiatives to limit the rise of global temperatures.

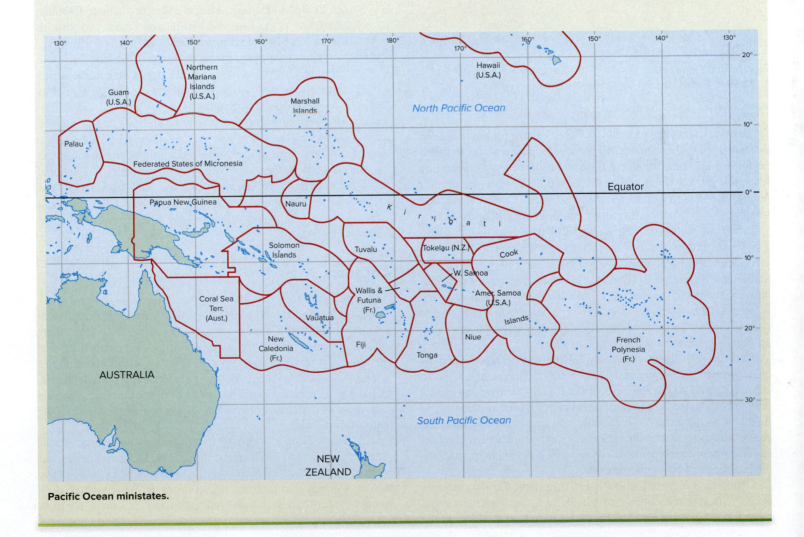

Pacific Ocean ministates.

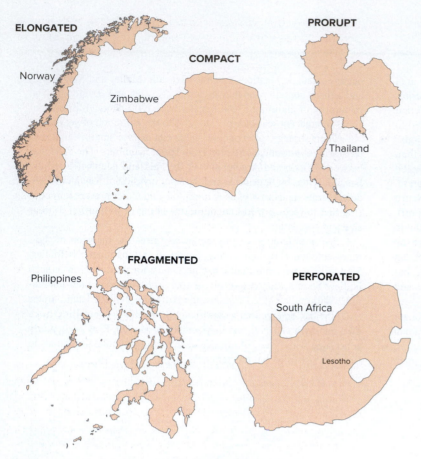

ELONGATED

COMPACT

PRORUPT

Norway

Zimbabwe

Thailand

FRAGMENTED

Philippines

PERFORATED

South Africa

Lesotho

Figure 8.7   **Shapes of states.** The sizes of the countries should not be compared. Each is drawn on a different scale.

the most efficient form would be a circle, with the capital located in the center. In such a country, all places could be reached from the center in a minimal amount of time and with the least expenditure for roads, railway lines, and so on. It would also have the shortest possible borders to defend. Zimbabwe, Uruguay, and Poland have roughly circular shapes, forming a **compact state** (**Figure 8.7**).

Prorupt states are nearly compact but possess one or sometimes two narrow extensions of territory. Proruption may simply reflect peninsular elongations of land area, as in the case of Myanmar (Burma) and Thailand. In other cases, the extensions have an economic or strategic significance, having been designed to secure state access to resources or to establish a buffer zone between states that would otherwise adjoin. Whatever their origin, proruptions tend to isolate a portion of a state.

The least-efficient shape administratively is represented by countries such as Norway and Chile, which are long and narrow. In such **elongated states,** the parts of the country far from the capital are likely to be isolated because great expenditures are required to link them to the core. These countries are also likely to encompass greater diversity of climate, resources, and peoples than compact states, perhaps to the detriment of national cohesion or, perhaps, to the promotion of economic strength.

The fourth class of shapes, that of **fragmented states,** includes countries composed entirely of islands (e.g., the Philippines and Indonesia), countries that are partly on islands and partly on the

mainland (Italy and Malaysia), and those that are chiefly on the mainland but whose territory is separated by another state (the United States). Fragmentation makes it harder for the state to impose centralized control over its territory, particularly when the parts of the state are far from one another. This is a problem in Indonesia, which is made up of more than 13,000 islands, stretched out along a 5100-kilometer (3200-mi) arc. Fragmentation helped lead to the disintegration of Pakistan. It was created in 1947 as a fragmented state, but East and West Pakistan were 1610 kilometers (1000 mi) from each other. That distance exacerbated economic and cultural differences between the two, and when the eastern part of the country seceded in 1971 and declared itself the independent state of Bangladesh, West Pakistan was unable to impose its control.

A special case of fragmentation occurs when a territorial outlier of one state, an **exclave,** is located within another state. Before German unification, West Berlin was an outlier of West Germany within East Germany (the German Democratic Republic). Europe has many such outlying bits of one country inside another. Kleinwalsertal, for example, is a piece of Austria accessible only from Germany. Baarle-Hertog is a fragment of Belgium inside Holland. Llivia is a Spanish town just inside France. Exclaves are not limited to Europe: African examples include Cabinda, an exclave of Angola, and Melilla and Ceuta, two Spanish exclaves in Morocco (**Figure 8.8**).

The counterpart of an exclave, an **enclave,** helps to define the fifth class of shapes, the **perforated state.** A perforated state completely surrounds a territory that it does not rule, as the Republic of South Africa surrounds Lesotho. The enclave, the surrounded territory, may be independent or may be part of another state. Two of Europe's smallest independent states, San Marino and Vatican City, are enclaves that perforate Italy. As an *exclave* of West Germany, West Berlin perforated the national territory of former East Germany and was an *enclave* in it. The stability of the perforated state can be weakened if the enclave is occupied by people whose value systems differ from those of the surrounding country.

Location

The significance of size and shape as factors in national well-being can be modified by a state's location, both absolute and relative. Although both Canada and Russia are extremely large, their *absolute* location in the upper-middle latitudes reduces their size advantages when agricultural potential is considered. For another example, Iceland has a reasonably compact shape, but its location in the North Atlantic Ocean, just south of the Arctic Circle, means that most of the country is barren. Settlement is confined to the rims of the island.

A state's *relative* location, its position compared to that of other countries, is as important as its absolute location. *Landlocked* states, those lacking ocean frontage and surrounded by other states, are at a geographic disadvantage (**Figure 8.9**). They lack easy access to maritime (seaborne) trade and to the resources found in coastal waters and submerged lands. Bolivia gained 480 kilometers

Figure 8.8   Spanish exclaves in North Africa and France.
Although Spanish troops seized the garrison towns of Melilla and Ceuta almost 500 years ago, and a majority of the exclaves's residents are of Spanish descent, Morocco still claims sovereignty over the towns. In recent years, Ceuta and Melilla have become the temporary stopping points for tens of thousands of would-be migrants to Spain. Emigrants from Mali, Nigeria, and as far away as Kashmir and Iraq enter Ceuta and Melilla requesting political asylum and seeking work permits or visas to enter Europe. Today the enclaves are surrounded by fences. Llivia became an exclave in 1660 when Spain ceded the surrounding area to France in the Treaty of the Pyrenees. Gibraltar is a British colony, and Andorra is an independent ministate.

(300 mi) of sea frontier along with its independence in 1825, but lost its ocean frontage by conquest to Chile in 1879. Its annual Day of the Sea ceremony reminds Bolivians of their loss and of continuing diplomatic efforts to secure an alternate outlet. There are about 40 landlocked states.

In a few instances, a favorable relative location constitutes the primary resource of a state. Singapore, a state of only 697 square kilometers (269 sq mi) and 5.4 million people, is located at the crossroads of world shipping and commerce. Based on its port and commercial activities, and buttressed by its more recent industrial development, Singapore has become a notable Southeast Asian economic success. In general, history has shown that countries benefit from a location on major trade routes, not only from the economic advantages such a location carries, but also because they are exposed to the diffusion of new ideas and technologies.

Cores and Capitals

Many states have come to assume their present shape, and thus the location they occupy, as a result of growth over centuries. They grew outward from a central region, gradually expanding into surrounding territory. The original nucleus, or **core area,** of a state usually contains its densest population and largest cities, the most highly developed transportation system, and the most developed economic base. All of these elements become less intense away from the national core. Urbanization ratios and city sizes decline, transport networks thin, and economic development is less intensive on the periphery than in the core.

Easily recognized and unmistakably dominant national cores include the Paris Basin of France; London and southeastern England; Moscow and the major cities of European Russia; northeastern United States and southeastern Canada; and the Buenos Aires megalopolis in Argentina. Not all countries have such clearly

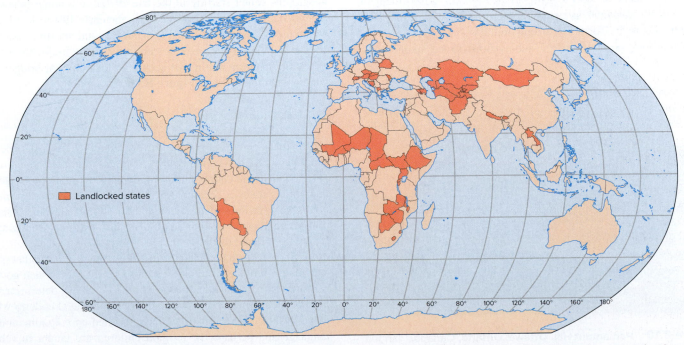

Figure 8.9   Landlocked states. Landlocked states are at a commercial and strategic disadvantage, compared to countries that have ocean frontage.

defined cores, and some may have two or more rival core areas. Chad, Mongolia, and Saudi Arabia have no clearly defined core, for instance, whereas Ecuador, Nigeria, Democratic Republic of the Congo, and Vietnam are examples of multicore states.

The capital city of a state is usually within its core region and frequently is the very focus of it, dominant not only because it is the seat of central authority but also because of the concentration of population and economic functions as well. That is, in many countries, the capital city is also the largest, or *primate,* city, dominating the structure of the entire country. Paris in France, London in the United Kingdom, and Mexico City are all examples of that kind of political, cultural, and economic primacy.

This association of capital with core is common in what have been called the *unitary states,* countries with highly centralized governments, relatively few internal cultural contrasts, a strong sense of national identity, and borders that are clearly cultural as well as political boundaries. Many European cores and capitals are of this type. This association is also found in many newly independent countries whose former colonial occupiers established a primary center of exploitation and administration and developed a functioning core in a region that lacked an urban structure or organized government. With independence, the new states retained the established infrastructure, added new functions to the capital, and, through lavish expenditures on government, public, and commercial buildings, sought to create prestigious symbols of nationhood.

In *federal states,* associations of more or less equal provinces or states with strong regional government responsibilities, the capital city may have been newly created to serve as the administrative center. Although part of a generalized core region of the country, the designated capital was not its largest city and acquired few of the additional functions to make it so. Ottawa, Canada; Washington, D.C.; and Canberra, Australia, are examples (**Figure 8.10**).

A new form of state organization, *regional* government or *asymmetric federalism*, is emerging in Europe. Asymmetric federalism occurs as formerly strong unitary states acknowledge the aspirations of their several subdivisions to be autonomous, grant to them varying degrees of local administrative control, but retain central authority over matters of nationwide concern, such as monetary policy, defense, and foreign relations. Autonomy is most likely to be granted to regions with the most outspoken residents, who claim that their region is different from the national whole. Such claims are mostly based on differences in religions, languages, or economic centers and interests. National governments may recognize regional capitals, legislative assemblies, and administrative bureaucracies. The asymmetric federalism of the United Kingdom, for example, now involves separate status for Scotland, Wales, and Northern Ireland, with their own capitals at Edinburgh, Cardiff, and Belfast, respectively. That of Spain recognizes Catalonia and the Basque country, with capitals in Barcelona and Vitoria, respectively.

All other things being equal, a capital located in the center of the country provides equal access to the government, facilitates communication to and from the political hub, and enables the government to exert its authority easily. Many capital cities, such as Washington, D.C., were centrally located when they were designated as seats of government, but lost their centrality as the state expanded.

Some capital cities have been relocated outside of peripheral national core regions, at least in part to achieve the presumed advantages of centrality. Two examples of such relocation are from Karachi inland to Islamabad in Pakistan and from Istanbul to Ankara, in the center of Turkey's territory. A particular type of relocated capital is the *forward-thrust capital* city, one that has been deliberately sited in a state's frontier zone to signal the government's awareness of regions away from the core and its interest in encouraging more-uniform development. In the late 1950s, Brazil moved its capital from Rio de Janeiro to the new city of Brasília to demonstrate its intent to develop the vast interior of the country. The West African country of Nigeria built the new capital of Abuja near its geographic center starting in the late 1970s, relocating its government offices and foreign embassies in the early 1990s. The British colonial government relocated Canada's capital six times between 1841 and 1865, in part seeking centrality to the mid-19th-century population pattern and in part seeking a location that bridged that country's cultural divide (**Figure 8.11**).

Boundaries: The Limits of the State

Recall that no portion of the Earth's land surface is outside the claimed control of a national unit, that even uninhabited Antarctica has had territorial claims imposed upon it (see Figure 8.3). Each of the world's states is separated from its neighbors by *international boundaries,* or lines that establish the limit of each state's jurisdiction and authority. Boundaries indicate where the sovereignty of one state ends and that of another begins.

Within its own bounded territory, a state administers laws, collects taxes, provides for defense, and performs other such government functions. Thus, the location of the boundary determines the kind of money people in a given area use, the legal code to which they are subject, the army they may be called upon to join, and the language and perhaps the religion children are taught in school. These examples suggest how boundaries serve as powerful reinforcers of cultural variation over the Earth's surface.

Figure 8.10 **Parliament Hill, Ottawa, Ontario, Canada.** Planned capitals are often architectural showcases, providing a focus for national pride. © *aimintang/Getty Images RF.*

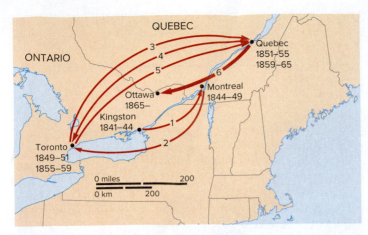

Figure 8.11 **Canada's migratory capital.** Kingston was chosen as the first capital of the united Province of Canada in preference to either Quebec, capital of Lower Canada, or Toronto, that of Upper Canada. In 1844, government functions were relocated to Montreal, where they remained until 1849, after which they shifted back and forth—as the map indicates—between Toronto and Quebec. An 1865 session of the provincial legislature was held in Ottawa, the city that became the capital of the Confederation of Canada in 1867. *Data: David B. Knight,* A Capital for Canada. *Chicago: University of Chicago, Department of Geography, Research Paper no. 182, 1977, Fig. 1, p. vii.*

Territorial claims of sovereignty are three-dimensional. International boundaries not only mark the outer limits of a state's claim to land (or water) surface but also are projected downward to the center of the Earth in accordance with international consensus allocating rights to subsurface resources. States also project their sovereignty upward, but with less certainty because of a lack of agreement on the upper limits of territorial airspace. Properly viewed, then, an international boundary is a line without breadth; it is a vertical interface between adjacent state sovereignties.

Before boundaries were delimited, nations or empires were likely to be separated by *frontier zones,* ill-defined and fluctuating areas marking the effective end of a state's authority. Such zones were often uninhabited or only sparsely populated and were liable to change with shifting settlement patterns. Many present-day international boundaries lie in former frontier zones, and in that sense, the boundary line has replaced the broader frontier as a marker of a state's authority.

Natural and Artificial Boundaries

Geographers have traditionally distinguished between "natural" and "artificial" boundaries. **Natural** (or *physical*) **boundaries** are those based on recognizable physiographic features such as mountains, rivers, and lakes. Although they might seem to be attractive as borders because they actually exist in the landscape and are visible dividing elements, many natural boundaries have proved to be unsatisfactory because they do not effectively separate states.

Many international boundaries lie along mountain ranges— for example, in the Alps, Himalayas, and Andes. Some have proved to be stable; others have not. Mountains are rarely total barriers to interaction. Although they do not invite movement,

they are crossed by passes, roads, and tunnels. High pastures may be used for seasonal grazing, and a mountain region may be a source of water for hydroelectric power. Nor is the definition of a boundary along a mountain range a simple matter. Should it follow the crests of the mountains or the *water divide* (the line dividing two drainage areas)? The two are not always the same. Border disputes between China and India are, in part, the result of the failure of mountain crests and headwaters of major streams to coincide. Recently, glaciers shrinking in the Alps have forced Italy to renegotiate its borders with Switzerland, France, and Austria, as the original borders were drawn along the ridge of the glaciers (**Figure 8.12**).

Rivers can be even less satisfactory as boundaries. In contrast to mountains, rivers foster interaction. River valleys are likely to be agriculturally or industrially productive and to be densely populated. For example, for hundreds of miles, the Rhine River serves as an international boundary in Western Europe. It is also a primary traffic route lined by chemical plants, factories, and power stations and dotted by the castles and cathedrals that make it one of Europe's major tourist attractions. It is more a common, intensively used resource than a barrier in the lives of the states it borders.

The alternative to natural boundaries are **artificial,** or **geometric, boundaries.** Frequently delimited as sections of parallels of latitude or meridians of longitude, they are found chiefly in Africa, Asia, and the Americas. The western portion of the United States–Canada border, which follows the 49th parallel, is an example of a geometric boundary. Many such boundaries were established when the areas in question were colonies, the land was only sparsely settled, and detailed geographic knowledge of the frontier region was lacking.

Boundaries Classified by Settlement

Boundaries can also be classified according to whether they were laid out before or after the principal features of the cultural landscape developed. An **antecedent boundary** is one drawn across an area before it is well populated—that is, before most of the cultural landscape features developed. To continue our earlier

Figure 8.12 **Several international borders run through the jumble of the Alps.** © *Chase Jarvis/Getty Images RF.*

example, the western portion of the United States–Canada boundary is such an antecedent line, having been established by a treaty between the United States and Great Britain in 1846.

Boundaries drawn after the development of the cultural landscape are termed **subsequent boundaries.** One type of subsequent boundary is a **consequent** (also called *ethnographic*) **boundary,** a border drawn to accommodate existing religious, linguistic, ethnic, or economic differences between countries. An example is the boundary drawn between Northern Ireland and Eire (Ireland). Subsequent **superimposed boundaries** may also be forced upon existing cultural landscapes, a country, or a people by a conquering or colonizing power that is unconcerned about preexisting cultural patterns. The colonial powers in 19th-century Africa superimposed boundaries upon established African cultures without regard to the tradition, language, religion, or ethnic affiliation of those whom they divided (see Figure 8.5).

When Great Britain prepared to leave the Indian subcontinent after World War II, it was decided that two independent states would be established in the region: India and Pakistan. The boundary between the two countries, defined in the partition settlement of 1947, was thus both a *subsequent* and a *superimposed* line. As millions of Hindus migrated from the northwestern portion of the subcontinent to seek homes in India, millions of Muslims left what would become India for Pakistan. In a sense, they were attempting to ensure that the boundary would be *consequent*—that is, that it would coincide with a division based on religion.

If a former boundary line that no longer functions as such is still marked by some landscape features or differences on the two sides, it is termed a *relic boundary.* The abandoned castles dotting the former frontier zone between Wales and England are examples of a relic boundary. They are also evidence of the disputes that sometimes attend the process of boundary making. A more recent example is the Berlin Wall, built by communist East Germany in 1961 to seal off the border between East and West Berlin. With the reunification of Germany in 1990, the wall was mostly dismantled. Berliners have chosen to preserve some pieces of the wall as historical monuments; elsewhere, the course of the wall is marked by a double row of paving stones, another form of relic boundary (**Figure 8.13**).

Boundaries as Sources of Conflict

Boundaries create many possibilities for conflict between countries. Although the causes of conflict are varied, geographic considerations underlie many of them. **Figure 8.14** shows for an imaginary state, Hypothetica, the spatial conditions that could give rise to conflict with its neighbors. Each condition is identified by number, and each is illustrated by real-world examples in the discussion that follows.

Landlocked States (Potential Trouble Spot #1) Hypothetica is a landlocked country, as are about one-fifth of the world's states (see Figure 8.9). To trade with overseas markets, landlocked states have to import and export their goods by land-based modes of transportation. They must cooperate with neighboring states and arrange for the goods to travel across a foreign country, and cooperation may be difficult.

Figure 8.13 **Dolbadarn Castle, in Wales.** Built in the 13th century, the stronghold served a military purpose and as a symbol of power and authority. © *Ingram Publishing/AGE Fotostock RF.*

Access to the sea usually is gained in one of two ways. Typically, the landlocked country arranges to use facilities at a foreign port and to have the right to travel to that port, but such arrangements are not without their problems. For centuries, landlocked countries have had to contend with restrictions, tolls, high fees for transit and storage, complicated customs formalities, the risk that the goods will be lost or damaged, and other obstructions to the easy movement of goods to and from the sea. In addition, they have had little or no control over the availability and efficiency of the transport and port facilities outside their borders, and they face the possibility that war will close their access to the sea. After seceding from Sudan, the newly independent state of South Sudan, recognized by the UN in 2011, is in the difficult position of having highly productive oil fields whose pipelines run to the Red Sea through Sudan. War with Sudan and conflict between ethnic groups within South Sudan have shut down the oil transportation. South Sudan proposes three possible solutions, all premised on building new pipelines: through Ethiopia to Djibouti on the Gulf of Aden; through Kenya to Lamu on the Indian Ocean; or through Uganda and Kenya to Mombasa

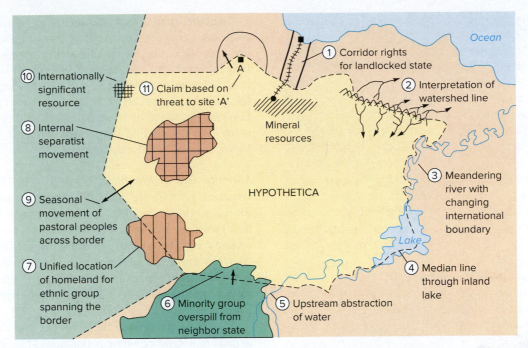

Figure 8.14 **Geographic sources of international stress.** To illustrate the conditions that can give rise to conflicts between states, eminent British geographer Peter Haggett drew this map of a hypothetical state and identified potential trouble spots. Real-world examples of the stress points and disputes shown in this map are discussed in the text. *Source: Peter Haggett,* Geography, a Global Synthesis *(Prentice Hall, 2001). Figure 17.10, p. 522.*

on the Indian Ocean. Any of these will take large investments and agreements with neighboring countries to allow the oil to flow to the ports (**Figure 8.15**).

Rather than depend on another country's port and goodwill, some landlocked states have gained access to the sea through a narrow corridor of land that reaches either the sea or a navigable river. Examples include the Congo Corridor of the Democratic Republic of Congo and the Caprivi Strip of Namibia, which was designed by the Germans to give what was then their colony of Southwest Africa access to the Zambezi River and Indian Ocean. Although these corridors have endured, others—such as the Polish and Finnish Corridors established after World War I—were short-lived.

Waterbodies as National Boundaries (#2, #3, #4, #5) As we noted earlier (p. 225), although rivers and lakes form parts of many national borders, they create many opportunities for conflict. Any body of water that forms part of a border requires agreement on where the boundary line should lie: along the right or left bank or shore, along the center of the waterway, or perhaps along the middle of the navigable channel. Soviet insistence that its sovereignty extended to the Manchurian (Dongbei) bank of the Amur and Ussuri Rivers was a long-standing matter of dispute and border conflict between the USSR and the People's Republic of China, only resolved with Russian agreement in 1987 that the boundary should pass along the main channel of the rivers. Even an agreement in accordance with international custom that the boundary be drawn along the main channel may be impermanent if the river changes its course, floods, or dries up.

Potential trouble spot #2 relates to the use of a watershed boundary in defining an international dividing line—that is, one that runs along a ridge or crest dividing two drainage areas. Disputes

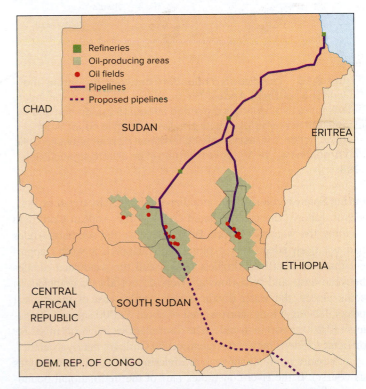

Figure 8.15 Like many other landlocked countries, the **new country of South Sudan** will have to negotiate with other countries to gain access to the sea. Its proposed pipelines might traverse as many as three different countries.

occur when states disagree about the interpretation of documents that define a boundary and/or the way the boundary was delimited. The boundary between Argentina and Chile, originally defined

Figure 8.16   The disputed boundary between Argentina and Chile in the southern Andes. The treaty establishing the boundary between the two countries preceded adequate exploration and mapping of the area, leaving its precise location in doubt. After years of friction, the last remaining territorial dispute between Chile and Argentina in the Andes was settled in an accord signed in late 1998.

during Spanish colonial rule and then established by treaty in 1881, was to follow "the most elevated crests of the Andean Cordillera dividing the waters" (**Figure 8.16**). Because the southern Andes had not been adequately explored and mapped, it wasn't apparent that the crestlines (highest peaks) and the watershed divides between east- and west-flowing rivers do not always coincide. In some places, the water divide is many miles east of the highest peaks, leaving a long, narrow area of about 52,000 square kilometers (20,000 sq mi) in dispute. The discrepancies in claims made for difficult relations between Argentina and Chile for nearly a century.

Pressure point #3 shows a meandering river (one that changes its course). If the river constitutes part of the international boundary, as it does here, the border will change over time. The boundary between the United States and Mexico, for example, which runs along the main channel of the Rio Grande, has changed as the river has altered its course. Similarly, a lake (#4) requires agreement on where the boundary between two countries lies. In the case of the United States and Canada, the two countries agreed that a line equidistant from the shores of Lake Erie and Ontario would form part of the international border.

Another trouble spot (#5) relates to Hypothetica's use of the river that flows downstream into it from another country. Water use is an increasing source of conflict among countries, particularly in arid or semiarid regions. When water is scarce, its abstraction, diversion, or pollution by one country can significantly affect the quantity and quality of water available to those downstream. Growing shortages of fresh water are leading to tensions along many rivers, including the Jordan, Tigris and Euphrates, Nile, Indus, Ganges, and Brahmaputra.

Minority Group Identification (#6, #7, #8)   Like nearly all countries, Hypothetica contains more than one culture group. In the real world, the locations of minority groups have led to international tensions, civil wars, wars of liberation, and international strife around the globe. As one of the most difficult problems with which countries must deal, minority group identification is discussed in greater detail in the section "Centrifugal Forces," but brief examples of pressure points #6, #7, and #8 are provided in the following paragraphs.

Conflicts can arise if the people of one state claim and seek to acquire a territory whose population is historically or ethnically related to that of the state but is now subject to a foreign government. This condition is represented by pressure point #6, minority group overspill from a neighboring state. Under these conditions, the desire to expand the country's borders is called **irredentism,** from the Italian word for "unredeemed." Since 1950, both India and Pakistan have claimed Jammu and Kashmir as part of their own national territory, based upon the fact that the area contains substantial Muslim and Hindu populations.

Hungary's claims to Transylvania, a Romanian province, are based on both historical and ethnic ties. The two countries have quarreled over it for centuries. Transylvania was under Hungarian control from 1649 until 1920, when—as part of the reordering of the political map of Europe that followed World War I—it became part of Romania. In 1940, Germany and Italy forced Romania to give the province back to Hungary, but the country had to surrender it again after World War II.

Earlier we discussed the case of the stateless nation (p. 217), a people without a state, and cited as examples Kurds, Roma, Basques, and Palestinians. Pressure point #7 shows a distinct ethnic group or nation located in both Hypothetica and a neighboring state. Conflict occurs when nations seek to govern themselves in their own state and try to carve out a new nation-state from portions of existing countries. As the example of the Basques indicates, they need not represent a majority of their residents in order to foment discord.

Basques live in a region overlapping France and Spain (**Figure 8.17**). In an attempt to dampen the separatist fires that had burned there since the 1960s, Spain granted its three Basque provinces a significant degree of self-rule in 1978, but that has failed to satisfy the extreme separatist movement Euskadi ta Askatasuna (ETA), which means "Basque Homeland and Liberty." The separatists contend that the Spanish state has attempted to destroy the Basques's unique cultural identity and to suppress their language, Euskadi, which is unrelated to any other language on Earth. They demand an independent, unified Basque state, not only for the Basque region of Spain but for a portion of southern France as

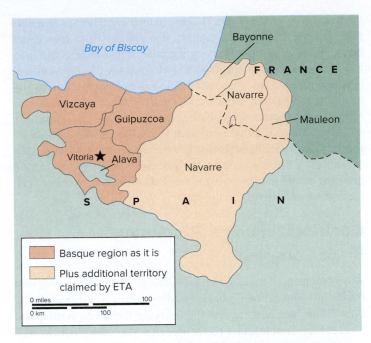

Figure 8.17   The Basque region straddles the border between Spain and France. Although the Basques have been granted a measure of self-rule for their region, militant separatists in the Euskadi ta Askatasuna (ETA) want to see the establishment of an independent state for the Basque region of Spain and a portion of southern France.

in this instance the influence of place—in addition to the stronger nation-state history of France—has resulted in differential identification with the proposed state. The ETA observed a unilateral cease-fire for 14 months following the 1998 Good Friday accords in Ireland; but after Spain cracked down on the group militarily, and then a second cease-fire, the group has taken up sporadic violence again, mainly aimed at hurting Spain's tourist economy.

In the case of pressure point #8, an internal separatist movement, the group seeking independence is totally contained within Hypothetica. The Civil War in the United States in the 19th century is one example of a secessionist conflict, but wars of secession were numerous in the 20th century, particularly in Africa and Asia.

Resource Disputes (#9, #10, #11)   Neighboring states are likely to covet the resources—whether they be valuable mineral deposits, rich fishing grounds, or a cultural resource, such as a site of religious significance—lying in border areas and to disagree over their use. In recent years, for example, the United States has been involved in resource disputes with both its immediate neighbors: with Mexico over the shared resources of the Colorado River and Gulf of Mexico and with Canada over the Georges Bank fishing grounds in the Atlantic Ocean.

Conflicts arise when neighboring states disagree over policies to be applied along a border. Such policies may concern the movement of traditionally nomadic groups (#9), immigration, customs regulations, and the like. U.S. relations with Mexico, for example, have been affected by the large number of illegal aliens and the flow of drugs entering the United States from Mexico (**Figure 8.18**).

well. Even though the Basques in Spain already possess far more autonomy than those in France, Spain has been the site of greater agitation and violence than France, leading scholars to theorize that

Figure 8.18   Border fence between Nogales, Arizona, and Nogales, Sonora, Mexico, as seen from the Arizona side. To stem the flow of undocumented migrants entering the United States via Mexico, the United States has constructed a fence along much of the border between the two countries. © *Dave Moyer RF.*

Figure 8.19   The Rumaila oil field. One of the world's largest oil reservoirs, Rumaila straddles the Iraq-Kuwait border. Iraqi grievances over Kuwaiti drilling were partly responsible for Iraq's invasion of Kuwait in 1990.

The location of an internationally significant resource in a border region (#10) provides another opportunity for conflict. One of the causes of the 1990–1991 war in the Persian Gulf was the huge oil reservoir known as the Rumaila field, which lies mainly in Iraq, with a small extension into Kuwait (**Figure 8.19**). Because the two countries were unable to agree on a formula for sharing production costs and revenues, Kuwait pumped oil from Rumaila without any international agreement. Iraq helped justify its invasion of Kuwait by contending that the latter had been stealing Iraqi oil in what amounted to economic warfare.

As global warming shrinks polar ice in the Arctic, a new area of conflict develops. Not only might the melting ice provide a quicker sea route from the Atlantic to the Pacific than the Panama Canal, but scientists estimate that the seabed might also contain as much as one-quarter of the world's undiscovered oil, gas, and minerals. Countries bordering the region, including Russia, Denmark, the United States, and Canada, are racing to establish claims to the area. Recently, a Russian submarine planted a titanium Russian flag on the seabed under the North Pole; a Danish expedition used crews of icebreakers to map the seabed; and Canada established two new military bases in the Canadian Arctic.

A final potential trouble spot is represented by site A on the map in Figure 8.14 (#11), the location of a resource that the state believes is crucial to its survival and that must be defended, even if it means claiming an adjacent piece of land in a neighboring state. The resource might be physical, such as a military post, or cultural, such as a holy city. Syria and Israel, for example, dispute the ownership of the Golan Heights, which contains water and is high ground. It allows the country that controls it to look down upon and listen to the country on the other side—Israelis can literally look at Damascus from Mount Hermon. A disputed cultural resource is represented by Jerusalem, a site of great religious significance to Christians, Jews, and Muslims. It has been the source of conflicts since at least the beginning of the first crusade in A.D. 1096. Believing that Jerusalem is vital to its own identity, in recent years Israel has effectively annexed much of Muslim East Jerusalem. Currently, one of the principal points of contention between the Israeli government and Palestinians is access to and the control of holy sites in the city.

Centripetal Forces: Promoting State Cohesion

At any moment in time, a state is characterized by forces that promote unity and national stability and by others that disrupt them. Political geographers refer to the unifying factors as **centripetal forces;** they bind together the people of a state, enable it to function, and give it strength. **Centrifugal forces,** on the other hand, destabilize and weaken a state. If centrifugal forces are stronger than those promoting unity, the very existence of the state will be threatened. In the sections that follow, we examine four forces—nationalism, unifying institutions, effective organization and administration of government, and systems of transportation and communication—to see how they can promote cohesion.

Nationalism

One of the most powerful of the centripetal forces is **nationalism,** an identification with the state and the acceptance of national goals. Nationalism is based on the concept of allegiance to a single country and the ideals and the way of life it represents. It is an emotion that provides a sense of identity and loyalty and of collective distinction from all other peoples and lands.

States purposely try to instill feelings of allegiance in their constituents, for such feelings give the political system strength. People who have such allegiance are likely to accept the rules governing behavior in the area and to participate in the decision-making process establishing those rules. In light of the divisive forces present in most societies, not everyone, of course, will feel the same degree of commitment or loyalty. The important consideration is that the majority of a state's population accept its ideologies, adhere to its laws, and participate in its effective operation. For many countries, such acceptance and adherence has come only recently and partially; in some, it is frail and endangered.

Recall that true nation-states are rare; in only a few countries do the territory occupied by the people of a particular nation and the territorial limits of the state coincide. Most countries have more than one culture group that considers itself separate in an important way from other citizens. In a multicultural society, nationalism helps integrate different groups into a unified population. This kind of consensus nationalism has emerged in countries such as the United States and Switzerland, where different culture groups have joined together to create political entities commanding the loyalties of all of their citizens.

Figure 8.20   **The ritual of the Pledge of Allegiance** is just one way in which schools in the United States seek to instill a sense of national identity in students. © BananaStock/PunchStock RF.

States promote nationalism in a number of ways. *Iconography* is the study of the symbols that help unite people. National anthems and other patriotic songs, flags, national sports teams, rituals, and holidays are all developed as symbols of a state in order to promote nationalism and attract allegiance (**Figure 8.20**). By ensuring that all citizens, no matter how diverse the population, will have at least these symbols in common, they impart a sense of belonging to a political entity, called, for example, Japan or Canada. In some countries, certain documents, such as the Magna Carta in England or the Declaration of Independence in the United States, serve the same purpose. Royalty may fill the need: in Sweden, Japan, and Great Britain, the monarchy functions as the symbolic focus of allegiance. Symbols and beliefs are major components of every culture. When a society is very heterogeneous, composed of people with different customs, religions, and languages, belief in the national unit can help weld them together.

Unifying Institutions

Institutions as well as symbols help develop the sense of commitment and cohesiveness essential to the state. Schools, particularly elementary schools, are among the most important of these. Children learn the history of their own country and relatively little about other countries. Schools are expected to instill the society's goals, values, and traditions; to teach the common language that conveys them; and to guide youngsters to identify with their country.

Other institutions that promote nationalism are the armed forces and, sometimes, a state church. The armed forces are, of necessity, taught to identify with the state. They see themselves as protecting the state's welfare from what are perceived to be its enemies.

In about one-quarter of the world's countries, the religion of the majority of the people has by law been designated a state church. In such cases, the church sometimes becomes a force for cohesion,

helping unify the population. This is true of Islam in Pakistan, Judaism in Israel, Buddhism in Thailand, and Hinduism in Nepal. In countries such as these, the religion and the church are so identified with the state that belief in one is transferred to allegiance to the other.

The schools, the armed forces, and the church are just three of the institutions that teach people to be members of a state. As institutions, they operate primarily on the level of the sociological subsystem of culture, helping structure the outlooks and behaviors of the society. But by themselves, they are not enough to give cohesion, and thus strength, to a state. Indeed, each of the institutions we have discussed can be a destabilizing centrifugal force.

Organization and Administration

A further bonding force is public confidence in the effective organization of the state. Can it provide security from external aggression and internal conflict? Are its resources distributed and allocated in such a way as to be perceived to promote the economic welfare of all its citizens? Are all citizens afforded equal opportunity to participate in government affairs (see "Legislative Women," p. 233)? Do institutions that encourage consultation and the peaceful settlement of disputes exist? How firmly established are the rule of law and the power of the courts? Is the system of decision making responsive to the people's needs?

The answers to these questions will vary from country to country, but they and similar ones are implicit in the expectation that the state will, in the words of the Constitution of the United States, "establish justice, insure domestic tranquillity, provide for the common defense, [and] promote the general welfare. . . ." If those expectations are not fulfilled, the loyalties promoted by national symbols and unifying institutions may be weakened or lost.

Transportation and Communication

A state's transportation network fosters political integration by promoting interaction between areas and by joining them economically and socially. The role of a transportation network in uniting a country has been recognized since ancient times. The saying that all roads lead to Rome had its origin in the impressive system of roads that linked Rome to the rest of the empire. Centuries later, a similar network was built in France, linking Paris to the various departments of the country. Often, the capital city is better connected to other cities than the outlying cities are to one another. In France, for example, it can take less time to travel from one city to another by way of Paris than by direct route.

Roads and railroads have played a historically significant role in promoting political integration. In the United States and Canada, they not only opened up new areas for settlement but also increased interaction between rural and urban areas. Because transportation systems play a major role in a state's economic development, it follows that the more economically advanced a country is, the more extensive its transport network is likely to be. At the same time, the higher the level of development is, the more the money will be invested in building transport routes. The two reinforce each other.

Transportation and communication, although encouraged within a state, are frequently curtailed or at least controlled between

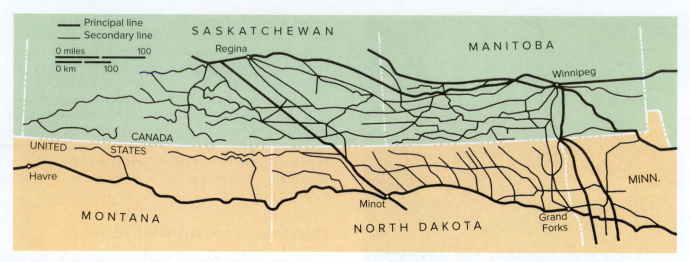

Figure 8.21 Canadian-U.S. railroad discontinuity. Canada and the United States developed independent railway systems connecting their respective prairie regions with their separate national cores. Despite extensive rail construction during the 19th and early 20th centuries, the pattern that emerged even before recent track abandonment was one of discontinuity at the border. Note how the political boundary restricted the ease of spatial interaction between adjacent territories. Many branch lines approached the border, but only eight crossed it. In fact, for more than 480 kilometers (300 mi), no railway bridged the boundary line. The international border—and the cultural separation it represents—inhibits other expected degrees of interaction. Telephone calls between Canadian and U.S. cities, for example, are far less frequent than would be expected if distance alone were the controlling factor.

states as a conscious device for promoting state cohesion through limitation on external spatial interaction (**Figure 8.21**). The mechanisms of control include restrictions on trade through tariffs and embargoes, legal barriers to immigration and emigration, and limitations on travel through passports and visa requirements.

Centrifugal Forces: Challenges to State Authority

State cohesion is not easily achieved or, once gained, invariably retained. Destabilizing centrifugal forces are ever-present, sowing internal discord and challenges to the state's authority (see "Terrorism and Political Geography," p. 234). Transportation and communication may be hindered by a country's shape or great size, leaving some parts of the country not well integrated with the rest. A state that is not well organized or administered stands to lose the loyalty of its citizens. Institutions that in some states promote unity can be a divisive force in others.

Religion, for example, can be a potent centrifugal force. It may compete with the state for people's allegiance—one reason the former USSR and other communist governments suppressed religion and promoted atheism. Conflict between majority and minority faiths within a country—such as between Catholics and Protestants in Northern Ireland or Hindus and Muslims in Kashmir and Gujarat State in India—can destabilize social order. Opposing sectarian views within a single, dominant faith can also promote civil conflict. Recent years have seen Muslim militant groups attempt to overturn official or constitutional policies of secularism or replace a government deemed insufficiently ardent in its imposition of religious laws and regulations. Islamic fundamentalist militancy has been a destabilizing force in Afghanistan, Egypt, and Saudi Arabia, among other countries.

Nationalism, in contrast to its role as a powerful centripetal agency, is also a potentially very disruptive centrifugal force. We previously identified four types of relationships between states and nations: a nation-state, a multinational state, a part-nation state, and a stateless nation (see Figure 8.4). The idea of the nation-state is that states are formed around and coincide with nations. It is a small step from that to the presumption that every nation has the right to its own state or territory.

Centrifugal forces are particularly strong in countries containing multiple nationalities and unassimilated minorities, racial or ethnic conflict, contrasting cultures, and a multiplicity of languages or religions. Such states are susceptible to nationalist challenges from within their borders: a country whose population is not bound by a shared sense of nationalism but is split by several local primary allegiances suffers from **subnationalism.** That is, many people give their primary allegiance to traditional groups or nations that are smaller than the population of the entire state.

Any country that contains one or more important national minorities is susceptible to challenges from within its borders if the minority group has an explicit territorial identification and believes that its right to *self-determination*—the right of a group to govern itself in its own state or territory—has not been satisfied. In its intense form, **regionalism,** a strong minority group self-awareness and identification with a region rather than with the state, can be expressed politically as a desire for greater autonomy (self-government) or even separation from the rest of the country. It is prevalent in many parts of the world today and has created currents of unrest within many countries, even long-established ones.

Canada, for example, houses a powerful secessionist movement in French-speaking Quebec, the country's largest province. In October 1995, a referendum to secede from Canada and become a sovereign country failed by a razor-thin margin (49% yes, 51% no). Quebec's nationalism is fueled by strong feelings of collective

LEGISLATIVE WOMEN

Women, a majority of the world's population, in general fare poorly in the allocation of such resources as primary and higher education, employment opportunities and income, and health care. That their lot is improving is encouraging. In nearly every developing country, women have been closing the gender gap in literacy, school enrolment, and acceptance in the job market.

But in the political arena, where power ultimately lies, women's share of influence is increasing only slowly and selectively. In 2016, out of a world total of some 200 countries, 23 had women as heads of government: presidents or prime ministers. In other words, more than 88% of the world's countries were led by men. Nor did women fare much better as members of parliaments. In that year, women held just 22% of all the seats in the world's legislatures.

In only 61 countries did women occupy one-quarter or more of the seats in the legislative house in 2016. Of these 61 countries, 21 were European (see the figure), 17 were African, 7 were Asian, and 14 were in Latin America. Rwanda was the most feminist, with 57% of its legislative members female. In no other country were women a legislative majority, and a number of countries had no female representatives at all. Twenty-five of those countries are in Europe, where women occupy about 25% of legislative seats in Europe overall. In some European countries, women occupy a tiny minority of legislative seats. This includes both established democracies of northern and Western Europe and many of the countries of southern and Eastern Europe.

Many countries are witnessing increased discontent with the proportion of women in legislatures. In the 1990s, women's legislative representation began to expand materially in many developed and developing democracies, and their "fair share" of political power began to be formally recognized or enforced. In Western countries, particularly, improvement in female parliamentary participation has become a matter of plan and pride for political parties and, occasionally, for governments themselves. Political parties from Mexico to China have tried to correct female under-representation, usually by setting quotas for women candidates, and a few governments—including Belgium and Italy—have tried to require their political parties to improve their balance.

France went further than any other country in acknowledging the right of women to equal access to elected office when in 1999 it passed a constitutional amendment requiring *parité*—parity, or equality. A year later, the National Assembly enacted legislation requiring the country's political parties to fill 50% of the candidacies in all elections in the country (municipal, regional, and European Parliament) with women or pay a fine. All poltical parties must put forward equal numbers of female and male candidates.

Quotas are controversial, however, and often are viewed with disfavor even by avowed feminists. Some argue that quotas are demeaning because they imply that women cannot match men on merit alone. Others fear that other groups (e.g., religious groups, ethnic minorities) would also seek quotas to ensure their fair representation in legislatures.

Notice that the United States was not among the countries where women held 25% or more of national legislative seats in 2016, unlike its neighbors Canada (28%) and Mexico (41%). In the 114th Congress (2015–2017), there were only 20 women serving in the Senate and 84 in the House of Representatives, making up 19% of both House and Senate.

American women have made greater electoral gains in state legislatures, where their percentage rose steadily over the past several decades, from 4% in 1969 to 24% in 2016. In 2016 there were 7383 state legislators, of which 1800 were women. Wide disparities exist among the states, however. Colorado had the most women in its legislature, with 42 out of 100 seats filled by women in 2016. At the other end of the spectrum were Wyoming (13%) and South Carolina (14%). Between 2005 and 2016, South Carolina fluctuated in its number of female legislators, with a low of 8.8% for the 2005–2008 period and has currently returned to the percentages seen in 1993.

A significant presence of women in legislative bodies makes a difference in the kinds of bills that get passed and the kinds of programs that receive government emphasis. Regardless of party affiliation, women tend to have somewhat different priorities than male lawmakers. For example, women are more apt than their male counterparts to sponsor bills and vote for measures in such policy areas as equal pay, child care, long-term care for the elderly, affordable health insurance, women's health issues, and women's rights, including divorce and spousal-abuse laws.

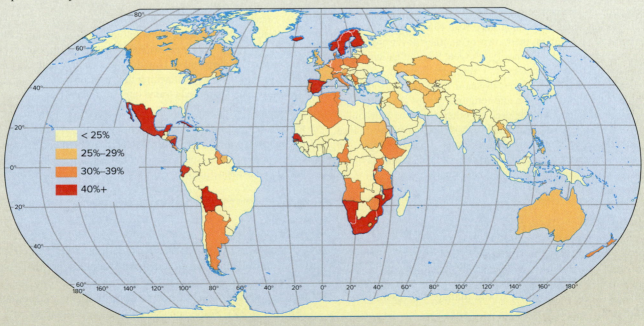

Countries where women held 25% or more of the legislative seats.

TERRORISM AND POLITICAL GEOGRAPHY

It has been more than 10 years since the terrorist attacks on the World Trade Center in New York City on September 11, 2001. For millions of people, though, that event is seared in their memories. It felt, as one singer put it, as though the "world stopped turning."

What is terrorism? How does it relate to political geography? Do all countries experience terrorism? Is terrorism new? Is there a way to prevent it? Attempting to answer these questions, difficult as they are, may help us understand the phenomenon.

Terrorism is the calculated use of violent acts against civilians and symbolic targets to publicize a cause, intimidate or coerce a civilian population, or affect the conduct of a government. *International terrorism,* such as the attacks of September 11, 2001, includes acts that transcend national boundaries. International terrorism is intended to intimidate people in other countries. *Domestic terrorism* consists of acts by individuals or groups against the citizens or government of their own country. *State terrorism* is committed by the agents of a government. *Subnational terrorism* is committed by groups outside a government.

Terrorism, thus, is a weapon. It is a weapon whose aim is intimidation and whose victims are usually civilians.

State terrorism is probably as old as the concept of a state. As early as 146 B.C., for example, Roman forces sacked and completely destroyed the city of Carthage, burning it to the ground, killing men, women, and children, and sowing salt on the fields so that no crops could grow. Governments have used systematic policies of violence and intimidation to further dominate and control their own populations. Nazi Germany, the Pol Pot regime in Cambodia, and Stalinist Russia are 20th-century examples of state terrorism. Heads of state ordered the murder, imprisonment, or exile of enemies of the state—politicians, intellectuals, dissidents—anyone who dared to criticize the government. In Rwanda, the former Yugoslavia, and Saddam Hussein's Iraq, state terrorism aimed against ethnic and religious minorities provided the government with a method of consolidating power; in each case, genocide, or mass murder of ethnic minority groups, was the result. The government or its agencies waged full-fledged military campaigns against minority groups.

Subnational terrorism began much later, at the same time as the rise of the nation-state. Subnational terrorism can be perpetrated by those who feel wronged by their own or another government. For example, ethnic groups in a minority who feel that the national government has taken their territory and absorbed them into a larger political entity, such as the Basques in Spain, have used terrorist acts to resist the government. Ethnic and religious groups that have been split by national boundaries imposed by others, such as Palestinian Arabs in the Middle East, have used terrorism to make governance impossible. Political, ethnic, or religious groups that feel oppressed by their own government, such as the Oklahoma City bombers in the United States, have committed acts of domestic terrorism.

Nearly every country has experienced some form of terrorism at some point since the mid-19th century. These acts have been as various as the anarchist assassinations of political leaders in Europe during the 1840s and in the United States in the late 19th century, the abduction of Canadian government officials by the Front Liberation du Quebec (FLQ) in 1970, and the release of sarin gas in the Tokyo subways in 1995 by the group Aum Shinrikyo.

The political and religious aims of these attackers, however, can cause confusion on the world stage. In 2001, the Reuters News Agency told its reporters to stop using the word *terrorism,* because "one person's terrorist is another's freedom fighter." The definition of *terrorism* rests on the ability to identify motives.

Although it may be difficult to distinguish among types of terrorism, it is even more difficult to prevent it. Generally speaking, there are four common responses to terrorism on the part of governments and international bodies:

1. Reducing or addressing the causes of terrorism. In some cases, political change can reduce a terrorist threat. For example, the 1998 Good Friday Agreement in Northern Ireland led to a reduction in terrorist acts; the Spanish government's granting of some regional autonomy to the Basques helped quiet the actions of the ETA and reduced the support of many Basque people for such acts.
2. Increasing international cooperation in the surveillance of subnational groups. Spurred by terrorist crimes in Bahrain and Saudi Arabia, the Arab Gulf States agreed in 1998 to exchange intelligence regarding terrorist groups, to share intelligence regarding the prevention of an anticipated terrorist act, and to assist one another in investigating terrorist crimes.
3. Increasing security measures in a country. In the United States, following September 11, 2001, the government organized a Department of Homeland Security, federalized air traffic screening, and increased efforts to reduce financial support for foreign terrorist organizations. In concert, the European Union froze the assets of any group on its list of terrorist organizations.
4. Using military means either unilaterally or multilaterally against terrorists or governments that sponsor terrorists. Following the September 11 attacks, the United States led a coalition of countries in attacking the government of Afghanistan, which had harbored Osama bin Laden's al-Qaeda terrorist organization.

Each response to terrorism is expensive, politically difficult, and/or potentially harmful to the life and liberty of civilians. Governments must decide which response or combination of responses is likely to have the most beneficial effect.

identity and distinctiveness, as well as by a desire to protect its language and culture. Additionally, separatists believe that the province, which has ample resources and one of the highest standards of living in the industrialized world, would manage successfully as a separate country.

In Western Europe, five countries (the United Kingdom, France, Belgium, Italy, and Spain) contain political movements whose members reject total control by the existing sovereign state and who claim to be the core of a separate national entity (**Figure 8.22**). Some separatists would be satisfied with *regional*

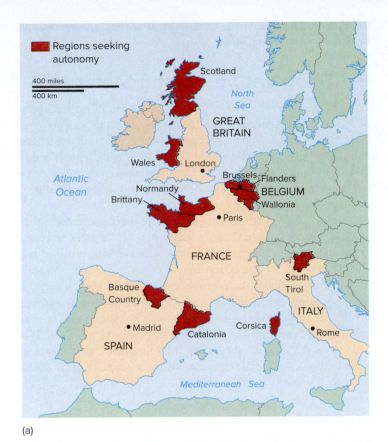

(a)

(b)

Figure 8.22   (a) Regions with active autonomous movement. Despite long-standing state attempts to culturally assimilate these historic nations, each contains a political movement that has recently sought or is currently seeking a degree of self-rule that recognizes its separate identity. Separatists on the island of Corsica, for example, want to secede from France, and separatists in Catalonia demand independence from Spain. The desires of nationalist parties in both Wales and Scotland were partially accommodated by the creation in 1999 of their own parliaments and a degree of regional autonomy, an outcome labeled "separation but not a divorce" from the United Kingdom. **(b)** Demonstrators carry a giant Basque flag during a march to call for independence for the **Basque region**. Spain has granted limited independence to the Basque region in an example of asymmetric federalism. *(b) © Aranberri/AP Images.*

autonomy, usually in the form of self-government, or "home rule"; others seek complete independence for their regions.

In an effort to defuse these separatist movements and to accommodate politically or culturally diverse peoples within their own borders, several European governments have moved in the direction of regional recognition and **devolution** (decentralization) of political control. Britain, France, Spain, Portugal, and Italy are among the countries that have recognized the need for administrative structures that reflect regional concerns and have granted a degree of political autonomy to recognized political subunits, giving them a measure of self-rule short of complete independence. In 1999, for example, voters in Scotland and Wales elected representatives to two newly created legislatures—the Scottish Parliament and the National Assembly of Wales. The legislatures have authority over such domestic matters as local government, housing, health, education, culture, transportation, and the environment. The British Parliament kept its powers over broad national policy concerns: defense, foreign policy, the economy, and monetary policy.

Nationalist challenges to state authority affect many countries outside of Western Europe. Many countries containing disparate groups that are more motivated by enmity than by affinity have powerful centripetal tendencies. The Basques of Spain and the Bretons of France have their counterparts in the Palestinians in Israel, the Sikhs in India, the Tamils in Sri Lanka, the Moros in the Philippines, and many others.

The countries of Eastern Europe and the republics of the former Soviet Union have seen many instances of regionally rooted nationalist feelings. Now that the forces of ethnicity, religion, language, and culture are no longer suppressed by communism, ancient rivalries are more evident than at any time since World War II. The end of the Cold War aroused hopes of decades of peace. Instead, the collapse of communism and the demise of the USSR spawned many smaller wars. Numerous ethnic groups, large and small, are asserting their identities and what they perceive to be their right to determine their own political status.

The national independence claimed in the early 1990s by the 15 former Soviet republics did not ensure the satisfaction of all separatist movements within them. Many of the new individual countries are subject to strong destabilizing forces that challenge their territorial integrity and survival. The Russian Federation itself, the largest and most powerful remnant of the former USSR, has 89 components, including 21 "ethnic republics" and a number of other nationality regions. Many are rich in natural resources, have non-Russian majorities, and seek greater autonomy within the federation. Some, indeed, want total independence.

As the USSR declined and eventually disbanded, it lost control of its communist satellites in Eastern Europe. That loss and resurgent nationalism led to a dramatic reordering of the region's political map. East Germany was reunified with West Germany in 1990, and 3 years later, the people of Czechoslovakia agreed to split their country into two separate, ethnically based states: the Czech Republic and Slovakia. More violently, Yugoslavia shattered into five pieces in 1991–1992, but with the exception of Slovenia, the boundaries of the five new republics did not match the territories occupied by nationalities, a situation that plunged the region into war as nations fought to redefine the boundaries of their countries. One tactic used to transform a multinational area into one containing only one nation is **ethnic cleansing,** the killing or forcible relocation of less-powerful minorities. It occurred in Croatia, Bosnia-Herzegovina, and the Kosovo province of southern Serbia. After a dozen years of efforts by NATO peace-keeping forces, these areas seem to be making progress toward long-term political stability.

Nationalist challenges to state authority have common elements. The two preconditions necessary to all separatist movements are *territory* and *nationality*. First, the group must be concentrated in a core region that it claims as a national homeland. It seeks to regain control of land and power that it believes were unjustly taken by the ruling party. Second, certain cultural characteristics must provide a basis for the group's perception of separateness, identity, and cultural unity. These might be language, religion, or distinctive group customs, which promote feelings of group identity at the same time that they foster exclusivity. Normally, these cultural differences have persisted over several generations and have survived despite strong pressures toward assimilation.

Other characteristics common to many separatist movements are a *peripheral location* and *social and economic inequality*. Troubled regions tend to be peripheral, often isolated in rural pockets, and their location away from the seat of central government engenders feelings of alienation, exclusion, and neglect. In addition, the dominant culture group is often seen as an exploiting class that has suppressed the local language, controlled access to the civil service, and taken more than its share of wealth and power. Poorer regions complain that they have lower incomes and greater unemployment than prevail in the rest of the state and that "outsiders" control key resources and industry. Separatists in relatively rich regions believe that they could exploit their resources for themselves and do better economically without the constraints imposed by the central state.

8.2 Cooperation Among States

The modern state is fragile and, as we have seen, its primacy may be less assured in recent years. In many ways, countries are now weaker than ever before. Many are economically frail, others are politically unstable, and some are both. Strategically, no state is safe from military attack, for technology now enables countries to shoot weapons halfway around the world. Is national security possible in the nuclear age?

Recognizing that a country cannot by itself guarantee either its prosperity or its own security, many states have opted to cooperate with others. These cooperative ventures are proliferating quickly, and they involve countries everywhere. They are also adding a new dimension to the concept of political boundaries, because the associations of states have borders of a higher spatial order than those between individual states. Such boundaries as the current division between North Atlantic Treaty Organization (NATO) and non-NATO states, or between the European Union area and other European countries, represent a different scale of the political ordering of space.

Supranationalism

Associations among states represent a new dimension in the ordering of national power and national independence. Recent trends in economic globalization and international cooperation suggest to some that the sovereign state's traditional responsibilities and authorities are being diluted by a combination of forces and partly delegated to higher-order political and economic organizations. Corporations and even nongovernmental agencies often operate in controlling ways outside of nation-state jurisdiction.

The rise of transnational corporations dominant in global markets, for example, limits the economic influence of individual countries. Cyberspace and the Internet are controlled by no one and are largely immune to the state restrictions on the flow of information exerted by many governments. And increasingly, individual citizens of any country have their lives and actions shaped by decisions not only of local and national authorities but also of regional economic associations (e.g., the North American Free-Trade Agreement), of multiparty military alliances (e.g., NATO), and of global political agencies (e.g., the United Nations).

The roots of such multistate cooperative systems are ancient—for example, the leagues of city-states in the ancient Greek world or the Hanseatic League of free German cities in Europe's medieval period. New cooperative systems have proliferated since the end of World War II. They represent a world trend toward a **supranationalism** composed of associations of three or more states created for mutual benefit and the achievement of shared objectives. Although many individuals and organizations decry the loss of national independence that supranationalism entails, the many supranational associations in existence early in the 21st century are evidence of their attraction and pervasiveness. Nearly all countries, in fact, are members of at least one—and most are members of many—supranational groupings.

The United Nations and Its Agencies

The United Nations (UN) is an organization that tries to be universal. Its membership has expanded from 51 countries in 1945 to 193 in 2013. The most recent state to be recognized as a member is South Sudan (2011).

The UN is the most ambitious attempt ever undertaken to bring together the world's countries in international assembly and to promote world peace. Stronger and more representative than its predecessor, the League of Nations, it provides a forum where countries can discuss international problems and regional concerns and a mechanism, admittedly weak but still significant, for forestalling disputes or, when necessary, for ending wars (**Figure 8.23**). The United Nations also sponsors 40 programs and agencies aimed at fostering international cooperation with respect to specific goals. Among these are the World Health Organization (WHO); the Food and Agriculture Organization (FAO); and the United Nations Educational, Scientific, and Cultural Organization (UNESCO). Many other UN agencies and much of the UN budget are committed to assisting member states with matters of economic growth and development.

Member states have not surrendered sovereignty to the UN, and the world body is legally and effectively unable to make or enforce a world law. Nor is there a world police force. Although there is recognized international law adjudicated by the International Court of Justice, rulings by this body are sought only by countries agreeing beforehand to abide by its arbitration. Finally, the United Nations has no authority over the military forces of individual countries.

Figure 8.23   UN Peacekeepers in Haiti after the earthquake of January 2010. A United Nations mission to Haiti was established in 2004 to stabilize the political situation after the exile of President Bertrand Aristide. After the earthquake, the UN Security Council increased the overall force levels to support the recovery, reconstruction, and stability efforts in Haiti. The UN continues to work to bring security and stability to the country. © Natalie Roeth RF.

A pronounced change both in the relatively passive role of the United Nations and in traditional ideas of international relations has begun to emerge, however. Long-established rules of total national sovereignty that allowed governments to act internally as they saw fit, free of outside interference, are fading as the United Nations increasingly applies a concept of "interventionism." The Persian Gulf War of 1991 was UN-authorized under the old rules prohibiting one state (Iraq) from violating the sovereignty of another (Kuwait) by attacking it. After the war, the new interventionism sanctioned UN operations within Iraq to protect Kurds within that country. Later, the UN intervened with troops and relief agencies in Somalia, Bosnia, and elsewhere, invoking an "international jurisdiction over inalienable human rights" that prevails without regard to state frontiers or sovereignty considerations. A UN effort in Afghanistan has the goal of helping the country establish the foundations for sustainable peace and development and aid in reconstruction and development; the UN has overseen elections there since 2002.

Whatever the long-term prospects for interventionism replacing absolute sovereignty, for the short term the UN remains the only institution where the vast majority of the world's countries can collectively discuss international political and economic concerns and attempt peacefully to resolve their differences. It has been particularly influential in formulating a law of the sea.

Maritime Boundaries

Boundaries define political jurisdictions and areas of resource control, but claims of national authority are not restricted to land areas alone. Water covers about two-thirds of the Earth's surface, and increasingly countries have been projecting their sovereignty seaward to claim adjacent maritime areas and resources. A basic question involves the right of states to control water and the resources it contains. The inland waters of a country, such as rivers and lakes, have traditionally been regarded as being within the sovereignty of that country. Oceans, however, are not within any country's borders. Are they, then, to be open to all states to use, or may a single country claim sovereignty and limit access and use by other countries?

For most of human history, the oceans remained effectively outside individual national control or international jurisdiction. The seas were a common highway for those daring enough to venture on them, an inexhaustible larder for fishermen, and a vast refuse pit for the muck of civilization. By the end of the 19th century, however, most coastal countries claimed sovereignty over a continuous belt 3 or 4 nautical miles wide (1 *nautical mile,* or 1 nm, equals 1.15 statute miles, or 1.85 km). At the time, the 3-nm limit represented the farthest range of artillery and thus the effective limit of control by the coastal state. Though recognizing the rights of others to innocent passage, such sovereignty permitted the enforcement of quarantine and customs regulations, allowed national protection of coastal fisheries, and made claims of neutrality effective during other people's wars. The primary concern was with security and unrestricted commerce. No separately codified law of the sea existed, however, and none seemed to be needed until after World War I.

A League of Nations Conference for the Codification of International Law, convened in 1930, inconclusively discussed maritime legal matters and identified areas of concern that were to become increasingly pressing after World War II. Important among these was an emerging shift from interest in commerce and national security to a preoccupation with the resources of the seas, an interest fanned by the *Truman Proclamation* of 1945. Motivated by a desire to exploit offshore oil deposits, the U.S. federal government, under this doctrine, laid claim to all resources on the continental shelf contiguous to its coasts. Other states, many claiming even broader areas of control, hurried to annex marine resources. Within a few years, one-quarter of the Earth's surface was appropriated by individual coastal countries.

An International Law of the Sea

Unrestricted extensions of jurisdiction and territorial disputes over proliferating claims to maritime space and resources led to a series of United Nations conferences on the Law of the Sea. Meeting over a period of years, delegates from more than 150 countries attempted to achieve consensus on a treaty that would establish an internationally agreed-upon "convention dealing with all matters relating to the Law of the Sea." The meetings culminated in a draft treaty in 1982, the **United Nations Convention on the Law of the Sea (UNCLOS).**

The convention delimits territorial boundaries and rights by defining four zones of diminishing control (**Figure 8.24**).

1. A *territorial sea* of up to 12 nautical miles (19 km) in breadth, over which coastal states have sovereignty, including exclusive fishing rights. Vessels of all types normally have the right of innocent passage through the territorial sea, although under certain circumstances noncommercial vessels (primarily military and research) can be challenged.
2. A *contiguous zone* of up to 24 nautical miles (38 km). Although a coastal state does not have complete sovereignty in this zone, it can enforce its customs, immigration, and sanitation laws and has the right of hot pursuit out of its territorial waters.
3. An **exclusive economic zone (EEZ)** of up to 200 nautical miles (370 km), in which the state has recognized rights to explore, exploit, conserve, and manage the natural

resources, both living and nonliving, of the seabed and waters (**Figure 8.25**). Countries have exclusive rights to the resources lying within the continental shelf when this extends farther, up to 350 nautical miles (560 km) beyond their coasts. The traditional freedoms of the high seas are to be maintained in this zone.
4. The *high seas* beyond the EEZ. Outside any national jurisdiction, they are open to all states, whether coastal or landlocked. Freedom of the high seas includes the right to sail ships, fish, fly over, lay submarine cables and pipelines, and pursue scientific research. Mineral resources in the international deep seabed area beyond national jurisdiction are declared the common heritage of humankind, to be managed for the benefit of all the peoples of the earth.

By the end of the 1980s, most coastal countries, including the United States, had used the UNCLOS provisions to proclaim and reciprocally recognize jurisdiction over 12-nautical-mile (19-km) territorial seas and 200-nautical-mile (370-km) economic zones. Despite reservations held by the United States and a few other industrial countries about the deep seabed mining provisions, the convention received the necessary ratification by 60 states and became international law in 1994. It has since been ratified by more than 100 other nations and has generally been respected.

UN Affiliates

Other fully or essentially global supranational organizations with influences on the economic, social, and cultural affairs of states and individuals have been created. Most are specialized international agencies, autonomous and with their own memberships but with affiliated relationships with the United Nations and operating under its auspices. Among them are the Food and Agriculture Organization (FAO), the World Bank, the International Labor Organization (ILO), the United Nations Children's Fund (UNICEF), the World Health Organization (WHO), and—of growing economic importance—the World Trade Organization (WTO).

The World Trade Organization, which came into existence in 1995, has become one of the most significant of the global expressions of supranational economic control. It is charged with enforcing the global trade accounts that grew out of years of international negotiations under the terms of the General Agreement on Tariffs

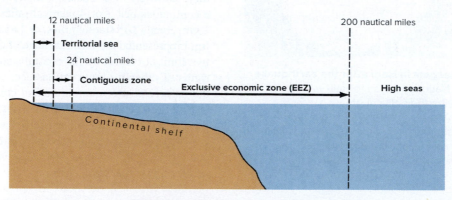

Figure 8.24   **Territorial claims permitted by the 1982 United Nations Convention on the Law of the Sea (UNCLOS).**

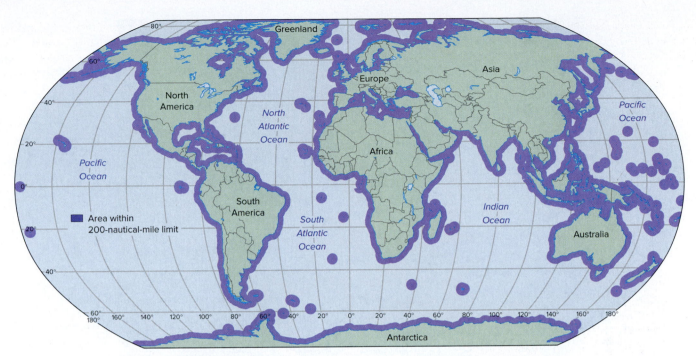

Figure 8.25   **The 200-nautical-mile exclusive economic zone (EEZ) claims of coastal states.** The provisions of the Law of the Sea Convention, in effect, changed the maritime map of the world. Three important consequences flowed from the 200-nautical-mile EEZ concept: (1) islands gained a new significance; (2) countries gained a host of new neighbors; and (3) the EEZ lines resulted in overlapping claims. EEZ lines are drawn around a country's possessions as well as around the country itself. Every island, no matter how small, has its own 200-nautical-mile EEZ. This means that although the United States shares continental borders with only Canada and Mexico, it has maritime boundaries with countries in Asia, South America, and Europe. All told, the United States may have to negotiate some 30 maritime boundaries, which is likely to take decades. Other countries, particularly those with many possessions, will have to engage in similar lengthy negotiations.

and Trade (GATT). The basic principle behind the WTO is that the 157 (as of 2013) member countries should work to cut tariffs, dismantle nontariff barriers to trade, liberalize trade in services, and treat all other countries uniformly in matters of trade. Any preference granted to one should be available to all.

Increasingly, however, regional rather than global trade agreements are being struck, and free-trade areas are proliferating. Only a few WTO members are not already part of another regional trade association. Some argue that such regional alliances make world trade less free by scrapping tariffs on trade among member states but retaining them on exchanges with nonmembers.

Regional Alliances

In addition to their membership in international agencies, countries have shown themselves willing to relinquish some of their independence to participate in smaller, multinational systems. These groupings can be economic, military, or political. Cooperation in the economic sphere seems to come more easily to states than does political or military cooperation.

Economic Alliances

Among the most powerful and far-reaching of the economic alliances are those that have evolved in Europe, particularly the **European Union (EU).** The EU grew out of the *Common Market,* which was established in 1957 and was composed at first of only

six states: France, Italy, West Germany, Belgium, the Netherlands, and Luxembourg. It added new members slowly at the outset, as the United Kingdom, Denmark, and Portugal joined the organization between 1973 and 1986; Greece, Spain, and Portugal joined during the 1980s; Austria, Finland, and Sweden joined in 1995. As it gathered momentum, more countries were admitted to the EU during the 2000s, including eight former Soviet bloc nations, from Estonia in the north to Croatia in the south (**Figure 8.26**). These additions brought the number of member nations to 28, increased the EU's landmass by 25%, raised its total population to more than 508 million people, and expanded its economy to rival that of the United States. The EU is the world's largest and richest bloc of countries.

Over the years, members of the European Union have taken many steps to integrate their economies and coordinate their policies in such areas as transportation, agriculture, and fisheries. A council of ministers, a commission, a European parliament, and a court of justice give the European Union supranational institutions with effective ability to make and enforce laws. By 1993, the EU had abolished most remnant barriers to free trade and the free movement of capital and people among its members, creating a single European market. In another step toward economic and monetary union, the EU's single currency, the euro, replaced separate national currencies in 1999. Notes and coins in 17 national currencies—such as the Portuguese escudo and the deutsche Mark—were withdrawn. Some countries within the EU such as Sweden and the United Kingdom have elected to delay moving to the single currency.

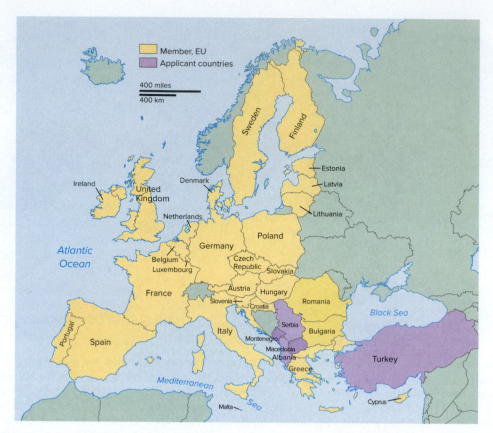

Figure 8.26   **The members of the European Union (EU) as of July 2016.** Croatia has recently (2013) been admitted to the EU while Iceland recently (2013) withdrew its application. Voters in the United Kingdom approved a 2016 referendum to leave the European Union. As of late 2016, no timetable has been established for the United Kingdom's departure. Candidate countries in talks to join the EU include Albania, Montenegro, Serbia, Turkey, and the former Yugoslav Republic of Macedonia. The EU has stipulated that in order to join, a country must have stable institutions guaranteeing democracy, the rule of law, human rights, and protection of minorities; a functioning market economy; and the ability to accept the obligations of membership, including the aims of political, economic, and monetary union.

The recent economic crisis has led many observers to question the euro's long-term viability.

Countries outside of Europe join together in regional alliances as well. In a fluid process, countries come together in an association to achieve economic and political goals. Sometimes members drop out, while others join. New treaties are made, and new coalitions emerge. Indeed, a number of such regional economic and trade associations have been added to the world supranational map. None are as encompassing in power and purpose as the EU, but all represent a cession of national independence to achieve broader regional goals.

NAFTA, the North American Free-Trade Agreement, was launched in 1994 and links Canada, the United States, and Mexico in an economic community aimed at lowering or removing trade and movement restrictions between the countries. A new agreement, CAFTA (United States–Central America–Dominican Republic Free-Trade Agreement), is a comprehensive trade agreement among Costa Rica, the Dominican Republic, El Salvador, Guatemala, Honduras, Nicaragua, and the United States. Free trade is not the only reason states cooperate. The Americas have other, similar associations with comparable trade enhancement objectives, though frequently they also have social, political, and cultural interests in mind. Caribbean Community and Common Market (CARICOM), for example, was established in 1974 to further

cooperation among its 15 members in economic, health, cultural, and foreign policy arenas. The Common Market of the Southern Cone (MERCOSUR), which unites Brazil, Argentina, Uruguay, and Paraguay (and associate members Bolivia and Chile) in the creation of a customs union to eliminate levies on goods moving among them, is a South American example.

A similar interest in promoting economic, social, and cultural cooperation and development among its members underpins the Association of Southeast Asian Nations (ASEAN). A less-wealthy African example is the Economic Community of West African States (ECOWAS). The Asia Pacific Economic Cooperation (APEC) forum includes China, Japan, Australia, Canada, and the United States among its 21 members and has a grand plan for "free trade in the Pacific" by 2020. More-restricted bilateral and regional preferential trade arrangements have also proliferated, creating a maze of rules, tariffs, and commodity agreements that result in trade restrictions and preferences contrary to the free-trade intent of the World Trade Organization.

Economic interests, then, motivate the establishment of most international alliances, though political, social, and cultural objectives also figure largely in many. Although the alliances themselves will change, the idea of supranational associations appears to have been permanently added to the national political and global realities of the 21st century.

Figure 8.27   **Western Hemisphere economic unions in 2013.** The number of international trade organizations has risen rapidly since the 1960s.

Three further points about regional international alliances are worth noting. The first is that the formation of a coalition in one area often stimulates the creation of another alliance by countries left out of the first. Thus, the creation of NAFTA was a precedent for CAFTA.

Second, the new economic unions tend to be composed of contiguous states (**Figure 8.27**). This was not the case with the recently dissolved empires, which included far-flung territories. Contiguity facilitates the movement of people and goods. Communication and transportation are simpler and more effective among adjoining countries than among those far removed from one another; common cultural, linguistic, and political traits and interests are more to be expected in countries adjacent to one another.

Finally, it does not seem to matter whether countries are alike or distinctly different in their economies, as far as joining economic unions is concerned. There are examples of both. If the countries are dissimilar, they may complement each other. This was one basis for the European Common Market, which preceded the EU. Dairy products and furniture from Denmark are sold in France, freeing France to specialize in the production of machinery and clothing. On the other hand, countries that produce the same raw materials hope that, by joining together in an economic alliance, they might be able to enhance their control of markets and prices for their products. The Organization of Petroleum Exporting Countries (OPEC) is a case in point. Other attempts to form commodity cartels and price agreements between producing and consuming countries are represented by the International Tin Agreement, the International Coffee Agreement, and others.

Military and Political Alliances

As we have seen, countries form alliances for other than economic reasons. Strategic, political, and cultural considerations may also foster cooperation. *Military alliances* are based on the principle that unity brings strength. Such pacts usually provide for mutual assistance in the case of aggression. Once again, action breeds reaction when such an association is created. The formation of the North Atlantic Treaty Organization (NATO), a defensive alliance of many European countries and the United States, was countered by the establishment of the Warsaw Treaty Organization, which joined the USSR and its satellite countries of Eastern Europe. Both pacts allowed the member states to base armed forces in one another's territories, a relinquishment of a certain degree of sovereignty uncommon in the past.

Military alliances depend on the perceived common interests and political goodwill of the countries involved. As political realities change, so, too, do the strategic alliances. NATO was created to defend Western Europe and North America against the Soviet military threat. When the dissolution of the USSR and Warsaw Pact removed that threat, the purpose of the NATO alliance became less clear. Since the 1990s, however, the organization has added seven members and has taken on a greater role in peacekeeping activities (**Figure 8.28**).

All international alliances recognize communities of interest. In economic and military associations, common objectives are clearly seen and described, and joint actions are agreed upon with respect to the achievement of those objectives. More generalized mutual concerns or appeals to historical interest may be the basis for primarily *political alliances*. Such associations tend to be rather loose, not requiring their members to yield much power to the union. Examples are the Commonwealth of Nations (formerly the British Commonwealth), composed of many former British colonies and dominions, and the Organization of American States (OAS), both of which offer economic as well as political benefits.

There are many examples of abortive political unions that have foundered because the individual countries could not agree on questions of policy and were unwilling to subordinate individual interests to make the union succeed. The United Arab Republic, the Central African Federation, the Federation of Malaysia and Singapore, and the Federation of the West Indies fall within this category.

Although many such political associations have failed, observers of the world scene speculate about the possibility that "superstates" will emerge from one or more of the international alliances that now exist. Will a "United States of Europe," for example, under a single common government, be the logical outcome of the expansion of the EU? No one knows, but as long as the individual state is regarded as the highest form of political and social organization (as it is now) and as the body in which sovereignty rests, such total unification is unlikely.

8.3 Local and Regional Political Organization

The most profound contrasts in cultures tend to occur between, rather than within, states, one reason political geographers traditionally have been interested primarily in country units. The emphasis

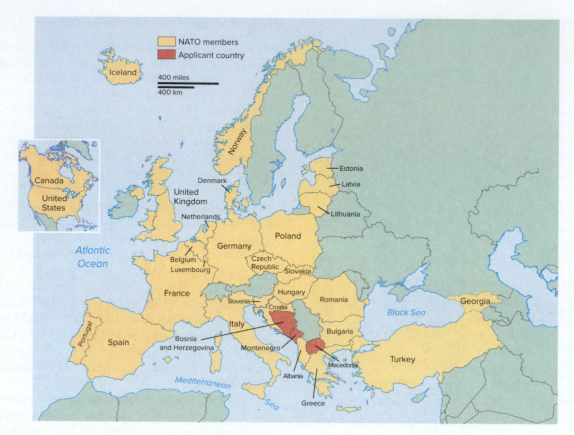

Figure 8.28   **The NATO military alliance** as of 2016 had 28 members. Current membership applicant countries include Bosnia and Herzegovina, Georgia, Montenegro, and the Former Yugoslav Republic of Macedonia.

on the state, however, should not obscure the fact that, for most of us, it is at that local level that we find our most intimate and immediate contact with government and its influence on the administration of our affairs. In the United States, for example, an individual is subject to the decisions and regulations made by the local school board, the municipality, the county, the state, and perhaps a host of special-purpose districts—all in addition to the laws and regulations issued by the federal government and its agencies. Among other things, local political entities determine where children go to school, the minimum size lot on which a person may build a house, and where one may legally park a car. Adjacent states of the United States may be characterized by sharply differing personal and business tax rates; differing controls on the sale of firearms, alcohol, and tobacco; variant administrative systems for public services; and different levels of expenditures for them (**Figure 8.29**).

All of these government entities are *spatial systems*. Because they operate within defined geographic areas, and because they make behavior-governing decisions, they are topics of interest to political geographers. In the concluding sections of this chapter, we will examine two aspects of political organization at the local and regional level. Our emphasis will be on the United States and Canadian scene simply because their local political geography is familiar to most of us. We should remember, however, that the North American structure of municipal governments, minor civil divisions, and special-purpose districts has counterparts in other regions of the world.

The Geography of Representation: The Districting Problem

There are more than 85,000 local government units in the United States. Slightly more than one-half of these are municipalities, townships, and counties. The remainder are school districts, water control districts, airport authorities, sanitary districts, and other special-purpose bodies. Boundaries have been drawn around each of these districts. Although the number of districts does not change greatly from year to year, many boundary lines are redrawn in any single year. Such *redistricting,* or *reapportionment,* is made necessary by shifts in population, as areas gain or lose people.

Every 10 years, following the U.S. census, updated figures are used to redistribute the 435 seats in the House of Representatives among the 50 states. Redrawing the congressional districts to reflect population changes is required by the Constitution, the intention being to make sure that each legislator represents roughly the same number of people. Since 1964, Canadian provinces and territories have entrusted redistricting for federal offices to independent electoral boundaries commissions. Although a few states in the United States also have independent, nonpartisan boards or commissions draw district boundaries, most rely on state legislatures for the task. Across the United States, the decennial census data are also used to redraw the boundaries of legislative districts within each state as well as those for local offices, such as city councils and county boards.

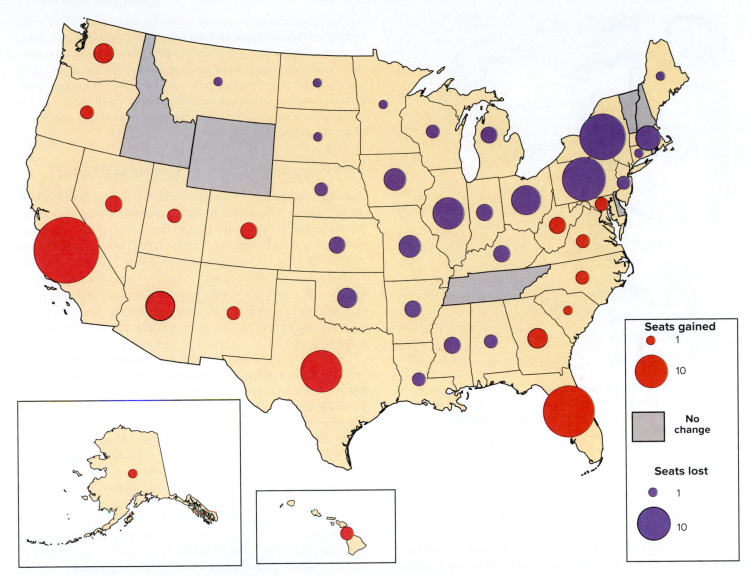

Figure 8.29   **Geographic shifts in congressional apportionment between 1930 and 2010** illustrate dramatic population movements to the South and West. Seats in the U.S. House of Representatives are reapportioned after each Census with the goal of achieving an equitable distribution based on population while maintaining at least one member for each state. Since 1930, New York has lost 18 seats and Pennsylvania 16, while California has gained 33, Florida gained 22, and Texas gained 15. There were no changes in seven states. After the 2010 Census, 12 seats were shifted from one state to another. *Source: Office of the Clerk, U.S. House of Representatives.*

The analysis of how boundaries are drawn around voting districts is one aspect of **electoral geography,** which also addresses the spatial patterns yielded by election results and their relationship to the socioeconomic characteristics of voters. In a democracy, it might be assumed that election districts should contain roughly equal numbers of voters, that electoral districts should be reasonably compact, and that the proportion of elected representatives should correspond to the share of votes cast by members of a given political party. Problems arise because the way in which the boundary lines are drawn can maximize, minimize, or effectively nullify the power of a group of people.

Gerrymandering is the practice of drawing the boundaries of legislative districts so as to unfairly favor one political party over another, to fragment voting blocs, or to achieve other nondemocratic objectives (**Figure 8.30**). A number of strategies have been employed over the years for that purpose. *Stacked* gerrymandering involves drawing circuitous boundaries to enclose pockets of strength or weakness of the group in power; it is what we usually think of as gerrymandering. The *excess vote* technique concentrates the support of the opposition in a few districts, which it can win easily, but leaves it few potential seats elsewhere. Conversely, the *wasted vote* strategy dilutes the opposition's strength by dividing its votes among a number of districts.

Assume that X and O represent two groups with an equal number of voters but different policy preferences. Although there are equal numbers of Xs and Os, the way electoral districts are drawn affects voting results. In **Figure 8.31a**, the Xs are concentrated in one district and will probably elect only one representative of four. The power of the Xs is maximized in Figure 8.31b, where they may control three of the four districts. The voters are evenly divided in Figure 8.31c, where the Xs have the opportunity to elect two of the four representatives. Finally, Figure 8.31d shows how both political

Figure 8.30 **The original gerrymander.** The term *gerrymander* originated in 1811 from the shape of an electoral district formed in Massachusetts while Elbridge Gerry was governor. When an artist added certain animal features, the district resembled a salamander and quickly came to be called a gerrymander. © *Bettmann/ Getty Images.*

parties may agree to delimit the electoral districts to provide "safe seats" for incumbents. Such a partitioning offers little chance for change.

Figure 8.31 depicts a hypothetical district, compact in shape with an even population distribution and only two groups competing for representation. In actuality, American municipal voting districts are often oddly shaped because of such factors as the city limits, historic settlement patterns, current population distribution, and transportation routes—as well as past gerrymandering. Further, in any large area, many groups vie for power. Each electoral interest group promotes its version of "fairness" in the way boundaries are delimited. Minorities seek representation in proportion to their numbers, so that they will be able to elect representatives who are concerned about and responsive to their needs.

Gerrymandering is not automatically successful. First, a districting arrangement that appears to be unfair may be appealed to the courts. In addition, many factors other than political party affiliation influence voting decisions. Key issues may cut across party lines, scandal may erode (or personal charm increase) votes unexpectedly, and the amount of candidate financing or the number of campaign workers may determine election outcomes if compelling issues are absent.

The Fragmentation of Political Power

Boundary drawing at any electoral level is never easy, particularly when political groups want to maximize their representation and minimize that of opposition groups. Furthermore, the boundaries that we may want for one set of districts may *not* be those that we want for another. For example, sewage districts must take natural drainage features into account, whereas police districts may be based on the distribution of the population or the number of miles of street to be patrolled. And school attendance zones must consider the numbers of school-aged children and the capacities of individual schools.

As these examples suggest, the United States is subdivided into great numbers of political administrative units whose areas of control are spatially limited. The 50 states are partitioned into more than 3000 counties ("parishes" in Louisiana), most of which are further subdivided into townships, each with a still lower level of governing power. This political fragmentation is further increased by the existence of nearly innumerable special-purpose districts whose boundaries rarely coincide with the standard major and minor civil divisions of the country, or even with one another (**Figure 8.32**). Each district represents a form of political allocation of territory to achieve a specific aim of local need or legislative intent.

Canada, a federation of 10 provinces and 3 territories, has a similar pattern of political subdivision. Each of the provinces contains minor civil divisions—municipalities—under provincial control, and all (cities, towns, villages, and rural municipalities) are governed by elected councils. Ontario and Quebec also have counties that group smaller municipal units for certain purposes. In general, municipalities are responsible for police and fire protection, local jails, roads and hospitals, water supply and sanitation, and schools, duties that are discharged either by elected agencies or by appointed commissions.

Most North Americans live in large and small cities. In the United States, these, too, are subdivided, not only into wards or

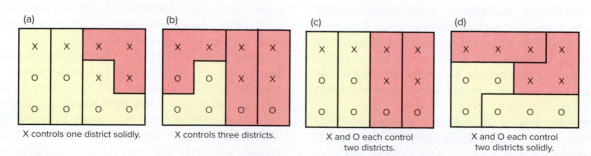

(a) X controls one district solidly.

(b) X controls three districts.

(c) X and O each control two districts.

(d) X and O each control two districts solidly.

Figure 8.31 **Alternative districting strategies.** Xs and Os might represent Republicans and Democrats, urban and rural voters, blacks and whites, or any other distinctive group.

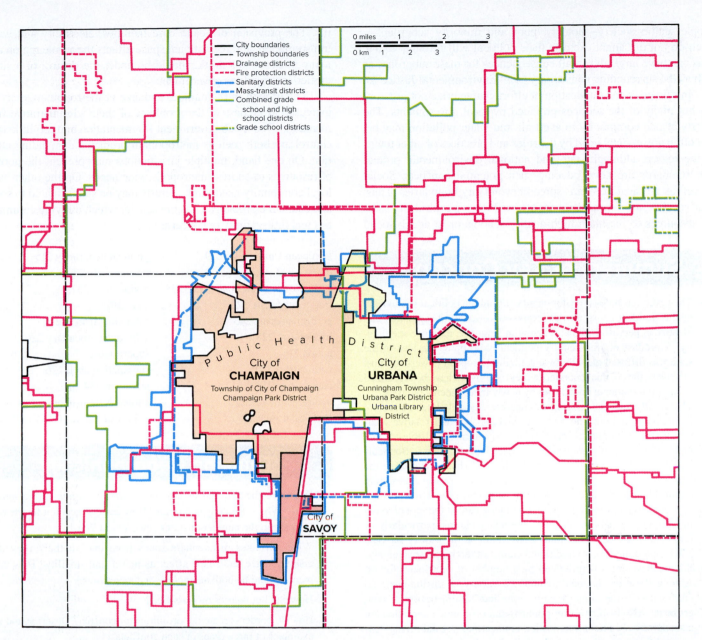

Figure 8.32   Political fragmentation in Champaign County, Illinois. The map shows a few of the independent administrative agencies with separate jurisdictions, responsibilities, and taxing powers in a portion of a single Illinois county. Among the other such agencies forming the fragmented political landscape are Champaign County itself, a forest preserve district, a public health district, a mental health district, the county housing authority, and a community college district.

precincts for voting purposes but also into special districts for such functions as fire and police protection, water and electricity supply, education, recreation, and sanitation. These districts almost never coincide with one another, and the larger the urban area, the greater the proliferation of small, special-purpose governing and taxing units. Although no Canadian community has quite the multiplicity of government entities plaguing many U.S. urban areas, major Canadian cities may find themselves with complex and growing systems of a similar nature. Even before its major expansion on January 1, 1998, for example, metropolitan Toronto had more than 100 authorities that could be classified as "local governments."

The existence of such a great number of districts in metropolitan areas may cause inefficiency in public services and hinder the

orderly use of space. *Zoning ordinances,* for example, are determined by each municipality. They are intended to allow the citizens to decide how land is to be used and, thus, are a clear example of the effect of political decisions on the division and development of space. Zoning policies dictate the areas where light and heavy industries may be located, the sites of parks and other recreational areas, the location of business districts, and the types and location of housing. Unfortunately, in large urban areas, the efforts of one community may be hindered by the practices of neighboring communities. Thus, land zoned for an industrial park in one city may abut land zoned for single-family residences in an adjoining municipality. Each community pursues its own interests, which may not coincide with those of its neighbors or of the larger region. In addition, some people within society—notably, poor and minority subgroups—often do not or cannot exercise their political will through voting. This can have large and small consequences for these subgroups as well as the surrounding community (see "Environmental Justice").

Inefficiency and duplication of effort characterize not just zoning but many of the services provided by local governments. The efforts of one community to avert air and water pollution may be, and often are, counteracted by the rules and practices of other towns in the region, although state and national environmental protection standards are now reducing such potential conflicts. Social as well as physical problems spread beyond city boundaries. Thus, nearby suburban communities are affected when a central city lacks the resources to maintain high-quality schools or to attack social ills. The provision of health care facilities, electricity and water, transportation, and recreational space affects the whole region and, many professionals think, should be under the control of a single unified metropolitan government.

The growth in the number and size of metropolitan areas has increased awareness of the problems of their administrative fragmentation. Too much government fragmentation and too little local control are both seen as metropolitan problems demanding attention. On one hand, multiple jurisdictions may prevent the pooling of resources to address metropolis-wide needs. On the other hand, local community needs and interests may be subordinated to social and economic problems of a core city for which outlying communities feel little affinity or concern.

Summary of Key Concepts

- The sovereign state is the dominant entity in the political subdivision of the world. It constitutes an expression of cultural separation and identity as pervasive as that inherent in language, religion, or ethnicity. A product of 18th-century political philosophy, the idea of the state was diffused globally by colonizing European powers. In most instances, the colonial boundaries they established have been retained as their international boundaries by newly independent countries.
- The greatly varying geographic characteristics of states contribute to national strength and stability. Size, shape, and relative location influence countries' economies and international roles, while national cores and capitals are the heartlands of states. Boundaries, the legal definition of a state's size and shape, determine the limits of its sovereignty. They may or may not reflect preexisting cultural landscapes and, in any given case, may or may not prove to be viable. Whatever their nature, boundaries are at the root of many international conflicts. Maritime boundary claims, particularly as reflected in the United Nations Convention on the Law of the Sea, add a new dimension to traditional claims of territorial sovereignty.
- State cohesiveness is promoted by a number of centripetal forces. Among these are national symbols; a variety of institutions; and confidence in the aims, organization, and administration of government. Also helping foster political and economic integration are transportation and communication connections. Destabilizing centrifugal forces, particularly ethnically based separatist movements, threaten the cohesion and stability of many states.
- Although the state remains central to the partitioning of the world, a broadening array of political entities affects people individually and collectively. Recent decades have seen a significant increase in supranationalism, in the form and variety of global and regional alliances, to which states have surrendered some sovereign powers. At the other end of the spectrum, expanding urban areas and government responsibilities raise questions of fairness in districting procedures and of effectiveness when political power is fragmented.

Key Words

antecedent boundary 225	core area 223
artificial (geometric) boundary 225	devolution 235
centrifugal force 230	electoral geography 243
centripetal force 230	elongated state 222
compact state 222	enclave 222
consequent boundary 226	ethnic cleansing 236

European Union (EU) 239	political geography 215
exclave 222	prorupt state 222
exclusive economic zone (EEZ) 238	regionalism 232
fragmented state 222	state 216
gerrymandering 243	subnationalism 232
irredentism 228	subsequent boundary 226
nation 216	superimposed boundary 226
nationalism 230	supranationalism 236
nation-state 216	terrorism 234
natural boundary 225	United Nations Convention on the
perforated state 222	Law of the Sea (UNCLOS) 238

Thinking Geographically

1. What are the differences among a *state*, a *nation*, and a *nation-state?* Why is a colony not a state? How can one account for the rapid increase in the number of states since World War II?

2. What attributes differentiate states from one another? How do a country's size and shape affect its power and stability? How can a piece of land be both an *enclave* and an *exclave?*

3. How can boundaries be classified?

4. How do borders create opportunities for conflict? Describe and give examples of three types of such conflicts.

5. Distinguish between *centripetal* and *centrifugal* political forces. Why is nationalism both a centripetal and a centrifugal force? What are some of the ways national cohesion and identity are achieved?

6. What characteristics are common to all or most separatist movements? Where are some of these movements active? Why do they tend to be located on the periphery rather than at the national core?

7. What types of international organizations and alliances can you name? What were the purposes of their establishment? What generalizations can you make regarding economic alliances?

8. How does the *United Nations Convention on the Law of the Sea* define zones of diminishing national control? What are the consequences of the concept of the 200-nautical-mile *exclusive economic zone?*

9. Why does it matter how boundaries are drawn around electoral districts? Theoretically, is it always possible to delimit boundaries "fairly"? Support your answer.

10. What reasons can you suggest for the great political fragmentation of the United States? What problems stem from such fragmentation?

CHAPTER 9

Economic Geography: Agriculture and Primary Activities

Wheat accounts for more than 20% of total calories consumed by humans. Wheat is a global commodity often produced on large farms. Contract harvesters follow the ripening wheat crop northward through the plains of the United States and Canada. © *Corbis RF.*

CHAPTER OUTLINE

The crop bloomed luxuriantly that summer of 1846. The disaster of the preceding year seemed over, and the potato, the sole sustenance of some 8 million Irish peasants, would again yield the bounty needed. However, within a week, wrote Father Mathew, "I beheld one wide waste of putrefying vegetation. The wretched people were seated on the fences of their decaying gardens . . . bewailing bitterly the destruction that had left them foodless." Colonel Gore found that "every field was black," and an estate steward noted that "the fields . . . look as if fire has passed over them." The potato was irretrievably gone for a second year; famine and pestilence were inevitable.

Within 5 years, the settlement geography of the most densely populated country in Europe was forever altered. The United States received a million immigrants, who provided the cheap labor needed for the canals, railroads, and mines that it was creating in its rush to economic development. New patterns of commodity flows were initiated as American maize for the first time found an Anglo-Irish market—as part of Poor Relief—and then entered a wider European market that had also suffered general crop failure in that bitter year. Within days, a microscopic organism, the cause of the potato blight, had altered the economic and human geography of two continents.

That alteration resulted from a complex set of intertwined causes and effects that demonstrates once again our repeated observation that apparently separate physical and cultural geographic patterns are really interconnected parts of a single reality. Central among those patterns are the ones the economic geographer isolates for special study. In Chapters 9 and 10, our attention is directed to the location of economic activities as we seek to answer the question of why they are distributed as they are.

Simply stated, **economic geography** is the study of how people earn their living, how livelihood systems vary from place to place, and how economic activities are spatially interrelated and linked. Economic geographers seek to understand what factors make some regions extremely productive and others less so, or some enterprises successful and others not. Of course, we cannot really comprehend the totality of the economic pursuits of approximately 7 billion human beings. We cannot examine the infinite variety of production and service activities found everywhere on the Earth's surface; nor can we trace all their innumerable interrelationships, linkages, and flows. Even if that level of understanding were possible, it would be valid for only a fleeting instant of time, for economic activities are constantly undergoing change.

Economic geographers seek consistencies. They attempt to develop generalizations that will help us comprehend the maze of economic variations characterizing human existence. From their studies emerges a deeper awareness of the dynamic, interlocking diversity of human enterprise, of the impact of economic activity on all other facets of human life and culture, and of the increasing interdependence of differing national and regional economic systems. The potato blight, although it struck only one small island, ultimately affected the economies of continents. In like fashion, the geographic distribution of commodities such as oil influences the relative wealth of countries, employment patterns, flows of international trade, political alliances, wars, and more (**Figure 9.1**).

9.1 The Classification of Economic Activity and Economies

Understanding livelihood patterns is made more difficult by five major environmental and cultural realities controlling the economic activities of humans. First, many production patterns are rooted in the spatially variable circumstances of the *physical environment*. The staple crops of the humid tropics, for example, are not part of the agricultural systems of the midlatitudes; livestock types that thrive in American feedlots or on western ranges are not adapted to the Arctic tundra or the margins of the Sahara Desert. The unequal distribution of useful mineral deposits gives to some regions and countries the economic prospects and employment opportunities denied to others. Forestry and fishing depend on still other natural resources unequal in occurrence, type, and value.

Second, within the bounds of the environmentally possible, economic or production decisions may be conditioned by *cultural considerations*. For example, culturally based food preferences rather than environmental limitations may dictate the choice of crops or livestock. Maize is a preferred grain in Africa and the Americas; wheat in North America, Australia, Argentina, southern Europe, and Ukraine; and rice in much of Asia. Pigs are not produced in Muslim areas. Third, the level of *technological development* of a culture will affect its recognition of resources or its ability to exploit them. Preindustrial societies do not know of, or need, iron ore or coal deposits that may underlie their hunting, gathering, or gardening grounds. Fourth, *political decisions* may encourage or discourage—through subsidies, protective tariffs, or production restrictions—patterns of economic activity. Fifth, ultimately production is controlled by *economic factors* of demand, whether that demand is expressed through a free market mechanism, through government intervention, or through the consumption requirements of a single family producing for its own needs.

Categories of Activity

One approach to categorize the world's productive work is to view economic activity as ranged along a continuum of both increasing complexity of product or service and increasing distance from the natural environment. Seen from that perspective, three distinct stages of economic activities may be distinguished: primary, secondary, and tertiary (**Figure 9.2**).

Figure 9.1 This oil tanker is part of a world of increasing **economic interdependence**. Oil is a global commodity with its price set by global markets. © *Malcolm Fife/Getty Images RF.*

3. TERTIARY (SERVICE) ACTIVITIES
- Wholesale and Retail Trade
- Transportation and Communication Services
- Business Services
  - Finance, Insurance, Real Estate
  - Accounting, Advertising, Architecture, Engineering, Legal Services
- Consumer Services
  - Eating and Drinking Establishments, Personal Services, Tourism
- Education, Fire Protection, Health Care, Nonprofit Organizations, Police

2. SECONDARY ACTIVITIES
- Manufacturing
- Processing
- Construction
- Power Production

1. PRIMARY ACTIVITIES
- Agriculture
- Gathering Industries
- Extractive Industries

Figure 9.2 **The categories of economic activity.** The three main sectors of the economy do not stand alone. For example, the primary sector produces the raw materials that are converted into finished products by the secondary sector, which are then distributed and marketed by the tertiary sector.

Primary activities are those that harvest or extract something from the earth. They are at the beginning of the production cycle, where humans are in closest contact with the resources and potentialities of the environment. Such primary activities involve basic food and raw material production. Hunting and gathering, grazing, agriculture, fishing, forestry, and mining and quarrying are examples (**Figure 9.3a**).

Secondary activities are those that add value to materials by changing their form or combining them into more useful, and therefore more valuable, commodities. That processing of raw materials into finished products ranges from simple handicraft production of pottery or woodenware to the delicate assembly of electronic goods or space vehicles (**Figure 9.3b**). Copper smelting, steelmaking, metalworking, automobile production, the textile and chemical industries—indeed, the full array of *manufacturing* and *processing industries*—are included in this phase of the production process. Also included are the production of *energy* (the "power company") and the *construction* industry.

Tertiary activities consist of those business and labor specializations that provide *services* to the primary and secondary sectors and *goods* and *services* to the general community and to the individual. The service sector includes wholesale and retail trade, which constitutes the vital link between producers and consumers. Business services include accounting, advertising, financial services, insurance, legal services, and real estate. Consumers may use some of these same services, although often from different providers. Other examples of consumer services include restaurants and pubs, repair and maintenance providers, and personal service establishments such as barbers and hair salons. Education, health care, transportation, and communication services are also included in the tertiary category.

These categories of production and service activities help us see an underlying structure to the nearly infinite variety of things people do to earn a living and to sustain themselves. But they tell us little about the organization of the larger economy of which the individual worker or enterprise is a part. For that wider organizational

(a)

(b)

Figure 9.3 (a) Harvesting these trees is a **primary activity**. (b) Processing them into paper products in this paper mill is a **secondary activity** that increases their value by altering their form. The products of many secondary industries—sheet steel from steel mills, for example—constitute "raw materials" for other manufacturers. *(a) © Glowimages RF (b) © Corbis RF.*

understanding of world and regional economies, we look to *systems* rather than *components* of economies.

Types of Economic Systems

Broadly viewed, national economies in the early 21st century fall into one of three major types of system: *subsistence, commercial,* or *planned.* None of these economic systems is or has been "pure." That is, none exists in isolation in an increasingly interdependent world. Each, however, displays certain underlying characteristics based on its distinctive forms of resource management and economic control.

In a **subsistence economy,** goods and services are created for the use of the producers and their kinship groups. Therefore, there is little exchange of goods and only limited need for markets. In the **market (commercial) economies** that have become dominant in nearly all parts of the world, producers or their agents, in theory, freely market their goods and services, and market competition is the primary force shaping production decisions and distributions. The laws of supply and demand determine the equilibrium market price and quantity (**Figure 9.4**). By contrast, in **planned economies** associated with communist-controlled societies, producers or their

agents dispose of goods and services through governmental agencies that control both supply and price. The quantities produced and the locational patterns of production are tightly programmed by central planning departments.

With a few exceptions—such as Cuba and North Korea—rigidly planned economies no longer exist in their classical form; they have been modified or dismantled in favor of free market structures or are only partially retained in the lesser degree of economic control associated with governmental supervision or ownership of selected sectors of increasingly market-oriented economies. Countries such as China, Russia, and the formerly communist countries of Central and Eastern Europe are now classified as **transition economies,** making the change from a centrally planned economy to a market-based economy. Nevertheless, the landscape evidence of centrally planned economies lives on. Their physical structures, patterns of production, and imposed regional interdependencies continue to influence the economic decisions of successor societies.

In actuality, few people are members of only one of these systems, although one may be dominant. A farmer in India may produce rice and vegetables primarily for the family's consumption but also save some of the produce to sell. In addition, members of the family may market cloth or other handicrafts they make. With the money derived

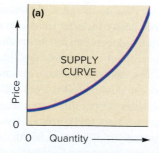

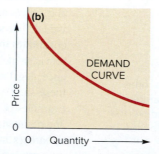

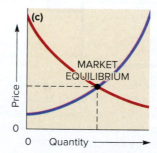

Figure 9.4 **Supply, demand, and market equilibrium.** The regulating mechanism of the market may be visualized graphically. **(a)** The *supply curve* tells us that, as the price of a good increases, more of that good will be made available for sale. Countering any tendency for prices to rise to infinity is the market reality that the higher the price, the smaller the demand as potential customers find other purchases or products more cost-effective. **(b)** The *demand curve* shows how the market will expand as prices drop and the good becomes more affordable and attractive to more customers. **(c)** *Market equilibrium* is marked by the point of intersection of the supply and demand curves and determines the price of the good and the quantity bought and sold.

from those sales, the Indian peasant is able to buy, among other things, clothes for the family, tools, or fuel. Thus, that Indian farmer is a member of at least two systems: subsistence and commercial.

In the United States, government subsidies or production controls for various types of goods and services (such as producing ethanol as a gasoline additive, growing wheat or sugarcane, constructing and operating nuclear power plants, or engaging in licensed personal and professional services) mean that the country does not have a purely market economy. To a limited extent, its citizens participate in a controlled and planned as well as in a free market environment. Many African, Asian, and Latin American market economies have been decisively shaped by governmental policies encouraging or demanding the production of export commodities rather than food for domestic consumers or promoting through import restrictions the development of domestic industries not readily supported by the national market alone. Example after example would show that there are very few people in the world who are members of only one type of economic system.

Inevitably, spatial patterns, including those of economic systems and activities, are subject to change. For example, the commercial economies of Western European countries, some with sizable infusions of planned economy controls, are being restructured by both increased free market competition and supranational regulation under the World Trade Organization and the European Union (see pp. 238–241). Many of the countries of Latin America, Africa, Asia, and the Middle East that traditionally were dominated by subsistence economies are now benefiting from technology transfer from advanced economies and integration into expanding global production and exchange patterns. The phenomenal growth of the Chinese economy is rewriting the map of economic activity and shifting the balance of economic power toward East Asia. Economic globalization increases linkages between distant regions and spreads wealth more widely, while also undermining the stability of established production locations. In short, the creative destruction of capitalism produces a constantly changing economic landscape.

Stages of Development

Despite such changes and global convergences, disparities between regions and countries in observable economic and social conditions obviously exist and, since the middle of the 20th century, have been the subject of theories and measurements of development and underdevelopment. We noted in Chapter 6 (pp. 144–148) that development implies the full productive use of a country's natural and human resources, and we traced the emergence of such comparative labels as *developed, less-developed, developing, newly industrializing,* and similar descriptors. We observed contrasts in the percentage of workers engaged in agriculture, industry, and services (see Figure 6.5) and in the availability and use of technologies such as the Internet (see Figure 6.6). A consequence of differences in the stage of development is apparent in the global map of per capita gross national incomes, or GNI (which measures the total domestic and foreign value added of all goods and services claimed by residents of a country during a year). Per capita incomes are dramatically higher in the developed states, even when corrected for price differences (**Figure 9.5**).

Thus, the United Nations recognizes a global contrast between an economically advanced industrialized "North" with relatively high per capita incomes and a "South" with little or no industrialization and low income levels (see Figure 6.7). In that contrast, a key indicator was the degree of an economy's industrialization and progression beyond a largely subsistence livelihood system.

Like any development statistic, GNI tells only part of a complex story. GNI measures miss the activities in the **informal economy** that are particularly important in developing countries. The informal economy is composed of activities that are unlicensed, lack formal contracts, and generate unreported earnings. Examples include raising one's own food, bartering, home sewing businesses, waste picking, sex work, shoeshining, and some forms of street vending (**Figure 9.6**). The informal economy is a growing proportion of the workforce and is vital to the livelihoods of a significant percentage of workers in Africa, Asia, Latin America, and the Caribbean (**Table 9.1**). While the informal economy is vital to the livelihoods of many households, for governments it represents lost tax revenues that could have funded schools, road improvements, and other public services.

In the 1960s, a dominant theory described normal development as a progression from the limitations of a traditional society of subsistence agriculture, low technology levels, and poorly developed commercial exchanges through "takeoff" stages of increasing investment in infrastructure and human capital, application of modern technology to resource exploitation, and a gradually expanding and maturing industrial base. Eventually, the developing state would achieve an "age of mass production and consumption" and, ultimately, the postindustrial status of the most-advanced, Western economies.

That theoretically expected progression proved illusory; many less- and least-developed countries remained locked in the pretakeoff stage, despite infusions of loans, investments, and technology transfers from the more-developed states. The 1960s, 1970s, and 1980s—all proclaimed by the United Nations as "Development Decades"—proved instead to be decades of disappointment, at least by economic measures. In the pursuit of development, many poor countries borrowed heavily during the 1960s and 1970s. Money was spent on hydroelectic dams, power plants, ports, and other large government-directed development projects. Unfortunately, many projects did not generate sufficient revenues to repay the loans, sparking a debt crisis in many developing countries. Debt burdens in many of the world's poorest countries exacerbated the already existing severe poverty and became so unmanageable that the World Bank and the International Monetary Fund created a debt relief program for the *heavily indebted poor countries*. Despite many challenges and missteps, substantial progress has been made in closing the gap between the Global South and the developed world.

A conspicuous difference between the economies of the developed and less-developed countries is the allocation of the labor force across the primary, secondary, and tertiary sectors and the relative contribution of agriculture, industry, and services to their gross domestic product (GDP). Degree of development, it is held, can be traced through the reallocation of labor resources from the basic primary (largely agricultural) activities to a rising share of workers engaged in secondary and then tertiary endeavors.

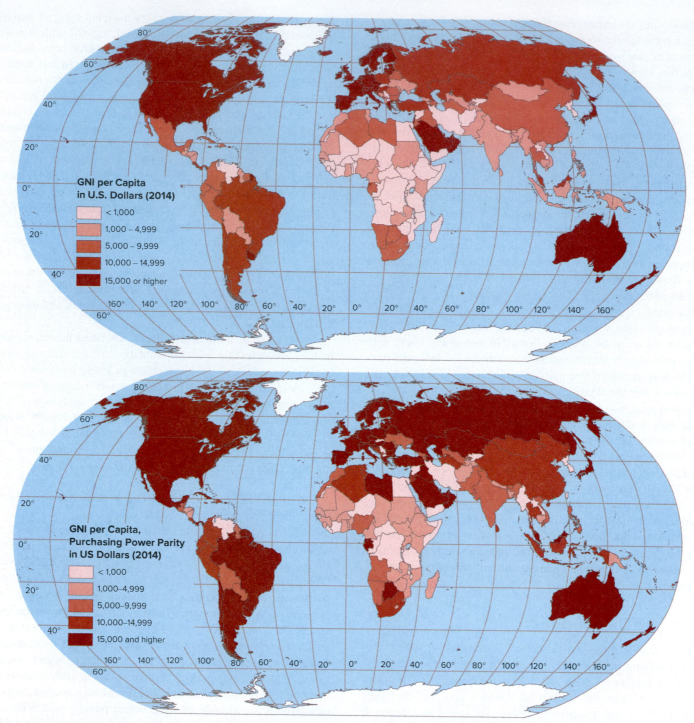

Figure 9.5   Two Contrasting Views of Income. (a) Gross national income per capita, 2014. GNI per capita is expressed in U.S. dollars at official exchange rates. This measure shows great contrasts between more- and less-advanced economies. **(b)** Purchasing power parity, 2015, takes account of variations in prices between countries. For example, travelers often notice that prices for the same goods and services are much higher in high-income countries than in low-income countries. Rather than use official exchange rates, purchasing power parity is calculated using a conversion factor based on the prices in the local currency for a set of identical goods. Neither measure accounts for the distribution of incomes within a country. *Source: Data from World Bank, 2016.*

The U.S. economy is among the most advanced in the world and has clearly gone through a transition from largely agricultural to industrial and now postindustrial (**Figure 9.7**).

The economy of the Central African Republic is typical of the world's least-developed countries with more than half of its labor force in agriculture. That stands in obvious contrast to highly developed Australia, with only 3% of its labor force involved in

agriculture and nearly 70% in services. In a similar way, using the relative contributions of the different categories of activity to GDP reveals meaningful contrasts between levels of national development. A sizable contribution to GDP from agriculture, for example, coupled with only a small input from services suggests a dependence on subsistence agriculture and a paucity or only low level of consumer services and business services—unmistakable

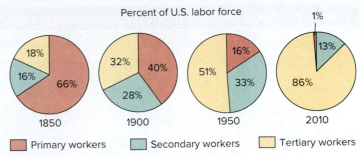

Percent of U.S. labor force

1850 · 1900 · 1950 · 2010

■ Primary workers   ■ Secondary workers   □ Tertiary workers

Figure 9.7   The changing sectoral allocation of labor in the United States demonstrates the industrialization of a rural, agricultural economy, followed by the transition to a postindustrial economy dominated by business services and consumer services.

services, marking their highly integrated knowledge-based exchange economies.

No matter what economic system locally prevails, in all instances transportation is a key variable. No advanced economy can flourish without a well-connected transport network. All subsistence societies—and subsistence areas of developing countries—are characterized by their isolation from regional and world routeways (**Figure 9.8**), and that isolation restricts their progression to more-advanced forms of economic structure.

Once-sharp contrasts in economic organization are becoming blurred, and countries' economic orientations are changing as globalization reduces structural contrasts in national economies. Still, both approaches to economic classification—by type of activity and by organization of economies—help us visualize and understand world economic geographic patterns. In this chapter, we begin with the primary sector, emphasizing the technologies, spatial patterns, and organizational systems for cultivating, raising, harvesting, and extracting resources from the earth.

9.2 Primary Activities: Agriculture

Humanity's basic economic concern is producing or securing food resources sufficient to meet individual daily energy requirements and balanced to satisfy nutritional needs. Those supplies may be acquired by the consumer directly through the primary economic activities of hunting, gathering, farming, and fishing (a form of "gathering") or indirectly through the performance of other primary, secondary, or tertiary economic activities that yield income sufficient to purchase needed daily sustenance.

Since the middle of the 20th century, a recurring but unrealized fear has been that the world's steadily increasing population would exceed available or potential food supplies. Instead, although global population has more than doubled since 1950, the total amount of human food produced worldwide since then has also more than doubled. The Food and Agriculture Organization (FAO) of the United Nations has set the minimum daily requirement for caloric intake per person at 2350. By that measure, annual food supplies are more than sufficient to meet world needs. That is, if total food resources were evenly distributed, everyone would have access to amounts sufficient for adequate daily nourishment. In reality, however, the number of undernourished people worldwide is unconscionable (**Figure 9.9**). Rising food prices, often due to increasing energy prices, combine with poverty as the main cause for

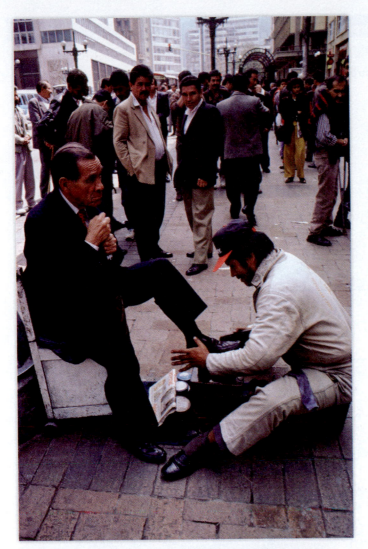

Figure 9.6   The informal economy includes this shoeshine stand in Bogota, Colombia. As commercialization of agriculture reduced rural employment, many workers migrated to cities and found work in the informal economy, the source of 40% of jobs in Latin American cities. They work as errand runners, street vendors, and open-air barbers, as well as unregistered workers in small construction and repair businesses. © *Flat Earth Images RF*.

Table 9.1   Size of the Informal Economy

	Africa	Latin America and the Caribbean	Asia
Nonagricultural Employment	78%	57%	45–85%
Urban Employment	61%	40%	40–60%
New Job Creation	93%	83%	NA

Source: J. Charmes, Informal Sector, Poverty and Gender: A Review of the Evidence, World Bank, 2000; and International Labour Organization, Women and Men in the Informal Economy: A Statistical Portrait, 2002.

indicators of economic underdevelopment (**Table 9.2**). Highly advanced, postindustrial societies, on the other hand, have relatively small inputs from primary production (including agriculture) and a dominating GDP input from the consumer and business

Table 9.2   Stage of Economic Development and the Structure of Output

Country Category	Value added as a Percentage of Gross Domestic Product, 2015 [a]			
	Agriculture	Mining, Construction, and Utilities	Manufacturing	Services
Least-Developed				
Central African Republic	58	6	6	30
Newly Industrialized				
Thailand	11	8	28	53
Industrial				
South Korea	2	9	30	59
Postindustrial				
United States	1	9	12	78

Source: The World Bank, *World Development Indicators,* 2016.

[a]Some values are from earlier years where 2010 was missing.

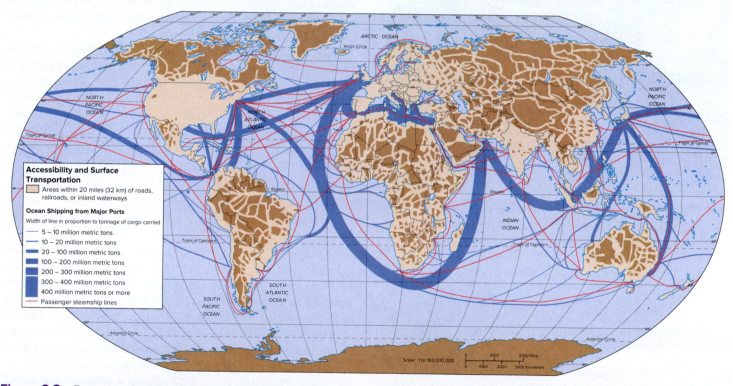

Figure 9.8   **Patterns of surface transportation and accessibility.** Accessibility is a key measure of economic development and of the degree to which a world region can participate in interconnected market activities. Isolated areas of countries with advanced economies suffer a price disadvantage because of high transportation costs. Lack of accessibility in subsistence economic areas slows their modernization and hinders their participation in the world market.

undernourished people. One of the Millennium Development Goal targets was to reduce undernourishment by one half. During the monitoring period of 1990 to 2015, the number of undernourished people dropped from 990 million to 800 million. The least progress occurred in countries suffering disasters or political instability.

This stark contradiction between sufficient worldwide food supplies and widespread malnutrition reflects inequalities in income; population growth rates; lack of access to fertile soils, credit, and education; local climatic conditions or catastrophes; and lack of transportation and storage facilities, among other reasons. By mid-century, the increasingly interconnected world population will expand to nearly 10 billion, and concerns with individual states' food supplies will inevitably continue and remain a persistent international issue.

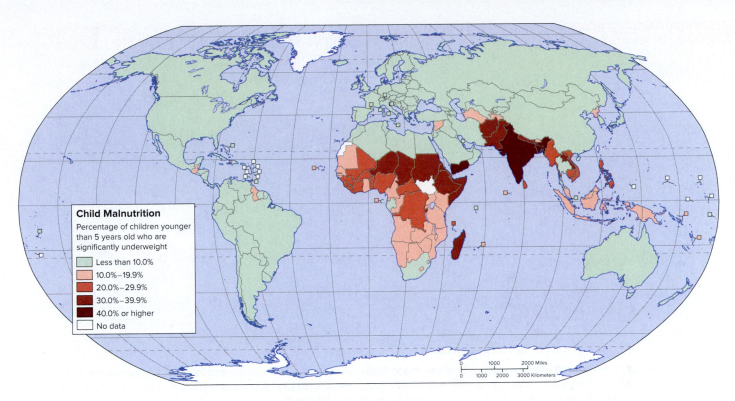

Figure 9.9   Percentage of children under 5 who were undernourished, 2009. In 2010, according to the Food and Agriculture Organization of the United Nations (FAO), there were about 925 million undernourished people worldwide. In contast, FAO data indicate that all industrialized countries have an average per capita caloric intake well above physiological requirements, and many face an obesity epidemic. *Source: World Bank, World Development Indicators, 2010.*

Before there was farming, *hunting* and *gathering* were the universal forms of primary production. These preagricultural pursuits are now practiced by at most a few thousand persons worldwide, primarily in isolated and remote pockets within the low latitudes and among the sparse populations of very high latitudes. The interior of New Guinea, rugged areas of interior Southeast Asia, diminishing segments of the Amazon rain forest, a few districts of tropical Africa and northern Australia, and parts of the Arctic regions still contain such preagricultural people. Their numbers are few and declining, and wherever they are brought into contact with more-advanced cultures, their way of life is eroded or lost.

Agriculture, defined as the growing of crops and the tending of livestock, has replaced hunting and gathering as the most significant of the primary activities. It is spatially the most widespread, found in all world regions where environmental circumstances—including adequate moisture, good growing season length, and productive soils—permit (**Figure 9.10**). Crop farming alone covers some 15 million square kilometers (5.8 million sq mi) worldwide, about 10% of the Earth's total land area. In many developing economies, at least half of the labor force is directly involved in farming and herding. Overall, however, employment in agriculture is steadily declining in developing countries, echoing but trailing the trend in highly developed economies, where direct employment in agriculture involves only a small fraction of the labor force (see Figure 6.5).

It has been customary to classify agricultural societies based on the importance of off-farm sales and the level of mechanization

and technological advancement. *Subsistence, traditional* (or *intermediate*), and *advanced* (or *modern*) are the terms usually employed to recognize both aspects. These are not mutually exclusive but, rather, are stages along a continuum of farm economies. At one end lies production solely for family sustenance, using simple tools and native plants. At the other end is the specialized, highly capitalized, industrialized agriculture for off-farm delivery that marks advanced economies. Between these extremes is the middle ground of traditional agriculture, where farm production is in partly for home consumption and in part oriented toward off-farm sale, either locally or in national and international markets. Complicating this contrast between subsistence and advanced agriculture is the recent trend towards gardening for home consumption and shopping at local farmers' markets in advanced countries. We can most clearly see the variety of agricultural activities and the diversity of controls on their spatial patterns by examining the subsistence and advanced ends of the agricultural continuum.

Subsistence Agriculture

A *subsistence* economic system involves nearly total self-sufficiency on the part of its members. Production for exchange is minimal and any exchange is noncommercial. Each family or close-knit social group relies on itself for its food and other most-essential requirements. Farming for the immediate needs of the family is, even today, the predominant occupation of humankind. In most of Africa, much of Latin America, and most of southern and eastern

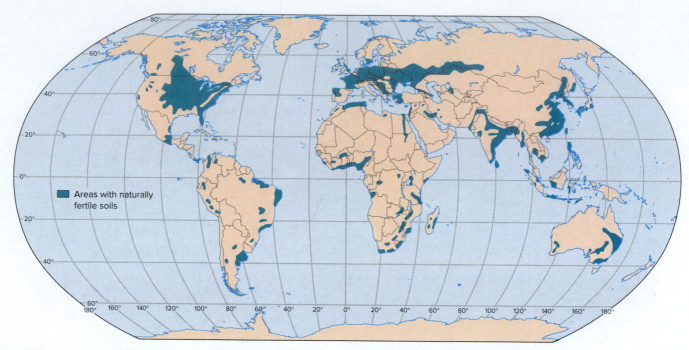

Figure 9.10 **Areas with naturally fertile soils** account for much of the world's grain production and include the corn and wheat belts of North America, Ukraine, and southern Siberia and the rice-producing regions of India and Southeast Asia. A comparison of this map with Figure 9.9 will help explain the prevalence of undernourishment in sub-Saharan Africa.

Asia, the majority of the people are primarily concerned with feeding themselves from their own land and livestock.

Two chief types of subsistence agriculture can be recognized: *extensive* and *intensive*. Although each type has several variants, the essential contrast between them is yield per unit of land used and, therefore, population-supporting potential. **Extensive subsistence agriculture** involves large areas of land and minimal labor input per hectare. Both product per land unit and population densities are low. **Intensive subsistence agriculture** involves the cultivation of small land holdings through the expenditure of great amounts of labor per acre. Yields per unit area and population densities are both high (**Figure 9.11**).

Extensive Subsistence Agriculture

Of the several types of *extensive subsistence* agriculture—varying one from another in their intensities of land use—two are of particular interest: nomadic herding and shifting cultivation.

Nomadic herding, the wandering but controlled movement of livestock solely dependent on natural forage, is the most extensive type of land use system (Figure 9.11). That is, it requires the greatest amount of land area per person sustained. Over large portions of the Asian semidesert and desert areas, in certain highland areas, and on the fringes of and within the Sahara, a relatively small number of people graze animals for consumption by the herder group, not for market sale. Sheep, goats, and camels are most common, while cattle, horses, and yaks are locally important. The reindeer of Lapland were formerly part of the same system.

Whatever the animals involved, their common characteristics are hardiness, mobility, and an ability to subsist on sparse forage. The animals provide a variety of products: milk, cheese, yogurt, and meat for food; hair, wool, and skins for clothing; skins for shelter; and excrement for fuel. For the herder, they represent primary subsistence. Nomadic movement is tied to sparse and seasonal rainfall or to cold-temperature regimes and to the areally varying appearance and exhaustion of forage. Extended stays in a given location are neither desirable nor possible. *Transhumance* is a special form of seasonal movement of livestock to exploit specific locally varying pasture conditions. Used by permanently or seasonally sedentary pastoralists and pastoral farmers, transhumance involves either the regular vertical movement from mountain to valley pastures between summer and winter months or horizontal movement between established lowland grazing areas to reach pastures temporarily lush from monsoonal (seasonal) rains.

As a type of economic system, nomadic herding is declining. Many economic, social, and cultural changes are causing nomadic groups to alter their way of life or to disappear entirely. On the Arctic fringe of Russia, herders under communism were made members of state or collective herding enterprises. In northern Scandinavia, Lapps (Saami) are engaged in commercial more than in subsistence livestock farming. In the Sahel region of Africa on the margins of the Sahara, oases once controlled by herders have been taken over by farmers, and the great droughts of recent decades have forever altered the formerly nomadic way of life of thousands.

A much differently based and distributed form of extensive subsistence agriculture is found in all of the warm, moist, low-latitude areas of the world. There, many people engage in a kind of nomadic farming. In these warm, wet tropical climates, once a field is put into production, dead plant matter decomposes rapidly and soil nutrients are quickly leached—washed away by rain or groundwater. Thus, soil fertility declines rapidly and, after harvesting several crops, the farmer is forced to move on. In a sense, they

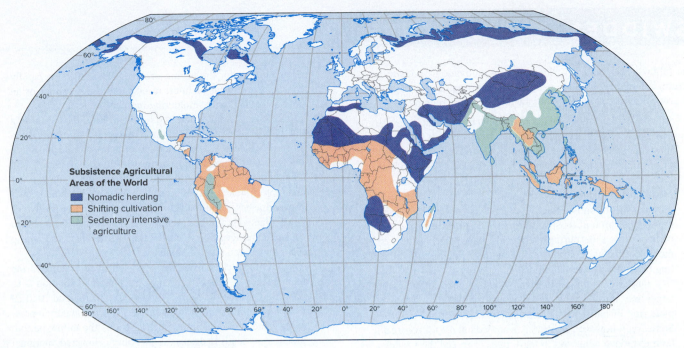

Figure 9.11   Subsistence agricultural areas of the world. Nomadic herding, supporting relatively few people, was the age-old way of life in large parts of the dry and cold world. Shifting cultivation, or swidden agriculture, maintains soil fertility by tested traditional practices in tropical wet and wet-and-dry climates. Large parts of Asia support millions of people engaged in sedentary intensive cultivation, with rice and wheat the chief crops.

rotate fields rather than crops to maintain productivity. This type of **shifting cultivation** has a number of names, the most common of which are *swidden* (an English localism for "burned clearing") and *slash-and-burn*. Each region of its practice has its own name—for example, *milpa* in Central and South America, *chitemene* in Africa, and *ladang* in Southeast Asia.

Characteristically, the farmers hack down the natural vegetation and burn the cuttings, which puts nutrients into the soil. Then they plant such crops as maize (corn), millet (a cereal grain), rice, manioc or cassava, yams, and sugarcane (**Figure 9.12**). Increasingly included in many of the crop combinations are such high-value, labor-intensive commercial crops as coffee, providing the

(a)

(b)

Figure 9.12   Preparation of a swidden plot in Liberia, Africa. (a) First the vegetation is hacked down. **(b)** Then the field is planted. Stumps and trees left in the clearing will remain after the burn. © *Albert Swingle.*

SWIDDEN AGRICULTURE

The following account describes shifting cultivation (swidden agriculture) among the Hanunóo people of the Philippines. Nearly identical procedures are followed in all swidden farming regions.

When a garden site of about one-half hectare (a little over one acre) has been selected, the swidden farmer begins to remove unwanted vegetation. The first phase of this process consists of slashing and cutting the undergrowth and smaller trees with bush knives. The principal aim is to cover the entire site with highly inflammable dead vegetation so that the later stage of burning will be most effective. Because of the threat of soil erosion the ground must not be exposed directly to the elements at any time during the cutting stage. During the first months of the agricultural year, activities connected with cutting take priority over all others.

Once most of the undergrowth has been slashed, the larger trees must be felled or killed by girdling (cutting a complete ring of bark) so that unwanted shade will be removed. Some trees, however, are merely trimmed but not killed or cut, both to reduce labor and to leave trees to reseed the swidden during the subsequent fallow period.

The crucial and most important single event in the agricultural cycle is swidden burning. The main firing of a swidden is the culmination of many weeks of preparation in spreading and leveling chopped vegetation, preparing firebreaks to prevent flames escaping into the jungle, and allowing time for the drying process. An ideal burn rapidly consumes every bit of litter; in no more than an hour or an hour and a half, only smoldering remains are left.

Swidden farmers note the following as the benefits of a good burn: (1) removal of unwanted vegetation, resulting in a cleared swidden; (2) extermination of many animal and some weed pests; (3) preparation of the soil for dibble (any small hand tool or stick to make a hole) planting by making it softer and more friable; (4) provision of an evenly distributed cover of wood ashes, good for young crop plants and protective of newly planted grain seed. Within the first year of the swidden cycle, an average of between 40 and 50 different types of crop plants have been planted and harvested.

The most critical feature of swidden agriculture is the maintenance of soil fertility and structure. The solution is to pursue a system of rotation of 1 to 3 years in crop and 10 to 20 in woody or bush fallow regeneration. When population pressures mandate a reduction in the length of the fallow period, productivity tends to drop as soil fertility is lowered, marginal land is utilized, and environmental degradation occurs. The balance is delicate.

Source: Adapted from Hanunoo Agriculture, *by Harold C. Conklin (FAO Forestry Development Paper No. 12), 1957.*

---

cash income that is evidence of the growing integration of all peoples into exchange economies. Initial yields—the first and second crops—may be very high, but they quickly become lower with each successive planting on the same plot. As that occurs, cropping ceases, native vegetation is allowed to reclaim the clearing, and gardening shifts to another newly prepared site. The first clearing will ideally not be used again for crops until, after many years, natural fallowing replenishes its fertility (see "Swidden Agriculture").

Less than 3% of the world's people are still predominantly engaged in tropical shifting cultivation on about one-seventh of the world's land area (see Figure 9.11). Because the essential characteristic of the system is the intermittent cultivation of the land, each family requires a total area equivalent to the garden plot in current use plus all land left fallow for regeneration. Population densities are low because much land is needed to support few people. It may be argued that shifting cultivation is a highly efficient cultural adaptation where land is abundant in relation to population and levels of technology and capital availability are low. Fossil fuels, chemical pesticides, and synthetic fertilizers are not required. However, as population densities increase, the system becomes less viable.

Intensive Subsistence Agriculture

About 45% of the world's people are engaged in intensive subsistence agriculture, which predominates in the areas shown in Figure 9.11. As a descriptive term, *intensive subsistence* is no longer fully applicable to a changing way of life and economy in which the distinction between subsistence and commercial is decreasingly valid. Although families may still be fed primarily with the produce of their individual plots, the exchange of farm commodities within the system is considerable. Production of foodstuffs for sale in rapidly growing urban markets is increasingly vital for the rural economies of subsistence farming areas and for the sustenance of the growing proportion of national and regional populations no longer themselves engaged in farming.

Nevertheless, hundreds of millions of Indians, Chinese, Pakistanis, Bangladeshis, and Indonesians, plus further millions in other Asian, African, and Latin American countries, remain small-plot, mainly subsistence producers of rice, wheat, maize, millet, or pulses (peas, beans, and other legumes). Most live in monsoon Asia, where warm, moist districts are well suited to the production of rice, a crop that under ideal conditions can provide large amounts of food per unit of land. Rice also requires a great deal of time and attention, for planting rice shoots by hand in standing fresh water is a tedious art (**Figure 9.13**). In the cooler and drier portions of Asia north of 20°N, wheat is grown intensively, along with millet and, less commonly, upland rice.

Intensive subsistence farming is characterized by large inputs of labor per unit of land; by small plots; by the intensive use of fertilizers, mostly animal manure (see "The Economy of a Chinese Village"); and by the promise of high yields in good years. For food security and dietary custom, *polyculture*—production of several different crops, often in the same field—is practiced. Vegetables and

THE ECONOMY OF A CHINESE VILLAGE

The village of Nanching is in subtropical southern China on the Zhu River delta near Guangzhou (Canton). Its traditional subsistence agricultural regime was described by a field investigator, whose account is condensed here. The system is still followed in its essentials in other rice-oriented societies.

In this double-crop region, rice was planted in March and August and harvested in late June or July and again in November. March to November was the major farming season. Early in March the earth was turned with an iron-tipped wooden plow pulled by a water buffalo. The very poor who could not afford a buffalo used a large iron-tipped wooden hoe for the same purpose.

The plowed soil was raked smooth, fertilizer was applied, and water was let into the field, which was then ready for the transplanting of rice seedlings. Seedlings were raised in a seedbed, a tiny patch fenced off on the side or corner of the field. Beginning from the middle of March, the transplanting of seedlings took place. The whole family was on the scene. Each took the seedlings by the bunch, 10 to 15 plants, and pushed them into the soft inundated soil. For the first 30 or 40 days the emerald green crop demanded little attention except for keeping the water at a proper level. But after this period came the first weeding; the second weeding followed a month later. This was done by hand, and everyone old enough for such work participated. With the second weeding went the job of adding fertilizer. The grain was now allowed to stand to "draw starch" to fill the hull of the kernels. When the kernels had "drawn enough starch," water was let out of the field, and both the soil and the stalks were allowed to dry under the hot sun.

Then came the harvest, when all the rice plants were cut off a few inches above the ground with a sickle. Threshing was done on a threshing board. Then the grain and the stalks and leaves were taken home with a carrying pole on the peasant's shoulder. The plant was used as fuel at home.

As soon as the exhausting harvest work was done, no time could be lost before starting the chores of plowing, fertilizing, pumping water into the fields, and transplanting seedlings for the second crop. The slack season of the rice crop was taken up by chores required for the vegetables which demanded continuous attention, since every peasant family devoted a part of the farm to vegetable gardening. In the hot and damp period of late spring and summer, eggplant and several varieties of squash and beans were grown. The green-leafed vegetables thrived in the cooler and drier periods of fall, winter, and early spring. Leeks grew year-round.

When one crop of vegetables was harvested, the soil was turned and the clods were broken up by a digging hoe and leveled with an iron rake. Fertilizer was applied, and seeds or seedlings of a new crop were planted. Hand weeding was a constant job; watering with the long-handled wooden dipper had to be done an average of three times a day, and in the very hot season when evaporation was rapid, as frequently as six times a day. The soil had to be cultivated with the hoe frequently as the heavy tropical rains packed the earth continuously. Instead of the two applications of fertilizer common with the rice crop, fertilizing was much more frequent for vegetables. Besides the heavy fertilizing of the soil at the beginning of a crop, usually with city garbage, additional fertilizer, usually diluted urine or a mixture of diluted urine and excreta, was given every 10 days or so to most vegetables.

Adapted from C. K. Yang, A Chinese Village in Early Communist Transition. The MIT Press, Cambridge, Mass. Copyright © 1959 by the Massachusetts Institute of Technology.

Figure 9.13   Intensive subsistence rice farming. Transplanting rice seedlings requires arduous hand labor by all members of the family. The newly flooded diked fields, previously plowed and fertilized, will have their water level maintained until the grain is ripe. The scene in this photograph is repeated wherever subsistence wet-rice agriculture is practiced. © frans lemmens/ Alamy Stock Photo.

some livestock are part of the agricultural system, and fish may be reared in rice paddies and ponds. Food animals include swine, ducks, and chickens, but since Muslims eat no pork, hogs are absent in their areas of settlement. Hindus generally eat little meat, mainly goat and lamb but not pork or beef. The large numbers of cattle in India are vital for labor and are a source of milk and cheese, as well as producers—through excrement—of fertilizer and fuel.

Not all of the world's subsistence farming is based in rural areas. Urban agriculture is a rapidly growing activity, with city farmers providing, according to United Nations figures, a significant fraction of the world's total food production. Occurring in all regions of the world, developed and underdeveloped, but most prevalent in Asia, urban agricultural activities range from small garden plots to backyard livestock breeding to fish raised in ponds and streams. Using the garbage dumps of Jakarta, the rooftops of Mexico City, and meager dirt strips along roadways in Kolkata or Kinshasa, millions of people are feeding their own families and supplying local markets with vegetables, fruit, fish, and even meat—all produced within the cities themselves and all without the expense and spoilage of storage or long-distance transportation.

In Africa, a fifth of the urban nutritional requirement is produced in the towns and cities; two of three Kenyan and Tanzanian urban families engage in farming, for example, and in Accra, Ghana's capital, urban farming provides the city with most of its fresh vegetables. In all parts of the developing world, urban subsistence farming has reduced the incidence of adult and child malnutrition in rapidly expanding cities.

Expanding Crop Production

Continuing population pressures on existing resources are a constant spur for seeking ways to expand the food supply available both to the subsistence farmers of the developing economies and

to the wider world as a whole. Two paths to promoting increased food production are apparent: (1) expand the land area under cultivation and (2) increase crop yields from existing farmlands.

The first approach—increasing cropland area—is not a promising strategy. Approximately 70% of the world's land area is agriculturally unsuitable, being too cold, too dry, too steep, or totally infertile. Of the remaining 30%, most of the area well suited for farming is already under cultivation, and of that area, millions of hectares annually are being lost through soil erosion, salinization, desertification, and the conversion of farmland to urban, industrial, and transportation uses in all developed and developing countries. Only the rain forests of Africa and the Amazon Basin of South America retain sizable areas of potentially farmable land. The soils of those regions, however, are fragile, are low in nutrients, have poor water retention, and are easily eroded or destroyed following deforestation. By most accounts, then, world food output cannot reasonably be increased by simple expansion of cultivated areas.

Intensification: the Green Revolution and Beyond

Increased productivity of existing cropland rather than expansion of cultivated area has been the key to the growth of agricultural output over the past few decades. Since the 1970s, world total grain production has increased faster than world population. The vast majority of that production growth was due to increases in yields rather than expansions in cropland. For Asia as a whole, cereal yields grew by more than 40% between 1980 and the early 21st century, accounted for largely by increases in China and India; they increased by more than 35% in South America. Two interrelated approaches to those yield increases mark recent farming practices.

First, throughout much of the developing world, production inputs such as water, fertilizer, pesticides, and labor have been increased to expand yields on a relatively constant supply of cultivable land. Irrigated area, for example, has nearly doubled since 1960. Global use of pesticides and herbicides has dramatically increased since the 1950s. Traditional practices of leaving land fallow (uncultivated) to renew its fertility have been largely abandoned, and double and triple cropping of land where climate permits has increased in Asia and even in Africa, where marginal land is put to near-continuous use to meet growing food demands. This intensive use of agricultural land has been made possible by the heavy use of chemical fertilizers to maintain fertility.

Many of these intensification practices are part of the second approach, linked to the **Green Revolution**—a complex of seed and management improvements adapted to the needs of intensive agriculture and designed to bring larger harvests from a given area of farmland. Genetic improvements in rice and wheat have formed the basis of the Green Revolution. Dwarfed varieties have been developed that respond dramatically to heavy applications of fertilizer, resist plant diseases, and can tolerate much shorter growing seasons than traditional native varieties. Adopting the new varieties and applying the irrigation, mechanization, fertilization, and pesticide practices they require have created a new "high-input, high-yield" agriculture.

WOMEN AND THE GREEN REVOLUTION

Tasks in traditional agriculture are often divided by gender as illustrated in Figure 9.12 where men prepare the swidden plot and women, often while carrying children, plant the crops. Women farmers grow at least one-half of the world's food and up to 80% in some African countries. They are responsible for an even larger share of food consumed by their own families: 80% in sub-Saharan Africa, 65% in Asia, and 45% in Latin America and the Caribbean. Further, women comprise between one-third and one-half of all agricultural laborers in developing countries.

Women's agricultural dominance in developing states is increasing as male family members continue to leave for cities in search of paid urban work. In Mozambique, for example, for every 100 men working in agriculture, there are 153 women. In nearly all other sub-Saharan countries, the female component runs between 120 and 150 per 100 men. The departure of men for near or distant cities means, in addition, that women must assume the effective management of their families' total farm operations.

Despite their fundamental role, however, women do not share equally with men in the rewards from agriculture or benefit equally from new agricultural techniques. First, most women farmers are involved in subsistence farming and food production for the local market, which yields little cash return. Second, they have far less access than men to credit at bank or government-subsidized rates that would make it possible for them to acquire the Green Revolution technology, such as hybrid seeds and fertilizers. Third, in some cultures women cannot own land and so are excluded from agricultural improvement programs and projects aimed at landowners. For example, many African agricultural development programs are based on the conversion of communal land, to which women have access, to private holdings, from which they are excluded. In Asia, inheritance laws favor male over female heirs, and female-inherited land is managed by husbands.

At the same time, the Green Revolution and its greater commercialization of crops have generally required an increase in labor per hectare, particularly in tasks typically reserved for women, such as weeding, harvesting, and postharvest work. If women are provided no relief from their other daily tasks, the Green Revolution for them may be more burden than blessing. But when mechanization is added to the new farming system, women tend to be losers. Frequently, such predominantly female tasks as harvesting or dehusking and polishing of grain—all traditionally done by hand—are given over to machinery, displacing rather than employing women. Even the application of chemical fertilizers ("a man's task") instead of cow dung ("women's work") has reduced the female role in agricultural development programs. The loss of those traditionally female wage jobs means that already poor rural women and their families have insufficient income to improve their diets even in light of substantial increases in food availability through Green Revolution improvements.

If women are to benefit from the Green Revolution, new cultural norms—or culturally acceptable accommodations within traditional household, gender, and customary legal relations—will be required. These must permit or recognize women's landowning and other legal rights not now clearly theirs, access to credit at favorable rates, and admission on equal footing with males to government assistance programs.

The Food and Agriculture Organization of the United Nations is working to enhance the role of women as contributors and beneficiaries of economic, social, and political development. Objectives include promoting gender-based equity in control of productive resources; enhancing women's participation in decision- and policy-making processes at all levels, local and national; and providing women with access to credit to enable them to engage as creators and owners of small-scale manufacturing, trade, or service businesses.

The model for that credit access is the Grameen Bank, established by a Bangladeshi economist Muhammad Yunus who went on to win a Nobel peace prize. Based on the conviction that access to credit should be a basic human right, the bank extends "microcredit"—the average loan is U.S. $160—for "microenterprises," with women being the primary borrowers. The Grameen concept has spread from Bangladesh across Asia and to Eastern Europe, Latin America, and Africa. The microloan concept has made a powerful difference in the financial security and resilience of some of the world's poorest people.

But a price has been paid. The Green Revolution is commercially oriented and demands high inputs of costly hybrid seeds, mechanization, irrigation, fertilizers, and pesticides. As the Green Revolution is adopted, traditional agriculture and subsistence agriculture are being displaced. Lost, too, are the food security that distinctive locally adapted native crop varieties provide and the nutritional diversity and balance that multiple-crop intensive gardening ensures. Subsistence farming, wherever practiced, is oriented toward risk minimization. Many differentially hardy varieties of a single crop guarantee some yield regardless of adverse weather, disease, or pest problems that might occur. Commercial agriculture, however, aims at profit maximization, not food security.

There are other costs for Green Revolution successes. For example, irrigation, responsible for an important part of increased crop yields, has destroyed large tracts of land; excessive salinity of soils resulting from poor irrigation practices can destroy soil productivity. And the huge amount of water required for Green Revolution irrigation has led to serious groundwater depletion; to conflict between agricultural and growing urban and industrial water needs in developing countries, many of which are in semiarid climates; and to worries about scarcity and future wars over water.

The presumed benefits of the Green Revolution are not available to all subsistence agricultural areas or advantageous to everyone engaged in farming. Most poor farmers on marginal and rain-fed (nonirrigated) lands, for example, have not benefited from the new plant varieties requiring irrigation and high chemical inputs, and women have often been losers in the transition from traditional to the more-industrialized farming practices of the Green Revolution (see "Women and the Green Revolution"). Africa is a case in point. Green Revolution crop improvements have concentrated on wheat, rice, and maize. Of these, only maize is important in Africa, where principal food crops include millet, sorghum, cassava, manioc, yams, cowpeas, and peanuts.

EATING LOCALLY ON THE COLLEGE CAMPUS

Stereotypes associate college life with a diet of fast-food hamburgers and greasy pizza. However, many college students are starting to pay close attention to their food choices. They are asking where their food comes from and how it impacts human health, animal welfare, and the environment. Some are even getting involved in producing their own food on campus gardens. At many colleges and universities students have succeeded in pressuring the college dining services to offer healthier, locally produced food by growing it on campus.

Several Midwestern colleges and universities have made commitments to growing more of their own food. A couple of generations ago, many of their students came from farms and had experience with fieldwork, milking cows, gathering eggs, butchering chickens, gardening, and canning. Back then, farms were small, nonspecialized general farming operations. Today, cash grain farmers plant thousands of hectares of just two crops—corn and soybeans. Livestock farmers raise thousands of hogs, turkeys, or chickens and dairies may house thousands of milk cows. While the Midwest is among the most productive of agricultural landscapes, very little is produced for direct local consumption. Meat and milk are shipped to distant markets, soybeans are refined for industrial processes, and corn goes to animal feed, to Asian or European markets, or to produce ethanol fuel. And no longer do Midwest colleges and universities get their students from farms or supply their cafeterias from local farms. Instead, the students come from cities and suburbs, and cafeteria food is delivered by large refrigerated trucks.

Many students have pressured their colleges and universities to provide space for organic campus gardens. At Calvin College in Grand Rapids, Michigan, the dining services has altered menus to incorporate the abundant greens produced by the campus farm. Students also collect and process hundreds of bottles of maple syrup from campus trees as well as collecting honey from bee hives located in the campus farm. At Gustavus Adolphus College in Saint Peter, Minnesota, the Big Hill Farm grew out of an undergraduate senior project

and is an important piece in the college's commitment to environmental sustainability. The college supports the farm with land, tillage equipment, and paid student internships. A grant paid for a greenhouse that extends the short Minnesota growing season. The farm and the student dining service have a mutually beneficial relationship. Food waste from the cafeteria is composted and used as fertilizer and soil amendment on the farm while the farm produces lettuce, tomatoes, peppers, beans, onions, melons, pumpkins, berries, and much more for the cafeteria. Where possible, the farm uses rare heirloom seed varieties to promote crop diversity. Extra produce is sold to the local food cooperative and at the local farmers' market. Students are now eating healthy, locally grown food, and the college is reducing its own impact on the environment. Students who work on the farm learn lifetime skills and describe their labors as both exhausting and deeply rewarding.

© Alicia De Jong

In many areas showing greatest past successes, Green Revolution gains are leveling off. The UN's Food and Agriculture Organization now considers the productivity gains of Green Revolution technologies "almost exhausted." Little prime land and even less water remain to expand farming in many developing countries, and the adverse ecological and social consequences of industrial farming techniques arouse growing resistance.

Hailed as a promising second Green Revolution, genetically modified (GM) crops have been widely adopted. GM seeds are created by combining the genetic material of different species. Since their introduction, the production of engineered crops is spreading rapidly. In 1996, the first year GM crops were commercially available, about 1.7 million hectares (4.3 million acres) were placed in biotechnology cultivation. Today, hundreds of millions of hectares are production with GM crops. Initially, GM crops were primarily grown in the most industrialized countries. However, due to European concerns about the health of GM foods and ecological fears about the potential for GM crops to cross-pollinate with wild plants, producing "superweeds," GM technology is now being

more rapidly adopted in developing countries. Almost one-half the global area planted with GM crops was in developing countries, notably Argentina, Brazil, India, and South Africa. The crops for which GM seeds are most prevalent are soybeans, cotton, corn, and canola. Herbicide resistance (Roundup Ready soybeans) and insect resistance (Bt corn and cotton) have been the most important of the genetic crop modifications. GM crop supporters argue that crops can be engineered to resist droughts or include added nutrients or other attributes that could be the solution to malnutrition in the world's poorest countries. Opponents point out that GM crops have led to higher herbicide use instead of the promised reduction. They also fear that GM crops shift control of the food system from local farmers to transnational chemical and biotechnology companies.

Commercial Agriculture

Few people or areas still retain the isolation and self-sufficiency of pure subsistence economies. Nearly all have been touched by a modern world of trade and exchange and, in response, have adjusted their traditional economies. Modifications of subsistence agricultural systems have inevitably made them more complex by imparting at least some of the diversity and linkages that mark the advanced economic systems of the more-developed world. Farmers in those systems produce not for their own subsistence but primarily for a market off the farm itself. They are part of the integrated exchange economies in which agriculture is but one element in a complex structure that includes employment in mining, manufacturing, processing, and the service activities of the tertiary sector. In those economies, farm production responds to market demand expressed through prices and is related to the consumption requirements of the larger society, rather than the need of farmers to feed themselves.

Production Controls

Agriculture within modern, developed economies is characterized by specialization—by enterprise (farm), by area, and even by country; by *off-farm sale* rather than subsistence production; and by *interdependence* of producers and buyers linked through markets. Farmers in a free market economy supposedly produce the crops that their estimates of market price and production cost indicate will yield the greatest return. Supply, demand, and the market price mechanism are the primary controls on agricultural production in commercial economies.

Total production costs, eventual harvest yields, and future market prices are among the many uncertainties that individual farmers must face. Beginning in the 1950s in the United States, specialist farmers and corporate purchasers developed strategies for minimizing those uncertainties. Processors sought uniformity of product quality and timing of delivery. Vegetable canners—of tomatoes, sweet corn, and the like—required volume delivery of raw products of uniform size, color, and ingredient content on dates that accorded with cannery and labor schedules. And farmers wanted the support of a guaranteed market at an assured price to minimize the uncertainties of their specialization and to stabilize the return on their investment.

The solution was contractual arrangements or vertical integration (where production, processing, and sales are all coordinated within one firm) uniting contracted farmer with purchaser-processor. Broiler chickens of specified age and weight, cattle fed to an exact weight and finish, wheat with a minimum protein content, popping corn with prescribed characteristics, potatoes of the kind and quality demanded by particular fast-food chains, and similar product specifications became part of production contracts between farmer and buyer-processor. In the United States, the percentage of total farm output produced under contractual arrangements or by vertical integration rose from 19% in 1960 to well over one-third by the turn of the century. For example, the vast majority of hogs are sold under some form of contract; in 1980 only 5% were sold that way. The term *agribusiness* is applied to the growing merging of the older, farm-centered crop economy and newer patterns of more integrated production and marketing systems.

Even for family farmers not bound by contractual arrangements to suppliers and purchasers, the older assumption that supply, demand, and the market price mechanism are effective controls on agricultural production is not wholly valid. In reality, those theoretical controls are joined by a number of nonmarket government influences that may be as decisive as market forces in shaping farmers' options and spatial production patterns. If there is a glut of wheat on the market, for example, the price per ton will come down and the area sown to it should diminish. It will also diminish regardless of supply if governments, responding to economic or political considerations, impose acreage controls.

Distortions of market control may also be introduced to favor certain crops or commodities through subsidies, price supports, market protections, and the like. The political power of farmers in the European Union, for example, secured generous product subsidies and, for the Union, immense unsold stores of butter, wine, and grains until 1992, when reforms began to reduce the surplus stockpiles even while increasing total farm spending. In Japan, the home market for rice is largely protected and reserved for Japanese rice farmers, even though their production efficiencies are low and their selling price is high by world market standards. In the United States, programs of farm price supports, acreage controls, financial assistance, and other government involvement in agriculture have given recurring and equally distorting effects.

A Model of Agricultural Location

Early in the 19th century, when transportation systems were less efficient and before such government influences were the norm, Johann Heinrich von Thünen (1783–1850) observed that uniformly fertile farmlands were used differently. Around each major urban market center, he noted, developed a set of concentric rings of different farm products (**Figure 9.14a**). The ring closest to the market featured intensive agriculture producing perishable commodities that were both expensive to ship and in high demand. The high prices they could command in the urban market made their production an appropriate use of high-value land near the city. Rings of farmlands farther away from the city were used for less-perishable commodities with lower transport costs, reduced demand, and lower market prices. General farming and grain farming replaced the market gardening of the inner ring. At the outer margins of profitable agriculture, farthest from the single central market, livestock grazing and similar extensive land uses were found.

To explain why this should be so, von Thünen proposed a formal spatial model, perhaps the first developed to analyze human activity patterns. He concluded that the uses to which parcels were put were a function of the differing "rent" values placed on seemingly identical lands. Those differences, he claimed, reflected the cost of overcoming the distance separating a given farm from a central market town. ("A portion of each crop is eaten by the wheels," he observed.) The greater the distance, the higher was the operating cost to the farmer, since transport charges had to be added to other expenses. When a commodity's production costs plus its transport costs just equaled its value at the market, a farmer was at the economic margin of its cultivation. Because in this model transport costs are uniform in all directions away from the center, the concentric zonal pattern of land use called the **von Thünen rings** results (Figure 9.14a).

The von Thünen model may be modified by introducing ideas of differential transport costs, variations in topography or soil

1. Dairying and market gardening
2. Specialty farming
3. Cash grain and livestock
4. Mixed farming
5. Extensive grain farming or stock raising

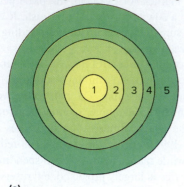

(a)

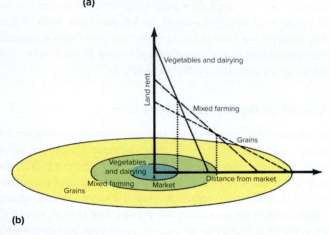

(b)

Figure 9.14 **(a) von Thünen's model.** Recognizing that, as distance from the market increases, the value of land decreases, von Thünen developed a descriptive model of intensity of land use that holds up reasonably well in practice. The most intensively produced agricultural crops are found on land close to the market; the less intensively produced commodities are located at more-distant points. The numbered zones of the diagram represent modern equivalents of the theoretical land use sequence von Thünen published in 1826. As the metropolitan area at the center increases in size, the agricultural specialty areas are displaced outward, but the relative position of each is retained. **(b) Transport gradients and agricultural zones.** In the model, perishable commodities such as fruits and vegetables have high transport rates per unit of distance; grain and other crops have lower rates. Land rent for any farm commodity decreases with increasing distance from the central market, and the rate of decline is determined by the *transport gradient* for that commodity. Crops that have both the highest market price and the highest transport costs will be grown nearest to the market. Less-perishable crops with lower production and transport costs will be grown at greater distances away.

fertility, or changes in commodity demand and market price. With or without such modifications, von Thünen's analysis helps explain the changing crop patterns and farm sizes evident on the landscape at increasing distance from major cities, particularly in regions dominantly agricultural in economy. Farmland close to markets takes on high value, is used *intensively* for high-value crops, and is

subdivided into relatively small units. Land far from markets is used *extensively* and in larger units (**Figure 9.14b**).

With improved transportation systems for perishable products and higher standards of living in industrial and postindustrial economies, the von Thünen regularities have undergone change. Today, for example, the presence of the nearby city may be felt in the form of market gardens and orchards (often community-supported organic agriculture or pick-your-own), nursery and bedding plant operations, hobby farms (frequently with horses), and land withheld from farming by developers who have purchased the land in anticipation of future subdivision.

Intensive Commercial Agriculture

Following World War II, agriculture in the developed world's market economies turned increasingly to concentrated methods of production. Machinery, chemicals, irrigation, and dependence on a restricted range of carefully selected and bred plant varieties and animal breeds all were employed in a concerted effort to wring greater production from each unit of farmland.

The goal, of course, was to increase off-farm sales as American agriculture increasingly shifted from an objective of self-sufficiency to total commitment to a fully commercial, exchange-economy stance. Prior to 1950 most U.S. farms had a significant subsistence orientation; they were "general farms" growing a variety of crops, some for sale and others for feed for farmstead livestock—a milk cow or two, chickens for the pot and for household eggs, a few hogs and steers, partly for farm slaughter and use. Their extensive kitchen gardens supplied vegetables and fruits for farm family seasonal consumption and home canning. In 1949 the average American farm sold only $4100 worth of products. By 2007, however, most farms had a full commitment to the market, average off-farm sales rose to more than $135,000, and farm families—like other Americans—shopped the supermarkets for their food needs. With the increases in capital investment and the need for larger farms to maximize return on that investment, many inefficient small farms have been abandoned. Consolidation has reduced the number and enlarged the size of farms still in production. From a high of 6.8 million in 1934, the number of U.S. farms of all sizes dropped to 2.1 million in 2002 and then increased to 2.2 million in 2007. The recent increase in the number of farms was due to more small farms, many of which were classified as lifestyle or retirement farms, in which farm products were not the primary source of income for the farmer. Small farms have also increased in part as a response to consumers' desire for organic and local food.

The reorientation of farm production goals in the United States and in most other highly developed market economies has led to significant changes in the scale of commercial agriculture. Reflecting the drive for enhanced and more-specialized output and the investment of large amounts of capital (for machinery, fertilizers, and specialized buildings, for example), all modern agriculture is "intensive." But the several types of farm specialization differ in how much capital is invested per hectare of farmed land (and, of course, in the specifics of those capital inputs). Those differences underlie generalized distinctions between traditional intensive and extensive commercial agriculture.

(a)

(b)

Figure 9.15 **Industrial poultry and livestock farming** are part of the changing scale of American agriculture. **(a)** Thousands of broiler chickens are raised in a single barn for 45 days before slaughtering. **(b)** This concentrated animal-feeding operation in Kansas is one of many on the Great Plains. Intensive commercial agriculture uses such methods to produce large volumes of a uniform product at the lowest possible price. *(a) © Digital Vision/PunchStock RF; (b) © Cathryn Dowd.*

The term **intensive commercial agriculture** is now usually understood to refer specifically to the production of crops that give high yields and high market value per unit of land. These include fruits, vegetables, and dairy products, all of which are highly perishable, as well as some "factory farm" production of livestock. Dairy farms and *truck farms* (horticultural, or "market garden," farms that produce a wide range of vegetables and fruits) are found near most medium-size and large cities. Because their products are perishable, transport costs increase because of the special handling that is needed, such as the use of refrigerated trucks and custom packaging. This is another reason for locations close to market.

Livestock-grain farming involves the growing of grain to be fed on the producing farm to livestock, which constitute the farm's cash product. In Western Europe, three-fourths of cropland is devoted to production for animal consumption; in Denmark, 90% of all grains are fed to livestock for conversion into meat, butter, cheese, and milk. Although livestock-grain farmers work their land intensively, the value of their product per unit of land is usually less than that of the truck farm. Consequently, in North America at least, livestock-grain farms are farther from the main markets than are horticultural and dairy farms. In general, the livestock-grain belts of the world are close to the great coastal and industrial zone markets. The Corn Belt of the United States and the livestock region of Western Europe are two examples.

In the United States—and commonly in all developed countries—the traditional livestock and grain operations of small and family farms have been largely supplanted by very large-scale, concentrated animal-feeding operations involving thousands or tens of thousands of closely quartered animals (**Figure 9.15**). From its inception in the 1920s, the intensive, industrialized rearing of livestock, particularly beef and dairy cattle, hogs, and poultry, has grown to dominate meat, dairy, and egg production. To achieve their objective of producing a large-volume, uniform product at the lowest possible cost, operators of livestock factory farms confine animals to pens or cages, treat them with antibiotics and vitamins to maintain health and speed growth, provide processed feeds that often contain low-cost animal by-products or crop residue, and deliver them under contract to processors, packers, or their parent company. Although serious concerns have been voiced about animal waste management and groundwater, stream, and atmospheric pollution, contract-based concentrated feeding operations now provide almost all supermarket meat and dairy products. The location of this form of intensive commercial farming is often determined not by land value or proximity to market but by land use restrictions and environmental standards imposed by state and county governments.

Extensive Commercial Agriculture

Farther from the market, on less-expensive land, there is less need to use the land intensively. Cheaper land gives rise to larger farm units. **Extensive commercial agriculture** is typified by large wheat farms and livestock ranching.

Large-scale wheat farming requires sizable capital inputs for planting and harvesting machinery, but the inputs per unit of land are low; wheat farms are very large. Nearly one-half of the farms in Saskatchewan, for example, are more than 400 hectares (1000 acres). The average farm in Kansas is more than 300 hectares (750 acres); in North Dakota, more than 525 hectares (1300 acres). In North America, the spring wheat (planted in spring, harvested in autumn) region includes the Dakotas, eastern Montana, and the southern parts of the Prairie Provinces of Canada. The winter wheat (planted in fall, harvested in midsummer) belt focuses on Kansas and includes adjacent sections of neighboring states (**Figure 9.16**). Argentina is the only South American country to have comparable large-scale wheat farming. In the Eastern Hemisphere, the system is fully developed only east of the Volga River in northern Kazakhstan and the southern part of West Siberia, and in southeastern and western Australia.

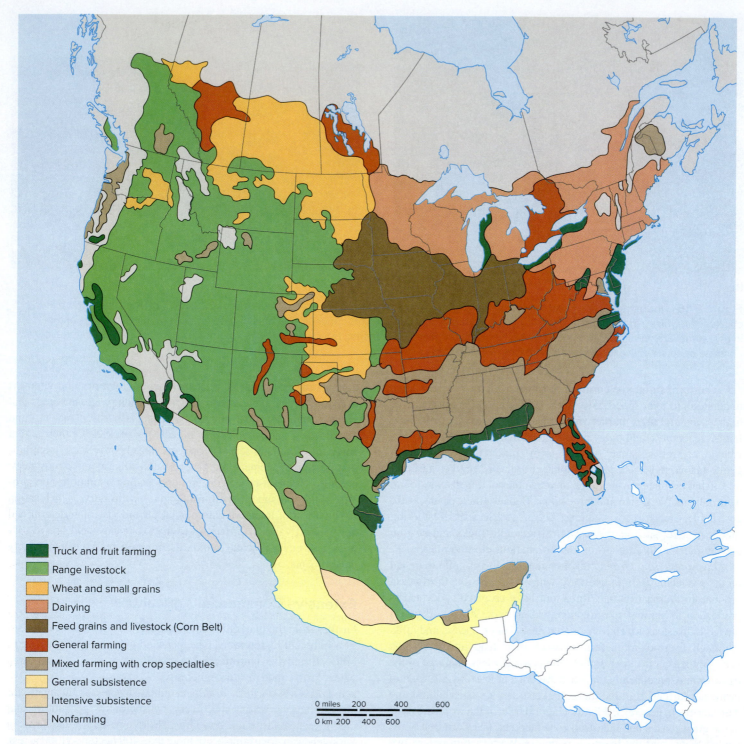

Figure 9.16 **Generalized agricultural regions of North America.** *Sources: U.S. Bureau of Agricultural Economics; Agriculture Canada; Mexico, Secretaría de Agricultura y Recursos Hidráulicos.*

Legend:
- Truck and fruit farming
- Range livestock
- Wheat and small grains
- Dairying
- Feed grains and livestock (Corn Belt)
- General farming
- Mixed farming with crop specialties
- General subsistence
- Intensive subsistence
- Nonfarming

Livestock ranching differs significantly from livestock-grain farming and, by its commercial orientation and distribution, from the nomadism it superficially resembles. A product of the 19th-century growth of urban markets for beef and wool in Western Europe and the northeastern United States, ranching has been primarily confined to areas of European settlement. It is found in the western United States and adjacent sections of Mexico and Canada (Figure 9.16); the grasslands of Argentina, Brazil,

Uruguay, and Venezuela; the interior of Australia; the uplands of South Island, New Zealand; and the Karoo and adjacent areas of South Africa (**Figure 9.17**). All except New Zealand and the humid pampas of South America have semiarid climates. All, even the most remote from markets, were a product of improvements in transportation by land and sea, the refrigeration of carriers, and meat-canning technology. In all, introduced beef cattle or sheep replaced original native fauna, such as bison on North America's

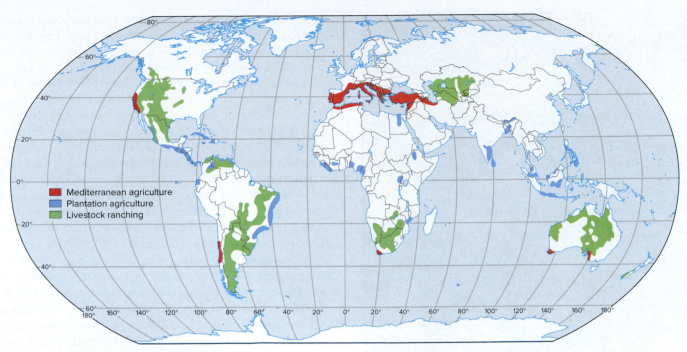

Figure 9.17   *Livestock ranching and special crop agriculture.* Livestock ranching is primarily a midlatitude enterprise catering to the urban markets of industrialized countries. Mediterranean agriculture and plantation agriculture are similarly oriented to the markets provided by advanced economies of Western Europe and North America. Areas of Mediterranean agriculture—all of roughly comparable climatic conditions—specialize in similar commodities, such as grapes, oranges, olives, peaches, and vegetables. The specialized crops of plantation agriculture are influenced by both physical geographic conditions and present or, particularly, former colonial control of areas.

Great Plains, almost always with eventual severe environmental deterioration. More recently, the midlatitude demand for beef has been blamed for expanded cattle production and the extensive destruction of tropical rain forests in Central America and the Amazon Basin, though in recent years Amazon Basin deforestation has reflected more the expansion of soybean farming than that of beef production.

In all of the ranching regions, livestock range (and the area exclusively in ranching) has been reduced as crop farming has encroached on its more-humid margins, as pasture improvement has replaced less-nutritious native grasses, and as grain fattening has supplemented traditional grazing. Since ranching can be an economic activity only where alternative land uses are nonexistent and land quality is low, ranching regions of the world characteristically have low population densities, low capitalization per land unit, and relatively low labor requirements.

Special Crops

Proximity to the market does not guarantee the intensive production of high-value crops, should terrain or climatic circumstances hinder it. Nor does great distance from the market inevitably determine that extensive farming on low-priced land will be the sole agricultural option. Special circumstances, most often climatic, make some places far from markets intensively developed agricultural areas. Two special cases are agriculture in Mediterranean climates and in plantation areas (Figure 9.17).

Most of the arable land in the Mediterranean Basin itself is planted to grains, and much of the agricultural area is used for grazing. *Mediterranean agriculture* as a specialized farming economy,

however, is known for grapes, olives, oranges, figs, vegetables, and similar commodities. These crops need warm temperatures all year-round and a great deal of sunshine in the summer. The Mediterranean agricultural lands indicated in Figure 9.17 are among the most productive agricultural lands in the world. The precipitation regime of Mediterranean climate areas—winter rain and summer drought—lends itself to the controlled use of water. Of course, much capital must be spent for the irrigation systems, another reason for the intensive use of the land for high-value crops that are, for the most part, destined for export to industrialized countries or areas outside the Mediterranean climatic zone and even, in the case of Southern Hemisphere locations, to markets north of the equator.

Climate is also considered the vital element in the production of what are commonly but imprecisely known as *plantation crops*. The implication of **plantation** is the introduction of a foreign element—investment, management, and marketing—into an indigenous culture and economy, often employing an introduced alien labor force. The plantation itself is an estate whose resident workers produce one or two specialized crops. Those crops, although native to the tropics, were frequently foreign to the areas of plantation establishment: African coffee and Asian sugar in the Western Hemisphere and American cacao, tobacco, and rubber in Southeast Asia and Africa are examples (**Figure 9.18**). Entrepreneurs in Western countries such as England, France, the Netherlands, and the United States became interested in the tropics partly because they afforded them the opportunity to satisfy a demand in temperate lands for agricultural commodities not producible in the market areas. Custom and convenience usually retain the term *plantation,* even where native producers of local crops dominate, as they do in cola nut production in Guinea, spice growing in India and

Figure 9.18 **Women picking coffee beans in Kenya.** Coffee is a classic plantation crop, with operations typically established by foreign capital in locations offering a suitable physical environment (climate and soils) and producing an introduced crop for a distant market. While coffee was first domesticated in East Africa, it is grown in many tropical countries, mostly for export to developed countries located in the midlatitudes. Note that the women laborers are also responsible for child care even as they work in the fields.
© Christopher Pillitz/Getty Images.

Sri Lanka, and sisal production in the Yucatán. As Figure 9.17 suggested, for ease of access to shipping most plantation crops are cultivated along or near coasts, since production for export rather than for local consumption is the rule.

Sustainable Agriculture

The adoption of large-scale, highly industrialized commercial agriculture has increased overall agricultural productivity and benefited successful farmers and agribusinesses. But it has created some social and environmental concerns. The negative effects of industrialized agriculture on the health of rural communities, ecosystems, and food systems have convinced some of the need for a more-sustainable mode of agriculture. As farms have grown larger and replaced human labor with machinery, the population involved in farming has dramatically declined. This has led to depopulation of rural areas and struggles to maintain the institutions and basic services necessary for a high quality of life.

Industrialized agriculture relies upon heavy inputs of fertilizers, pesticides, and herbicides, each of which has had a negative effect on the health of people, wildlife, surface waterways, and coastal systems that receive agricultural runoff (see Chapter 13). Industrialized agriculture relies on large quantities of fossil fuels to fuel the machinery, manufacture petrochemical-based fertilizers, and distribute the food around the world. Finally, there are serious concerns about the quality of the food supply and its relationship to human health, the obesity epidemic, and diet-related diseases such as cancer and diabetes. Concerns range from the relative lack of fresh fruits and vegetables in the diets of rich countries, to the safety of foods with pesticide and hormone residues, to the growth of antibiotic resistance due to the overuse of antibiotics in livestock feed.

The sustainable agriculture movement is a collection of alternative approaches to agriculture that seek to enhance social, ecological, economic, and individual health. Sustainable agriculture advocates emphasize local knowledge, local markets, smaller-size operations, and farm diversification. Sustainable agriculture relies upon local knowledge of soil and plant conditions and uses traditional agriculture methods adapted to particular places. As consumers show concern about the health of their food, where it comes from, and how far it has traveled to their table, they've turned to organic foods, farmers' markets, urban gardening, and community-supported agriculture (see "Eating Locally on the College Campus," p. 264). Where industrial agriculture creates monocultures and specialized producers, sustainable agriculture advocates believe in farm diversification and biodiversity. They argue that a region with many smaller producers growing many different crops and raising different livestock is more resilient and creates a healthier local economy and environment. In many ways, sustainable agriculture is a return to methods of agriculture that were widely used prior to World War II. Thus, critics question whether sustainable agriculture will be able to maintain the productivity gains of industrial agriculture and feed the world's growing population.

9.3 Other Primary Activities

In addition to agriculture, primary economic activities include fishing, forestry, and the mining and quarrying of minerals. These industries involve the direct exploitation of natural resources that are unequally available in the environment and differentially evaluated by different societies. Their development, therefore, depends on the occurrence of perceived resources, the technology to exploit their natural availability, and the cultural awareness of their value. (The definition, perception, and utilization of resources are explored in depth in Chapter 12.)

Two of them—fishing and forestry—are **gathering industries** based on harvesting the natural bounty of renewable resources, though once in serious danger of depletion through overexploitation. Mining and quarrying are **extractive industries,** removing nonrenewable metallic and nonmetallic minerals, including the mineral fuels, from the earth's crust. They are the initial raw material phase of modern industrial economies.

Fishing

In both the developed and less-developed countries, fish are an important part of diets. Fish and shellfish account for over 15% of all human animal protein consumption, and an estimated 3 billion people depend on fish for at least 20% of their protein. The majority of the world's annual fish harvest is consumed by humans, although a portion is processed into fish meal to be fed to livestock or used as fertilizer. Those two quite different markets have increased both the demand for and the annual harvest of fish. Indeed, so rapidly have pressures on the world's fish stocks increased that evidence is unmistakable that, at least locally, their *maximum sustainable yield* is being exceeded. The **maximum sustainable yield** of a resource is the largest volume or rate of use that will not impair its ability to be renewed or to maintain the same future productivity. For fishing, that level is marked by a catch or harvest equal to the net growth of the replacement stock.

The annual fish supply comes from three sources:

1. the *inland catch,* from ponds, lakes, and rivers
2. *fish farming (or aquaculture),* where fish are produced in a controlled and contained environment
3. the *marine catch,* all wild fish harvested in coastal waters or on the high seas

The inland catch supplies a modest fraction of the global fish catch, while fish farming continues to grow, producing nearly 40% of the world's fish harvest (**Figure 9.19**). While the world's ocean catch is still the most important source of fish, its annual harvest has been stable or declining since the late 1980s. Inland waters supply less than 10% of that catch. Fish farming, both inland and marine, accounts for about one-third of total global production. Most of the marine capture is made in coastal wetlands, estuaries, and the relatively shallow coastal waters above the continental shelf. Near shore, shallow embayments and marshes provide spawning grounds, and river waters supply nutrients to an environment highly productive of fish. Increasingly, these are also areas seriously affected by pollution from runoff and ocean dumping, an environmental assault so devastating in some areas that fish and shellfish stocks have been destroyed with little hope of revival.

Commercial marine fishing is largely concentrated in northern waters, where warm and cold currents join and mix and where such familiar food species as herring, cod, mackerel, haddock, and flounder congregate, or "school," on the broad continental shelves (**Figure 9.20**). Two of the most heavily fished regions are the Northeast Pacific and Northwest Atlantic. Tropical fish species tend not to school and, because of their high oil content and unfamiliarity, are less acceptable in the commercial market. They are, however, of great importance for local consumption. Only a very small percentage of total marine catch comes from the open seas that make up more than 90% of the world's oceans.

Modern technology and more-aggressive fishing fleets of more countries greatly increased annual marine capture in the years after 1950. That technology included the use of sonar, radar, helicopters, and satellite communications to locate schools of fish; more efficient nets and tackle; and factory trawlers to follow fishing fleets to prepare and freeze the catch. In addition, more nations granted ever-larger subsidies to expand and reward their marine trawler

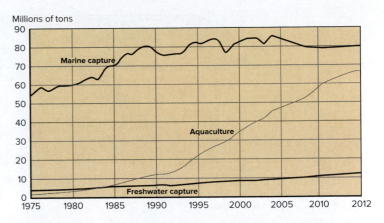

Figure 9.19 Officially recorded annual fish harvests, 1975–2012, rose irregularly from 66 million tons in 1975 to some 158 million tons in 2012. The 1993 and 1998 dips are associated with ocean temperature changes produced by El Niño. The FAO estimates that 20 million to 40 million tons per year of unintended marine capture of juvenile or undersized fish and nontarget species are discarded each year. *Source: Food and Agriculture Organization (FAO).*

operations. The rapid rate of increase led to inflated projections of continuing or growing fisheries productivity and to optimism that the resources of the oceans were inexhaustible.

Quite the opposite has proved to be true. In fact, in recent years, the productivity of many major marine fisheries has declined because *overfishing* (catches above reproduction rates) and the pollution of coastal waters have seriously endangered the supplies of traditional and desirable food species. The decline, coupled with a growing world population, has caused a serious drop in the average per capita marine catch. The UN reports that all 17 of the world's major oceanic fishing areas are being fished at or beyond capacity; 13 are in decline. The plundering of coastal waters off North America has imperiled a number of the most desirable fish species; in 1993, Canada shut down its Atlantic cod industry to allow stocks to recover, and on the west coast, the Pacific salmon are in peril. The U.S. authorities report that 67 North American species are overfished and 61 harvested to capacity. Pacific bluefin tuna, popular in sushi, are reportedly near collapse due to overfishing of juvenile fish that have yet to reproduce.

Overfishing is partly the result of the accepted view that the world's oceans are common property, a resource open to anyone's use, with no one responsible for their maintenance, protection, or improvement. The result of this "open seas" principle is but one expression of the so-called **tragedy of the commons**[a]— the economic reality that, when a resource is available to all, each user, in the absence of collective controls, thinks he or she is best served by exploiting the resource to the maximum, even though this means its eventual depletion. In 1995, more than 100 countries adopted a treaty to regulate fishing on the open oceans outside territorial waters. Applying to such species as cod, pollock, and tuna— that is, to migratory and high-seas species—the treaty requires fishermen to report the size of their catches to regional organizations

[a] The *commons* refers to undivided land available for the use of everyone; usually, it meant the open land of a village that all used as pasture. The *Boston Common* originally had this meaning.

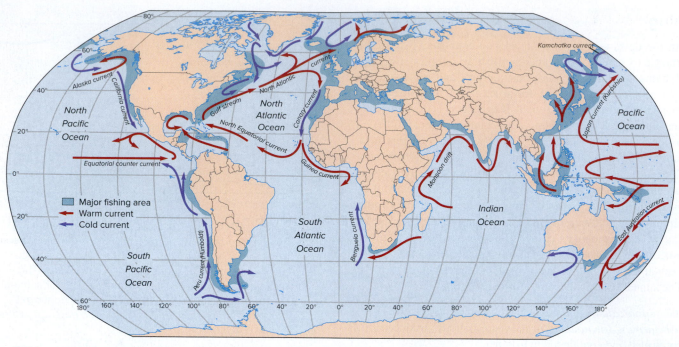

Figure 9.20   **The major commercial marine fisheries of the world.** The waters within 325 kilometers (200 miles) of the U.S. coastline account for almost one-fifth of the world's annual fish and shellfish harvests. Overfishing, urban development, and the contamination of bays, estuaries, and wetlands have contributed to the depletion of the fish stocks in those coastal waters.

that would set quotas and subject vessels to boarding to check for violations. These and other fishing control measures could provide the framework for the future sustainability of important food fish stocks, though they appear to be too late to save or revive the marine food chain in such formerly important fisheries as the Atlantic coast of Canada.

One approach to increasing the fish supply is through *fish farming,* or **aquaculture,** the breeding of fish in freshwater ponds, lakes, and canals or in fenced-off coastal bays and estuaries or enclosures. Aquaculture production has provided a growing share of the total fish harvest in recent years; its contribution to the human food supply is even greater than raw production figures suggest. Whereas one-third of the conventional fish catch is used to make fishmeal and fish oil, virtually all farmed fish are used as human food. Fish farming has long been practiced in Asia (**Figure 9.21**), where fish are a major source of protein, but now takes place on every continent. Critical environmental problems associated with marine aquaculture exist: pollution from fish wastes and chemicals

Figure 9.21   **Fish farming in China.** Fish farming is one of the fastest-growing sectors in world food production with Asian countries supplying the vast majority of the total fish farming harvest. As shown here, fish farming can be done in flooded rice paddies. Fish wastes enhance soil fertility and the fish eat insects that damage rice crops.
© *Greg Baker/AP Images.*

and drugs; transference of disease to wild fish stocks; depletion of wild stock to provide feed for farmed fish; genetic damage to wild stock from escaped alien or genetically altered farmed fish; and more. Despite concerns about its potential adverse consequences, its rapid and continuing production increase makes aquaculture the fastest-growing sector of the world food economy.

Forestry

Forests, like fish, are a heavily exploited renewable resource. Even after millennia of land clearance for agriculture and, more recently, commercial lumbering, cattle ranching, and fuelwood gathering, forests still cover nearly one-third of the world's land area. Livelihoods based on forest resources are spatially widespread and parts of both subsistence and commercial economies. A full treatment of forest resources and their nature, distribution, and exploitation appears in Chapter 12.

Mining and Quarrying

Societies at all stages of economic development engage in agriculture, fishing, and forestry. The extractive industries—mining and drilling for nonrenewable mineral wealth—emerged only when cultural advancement and economic necessity made possible a broader understanding of the Earth's resources. Now those extractive industries provide the raw material and energy base for the way of life experienced by people in the advanced economies and are the basis for an important part of the international trade connecting the developed and developing countries of the world.

Extractive industries depend on the exploitation of minerals that are unevenly distributed in the Earth's crust. The locations of mineral deposits are determined by past geologic events, not by contemporary market demand. Thus, mines are often located in remote locations (**Figure 9.22**). Because usable deposits are the result of geological processes operating in the distant past, there is little relationship between the development status of a country and its mineral resources. The laws of probability predict that larger states are more likely to be the beneficiaries of such "accidents." Still, many smaller developing countries, especially in Africa, are major sources of one or more-critical raw materials and, therefore, are important participants in the growing international trade in minerals.

Transportation costs play a major role in determining where *low-value minerals* will be mined. Materials such as gravel, limestone for cement, and aggregate are found in such abundance that they have value only when they are near the site where they are to be used. For example, gravel for road building has value if it is near the road-building site, not otherwise. Transporting gravel hundreds of miles is an unprofitable activity.

The production of other minerals—especially *metallic minerals* such as copper, lead, and iron ore—is affected by a balance of three forces: the quantity available, the richness of the ore, and the distance to markets. Also important are labor costs and land acquisition and royalty costs which may equal or exceed other considerations in mine development decisions (see "Public Land, Private Profit," p. 272). Even if these conditions are favorable, mines

Figure 9.22 **Molybdenum mine and processing facilities at the Climax mine near Leadville, Colorado.** Mines are often located in remote areas such as this mountain pass at an elevation of about 3500 meters (11,400 feet) above sea level. Concentrating mills crush the ore, separating molybdenum-bearing material from the rocky mass containing it. The great volume of waste material removed assures that most concentrating operations are found near the ore bodies. © *National Geographic Creative/Alamy Stock Photo RF*

may not be developed or even remain operating if supplies from competing sources are more cheaply available in the market. The developed industrial countries, whatever their former or even present mineral endowment, frequently find themselves at a competitive disadvantage against producers in developing countries with lower-cost labor and state-owned mines with abundant, rich reserves. As a consequence, North American iron ore, copper, nickel, zinc, lead, and molybdenum mines have experienced periodic shutdowns when cyclical prices dropped below domestic production costs.

Mining companies are frequently global in their scope of operations, and many operate mines on all continents. This has brought investment to developing countries, but also loss of local control. Some developing countries have responded by imposing higher taxes on mines and imposing minimum requirements for local ownership.

When the ore is rich in metallic content, it is profitable (in the case of iron and aluminum ores) to ship it directly to the market for refining. But, of course, the highest-grade ores tend to be mined first. Consequently, the demand for low-grade ores has been increasing in recent years as richer deposits have been depleted. Low-grade ores are often upgraded by various types of separation treatments at the mine site to avoid the cost of transporting waste materials not wanted at the market. Concentration of copper is nearly always mine-oriented while refining takes place near areas of consumption.

The large amount of waste in copper (98% to 99% or more of the ore) and in most other industrially significant ores should not be considered the mark of an unattractive deposit. Indeed, the opposite may be true. Many higher-content ores are left unexploited— because of the cost of extraction or the smallness of the reserves—in favor of the use of large deposits of even very low-grade ore. The

Public Land, Private Profit

When President Ulysses S. Grant signed the Mining Act of 1872, the presidential and congressional goal was to encourage Western settlement and development by allowing any "hard-rock" miners (including prospectors for silver, gold, copper, and other metals) to mine federally owned land without royalty payment. It further permitted mining companies to gain clear title to publicly owned land and all subsurface minerals for no more than $12 a hectare ($5 an acre). Under those liberal provisions, mining firms have bought 1.3 million hectares (3.2 million acres) of federal land since 1872 and each year remove some $1.2 billion worth of minerals from government property. In contrast to the royalty-free privileges granted to metal miners, oil, gas, and coal companies pay royalties of as much as 12.5% of their gross revenues for exploiting federal lands.

Whatever the merits of the 1872 law in encouraging the economic development of lands otherwise unattractive to homesteaders, modern-day mining companies throughout the Western states have secured enormous actual and potential profits from the law's generous provisions. In Montana, a company claim to 810 hectares (2000 acres) of land would cost it less than $10,000 for an estimated $4 billion worth of platinum and palladium; in California, a gold mining company in 1994 sought title to 93 hectares (230 acres) of federal land containing a potential of $320 million of gold for less than $1200. Foreign as well as domestic firms may be beneficiaries of the 1872 law. In 1994, a South African firm arranged to buy 411 hectares (1016 acres) of Nevada land with a prospective $1.1 billion in gold from the government for $5100. And a Canadian firm in 1994 received title to 800 hectares (nearly 2000 acres) near Elko, Nevada, that cover a likely $10 billion worth of gold—a transfer that Interior Secretary Bruce Babbitt dubbed "the biggest gold heist since the days of Butch Cassidy." And in 1995 Mr. Babbitt conveyed about $1 billion worth of travertine (a mineral used in whitening paper) under 45 hectares (110 acres) of Idaho to a Danish-owned company for $275.

What was fitting in 1872 feels like a "gold heist" to today's congress and to the American public who own the land. In part, that feeling results from the fact that mining companies create environmental damage that requires public funding to restore. The mining firms may destroy whole mountains to gain access to low-grade ores and leave toxic mine tailings, surface water contamination, and open-pit scarring of the landscape as they move on or disappear. The projected public costs of cleaning up more than 50,000 abandoned mine sites, thousands of miles of damaged or dead streams, and several billion tons of contaminated waste are estimated at a minimum of $35 billion.

A congressional proposal introduced, but defeated, in 1993 would have required mining companies to pay royalties of 8% on gross revenues for all hard-rock ores extracted and would have prohibited them from outright purchase of federal land. The royalty provision alone would have yielded nearly $100 million annually at 1994 levels of company income. Mining firms claim that an imposition of royalties might well destroy America's mining industry. They stress both the high levels of investment they must make to extract and process frequently low-grade ores and the large number of high-wage jobs they provide as their sufficient contribution to the nation. The Canadian company involved in the Elko site, for example, reports that, since it acquired the claims in 1987 from their previous owner, it has expended more than $1 billion, plus has made additional donations for town sewer lines and schools and has created 1700 jobs. The American Mining Congress estimates that the proposed 8% royalty charge would have cost 47,000 jobs out of 140,000, and even the U.S. Bureau of Mines assumes a loss of 1100 jobs.

Public resistance to Western mining activities is taking its toll. State and federal regulatory procedures, many dragging on for a decade or more, have discouraged opening new mines; newly enacted environmental regulations restricting current mining operations (for example, banning the use of cyanide in gold and silver refining) reduce their economic viability. In consequence, both investment and employment in U.S. mining are in steady decline, eroding the economic base of many Western communities.

Considering the Issues

1. Do you believe the 1872 Mining Act should be repealed or amended? If not, what are your reasons for arguing for retention? If so, would you advocate the imposition of royalties on mining company revenues? At what levels, if any, should royalties be assessed? Should mining and energy companies be treated equally for access to public land resources? Why or why not?

2. Would you propose to prohibit outright land sales to mining companies? If not, should sales prices be determined by the surface value of the land or by the estimated (but unrealized) value of mineral deposits it contains?

3. Do you think that cleanup and other charges now borne by the public are acceptable in view of the capital investments and job creation of hard-rock mining companies? Do you accept the industry's claim that an imposition of royalties would destroy American metal mining? Why or why not?

4. Do you favor continued state and federal restrictions on mining operations, even at the cost of jobs and community economies? Why or why not?

---

attraction of the latter is a size of reserve sufficient to justify the long-term commitment of development capital and, simultaneously, to ensure a long-term source of supply. At one time, high-grade magnetite iron ore was mined and shipped from the Mesabi area of Minnesota. Those deposits are now exhausted. However, immense amounts of capital (including, since 2003, Chinese capital) have been invested in the mining and processing into high-grade iron ore pellets of the virtually unlimited supplies of low-grade iron-bearing rock (taconite) still remaining.

Such investments do not ensure the profitable exploitation of the resource. The metals market is highly volatile. Rapidly and widely fluctuating prices can quickly change profitable mining and refining ventures to losing undertakings. Marginal gold and silver deposits are opened or closed in reaction to trends in precious metals prices. Taconite beneficiation (waste material removal) in the Lake Superior region has cycled through booms and slowdowns in response to the decline of the U.S. steel industry and the growing demand from overseas. In market economies, cost and market controls dominate economic decisions. In planned economies, cost may be a less-important consideration than goals of national development and resource independence.

Figure 9.23 **Grain being loaded for export** at the port of New Orleans, Louisiana. International trade in agricultural commodities is an important part of the global economy. Agricultural subsidies in the United States often make it difficult for producers in other countries to compete economically. © *Glowimages RF.*

The advanced economies have reached that status through their control and use of fossil fuels. Domestic supplies of fuels, therefore, are often considered basic to national strength and independence. When those supplies are absent, developed countries are concerned with the availability and price of coal, oil, and natural gas in international trade. Fossil fuels and other energy sources are given extended discussion in Chapter 12.

9.4 Trade in Primary Products

With the liberalization of international trade policies that began in the 1980s and the growth of bulk shipping fleets, international trade has undergone massive expansion. International trade accounts for a significant and growing fraction of the world's economic activity (**Figure 9.23**). Agricultural products, minerals, and fuels account for just under 30% of the total value of international trade. During the first half of the 20th century, the world distribution of supply and demand resulted in a simple pattern of commodity flow: from raw material producers located within less-developed countries to the processors, manufacturers, and consumers in the more-developed countries. The reverse flow carried manufactured goods processed in the industrialized states back to the developing countries. That two-way trade benefited the developed states by providing access to industrial raw materials and food not available domestically and gave them access to markets for their manufactured goods. Developing countries received needed investments but were in a less-powerful position because manufactured goods are generally more valuable.

By the late 20th century, however, world trade flows and export patterns of the emerging economies had radically changed. Raw materials greatly decreased and manufactured goods increased in the export flows from developing states. Even with the overall decline in raw material exports, however, trade in unprocessed goods remains dominant in the economic well-being of many of the world's poorer economies. Increasingly, the terms of the traditional

trade flows on which they depend have been criticized as unequal and damaging to commodity-exporting countries.

United Nations data indicate that 95 developing countries depend on primary commodities for more than one-half of their export earnings. As a group, African countries are the most dependent on primary commodities, with over 80% of the continent's export earnings coming from that category. Countries that depend on primary commodities for the bulk of their export earnings are vulnerable to price volatility and technological substitutes. Sometimes prices fluctuate by 50% within a single year. Prices rise sharply in periods of product shortage or international economic growth and swiftly decline when the world economy slows.

Whatever the current world prices of raw materials may be, as a group, raw material–exporting states have long expressed resentment at what they perceive as commodity price manipulation by rich countries and corporations to ensure low-cost supplies. Although collusive price fixing has not been demonstrated, technological substitutes can reduce demand for primary commodities. Glass fibers replace copper wire in telecommunications applications; synthetic rubber replaces natural rubber; glass and carbon fibers provide the raw material for rods, tubes, and sheet panels and other products superior in performance and strength to the metals they replace. For example, the invention of synthetic fibers decreased the demand for wool from Uruguay, damaging the sheep and wool industries upon which its economy depended. Thus, even as the world economy expands, demands and prices for traditional raw materials may remain depressed.

While prices paid for developing-country commodities tend to be low, prices charged for the manufactured goods offered in exchange by the developed countries tend to be high. To capture processing and manufacturing profits for themselves, some developing states have placed restrictions on the export of unprocessed commodities. Malaysia, the Philippines, and Cameroon, for example, have limited the export of logs in favor of increased domestic processing of sawlogs and exports of lumber. Some developing countries have also encouraged domestic manufacturing to reduce imports and to diversify their exports. Frequently, however, such exports meet with tariffs and quotas protecting the home markets of the industrialized states.

In 1964, in reaction to the whole range of perceived trade inequities, developing states promoted the establishment of the United Nations Conference on Trade and Development (UNCTAD). Its central constituency—the "Group of 77," which had expanded to 132 developing states today—continues to press for a new world economic order based, in part, on an increase in the prices and values of exports from developing countries, a system of import preferences for their manufactured goods, and a restructuring of international cooperation to stress trade promotion and recognition of the special needs of poor countries.

The World Trade Organization, established in 1995 (and discussed in detail in Chapter 8), was designed, in part, to reduce trade barriers and inequities. It has, however, been judged by its detractors as ineffective on issues of importance to developing countries. Chief among the complaints is the continuing failure of the industrial countries to significantly (or at all) reduce protections for their own agricultural and mineral industries. The European Union, the United States, and Japan together spend over

$380 billion per year on agricultural subsidies, about six times the amount they contribute on foreign assistance. Those subsidies prevent poor countries from strengthening their own agricultural sector and reduce incomes in poor countries.

In 2001, members of the World Trade Organization met in Doha, Qatar, to begin negotiations on opening world markets. Low-income, developing countries argued for the elimination of agricultural subsidies and protectionist policies in the European Union and the United States. In turn, the rich economies insisted on significant concessions from poor countries on trade in both manufactured goods and, particularly, services. The "Doha Round" round of trade negotiations continued through 2008 when trade negotiations broke down. Despite the fact that agriculture makes up less than 10% of world trade, it has been the roadblock in all global trade talks. The goals of greater fairness and openness in world trade, while maintaining special consideration for the economic development needs of poor countries, have yet to be achieved to the satisfaction of all parties. Finding a solution to these disputes requires consideration of the unique needs and economic specializations of each world region. The secondary and tertiary sectors of the economy, which traditionally were dominated by the advanced countries, are the subject of the following chapter.

Summary of Key Concepts

- Three recognized types of economic systems are *subsistence, commercial,* and *planned.* The first is concerned with production for the immediate consumption of individual producers and family members. In the second, economic decisions respond to impersonal market forces and reasoned assessments of monetary gain. In the third, at least some nonmonetary social or political goals influence production decisions. Most planned economies are now considered transition economies as they introduce market mechanisms.
- Economic activities may be grouped into three broad categories based on their role in the stages of production: *primary* activities (food and raw material production), *secondary* industries (processing and manufacturing), and *tertiary* activities (consumer services, business services, and distribution and sales).
- Different countries and regions are at different stages in their economic development; there is no single, inevitable pattern of progression from underdeveloped subsistence status to the advanced integration of secondary and tertiary activities that marks a modern market economy.
- Agriculture, the most extensively practiced of the primary industries, is part of the spatial economy of both subsistence and advanced societies. In the first instance, it is responsive to the immediate consumption needs of the producer group and reflective of environmental conditions. In the second, agriculture reacts to consumer demand expressed through free or controlled markets.
- Agricultural land uses and land values are influenced by the land's location relative to markets, as first theorized by von Thünen. Transportation costs for different commodities result, in theory, in concentric rings of progressively less-intensive agricultural land use around cities.
- No national economy exists in isolation; each is an interconnected part of a world system, of economic and cultural integration. Trade in primary commodities is an important and sometimes controversial part of the world economy. Many developing countries depend on primary commodities for the majority of their export earnings.

Key Words

agriculture 255	market (commercial)
aquaculture 270	economy 250
economic geography 248	maximum sustainable yield 269
extensive commercial	nomadic herding 256
agriculture 265	planned economy 250
extensive subsistence	plantation 267
agriculture 256	primary activity 249
extractive industry 268	secondary activity 249
gathering industry 268	shifting cultivation 257
Green Revolution 260	subsistence economy 250
informal economy 251	tertiary activity 249
intensive commercial	tragedy of the commons 269
agriculture 265	transition economy 250
intensive subsistence	von Thünen rings 263
agriculture 256	

Thinking Geographically

1. What are the distinguishing characteristics of the economic systems labeled *subsistence, commercial, planned,* and *transitional?* Are they mutually exclusive, or can they coexist within a single political unit?

2. How is *intensive subsistence* agriculture distinguished from *extensive subsistence* cropping? Why, in your opinion, have such different land use forms developed in separate areas of the warm, moist tropics?

3. Briefly summarize the assumptions and dictates of von Thünen's agricultural model. How might the land use patterns predicted by the model be altered by an increase in the market price of a single crop? A decrease in the transportation costs of one crop but not of all crops?

4. What economic or ecological problems affect the gathering industries of forestry and fishing? What is maximum sustainable yield? Is that concept related to the problems you discerned?

CHAPTER 10

Economic Geography: Manufacturing and Services

Digital communication technologies have allowed the globalization of service industries. These call center workers in India work night shifts to cater to customers in Europe and North America. © Moodboard/Thinkstock RF.

CHAPTER OUTLINE

LEARNING OBJECTIVES

After studying this chapter you should be able to:

10.1 Explain Weber's least-cost theory for the optimim location of different manufacturing industries.

10.2 Compare the locational strategies for bulk-reducing and bulk-gaining industries.

10.3 Explain how contemporary locational factors are creating a new international division of labor.

10.4 Explain the role of transnational corporations in the global economy.

10.5 Identify major centers of manufacturing and high-technology innovation.

10.6 Distinguish the locational patterns for consumer services and business services.

Route 837 connects the four U.S. Steel plants stretched out along the Monongahela River south of Pittsburgh. In the late 1960s, 50,000 workers labored in those mills, and Route 837 was choked with the traffic of steelworkers on their way to and from shifts and trucks loaded with steel. By 1979, blast furnaces in the aging mills were being shut down as lower-cost imports from Asia and Europe gained a foothold in the U.S. market. By the mid-1980s, with employment in the steel plants of the "Mon" Valley well below 5000, the highway was only lightly traveled and the mills were closed and deserted. From Massachusetts to Wisconsin, competition with Japanese imports and the opening of new assembly plants in lower-wage countries such as Mexico led to manufacturing job losses and empty factories. Derelict factories, once the economic heart of communities across the Northeast and Midwest, were left to rot.

At the same time, traffic was building along many highways in the northeastern part of the country. Four-lane Route 1 was clogged with traffic along the 42 kilometers (26 m) of the "Princeton Corridor" in central New Jersey as that stretch of road in the 1980s added more new office space, research laboratories, hotels, conference centers, and residential subdivisions than anywhere else in the eastern United States. Farther to the south, in the Virginia and Maryland suburbs of Washington, D.C., traffic grew heavy along the Capital Beltway and Dulles Toll Road, where vast office building complexes, defense-related industries, and commercial centers were converting rural land to urban uses. And east of New York City, traffic jams were monumental around Stamford, Connecticut, in Fairfield County, as it became a leading corporate headquarters town with 150,000 daily in-commuters.

By the early 1990s, traffic in Fairfield County had thinned as corporate takeovers, leveraged buyouts, and "downsizing" reduced the number and size of companies and their need for both employees and office space. Vacancies exceeded 25% among the office buildings and research parks so enthusiastically built during the 1970s

and 1980s, and vacant "corporate campuses" lined stretches of formerly clogged highways. Starting in the late 1990s, the explosive growth of China's manufacturing exports led to factory closures in a wide swath of the world including textile and garment factories in the Carolinas. At the same time, elsewhere in North America, economic prosperity induced rising traffic levels and rapid housing and commercial development. In California's "Silicon Valley" it was the computer software industry while in other "high tech" hot spots it was aerospace, biomedical devices, biotechnology, or energy technologies that fueled economic growth. Other urban regions prospered based on corporate headquarters and advanced business services such as finance and insurance.

These contrasting and fluctuating patterns of traffic flow symbolize the ever-changing structure of the North American economy. The smokestack industries of the 19th and early 20th centuries have declined, replaced by research park industries, shopping centers, and office building complexes that in their turn experience cyclical prosperity and adversity. The continent's economic landscape and employment structure are continually changing (**Figure 10.1**). And North America is not alone. Change is the ever-present condition of contemporary economies, whether of the already industrialized, advanced countries or of those newly developing in an integrated world marketplace. Resources are exploited and exhausted; markets grow and decline; patterns of economic advantage undergo alteration and reversal; global competition and technological innovations upset established hierarchies and create new patterns of growth, prosperity, and decline. In this chapter, our attention is directed to the changing locations of secondary and tertiary economic activities as we explore why they are distributed as they are and what the future might hold.

10.1 Industrial Location Theory

Since the Industrial Revolution and introduction of the factory system in late 18th-century England, manufacturing and heavy industry have been seen as the measures and symbols of economic development. Wherever introduced, manufacturing has been the catalyst for the whole range of economic and social changes recognized as modernization and the eagerly sought escape from traditional subsistence economies. The factory system and mass production of standardized, lower-cost goods were the spur to inventions and improvements in transportation and to the urbanization of populations freed from peasant agricultural drudgery. Wages and a money economy became the norm. Wholesale and retail trade expansion, increased labor specialization, and the rise of an economic middle class accompanied the introduction and expansion of manufacturing—first in Europe and North America and now throughout the world. And everywhere the new industrial economy was introduced, it prospered or faltered to the extent that it observed the new factors influencing industrial location and spatial behavior.

Although primary industries are locationally tied to the natural resources they gather or exploit, secondary and tertiary economic activities are less concerned with conditions of the physical environment. For them, location is more closely related to cultural and economic conditions. They are movable rather than spatially tied

Figure 10.1   Changing economic landscapes. The shining financial district of Shanghai, China, along the Huangpu River symbolizes some of the dramatic changes taking place in the geography of secondary and tertiary activites. China's dramatic growth as a manufacturing exporter has led to a growing role in financial services and other advanced business services. © *Elysee Shen/DigitalVision/Getty Images RF.*

and are assumed to respond to recurring locational requirements and controls.

Those controls are derived from human spatial behavior in general and economic behavior in particular. We have already explored some of those assumptions in earlier discussions. We noted, for example, that the intensity of spatial interaction decreases with distance—distance decay, we called it. Recall that von Thünen's model of agricultural land use was rooted in conjectures about transportation cost and land value relationships.

Such simplifying assumptions help us understand an accepted common set of controls and motivations guiding human economic behavior. We assume, for example, that people are *economically rational;* that is, given the information at their disposal, they make locational, production, or purchasing decisions in light of a perception of what is most cost-effective and advantageous. From the standpoint of producers or sellers of goods or services, it is further assumed that each is intent on *maximizing profit.* To reach that objective, a host of production and marketing costs and political, competitive, and other limiting factors may be considered, but the ultimate goal of profit-seeking remains clear. Although increasingly challenged as too rigid and unrealistic as explanations of actual human behavior and economic decision making, those assumptions still underlie most current analyses of the spatial patterns of industry.

Weber's Least-Cost Industrial Location Model

When market principles are controlling, entrepreneurs seek to maximize profits by locating manufacturing activities at sites of lowest total input costs (and high revenue yields). In order to

assess the advantages of one location over another, industrialists must evaluate the most important **variable costs**. They subdivide their total costs into categories and note how each cost will vary from place to place. In different industries, transportation charges, labor rates, power costs, plant construction or operation expenses, the interest rate of money, or the price of raw materials may be the major variable cost. The industrialist must look at each of these and, by a process of elimination, select the lowest-cost site. If the producer then determines that a large enough market can be reached cheaply enough, the location promises to be profitable.

In the economic world, nothing remains constant. Because of a changing mix of input costs, production techniques, and marketing activities, many initially profitable locations do not remain advantageous. Migrations of population, technological advances, and changes in the demand for products affect industrialists and industrial locations greatly. The abandoned mills and factories of New England or steel towns of Pennsylvania, even the "deindustrialization" of America itself in the face of a changing domestic economy and foreign industrial cost advantages and competition, are testimony to the impermanence of the "best" locations (**Figure 10.2**). Similarly, the spread of manufacturing from the more-developed to the less-developed regions of the world since World War II testifies to changing perceptions of manufacturing costs and locational advantages.

The concern with variable costs as a determinant in industrial location decisions has inspired an extensive theoretical literature. Much of it is based on industrial patterns and economic assessments seen as controlling during the later 19th and early 20th centuries and extends the **least-cost theory** proposed by German economist Alfred Weber (1868–1958), and sometimes called *Weberian analysis.* Weber explained the optimum location of a manufacturing

Figure 10.2   Deindustrialization. This derelict, abandoned farm equipment factory in Brantford, Ontario, typifies the structural changes occurring in postindustrial economies. © *Mark Bjelland.*

establishment in terms of the minimization of three basic expenses: relative transport costs, labor costs, and agglomeration costs. **Agglomeration** refers to the clustering of productive activities and people for mutual advantage. Such clustering can produce *agglomeration economies* through shared facilities and services. Diseconomies such as higher rents or wage levels resulting from competition for these localized resources may also occur.

Weber concluded that transport costs were the major consideration determining location. That is, the optimum location would be found where the costs of transporting raw materials to the factory and finished goods to the market were at their lowest (**Figure 10.3**). He noted, however, that, if variations in labor or agglomeration costs were sufficiently great, a location determined solely on the basis of transportation costs might not, in fact, be the optimum one. In later sections, we will revisit the importance of agglomeration and labor costs.

An example of such considerations is found in the location of steel mills in the United States and Canada. The iron and steel industry required multiple inputs from different locations: limestone, coking coal from Appalachia, and iron ore from mines in northern Minnesota and Michigan's Upper Peninsula. Steel mills located near the southern Great Lakes minimized the total costs of assembling the different raw materials and became the centers for other heavy industries (**Figure 10.4**).

Assuming, however, that transportation costs determine the "balance point," the optimum location will depend on distances, the respective weights of the raw material inputs, and the final weight of the finished product. The production process may be either *material-oriented* or *market-oriented*. Material orientation reflects a sizable weight loss during the production process; market orientation indicates a weight gain (**Figure 10.5**).

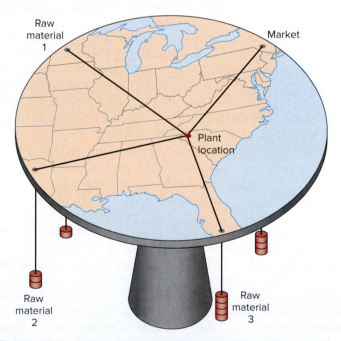

Figure 10.3   Plane table solution to a plant location problem. This mechanical model, suggested by Alfred Weber, uses weights to demonstrate the least-transport-cost point where there are several sources of raw materials. When a weight is allowed to represent the "pull" of raw material and market locations, an equilibrium point is found on the plane table. That point is the location at which all forces balance one another and represents the least-cost plant location.

For many theorists, Weber's least-cost analysis is unnecessarily rigid and restrictive. They propose instead a *substitution principle* which recognizes that in many industrial processes it is possible to

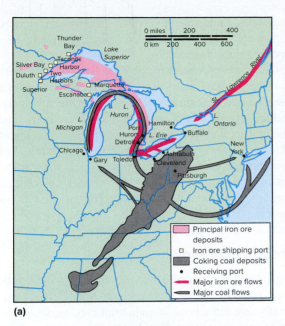

(a)

(b)

Figure 10.4   **Industrial location considerations in the steel industry.** (a) When an industrial process requires several heavy or bulky raw materials, an intermediate production site is often the least-cost location. (b) Steel mill in Gary, Indiana. The city of Gary, at the southern tip of Lake Michigan, was founded by the U.S. Steel Corporation in the early 1900s for the sole purpose of making steel at the lowest-cost location. Coal from Appalachia is transported by rail or water and iron ore from northern Minnesota and upper Michigan is shipped by boat on the Great Lakes. Other steelmaking locations such as Hamilton, Toledo, Cleveland, and Pittsburgh share similar locational advantages. © *Stockbyte/Thinkstock Images RF.*

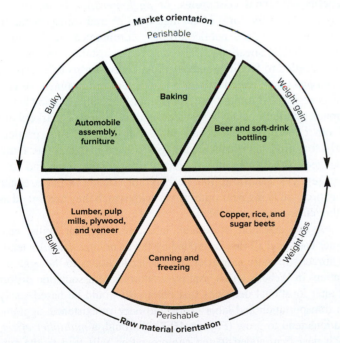

Figure 10.5   **Spatial orientation tendencies.** *Raw material orientation* exists when there are limited alternative material sources, when the material is perishable, or when, in its natural state, it contains a large proportion of impurities or nonmarketable components. *Market orientation* represents the least-cost solution when manufacturing uses commonly available materials that add weight to the finished product, when the manufacturing process produces a commodity much bulkier or more expensive to ship than its separate components, or when the perishable nature of the product demands processing at individual market points. *Source: Data: Truman A. Hartshorn, Interpreting the City, 1980 John Wiley & Sons, Inc., New York, NY..*

replace a declining amount of one input (e.g., labor) with an increase in another (e.g., capital for automated equipment) or to increase transportation costs while simultaneously reducing land rent. With substitution, a number of different points may be optimal manufacturing locations. Further, they suggest, a whole series of points may exist where total revenue of an enterprise just equals its total cost of producing a given output. These points, connected, mark the *spatial margin of profitability* and define the area within which profitable operation is possible (**Figure 10.6**). Location anywhere within the margin ensures some profit and tolerates both imperfect knowledge and personal (rather than economic) considerations.

Other Locational Considerations

The behavior of individual firms seeking specific production sites under competitive conditions forms the basis for most classical industrial location theory. But such theory no longer fully explains world or regional patterns of industrial localization or specialization; nor does it account for locational behavior that is uncontrolled by objective "factors," which are influenced by new production technologies and corporate structures or are directed by noncapitalistic planning goals.

Traditional theories sought to explain locational decisions for plants engaged in mass production for mass markets where transportation lines were fixed and transport costs were relatively high. Both conditions began to change significantly during the last years of the 20th century. Assembly line production of identical commodities by a rigidly controlled and specialized labor force for generalized mass markets—known as **Fordism** to recognize Henry Ford's pioneering development of the system—shifted toward less-developed regions. In developed regions, the surviving manufacturing shiefted toward post-Fordist *flexible manufacturing* processes based on smaller production runs of a greater variety of goods aimed at smaller,

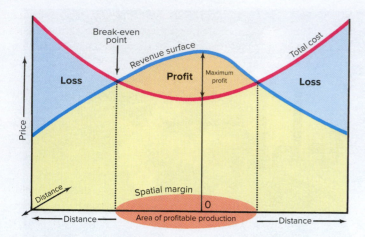

Figure 10.6   **The spatial margin of profitability.** In the diagram, 0 is the single optimal profit-maximizing location, but location anywhere within the area defined by the intersections of the total-cost and total-revenue surfaces will permit profitable operation. Some industries will have wide margins; others will be more spatially constricted. Skilled entrepreneurs may be able to expand the margins farther than less-able industrialists. Importantly, a *satisficing* location may be selected by reasonable estimate even in the absence of the totality of information required for an *optimal* decision.

niche markets. At the same time, information technology applied to machines and operations, increasing flexibility of labor, declining costs for transportation, and a greater emphasis on speed rather than cost of shipment have altered the underlying assumptions of Weber's classical theories.

Transport Characteristics

Within both national and international economies, the type, cost, and efficiency of transport modes have been central to the spatial patterns of production, explaining the location of a large variety of economic activities. Waterborne transportation is nearly always cheaper than any other mode of conveyance, and the enormous amount of commercial activity that takes place on coasts or on navigable rivers leading to coasts is an indication of that cost advantage. When railroads were developed and the commercial exploitation of inland areas could begin, coastal sites continued to be important as more and more goods were transferred there between low-cost water and land modes. The advent of highway transportation vastly increased the number of potential "satisficing" manufacturing locations by freeing factories from fixed-route corridors. Every change in carrier mode, efficiency, or cost structure has direct implications for locations of economic activity.

In the rare instance when transportation costs become a negligible factor in production and marketing, an economic activity is said to be *footloose*. Some manufacturing facilities are located without reference to raw materials; for example, the raw materials for electronic products such as computers are so valuable, light, and compact that transportation costs have little bearing on where production takes place. Others are inseparable from the markets they serve and are so widely distributed that they are known as

ubiquitous industries. Newspaper publishing, bakeries, and dairies, all of which produce a highly perishable commodity designed for immediate consumption, are examples.

Overall, transportation costs have been declining and efficiencies have been increasing. The advent of near-universal commercial jet aircraft service, the development of large ocean-going superfreighters, and the introduction of containerization and intermodal transfers in both water and land shipments of goods have reduced the costs and increased the speed of freight services. As those costs have decreased, manufacturing location has become more influenced by nontransport locational factors. To that extent, Weberian location theories have reduced applicability to the modern global space economy.

Agglomeration Economies

The geographic concentration of economic, including industrial, activities is the norm at the local or regional scale. The cumulative and reinforcing attractions of industrial concentration and urban growth are recognized locational factors, but ones not easily quantified. While Weber's theory emphasized transport costs, he recognized the importance of *agglomeration*, the spatial concentration of people and activities for mutual benefit. That is, the clustering of industrial activities often produces benefits for individual firms that they could not experience in isolation. Those benefits—**external economies**, or *agglomeration economies*—may accrue in the form of savings from shared transport facilities, social services, public utilities, communication facilities, and the like. Collectively, these and other installations and services needed to facilitate industrial and other forms of economic development are called **infrastructure**. Clustering may also create pools of skilled and ordinary labor, specialized knowledge, capital, ancillary business services, and, of course, a market built of other industries and urban populations. New firms, particularly, may find significant advantages in locating near other firms engaged in the same activity, for labor specializations and support services specific to that activity are already in place. Some may profit in being near other firms with which they are linked as either customers or suppliers.

A concentration of capital, labor, management skills, customer base, and all that is implied by the term *infrastructure* tends to attract still more industries from other locations to the agglomeration. In Weber's terms, that is, economies of association distort or alter locational decisions that otherwise would be based solely on transportation and labor costs, and once in existence, agglomerations tend to grow (**Figure 10.7**). Through a *multiplier effect,* each new firm added to the agglomeration will lead to the further development of infrastructure and linkages. As we will see in Chapter 11, the multiplier effect also implies total (urban) population growth and thus the expansion of the labor pool and the localized market that are part of agglomeration economies.

Just-in-Time and Flexible Production

Agglomeration economies and tendencies are also encouraged by newer manufacturing policies practiced by both older, established industries and newer, post-Fordist plants.

Figure 10.7   Agglomeration economies. On a small scale, the planned industrial park furnishes to its tenants external agglomeration economies similar to those offered by large urban concentration of industry. An industrial park provides a subdivided tract of land developed according to a comprehensive plan for the use of multiple firms. Since the park developers, whether private companies or public agencies, supply the basic infrastructure of streets, water, sewage disposal, power, transport facilities, and, perhaps, private police and fire protection, park tenants are spared the additional cost of providing these services themselves. In some instances, factory buildings are available for rent, still further reducing firm development outlays. Counterparts of industrial parks for manufacturers are the office parks, research parks, science parks, and the like for "high-tech" firms and for enterprises in the tertiary sector. © *Mark Bjelland.*

Traditional Fordist industries required the on-site storage of large lots of materials and supplies ordered and delivered well in advance of their actual need in production. That practice permitted cost savings through infrequent ordering and reduced transportation charges, and it made allowances for delayed deliveries and for the inspection of received goods and components. The assurance of supplies on hand for long production runs of standardized outputs was achieved at high inventory and storage costs.

Just-in-time (JIT) manufacturing, in contrast, seeks to reduce inventories through the production process by purchasing inputs for arrival just in time to use and producing output just in time to sell. Rather than the costly accumulation and storage of supplies, JIT requires frequent ordering of small lots of goods for precisely timed arrival and immediate deployment to the factory floor. Such *lean manufacturing* based on the frequent purchasing of immediately needed goods requires rapid delivery by suppliers and encourages them to locate near the buyer. Recent manufacturing innovations thus reinforce and augment the spatial agglomeration tendencies evident in the older industrial landscape and deemphasize the applicability of older, single-plant location theories.

JIT is one expression of a transition from mass production Fordism to more **flexible production** systems. That flexibility is designed to allow producers to shift quickly and easily between different levels of output and, importantly, to move from one factory process or product to another as market demand dictates. Flexibility of that type is made possible by new technologies of easily reprogrammed computerized machine tools and by computer-aided design and computer-aided manufacturing systems. These technologies permit small-batch, just-in-time production and distribution responsive to current market demand as monitored by computer-based information systems.

Flexible production, to a large extent, requires significant acquisition of components and services from outside suppliers rather than from in-house production. In manufacturing, *outsourcing*—subcontracting production work to outside companies—has become an important element in just-in-time acquisition of preassembled components for snap-together fabrication of finished products. For example, modular assembly, in which many subsystems of a complex final product enter the plant already assembled, reduces factory space and worker requirements. The premium that flexibility places on proximity to component suppliers adds still another dimension to industrial agglomeration tendencies.

"Flexible production regions" have, according to some observers, emerged in response to the new flexible production strategies and interfirm dependencies. Those regions, it is claimed, are usually some distance—spatially or socially—from established concentrations of Fordist industrialization. In part, this is so because outsourcing firms often rely upon flexible, nonunion labor rather than the unionized labor associated with Fordism.

Comparative Advantage and the New International Division of Labor

The principle of *comparative advantage* and the practice of *offshoring* are of growing international importance in decisions regarding industrial location. They are interconnected in that each reflects cost advantages of specialization and each is dependent on free trade policies and the free flow of information. **Comparative advantage** is an economic concept that suggests areas and countries can best improve their economies and living standards through specialization and trade. These benefits will follow if each area or country concentrates on the production of those items for which it has the greatest relative advantage over other areas or for which it has the least relative disadvantage and imports all other goods. This principle is basic to the understanding of regional specialization, and it applies as long as areas have different relative advantages for two or more goods and free trade exists between them.

The logic of comparative advantage was recognized by economists in the 19th century when specialization and exchange involved shipments of grain, coal, or manufactured goods whose relative costs of production in different areas were clearly evident. Today, when the comparative advantages of other countries may reflect lower costs for labor or less-stringent environmental regulations, the application of the principle is seen in a much less-favorable light by some critics. They observe that manufacturing activities may relocate from higher-cost market-country locations to lower-cost foreign production sites, taking jobs and income away from the consuming country to the apparent detriment of that country's prosperity.

Outsourcing involves subcontracting production or service tasks to an outside company rather than performing them in-house. Outsourcing gives companies greater flexibility and can save costs. **Offshoring** is the practice of either hiring foreign workers or, commonly, contracting with a foreign third-party service provider to take over particular manufacturing operations. Offshoring has become an increasingly standard cost-containment strategy, reflecting the recent steep decline in shipping costs, the ease of long-distance communications, and the growing manufacturing capabilities of developing countries.

The motivation for offshoring is obvious after examining the wide spatial variation in hourly compensation (wages plus benefits) for manufacturing work. According to the U.S. Bureau of Labor Statistics in 2012, costs varied from $58 per hour in Norway or $35 per hour in the United States to just $6 per hour in Mexico and $1.90 per hour in the Philippines. Defenders of free trade and comparative advantage argue that the increased efficiencies of such industrial offshoring increases overall prosperity.

A distinctive regional illustration of the changing geography of industrial production is found along the northern border of Mexico. In the 1960s, Mexico enacted legislation permitting foreign (specifically, American) companies to establish "sister" plants, called *maquiladoras,* within 20 kilometers (12 mi) of the U.S. border for the duty-free assembly of products destined for re-export. By the early 20th century more than 3000 such assembly and manufacturing plants had been established to produce a diversity of goods, including electronic products,

textiles, furniture, leather goods, toys, and automotive parts. The plants generated direct and indirect employment for more than a million Mexican workers (**Figure 10.8**) and for large numbers of U.S. citizens, employees of growing numbers of American-side *maquila* suppliers and of diverse service-oriented businesses spawned by the "multiplier effect."

On the broader world scene, offshoring often involves production of consumer goods by developing countries that have benefited from the transfer of technology and capital from industrialized states. For example, electrical and electronic goods from China and Southeast Asia compete with and replace similar goods formerly produced by Western firms. Such replacements, multiplied many times over, have resulted in new global patterns of industrial regions and specializations. They have also strikingly changed the developing world's share of gross global output from an estimated 20% in the mid-20th century to about half today. That improvement reflected, in part, growth in manufacturers' share of their exports. For some observers, that change is ample proof of the beneficial impact of comparative advantage on the world economy.

The exploitation of comparative advantage by offshoring manufacturing activities to less-developed regions and transferring technology from economically advanced to underdeveloped economies has introduced a **new international division of labor (NIDL)**. In the 19th century and the first half of the 20th century,

Figure 10.8 **Offshoring.** Seeking lower labor costs, a large proportion of American electronics, small appliance, auto, and garment industries has been moved to offshore subsidiaries or contractors in Asia and Latin America. More recently, higher-skilled work such as software design and engineering has been shifted to offshore locations such as this Google facility in China. © *Renaud Rebardy/Alamy Stock Photo.*

WHERE DO YOUR CLOTHES COME FROM?

One of the distinguishing features of humans is that, unlike other animals, we (almost always) clothe our bodies. The clothing we wear expresses our culture, values, social status, and self-identity. In the United States, at least, clothes are required to carry a label indicating the country of origin. A quick check through your closet is likely to reveal the international nature of the clothing industry. Clothing is second only to agriculture as the leading product in international trade. Textiles and clothing production were at the heart of the original Industrial Revolution and are one of the leading ways for developing countries to industrialize. The clothing industry was the leader in globalizing production and creating a new international division of labor. The technological requirements are fairly simple, and the production process is labor-intensive, offering a comparative advantage to low-wage countries. After World War II, Japan used clothing production to jump-start its manufacturing sector. The newly industrializing countries, and in particular China, are following that pattern. China is now the world's leading clothing manufacturer and is gaining share at the expense of most other countries. The geography of clothing production is changing rapidly. In a vintage clothing store, many of the clothes will have been manufactured in the United States.

Broader trends in manufacturing are also evident in the clothing industry. Export-processing zones are common for garment manufacturing. Maquiladoras along the Mexico-U.S. border assumed an important role after the passage of the North American Free Trade Act. Just-in-time manufacturing and lean, flexible manufacturing have become increasingly important along with more-rapid turnover of styles with the advent of fast-fashion as pioneered by global brands such as Swedish retailer H and M and Spanish brand Zara. While most mass-market production has moved to developing countries, most major brands and customers are based in the developed countries. High-end fashion production must be closely connected to the designers and shows in the major fashion centers of New York, London, Paris, and Milan. Thus, most production of high-priced, specialty fashion remains near those cities. Lower-cost fashion, especially for discount retailers, has moved most relentlessly to the lowest-wage countries.

Working conditions in clothing factories, or "sweatshops" as they sometimes known, remain a major concern of human rights watch groups. Workers are mostly female, may be subject to long hours and unsafe working conditions, and may have little recourse to file complaints.

the international division of labor invariably involved exports of manufactured goods from the "industrial" countries and of raw materials from the "colonial" or "undeveloped" economies. Roles have now altered. Manufacturing no longer is the mainstay of the economy or the employment structure of Europe or North America, and the world pattern of industrial production is shifting to reflect the growing dominance of countries once regarded as subsistence peasant societies that are now major manufacturing exporters for the world market (see the feature "Where Do Your Clothes Comes From?" p. 283). In recognition of that shift, the NIDL builds on the current trend toward the increased subdivision of manufacturing processes into smaller steps. That subdivision permits multiple outsourcing and offshoring opportunities based on differential land and capital costs and skill levels available in the globalized world economy.

Political Considerations

Location theories dictate that, in a pure, competitive economy, the costs of material, transportation, labor, and plant should be dominant in locational decisions. Obviously, neither in the United States nor in any other market economy do the idealized conditions exist. Other constraints—some representing cost considerations, others political or social impositions—also affect, perhaps decisively, the locational decision process. Land use and zoning controls, environmental quality standards, government regional development inducements, local tax abatement provisions or developmental bond authorizations, noneconomic pressures on quasi-government corporations, and other considerations

constitute attractions or repulsions for industry outside of the context and consideration of pure theory (see "Incentives or Bribery?" p. 293). If these political considerations become compelling, the assumptions of classical industrial location theory no longer dominate, and locational decisions resemble those made in centrally planned economies.

No other imposed considerations were as pervasive as those governing industrial location in planned economies. The theoretical controls on plant locational decisions that apply in commercial economies were not, by definition, determinant in the centrally planned Marxist economies of Eastern Europe and the former Soviet Union. In those economies, plant locational decisions were made by government agencies rather than by individual firms.

Bureaucratic rather than company decision making did not mean that locational assessments based on factor cost were ignored; it meant that central planners were more concerned with other than purely economic considerations in the creation of new industrial plants and concentrations. Important in the former Soviet Union, for example, was a controlling policy of the *rationalization* of industry through full development of the resources of the country wherever they were found and without regard to the cost or competitiveness of such development. Inevitably, although the factors of industrial production are identical in capitalist and noncapitalist economies, the philosophies and patterns of industrial location and areal development will differ between them. Because major capital investments are relatively permanent additions to the landscape, the results of their often noneconomic political or philosophical decisions are fixed and will long remain to influence industrial regionalism and competitive efficiencies into the postcommunist present and future.

Those same decisions and rigidities continue to inhibit the transition by the formerly fully planned economies to modern capitalist industrial techniques and flexibilities.

10.2 World Manufacturing Patterns and Trends

Whether locational decisions are made by private entrepreneurs or by central planners—and on whatever considerations those decisions are based—the results over many years have produced a distinctive world pattern of manufacturing. **Figure 10.9** suggests the striking prominence of a relatively small number of major industrial concentrations localized within relatively few countries primarily but not exclusively parts of the "industrialized" or "developed" world. These can be roughly grouped into four commonly recognized major manufacturing regions: *Northastern and Midwestern United States and Eastern Canada, Western and Central Europe, Eastern Europe,* and *Eastern Asia.* Together, the industrial plants within these established concentrations account for an estimated three-fifths of the world's manufacturing output by volume and value.

Their continuing dominance is by no means assured. The first three—those of the United States, Canada, and Europe—were the beneficiaries of an earlier phase in the development and spread of manufacturing following the Industrial Revolution of the 18th century and lasting until after World War II. The countries within them now increasingly exhibit postindustrial economies in which traditional manufacturing and processing are of declining relative importance.

The fourth—the East Asian industrial region—is a part and forerunner of the wider, new pattern of world industrialization that has emerged in recent years, the result of international cultural convergence and technology transfers since World War II, as briefly reviewed in Chapter 6. The older, rigid industrial "North–South" split between the developed and developing worlds has rapidly weakened as the full range of industrial activities from primary metal processing (e.g., the iron and steel industries) through advanced electronics assembly has diffused to an ever-expanding list of countries.

Such states as Mexico, Brazil, China, and others of the developing world have created industrial regions of international significance, and the contribution to world manufacturing activity of the smaller, newly industrializing countries (NICs) has been growing significantly. The list of NICs includes the four Asian tigers: South Korea, Hong Kong, Singapore, and Taiwan. They each developed by combining well-educated workforce, infrastructure investments, and policies supporting export-oriented industrialization. Following them came the Asian dragons: Malaysia, Indonesia, and Thailand. The largest of all NICs is China whose economy has witnessed tremendous growth since the 1970s, passing Japan in 2010 to become the world's second-largest economy after the United States.

The adoption of efficient and sercure containerized shipment of high-value goods has helped NICs succeed as manufacturing exporters (**Figure 10.10**) Even economies that until recently were overwhelmingly subsistence or dominated by agricultural or mineral exports have become important players in the changing world manufacturing scene. Foreign branch plant investment in low-wage Asian, African, and Latin American states has not only built up their industrial infrastructure, but also increased their gross national products and per capita incomes sufficiently to permit expanded production for growing domestic—not just export—markets.

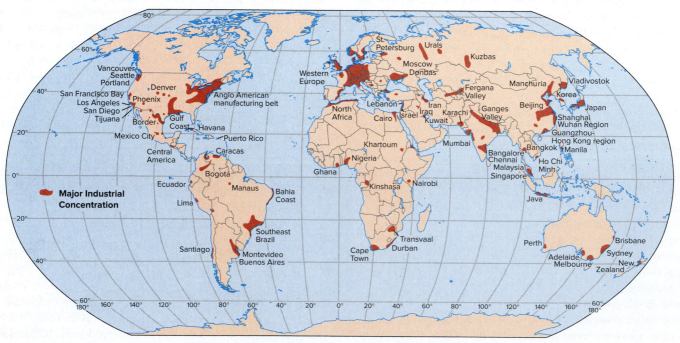

Figure 10.9 World industrial regions. Industrial districts are not as continuous or "solid" as the map suggests. Manufacturing is a relatively minor user of land even in areas of its greatest concentration.

Figure 10.10   **Standardized cargo containers have revolutionized shipping.** The Port of Vancouver, Canada, is a major terminal for goods exported from Asia to North America. The containers will be loaded onto trucks and railcars for delivery to their final destinations. © Mark Bjelland.

10.3 High-Tech Innovation

Major industrial districts of the world developed over time as entrepreneurs and planners established traditional secondary industries according to the factors found in classical location theories. The location of raw materials was a critical factor as were transportation costs. Those theories are less applicable in explaining the location of the latest generation of manufacturing activities: the high-technology—or *high-tech*—processing and production that are increasingly essential to the advanced economies. For high-technology firms, new and different patterns of locational orientation and competitive advantage have emerged. Often the new factors have to do with the clustering of talented, creative workers.

High technology is more an ambiguous concept than a precise definition. Today's high-tech fields may become standard in just a few years and old, obsolete fields within a few decades. It probably is best understood as the application of intensive research and development efforts to the creation and manufacture of new products of an advanced scientific and engineering character. Professional—"white-collar"—workers make up a large share of the total workforce. They include research scientists, engineers, and skilled technicians. When these high-skill specialists are added to administrative, supervisory, marketing, and other professional staffs, they may greatly outnumber actual production workers in a firm's employment structure.

Although only a few types of industrial activity are generally reckoned as exclusively high-tech—electronics, communications, computers, software, pharmaceuticals and biotechnology, aerospace, and the like—advanced technology is increasingly a part of the structure and processes of all forms of industry. Robotics on the assembly line, computer-aided design and manufacturing, electronic controls of smelting and refining processes, and the constant development of new products of the chemical and pharmaceutical industries are cases in point. In the world of high technology, manufacturing work is often knowledge-intensive, involving the operation of expensive, sophisticated equipment.

The impact of high-tech industries on patterns of economic geography is expressed in several ways. First, high-tech activities are major factors in employment growth and manufacturing output in the advanced and newly industrializing economies. Relatively high wages in high-tech occupations reflect the level of training required and the high worker productivity. Second, high-tech industries have become regionally concentrated in centers of innovation, frequently forming self-sustaining, highly specialized agglomerations (**Figure 10.11**). Third, the growth of high-tech industries is often accompanied by the offshoring of less-skilled production and assembly tasks, spurring the growth of NICs.

Agglomeration economies are particularly important to high-tech innovation. In certain places around the globe, talented, creative people come together so that knowledge of how to make a particular product seems to be "in the air." Concentrations of high-tech employment include California, the Pacific Northwest (including British Columbia), New England, New Jersey, Texas, and Colorado. Within these and other states or regions of high-tech concentration, specific locales have achieved prominence. Silicon Valley in Santa Clara County near San Francisco is probably the most important. The Silicon Valley region boasts institutions such as Stanford University and government research labs, a high quality of life, and a remarkable concentration of skilled workers who can share ideas in formal or informal social settings (**Figure 10.12**).

Other clusters of high-tech employment include Irvine and Orange County south of Los Angeles; the Silicon Forest near Seattle; North Carolina's Research Triangle; Utah's Software Valley; Routes 128 and 495 around Boston; Silicon Swamp of the Washington, D.C., area; Ottawa, Canada's, Silicon Valley North; and the Canadian Technology Triangle west of Toronto.

Within such concentrations, specialization is often the rule: biomedical technologies in Minneapolis and Philadelphia; biotechnology around San Antonio; computers and semiconductors in eastern Virginia and at "Silicon Hills" in Austin, Texas; biotechnology and telecommunications in New Jersey's Princeton Corridor; and telecommunications and Internet industries near Washington, D.C. Elsewhere, Scotland's Silicon Glen, England's Sunrise Strip and Silicon Fen, Wireless Valley in Stockholm, Zhong Guancum in suburban Beijing, and the high-tech industries zone in Xian, and High-Tec city, Pune, and Bangalore, India, are

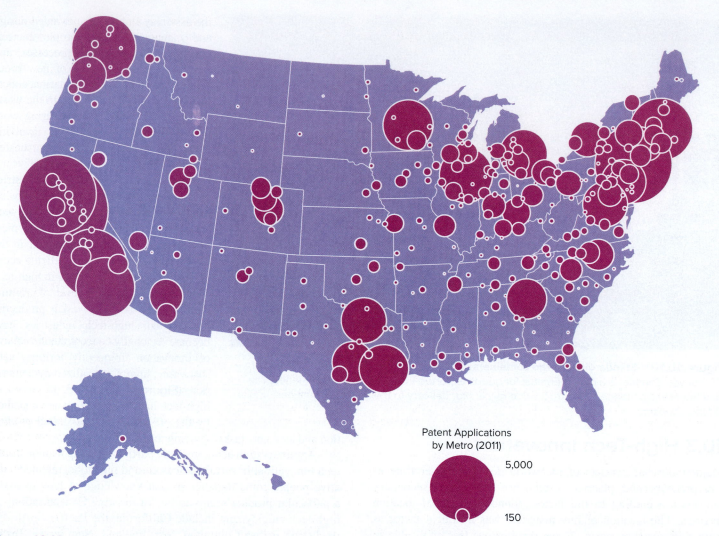

Figure 10.11   Patent applications are clustered in hubs of technological innovation. Technological innovation is geographically uneven. Hubs of innovation tend to be home to highly educated residents, research universities, research laboratories, established technology companies, venture capital investors, and an air of innovation, idea exchange, and entrepreneurship. *Source: http://www.citylab .com/work/2013/10/where-americas-inventors-ara/7069/.*

other examples of industrial landscapes characterized by low, modern, dispersed office-plant-laboratory buildings rather than by massive factories, mills, or assembly structures, freight facilities, and storage areas. Planned business parks catering to the needs of smaller companies are increasingly a part of regional and local economic planning.

The spatial patterns of high-tech industries suggest that they respond to different locational factors than those controlling heavy manufacturing industries. At least five locational tendencies have been recognized: (1) proximity to major research universities or government research laboratories and to a large pool of scientific and technical labor skills; (2) the avoidance of areas with strong labor unionization where contract rigidities might slow innovation and workforce flexibility; (3) locally available venture capital and entrepreneurial daring; (4) location in regions and major metropolitan areas with favorable "quality of life" reputations—climate, scenery, recreation, good universities, artistic and cultural opportunities, and large enough to provide job opportunities for professionally trained partners; (5) the availability of first-quality communication

and transportation facilities to unite separated stages of research, development, and manufacturing and to connect the firm with suppliers, markets, finances, and the government agencies so important in supporting research.

Nearly all the major high-tech agglomerations have developed on the outer suburban peripheries of metropolitan areas far from inner-city problems and disadvantages. Many have emerged as self-sufficient areas of subdivisions, shopping centers, schools, and parks in close proximity to the company locations and business parks that form their core. Although the New York metropolitan area is a major high-tech concentration, most of the technology jobs are suburban, not in Manhattan, which is more attractive to financial services and creative service industries, such as advertising and Web design.

The formation of new firms is frequent and rapid in industries where discoveries are constant and innovation is continuous. Because many are "spin-off" firms founded by skilled employees leaving established local companies, areas of existing high-tech concentration tend to spawn new entrants. The large pool of skilled

Figure 10.12 **Silicon Valley.** The area around San Jose, California, south of San Francisco, is the world's leading hub for high-tech innovation. The valley was once known for its orchards but is now home to leading technology companies such as Apple, Advanced Micro Devices, Cisco, eBay, Facebook, Google, Hewlett-Packard, Intel, Oracle, and Tesla Motors. It is named for the silicon chips used in computer and electronics semiconductors. © *David McNew/Hulton Archive/Getty Images.*

information-technology goods, such as laptop computers, mobile phones, and digital cameras. With rising education levels, countries such as China, India, Singapore, and South Korea are producing large numbers of highly trained scientists and engineers capable of doing much more than assembly work. Thus, computer software companies have begun taking advantage of India's strengths in engineering and computer science, making Bangalore and Hyderabad major world players in software development.

10.4 Transnational Corporations

Outsourcing is one piece of the growing international structure of manufacturing and service enterprises. Business and industry are increasingly stateless and economies borderless as giant **transnational corporations (TNCs)**—private firms that have established branch operations in nations foreign to their headquarters' country—become an ever-more important driving force in the globalizing world space economy.

workers and availability of supporting industries improve the chances of success for new spin-off firms. Agglomeration, therefore, is both a product and a cause of spatial concentrations.

Not all phases of high-tech production, however, must be concentrated. The professional, scientific, and knowledge-intensive aspects of the high-tech economy are often located far from the component manufacturing and assembly operations. Highly automated or low-skill assembly tasks are "foot-loose"; they require highly mobile capital and technology investments but may be performed at a lower cost in lower-wage countries such as China, Taiwan, Singapore, Malaysia, or Mexico. Major high-tech companies such as Apple, Sony, and Microsoft outsource the assembly of electronic devices to independent contract manufacturers. Most often the same factory produces similar or identical products under a number of different brand names (**Figure 10.13**). Contract electronics manufacturers are sometimes under extreme pressure to meet production deadlines because of the highly competitive, time-sensitive nature of their products.

High-tech products often have complex, highly international **commodity chains**—steps in the production and distribution process. The iPhone was designed by U.S. engineers, but its manufacture uses rare metals from Asia and Africa and specialized components manufactured in Germany, Korea, Taiwan, and Japan, all of which come together in a Chinese assembly plant.

Through such outsourcing and technology transfers, high-tech activities are spread to NICs. This globalization through geographic transfer and diffusion represents an important impact of high-tech activities on world economic geographic patterns. For example, China has surpassed the United States in exporting

Figure 10.13 **Contract electronics manufacturing.** Assembly work that can be broken down into multiple, repetitive steps is often outsourced to overseas contractors. The Taiwanese company FoxConn Technology Group operates large assembly plants in China, Brazil, Mexico, the Czech Republic, and many other countries, employing 1.2 million workers in 2012. It manufactures many popular computer, consumer electronics, and communication devices such as the iPad, iPhone, PlayStation, XBox 360, Wii, and Amazon Kindle. Its largest plant in Shenzhen, China, nicknamed FoxConn City, is a walled, self-contained campus complete with factories, dormitories, and all the services needed by its estimated 250,000 workers. © *STR/Getty Images.*

The total annual revenue of the world's largest TNCs—also known as multinationals—rivals the gross domestic product of entire countries. For example, Wal-Mart Stores in 2015 had $480 billion in revenues which, if it were a country, would have placed it in close competition with Argentina, Belgium, and Poland. The great majority of large TNCs are headquartered in East Asia, Europe, and the United States. With the exception of Wal-Mart, most of the largest TNCs are involved in manufacturing or petroleum extraction and refining. Ranking companies by total revenues in 2014 would put Walmart, Sinopec Group, Royal Dutch Shell, China National Petroleum, and Exxon Mobil in the top five positions. Although tertiary activities have also become international in scope and transnational in corporate structure, the locational and operational advantages of multicountry operation were first discerned and exploited by manufacturers.

The direct impact of TNCs is limited to relatively few countries and regions. **Foreign direct investment (FDI)**—the purchase or construction of factories and other fixed assets by TNCs—has been a significant engine of globalization. While over one-half of FDI goes from one developed country to another developed country, a growing proportion is invested in less-developed economies, potentially stimulating their economic growth. The three main sources for outward FDI are the countries or regions that are home to the largest TNCs—the United States, Europe, Hong Kong, and Japan. Within Europe, the United Kingdom, Germany, and France are the leaders in foreign direct investment. The leading destinations for inward FDI are Hong Kong, China, Singapore, Mexico, Brazil, and India. Distance and proximity influence where FDI flows go. For example, FDI from the United States is more likely to go to Latin America while Asian countries are more likely to invest in other Asian countries.

The portion of FDI going to the 50 least-developed countries as a group—including nearly all African states—remains less than 5%. Despite poor countries' hopes for foreign investment to spur their economic growth, critics argue it is counterproductive. Economic control is lost to a foreign firm and may undermine political sovereignty as the TNC demands subsidies and tax breaks. TNCs may rely on foreign suppliers instead of local firms, bankrupt local competitors who lack the capital to compete, and then return their profits to the home country rather than reinvest them in the host country.

Investment outflows from companies based in India, Brazil, South Africa, Malaysia, and China (among others) have swelled, with an increasing share going to other developing countries. Because over 80% of the world's 7 billion consumers live in the expanding less-developed nations, TNCs based in newly industrializing countries have the strength of familiarity with those markets and have an advantage in supplying them with goods and services that are usually cheaper and more effectively distributed than those of many Western TNCs.

The advanced-country destination of over one-half of FDI capital flows is understandable: TNCs are actively engaged in merging with or purchasing competitive established firms in already developed foreign market areas. Because most TNCs operate in only a few industries—computers, electronics, petroleum and mining, motor vehicles, chemicals, and pharmaceuticals—the worldwide impact of their consolidations is significant. Some dominate the marketing and distribution of basic and specialized commodities. In raw materials, a few TNCs account for most world trade in wheat, maize, coffee, cotton, iron ore, and timber, for example. Because they are international in operation with multiple markets, plants, and raw material sources, TNCs actively exploit the principle of comparative advantage and seize opportunities for outsourcing and offshoring. In manufacturing they have internationalized the plant-siting decision process. TNCs produce in the country or region where costs of materials, labor, or other production inputs are minimized. At the same time, they can shift their official headquarters location and pay taxes where the rates are lowest. Research and development, accounting, and other corporate activities are placed wherever it is economical and convenient.

TNCs have become global entities because global communications make it possible (**Figure 10.14**). Many have lost their original national identities and are no longer closely associated with or controlled by the cultures, societies, and legal systems of a nominal home country. At the same time, their multiplication of economic activities has reduced any earlier identifications with single products or processes and given rise to "transnational integral conglomerates" spanning a large spectrum of both service and industrial sectors.

10.5 Tertiary Activities

Primary activities, you will recall, gather, extract, or grow things. *Secondary* industries add value to the products of primary industry through manufacturing and processing. A major and growing segment of both domestic and international economic activity, however, involves *services* rather than the production of commodities. These *tertiary* activities consist of those business and labor specializations that provide services to the primary and secondary sectors, to the general community, and to individuals. They imply pursuits other than the actual production of tangible commodities.

As we saw in Chapter 9, regional and national economies undergo fundamental changes in emphasis in the course of their development. Subsistence societies exclusively dependent on primary industries may progress to secondary-stage processing and manufacturing activities. In that progression, the importance of agriculture, for example, as an employer of labor or as a contributor to national income declines as that of manufacturing expands. As economic growth continues, secondary activities in their turn are replaced by service, or tertiary, functions as the main support of the economy.

Many of the economically advanced countries that originally dominated world manufacturing experienced deindustrialization starting in the 1970s. Rising labor costs in advanced economies, space-shrinking technologies for communications and transportation, the growth of transnational corporations, technology transfer to developing countries, and outsourcing have produced a new international division of labor. The earlier competitive manufacturing advantages of the developed countries could no longer be maintained and were replaced by a new focus on service activities. Based on the contribution of each sector to their gross domestic products, it is the advanced economies that have most completely made that transition and are often referred to as "postindustrial" (**Table 10.1**).

Perhaps more than any other major country economy, the United States has reached postindustrial status. Its primary sector

Figure 10.14 **Transnational corporations (TNCs) are engines of globalization.** The number of TNCs has grown rapidly since the 1970s. The headquarters of top TNCs are dominated by the United States, Europe, China, Japan. Their name recognition and impact, however, is global as suggested by these scenes from the developing world. McDonald's—USA and BP—U.K. shown in Russia; Wal-Mart—U.S.A. in China and Pepsi—U.S.A. in Vietnam. *(a) © Bloomberg/Getty Images; (b) © Cancan Chu/Getty Images News/Getty Images; (c) © Peter Charlesworth/LightRocket/Getty Images.*

Table 10.1	Contribution of the Service Sector to Gross Domestic Product		
	Percentage of GDP		
Country Group	**1960**	**1980**	**2013**
Low-income	32	29	47
Middle-income	47	48	56
High-income	54	61	74
United States	58	63	78
World		55	71

Source: Data from World Bank.

component fell from 66% of the labor force in 1850 to 1% in 2010, and the service sector rose from 18% to 86% (see Figure 9.7). Virtually all net job growth in the United States over the past two decades occurred in services, leaving behind many resource-dependent and manufacturing-dependent communities. Comparable changes have occurred in Japan, Canada, Australia, Israel, and all major Western European countries.

The significance of tertiary activities to national economies and the contrast between more-developed and less-developed states are made clear not just by employment but also by the differential contribution of services to the gross domestic products of states. The relative importance of services displayed in **Figure 10.15** shows a marked contrast between advanced and subsistence societies. The greater the service share of an economy, the greater the average incomes and economic complexity of that society. That share has grown over time among most regions, and all national income categories as all economies have shared to some degree in economic growth and integration into the world economy Indeed, the expansion of the tertiary sector in modernizing East Asia, South Asia, and the Pacific has exceeded the world average in recent decades as these regions catch up.

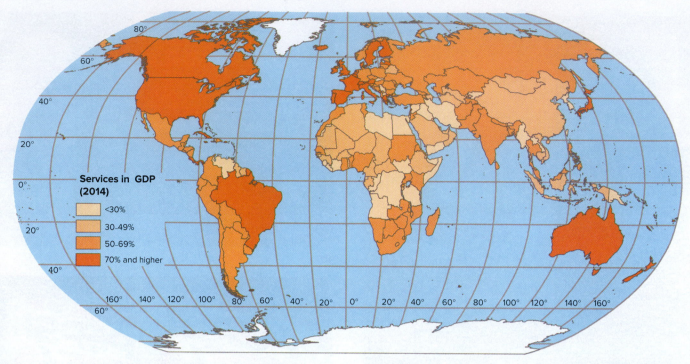

Figure 10.15   Services in GDP. Services accounted for over 70% of global GDP in 2013, up sharply from about 55% in 1980. As the map documents, the contribution of services to individual national economies varied greatly; Table 10.1 indicates all national income categories shared to some degree in the expansion of service activities. *Source: World Bank, World Development Indicators 2015.*

Just as service activities have been major engines of national economic growth, so too have they become an increasing factor in international trade flows and economic interdependence. Between 1980 and 2010, services increased from 15% of total world trade to 20%. Rapid advances and reduced costs in information and communications technology have been central elements in the internationalization of services, as wired and wireless communication and data transmission costs have dropped to negligible levels. Many services considered nontradable until recently are now actively exchanged at long distance, as the growth of services offshoring clearly shows. When manufacturing jobs were first offshored, most services workers from advanced countries assumed their jobs would remain secure. Increasingly that assumption is being challenged for certain types of service sector activities.

Types of Service Activities

Tertiary and *service* are broad and imprecise terms that cover activities ranging from the neighborhood barber to the president of the World Bank. As a category, services takes in low-order personal and retail activities as well as higher-order, knowledge-based professional services performed primarily for other businesses, not for individual consumption. Based on who purchases the services, we distinguish between consumer services and business services. **Consumer services** are performed for individuals and include entertainment, tourism, restaurants, hotels, bars, maintenance services, education, health care, and the vast array of personal services. **Business services** are performed for companies and include finance, insurance, real estate, legal services, accounting, architecture, and engineering consulting services.

Wholesale and retail trade are categories of services that link producers and consumers. Transportation and communication services also serve both producers and consumers. Government and nonprofit service providers are also important components of the service economy.

Growth in the tertiary sector has numerous explanations. It reflects the development of ever more-complex social, economic, and administrative structures, the effects of rising personal incomes and changes in family structure and individual lifestyles. For example, in subsistence economies, families care for their own children, produce and prepare their own food, and repair their own houses. In postindustrial societies, people hire child-care workers to care for the children, send their children to formal schools and universities, purchase prepared meals in restaurants, and hire contractors to repair their houses. Similar needs are met, but with very different employment structures.

As personal incomes rise, a greater proportion of income is spent on services rather than primary products or durable goods. If a person gets a raise, she or he might take a cruise ship vacation or dine out at restaurants more often, but isn't likely to add a second washing machine. Growth in the health care industry is driven by both rising incomes and the aging of society that inevitably occurs when a society completes the demographic transition. Growing complexity in the economy translates into the need for higher levels of education and training as well as more government employees to collect taxes, control borders, alleviate poverty, ensure public safety, plan community development, monitor commerce, protect the environment, and maintain safe workplaces.

Part of the growth in the tertiary component is statistical, rather than functional. We saw in our discussion of manufacturing that outsourcing was increasingly used to reduce costs and improve efficiency. In the same way, outsourcing of services formerly provided in-house is also characteristic of current business practice. Cleaning and maintenance of factories, shops, and offices—formerly done by the company itself as part of internal operations—now are subcontracted to specialized service providers. The jobs are still done, perhaps even by the same personnel, but worker status has changed from secondary (as employees of a manufacturing plant, for example) to tertiary (as employees of a service company).

Locational Interdependence Theory for Services

The locational controls for services are simpler than those for the manufacturing sector. Service activities are by definition market-oriented. Those dealing with transportation and communication are concerned with the location of people and commodities to be connected or moved; their locational determinants are therefore the patterns of population distribution and the spatial structure of production and consumption. Just as Weber offered a classic location theory for manufacturing enterprises, economist Harold Hotelling (1895–1973) used simplifying assumptions to create the locational interdependence model for retail services. In the locational interdependence model, the location decisions of firms are influenced by those of its competitors. Firms choose locations that give them a measure of spatial monopoly to maximize revenues, rather than minimize costs as in Weber's model.

Imagine the location decisions of two firms in competition with each other, each selling identical goods to customers evenly spaced along a linear market. The usual example cited is of two ice cream vendors, each selling the same brand at the same price along a stretch of beach with a uniform distribution of people. Beachgoers will purchase the same amount of ice cream (that is, demand is inelastic—is not sensitive to a change in the price) and will patronize the store closest to them. **Figure 10.16** suggests that the two sellers would eventually cluster at the midpoint of the linear market (the beach) so that each vendor could supply customers at the extremities of the market without yielding locational advantage to the other competitor.

This is a spatial solution that maximizes revenues for sellers but does not minimize costs for customers. The lowest total cost location would be for each vendor to locate at the midpoint of his or her half of the beach, as shown at the top of Figure 10.16, where the total effort expended by customers walking to the ice cream stands (or cost by sellers delivering the product) is least. To maximize market share, however, one seller might decide to relocate immediately next to the competitor (Figure 10.16b), capturing three-fourths of the market. The logical retaliation would be for the second vendor to jump back over the first to recapture market share. Ultimately, side-by-side locations (Figure 10.16c) at the centerline of the beach are inevitable, and a stable placement is achieved since neither seller can gain any further advantage from moving. But now the average customer has to walk farther to satisfy the desire for ice cream than she or he did initially; that is, the total cost or delivered price (ice cream purchase plus effort expended) has increased. If a third

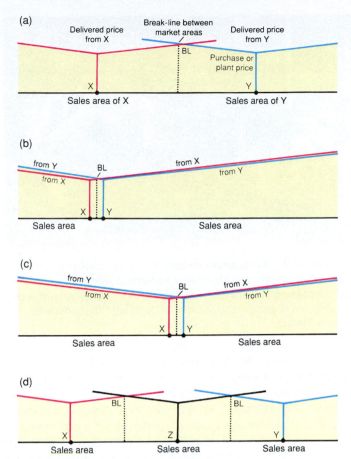

Figure 10.16 **Locational interdependence in retail location.** The Hotelling model predicts where retail outlets will locate. The model assumes customers are evenly spread in a linear market. An example would be vacationers at a beach. The initially socially optimal locations shown in (**a**) minimize total travel costs. However they will be vacated by sellers in search of a market advantage as shown in (**b**). After multiple moves, sellers will reach the competitive equilibrium shown in (**c**). This competitive equilibrium poorly serves customers at the periphery. Spatial dispersion may occur if another competitor enters the market or the sellers agree to subdivide the market by agreement as shown in (**d**).

vendor enters the market, the optimal locations for each vendor change to a more dispersed pattern (Figure 10.16d).

The locational interdependence model offers some simple lessons. First, the locational controls for services depend on the locations of both customers and competitors. Second, under one set of conditions they may produce a clustered pattern and another set, a dispersed pattern. Third, the Hotelling model suggests that a location solution that optimizes revenue for sellers may not be optimal for customers.

10.6 Consumer Services

The supply of consumer services must match the spatial distribution of effective demand—wants backed up by dollars, that is, made effective through purchasing power. Retailers, restaurants, and personal service providers are savvy about locating close to their customers, and the most successful chains use

Figure 10.17 **Consumer services.** Low-level consumer services are most efficiently and effectively performed where demand and purchasing power are concentrated, as this garment repairer working in a street in central Kathmandu, Nepal, demonstrates. © Mark Bjelland.

geodemographic analysis to find optimal locations within cities. Retailers and consumer service providers tend to locate, therefore, where population density, buying power, and traffic are most concentrated (**Figure 10.17**). Prior to the 1960s, shopping for clothes, furniture, or housewares meant a trip downtown where nearly all the stores were clustered. However, as middle-class residents left central cities for the suburbs, the department stores quickly chased their customers into newly developed suburban shopping malls. The location of retail services is an important topic in urban geography and receives greater attention in Chapter 11. In the following sections we give special attention to two special types of consumer services: tourism and gambling.

Tourism

Tourism—travel undertaken for purposes of recreation rather than business—has become the most important single tertiary sector activity and the world's largest industry in jobs and total value generated. On a worldwide basis, in 2010 travel and tourism accounted for some 250 million jobs and about 9% of the world's gross domestic product. Domestic tourism leads to spending on transportation, roadside services, lodging, meals, entertainment, theme parks, and national parks. International tourism, on the other hand, generates new income and jobs in developing states as they are "discovered" as tourist destinations, whether for their climate, unspoiled character, or unique

culture and cultural landscapes (**Figure 10.18**). For one-half of the world's 50 poorest countries, tourism has become the leading service export sector.

The growth of tourism is part of a broader shift in emphasis from production to consumption that accompanies rising standards of living. Like manufacturing, the tourism industry has experienced a post-Fordist transition away from one-size-fits-all mass-produced tourist destinations to numerous consumer niches. Tourists with enough money can choose between cruise ship vacations, beach vacations, African safaris, ecotourism to "unspoiled" wilderness areas, adventure tourism such as helicopter skiing in the Canadian Rockies or kayaking in Alaska, cultural tourism to exotic places such as Bali or Guatemala, heritage tourism in historic villages and cities, sex tourism, bicycle tours through Europe, wine tourism, gambling, and more. **Ecotourism**—travel to wild and scenic locations in ways that are sensitive to local social and environmental concerns—offers an alternative approach. Ecotourism providers attempt to raise the environmental and cultural awareness of tourists, while supporting goals of natural conservation and sustainable local economic development (**Figure 10.19**).

The geographic pattern of tourist destinations is highly uneven. Just imagine the different challenges facing a tourism and convention bureau working to promote Iowa versus one working on behalf of Hawaii. The geographic features of the destination matter, but so does the level of tourism infrastructure and proximity to potential customers. Modest slopes that would hardly qualify as foothills in the mountainous western United States have been turned into downhill ski resorts in the Midwest. The important factors are proximity to major cities such as Chicago or Detroit and the developer's willingness to add the necessary infrastructure such as snowmaking and lifts.

Figure 10.18 **Tourism is the world's largest industry.** Cruise ships at port in St. Thomas, U.S. Virgin Islands. Tourism totally dominates the economy of the U.S. Virgin Islands. About 2 million tourists visit these islands each year, dwarfing the local population of just over 100,000 residents. © Al Rod/AGE Fotostock RF.

GEOGRAPHY
& PUBLIC POLICY

Incentives or Bribery?

When Boeing decided to build its 787 Dreamliner aircraft in North Charleston, South Carolina it was not based on traditional concerns for transportation costs or agglomeration economies. Instead, the decision was influenced by a $900 million incentive package that included state borrowing to help build the facility, state funding for worker training programs, property tax breaks, and corporate income tax credits. In 1985 it cost Kentucky more than $140 million in incentives—about $47,000 a job—to induce Toyota to locate an automobile assembly plant in Georgetown, Kentucky. That was cheap. In 1993, Alabama spent $169,000 per job to lure Mercedes-Benz to that state; Mississippi agreed to give $400 million in spending and tax rebates to Nissan in 2001; and in 2002 Georgia gave DaimlerChrysler $320 million in incentives in successful competition with South Carolina to secure the company's proposed new factory. Earlier, Kentucky bid $350,000 per job in tax credits to bring a Canadian steel mill there.

The spirited auction for jobs is not confined to manufacturing. Minnesota spent $500,000 for each of the 1500 jobs created by Northwest Airlines at new reservation and maintenance facilities. Illinois gave $240 million in incentives ($44,000 per job) to keep 5400 Sears, Roebuck employees within the state, and New York City awarded $184 million to the New York Mercantile Exchange and more than $30 million each to financial firms Morgan Stanley and Kidder, Peabody to induce them to stay in the city. For some, the bidding among states and locales to attract new employers and employment gets too fierce. Kentucky withdrew from competition for a United Airlines maintenance facility, letting Indianapolis have it when Indiana's offered package exceeded $450 million. In 2003, following a slowdown in air travel, United walked away from a fully completed operational facility, leaving the city and state with $320 million of bonded debt and a complex of empty hangars and office buildings.

Inducements to lure companies are not just in cash and loans—though both figure in some offers. For manufacturers, incentives may include workforce training, property tax abatement, subsidized costs of land and building or their outright gifts, below-market financing of bonds, and the like. Similar offers are regularly made by states, counties, and cities to wholesalers, retailers, and major office worker and other service activity employers. The total annual loss of city and state tax revenue through abatements, subsidies, grants, and the like to benefit retained or attracted firms has been estimated at $30 billion to $40 billion. The objective, of course, is not just to secure the new jobs represented by the attracted firm but to benefit from the general economic stimulus and employment growth that those jobs—and their companies—generate. Auto parts manufacturers are presumably attracted to new assembly plant locations; cities grow and service industries of all kinds—doctors, department stores, restaurants, food stores—prosper from the investments made to attract new basic employment.

Not everyone is convinced that those investments are wise, however. The Council for Urban Economic Development has actively lobbied against incentives, and many academic observers note that industrial attraction amounts to a zero sum game: unless the attracted newcomer is a foreign firm, whatever one state achieves in attracting an expanding U.S. company comes at the expense of another state. Economic geographers have observed that the communication and transportation revolutions that have given us a more globalized economy have also weakened the ties between businesses and communities. Corporate investments are highly mobile forcing cities and states to adopt an entrepreneurial posture in order to attract investment and jobs.

Some doubt that inducements matter much, anyway. Although companies seeking new locations will shop around and solicit the lowest-cost, best deal possible, their site choices are apt to be determined by conventional business considerations: access to labor, suppliers, and markets; transportation and utility costs; the nature of the workforce; and overall costs of living. Only when two or more similarly attractive locations have essentially equal cost structures might such special incentives as tax reductions or abatements be a determining factor in a locational decision.

Considering the Issues

1. As a citizen and taxpayer, do you think it is appropriate to spend public money to attract new employment to your state or community? If not, why not? If yes, what kinds of incentives and what total amount offered per job seem appropriate to you? What reasons support your opinion?
2. If you believe that "best locations" for the economy as a whole are those determined by pure location theory, what arguments would you propose to discourage locales and states from making financial offers designed to circumvent decisions clearly justified on abstract theoretical grounds?

Tourism is an important tool in economic development, but geographers have raised a number of critical questions about the industry. Many of the jobs in the tourism industry are low-skill, low-wage positions such as hotel maids, and profits often return to developed countries that are the home of the transnational corporations that own and operate the resorts. Tourism can be exploitative, particularly sex tourism. Tourist visits are often highly seasonal, and this creates stresses on the destination, both during the on-season and off-season. Tourist destinations can go in and out of style, destabilizing the local economy. Tourism transforms places, often dramatically, and in some cases undermines the original tourist attraction. In the United States, the areas just outside designated wilderness areas have become magnets for new housing, hotels, amusement parks, and other development. Cultural tourism inevitably changes the culture and cultural landscapes upon which it is based. In response, ecotourism has emerged as an ethical form of tourism that focuses on education, minimizing environmental impacts, and using locally owned service providers.

Gambling

Gambling is a fast-growing industry that draws large numbers of tourists and in the process remakes places and local economies. In the United States, the gambling industry attracts almost 15% of all entertainment or recreational spending, and it generates more revenue than professional sports, museums, performing arts, fitness

293

Figure 10.19   Ecotourism. Ecotourism seeks to empower local communities while preserving natural habitat, and environmental quality. © *Brand X Pictures/SuperStock RF.*

centers, golf courses, or amusement parks. The geography of gambling is determined by legal structures, political boundaries, and proximity to consumers. Gambling was once concentrated in a few, select locations where it was permitted: Las Vegas, Nevada, Atlantic City, New Jersey, cruise ships (some of which never left shore), Monte Carlo, and Macau, the only Chinese territory where gambling is permitted. The dominance of those gambling centers is being challenged by the rise of lotteries and Internet gambling. The Indian Gaming Regulatory Act of 1988 allowed Indian reservations to operate casinos, and today there are more than 400 Indian casinos in the United States (**Figure 10.20**). Indian reservations located near major population centers or interstate highways are major beneficiaries, and they have often funneled their substantial profits into improving conditions on the reservation. Reservations in Florida, California, and Connecticut are among the most profitable, although many of the jobs go to outsiders. Unfortunately, reservations in remote locations rarely benefit from casinos.

10.7 Business Services

Business services are specialized activities performed for other businesses. They allow producers to realize cost savings by outsourcing specialized tasks when they are needed without the expense of adding to their own labor force. One difference between consumer services and business services is that knowledge and skill-based business service establishments can be spatially divorced from their clients; they are not tied to resources, affected by the environment, or necessarily localized by market. Of course, when high-level personal, face-to-face contacts are required, service firms will often locate close to their clients, the primary, secondary, or tertiary industries they serve. But the transportability of producer services

also means that many of them can be spatially isolated from their client base.

As with other industries, the trade-offs between costs and proximity are one of the central tensions when a business services firm chooses office space. The clients for business services firms are the major companies, many of which are headquartered in the largest cities where real estate and labor costs are highest.

Business services are "knowledge" activities that are highly dependent on communication. Sometimes, knowledge-based activities are referred to as the *quaternary sector* of the economy. The spatial dispersion of some kinds of tasks has been facilitated by innovations in information and communication technologies. Satellite and fiber-optic cables, wireless communications, and the Internet permit the spatial separation of office work into "front-office" and "back-office" tasks. Front-office work involves the creation and exchange of new ideas, while back-office work is often repetitive and requires fewer specialized skills. Front-office tasks involve face-to-face interactions with clients where projecting the correct corporate image is imperative (**Figure 10.21**). Front-office work requires and can bear the high costs of the most prestigious commercial real estate—in high-quality office buildings with prestigious addresses (Park Avenue, Wall Street) or well-known signature office buildings (Transamerica Tower, Seagram Building).

Different types of service sector professionals have different locational needs and preferences. Political lobbyists and companies that do consulting work for federal government agencies need to be in Washington, D.C., and often cluster along the famed Capital Beltway. Investment and law firms prefer the most prestigious addresses in downtowns. Engineering consulting firms prefer suburban office parks or research campuses. Advertising, architecture, and other design professions often try to project a more relaxed, creative image, frequently choosing 19th century or early 20th century factories or warehouse buildings that have been converted to office space.

The back office was once literally in the back of the same office building, but now may be spatially distant from the headquarters of either the service or client firms. Insurance claims, credit card charges, stock market transactions, and call centers are handled more cost-effectively in low-rent, low-labor-cost locations—often in suburbs or small towns in rural states—than in the financial districts of major cities. While New York City remains the center for front-office financial services work in the United States, the relatively small city of Sioux Falls, South Dakota, has several thousand employees engaged in back-office work for major banks and credit card companies.

Developing countries have been particular beneficiaries of space-shrinking digital communications technologies such as fiber-optic cables and the Internet. The increasing tradability of services has expanded the international comparative advantage

Figure 10.20 **Indian casinos.** Casinos proliferated across the United States after passages of the 1988 Indian Gaming Regulatory Act. The Act gave tribes the right to operate casinos as a way of addressing unemployment and poverty on reservations. As in all service industries, location is essential, and reservations in or near major metropolitan areas or major highway corridors have profited most. © *Mark Bjelland.*

Figure 10.21 **Front-office workplaces.** The Chicago Loop is home to the Chicago Board of Trade, the world's oldest commodity futures market, and many corporate headquarters and financial and legal services firms. Businesses that place a premium on face-to-face interactions and prestigious office space often chose locations in central business districts such as the Chicago Loop. © *Pawel.gaul/Getty Images RF.*

of developing states in relatively labor-intensive service activities such as mass data processing and computer software development. At the same time, developing countries have benefited from increased access to efficient, state-of-the-art equipment and techniques transferred from advanced economies. Thus, increasing volumes of back-office work for Western insurance, finance, accounting, and airline companies are being performed abroad.

With an ever-increasing portion of the developing world acquiring the education and experience to provide skilled business services at a level comparable to that formerly available only in advanced countries, traditional notions of comparative advantage are disappearing in the face of a new era of hypercompetition, at least in business services. Thus, service outsourcing has broadened to include higher-level tasks such as paralegal and legal services, accountancy, medical analysis and technical services, architectural and engineering design, and research and development. When the practice employs highly educated and talented specialists receiving developing-world salaries, the cost attractions for companies are strong.

Wired and wireless transmission of data, documents, medical and technical records, charts, and X-rays makes distant consumer and business services immediately and efficiently accessible. Further, many higher-level services are easily subdivided and performable in sequence or simultaneously in multiple locations. The well-known "follow the sun" practices of software developers who finish a day's tasks only to pass on work to colleagues elsewhere in the world are now increasingly used by professionals in many other fields. As transnational corporations use computers around the clock for data processing, they can exploit or eliminate time zone differences between home office countries and host countries of their affiliates. Such cross-border intrafirm service transactions are not usually recorded in trade statistics, but are part of the growing volume of international services flows.

With its large population of well-educated English speakers, India has emerged as the dominant competitor and beneficiary of services outsourcing, echoing China's position as the preferred destination of production outsourcing. The concentration of computer software development around Bangalore, Pune, and Hyderabad has made India a major world player in software innovation, for example, whereas elsewhere in that country increasing volumes of back-office work for Western insurance companies and airlines is being performed. Customer interaction services ("call centers") formerly based in the United States are now increasingly relocated to India, employing workers trained to use an American nickname and speak in perfect American English. Claims processing for U.S.-based life and health insurance firms formerly was concentrated in English-speaking Caribbean states to take advantage of lower wages and the availability of a large pool of educated workers there. Increasingly, such business process outsourcing has shifted to India, Eastern Europe, and in some cases China. In all such cases, the result is accelerated technology transfer in such key areas as information and telecommunications services.

Table 10.2 Shares of World Trade in Commercial Services (Exports, 2014)	
Country or Category	**Percent of World Trade in Services**
United States	13.7
United Kingdom	7.1
China (incl. Hong Kong)	6.7
France	5.4
Germany	5.4
Japan	3.1
India	3.1
Netherlands	3.1
High-income states	80.7
Sub-Saharan Africa	1.1

Source: Data from International Monetary Fund, *Balance of Payments Statistics Yearbook*, 2015.

Despite the increasing share of global services trade held by developing countries, world trade—imports plus exports—in services is still overwhelmingly dominated by a very few of the most advanced states (**Table 10.2**). The country contrasts are great, as a comparison of the data for the "high-income" countries with sub-Saharan Africa documents. The single small island state of Singapore has a larger share of world services trade than all of sub-Saharan Africa.

The same cost and skill advantages that enhance the growth and territorial expansion of domestic service sector firms also operate internationally. Principal banks of all advanced countries have established foreign branches, and the world's leading banks have become major presences in the primary financial capitals. In turn, a relatively few world cities have emerged as international business and financial centers whose operations and influence are continuous and borderless (**Figure 10.22**). The world's key cities for banking, securities firms, and stock exchanges are spread across the globe, allowing almost continuous 24-hours per day trading (**Figure 10.23**). Meanwhile a host of offshore banking havens have emerged to exploit gaps in regulatory controls and tax laws (**Figure 10.24**).

Accounting firms, advertising agencies, management consulting companies, and similar establishments of primarily North American or European origin have increasingly established their international presence, with main branches located in principal business centers worldwide. Those advanced and specialized service components help swell the dominating role of the United States and European Union in the structure of world trade in services.

The list of services employment is long. Its diversity and familiarity remind us of the complexity of modern life and of how far removed we are from the subsistence economies. As societies advance economically, the share of employment and national income generated by the primary, secondary, and tertiary sectors continually changes; the spatial patterns of human activity reflect those changes. The shift is steadily away from production and processing and toward the trade, personal, and professional services of the tertiary sector. That transition is the essence of the now-familiar term *postindustrial*.

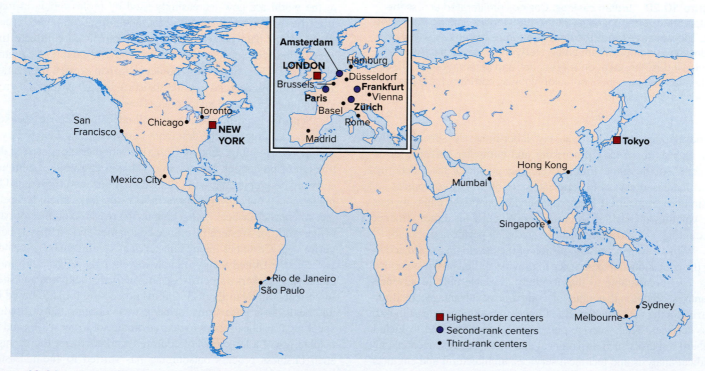

Figure 10.22   **The hierarchy of international financial centers,** topped by New York and London, indicates the tendency of highest-order quaternary activities to concentrate in a few world and national centers. *Source: Peter Dicken. Global Shift. 4th ed. Guilford Press, 2003, Figure 13.8.*

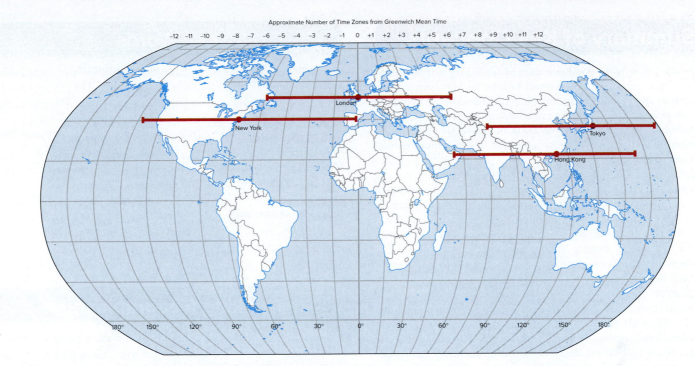

Figure 10.23 **The world's major stock exchanges.** Linked together by telecommunications and electronic funds transfer systems, traders at the major exchanges can potentially operate 24-hours per day. Based on market capitalization in December 2011, New York, London and Tokyo had the largest exchanges, followed by Shanghai, Hong Kong, Toronto, São Paulo, Sydney, Frankfurt, and Shenzhen. The red bars indicate the length of daytime trading hours.

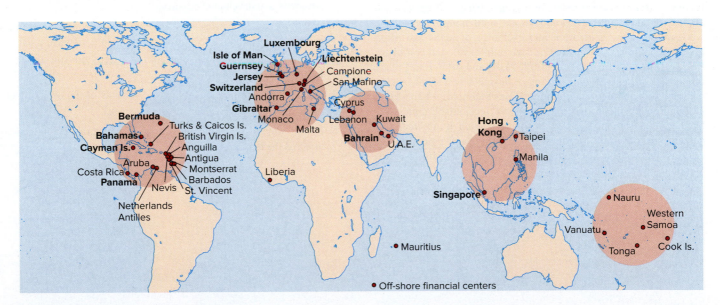

Figure 10.24 **Offshore banking.** Offshore financial centers, mostly in small island states and micro-states, allow "furtive money" to avoid taxation and regulatory scrutiny. These financial havens have low tax rates and relaxed financial regulations. They are spread around the world to offer proximity to major financial centers and 24-hour trading. International pressure has led most of the tax havens to agree to greater openness and less protective secrecy. *Source: Peter Dicken. Global Shift. 4th ed. Guilford Press, 2003, Figure 13.10.*

Summary of Key Concepts

- Manufacturing is the dominant form of secondary activity and is evidence of economic advancement beyond the subsistence level. Location theories help explain observed patterns of industrial development. Those theories are based on simplifying assumptions about fixed and variable costs of production and distribution, including costs of raw materials, power, labor, market accessibility, and transportation.

- Agglomeration economies and the multiplier effect may make existing industrial locations more attractive while comparative advantage may influence production decisions of entrepreneurs. Just-in-time and flexible productions systems introduce different locational considerations than those shaping traditional theories. A growing number of transnational corporations (TNCs) with multiple markets, plants, and raw material sources actively exploit advantages of outsourcing and offshoring in making foreign direct investments away from their home country base.

- A large share of global manufacturing activity is found within a relatively small number of major industrial concentrations and multinational regions. The most-advanced countries within those regions are undergoing deindustrialization as newly industrializing countries with more favorable cost structures compete for markets. In the advanced economies, tertiary activities become more important as secondary sector employment and share of gross national product declines.

- Consumer services tend to locate based on the density and buying power of customers.

- Business services are divided into front-office and back-office activities, each with their own locational tendencies. Front-office work involves the generation and exchange of ideas and often takes places in central business districts in developed countries. Back-office work involves repetitive tasks and often takes place in lower-cost regions. Outsourcing of business services has been made possible by digital communications technologies and India has emerged as the leader in business services outsourcing.

- No national economy exists in isolation; each is an interconnected part of a world system, of economic and cultural integration. Events affecting one affect all. Despite differences in language, culture, or ideology, we are inseparably a single people economically, a unification increasingly controlled by the growing global urbanization discussed in the following chapter.

Key Words

agglomeration 278	infrastructure 280
business service 290	just-in-time (JIT)
commodity chain 287	manufacturing 281
comparative advantage 281	least-cost theory 277
consumer service 290	new international division of labor
ecotourism 292	(NIDL) 282
external economies 280	offshoring 282
flexible production 281	outsourcing 282
Fordism 279	transnational corporation
foreign direct investment	(TNC) 287
(FDI) 288	variable cost 277

Thinking Geographically

1. What simplifying assumptions did Weber make in his theory of plant location? In what ways does the Weberian search for the *least-cost location* differ from the recognition of the *spatial margin of profitability*? How does Weberian theory help explain the location of traditional heavy industry clusters?

2. How, in your opinion, do the concepts or practices of *comparative advantage* and *outsourcing* affect the industrial structure of advanced and developing countries?

3. As high-tech industries and employment become more important in the economic structure of advanced countries, what consequences for economic geographic patterns do you anticipate? Explain.

4. What have been the motivations and rewards of the outsourcing of service activities by developed country firms?

5. In what ways, do you think, has outsourcing favorably or unfavorably affected the home country economies of the outsourcing firms?

CHAPTER 11

An Urban World

San Francisco, California. ©Jan Hanus/Alamy Stock Photo RF.

CHAPTER OUTLINE

Cairo was a world-class city in the 14th century. Situated at the crossroads of Africa, Asia, and Europe, it dominated trade on the Mediterranean Sea. By the early 1300s, it had a population of half a million or more, with 10- to 14-story buildings crowding the city center. Cairo's chronicler of the period, Taqui-al-Din-Al-Maqrizi, recorded the construction of a huge building with shops on the first floors and apartments housing 4000 people above. One Florentine visitor estimated that more people lived on a single Cairo street than in all of Florence. Travelers from all over Europe and Asia made their way through Cairo, and the shipping at the port of Bulaq outdistanced those of Venice and Genoa combined. There were more than 12,000 shops, some specializing in luxury goods from all over the world—Siberian sable, chain mail, musical instruments, cloth, songbirds. Travelers marveled at the size, density, and variety of Cairo, comparing it favorably with cities such as Venice, Paris, and Baghdad.

Today, Cairo is a vast, sprawling metropolis representative of several recent trends in urbanization in developing countries where population growth far outstrips economic development. The 1970 population of Egypt was 33 million; today it is 83 million, a result of better health care, a big drop in infant mortality, and longer life expectancies. Some 11 million people reside in the Cairo greater metropolitan area. Greater Cairo now extends more than 450 square kilometers (175 sq mi) and has a population density of about 33,000 per square kilometer (12,000 per sq mi). Population trends are expected to continue, with a predicted population of around 13.5 million by 2015. And the city continues to grow, spreading onto valued farmland and decreasing food availability for the country's growing population.

A steady stream of migrants arrives in Cairo daily because it is the place where people think opportunities are available, where life will be better and brighter than in the crowded country-side. The city is the symbol of modern Egypt, a place where young people are willing to undergo deprivation for the chance to "make it." But real opportunities continue to be scarce. The poor, of whom there are millions, crowd into row after row of apartment houses, many of them poorly constructed. Tens of thousands more live in rooftop sheds or on small boats on the Nile; one-half million find shelter in the Northern and Southern Cemeteries—known as the Cities of the Dead—on Cairo's eastern edge. On occasion, buildings collapse; the earthquake of October 12, 1992, measured 5.9 on the Richter scale but caused enormous damage, leveling thousands of structures.

One's first impression when arriving in central Cairo is of opulence, a stark contrast to what lies outside the center. High-rise apartments, regional headquarters buildings of multinational corporations, and modern hotels stand amid clogged streets, symbols of the new Egypt. The well-to-do can eat at McDonald's, Pizza Hut, T.G.I. Friday's, Chili's, or KFC; new suburban developments and exclusive residential communities create enclaves for the wealthy. The plush apartments and expensive cars, however, are only a short distance from the slums that are home to masses of under-employed people, perhaps as much as 20% of Cairo's population (**Figure 11.1**).

Like cities nearly everywhere in the developing world, Cairo has experienced explosive growth that finds an increasing proportion of the population housed in urban areas without the facilities to support them all. Street congestion and idling traffic generate some of the world's worst air pollution. Both the Nile River and the city's treated drinking water show dangerous levels of lead and cadmium, by-products of the local lead smelter and whose health effects will be borne for years to come. Because of its rapid growth, Cairo has not been able to plan and create adequate infrastructure for its inhabitants. The 1992 earthquake prompted the city's adoption of its first development plan.

Cairo is a classic case of the urban explosion. Its cultural and economic influence and problems of population, poverty, planning, and infrastructure face all cities to one degree or another and are some of the topics we consider in this chapter.

11.1 An Urbanizing World

Cities today are growing at a phenomenal rate. In 1900 only 13 cities had a population of more than 1 million people; in 2010 there were 488 such cities (**Figure 11.2**). The UN Population Bureau projects that there will be 662 cities with populations exceeding 1 million by 2030.

In 1900, no cities had a population of more than 10 million people. Twenty-eight metropolises had 10 million or more people in 2014 and 41 are expected by 2030 (**Table 11.1** and **Figure 11.3**). The United Nations calls these *megacities*. Of course, as we saw in Chapter 5, the urban component of the world's population has greatly increased. Urbanization and metropolitanization have increased more rapidly than the growth of total population, however. In the year 1800, only 3% of the world's population lived in cities. In the last decade the percentage of the world's population living in cities has tipped the balance, with more than 54% now living in urban areas. The amount of urban growth differs from region to region and from country to country, but all countries have one thing in common: the proportion of the people living in cities is rising.

Table 11.2 shows world urban population by region. While Asia and Africa are the least-urbanized continents, cities are growing particularly quickly there. While some cities will grow

Figure 11.1   Cairo, Egypt. The population growth in the greater metropolitan area—from some 3 million in 1970 to 18 million today—has been mirrored in many developing countries. The rapid growth of urban areas brings with it housing shortages, inadequate transportation and other infrastructure development, unemployment, poverty, and environmental deterioration. © *McGraw-Hill Education/Barry Barker.*

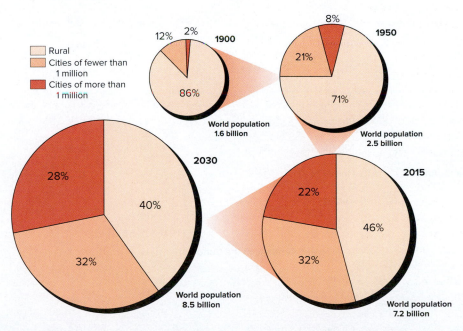

Figure 11.2   Trends in world urbanization. Note the steady decline in the proportion of people living in rural areas. The United Nations predicts that virtually all of the population growth during the next 15 years will be concentrated in the urban areas of the world. *Source: UN Population Division.*

into megacities, cities with less than 1 million residents will grow faster than the very largest cities. Industrialization spurred the earlier rapid urbanization in Western Europe and North America. In many of the still-developing countries, however, urban expansion is only partly the result of the transition from agricultural to industrial economies. Rather, in many of those areas people flee impoverished rural districts; by their numbers and high fertility rates they accelerate city expansion. Industrialization fosters urbanization, but in developing countries, urbanization has resulted only partly from industrialization. People flock to the cities seeking a better life than they can find in rural areas, but they often do not find it. The cities of sub-Saharan Africa are growing at a rapid rate, largely due to rural-to-urban migration. However, the urban growth is beyond the capability of the economic system to create

Table 11.1 Cities with 10 Million or More Inhabitants (Megacities), 2014 and Projections for 2030 (Population in Millions)

City	Country	2014 Population	2030 Population
Tokyo	Japan	37.8	37.2
Delhi	India	25.0	36.1
Shanghai	China	23.0	30.8
Mexico City	Mexico	20.8	23.9
São Paulo	Brazil	20.8	23.4
Mumbai (Bombay)	India	20.7	27.8
Osaka-Kobe	Japan	20.1	20.0
Beijing	China	19.5	27.7
New York-Newark	United States	18.6	19.9
Cairo	Egypt	18.4	24.5
Dhaka	Bangladesh	17.0	27.4
Karachi	Pakistan	16.1	24.8
Buenos Aires	Argentina	15.0	17.0
Kolkata (Calcutta)	India	14.8	19.1
Istanbul	Turkey	14.0	16.7
Chongqing	China	12.9	17.4
Rio de Janeiro	Brazil	12.8	14.2
Manila	Philippines	12.8	16.8
Lagos	Nigeria	12.6	24.2
Los Angeles-Long Beach-Santa Ana	United States	12.3	13.3
Moskva (Moscow)	Russia	12.1	12.2
Guangzhou, Guangdong	China	11.8	17.6
Kinshasa	Democratic Republic of the Congo	11.1	20.0
Tianjin	China	10.9	14.7
Paris	France	10.8	11.8
Shenzhen	China	10.7	12.7
London	United Kingdom	10.2	11.5
Jakarta	Indonesia	10.2	13.8

Source: The UN Population Division. World Urbanization Prospects, 2014 Revision.

employment, housing, and social services. Shantytowns and squatter settlements, in addition to unemployment and underemployment, are characteristic of cities such as Lagos, Nigeria, and Dakar, Senegal.

In this chapter, our first objective is to consider the major factors responsible for the development, location, and functions of urban areas. Our second goal is to examine the systems of urban areas—the relationships they bear to one another. Our third goal is

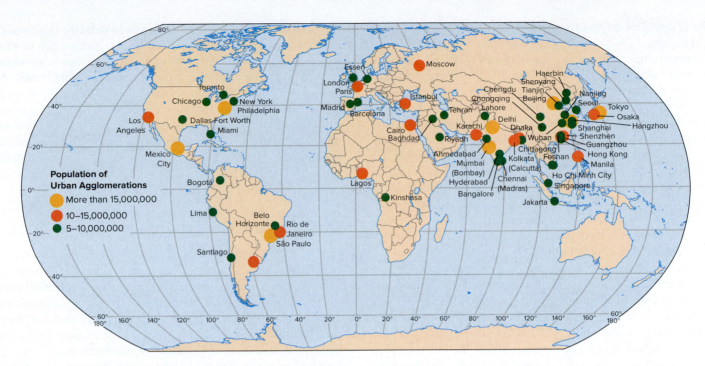

Figure 11.3   **Metropolitan areas of 5 million or more in 2010.** Massive urbanized districts are no longer characteristic only of the industrialized, developed countries. Note the clusters of large metropolitan areas in developing countries, such as Rio de Janeiro–São Paulo in Brazil, Beijing–Tianjin in China, and the Mumbai (Bombay) region of India. *Source: UN Population Division.*

Table 11.2	Estimated Urban Share of Total Population for Selected Regions, 1950 and 2014, with Projections to 2050		
Region	**1950**	**2014**	**2050**
North America	64	82	89
Latin America and the Caribbean	41	80	87
Europe	51	73	82
Oceania	62	71	73
Asia	18	48	64
Africa	14	40	58
World	**29**	**52**	**67**

Source: UN Population Division.

to identify the nature of land use patterns within those areas. Our fourth goal is to differentiate cities around the world by reviewing some of the factors that help explain their special nature.

11.2 Origins and Evolution of Cities

People need to be near one another: they gather together to form couples, families, groups, organizations, and towns. Beyond companionship, people need one another to sustain important support systems. The origins of towns lie in several factors: the existence of a settled community (not a hunter-gatherer society); a concentration of people; groups not directly engaged in agriculture; and

the existence and governance of an elite group. These factors are the basis for urban settlement and the underpinnings of civilization (note that *city* and *civilization* have the same Latin root, *civis*). Although civilization was necessary before the first towns and cities emerged, the development of cities depended on favorable circumstances, including fertile soil, the presence of water for transportation, building materials, and a defensible situation. Technology in the form of agricultural production and food transport and storage was also necessary for urban development. Given these conditions, how did cities begin in the ancient world?

First and foremost, the earliest cities depended on the existence of an agricultural surplus. Many early cities included farms within their walls, but the main distinction between the city and the countryside stemmed from the nonagricultural aspect of most urban dwellers. This meant that food had to be provided to the urban population by the **hinterland** surrounding the city. Advances in agriculture, stemming from fortunate circumstances and improvements in technology, created a food surplus in an area surrounding a population center. Those in the nascent town who were not farmers were free to specialize in other occupations—metalworking, pottery making, cloth weaving—producing goods for other town dwellers and the farm population on which they depended. Still others became scribes, merchants, priests, and soldiers, providing the services and refining the power structure on which the organized urban and rural society depended.

Social organization and power, as reflected in religious hierarchy and civil administration, were the second necessary precursor to urban development. Most ancient cities centered on a temple or a palace, which housed the priests, the granary, the schools, and often the ruler. Cities became the seats of central power, cementing the relationship to the hinterland and allowing the extraction of the agricultural surplus from the hinterland and its redistribution in the city.

Ancient cities grew in spots that were easy to defend. Positions along rivers aided in transportation, but hilltops offered defensive advantages. Often residents built walls to surround the town and aid in its defense. Building walls and sometimes moats was a strategy used in much of Europe, but these defensive frameworks could also limit city expansion. Some cities, such as Rome, went through multiple rounds of wall construction, with each addition extending the area in which residents could live.

The fourth factor in the emergence of urban areas was the development of a more-complex economy. Cities extracted their sustenance from the surrounding hinterland. As the city extended its power and organization, it could extend its hold over a wider hinterland. As agricultural technology improved, the hinterland could produce more food, and as methods of transportation and storage improved, cities could bring in more food and store it safely for eventual redistribution to the urban population. The size of a hinterland could limit urban growth, as cities could grow only if the agricultural surplus also grew.

In Europe and Asia, from about the 10th to the 18th centuries, shifts in economic relationships changed the simple extractive relationship between the city and its hinterland. As trade became the engine of the economy, urban merchants began to buy and sell. In general, they traded raw materials—wool, wood, spices—which were then used to produce finished goods, such as textiles, boats, and food.

With the Industrial Revolution, another shift in cities took place. The Industrial Revolution accelerated the rate of urban growth, initially in Europe. Powered by water or coal, new machinery in factories, operated by growing working populations, fostered mass production. Cities, once centered on the temple or the palace, once surrounded by walls, once focused on the marketplace and the waterfront wharves, changed utterly: their economic fortunes centered on the factories, the railroads, and housing for the factory workers.

This very brief overview of the development of urban areas only scratches the surface of a very complex phenomenon and a very long history, but it is important to understand the origins of cities before examining other aspects of cities. Today, cities in newly industrializing countries in Asia or Latin America are witnessing some of the same explosive growth and social polarization that historically accompanied industrialization. Meanwhile, in more-developed countries, the transition to a service economy has caused cities to take on a postindustrial character. Smokestacks have crumbled, and former factories and industrial areas have been redeveloped for parks, housing, and commercial uses. Consumption and service sector activities rather than heavy industry dominate the postindustrial city.

Defining the City Today

Urban areas are not of a single type, structure, or size. Their common characteristic is that they are nucleated, nonagricultural settlements. At one end of the size scale, urban areas are small towns with perhaps a single main street of shops; at the opposite end, they are complex, multifunctional metropolitan areas or megacities (**Figure 11.4**). The word *urban* is often used to describe such places as a town, city, suburb, or metropolitan area, but it is a general term, not used to specify a particular type of settlement. People use common terms differently. What a resident of rural Vermont or West Virginia calls a city may not at all be afforded that name and status by an inhabitant of California or New Jersey. In addition, one should keep in mind that the term *urban* differs the world over: in the United States, the Census Bureau describes *urban* places as having 2500 or more inhabitants; in Greece, *urban* is defined as municipalities in which the largest population center has 10,000 or more inhabitants; in Nicaragua, it denotes administrative centers of localities with streets, electric lights, and at least 1000 inhabitants. It is necessary in this chapter to agree on the meanings of terms commonly employed but interpreted in different ways.

The words *city* and *town* denote nucleated settlements, multifunctional in character, including an established central business district and both residential and nonresidential land uses. Towns are

(a)

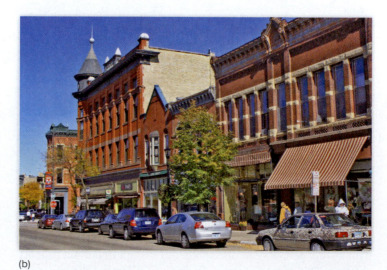
(b)

Figure 11.4 **The differences in size, density, and land use complexity** are immediately apparent between (a) a city (Chicago) and (b) a town (Northfield, MN). One is a city, the other is a town, but both are urban places. *(a) © dibrova/iStock/Getty Images RF, (b) © Arthur Getis.*

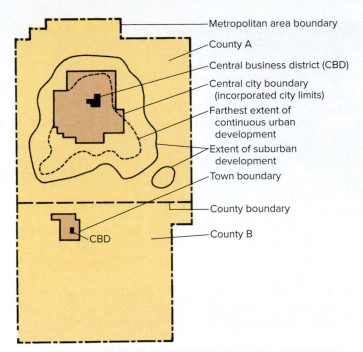

Figure 11.5   A hypothetical spatial arrangement of urban units within a metropolitan area. Sometimes the official limits of the central city are very expensive and contain areas commonly thought of as suburban or even rural. On the other hand, older eastern U.S. cities and some in the West, such as San Francisco, more often have official limits that contain only part of the high-density land uses and populations of their metropolitan areas.

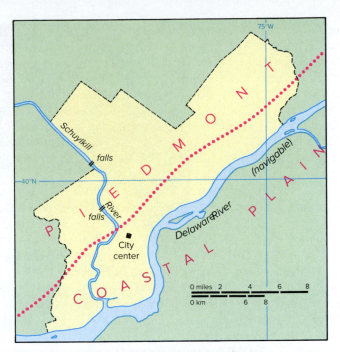

Figure 11.6   The site of Philadelphia.

smaller and have less functional complexity than cities, but they still have a nuclear business concentration. **Suburb** denotes a subsidiary area, a functionally specialized segment of a large urban complex, dependent on an urban area. It may be dominantly or exclusively residential, industrial, or commercial. Suburbs, however, can be independent political entities. The **central city** is the part of the urban area contained within the suburban ring; it usually has official boundaries.

Some or all of these urban types may be associated into larger units. An **urbanized area** is a continuously built-up landscape defined by building and population densities with no reference to political boundaries. It may contain a central city and many contiguous cities, towns, suburbs, and other urban tracts. A **metropolitan area**, on the other hand, refers to a large-scale *functional* entity, perhaps containing several urbanized areas, discontinuously built up but nonetheless operating as an integrated economic whole (**Figure 11.5**).

The Location of Urban Settlements

Urban centers are connected to other cities and rural areas. Cities exist to provide services for themselves and for others outside them. They rely on outside areas for goods and services not produced locally and as markets for their products and activities.

To perform the tasks that support it and to add new functions as demanded by the larger economy, the urban unit must be efficiently located. Efficiency may derive from its centrality to the area it serves, from the physical characteristics of its site, or from its

location relative to the resources, productive regions, and a transportation network connecting it to its markets.

In discussing urban settlement location, geographers frequently distinguish between site and situation, concepts introduced in Chapter 1. You will recall that **site** refers to the exact location of a settlement and can be described either in terms of latitude and longitude, or in terms of the physical characteristics of the site. For example, the site of Philadelphia is an area bordering and west of the Delaware River north of the intersection with the Schuylkill River in southeast Pennsylvania (**Figure 11.6**).

The description can be more or less exhaustive, depending on the purpose it is meant to serve. In the Philadelphia case, the fact that the city is partly on the Atlantic coastal plain, is partly in the piedmont (foothills), and is served by a navigable river is important if one is interested in the development of the city during the Industrial Revolution. As **Figure 11.7** suggests, water transportation and power were important localizing factors when major American cities were established on the East Coast.

Classifications of cities according to site characteristics have been proposed, recognizing special placement circumstances. These include *break-of-bulk* locations, such as river crossing points where cargoes and people must interrupt a journey; *head-of-navigation* or *bay head* locations where the limits of water transportation are reached; and *railhead* locations where a railroad ended. In Europe, security and defense—island locations or elevated sites—were considerations in earlier settlement locations. Water power sites and proximity to coal fields were prime considerations in selecting city sites during the Industrial Revolution.

Whereas *site* suggests absolute location, **situation** indicates relative location; it places a settlement in relation to the physical and cultural characteristics of the surrounding areas. Very often it is important to know what kinds of possibilities and activities exist in the area near a settlement, such as the distribution

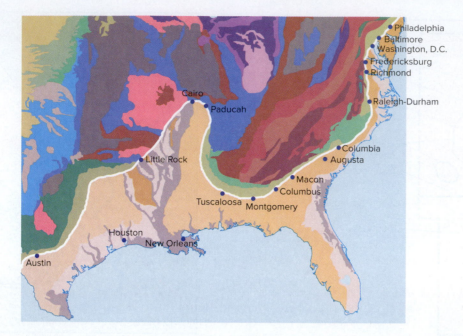

Figure 11.7   **The fall line separates hard Paleozoic metamorphic rocks to the west from the softer sedimentary rocks of the coastal plain.** Rivers flowing from the west move powerfully over this erosional scarp. The fall line hosted waterwheel-powered industries in colonial times and helped to determine the location of major cities such as Philadelphia, Richmond, and Baltimore. *Source: Data from U.S. Geological Survey.*

of raw materials, market areas, agricultural regions, mountains, and oceans.

The site of central Chicago is 41°52′N, 87°40′W, on a lake plain. More important, however, is its situation close to the deepest penetration of the Great Lakes system into the interior of the country, astride the Great Lakes–Mississippi waterways, and

near the western margin of the manufacturing belt, the northern boundary of the Corn Belt, and the southeastern reaches of a major dairy region. It is the most important railroad hub in the United States and is located central to ore fields and coal deposits (**Figure 11.8**). As a gateway to the West from the East and vice versa, Chicago's O'Hare International Airport is one of the

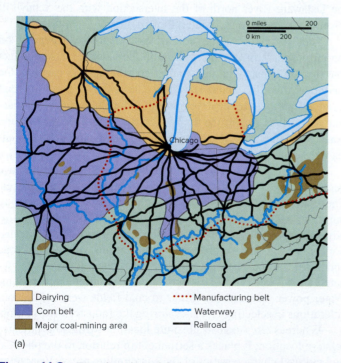

Dairying
Corn belt
Major coal-mining area
Manufacturing belt
Waterway
Railroad

(a)

(b)

Figure 11.8   **The situation of Chicago helps to suggest the reasons for its functional diversity and size. (a)** Chicago grew large because of its central location relative to different resources and transportation systems, and **(b)** became the nation's most important railroad center. *(b) © Christian B. Valle/Moment Open/Getty Images RF.*

busiest in the United States. From this description of Chicago's situation, implications relating to market, to raw materials, and to transportation centrality can be drawn.

The site or situation that originally gave rise to an urban unit may not remain the essential ingredient for its growth and development for very long. The already existing markets, labor force, and urban facilities of a successful city may attract people and activities totally unrelated to the initial localizing forces. For instance, although coal mining had much to do with Pittsburgh's early growth, other factors are more important to its well-being today.

The Economic Base

Every urban area has an **economic base**, the activities people do to support the urban population. These are activities such as manufacturing goods, repairing roads, managing stores, tending to the sick, and teaching children. The economic base of an urban area can be categorized in terms of basic and nonbasic sectors.

The **basic sector** of an urban area's economic structure is made up of the activities of people that bring in money from outside the community. People who produce goods or perform services bought by those outside the urban area are engaged in "export" activities. For example, people manufacture semiconductors in Hillsboro, Oregon. The semiconductors are bought by computer manufacturers all over the world, making the semiconductor manufacturers part of the basic sector of the economic structure of Hillsboro.

Other workers produce goods or services for residents of the city itself. They are not bringing new money into the community, as their goods and services are not being exported. This is the **nonbasic sector** of a city's economy, which is crucial for its internal functioning. It includes the operation of stores, offices, city government, local transit, and school systems.

The total economic structure of an urban area equals the sum of its basic and nonbasic activities. It is difficult to classify work as belonging exclusively to one sector or the other, however. Doctors, for example, may have mainly local patients and thus are members of the nonbasic sector, but the moment they treat someone from outside the community, they bring new money into the city and become part of the basic sector.

Most cities perform many export functions, and the larger the urban unit, the more multifunctional it becomes. Nonetheless, even in cities with a diversified economic base, one export activity, or a very small number of export activities, tends to dominate the structure of the community. Such functional specialization permits the classification of cities into categories: manufacturing, retailing, wholesaling, transportation, government, and so on.

Assuming it were possible to divide with complete accuracy the employed population of a city into totally separate basic and service (nonbasic) components, a ratio between the two employment groups could be established. This basic–nonbasic ratio shown in **Figure 11.9** indicates that as a settlement increases in size, the number of nonbasic personnel grows faster than the number of new basic workers. The graph suggests that service sector jobs, most of which are nonbasic, will be more common in larger cities. In cities with a population of 1 million, the ratio is about two nonbasic workers for every basic worker. This means that adding 10 new basic employees expands the labor force by 30 (10 basic,

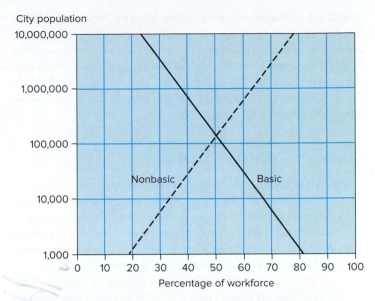

Figure 11.9 **A generalized representation of the proportion of the workforce engaged in basic and nonbasic activities.** As settlements become larger, a greater proportion of the workforce is employed in nonbasic activities. A city of 10 million will have about one-quarter of the workforce engaged in basic activities, whereas one with only 100,000 residents will have more than one-half of the workforce so employed.

20 nonbasic). The resultant increase in total population is equal to the added workers plus their dependents. Thus, a **multiplier effect** exists in which every new basic sector job creates additional nonbasic jobs. For example, if the computer semiconductor manufacturer in Hillsboro, Oregon, hires new technicians, they and their dependents will create a demand for more doctors, teachers, and baristas. When news media report that a new manufacturing plant will create a certain number of new jobs in addition to those at the plant, they are referring to the multiplier effect. The size of the multiplier effect is determined by the community's basic–nonbasic ratio.

11.3 Functions of Cities

Most modern cities take on multiple functions. These include manufacturing, retailing, transportation, public administration, military uses, housing cultural and educational institutions, and, of course, the housing of their own citizens. Yet most cities specialize in categories, and even for truly diversified cities it can help to consider cities in regard to their relationship to markets, sites of production, or administration and other services.

No matter what their size, urban settlements exist for the efficient performance of necessary functions. They have three main functions: (1) central place functions, or providing general goods and services for a surrounding area; (2) transport functions; and (3) specialized functions not necessarily geared to the local area. All towns provide the first two functions, but not necessarily the third.

Detroit, for example, grew to a population of 1,849,568 at its height in 1950 as it provided a point of general merchandise services, convenient transportation on the Great Lakes and the web of railroad networks crisscrossing the Midwest, and a specialized function of manufacturing automobiles. Now, there is only one automobile assembly plant in a declining Detroit. Other examples of specialized functions are those of Washington, D.C., which provides government services to the whole United States, and the Mayo Clinic in Rochester, Minnesota, which provides specialized health services to all who seek them.

Cities as Central Places

When people want to buy an item or purchase a service, they go to the city. For as long as cities have existed, they have served as marketplaces, not only for their own residents but also for the population beyond the city limits. Small cities provide a range of goods and services that suffice for most people's needs. But truly unique items or special services often can be found only in the biggest cities.

The geographer Walter Christaller developed **central place theory** (see "Central Place Theory," p. 309) to explain the size and location of settlements. He detected a pattern of small, medium, and larger settlements that were dependent on one another. Small towns, he postulated, would serve as marketplaces for the surrounding populations, while expensive luxury goods would be available in the large cities that served the surrounding small towns. People would have to travel short distances for basic goods, such as groceries, and longer distances for rarer items, such as luxury cars. Christaller's theories have been shown to be generally valid in widely differing areas within the commercial world. When varying incomes, cultures, landscapes, and transportation systems are taken into consideration, the theories hold up rather well. They are particularly applicable to agricultural areas, especially with regard to the size and spacing of cities and towns. If we combine a Christaller-type approach with the ideas that help us understand industrial location and transportation alignments (see Chapter 9), we have a fairly good understanding of the location of most cities and towns.

The interdependence of small, medium, and large cities can also be seen in their influence on one another. The sphere of influence of an urban unit is usually proportional to its size. A small city may influence a local region of, say, 65 square kilometers (25 sq mi) if, for example, its newspaper is delivered to that region. Beyond that area, another city may be the dominant influence through its banking services, television stations, and sports teams. **Urban influence zones** are the areas outside of a city that are still affected by it. As the distance away from a city increases, the city's influence on the surrounding countryside decreases (recall the idea of distance decay discussed in Chapter 7).

Intricate relationships and hierarchies are common among cities of all sizes. Consider Grand Forks, North Dakota, which for local market purposes dominates the rural area immediately surrounding it. However, Grand Forks is influenced by political decisions made in the state capital, Bismarck. For a variety of cultural, commercial, and banking activities, Grand Forks is influenced by Minneapolis. As a center of wheat production, Grand Forks and Minneapolis are subordinate to the grain market in Chicago. Of course, the pervasive agricultural and other political controls exerted from Washington, D.C., on Grand Forks, Minneapolis, and Chicago indicate how large and complex are the urban zones of influence.

Cities as Centers of Production and Services

Urban growth, particularly in the last 200 years, has been tied to the development of industries. Manufacturing of some sort was always an important part of cities, but before the Industrial Revolution it tended to be on a small scale. With the growth of mass production, manufacturing became the primary engine driving the urban economy. Industrial products are usually exported to other places and bring in money that is distributed throughout the urban economy.

Most cities, especially large ones, perform many export functions. Nonetheless, even in cities with a diversified economic base, a few export activities tend to dominate the structure of the community and identify its operational purpose within a system of cities. Recall that the term *multiplier effect* means that a city's employment and population grow with the addition of manufacturing workers and dependents as a supplement of new basic or manufacturing employment.

The growth of cities may be self-generating—"circular and cumulative"—acquiring new people and functions attracted by the existing markets, labor force, and urban facilities. Service activities such as banking and legal services, government services, and the like may generate basic and nonbasic additions to the labor force. In recent years, service industries have developed to a point where new service activities serve older ones. For example, computer systems firms aid banks in developing more-efficient computer-driven banking systems. A list of the largest U.S. metropolitan areas is given in **Table 11.3.**

Just as settlements grow in size and complexity, so do they decline. When the demand for the goods and services of an urban unit falls, fewer workers are needed, and thus both the basic and the nonbasic components of a settlement system are affected. There is, however, a resistance to decline that impedes the process and delays its impact. Whereas settlements can grow rapidly as migrants respond quickly to the need for more workers, under conditions of decline, many of those who have developed roots in the community are hesitant to leave or may be financially unable to move to another locale. **Figure 11.10** shows that in recent years urban areas in the South and West of the United States have been growing rapidly, while those in the Northeast and the North Central regions have grown more slowly.

Cities as Centers of Administration and Institutions

The earliest cities were marked by the presence of a temple, a granary, and the residence of the ruler. Cities have always been centers for administration. State and federal capitals are almost always in towns of some size, and these cities usually grow substantially once the government is in place, as governments are important employers.

In cities of all sizes, some proportion of the population is employed in government services. In addition to those in the government, whether it is federal, state, metropolitan, or local,

CENTRAL PLACE THEORY

In 1933, the German geographer Walter Christaller attempted to explain the size and location of settlements. He developed a framework, called *central place theory,* for understanding town interdependence. Christaller recognized that his theory would best be developed in rather idealized circumstances, such as the following.

1. Towns that provide the surrounding countryside with such fundamental goods as groceries and clothing would develop where farmers specialized in commercial agricultural production.
2. The farm population would be dispersed in a generally even pattern.
3. The people would possess similar tastes, demands, and incomes.
4. Each kind of product or service available to the population would have its own *threshold,* or minimum number of consumers needed to support its supply. Because such goods as luxury automobiles are either expensive or not in great demand, they would have a high threshold, whereas fewer consumers would be required to support a small grocery store.
5. Consumers would purchase goods and services from the nearest store.

When all the assumptions are considered simultaneously, they yield the following results.

1. A series of hexagonal market areas that cover the entire plain will emerge, as shown in the figure.
2. There will be a central place at the center of each of the hexagonal market areas.

The two A central places are the largest on this diagram of one of Christaller's models. The B central places offer fewer goods and services for sale and serve only the areas of the intermediate-sized hexagons. The many C central places, which are considerably smaller and more closely spaced, serve still smaller market areas. The goods offered in the C places are also offered in the A and B places, but the latter offer considerably more as well as more-specialized goods. Notice that the places of the same size are equally spaced. *Arthur Getis and Judith Getis, "Christaller's Central Place Theory,"* Journal of Geography, 1966. *Used with permission of the National Council for Geographic Education, Indiana, PA.*

3. The size of the market area of a central place will be proportional to the number of goods and services offered from that place.

In addition, Christaller reached two important conclusions. First, towns of the same size will be roughly evenly spaced, and larger towns will be farther apart than smaller ones. This means that many more small than large towns will exist. In the figure, the ratio of the number of small towns to towns of the next-larger size is 3 to 1. This distinct, steplike series of towns in size classes differentiated by both size and function is called a *hierarchy of central places.*

Second, the system of towns is interdependent. If one town were eliminated, the entire system would have to readjust. Consumers need a variety of products, each of which has a different minimum number of customers required to support it. The towns containing many goods and services become regional retailing centers, and the small central places serve just the people immediately in their vicinity. The higher the threshold of a desired product, the farther, on average, the consumer must travel to purchase it.

Consider, for example, fans of major league baseball teams. There are a relatively small number of major league baseball teams in the Midwest, where they are in urban areas separated by quite some distance. Christaller's model predicts that hexagonal market areas, centered on central places, will be proportional to the services offered there—in this case, a sport. Indeed, a map of baseball fan (self-reported) affiliations shows large, circular or hexagonal areas centered on cities with teams, bounded by the area of fans for the next team. To see the map and contribute your own fan information, see http://commoncensus.org/sports.php.

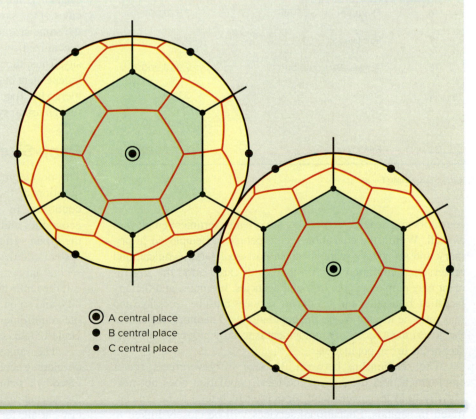

- ◉ A central place
- ● B central place
- ● C central place

Table 11.3	U.S. Metropolitan Area Populations of More Than 2.5 Million, 2015	
Rank	Metropolitan Area (largest principal city(s))	2015 Population
1	New York–Newark	20,180,000
2	Los Angeles	13,340,000
3	Chicago	9,550,000
4	Dallas-Fort Worth	7,100,000
5	Houston	6,660,000
6	Washington	6,100,000
7	Philadelphia	6,070,000
8	Miami	6,010,000
9	Atlanta	5,710,000
10	Boston	4,770,000
11	San Francisco	4,656,000
12	Phoenix–Mesa	4,570,000
13	Riverside	4,490,000
14	Detroit	4,300,000
15	Seattle	3,730,000
16	Minneapolis–St. Paul	3,520,000
17	San Diego	3,300,000
18	Tampa–St. Petersburg	2,980,000
19	Denver	2,810,000
20	St. Louis	2,810,000
21	Baltimore	2,800,000

Source: U.S. Census Bureau.

educators and health care workers are often government employees as well. Within the education sector, people are employed in primary and secondary schools and in postsecondary, technical, and professional education. Within the health care sector, the government may employ people in social agencies, hospitals, and therapy facilities. Education and health care services must be accessible and thus must be located where the population is. Therefore, the size of the government sector of employment is usually proportional to the size of the urban population.

The relationship between the size of government sector employment and the size of the local population does not necessarily hold for government services, which tend to be concentrated in capitals. Educational services will be closely linked to

distribution of population, but some cities have a disproportionate education sector population, especially cities that are home to a university or college. Madison, Wisconsin, for example, has large government sector employment numbers, as it is the state capital and home to a large public university.

11.4 Systems of Cities

Cities today are interdependent. Any given city may have few or multiple functions, and these will be influenced by its location and size. It will also be affected by the distance from, size of, and functions of other cities. Taken together, cities are part of a system of urban settlement.

The Urban Hierarchy

Perhaps the most effective way to recognize how systems of cities are organized is to consider the **urban hierarchy**, a ranking of cities based on their size and functional complexity. One can measure the numbers and kinds of services each city or metropolitan area provides. The hierarchy is then like a pyramid; a few large and complex cities are at the top, and many smaller ones are at the bottom. There are always more smaller cities than larger ones.

When a spatial dimension is added to the hierarchy, as in **Figure 11.11**, it becomes clear that an areal system of metropolitan centers, large cities, small cities, and towns exists. Goods, services, communications, and people flow up and down the hierarchy. The few high-level metropolitan areas provide specialized functions for large regions, while the smaller cities serve smaller districts. The separate centers interact with the areas around them, but because cities of the same level provide roughly the same services, those of the same size tend not to serve one another unless they provide a specialized service, such as a major hospital or a research university. Thus, the settlements of a given level in the hierarchy are not independent but interrelated with communities of other levels in that hierarchy. Together, all centers at all levels in the hierarchy constitute an urban system.

Rank-Size Relationships

The development of city systems on a global scale raises the question of the organization of city systems within regions or countries. In some countries, especially those with complex economies and a long urban history, the **rank-size rule** describes the urban system. It tells us that the nth-largest city of a national system of cities will be $1/n$ the size of the largest city. That is, the second-largest settlement will be one-half the size of the largest, the tenth-biggest will be one-tenth the size of the first-ranked city, and so on. Although no national urban system exactly meets the requirement of the rank-size rule, those of Russia and the United States closely approximate it.

The rank-size ordering is less applicable to countries with developing economies and those in which the urban system is dominated by a **primate city**, one that is far more than twice the size of the second-ranked city. In fact, there may be no obvious "second city" at all, for a characteristic of a primate city hierarchy is one

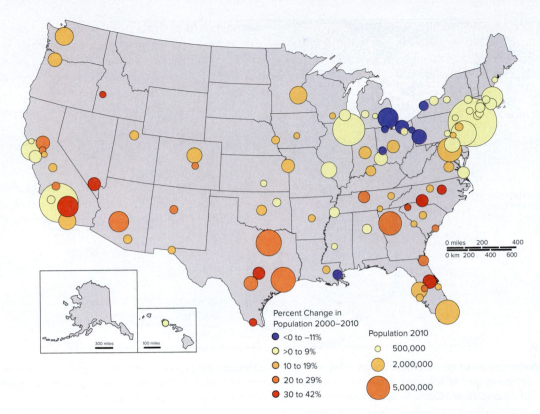

Figure 11.10 **The pattern of metropolitan growth and decline in the United States, 2000–2010.** Shown are the 100 largest metropolitan areas in 2000. The cities of the Sun Belt showed the greatest rate of growth. Modest growth marked most metropolitan regions of the Northeast and Midwest. Meanwhile, hurricane-racked New Orleans and deindustrializing regions of the Manufacturing Belt witnessed actual decline. *Source: Bjelland, Human Geography 12e, Fig.11.34, p. 375.*

very large city, few or no intermediate-size cities, and many subordinate smaller settlements. For example, metropolitan Seoul (with 9.8 million in 2010) contains more than 20% of the total population of South Korea. Bangkok is home to more than one-half of the urban residents of Thailand.

The capital cities of many developing countries display this kind of overwhelming primacy. In part, their primate city pattern is a heritage of their colonial past, when economic development, colonial administration, and transportation and trade activities were concentrated at a single point. Dakar (Senegal), Luanda (Angola), and many other capital cities of African countries are examples.

In other instances, including Cairo, development and population growth have tended to concentrate disproportionately in a capital city whose very size attracts further development and growth. Many European countries (e.g., the United Kingdom, France, and Austria) show a primate structure, often due to the historic concentration of economic and political power around the royal court in a capital city that was, perhaps, also the administrative and trade center of a larger colonial empire.

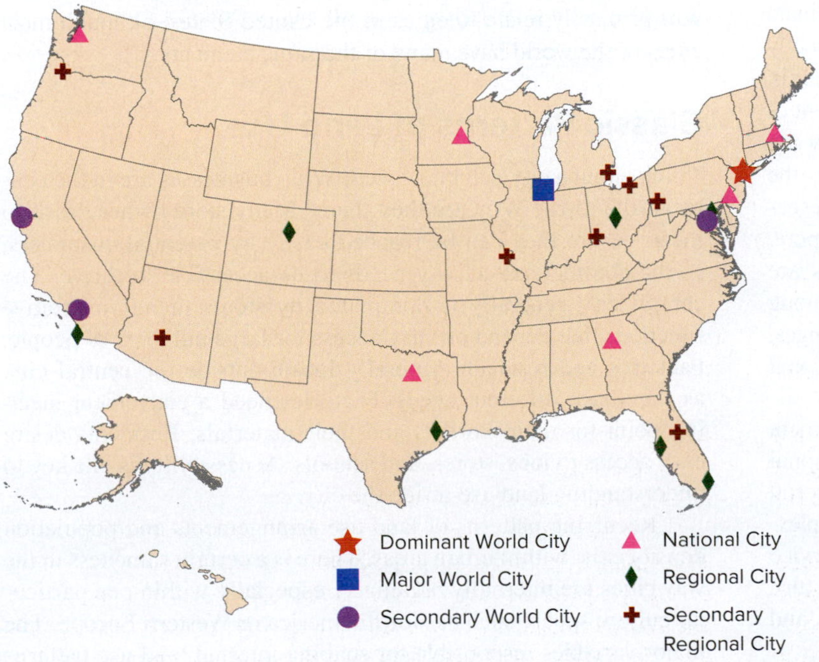

Figure 11.11 **A hierarchy of U.S. metropolitan areas.** Only major metropolitan areas are shown. This classification is based on the **(a)** total employment in each metropolitan area, **(b)** the total number of workers in business and financial operations, and **(c)** the location quotient of business and financial workers. *Source: Based on financial employment data from the U.S. Bureau of Labor Statistics.*

★ Dominant World City ▲ National City

■ Major World City ◆ Regional City

● Secondary World City ✚ Secondary Regional City

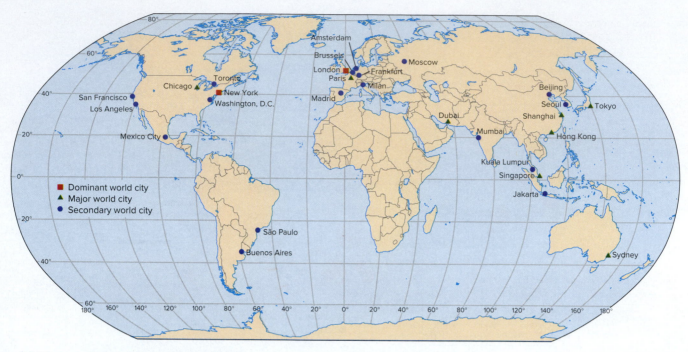

Figure 11.12 **A classification of world cities.** World cities are centers of international production, marketing, and finance. They are bound together in complex networks and all are interconnected in many different ways. *Source: Data are from "The World According to GaWC 2010" by the Globalization and World Cities Research Network, http://www.lboro.ac.uk/gawc/world2010t.html.*

World Cities

Standing at the top of national systems of cities are a relatively few centers that may be called **world cities**. These large urban centers are the "control and command" centers of the global economy. They are control points for international advanced producer services, including finance, advertising, banking, and law.

London and New York are the dominant world cities. They contain the highest number of transnational service offices and the headquarters of multinational corporations, and they dominate commerce in their respective parts of the world. Each is directly linked to a number of other world cities. All are bound together in complex networks that control the organization and management of the global system of finance, manufacturing, and trade. They are key to the globalization of the economy. **Figure 11.12** shows the links among the dominant centers and the suggested major and secondary world cities, which include Hong Kong, Paris, Singapore, Tokyo, Shanghai, Chicago, Dubai, and Sydney. These cities are all interconnected by advanced communications systems among governments, major corporations, stock and futures exchanges, securities and commodity markets, major banks, and international organizations.

Major international corporations spur world city development and dominance. The growing size and complexity of transnational corporations dictates their need to outsource central managerial functions to specialized service firms to minimize the complexity of control over dispersed operations. Those specialized service agencies—legal, accounting, financial, and so on—in their turn need to draw on the very large pools of expertise, information, and talent available only in world cities.

11.5 Inside The City

Urban areas have distinct physical and cultural landscapes, so an understanding of the nature of cities is incomplete without a knowledge of their internal characteristics. So far, we have explored the origins and functions of cities within hierarchical urban systems. Now we turn to the urban patterns of land use, changes in urban form, the social areas of cities, and the institutional controls that determine much of the character of an urban area. This discussion will primarily relate to cities in the United States, although most cities of the world have many of the same elements.

Classic Patterns of Land Use

Think about a city you know well. What businesses are in the center of the town? Why are they there? Many store owners wish to locate where they can be reached easily by potential customers, as the commercial activity needs to be accessible to thrive. The central city generally is dominated by stores or administrative functions that depend on easy access for large numbers of people. Factories and residents generally locate outside the central city, as they have different needs. Factories need a convenient meeting point for their workers and their materials. Residents desire easy access to jobs, stores, and schools. Accessibility is one key to understanding land use inside the city.

Recurring patterns of land use arrangements and population density exist within urban areas. There is a certain sameness to the way cities are internally organized, especially within one particular culture sphere such as North America or Western Europe. The major variables responsible for shaping internal land use patterns are accessibility, a competitive market in land, and the transportation technologies available during the periods of urban growth. Successive transit systems—first, walking, then mass transit systems, and later automobiles—have given rise to three sharply different urban land use patterns.

The Central Business District

In the United States, the first cities grew up around pedestrian movement and pack-animal haulage. Populations were small and the cities compact. During the late 19th and early 20th centuries, cities installed mass-transit systems, which were costly but efficient. Even with their introduction, however, only land within walking distance of the mass-transit routes could be incorporated successfully into the expanding urban structure. Within the older central city, locations at the city core, called the **central business district (CBD)**, had the highest accessibility and were therefore the most desirable for many functions. The central business district was located at the convergence of mass-transit lines (in European cities and large U.S. cities) or the central roads of the city (in smaller cities).

Land parcels at the city center had high value in addition to their high accessibility. Business owners who demanded the greatest accessibility for their establishments bid the most for land parcels within the CBD. The slightly less accessible CBD parcels generally became sites for tall office buildings (skyscrapers), the principal hotels, and similar land uses that helped produce the skyline of the commercial city. Public uses, such as parks and schools, were allocated land according to criteria other than ability to pay.

Land in the CBD was a scarce commodity, and its scarcity made it expensive, demanding intensive, high-density utilization. Because of its limited supply of usable land, the industrial city of the mass-transit era was compact. It was characterized by high residential and structural densities, and at its margins there was a sharp break between urban and nonurban uses. The older central cities of the northeastern United States and southeastern Canada display this pattern.

Outside the Central Business District

Just outside the core area of the city, industry controlled land adjacent to essential cargo routes: rail lines, waterfronts, rivers, or canals. Commercial aggregations developed at the outlying intersections—transfer points—of the mass-transit network. Strings of stores, light industries, and high-density apartment structures could afford and benefit from location along high-volume transit routes. The least-accessible locations within the city were left for the least-competitive users: low-density residences. A diagrammatic summary of this repetitive allocation of space among competitors for urban sites is shown in **Figure 11.13.**

The competitive bidding for land should yield—in theory, at least—two separate but related distance-decay patterns. Both land value and population density decrease as distance from the CBD increases. Land value declines in a distinct pattern: there is a sharp drop in land value a short distance from the *peak land value intersection,* the most accessible and costly parcel of the central business district, and then the value declines less steeply to the

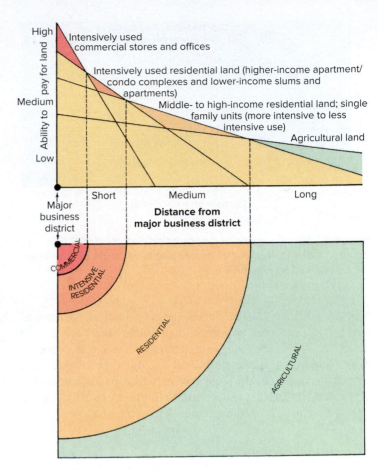

Figure 11.13   A generalized urban land use pattern. The model depicts the location of various land uses in an idealized urban area where the highest bidder gets the most-accessible land.

margins of that built-up area. The population density pattern of the central city shows a comparable distance-decay arrangement, as suggested by **Figure 11.14.**

Automobile-Based Patterns

Starting in the 1940s, automotive transportation became dominant in the movement of people and goods, and streetcar systems lost riders and were often converted to bus systems. As highway systems were extended outward after World War II, vast areas of lower-priced land on the urban fringe were opened up for development. As the wealthy and middle-class families moved away from the city center, the zones shifted outward, flattening the density–distance curve. The compact older mass-transit city created prior to World War II was fundamentally changed and succeeded by the low-density unfocused urban and suburban sprawl of the automobile city. The automobile made vast areas accessible, creating the possibility of multiple business districts rather than a single central business district. The peak land value intersection was now likely to be the intersection of a major radial highway with a circumferential (beltway) highway or even an entire highway corridor. Still, the concepts of accessibility and competitive bidding for land shown in Figure 11.14 apply. In newer automobile-based development, major commercial uses occupy the most-accessible and most-expensive land along major

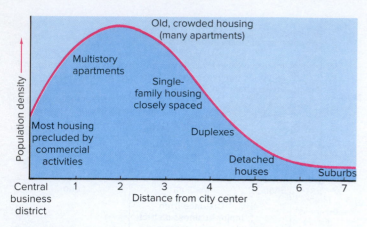

Figure 11.14 **A generalized population density curve.** As distance from the area of multistory apartment buildings increases, the population density declines.

highway corridors. Higher-density housing such as apartments, townhouses, and condominiums often border these commercial districts, and lower-density, single-family housing is found in more-secluded, less-accessible locations. In most communities, these patterns are not the product of pure free market bidding but are dictated by land use and zoning plans that try to anticipate the results of competitive bidding for each piece of land.

Regional Differences

The timing of an urban region's growth determines the relative mix of walking city, mass-transit city, and automobile city. Only the oldest parts of eastern cities such as Old Quebec and Boston's Beacon Hill still display remnants of the walking city. Cities in the East and Midwest such as Philadelphia and Chicago have large areas developed when mass-transit was dominant. The density and design of the newer cities of the West and Southwest, as well as the suburban growth areas of older centers, have been influenced primarily or exclusively by the automobile and motor truck, not by mass-transit and railroads. The land use contrast between regions is not absolute, of course, as older cities have adapted to the automobile and rapidly growing cities in the West and Southwest have added light rail transit systems. Even so, the different patterns have not been totally erased, since cities, like other cultural landscapes, are built up over time, layer upon layer. Thus, the ever-changing 21st-century American city shows the intermingling of influences from different eras of city building. What the future holds for our cities is hard to say, but many urban geographers and planners are arguing for a return to the transit-oriented pattern of urban growth for reasons of energy conservation and environmental sustainability.

Models of Urban Form

Generalized models—simplified graphic summaries—of urban growth and land use patterns were proposed during the 1920s and 1930s. While such models generalize what admittedly is a tremendous degree of variation in cities, they also help us understand some regularities in urban shape. More recently, urban geographers have begun to employ models that help us better understand a decentralized city.

The common starting point of the early models is the distinctive central business district found in every older central city. The core of the CBD displays intensive land use development: tall buildings, many stores and offices, and crowded streets. Framing the core is a fringe area of warehousing, transportation terminals, and light industries (as long as they require few raw materials and pollute very little). Just beyond the urban core, residential land uses begin.

The **concentric zone model** (**Figure 11.15a**) was developed by a sociologist, Ernest Burgess, in the 1920s. He described five zones, radiating outward from the first, the CBD. The second ring, the zone of transition, is characterized by stagnation and deterioration and contains high-density, low-income slums, rooming houses, and perhaps ethnic ghettoes. The third zone is a zone of workers' homes, usually smaller, older homes on modest lots; the fourth and fifth zones are areas of middle-class and well-to-do houses and apartments.

The concentric zone model is dynamic. Each type of land use and each group tend to move outward into the next outer zone. The movement was seen as part of a ceaseless process of invasion and succession that yielded a restructured land use pattern and population segregation by income level. The development of Chicago (**Figure 11.16**) shows some accordance with this model.

The **sector model** (see **Figure 11.15b**), developed in the 1930s, focuses on transportation arterials. It posits that high-rent residential areas expand outward from the city center along major transportation routes such as suburban commuter rail lines. As cities grow, this model proposes, the highest-income groups move into new homes in new neighborhoods that are located along existing transportation routes radiating outward from the center of the city. Middle-income housing clusters around the housing for the wealthy, and low-income housing occupies land adjacent to the areas of industry and associated transportation, such as freight railroad lines.

The sector model is also dynamic, marked by a filtering-down process as older areas are abandoned by the outward movement of their original inhabitants, with the lowest-income populations (closest to the center of the city and farthest from the current location of the wealthy) becoming the dubious beneficiaries of the least-desirable vacated areas. The expansion of the city is radial, not zonal, as in the concentric zone model. The accordance of the sector model with the actual pattern that developed in Calgary, Canada, is suggested in **Figure 11.17**.

The third model of urban land use patterns, the **multiple-nuclei model** (see **Figure 11.15c**), counters the central assumption of the concentric zone and sector models—that urban growth and development spreads outward from a single central core. This model states that large cities develop by peripheral spread from several nodes of growth, not just one. Certain activities are limited to specific sites based on their needs: the retail district needs accessibility, whereas a port needs a waterfront location, for example. Peripheral expansion of the separate nuclei eventually leads to coalescence and the meeting of incompatible land uses along the lines of juncture. The urban land use pattern, therefore, is not regularly structured from a single center in a sequence of circles or a

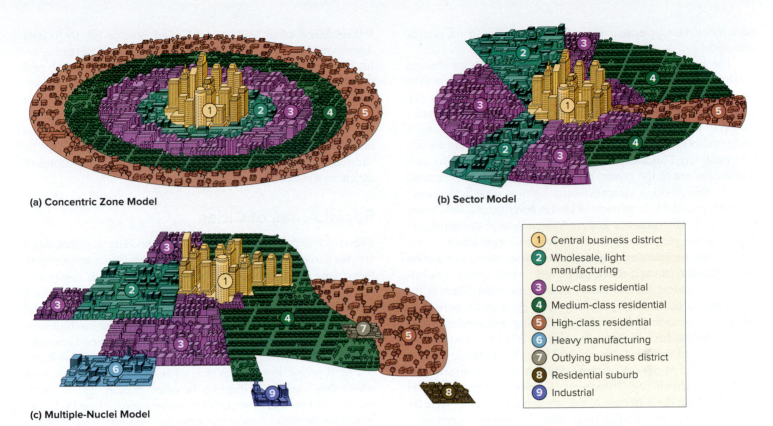

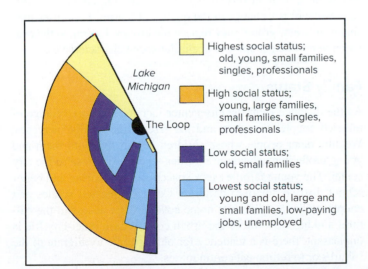

Figure 11.15   Three classic models of the internal structure of cities: (a) the concentric zone model, (b) the sector model, and (c) the multiple-nuclei model. *Source: Data: "The Nature of Cities" by C. D. Harris and E. L. Ullman in volume no. 242 of The Annals of the American Academy of Political and Social Science. The American Academy of Political and Social Science, Philadelphia, PA, 1945.*

Figure 11.16   A diagrammatic representation of the major social areas of the Chicago region. The central business district is known as the Loop. *Source: Data: Philip Rees, "The Factorial Ecology of Metropolitan Chicago," M. A. Thesis, University of Chicago, 1968.*

Figure 11.17   The land use pattern in and around Calgary, Alberta, Canada, in 1981. Physical and cultural barriers and the evolution of urban areas over time tend to result in a sectoral pattern of similar land uses. Calgary's central business district is the focus for many of the sectors. *P. J. Smith, "Calgary: A Study in Urban Patterns," Economic Geography, 38, no. 4, p. 328, Clark University, Worcester, MA, 1962.*

series of sectors but is based on separately expanding clusters of contrasting activities.

Although the culture, society, economy, and technology that these three models summarized have now been superseded, the physical patterns they explained remain as vestiges on the current landscape. North American cities up until 1950 resembled the

concentric zone or the sector models in that they had a clearly defined and dominant CBD; but cities grew more and more complex in the years following World War II. The multiple-nuclei model may more closely accord with urban development in the recent past, but it should be supplemented by a fourth visualization, the peripheral model.

The **peripheral model** takes into account the major changes in urban form that have taken place since World War II, especially the suburbanization of what were once central city functions. The peripheral model supplements rather than supplants the three earlier models of urban land use patterns (**Figure 11.18**). It focuses on the peripheral belt that lies within the metropolitan area, but outside the central city. Functions of the peripheral belt are defined by their relationship not to the center, but to other parts of the peripheral zone. In this model, a circumferential highway outside the center city makes available large tracts of land to development. Residences are relatively homogeneous and are located in large developments. Nodes on the peripheral belt are centers for employment or services. These include shopping malls, industrial parks, distribution and warehouse clusters, office parks, and airport clusters (including hotels, meeting facilities, and car rental agencies). Much of the life of the residents of the periphery takes place on the periphery, not in the city, as they shop for food, clothing, and services in the shopping malls; find recreation in country clubs and entertainment complexes; and find employment in the industrial or office parks.

Observers of the urban scene disagree over the direction of the U.S. peripheral city. Those who first proposed the peripheral pattern felt that the periphery of the city is not completely separate from the city; it is still a functional part of the metropolitan area. More recently, some scholars have argued that U.S. cities are developing complex suburban downtowns that compete with the CBD. They find that the metropolis is dividing into a set of increasingly self-contained urban realms, each served more or less by its own downtown. This trend relegates the influence of the CBD to the status of one of many urban centers within the metropolis. The new polycentric metropolis shows a CBD and a surrounding constellation of suburban downtowns, which are vitally important to the metropolitan, and indeed the national, economy.

The models of urban form discussed in this section aid our understanding of urban structure and development, but it must be stressed that a model is not a map and that many cities have some of the characteristics of several models.

Social Areas of Cities

The larger and more economically and socially complex cities are, the stronger is the tendency for city residents to segregate themselves into groups on the basis of *social status, family status,* and *ethnicity.* In a large metropolitan region with a diversified population, this territorial behavior may be a defense against the unknown or unwanted; a desire to be among similar kinds of people and close to preferred stores, restaurants, and bars; a response to income constraints; or a result of social and institutional barriers. Most people feel more at ease when they are near those with whom they can easily identify. In cities, people tend to group according to income or occupation (social status), stage in the life cycle (family status), and language or race (ethnic characteristics) (see "Birds of a Feather," p. 317).

Many of these groupings are fostered by the size and cost of available housing. Land developers, especially in cities, produce homes of similar quality in specific areas. Of course, as time elapses, there is a change in the quality of houses. Land uses may change and new groups may replace old groups, leading to the evolution of new neighborhoods of similar social characteristics.

Family Status

As the distance from the city center increases, the percentage of married couples increases and the size of the family increases. Wealthy older people whose children do not live with them and young professionals without families tend to live close to the city center. The young families seek space for child rearing and better school districts for their children. College-educated singles and couples without children at home covet greater access to the cultural and business life of the urban core. Where inner-city life is unpleasant, there is a tendency for older people to migrate to the suburbs or to retirement communities.

Within lower-status populations, the same pattern tends to emerge. Transients and single people are housed in the inner city, and families, if they find it possible or desirable, live farther from the center. The arrangement that emerges is a concentric-circle patterning according to family status (**Figure 11.19**).

Social Status

The social status of an individual or a family is determined by income, education, occupation, and home value. In the United States, high income, a college education, a professional or managerial position, and high housing value confer high status. High

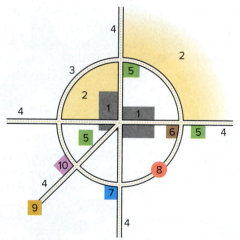

1. Central city
2. Suburban residential area
3. Circumferential highway
4. Radial highway
5. Shopping mall
6. Industrial district
7. Office park
8. Service center
9. Airport complex
10. Combined employment and shopping center

Figure 11.18   The peripheral model of urban form. This model supplements the concentric zone, sector, and multiple-nuclei models.

BIRDS OF A FEATHER . . . OR, WHO ARE THE PEOPLE IN YOUR NEIGHBORHOOD?

How does Starbucks or Subway decide where to open a new store? How does McDonald's or Burger King decide which 99-cent menu items to promote at a certain site, or whether it can profitably offer salads at that franchise? Are there enough families with children to justify building a play area?

Many businesses, large and small, base their decisions on marketing analysis systems. The media research firm Nielsen sells businesses geographically based on market analyses originally developed by Claritas, Inc. The system is based on census data and allows customers to map potential customers by multiple characteristics: Starbucks, for example, could use the system to pinpoint areas with substantial populations of 18- to 34-year-olds (a demographic whose coffee intake is predicted to increase) with higher incomes. An auto dealership could use the system to understand the driving preferences of residents of a certain zip code (77024, or Spring Valley, Texas, a suburb of Houston, for example, is home to people who prefer Acuras and Land Rovers).

Businesses can use these market analysis tools to pinpoint business locations and to understand local landscapes, traffic patterns, retail environment, and financial services available within a certain radius. Nielsen uses demographic characteristics to categorize Americans by their social groups, life stages, and household characteristics. In addition, the company has created dozens of market segments with catchy names, ranging from "Blueblood Estates" (wealthy, older, with kids) to "Suburban Sprawl" (45- to 64-year-olds, suburban, who are college graduates) to "Shotguns and Pickups" (young rural families who go to auto races and drive Ford F-series trucks). Some of the others include "Fast-Track Families" (numerous children, spacious homes, lots of disposable income to spend on the latest technology); "Urban Achievers" (children of immigrants in multilingual neighborhoods in coastal cities); and "Young Digerati" (tech-savvy college graduates living on the urban fringe). Businesses can use these categories to understand potential customers in their chosen location.

The company realizes that the designations don't define the tastes and habits of every single person in a community, but they do identify the behavior that most people are apt to follow. In the market segment "Bohemian Mix" (mobile urbanites, ethnically diverse, younger than 55), for example, members are likely to rent foreign videos and watch soccer on TV, and they will be quick to check out the latest movie, club, and microbrewery. Each zip code across the country has been associated with one or more market segments. If you want to know how marketing analysts have categorized your neighborhood, go to https://segmentationsolutions.nielsen.com/mybestsegments/ click on "Zip Code Look-up," and enter your zip code.

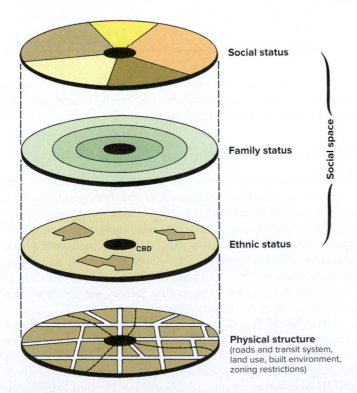

Figure 11.19 **The social geography of American and Canadian urban areas.** *Robert A. Murdie, "Factorial Ecology of Metropolitan Toronto," Research Paper 116, Department of Geography Research Series, University of Chicago, 1969.*

home value can mean an expensive rented apartment or a large house with extensive grounds. A good housing indicator of social status is people per room. A low number of people per room tends to indicate high status. Low status characterizes people with low-income jobs living in low-value housing. People tend to sort into neighborhoods where most of the heads of households are of similar status.

Patterns of social status divisions are in concord with the sector model. In most cities, people of similar social status are grouped in sectors that fan out from the innermost urban residential areas (Figure 11.19). The pattern in Chicago was illustrated in Figure 11.16. If the number of people within a given social group increases, they tend to move away from the central city along an arterial connecting them with the old neighborhood. Major transport routes leading to the city center are the usual migration routes out from the center.

Today, social status divisions are often perpetuated by political boundaries between separate municipalities or school districts. Communities on either side of the divide may differ greatly in relative income. Many residential developments are also income-segregated, because the houses are of similar value. To preserve the upscale nature of a development and protect land values, self-governing community associations enact conditions and restrictions (see "The Gated Community," p. 318). Pervasive and detailed, these conditions specify such things as the size, construction, and color of walls and fences; the size and permitted uses of rear and side yards; and the design of lights and mailboxes. Some go so far as to tell residents what trees they may plant, what

THE GATED COMMUNITY

Approximately one in six Americans lives in a *master-planned community.* Particularly characteristic of the fastest-growing parts of the country, most of these communities are in the South and West, but they occur everywhere. Master-planned communities date back to the 1960s, when Irvine, California, and Sun City, Arizona, were built.

A subset of the master-planned community is the **gated community,** a fenced or walled residential area where access is limited to designated individuals. More than 9 million Americans live in these middle- and high-income developments. Entry to these communities within communities is restricted; gates are staffed by security guards or are accessible only by computer key card or telephone. Some of the communities hire private security forces to patrol the streets. Surveillance systems monitor common recreational areas, such as community swimming pools, tennis courts, and health clubs. Houses are commonly equipped with security systems. Troubled by the high crime rates, drug abuse, and gangs that characterize many urban areas, people seek safety within their walled enclaves.

Gated and sheltered communities are not just an American phenomenon but are increasingly found in all parts of the world. More and more guarded residential enclaves have been sited in such stable Western European states as Spain, Portugal, and France. Developers in Indian cities have also used gated communities to attract wealthy residents. Trying to appeal to Indians returning to that country after years in areas such as the Boston high-tech corridor and Silicon Valley, developers have built enclaves, with names such as Regent Place and Golden Enclave, that boast two-story houses and barbecues in the backyards. Mexico has the largest percentage of its population living in gated communities, a result of the large income gap between wealthy and poor urbanites.

Gated community. © *Arthur Getis.*

Elsewhere, as in Argentina or Venezuela in South America or Lebanon in the Middle East—with little urban planning, unstable city administration, and inadequate police protection—not only rich but also middle-class citizens are opting for protected residential districts. In China and Russia, the sudden boom in private and guarded settlements reflects in part a new form of postcommunist social class distinction, while in South Africa gated communities serve as effective racial barriers.

pets they may raise, and where they may park their boats or recreational vehicles.

Ethnicity

For some groups, ethnicity is a more important factor in determining residential location than social or family status. Areas of homogeneous ethnic identification appear in the social geography of cities as separate clusters or nuclei reminiscent of the multiple-nuclei concept of urban structure. For some ethnic groups, cultural segregation is both sought and vigorously defended, even in the face of pressures for neighborhood change exerted by potential competitors for housing space. The durability of "Little Italys" and "Chinatowns" and of Polish, Greek, Armenian, and other ethnic neighborhoods in many American cities is evidence of the persistence of self-maintained segregation.

Certain ethnic or racial groups, especially African Americans, have been segregated by discrimination in housing markets. Every city in the United States has one or more black areas that, in many respects, may be considered cities within a city. **Figure 11.20** illustrates the concentration of blacks, Hispanics, and other racial/ethnic groups in distinct neighborhoods in Los Angeles. The social and economic barriers to movement outside the area have always been high. In many American cities, the poorest residents are the blacks,

who are often relegated to the lowest-quality housing in the least-desirable areas of the city. Similar restrictions have been placed on Hispanics and other non-English-speaking minorities.

Changes in Urban Form

The 20th century saw massive change in U.S. urban patterns. This was the consequence of the creation of a technological, physical, and institutional structure that fostered change. First, the improvement of the automobile increased its reliability, range, and convenience, freeing its owner from dependence on fixed-route public transit for access to home, work, or shopping. Then the acceptance of a maximum 40-hour workweek in 1938 guaranteed millions of Americans the time for a commuting journey not possible when workdays of 10 or more hours were common.

After World War II, the United States witnessed a significant increase in the number of its citizens owning their own homes, from below 50% in 1945 to 60% in 1960. The government stimulated this boom by authorizing increased spending for home loans by the Federal Housing Administration (FHA) and the Veterans Administration (VA). These agencies offered much more generous terms than private bankers had before the war, when buyers had to put down large down payments (sometimes 50% or more) and repay their loans in a short time, often 10 years. The FHA and VA revolutionized home

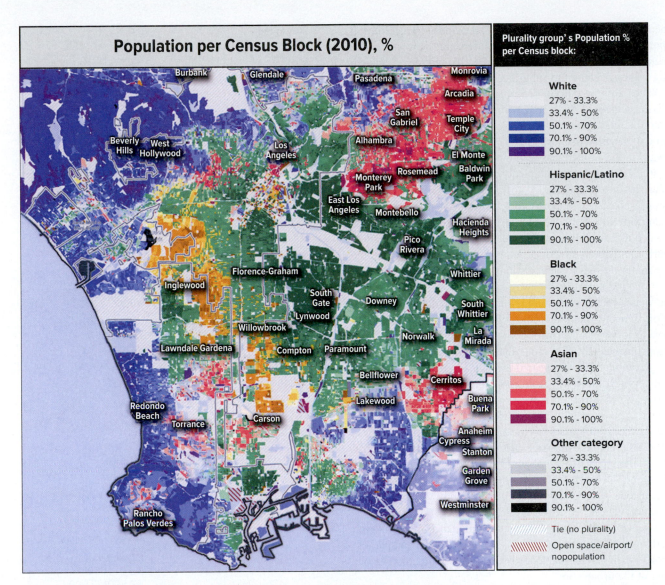

Figure 11.20   Racial/ethnic patterns in Los Angeles, 2010. Although Los Angeles County has an extremely diverse population, people tend to cluster in distinct neighborhoods by race and ethnicity. Between 2000 and 2010, the neighborhood of Inglewood became less black and more Hispanic, the neighborhoods of Rosemead and Monterey Park became less Hispanic and more Asian, and the mostly white Torrance neighborhood became more Asian. To see ethnicity shifts like these for New York, Detroit, Chicago, and other metropolitan regions visit http://www.urbanresearchmaps.org/comparinator/pluralitymap.htm *Source: urbanresearchmaps.org/comparinator/pluralitymap.htm.*

buying, offering mortgages of up to 90% of a home's value and up to 30 years to pay off loans. The VA permitted many veterans to purchase homes with virtually no down payment.

In addition, the development of the interstate highway system made commuting long distances more feasible. During the 1970s, the interstate highway system was substantially completed, and major metropolitan expressways were put in place, allowing sites 30 to 45 kilometers (20 to 30 mi) or more to be considered commuting distance between the workplace and home.

The combination of new transportation flexibility, widespread homeownership, shorter workweeks, and new roads opened up vast new acreages of nonurban land to urban development. These developments changed the prevailing patterns of population growth. Over the past 60 to 70 years, U.S. cities have experienced a depopulation at the center as residents, businesses, and industry moved

outward into suburbs; then a consequent decline at the city center; and recently a modest reversal of these trends as some people and economic functions have returned to the core areas of many cities.

Suburbanization and Edge Cities

Demands for housing, pent up by years of economic depression and wartime restrictions, were loosed in a flood after 1945, and a massive suburbanization altered the existing pattern of urban America. Between 1950 and 1970, the two most prominent patterns of population growth were the *metropolitanization* of people and, within metropolitan areas, their *suburbanization*.

Residential land uses led the rush to the suburbs. Typically, uniform but spatially discontinuous housing developments were built beyond the boundaries of older central cities. The new design was an unfocused sprawl because it was not tied to mass-transit lines. It also represented a massive relocation of purchasing power to which retail merchants were quick to respond. The planned major regional shopping center became the suburban counterpart of higher-order central places and the outlying commercial districts of the central city. Smaller shopping malls and strip shopping centers gradually completed the retailing hierarchy.

Faced with a newly suburbanized labor force, industry followed the outward move, attracted as well by the economies derived from modern single-story plants with plenty of parking space for employees. Industries no longer needed to locate near railway facilities; freeways presented new opportunities for lower-cost, more-flexible truck transportation. Service industries were also attracted by the purchasing power and large, well-educated labor force now living in the suburbs, and complexes of office buildings developed, like the shopping malls, at freeway intersections and along freeway frontage roads and major connecting highways.

The major metropolitan areas rapidly expanded in area and population. During the 1980s and 1990s, cities that were growing showed most of their growth about 16 kilometers (10 mi) from the central business district; cities that shrank during these years had a large population drop near the center city, but growth 32 to 48 kilometers (20 to 30 mi) from the central business district. In other words, cities during these years grew more in their suburban fringe than in the central business district.

During the 2000s, patterns changed. Cities on the increase in these years, such as Atlanta, saw greater growth in the central city and smaller growth in the suburbs. Declining cities, such as Cleveland and Detroit, show continued suburbanization and a small trend toward repopulating the urban center. **Figure 11.21** compares population density in Atlanta in 1980 and 2010, showing the extent of suburbanization and maintenance of population in the urban core. **Figure 11.22** performs the same comparison for Detroit, which is remarkable for both its suburbanization and its hollowing of the urban core.

A new urban America has emerged on the perimeters of the major metropolitan areas. With increasing sprawl and the rising costs implicit in the ever-greater spatial separation of the functional segments of the fringe, peripheral expansion slowed, the supply of developable land was reduced, and the intensity of land development grew. No longer dependent on the central city, the suburbs were reborn as vast, collectively self-sufficient outer cities, marked by landscapes of industrial parks, skyscraper office parks, massive retailing complexes, and a proliferation of gated communities and apartment complexes.

The new suburbia began to rival the older central business district in size and complexity. Collectively, the new centers surpassed the central cities as generators of employment and income. Together with the older CBDs, the suburbs perform many services that mark the postindustrial metropolis. During the 1980s, more office space was created in the suburbs than in the central cities of America. Tysons Corner, Virginia (between Arlington and Reston), for example, became the ninth-largest central business district in the United States. Regional and national headquarters of leading corporations, banking, professional services of all kinds, major hotel complexes, and recreational centers became parts of the new outer cities. **Edge cities** are defined by their large nodes

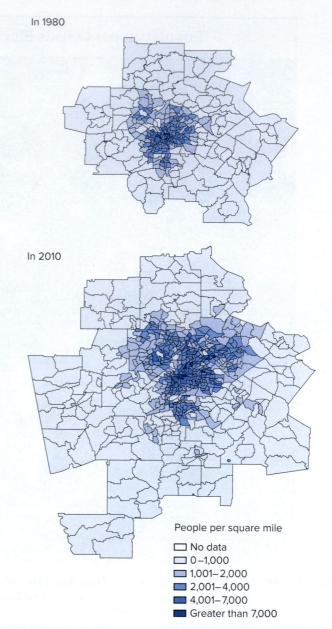

In 1980

In 2010

People per square mile

- ☐ No data
- ☐ 0–1,000
- ☐ 1,001–2,000
- ☐ 2,001–4,000
- ☐ 4,001–7,000
- ☐ Greater than 7,000

Figure 11.21 **Population density, Atlanta, 1980 and 2010.** Note that, while the suburbs have grown, the urban core remains densely populated. *Source: Kyle Fee, "Urban Growth and Decline: The Role of Population Density at the City Core," Federal Reserve Bank of Cleveland, http://www.clevelandfed.org/research /commentary/2011/2011-27.cfm accessed on 1/6/13.*

of office and commercial buildings and characterized by having more jobs than residents within their boundaries.

Edge cities now exist in all regions of the United States. The South Coast Metro Center in Orange County, California; the City Post Oak–Galleria center on Houston's west side; King of Prussia and the Route 202 corridor northwest of Philadelphia; the Meadowlands, New Jersey, west of New York City; and Schaumburg, Illinois, in the western Chicago suburbs are a few examples of this new urban form. There is even a suburb of Atlanta named Periphery Center. Location factors for edge cities include proximity to major highway corridors, international airports, and areas of high social status.

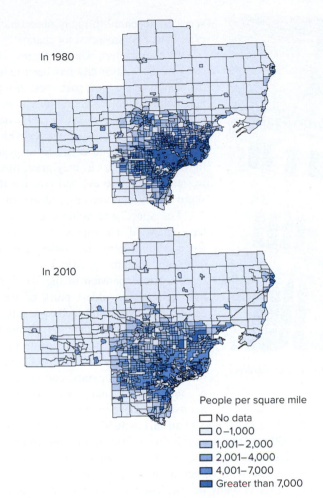

In 1980

In 2010

People per square mile

☐ No data
☐ 0–1,000
☐ 1,001–2,000
☐ 2,001–4,000
■ 4,001–7,000
■ Greater than 7,000

Figure 11.22   Population density, Detroit, 1980 and 2010. Suburbs around Detroit have grown and the central business district has lost population, creating what some term "the urban doughnut." *Source: Kyle Fee, "Urban Growth and Decline: The Role of Population Density at the City Core," Federal Reserve Bank of Cleveland, http://www.clevelandfed.org/research /commentary/2011/2011-27.cfm accessed on 1/6/13.*

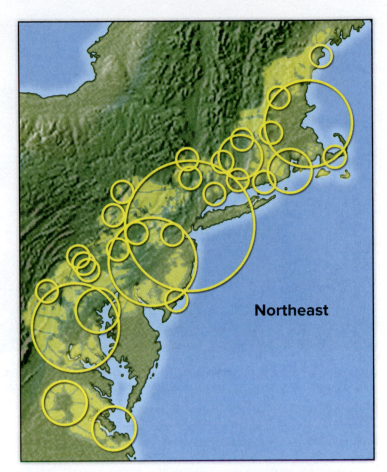

Northeast

Figure 11.23   A megalopolis in the northeast corridor. With its principal cities of Boston, New York, Philadelphia, Baltimore, and Washington, D.C., the northeast corridor has a 2010 population of 52 million, or 17% of the U.S. population. The area is important economically, as it produces 20% of the country's gross domestic product (GDP). It is projected to grow by 18 million people by 2050. *Source: All information from America 2050, America's Regional Plan Association, http://www.america2050.org/northeast.html accessed 1/2013.*

In recent years, suburban areas have expanded to the point where metropolitan areas are coalescing. The Boston-to-Washington corridor, often called a **megalopolis**, is a continuously built-up region that stretches from north of Boston to south of Washington, D.C. (**Figure 11.23**). The megalopolis' tightly interwoven suburban and urban areas encroach on rural landscapes. New growth centers compete with central cities: Virginia suburbs specialize in defense-related industries; Maryland suburbs specialize in health, space, and communications firms. The area's population is estimated around 50 million and expected to grow to 58 million by 2025.

Decline of the Central City

Changing patterns of accessibility have had a tremendous effect on the vitality of the central business district and of the inner city surrounding it. The CBD once enjoyed high levels of accessibility, at the center of streetcar and intercity rail lines. With the coming of the interstate highway system and widespread air transportation, central cities were increasingly viewed as overly congested and relatively inaccessible. The dynamic that provided economic advantages to central cities worked to their detriment. Many jobs moved from the central city to the urban periphery, along with much of the prosperous population.

Suburbanization redistributed jobs and population, resulting in spatial and political separation of social groups. The upwardly mobile residents of the city—younger, wealthier, and better educated—took advantage of the automobile and the freeway to leave the central city. The poorer, older, least-advantaged urbanites were left behind and decay and abandonment set in (**Figure 11.24**). The central cities and suburbs became increasingly differentiated. Large areas within the cities now contain only the poor and minority groups, a population barely able to pay the rising costs of social services that their numbers, neighborhoods, and condition require.

The services needed to support the poor include welfare payments, social workers, police and fire protection, health delivery systems, homeless shelters, and subsidized housing. Central cities,

Figure 11.24   **Derelict buildings in the central business district of Gary, Indiana.** Some areas of large cities have been abandoned and now house only the poor. © *Mark Bjelland.*

by themselves, are unable to raise the taxes needed to support such an array and intensity of social services because they have lost the tax base represented by suburbanized commerce, industry, and upper-income residential uses. Lost, too, were the job opportunities that were once a part of the central-city structure. Increasingly, the poor and minorities are trapped in a central city without the possibility of nearby employment and are isolated by distance, immobility, and unfamiliarity from the few remaining low-skill jobs, which are now largely in the suburbs. This unfortunate circumstance is often called a *spatial mismatch*. Many large urban areas have meager public transportation systems that do not provide access to all parts of the urban region. In Detroit, it can take 3 hours and three buses to travel from the urban core to a job in the suburbs. This particularly affects those without the means to own or maintain an automobile, especially women and minorities.

In an effort to help struggling cities and their residents, the federal government initiated urban renewal programs. Spurred by the Housing Act of 1949, governments cleared slums, reconstructed center cities, and built public housing, cultural complexes, and industrial parks. Unfortunately, these public housing projects often became places of concentrated poverty and high crime rates. Many have since been torn down (**Figure 11.25**).

In cities with a declining industrial base and concentrated poverty, the battle to maintain or revive the central city is frequently judged to be a losing one. Cities such as Detroit, Michigan; Toledo, Ohio; and Bridgeport, Connecticut, witnessed multiple failed attempts at urban renewal. In recent years, the central city has been the destination of thousands of homeless people (see "The Homeless," p. 324). Many live in public parks, in doorways, by street-level warm-air exhausts of subway trains, and in subway stations. Central-city economies, with their high land and housing values, limited job opportunities for the unskilled, and inadequate resources for social

services, appeared to many observers to offer few or no prospects for change.

In the western United States, the story of the central city has been rather different. For the most part, these newer, automobile-oriented metropolises were able to expand physically to keep the new growth areas on their peripheries within the central-city boundaries. By expanding their boundaries as they grew, many western cities have avoided creating the sharp divides between city and suburb.

The speed and volume of growth have spawned a complex of concerns, some reminiscent of older eastern cities and others specific to areas of rapid urban expansion in the West. As in the East, the oldest parts of western central cities tend to be pockets of poverty, racial conflict, and abandonment. In addition, western central-city governments face all the economic, social, and environmental consequences of unrestricted marginal expansion. Scottsdale, Arizona, for example, covered 2.6 square kilometers (1 sq mi) in 1950; today it has grown to nearly 500 square kilometers (about 200 sq mi), four times the physical size of San Francisco. Phoenix, with which Scottsdale has now coalesced, is larger in area than Los Angeles, which has three times as many people. The phenomenal growth of Las Vegas, Nevada, has similarly converted vast areas of desert landscape to low-density urban use (**Figure 11.26**).

Revitalizing the Urban Core

Central cities in the United States hit their low point in the late 1970s and early 1980s: fiscal crises, high crime, and depopulation seemed irreversible. By the late 1980s, the arrival of the digital age led some to proclaim that new communication devices would eliminate the need for the face-to-face interaction intrinsic to cities. Instead, digital communications have become centralizing forces, facilitating the growth of knowledge-based industries and activities such as finance, entertainment, health care, and corporate management. These industries depend on dense, capital-intensive information technologies concentrated in centralized locations. Cities—particularly large metropolitan cores—provide the first-rate telecommunications and fiber-optic infrastructures and the access to skilled workers, customers, investors, and research, educational, and cultural institutions needed by the modern, postindustrial economy.

As a reflection of cities' renewed attractions, employment and gross domestic product in the country's 50 largest urban areas began to grow again in the 1990s, reversing patterns of stagnation and decline in the preceding decades. Demand for downtown office space was met by extensive new construction and urban renewal, and even manufacturing has revived in the form of small and mid-size companies providing high-tech equipment and processes.

Figure 11.25 **Faulty towers.** Many elaborate—and massive—public housing projects have been failures. Chicago's Robert Taylor Homes, completed in 1962, consisted of 28 identical 16-story buildings, the largest public housing unit in the world and the biggest concentration of poverty in America. The 4400 apartments occupied a 3-kilometer-long (2-mi-long) stretch of South State Street, a concrete curtain that concealed from passing traffic the vandalism and crime that wracked the project. The growing awareness that public high-rise developments intended to revive the central city do not meet the housing and social needs of their inhabitants led to the razing since 1990 of nearly 100,000 of the more than 1.3 million public housing units in cities around the country, including the Robert Taylor Homes. © Beth A. Keiser/ AP Images.

Figure 11.26 **Urban sprawl in the Las Vegas, Nevada, metropolitan area.** Like many western cities, Las Vegas has grown over great expanses of desert in order to keep pace with a rapidly growing population. The fastest-growing metropolitan area in the United States in the 1990s, Las Vegas increased from a little more than 850,000 people in 1990 to more than 1,560,000 by the end of the decade, an increase of 83%. *Source: USDA-Natural Resources Conservation Service, photo by Lynn Betts.*

Figure 11.27 **Gentrified housing in Charleston, South Carolina.** Gentrification tends to occur where there is a supply of older, architecturally-interesting buildings in close proximity to a concentration of advanced service sector jobs. In Charleston, gentrification has resulted in significant displacement of African Americans by incoming whites. © Aimintang/iStock/Getty Images RF

These, in turn, support a growing network of suppliers and specialized services with "circular and cumulative" growth the result.

Urban centers became "sexy" places of consumption, promoted by popular television shows and movies. Some of the new office workers chose to live in central-city neighborhoods that offered solidly built, character-laden housing stock in highly accessible locations. In some cities, **gentrification**, the rehabilitation of housing in the oldest and now deteriorated inner-city areas by middle- and high-income groups, renewed depressed neighborhoods surrounding the central business district (**Figure 11.27**). Welcomed by many as a positive, privately financed force in the renewal of depressed urban neighborhoods, gentrification can have serious negative social and housing impacts on the low-income, frequently minority families who are displaced. Gentrification is another expression of the continuous remaking of urban land use and social patterns in accordance with the rent-paying abilities of alternate potential occupants. The rehabilitation and replacement of housing stock that it implies yield inflated rents and prices that push out established residents, disrupting the social networks they have created. An influx of younger, wealthier professionals has helped revitalize inner-city zones. Nearly all North American cities have witnessed a significant increase in the downtown residential population.

The Homeless

Every large city in the United States and Canada has hundreds or even thousands of people who lack homes of their own. One sees them pushing shopping carts containing their worldly goods, lining up at soup kitchens or rescue missions, and sleeping in parks or doorways. Reliable estimates of their numbers simply do no exist; official counts place the numbers of homeless Americans anywhere between 2.3 and 3.5 million. In other words, approximately 1% of the U.S. population experiences homelessness in a given year. Nearly one-quarter of those people were children; less than 2% were older than 62. Homelessness strikes minorities disproportionately: African-Americans constitute about 12% of the population, but 45% of people who are homeless. Perhaps 25% of the homeless have disabilities such as severe mental illness and chronic substance abuse. Other studies estimate that 14% of the U.S. population has been homeless at some point in their lives.

The existence of the homeless raises a multitude of questions—the answers, however, are yet to be agreed upon. Who are the homeless, and why have their numbers increased? Who should be responsible for coping with the problems they raise? Are there ways to eliminate homelessness?

Some people believe the homeless are primarily the impoverished victims of a rich and uncaring society. They view them as ordinary people, but ones who have had a bad break and have been forced from their homes by job loss, divorce, domestic violence, or incapacitating illness. They point to the increasing numbers of runaway teenagers, families, women, and children among the homeless. The women and children are less visible than the "loners" (primarily men) because they tend to live in cars, emergency shelters, or doubled up in substandard buildings. Advocates of the homeless argue that federal government budget cuts and gentrification have led to a dire shortage of affordable housing. During the same period, city governments demolished low-income housing, especially single-room occupancy hotels, in the name of urban revitalization. In addition, many states reduced funding for mental hospitals, casting institutionalized people out onto the streets.

At the other end of the spectrum are those who see the homeless chiefly as people responsible for their own plight. In the words of one commentator, they are "deranged, pathological predators who spoil neighborhoods, terrorize passersby, and threaten the commonweal." They point to studies showing that a significant percentage of the homeless suffer from alcoholism, drug abuse, or mental illness and argue that people are responsible for the alcohol and drugs they ingest; they are not helpless victims of a disease.

Communities have tried a number of strategies to cope with their homeless populations. Many have set up temporary shelters, especially in cold weather; others subsidize permanent housing and/or group homes. They encourage private, nonprofit groups to establish soup kitchens and food banks. Others attempt to drive the homeless out of town, or at least to parts of town where they will not be visible. They forbid loitering in city parks or on beaches after midnight, install sleep-proof seats on park benches and bus stations, and outlaw aggressive panhandling.

Neither point of view appeals to those who believe that homelessness is more than simply a lack of shelter. What the homeless need, others say, is a "continuum of care"—an entire range of services that includes education, job training, treatment for drug and alcohol abuse and mental illness, as well as, first and foremost, affordable housing options.

Considering the Issues

1. What is the nature of the homeless problem in the community where you live or with which you are most familiar? Has your community done anything to make the homeless feel unwelcome?

2. Where should responsibility for the homeless lie: at the federal, state, or local government level? Or is it best left to private groups such as churches and charities? Why?

3. Some people argue that giving money, food, or housing but no therapy to street people makes one an "enabler," or accomplice, of addicts. Do you agree? Why or why not?

4. What do you think of solutions that focus on first providing stable, permanent housing for the homeless and then providing needed support services.

A tent camp of homeless persons in Los Angeles. © *Mark Bjelland.*

The reason for that expected and actual growth lies in demographics. Young professionals are marrying and having children later or, often, are divorced or never married. For them—a growing proportion of Americans—suburban life and shopping malls hold few attractions, while central-city residence offers high-tech and executive jobs within walking or biking distance and cultural, entertainment, and boutique shopping opportunities close at hand. Gay couples and families often choose to live in urban centers as

Figure 11.28   **New uses for industrial buildings.** This former tractor factory in Minneapolis, like most of the neighboring factory and warehouse buildings, has been converted to loft housing. © *Mark Bjelland*

well. The younger group has been joined by "empty-nesters," couples who no longer have children living at home and who find big houses on suburban lots no longer desirable. By their interests and efforts these two groups have largely or completely remade and upgraded such old urban neighborhoods and industrial districts as the Mill District of Minneapolis, the Armory District of Providence, Rhode Island, the Denny Regrade and Belltown of Seattle, Main St./Market Square District of Houston, and many others throughout the country (**Figure 11.28**).

Part of the new vigor of central cities comes from new immigrants who spread beyond the "gateway" cities. They have become deeply rooted in their new communities by buying and renovating homes in inner-city areas, spending money in neighborhood stores, and most importantly establishing their own business districts (**Figure 11.29**). They are also important additions to the general urban labor force, providing the skilled and unskilled workers needed in expanding office work, service, and manufacturing sectors.

Local, state, and federal governments are fostering the revival of urban areas by investing in slum clearance, park development, cultural center construction, sports facilities, and the like. Columbus, Ohio, for example, recently built a hockey arena downtown and a soccer stadium north of the CBD, promoted the rehabilitation of several neighborhoods near the downtown, and aided the development of art galleries that now dominate the northern part of the downtown. Indianapolis city officials are emptying housing projects in the Chatham Arch neighborhood and selling them to developers for conversion into apartments and condominiums. Renovation of an old cotton mill in the Cabbage-town district of Atlanta has produced 500 new lofts in a building-recycling project common to many older cities. And as whole areas are gentrified or redeveloped residentially, other investment flows into nearby commercial activities. For example, Denver's LoDo district, once a skid row, has been wholly transformed into a thriving area of shops, restaurants, and sports bars along with residential lofts, a baseball stadium, and basketball arena.

Increasingly, central cities and metropolitan areas of both the eastern and the western United States are seeking to restrain rather than encourage physical growth. Portland, Oregon, drew a "do not pass" line around itself in the late 1970s, prohibiting urban conversion of surrounding forests, farmlands, and open space. Rather than lose people and functions, it has added both while preserving and increasing parklands and other urban amenities.

Other cities, metropolitan areas, and states are also beginning to resist and restrict urban expansion. "Smart growth" programs have been adopted by Colorado, Delaware, Minnesota, and Washington. These programs encourage governments and private developers to build a range of housing types, create walkable neighborhoods, mix land uses, preserve open space, provide a variety of transportation options, and direct support toward existing communities rather than continue the city sprawl. Developers generally find it easier to build large developments on new land rather than to fill in or rehabilitate city neighborhoods. In both the West and the East, U.S. cities are beginning to tighten controls on unrestricted growth, but private-sector resistance to controlling sprawl is still an important factor.

Figure 11.29   **Immigrants have revitalized central city neighborhoods.** Many African, Asian, and Latin American immigrants have established their own businesses, fixing, making, or selling things, adding to the vitality of central cities. © *David Grossman/The Image Works*

Institutional Controls

The governments—local and national—of most Western urbanized societies have instituted innumerable laws to control all aspects of urban life, including the rules for using streets, the provision of sanitary services, and the use of land. In this section, we touch upon only land use.

Institutional and government controls have strongly influenced the land use arrangements and growth patterns of most cities in the world. Cities have adopted land use plans and enacted subdivision control regulations and zoning ordinances to realize those plans. They have adopted building, health, and safety codes to ensure legally acceptable urban development and maintenance. All such controls are based on broad applications of the police powers of municipalities and their rights to ensure public health, safety, and well-being even when private-property rights are infringed.

These nonmarket controls on land use are designed to minimize incompatibilities (residences adjacent to heavy industry, for example) and provide for the creation in appropriate locations of public uses (the transportation system, waste disposal, government buildings, prisons, and parks) and private uses (colleges, shopping centers, and housing) needed for and conducive to a balanced, orderly community. In theory, such careful planning should preclude the emergence of slums, so often the result of undesirable adjacent uses, and should stabilize neighborhoods by reducing market-induced pressures for land use change.

In the United States, zoning ordinances and subdivision regulations have been used to exclude "undesirable" uses—apartments, special housing for the aged, low-income housing, halfway homes—from upper-income areas. Bitter court battles have been waged over "exclusionary" zoning practices: while proponents contend that the undesirable uses will harm the city and, more importantly, property values, opponents argue that such uses are important to the city as a whole. In recent years, some newly incorporated suburbs have sought to reduce the costs of growth—such as developing infrastructure and building schools and playgrounds—by zoning out housing that would bring in children.

In most of Asia, there is no zoning, and it is quite common to have small-scale industrial activities operating in residential areas. Even in Japan, a house may contain several people doing piecework for a local industry. In both Europe and Japan, neighborhoods have been built and rebuilt gradually over time, and it is usual to have a wide variety of building types from various eras mixed together on the same street. In the United States and Canada, such mixing is much rarer and is often viewed as a temporary condition as areas are in transition toward total redevelopment. Perhaps the only exception to this in a large city is Houston, Texas, which has no zoning regulations.

11.6 Global Urban Diversity

The city is a global phenomenon. Cities' structures, forms, and functions differ between regions, reflecting diverse heritages and economies. Much of the discussion in this chapter has dwelt on models of the city in the United States. But the models and descriptions of the U.S. city do not generally apply to cities in other parts of the world. Those cities have developed different functional and structural patterns, some so radically different from the models we have explored that we would find them unfamiliar and uncharted landscapes, indeed. The city is universal; its characteristics are cultural and regional.

Western European Cities

Although each is unique historically and culturally, Western European cities as a group share certain common features. They have a much more compact form and occupy less total area than American cities of comparable population, and most of their residents are apartment dwellers. Residential streets of the older sections tend to be narrow, and front, side, or rear yards or gardens are rare.

European cities also enjoy a long historical tradition. A generally common heritage of medieval origins, Renaissance restructurings, and industrial-period growth has given the cities of Western Europe distinct features. Despite wartime destruction and postwar redevelopment, many still bear the impress of past occupants and technologies, even as far back as Roman times. An irregular system of narrow streets may be retained from the random street pattern developed in medieval times of pedestrian and pack-animal movement. Main streets radiating from the city center and cut by circumferential "ring roads" tell us the location of high roads leading into town through the gates in city walls now gone and replaced by circular boulevards. Broad thoroughfares, public parks, and plazas mark Renaissance ideals of city beautification and the aesthetic need for processional avenues and promenades.

European cities were developed for pedestrians and still retain the compactness appropriate to walking. The sprawl of American peripheral or suburban zones is generally absent. At the same time, compactness and high density do not mean skyscraper skylines. Much of urban Europe predates the steel-frame building and the elevator. City skylines tend to be low, three to five stories in height, sometimes (as in central Paris) held down by building ordinance or by prohibitions against private structures exceeding the height of a major public building, often the central cathedral (**Figure 11.30**). Where those restrictions have been relaxed, however, tall office buildings have been erected.

Compactness, high densities, and apartment dwelling encouraged the development and continued importance of public transportation, including well-developed subway systems. The private automobile has become much more common of late, though most central-city areas have not yet been significantly restructured with wider streets and parking facilities to accommodate it. For instance, private autos are charged a high fee to enter the central business district of London. The automobile is not the universal need in Europe that it has become in American cities, as home and work are generally more closely spaced. Even though most European cities have core areas that tend to be stable in population and attract, rather than repel, the successful middle class and the upwardly mobile, they are also affected by the processes of decentralization. Many residents choose to live in suburban locations as car ownership and use become more commonplace, a pattern also seen in Canada (see "The Canadian City," p. 328).

The old city fortifications may mark the boundary between the core and the surrounding transitional zone of substandard housing, 19th-century industry, and recent immigrants. European governments used the strategy of grouping industrial developments and

Figure 11.30 **Low-rise European cities.** Even in their central areas, many European urban centers show a low profile, like that of Paris, seen here. Although taller buildings—20, 30, and even 50 or more stories in height—have become more common in major urban areas since World War II, they are neither the universal mark of the central business districts they have become in the United States nor the generally welcomed symbols of progress and pride. © *Photov.com/Pixtal/AGE Fotostock RF.*

working-class homes in suburban areas outside the city core; these have been the areas in which immigrants, often from North Africa and Turkey, have found housing. Some of these areas have been neglected and suffer decline, especially when unemployment rates are high.

Eastern European Cities

Cities of Eastern Europe, including Russia and the former European republics of the Soviet Union, once part of the communist world, make up a separate urban class. The poscommunist city shares many of the traditions and practices of Western European cities, but it differs from them in the centrally administered planning principles that were, in the communist period (1945–1990), designed to shape and control both new and older settlements. The governments' goals were, first, to limit city size to avoid metropolitan sprawl; second, to ensure an internal structure of neighborhood equality and self-sufficiency; and third, to segregate land uses. The planned communist city fully achieved none of these objectives, but by attempting them it emerged as a distinctive urban form.

The postcommunist city is compact, with relatively high building and population densities reflecting the nearly universal apartment dwelling, and with a sharp break between urban and rural land uses on its periphery. It depends nearly exclusively on public transportation.

During the communist period, the Eastern European city differed from its Western counterpart in its purely governmental, rather than market, control of land use and functional patterns. This control dictated that the central area of cities be reserved for public use, not occupied by retail establishments or office buildings on the Western, capitalist model. Eastern European governments preferred a large central square ringed by administrative and cultural buildings with space nearby for a large recreational and commemorative

Figure 11.31 **Communist-era housing.** This scene from Bratislava, Slovakia, shows "superblocks" of self-contained apartment housing microdistricts that contain their own shopping areas, schools, and other facilities. © *Josie Elias/Photodisc/Getty Images RF.*

park. They emphasized industrial development of cities and marginalized consumer services.

Residential areas were made up of *microdistricts,* assemblages of uniform apartment blocks housing perhaps 10,000 to 15,000 people, surrounded by broad boulevards, and containing centrally sited nursery and grade schools, grocery and department stores, theaters, clinics, and similar neighborhood necessities and amenities (**Figure 11.31**). Residential areas were expected to be largely self-contained in the provision of goods and services, minimizing the need for a journey to centralized shopping locations. Plans called for the effective separation of residential quarters from industrial districts by landscaped buffer zones; but in practice many microdistricts were built by factories for their own workers and were located immediately adjacent to the workplace. Because microdistricts were most easily and rapidly constructed on open land at the margins of expanding cities, high residential densities have been carried to the outskirts of town.

This pattern is changing as market principles of land allocation are adopted. Now that private interests can own land and buildings, historic apartments and townhouses are being restored and becoming the fashionable places to live. Builders are constructing Western-style shopping malls, spacious apartments, and single-family homes. Many peripheral superblock apartment complexes have been depopulated as residents relocate based on market preferences. The weighting of the economy is changing, with less emphasis on industry and greater emphasis on the service sector. The socioeconomic divisions in cities are increasing as the newly rich move to higher-priced housing while housing shortages for the less well off have become acute.

Rapidly Growing Cities in the Developing World

The fastest-growing cities, and the fastest-growing urban populations, are found in the developing world, especially Asia and Africa. Industrialization has come to most of them only recently. Modern technologies in transportation and public facilities are sometimes lacking, and the structures of cities and the cultures

THE CANADIAN CITY

In 1900, only 37% of Canadians lived in urban areas; in 2010, some 81% of the population was urban. Cities such as Calgary and Edmonton, in Alberta, and Toronto, in Ontario, have seen high population growth rates recently; poorer regions have seen a decline similar to that of declining U.S. industrial cities.

While urban forms are similar, there are some subtle but significant differences between Canadian and U.S. cities. The Canadian city, for example, is more compact than its U.S. counterpart of equal population size, with a higher density of buildings and people and less suburbanization of populations and functions.

Space-saving multiple-family housing units are more prevalent in Canada, so a similar population size is housed on a smaller land area with much higher densities within the central area of cities. The Canadian city is better served by and more dependent on mass transportation than in the U.S. city. Because Canadian metropolitan areas have only one-fourth as many miles of expressway lanes per capita as U.S. metropolitan areas, suburbanization is less extensive north of the border than south of it.

The differences are cultural as well. Cities in both countries are ethnically diverse (Canadian communities, in fact, have a higher proportion of foreign-born), but U.S. central cities exhibit far greater internal distinctions in race, income, and social status and more pronounced contrasts between central-city and suurban residents. That is, there has been much less "flight to the suburbs" by middle-income Canadians. As a result, the Canadian city shows greater social stability, higher per capita average income, greater retention of shopping facilities, and more employment opportunities and urban amenities than its U.S. central-city counterpart. In particular, it does not have the rivalry from the well-defined, competitive, edge cities of suburbia that fragment U.S. metropolitan complexes.

High-density Canadian cities. The downtown peninsula of Vancouver has been filled with residential towers that are home to 70,000 residents. Vancouver's high density allows it to protect ample park, forest, and agricultural land. On average, Canadian metropolitan areas are almost twice as densely populated as metropolitan areas in the United States. On a per capita basis, Canadian urbanites are two and a half times more dependent on public transportation than are American city dwellers. That reliance gives form, structure, and coherence to the Canadian central city, qualities lost in the sprawling and fragmented U.S. metropolis. © *Mark Bjelland.*

of their inhabitants are far different from those familiar to North Americans. The developing world is vast in extent and diverse in its physical and cultural landscapes.

The backgrounds, histories, and current economies and administrations of developing-world cities vary so greatly that it is impossible to generalize about their internal structure. Some are ancient, having been established centuries before the more-developed cities of Europe and North America. Others are still preindustrial, with only a modest central commercial core; they lack industrial districts, public transportation, or any meaningful degree of land use separation. Still others, though increasingly Western in form, are only beginning to industrialize.

Despite the variety of urban forms found in such diverse regions as Latin America, Africa, the Middle East, and Asia, we can identify some features common to most of them. First, most of what are currently categorized as developing countries have a colonial legacy, and several major cities were established principally to serve the needs of the colonizing country. The second feature is that urban primacy and the tremendous growth that these cities are generally experiencing as their societies industrialize have left many of these cities with inadequate facilities and no way to keep up with population growth. Third, most cities in developing countries are now characterized by neighborhoods hastily built by new migrants, away from city services and often occupying land illegally. Such squatter settlements are a large and growing component of these cities and reflect both the city's relative opportunity and its poverty. Finally, in many cases, the government has responded with greater planning, sometimes going so far as to move the national capital away from the overcrowded primate city to a new location or to create entirely new cities to house industrial or transport centers.

Colonial and Noncolonial Antecedents

Cities in developing countries originated for varied reasons and continue to serve several functions based on their position as market, production, government, or religious centers. Their legacy and purpose influence their urban form.

Many cities in developing countries are products of colonialism, established as ports or outposts of administration and exploitation, built by Europeans on a Western model. For example, the British built Kolkata (Calcutta), New Delhi, and Mumbai (Bombay) in India and Nairobi and Harare in Africa. The French developed Hanoi and Ho Chi Minh City (Saigon) in Vietnam, Dakar in Senegal, and Bangui in the Central African Republic. The Dutch used Jakarta as their main outpost, Belgium had Kinshasa (formerly Leopoldville) in what is now the Democratic Republic of the Congo, and the Portuguese founded a number of cities in Angola and Mozambique. Colonialists controlled the economies of the regions from these urban centers.

Urban structure is a function not just of the time when a city was founded, or of who the founder was, but also of the role a city plays in its own cultural milieu. Land use patterns in capital cities reflect the centralization of government functions and the concentration of wealth and power in a single city of a country (**Figure 11.32a**). The physical layout of a religious center, or sacred city, is conditioned by the religion it serves, whether it is Hinduism, Buddhism, Islam, Christianity, or another faith. Typically, a monumental structure—such as a temple, mosque, or cathedral—and associated buildings rather than government offices occupy the city center (**Figure 11.32b**). Traditional market centers for a wide area (Timbuktu in Mali and Lahore in Pakistan) or cultural capitals (Addis Ababa in Ethiopia and Cuzco in Peru) have land use patterns that reflect their special functions. Likewise, port cities such as Dubai (United Arab Emirates), Haifa (Israel), and Shanghai (China) have a land use structure different from that of an industrial or mining center such as Johannesburg (South Africa). Adding to the complexity are the facts that cities with a long history reflect the changes wrought by successive rulers and/or colonial powers and that as some of the megacities in the developing world have grown, they have engulfed nearby towns and cities.

Urban Primacy and Rapid Growth

In many developing countries, the population is disproportionately concentrated in national and regional capitals. Few developing countries have mature, functionally complex small- and medium-size centers. The primate city dominates their urban systems. One-fifth of all Nicaraguans live in Managua, and Libreville contains one-third of the population of Gabon.

Many cities in developing countries with rapidly growing economies have a vibrant and modern city center. Such districts contain amenities that could be found in many cosmopolitan centers and are the places where the wealthiest members of society work and often live. This is also the part of the city that businesspeople, officials, tourists, and other visitors are most likely to see. Some cities have

(a)

(b)

Figure 11.32 **Developing-world cities.** Compare the monumental modernist government buildings in (**a**) the capital city of Brasília, Brazil, with (**b**) a state capital, Guanajuato, Mexico, which is dominated by religious structures. *(a) © Julia Waterlow/Corbis Documentary/Getty Images. (b) © Jose Fuste Raga/Corbis.*

Figure 11.33 **Prosperous, modern, city center in Dubai, United Arab Emirates.** Dubai grew rich from oil revenues but is now a diversified business and tourist center. It features the world's tallest building and largest indoor shopping mall. Contrast this photo with that of the slum in Manila in Figure 11.35. © *Barry Schofield/Alamy Stock Photo RF.*

Figure 11.34 **High-rise apartments in densely-populated Mumbai, India.** © *Image Source RF.*

poured enormous efforts into these city centers: for example, Dubai, in the United Arab Emirates, has many skyscrapers and is home to the tallest building in the world (**Figure 11.33**).

Yet the presence of gleaming downtowns cannot disguise the dramatic dualism of rich and poor. In fact most of these cities simply cannot keep pace with the massive growth they are experiencing. Many cities in Latin America, Africa, and Asia are adding population at terrific rates: in India, rural-to-urban migration is proceeding at such a pace that observers predict that 31 villagers will continue to arrive in an Indian city every minute over the next 43 years—700 million people altogether. They leave their villages to become their family breadwinners; they will become the cities'

underclass, driving taxis, parking cars, or selling tea on street corners and crowd the already densely populated cities (**Figure 11.34**).

Lagos, Nigeria, is the second most populous city in Africa, following Cairo, Egypt. It has a current population of 12.6 million and a growth rate that is 10 times faster than that of New York or Los Angeles. It adds 2000 people each day. Its population is expected to reach 24.2 million by 2030. Although most of the population of Lagos has access to electricity, only 10% of households are connected to a public drinking water source and only 2% are connected to a sewage system. Much of the city's human waste is disposed of through open ditches that discharge onto the city's waterfront, although the government has commited to boosting wastewater treatment capacity.

Nigeria and other developing countries are experiencing urban growth driven by rural-to-urban migration. In 2009, China had 145 million rural-to-urban migrants, more than 30 times the number of Irish immigrants to the United States from 1820 to 1930. The government allows such migration to promote cheap factory and construction labor and its long-term goal of turning its mostly rural economy into an urban economy.

Traffic in these rapidly growing cities is almost always a problem. Most people do not own cars and must rely on public transportation. Where the public transport system is limited, the result has been an overcrowded city centered on a single major business district in the old tradition. In Lagos, it takes an average of 2 to 3 hours to travel 10 to 20 kilometers (6 to 12 mi), as there are only three bridges connecting the city's four islands to the mainland. In cities with expanded automobile use, congestion and air pollution have become severe. Mexico City and Bangkok are especially plagued by traffic-choked streets and pollution.

For those who are well off, living in many of these cities can present a series of challenges. Unlike residents of cities in more-developed countries, they cannot take urban planning, a stable city administration, and adequate protection for granted.

Squatter Settlements

The developing countries have experienced massive in-migrations from rural areas as vast numbers of low-income residents have migrated to cities with the hope of finding jobs and improving their socioeconomic condition. That hope is rarely realized because the cities already have populations greater than their formal functions and employment bases can support. In all of them, large numbers of people support themselves in the "informal" sector—as food vendors, peddlers of cigarettes or trinkets, streetside barbers or tailors, errand runners, or package carriers—outside the usual forms of wage labor.

Most of the new urbanites have little choice but to pack themselves into shantytowns and squatter settlements on the fringes of the city. These informal communities—*favela* in Brazil, *barrio* in Mexico, *kampung* in Indonesia, *gecekondu* in Turkey, or *katchi abadi* in Pakistan—are isolated from sanitary facilities,

Figure 11.35   **A slum in Manila, Philippines.** Where states or municipalities provide little or no housing for the poor, people create their own. © *Marcus Lindstrom/E+/Getty Images RF.*

Table 11.4	Slum Dwellers' Share of Urban Population		
Region	1990	2000	2014
Sub-Saharan Africa	70	65	55
South Asia	57	46	31
East Asia	44	37	25
Western Asia	23	21	25
Latin America and Caribbean	34	29	20
North Africa	34	20	11
Oceania	24	24	24
Developing regions	*46*	*39*	*30*

Source: United Nations, The Millennium Development Goals Report, 2010 and 2015.

public utilities, and job opportunities found only at the center. In some cities, one-third to two-thirds of the population live in shantytowns, often in defiance of the law and with no legal title to their shacks.

Impoverished squatter districts exist around most major cities in Africa, Asia, and Latin America (**Figure 11.35**). In the vignette that opened this chapter, you read that one-half million people are estimated to live in Cairo's Northern and Southern Cemeteries. In the sprawling slum district of Nairobi, Kenya, called Mathare Valley, one-quarter million people are squeezed into 15 square kilometers (6 sq mi) with little or no access to public services such as water, sewers and drains, paved roads, and garbage removal. Similar shantytowns and squatter settlements house approximately one-fourth of the population of such Asian cities as Bangkok (Thailand), Kuala Lumpur (Malaysia), and Jakarta (Indonesia). The percentage is even higher in a number of Indian cities, including Chennai (Madras) and Kolkata (Calcutta). Crumbling tenements house additional tens of thousands, many of whom are eventually forced into shantytowns by the conversion of their residences into commercial property or high-income apartments.

Over the past decade the percentage of urban residents living in slums has decreased in most regions, while their absolute numbers have increased as the cities grow and grow (**Table 11.4**). Conflict-torn regions are an area where conditions have worsened. Sometimes residents of squatter settlements have successfully lobbied governments for water, sewers, roads, and other infrastructure and over time have become more established neighborhoods. One of the major steps in upgrading slums is to give residents some form of secure right to the land on which their dwelling is located. Over time as incomes stabilize, shacks can be improved into regular houses and slums can become stable neighborhoods. Unless the land is unsafe or unstable, slum upgrading is preferable to the demolition and relocation that displaces people and breaks apart dense social networks.

Planned Cities

Some capital cities have been relocated outside the core regions of their countries, either to achieve the presumed advantages of centrality or to encourage more-uniform national development; examples include Islamabad (Pakistan), Ankara (Turkey), Brasília (Brazil), and Abuja (Nigeria) (see Chapter 8, 224). A number of developing countries have also created or are currently building new cities that are intended to draw population away from overgrown metropolises and/or to house industrial or transport centers.

The huge numbers of rural-to-urban migrants in China have strained its cities' capability to house them all. In Shanghai, the demand for housing has spawned hundreds of new apartment developments of multiple skyscrapers with names such as Rich Gate and Home of the Tycoons. The city has more than 4000 buildings 18 stories high or taller, far more than New York, and new satellite towns are planned and established in a short time. An hour north of Beijing is Sun City, a gated community with California-style houses; connected to Beijing by two six-lane superhighways, the area is scheduled for rapid development. Urban populations are expected to grow exponentially, so planning for new developments and towns in China continues apace.

Planning can also affect the fate of older cities experiencing huge population growth. Hyderabad, India, has a population of 7.6 million, which is expected to grow to 9 million by 2015. To provide job opportunities for its growing population, provincial officials have attracted international technology, pharmaceutical, biotech, and banking and insurance companies to planned suburbs (**Figure 11.36**). They directed funds into cleaning and "greening" the city, planting trees, and creating parks and gardens. In addition, officials fought corruption and collected taxes via computer, opening transaction centers to make paying bills and banking simpler. The city opened a new international airport to rival Mumbai and Delhi in 2008. Hyderabad is not alone: examples of city planners trying to cope with growth, some more successfully than others, exist on every continent.

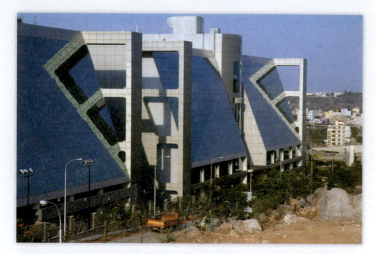

Figure 11.36   The Cyber Gateway Building in Hyderabad's Hitech City. The complex houses such firms as the multinational software companies Microsoft, IBM, and Toshiba, as well as such Indian companies as Wipro, which provides information technology services and product design. Hitech City also houses professional schools in business and information technology. © *Idealink Photography/Alamy Stock Photo.*

Summary of Key Concepts

- Cities developed in the ancient world under certain circumstances: in areas where there was a food surplus, social organization, a defensible situation, and a developing economy. Although there have been urban areas since ancient times, only recently have urban areas become home to the majority of people in industrialized countries and the commercial crossroads for uncounted millions in the developing world.

- All settlements growing beyond their village origins take on functions uniting them to the countryside and to a larger system of settlements. As they grow, they become functionally complex. Their economic base, composed of both *basic* and *nonbasic* activities, may become diverse. Basic activities represent the functions performed for the larger economy and urban system; nonbasic activities satisfy the needs of the urban residents themselves. Cities are marketplaces, centers of production, and centers of government administration.

- Systems of cities are reflected in the urban hierarchy and the rank-size rule. When a city is far larger than all others in the country, it is termed a *primate* city. Many countries have this dominant city as their main urban settlement, but there are only a few world cities, which dominate the global economy.

- Inside North American cities, urban areas often take on similar forms. At the core, the central business district has highest value and accessibility. Outside the CBD, lower-order commercial uses dominate. These patterns have inspired geographers to summarize urban form in the concentric zone, sector, multiple-nuclei, and peripheral models. The period following World War II brought massive changes in urban organization, with the decline and revitalization of the central city, accompanied by the rise and expansion of the suburbs. These changes have been accentuated by the tendency for urban dwellers to sort themselves by family status, social status, and ethnicity. In Western countries, these patterns are also influenced by government controls, which help determine land use.

- Urbanization is a global phenomenon, and the U.S. and Canadian models of city systems, land use, and social area patterns differ substantially from those of cities in the rest of the world, reflecting diverse heritages and economic structures. Western European cities differ from those in Eastern Europe, where land use reflects communist principles of city structure. Cities in the developing world are currently growing so quickly that they are unable to provide all their residents with employment, housing, safe water, sanitation, and other minimally essential services and facilities. In some cases, governments have been able to alleviate the situation somewhat by planning new cities or applying planning principles to rapidly growing older cities.

Key Words

basic sector 307	multiplier effect 307
central business district (CBD) 313	nonbasic sector 307
central city 305	peripheral model 316
central place theory 308	primate city 310
city 304	rank-size rule 310
concentric zone model 314	sector model 314
economic base 307	site 305
edge city 320	situation 305
gated community 318	suburb 305
gentrification 323	town 304
hinterland 303	urban hierarchy 310
megalopolis 321	urban influence zone 308
metropolitan area 305	urbanized area 305
multiple-nuclei model 314	world city 312

Thinking Geographically

1. Consider the city or town in which you live or attend school or with which you are most familiar. In a brief paragraph, discuss that community's site and situation; point out the connection, if any, between its site and situation and the basic functions that it performed earlier or now performs.

2. Describe the *multiplier effect* as it relates to the population growth of urban units.

3. Is there a hierarchy of retailing activities in the community with which you are most familiar? Of how many and of what kinds of levels is that hierarchy composed? What localizing forces affect the distributional pattern of retailing within that community?

4. Briefly describe the urban land use patterns predicted by the *concentric zone, sector, multiple-nuclei,* and *peripheral* models of urban development. Which one, if any, best corresponds to the growth and land use pattern of the community most familiar to you?

5. In what ways do social status, family status, and ethnicity affect the residential choices of households? What expected distribution patterns of urban social areas are associated with each? Does the social geography of your community conform to the predicted pattern?

6. How has suburbanization damaged the economic base and the financial stability of the central city?

7. In what ways does the Canadian city differ from the pattern of its U.S. counterpart?

8. Why are metropolitan areas in developing countries expected to grow larger than many Western metropolises by the year 2015?

9. What are *primate* cities? Why are primate cities in the developing countries overburdened? What can be done to alleviate the difficulties?

10. What are the significant differences in the generalized pattern of land uses of North American, Western European, and Eastern European cities?

11. How are cities in the developing world influenced by their colonial pasts? If you had to create a model for land use in such cities as Mumbai and Bangkok, what essential elements would you incorporate?

An oil well in Alberta, Canada with the Rocky Mountains in the distance. © *Michael Interisano/Design Pics/Getty Images RF.*

CHAPTER OUTLINE

"The world's richest nation" in terms of income per person was how an article in *National Geographic* described Nauru in 1976. A single-island country in the South Pacific Ocean, halfway between Hawaii and Australia, Nauru today is a wasteland whose barren and inhospitable landscape suggests a creation of science fiction. What happened between 1976 and now is that Nauru was almost totally stripped of the one natural resource that gave it value—high-grade phosphate rock for fertilizer—and most of the island is no longer inhabitable. Laid down over millions of years, the phosphates were the product of seagull droppings on coral.

Located just south of the equator, Nauru is tiny, with an area of just 21 square kilometers (8 sq mi). A narrow coastal belt surrounds a central plateau that rises to 65 meters (213 ft) above sea level, covers 80% of the land area, and contains valuable deposits of phosphate. There is just one paved road, a loop that circles the island.

For several thousand years, Nauruans lived within the island's limits despite frequent droughts. A population of about 1000 depended on food from the sea and from the tropical vegetation. The plunder of the island began after Nauru became a colony of Germany in 1888. At the end of World War I, the League of Nations gave Great Britain, Australia, and New Zealand the right to administer the island and set up the British Phosphate Commission to run the phosphate industry. Most of the phosphate was shipped to Australia for fertilizer. When Nauru received independence in 1968, the people chose to continue the extraction, which brought in tens of millions of dollars annually and gave the country one of the highest per capita incomes in the world.

Now, however, most of the phosphate has been mined. Deposits are expected to be exhausted in just a few years. Apart from some tropical fruits, no other resources exist; virtually everything must be imported for the 13,500 inhabitants: foodstuffs, fuels, manufactured goods, machinery, building and construction materials, and even water when the desalination plant fails. Poor investments and mismanagement of a billion-dollar trust fund have plunged the country into debt. The environmental damage from strip mining has been severe. The lush tropical forests were removed to facilitate mining. For thousands of years, they had provided refuge for native and migratory birds. The plateau is now a dry, jagged wasteland of tall, fossilized coral pinnacles, all that remain after the phosphate that lies between them is removed (**Figure 12.1**). A rise in sea level due to glacial melting and the thermal expansion of the oceans (see Chapter 4) would further impact Nauru, resulting in a major displacement of people from the coast to higher elevations.

Nauru exemplifies the difference between a sustainable society and one that squanders resources that have accumulated over millions of years. Population growth and economic development have magnified the extent and the intensity of human depletion of the treasures of the Earth. Resources of land, of ores, and of most forms of energy are finite, but the resource demands of an expanding, economically advancing population appear to be limitless. This imbalance between resource availability and use has been a concern at least since the time of Malthus and Darwin, but it wasn't until the 1970s that the rate of resource depletion and the associated environmental degradation became a major public policy issue.

Resources are unevenly distributed in kind, amount, and quality and do not match the distribution of population and demand. In this chapter, we survey the natural resources on which societies depend, their patterns of production and consumption, and the problem of managing those resources in light of growing demands and shrinking reserves.

We begin our discussion by defining some commonly used terms.

Figure 12.1 Environmental devastation. Intensive phosphate mining has left most of Nauru a wasteland. After mining, the land is unsuitable for agriculture or other human uses. *Lorrie Graham/AusAID.*

12.1 Resource Terminology

A **resource** is a naturally occurring, exploitable material that a society perceives to be useful to its economic and material well-being. Willing, healthy, and skilled workers constitute a valuable resource, but without access to materials such as fertile soil or petroleum, human resources are limited in their effectiveness. In this chapter, we devote our attention to physically occurring resources, or, as they are more commonly called, *natural* resources.

The availability of natural resources is a function of two things: the physical characteristics of the resources themselves and human economic and technological conditions. The physical processes that govern the formation, distribution, and occurrence of natural resources are determined by physical laws over which people have no direct control. We take what nature gives us. To be considered a resource, however, a given substance must be *understood* to be a resource. This is a cultural, not purely a physical, circumstance. Native Americans may have viewed the resource base of Pennsylvania as composed of forests for shelter and fuel, as well as the habitat of the game animals (another resource) on which they depended for food. European settlers viewed the forests as the unwanted covering of the resource that *they* perceived to be of value: soil for agriculture. Still later, industrialists appraised the underlying coal deposits, ignored or unrecognized as a resource by earlier occupants, as the item of value for exploitation (**Figure 12.2**).

Natural resources are usually recognized as falling into two broad classes: renewable and nonrenewable.

Renewable Resources

Renewable resources are materials that are replaced or replenished by natural processes. They can be used over and over; the supplies are not depleted. A distinction can be made, however, between those that are perpetual and those that are renewable only if carefully managed (**Figure 12.3**). **Perpetual resources** come from sources that are virtually inexhaustible, such as the sun, the wind, waves, tides, and geothermal energy.

Potentially renewable resources are renewable if left to nature but can be destroyed if people use them carelessly. These include groundwater, soil, plants, and animals. If the rate of exploitation exceeds that of regeneration, these renewable resources can be depleted. Groundwater extracted beyond the replacement rate in arid areas may be as permanently removed as if it were a nonrenewable ore. Soils can be totally eroded, and an animal species may be completely eliminated. Forests are a renewable resource only if people are planting at least as many trees as are being cut.

Nonrenewable Resources

Nonrenewable resources exist in finite amounts or are generated in nature so slowly that for all practical purposes the supply is finite. They include the fossil fuels (coal, crude oil, natural gas, oil shales, and oil sands), the nuclear fuels (uranium and thorium), and a variety of nonfuel minerals, both metallic and nonmetallic. Although the elements of which these resources are composed cannot be destroyed, they can be altered to less useful or available

Figure 12.2   The original hardwood forest covering these West Virginia hills was removed by settlers who saw greater resource value in the underlying soils. The soils, in turn, were stripped away for access to the still more valuable coal deposits below. Resources are as a culture perceives them, though exploitation may consume them and destroy the potential of an area for alternate uses. © C. Borland/PhotoLink/Getty Images RF.

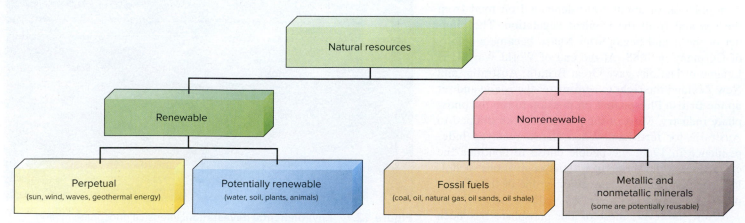

Figure 12.3   A classification of natural resources. Renewable resources can be depleted if the rate of use exceeds that of regeneration.

forms, and they are subject to depletion. The energy stored in a unit volume of the fossil fuels may have taken eons to concentrate in usable form; it can be converted to heat in an instant and be effectively lost forever.

Fortunately, many minerals can be *reused* even though they cannot be *replaced*. If they are not chemically destroyed—that is, if they retain their original chemical composition—they are potentially reusable. Aluminum, lead, zinc, and other metallic resources, plus many of the nonmetallics, such as diamonds and petroleum by-products, can be used time and again. However, many of these materials are used in small amounts in any given object, so that recouping them is economically unfeasible. In addition, many materials are now being used in manufactured products, so that they are unavailable for recycling unless the product is destroyed. Consequently, the term *reusable resource* must be used carefully. At present, all mineral resources are being mined much faster than they are being recycled.

Resource Reserves

Some regions contain many resources; others, relatively few. No industrialized country, however, has all the resources it needs to sustain itself. The United States has abundant deposits of many minerals, but it depends on other countries for such items as tin and manganese. The actual or potential scarcity of key nonrenewable resources makes it desirable to predict their availability in the future. We want to know, for example, how much petroleum remains in the Earth and how long we will be able to continue using it.

Any answer will be only an estimate, and for a variety of reasons such estimates are difficult to make. Exploration has revealed the existence of certain deposits, but we have no sure way of knowing how many remain undiscovered. Further, our definition of what constitutes a usable resource depends on current *economic* and *technological* conditions. If they change—if, for example, it becomes possible to extract and process ores more efficiently—our estimate of reserves also changes. Finally, the answer depends in part on the rate at which the resource is being used, but it is impossible to predict future rates of use with any certainty. The current rate could drop if a substitute for the resource in question were discovered, or it could increase if population growth or industrialization placed greater demands on it.

A useful way of viewing reserves is illustrated by **Figure 12.4**. Assume that the large rectangle includes the total stock of a particular resource, all that exists of it in or on the Earth. Some deposits of that resource have been discovered; they are shown in the left-hand column as "identified" amounts. Deposits that have not been located are called "undiscovered" amounts. Deposits that are economically recoverable with current technology are at the top of the diagram, whereas those labeled "subeconomic" are not attractive for any number of reasons (the concentration is not rich enough, it would require expensive treatment after mining, it is not accessible, and so on).

We can properly term **proved**, or **usable**, **reserves**—quantities of a resource that can be extracted profitably from known deposits—only the portion of the rectangle indicated by the pink tint. These are the amounts that have been identified and that can be recovered under existing economic and operating conditions. If new deposits of the resource are discovered, the reserve category will shift to the

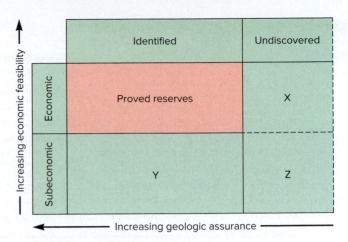

Figure 12.4   The variable definition of *reserves*. Proved, or usable, reserves consist of amounts that have been identified and can be recovered at current prices and with current technology. X denotes amounts that would be attractive economically but have not yet been discovered. Identified but not economically attractive amounts are labeled Y, and Z represents undiscovered amounts that would not be attractive now even if they were discovered. *Source: General Classification of Resources by the U.S. Geological Survey.*

right; improved technology or increased prices for the product can shift the reserve boundary downward. An ore that was not considered a reserve in 1950, for example, may become a reserve in the future if ways are found to extract it economically.

12.2 Energy Resources and Industrialization

Although people depend on a wide range of resources contained in the biosphere, energy resources are the "master" natural resources. We use energy to make all other resources available (see "What Is Energy?"). Without the energy resources, all other natural resources would remain in place, unable to be mined, processed, and distributed. When water becomes scarce, we use energy to pump groundwater from greater depths or to divert rivers and build aqueducts. Likewise, we increase crop yields in the face of poor soil management by investing energy in fertilizers, herbicides, farm implements, and so on. By the application of energy, the conversion of materials into commodities and the performance of services far beyond the capabilities of any single individual are made possible. Further, the application of energy can overcome deficiencies in the material world that humans exploit. High-quality iron ore may be depleted, but by massive applications of energy, the iron contained in rocks of very low iron content can be extracted and concentrated for industrial use.

Energy can be extracted in a number of ways. Humans themselves are energy converters, acquiring their fuel from the energy contained in food. Our food is derived from the solar energy stored in plants via photosynthesis. In fact, nearly all energy sources are really storehouses for energy originally derived from the sun. Among them are wood, water, wind, and fossil fuels. People have harnessed each of these energy sources, to a greater or lesser degree. Preagricultural societies

WHAT IS ENERGY?

People have built their advanced societies by using inanimate energy resources. **Energy**—the ability to do work—is either potential or kinetic. *Potential* energy is stored energy; when released, it is in a form that can be harnessed to do work. *Kinetic* energy is the energy of motion; all moving objects possess kinetic energy.

Assume that an elevated reservoir contains a large amount of stored water. The water is a storehouse of potential energy. When the gates of the dam holding back the water are opened, water rushes out. Potential energy has become kinetic energy, which can

be harnessed to do such work as driving a generator of electricity. No energy has been lost; it has simply been converted from one form to another.

Unfortunately, energy conversions are never complete. Not all of the potential energy of the water can be converted into electrical energy. Some potential energy is always converted to heat and dissipated to the surroundings. *Energy efficiency* is the measure of how well we can convert one form of energy into another without waste—that is, the ratio of energy that is produced to the amount of energy consumed in the production process.

---

depended chiefly on the energy stored in wild plants and animals for food, although people developed certain tools (such as spears) and customs to exploit the energy base. For example, they added to their own energy resources by using fire for heating, cooking, and clearing forest land.

Sedentary agricultural societies developed the technology to harness increasing amounts of energy. The domestication of plants and animals, the use of wind to power ships and windmills, and the use of water to turn waterwheels all expanded the energy base. For most of human history, wood was the predominant source of fuel, and even today at least one-half of the world's people depend largely on fuelwood for cooking and heating.

However, it was the shift from renewable resources to those derived from nonrenewable minerals, chiefly fossil fuels, that sparked the Industrial Revolution, made possible the population increases discussed in Chapter 5, and gave population-supporting capacity to areas far in excess of what would be possible without inanimate energy sources. The enormous increase in individual and national wealth in industrialized countries has been built in large measure on an economic base of coal, oil, and natural gas. They are used to provide heat, to generate electricity, and to run engines.

Energy consumption goes hand in hand with industrial production and increases in per capita income. In general, the greater the level of energy consumption, the higher the gross national income per capita. As people grow richer, they want better homes, more cars, and all the other goods that characterize developed countries—which means a large increase in the global demand for energy and industrial raw materials. Two of the questions now being

asked are whether the world has enough resources to meet the rising demand in the developing countries and what effects that demand will have on the supplies and prices of resources.

12.3 Nonrenewable Energy Resources

Crude oil, natural gas, and coal have formed the basis of industrialization. **Figure 12.5** shows past energy consumption patterns in the United States. Burning wood supplied most energy needs until about 1885, by which time coal had risen to prominence. The proportion of energy needs satisfied by burning coal peaked

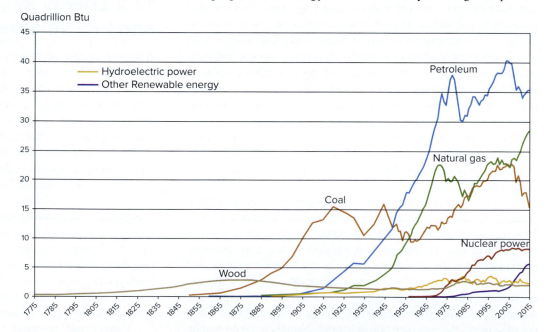

Figure 12.5   Sources of energy in the United States, 1771–2015. Wood was the primary source of energy until coal surpassed it in 1885. Petrolem and natural gas passed coal in the 1950s and remain the two most important sources of energy. Starting in the 1970s, nuclear energy emerged as an important means of generating electricity. Renewable energy has grown rapidly in the past decade. Hydroelectricity has been the dominant source of renewable energy since the 1930s, followed by biofuels, wood, wind, solar, waste to energy conversion, and geothermal. *Source: Data from United States Energy Information Administration,* Energy Perspectives 2011 and Annual Energy Overview 2015.

about 1910; from then on, oil and natural gas were increasingly substituted for coal. The use of new drilling techniques since 2005 has contributed to greater natural gas use and less coal use. Despite recent growth in renewable energy supplies, fossil fuels still dominate. In 2015 they accounted for about 90% of national energy consumption.

Crude Oil

Today, crude oil and its by-products account for about 33% of the commercial energy (excluding wood and other traditional fuels) consumed in the world. Some world regions and industrial countries have a far higher dependency. **Figure 12.6** shows the main producers and consumers of crude oil (also called petroleum). Notably, the four largest consumers, the United States, China, Japan, and India, all consume significantly more oil than they produce.

After it is extracted from the ground, crude oil must be refined. The hydrocarbon compounds are separated and distilled into waxes and tars (for lubricants, asphalt, and many other products) and various fuels. Petroleum rose to importance because of its combustion characteristics and its adaptability as a concentrated energy source for vehicles. One barrel (42 U.S. gallons) of crude oil produces about 20 gallons of gasoline, 10 gallons of diesel fuel and home heating oil (combined), 5 gallons of jet fuel, and other products. In the United States, transportation fuels account for two-thirds of all oil consumption.

As **Figure 12.7** shows, oil from a variety of production centers flows, primarily by water, to the industrially advanced countries. Note that the United States imports oil from a number of regions. The other major importers, Western Europe and Japan, import chiefly Middle Eastern oil.

The efficiency of pipelines, supertankers, and other modes of transport and the low cost of oil helped create a world dependence on that fuel, even though coal was still generally and cheaply available. The pattern is aptly illustrated by American reliance on foreign oil. For many years, U.S. oil production had remained at about the same level, 8 to 9 million barrels per day. Between 1970 and 1977, however, as domestic supplies became much more expensive to extract, consumption of oil from foreign sources increased dramatically, until almost one-half of the oil consumed nationally was imported. The dependence of the United States and other advanced industrial economies on imported oil gave the oil-exporting countries tremendous power, reflected in the soaring price of oil in the 1970s. During that decade, oil prices rose dramatically, largely as a result of the strong market position of the Organization of Petroleum Exporting Countries (OPEC).

Among the side effects of the oil "shocks" of 1973–1974 and 1979–1980 were worldwide recessions, large net trade deficits for oil importers, a reorientation of world capital flows, and a depreciation of the U.S. dollar against many other currencies. On the positive side, the soaring oil prices of the 1970s triggered energy conservation efforts, oil exploration in non-OPEC countries, improvements in oil-drilling technology, and a search for alternative energy sources. Higher energy prices diminished total energy consumption, partly because of weak demand during the recession and partly because the high prices fostered conservation. Industrial countries have learned to use much less oil for each unit of output. In general, cars, planes, and other machines are more energy-efficient than they were in the 1970s, as are industries and buildings constructed in recent years.

After 1985, however, both global production and global consumption of oil increased steadily. And the United States, which had satisfied more than two-thirds of its oil needs in the mid-1980s by domestic production, began to increase its imports due to widespread adoption of gas-guzzling vehicles combined with weakening domestic production. In 2007, the United States imported 66% of the oil it consumed annually (see "Fuel Economy and CAFE Standards," pp. 343–344). Since then, imports have decreased due to improved fuel economy and increased domestic production.

It is particularly difficult to estimate the size of oil reserves. Not only are estimates constantly revised as oil is extracted and new reserves are located, but many governments tend to maintain some secrecy about the sizes of reserves, understating official national estimates. Nonetheless, it is clear that oil is a finite resource and that oil reserves are very unevenly distributed among the world's countries (**Figure 12.8**). Slightly more than 1200 billion barrels are classified

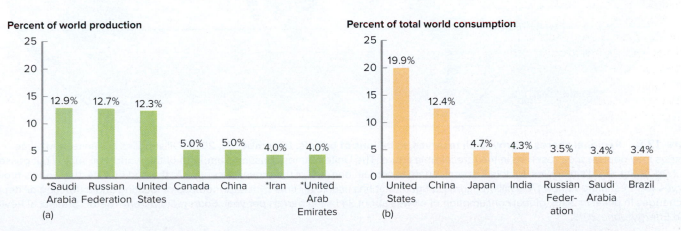

Figure 12.6 **(a) Leading producers of oil.** These seven countries produced 56% of the world's oil in 2014. Members of OPEC are marked with an asterisk. All told, OPEC countries accounted for 41% of oil production in 2014. **(b) Leading consumers of oil.** The seven countries shown here each consumed 3% or more of the world's oil in 2014. *Source: Data from* The BP Statistical Review of World Energy, *June 2015.*

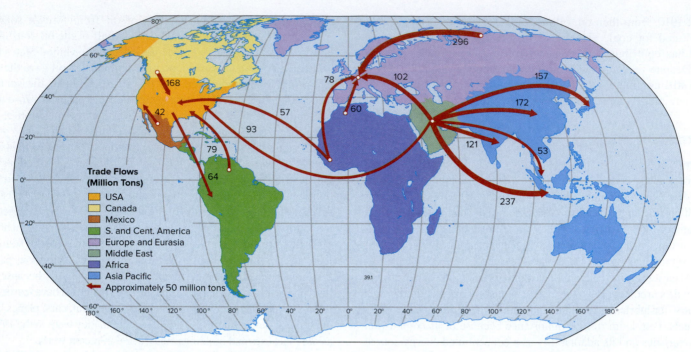

Figure 12.7   Major international crude oil shipments, 2014. Note the dominant position of the Middle East in terms of oil exports and the heavy reliance of the Asia Pacific region on imports. The arrows indicate origin and destination, not specific routes. The line widths are roughly proportional to the volume of movement. In 2014, the United States imported 54% of the oil it used, down from 66% in 2008. *Source: Data from* The BP Statistical Review of World Energy, *June 2015.*

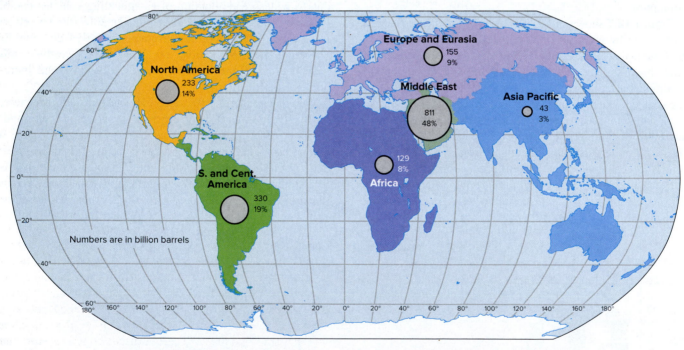

Figure 12.8   Regional shares of proved oil reserves, in billions of barrels, December 31, 2014. Oil supplies are finite, and some countries may deplete their reserves in the foreseeable future. The United States consumes one-fifth of the world's oil supply but possesses only 2.9% of the world's reserves. Middle Eastern countries contain one-half of the proved reserves. At the end of 2014, the world's proved oil reserves were estimated to be 1700 billion barrels, a figure that has risen over time with new methods of locating and extracting oil deposits and changes in price. In 2015, global consumption of oil was about 34 billion barrels per year. *Source: Data from* The BP Statistical Review of World Energy, *June 2015.*

as proved reserves, and another 900 billion are thought to exist in undiscovered reservoirs. If all the oil could be extracted from the known reserves, and if the current rate of production holds, the proved reserves would last only about 50 years. The ratio of production to reserves, however, gives some Middle Eastern countries more than a century of pumping at current rates before their oil fields run dry.

For half a century, there have been predictions that the world would soon run out of oil. Pessimists still believe that global oil production could peak this decade, whereas optimists think we will rely on oil for most, if not all, of this century. They argue that technological advances in exploration and production, such as horizontal drilling, hydraulic fracturing, deep-water drilling, and enhanced recovery, will so significantly increase the amount of oil that can be recovered from beneath the ground that we needn't worry about gas pumps running dry.

Horizontal drilling and **hydraulic fracturing** in tight oil-bearing shale formations have opened up new supplies, such as the Bakken formation in western North Dakota. The well is drilled vertically until it reaches the oil-bearing shale formation; then the shaft turns and drills laterally. Under high pressure, chemically treated water and sand are injected underground to break open tight oil-bearing rock formations. The sand serves to hold open the fractures so the oil can travel to the well. This hydraulic fracturing (or "fracking") technology has greatly expanded the economic reserves of petroleum. A 2008 article in *National Geographic* magazine portrayed North Dakota as "the vanishing prairie" and described the state's many dying towns and abandoned farmhouses. Today, jobs in the oil industry have created boom towns with some of the fastest population growth rates in the country. Housing shortages have replaced problems of abandoned housing.

Whereas it was once thought that offshore oil existed only in shallow waters, current thinking is that enormous amounts of oil lie thousands of feet below sea level off the Gulf of Mexico, Brazil, and West Africa. For several years now, a number of oil companies have pumped oil from the Gulf of Mexico at depths exceeding 1000 meters (3280 ft).

Even more promising is the possibility of recovering more oil from existing reservoirs. Currently, on average, only 30% to 35% of the oil in a reservoir is brought to the surface; most of the oil remains unrecovered. Oil industry optimists believe that enhanced recovery techniques (injecting water, gases, or chemicals into wells to force out more oil) will make 60% to 70% of the oil in a reservoir recoverable.

Finally, although most geologists agree that there are few huge oil fields left to be discovered, some of these have yet to be tapped in a major way. Industry analysts think significant increases in production are likely to come from Russia and a number of the countries of the former Soviet Union (Kazakhstan, Tajikistan, Uzbekistan, Turkmenistan, Kyrgyzstan, and Azerbaijan). Kazakhstan, for example, has developed large new oil fields and recently constructed a new pipeline from the Caspian Sea to China, and Russia has opened huge offshore oil fields in its far east, near Sakhalin Island in the Sea of Okhotsk.

Coal

Coal was the energy source that fueled the Industrial Revolution. From 1850 to 1910, the proportion of U.S. energy supplied by coal rose from 10% to almost 80%. Although the consumption of coal declined as the use of petroleum expanded, coal remained the single most important domestic energy source until 1950 (see Figure 12.5, p. 338).

Although coal is a nonrenewable resource, world supplies are so great that its resource life expectancy may be measured in centuries, not in the much shorter spans usually cited for oil and natural gas. The United States alone possesses nearly 240 billion tons of coal considered potentially minable on an economic basis with existing technology. At current production levels, these demonstrated reserves would be sufficient to meet the domestic demand for coal for another 2.5 centuries. The more pressing concern is the effect on the global climate of burning those abundant coal reserves (see Chapter 4).

Worldwide, the most extensive deposits are concentrated in the middle latitudes of the Northern Hemisphere, as shown in **Figure 12.9**. Two countries, China and the United States, have dominated world coal production in recent years. China's coal production and consumption has matched the rise of its economy. It now accounts for one-half of the coal produced and consumed in the world. Since 2000, the use of coal worldwide has grown by about 60%, with most of that growth taking place in Asia. Coal production has decreased slightly in the United States, Western Europe, and the countries of the former Soviet Union, as governments have discontinued subsidies to the industry. The use of coal continues to grow in many Asian countries (India, Indonesia, South Korea, Vietnam, and Japan). China has increased both its production and its consumption of coal by a factor of 2.5 since 2000, a result of its rapid economic growth. In the United States and other industrialized countries, coal is used chiefly for electric power generation and to make coke for steel production. In less-developed countries, coal is widely used for home heating and cooking, as well as to generate electricity and fuel factories.

Coal is not a resource of constant quality. It ranges from lignite (barely compacted from the original peat) through bituminous coal (soft coal) to anthracite (hard coal), each *rank* reflecting the degree to which organic material has been transformed. Anthracite has a fixed carbon content of about 90% and contains very little moisture. Conversely, lignite has the highest moisture content and the lowest amount of elemental carbon, and thus the lowest heat value. About one-half of the demonstrated reserve base in the United States is bituminous coal, concentrated primarily in the states east of the Mississippi River.

Besides rank, the *grade* of a coal, which is determined by its content of waste materials (particularly ash and sulfur), helps to determine its quality. Good-quality bituminous coals with the caloric content and the physical properties suitable for producing coke for the steel industry are decreasingly available and are increasing in cost. Anthracite, formerly a dominant fuel for home heating, is now much more expensive to mine and finds no ready industrial market. The value of a coal deposit is determined not only by its rank and grade but also by its accessibility, which depends on the thickness, depth, and continuity of the coal seam and its inclination to the surface. The two methods of extracting coal are surface mining and underground mining. In *surface mining* (also called *strip mining*), huge machines strip off the soil and rock above the vein of coal—the *overburden*—to get at the coal beneath. In general, surface mining is used if the overburden is less than 100 meters (328 ft) thick. The negative environmental consequences of strip mining are discussed in Chapter 13 (pp. 394–399). If the overburden is thicker, coal is extracted using *underground mining* (also called *shaft mining*) where operators sink two or more shafts down to the coal deposit. Underground mining is not only expensive but is among the most hazardous occupations in the world. In China, coal miners make up 4% of the workforce but 45% of the workplace fatalities.

In spite of their generally lower heating value, western U.S. coals are now attractive because of their low sulfur content. They

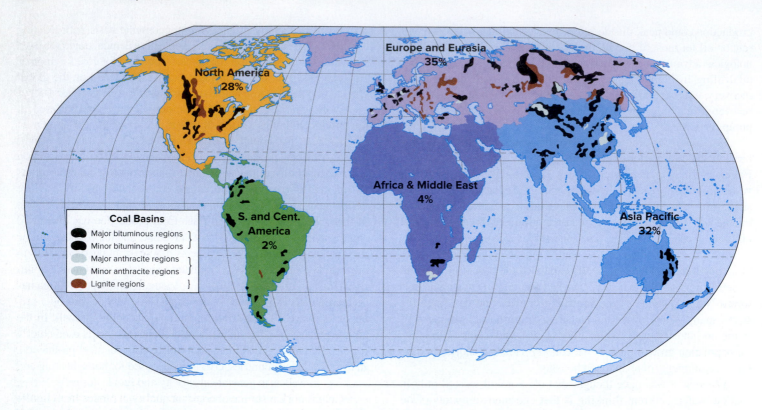

Figure 12.9 **Regional shares of proved coal reserves, December 31, 2014.** Major coal basins are concentrated in the Northern Hemisphere. Just five countries contain three-quarters of the world's coal reserves: the United States (28%), the Russian Federation (18%), China (13%), India (7%), and Australia (9%). Despite their large size, Africa and South America have very little coal. In 2011, China was the world's largest coal producer and consumer, accounting for one-half of the world's coal production and consumption. *Source: Reserve data from* The BP Statistical Review of World Energy, *June 2015.*

do, however, require expensive transportation to markets or high-cost transmission lines if they are used to generate electricity for distant consumers (**Figure 12.10**).

The ecological, health, and safety problems associated with the mining and combustion of coal must also be figured into its cost. The destruction of the original surface and the acid contamination of lakes and streams associated with the strip mining and burning of coal are partially controlled by environmental protection laws, but these measures add to the costs. Eastern U.S. coals have a relatively high sulfur content, and costly techniques for the removal of sulfur and other wastes from stack gases are now required by most industrial countries, including the United States.

The cost of moving coal influences its patterns of production and consumption. Coal is bulky and is not as easily transported as nonsolid fuels. Indeed, the high cost of transporting coal induced the development of major heavy industrial centers directly on coal fields—for example, Pittsburgh, the Ruhr, the English Midlands, and the Donets district of Ukraine.

Natural Gas

Coal is the most abundant fossil fuel, but natural gas has been called the nearly perfect energy source. It is a highly efficient, versatile fuel that requires little processing. Of the fossil fuels, natural gas (which is mainly methane) has the least impact on the environment.

It burns cleanly. The chemical products of burned methane are carbon dioxide and water vapor, which are not pollutants, although they are greenhouse gases.

Figure 12.10 **Unit trains for coal.** Long-distance transportation adds significantly to the cost of low-sulfur western U.S. coals because they are remote from eastern U.S. markets. To minimize these costs, unit trains carrying only coal engage in a continuous shuttle movement between western strip mines and eastern power plants. © *Mike Danneman/Getty Images RF.*

Fuel Economy and Cafe Standards

Some have likened the United States to a crude oil junkie, dependent on daily fixes of petroleum. On average, every person in the United States uses 2.3 gallons of oil each day. Transportation accounts for two-thirds of that consumption. What are some of the implications of this dependence?

Consider the data in the following table.

Country	2014 Proved Reserves (Billion Barrels)	2014 Production (Billion Barrels)	2014 Consumption (Billion Barrels)
United States	49	4.3	6.9
Canada	173	1.6	0.9
Mexico	11	1.0	0.7

© Comstock/Getty Images RF.

Notice the imbalance between yearly production and consumption for the United States. In contrast to its hemispheric neighbors, Canada and Mexico, the United States consumes far more oil than it produces. Americans continue to drive their cars, and their manufacturing plants continue to turn out a wide range of petroleum-based products, only because the country imports about 7 million barrels of oil a day. That is, the United States relies on foreign sources to meet more than one-half of its crude oil needs.

Slightly less than one-half of the oil used in the United States is burned as fuel in private automobiles, sport utility vehicles (SUVs), and light trucks. When the price of oil tripled in the mid-1970s, Congress set fuel efficiency standards, and Americans began buying smaller cars, helping bring down the per capita consumption of gasoline. Between 1973 and 1987, the average fuel efficiency of new American automobiles increased from 13.1 to 22.1 miles per gallon (mpg). That increase alone cut gasoline consumption by 20 billion gallons per year, lowering oil imports by 1.3 million barrels of oil per day.

As the price of gasoline declined in the 1990s, however, Americans reverted to past practices, buying large, non-fuel-efficient vehicles, such as sport utility vehicles and large pickup trucks with extended cabs. They became so popular in the United States that they soon accounted for about one-half of all new passenger vehicles sold in the country. Big SUVs are gas guzzlers, and their popularity has driven down the average gas mileage (fuel-economy) ratings of vehicles driven in America. The fuel efficiency of automobiles varies widely. The most efficient vehicles are gasoline-electric hybrids that get 50 mpg or better, while the least efficient average only 10 mpg. The high fuel consumption of large SUVs not only exacerbates U.S. dependence on foreign sources of oil but also has negative environmental impacts. If the average fuel efficiency of the vehicle fleet increased just 3 mpg, the United States would save 1 million barrels of oil per day, out of the roughly 10 million barrels it imports. Automobiles account for about 20% of the country's carbon dioxide emissions, and the more fuel a vehicle consumes, the greater its emissions. A car that gets 30 mpg has only one-half of the emissions of one that averages 15 mpg. Every gallon of gasoline burned puts about 13 pounds of carbon dioxide into the atmosphere, contributing to global warming. The nitrogen oxides and hydrocarbons emitted by vehicles contribute to acid rain and ozone smog (topics discussed in Chapter 13).

One way to address these environmental concerns and simultaneously reduce American dependence on imported oil would be to increase energy efficiency. Current federal *Corporate Average Fuel Economy (CAFE) standards* require each automaker's fleet of new passenger cars to average 27.5 mpg and its light truck fleet (pickups, SUVs, and minivans) 21.3 mpg. The less-stringent fuel-economy standard for light trucks originally was intended to avoid penalizing builders, farmers, and others who relied on pickup trucks for their work; but it opened the way for automakers to replace station wagons and sedans with vehicles that met the definition of a light truck: SUVs and minivans.

In 2007, Congress passed the Energy Independence and Security Act, the first overhaul of CAFE standards in decades. The Obama administration released new standards in 2012, requiring automakers to boost fleetwide gas mileage to at least 35.5 mpg by the year 2016 and to 54.5 mpg by 2025. In 2015, fuel efficiency standards were proposed for buses and medium and heavy duty trucks. Unlike in the past, U.S. automakers generally supported the changes while admitting that new technological breakthroughs will be required. Proponents of raising the CAFE standards significantly higher note that the technologies to boost automobile efficiency significantly without sacrificing safety already exist. They point out that Japanese standards mandate 30.3 mpg, and European regulations an even higher 33.0 mpg. Increased energy efficiency is the quickest and cheapest path to reduced fuel consumption and a cleaner and healthier world, they argue. Additionally, they contend that SUVs endanger other motorists because they limit visibility and because they cause greater damage in a crash.

Those opposing the change point out that the automobile industry offers a variety of vehicles, some more fuel-efficient, others less, and that people are free to choose the model they like. Most Americans, they say, are more concerned about automotive safety, comfort, and performance than about fuel economy. Some buyers cite the safety of SUVs. As one SUV owner noted, "It's a very safe vehicle. I bought this thing because it was a big iron tanker that wouldn't bend. You get hit by something and don't feel it." Others value the space a minivan or an SUV offers: "When you have a boat to haul and a family to transport, you don't have a lot of choice; you need large capacity."

Considering the Issues

1. With which of the following statements do you agree? Disagree? Why? "Allowing the American free-market system to meet America's energy needs hasn't worked. The automobile industry cannot

(continued)

be trusted to regulate itself. Automakers won't improve the fuel economy of their fleets unless the government forces them to."

"Government regulation isn't the way to go. Maybe $8-per-gallon gasoline will make people buy more fuel efficient vehicles."

"SUVs are safer in a crash than compact cars. If I want to protect my family by driving an SUV, I should have that right. We have to balance environmental concerns with safety and the customer's freedom of choice."

2. Do you think the Obama administration should have raised CAFE standards as high as it did for 2016 and 2025 models? Domestic auto dealers have opposed raising the standards, claiming they will increase the cost of a new vehicle by nearly $3000, shutting 7 million people out of the new-car market.

3. In the past when the federal government raised the CAFE fuel-economy standards, required seatbelts, or required air pollution equipment, the domestic automobile companies strongly opposed the changes, arguing that it would rob them of their profits and force them to lay off workers. Why do you think that the automobile companies agreed to work toward the new fuel economy standards?

4. Should SUVs and minivans continue to be subject to less-stringent regulations than other passenger cars? Why or why not?

5. Should automakers be allowed to pay penalties rather than comply with the CAFE standards?

As Figure 12.5 indicates, the 20th century saw an appreciable growth in the proportion of U.S. energy supplied by gas. In 1900, it accounted for about 3% of the national energy supply. By 1980, the figure had risen to 30%, but then it declined to 26% by 2011. The trend in the rest of the world has been in the opposite direction. Global production and consumption increased significantly after the oil shock of 1973–1974 and in 2014 had nearly doubled, accounting for almost 25% of global energy consumption.

Natural gas is used mostly for industrial and residential heating and for electricity generation. In fact, gas has overtaken both coal and oil as a house-heating fuel, and more than one-half of the homes in the United States are now heated by gas. Natural gas is the primary ingredient in creating the chemical fertilizers that were part of the Green Revolution in agriculture (see "Eating Fossil Fuels," pp. 344). A portion of the natural gas supply is also processed into plastics, synthetic fibers, and insecticides.

Very large natural gas fields were discovered in Texas and Louisiana as early as 1916. Later, additional large deposits were found in the Kansas–Oklahoma–New Mexico region. At that time, the south-central United States was too sparsely settled to make use of the gas, and in any case, it was oil, not gas, that was being sought. Many wells that produced only gas were capped. Gas found in conjunction with oil was vented or burned at the wellhead as an unwanted by-product of the oil industry. The situation changed only in the 1930s, when pipelines were built to link the southern gas wells with customers in Chicago, Minneapolis, and other northern cities. Like oil, natural gas flows easily and cheaply by pipeline. In the United States, the interstate and intrastate pipeline system is more than 500,000 kilometers (310,000 mi) long.

Unlike oil, however, gas does not move as freely in international trade (**Figure 12.11**). Transoceanic shipment involves costly equipment for liquefying the gas by cooling it to −126°C (−260°F), for double-hulled tankers that can contain the liquid under appropriate low temperature conditions, and for reheating facilities at the destination port, where it is injected into the local pipeline system. **Liquefied natural gas (LNG)** is extremely hazardous because mixtures of methane and air are explosive. Because an accident or a terrorist strike on an LNG tanker carrying millions of gallons of highly flammable fuel would generate an enormous fireball, onshore facilities are risky. Offshore facilities, from which the gas is piped underwater to the port, are less hazardous.

Like other fossil fuels, natural gas is nonrenewable; its supply is finite. Estimates of reserves are difficult to make because they depend on what customers are willing to spend for the fuel, and estimates have risen as the price of gas has increased. Further complicating the estimate of supplies is uncertainty about potential gas resources in unusual types of geologic formations. These include tight sandstone formations, deep (below 6000 m, or 20,000 ft) geologic basins, and shale and coal beds.

Worldwide, the Middle East dominates the proved gas reserves with 43% of world supplies (**Figure 12.12**). The gas in the proved reserves would last about 55 more years at current production rates, but new discoveries could add significantly to the life expectancy of world reserves if they were developed.

In the United States, the Texas–Louisiana and Kansas–Oklahoma–New Mexico regions traditionally dominated domestic natural gas output, but there are thought to be gas deposits beneath almost all states. In addition, many offshore areas are known to contain gas. Potential Alaskan reserves are estimated to be at least twice as large as today's proved reserves in the rest of the country.

As with petroleum extraction, horizontal drilling combined with hydraulic fracturing has led to dramatic growth in natural gas

EATING FOSSIL FUELS

The intensive commercial agricultural systems described in Chapter 9 depend heavily on fossil fuels. Use of fossil fuels began with mechanized farm equipment such as tractors. As populations urbanized and agriculture became more specialized, the distance from farm to table increased. Thus, transportation of food to market consumes significant energy resources. The Green Revolution that played such an important role in feeding the expanding global population relied upon commercial fertilizer inputs. Commercial fertilizers are produced using the Haber process which uses natural gas as the primary input. Together, mechanization, long-distance transportation, and the use of agricultural chemicals create a food system that converts stored fossil fuel energy to food energy.

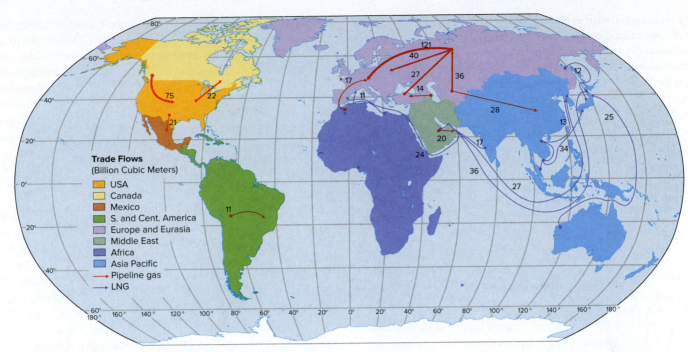

Figure 12.11   **Major worldwide trade flows of natural gas, 2014.** Russia exported gas to 17 other European countries, accounting for about 28% of the worldwide trade flows. Other major natural gas exporters are Qatar, Norway, Canada, United States, Netherlands, Malaysia, and Australia. Most natural gas flows by pipeline, but diminishing supplies of natural gas in many developed countries, the discovery of large reserves of natural gas in remote regions, and a reduction in the costs associated with constructing facilities for shipping liquefied natural gas (LNG) are combining to make LNG more attractive. The largest LNG importer is Japan, followed by South Korea. Only annual flows of 10 billion cubic meters or greater are shown. *Source: Data from* The BP Statistical Review of World Energy, *June 2015.*

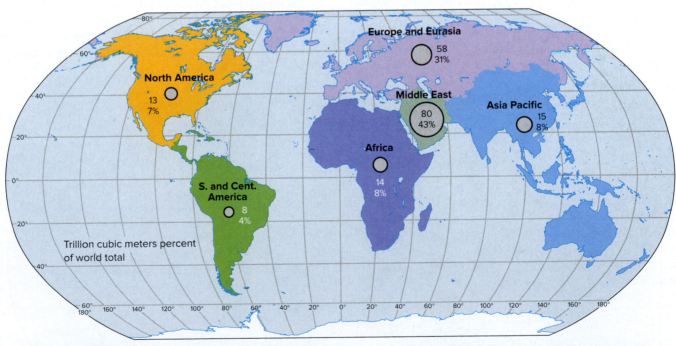

Figure 12.12   **Proved natural gas reserves, December 31, 2014.** Russia contains the largest natural gas reserves of any single country, about 21% of the world total (more than 5 times the size of U.S. reserves). Major reserves also exist in the countries of the Middle East, particularly Iran and Qatar. *Source: Data from* The BP Statistical Review of World Energy, *June 2015.*

production in the United States. **Shale gas** is natural gas trapped within sedimentary shale rocks, often in close proximity to petroleum deposits. Horizontal drilling in shale gas deposits has opened once inaccessible supplies in the Barnett Shale near Fort Worth, Texas, and Marcellus Shale in Pennsylvania. Pennsylvania's natural gas production quadrupled between 2009 and 2011 due to horizontal drilling in the Marcellus Shale. North Dakota's Bakken lacks the connection to natural gas pipelines, and about one-third of the natural gas vented during oil extraction is simply burned off in wasteful flares that are visible from space. With the new technologies of horizontal drilling and hydraulic fracturing, gas reserves may be sufficient for another century. Of course, these less-accessible supplies and unconventional drilling techniques will be more costly to develop, and hence more expensive.

Hydraulic fracturing has generated a wide range of environmental concerns. Of primary concern is the potential to contaminate groundwater with the injected chemicals. Another concern is storing, treating, and disposing the wastewater generated by the process. In some areas, the volumes of water used during hydraulic fracturing have led to lowered stream flows or lowered water table levels in wells.

Oil Shale and Oil Sands

A similar situation affects the prospects of oil extraction from unconventional fossil fuels: oil shale and oil sands. Both are stored in unconventional places, within rock and sand.

Oil shale should not be confused with tight oil-bearing shales such as the Bakken Formation that contain liquid oil (and natural gas) that is released through hydraulic fracturing. Neither shale nor oil, **oil shale** is fine-grained sedimentary rock rich in organic material called *kerogen*. A tremendous potential reserve of hydrocarbon energy, the rocks are calcium and magnesium carbonates more similar to limestone than to shale, and the hydrocarbon, kerogen, is not oil but a waxy, tarlike substance that adheres to the grains of carbonate material. The crushed rock is heated to a temperature high enough (more than 480°C, or 900°F) to decompose the kerogen, releasing a liquid oil product, *shale oil*.

World reserves of oil shale are enormous. Known deposits estimated to contain at least 800 billion barrels of recoverable shale oil are found in the United States, Brazil, Russia, China, and Australia

(**Figure 12.13**). The richest deposits in the United States are in the Green River formation, which lies beneath Colorado, Utah, and Wyoming. They contain enough oil to supply the needs of the United States for another century, and in the 1970s they were thought to be the answer to national energy self-sufficiency. Several billion dollars were invested in oil shale research and development operations in the Piceance Basin near Grand Junction, Colorado, but as oil prices fell in the 1980s, interest in the projects waned. The last plant, that at Parachute Creek, Colorado, was abandoned in 1992.

Another source of petroleum liquids is **oil sands.** Also called tar sands, oil sands are a mixture of sand, clay, and silt (85%); water (5%); and *bitumen* (10%), a very thick, tarlike, high-carbon petroleum. The crude bitumen is too thick to flow out of the rock, so the oil sand must be mined, crushed, and heated to extract the petroleum. It takes about 2 metric tons of oil sands to produce the equivalent of 1 barrel of crude oil. Global oil sand resources are thought to be many times larger than conventional oil resources, containing more than 2 trillion barrels of oil, much of it in the province of Alberta, Canada (**Figure 12.14**). Four major deposits in Alberta are estimated to contain as much as 1 trillion barrels of bitumen. The largest deposit, the Athabasca, is near Fort McMurray in northern Alberta. Oil sand deposits are also found in Venezuela, Trinidad, Russia, and Utah.

Producing oil from either oil shale or oil sands requires high capital outlays and carries substantial environmental costs. It requires a great deal of energy and fresh water, disturbs large areas of land, and produces enormous amounts of waste. The production process also emits greenhouse gases and pollutes air, water, and surrounding soil.

At present crude oil prices, oil shale development is relatively limited. Oil sands technology, however, has improved so that it is economical at current prices. Alberta's oil sands are estimated to contain 170 billion barrels of oil that can be economically recovered with current technology. The Canadian Province of Alberta

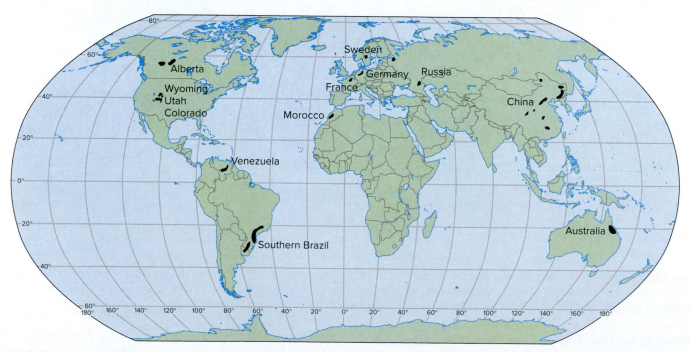

Figure 12.13   Oil shale deposits. The United States contains about two-thirds of the world's known supply of oil shale, with the richest deposits located in the Green River formation. Oil shale is different than oil-bearing shale that releases liquid oil through hydraulic fracturing. Oil shale is a sedimentary rock that requires expensive treatment and is unlikely to become an important resource until cheaper ways are found to process it and solve the waste disposal and land reclamation problems its mining poses.

(b)

Figure 12.14   **Canadian oil sands. (a)** Oil sand deposits in Alberta, Canada. **(b)** Production of synthetic oil from the oil sands involves four steps: (1) the removal of overburden; (2) mining and transportation to extraction units; (3) the addition of steam and hot water to separate the bitumen from the tailings residue; and (4) the refinement of the bitumen into coke and distillates. It takes about 2 tons of sand to yield 1 barrel of oil. *(b) © Blackfox Images/Alamy Stock Photo.*

now ranks third in oil reserves behind Venezuela and Saudi Arabia. In 2011, about one-fourth of total U.S. crude oil imports came from Alberta's oil sands. Further production for U.S. markets is hampered by a lack of pipeline capacity. Construction of a new pipeline to the United States is on hold because of serious environmental concerns about air, water, and land pollution during the processing of oil sands, the possibility of pipeline leaks, and the greenhouse gas emissions from using such a vast source of fossil fuels. While renewable energy sources are still a relatively small proportion of the total worldwide energy supply, nuclear power is a well-developed technology that does not emit greenhouses gases.

Nuclear Energy

Proponents consider nuclear power a major long-term solution to energy shortages. Assuming that the technical problems can be solved, proponents contend that nuclear fuels could provide a virtually inexhaustible source of energy. Other commentators, pointing to the dangers inherent in any system dependent on the use of radioactive fuels, argue that nuclear power poses technological, political, social, and environmental problems for which society has no solutions. Basically, energy can be created from the atom in two ways: nuclear fission and nuclear fusion.

Nuclear Fission

The conventional form of **nuclear fission** for power production involves the controlled "splitting" of an atomic nucleus of uranium-235, the only naturally occurring fissile isotope. When U-235 atoms are split, about one-thousandth of the original mass is converted to heat. The released heat is transferred through a heat exchanger to create steam, which drives turbines to generate electricity.

A single pound of U-235 contains the energy equivalent of nearly 5500 barrels of oil. More than 440 commercial nuclear reactors around the world tap that energy, generating about 16% of the world's electricity (**Figure 12.15**). Nearly one-fourth of those plants are in the United States.

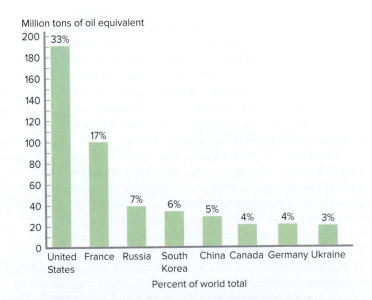

Figure 12.15   **Leading producers and consumers of nuclear energy, 2014.** These eight countries produce and use about three-fourths of the world's nuclear energy. Since the 2011 Fukushima disaster, Japan has progressively shut down its nuclear reactors, dropping nuclear energy production from 66 million tons of oil equivalent in 2010 to zero in 2014. Few nuclear power plants exist in developing countries. *Source: Data from* The BP Statistical Review of World Energy, *June 2015.*

Some countries are much more dependent than others on nuclear power. Public acceptance has varied depending upon the domestic availability of other fuel sources and upon public concerns about the safety of reactors and radioactive waste storage and disposal facilities. Nuclear power provides more than 75% of the electricity in France and Lithuania, and about 50% in Belgium and Sweden, but it accounts for only 20% or less of the electric power generated in both the United States and Canada. Several countries have rejected the nuclear option altogether. Denmark, Italy, Greece, Australia, and New Zealand are among the countries that decided to remain "nuclear-free" and never built nuclear power plants. A few countries, including Germany, Sweden, and the Philippines, plan to phase out and dismantle their reactors.

Many countries, however, have seen a recent revival of interest in nuclear power, stemming in part from the belief that nuclear energy is the best way to reduce carbon emissions and slow the rate of climate change. Worldwide, as of 2016, 65 reactors were under construction in 15 countries, with many more planned or proposed. In China, a soaring demand for electricity and a desire to reduce reliance on imported oil and natural gas and heavily polluting coal-fired power plants have prompted the government to continue building new reactors. Other countries with two or more reactors include Russia, India, United States, United Arab Emirates, South Korea, Pakistan, Belarus, Slovakia, Ukraine, and Japan.

With few fossil fuel resources, Japan once relied on 55 nuclear power plants to generate almost one-third of its electricity. However, a 2011 earthquake and tsunami led to disastrous radioactive releases from the Fukushima nuclear reactor. With the loss of electricity for its cooling systems, three reactor cores melted down and spent fuel rods overheated. Approximately 100,000 persons were evacuated from the surrounding region and as of this writing have not returned. The Fukushima disaster led to immediate shutdowns of Japan's other nuclear reactors and to political commitments to wean Japan off nuclear power. Nonetheless, Japan began restarting its existing nuclear reactors in 2015 as well as building new reactors.

Expansion of the nuclear power industry in the United States is less certain (**Figure 12.16**). The high costs of constructing, licensing, and operating the plants made nuclear power more expensive than energy derived from other sources. Public acceptance of nuclear energy has risen and fallen with the technology's safety record. The partial reactor meltdown in Three Mile Island, Pennsylvania, in 1979, the catastrophe at Chernobyl in Ukraine in 1986, and the 2011 Fukushima disaster each heightened safety concerns. The lack of safe storage sites for radioactive reactor waste has also eroded public support for nuclear energy. Finally, particularly after September 11, 2001, many view nuclear plants as tempting targets for terrorist attacks.

Despite these concerns, nuclear power may stage a modest comeback in the United States. Many of the older reactors will be closed in the coming years as their steam generators corrode and steel pressure vessels become too brittle to operate safely. Several

(a)

• Operating reactors

(b)

Figure 12.16   Operating nuclear power plants in the United States, 2016. (a) The Cook nuclear power plant on the shores of Lake Michigan. **(b)** Nuclear power plants are built next to rivers, lakes, or oceans because they need large quantities of cooling water for their waste heat. The 99 plants produce about 20% of the country's electricity. After 35 years with almost no new nuclear power plant construction, analysts expect up to nine new nuclear power plants to be operating by 2020. *(a) Source: NRC File Photo. (b) Source: Data from World Nuclear Association.*

companies have received approval from the Nuclear Regulatory Commission for relicensing or upgrades of existing facilities. The energy bill approved by Congress in 2005 included billions of dollars in incentives—including tax credits, subsidies, loan guarantees, and federal insurance—to encourage production of nuclear energy.

Nuclear Fusion

Unlike a fission reaction, which splits an atom, a **nuclear fusion** reaction forces forms of hydrogen known as deuterium and tritium to combine to form helium, releasing tremendous amounts of energy. Fusion is the process that makes the sun and other stars burn; it is also the basis of the hydrogen bomb, which uses a brief, uncontrolled thermonuclear fusion reaction.

More difficult to achieve than fission, nuclear fusion requires heating the atoms to extremely high temperatures, until their nuclei collide and fuse. One of the technological problems facing fusion researchers is to find material for the containment vessel that is resistant to radiation and temperatures 100,000,000°C (180,000,000°F) or higher. A seven-member consortium of the United States, European Union, India, Russia, China, South Korea, and Japan is collaborating on construction of the world's first large-scale nuclear fusion reactor in France. The International Thermonuclear Experimental Reactor (ITER) is intended to be a demonstration plant to prove that fusion can be harnessed as an economically viable source of energy. It will take at least a decade to build the plant.

If the developmental problems associated with nuclear fusion are solved, Earth's electricity requirements would presumably be satisfied for millions of years. One cubic kilometer of ocean water, the source of deuterium atoms, contains as much potential energy as the world's entire known oil reserves. Advocates of fusion cite other advantages. The radioactive processes are short-lived, and the waste products are benign. Unlike fission reactors, fusion reactors do not use uranium-235, a raw material that is in short supply; and unlike conventional power plants, they would not emit the pollutants carbon dioxide, sulfur dioxide, and nitric oxide.

Skeptics point out that, despite 50 years of research, scientists have not solved the problem of controlling fusion so that its released energy can be used. They argue that the costs of a fusion plant will be enormous and that electricity can be supplied more cheaply by conventional means. Finally, they say, fusion energy could be accompanied by health and environmental problems we have not conceived of yet.

12.4 Renewable Energy Resources

The safety problems posed by nuclear energy, the threatened depletion of some finite fossil fuels, the desire to be less dependent on foreign sources of energy, and concerns about global climate change have increased interest in *renewable* resources. One of the advantages of such resources is their ubiquity. Most places on Earth have an abundance of sunlight, rich plant growth, strong wind, or heavy rainfall. Another advantage is ease of use. It doesn't take advanced technology to utilize many of the renewable resources, one reason for their widespread use in developing countries. The most common renewable source of energy is plant matter.

Biomass Fuels

More than one-half of the people in the world depend on wood and other forms of biomass for their daily energy requirements. **Biomass fuel** is any organic material produced by plants, animals, or microorganisms that can be burned directly as a heat source or converted into a liquid or gas. In addition to wood, biomass fuels include leaves, crop residues, peat, manure, and other plant and animal material. In Ethiopia and Bangladesh, biomass supplies nearly all of the total energy consumed. In contrast, energy from the conversion of wood, grasses, and other organic matter is relatively minor in the developed world.

There are two major sources of biomass: (1) trees, grain and sugar crops, and oil-bearing plants such as sunflowers and (2) wastes, including crop residues, animal wastes, garbage, and human sewage. Biomass can be transformed into fuel in many ways, including direct combustion, gasification, and anaerobic digestion. Further, conversion processes can be designed to produce, in addition to electricity, solid (wood and charcoal), liquid (oils and alcohols), and gaseous (methane and hydrogen) fuels that can be easily stored and transported.

Wood

The great majority of the energy that is produced from biomass comes from *wood*. In 1850, the United States obtained about 90% of its energy needs from wood. Although wood now contributes only 3% to the energy mix of the country as a whole, the percentage varies by region, with wood fuel providing about 15% of the energy used in Maine and Vermont. In developing countries, wood is a key source of energy, used for space heating, cooking, water heating, and lighting. This dependence on wood is leading in places to severe depletion of forests, a subject discussed later in this chapter (see "Forest Resources," pp. 365–371).

A second biomass contribution to energy systems is *alcohol,* which can be produced from a variety of plants. After the oil shortages of the 1970s, Brazil, which is poorly endowed with fossil fuels, embarked on an effort to develop its indigenous energy supplies in order to reduce the country's dependence on imported oil. All gasoline sold in Brazil contains 25% grain alcohol (*ethanol*) made from sugarcane, a mixture on which conventional automobile engines can run without modifications. In 2003, "flex-fuel" cars made their debut in Brazil. The modified engines of these cars are designed to run on pure gasoline, alcohol, or any combination of the two. Buyers are attracted by the low price of alcohol and the fact that virtually all of the country's service stations sell the fuel. All new cars sold in Brazil have flex-fuel engines.

In the United States, ethanol is currently blended with gasoline at a ratio of 5% to 10% as an oxygenate. But the desire to reduce dependence on imported oil resulted in the Energy Independence and Security Act of 2007 which called for dramatic increases in biofuel production. In the United States, 95% of all ethanol is derived from corn, a fuel with significant disadvantages when compared with sugarcane.

- Sugarcane can be grown on marginal soils in a tropical climate. As grown in the United States, corn requires good land and heavy applications of nitrogen fertilizer, herbicides, and pesticides, which pollute water supplies.
- Corn takes a lot of land to grow. To replace an additional 5% of U.S. gasoline consumption, roughly 117 million acres (the size of Wisconsin and Nebraska combined) would have to be planted in corn.
- Because sugarcane is more energy-efficient, Brazil produces far more ethanol per hectare than the United States does with corn.
- The energy balance, or fossil fuel energy used to make the fuel (input) compared with the energy in the fuel (output), tilts

heavily in favor of sugarcane. Cane yields 8 units of ethanol for every 1 unit of fossil fuel; for corn, the energy ratio is just 1.3 units of ethanol for every 1 unit of fossil fuel.

- Cane is cheaper to process because it's already sugar and doesn't need converting before distillation. Corn, however, is ground up and combined with water and enzymes that convert the starch into sugars, which turn into alcohol during fermentation. Distillation separates the alcohol from the water.
- Rising prices for corn will encourage farmers to plant millions of acres of farmland now set aside for soil and wildlife conservation.
- Ethanol distilleries currently consume about one-fifth of the U.S. corn crop. Converting more corn to fuel will drive up the prices of livestock feed, meat, cereals, and other goods with the effects felt most profoundly by the world's poor who spend a large proportion of their budget on food.

A preferred way to produce biofuels without competing with food crops is by using waste. In fact, the 2007 Energy Independence Act requires that the majority of biofuels be produced from feedstocks other than corn. Research and development continues on producing ethanol from woody biomass in cellulosic ethanol plants.

Waste

Waste, including crop residues and animal and human refuse, represents the second broad category of organic fuels. Particularly in rural areas, energy can be obtained by fermenting such wastes to produce methane gas (also called *biogas*) in a process known as *anaerobic digestion* (**Figure 12.17**). A number of countries, including India, South Korea, and Thailand, have national biogas programs, but the largest effort to generate substantial quantities of methane gas for rural households has been undertaken in China. There, backyard fermentation tanks (biodigesters) supply as many as 35 million people with fuel for cooking, lighting, and heating. The technology has been kept intentionally simple. A stone fermentation tank is fed with wastes, which can include straw and other crop residues in addition to manure. These are left to ferment under pressure, producing methane gas that is later drawn through a hose into the farm kitchen. After the gas is spent, the remaining waste is pumped out and used in the fields for fertilizer.

Hydropower

The second most common source of renewable energy is **hydropower**, which exploits the energy present in falling or flowing water. Hydropower is generated when water falls or flows from one level to another, either naturally or over a dam. The falling water can then be used to turn waterwheels, as it was in ancient Egypt, or modern turbine blades, powering a generator to produce electricity. Hydropower is a clean source of energy. The water is neither polluted nor consumed during power generation, although in arid areas some water in the reservoir may be lost to evaporation. Generally speaking, as long as a stream continues to flow, hydropower is renewable.

Tied as it is to a source of water, hydroelectric power production is location-specific with regard to generation. The power that can be generated is a function of two variables: the elevation drop and the flow rate. Regions with mountains and high precipitation, such as the U.S. Pacific Northwest, are most suited to hydroelectric power. Although most countries generate some hydropower, just four countries account for more than one-half of the total world hydroelectric production: China, Canada, Brazil, and the United States, in decreasing order. In the United States, hydropower is generated at more than 1900 sites in 47 states. Still, most of the country's hydropower is generated in a few places with abundant water and elevation changes (**Figure 12.18**). The majority of the country's developed capacity is concentrated in just three areas: the northwest (Washington, Oregon, and Idaho), the multistate Tennessee River valley area of the southeast, and the Great Lakes where a few large pumped storage facilities are located. That pattern is a result of both the physical environment and the role that agencies such as the U.S. Bureau of Reclamation and the Tennessee Valley Authority (TVA) have played in hydropower development.

Transmitting hydroelectricity over long distances is costly, so it is generally consumed in the region where it is produced. This fact helps account for variations in the pattern of consumption and for the energy mix used in specific regions. Thus, although hydropower provides the United States as a whole with about 7% of its electricity, it supplies the majority of electricity used in Idaho, Oregon, and Washington.

Figure 12.17   A biogas generator in Nepal. Animal and vegetable wastes are significant sources of fuel in countries such as Nepal, Pakistan, India, and China. Wastes inserted into the tank in the foreground are mixed with water and decaying organic material. As the wastes decompose, gases are emitted. A chamber collects the gas and tubes direct it to the kitchen stove. © *Joerg Boethling/Alamy Stock Photo.*

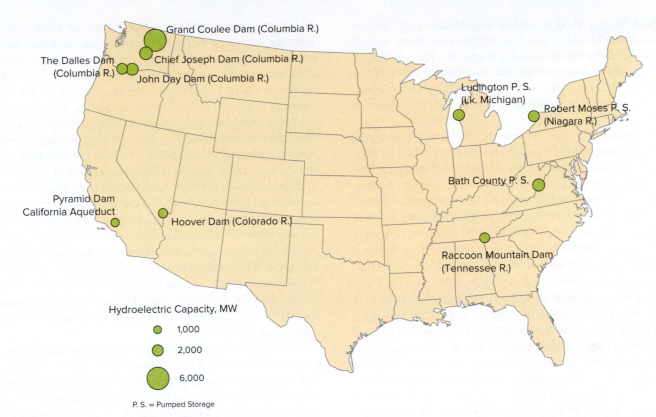

Figure 12.18 **Leading hydropower facilities in the United States.** Washington, the location of the country's largest hydroelectric facility—the Grand Coulee Dam—generates 32% of total U.S. hydropower. The United States has developed about one-half of its potential hydroelectric capacity. *Source of data:* United States Society on Dams.

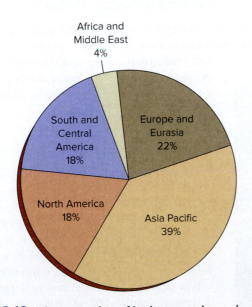

Figure 12.19 **Consumption of hydropower by region, 2014.** The percent of world hydropower consumption is indicated for each region. Hydropower's contribution to the electricity supply is not limited to industrialized countries. In South America, for example, hydropower provides about 70% of the electricity; the figure for the developing world as a whole is 44%. *Source: Data from* The BP Statistical Review of World Energy, *June 2015.*

Consumption patterns around the world are shown in **Figure 12.19**. Hydropower's contribution to a country's energy supplies varies greatly. Several countries—New Zealand, Switzerland,

and Brazil—obtain more than three-quarters of their electricity from hydropower. Hydroelectric power plants provide virtually all of the electricity in Paraguay and Norway. It is estimated that people use only 10% of the world's potential hydropower supply. Water resources that haven't yet been exploited for hydropower are still abundant in Central and South America, Africa, India, and China.

In addition to supplying electrical power, dams provide flood protection, irrigation water, and recreational opportunities. Despite these and other advantages, the exploitation of hydropower has significant environmental and social costs (see "Dammed Trouble," pp. 353–354). The dam itself can block fish migration that is necessary for spawning of certain species. Flooding valleys for water storage submerges forests, farmlands, and villages, sometimes resulting in the displacement of tens of thousands or even millions of people. The reservoirs flood natural wetlands and river habitats, alter stream-flow patterns, and trap silt that otherwise would flow downstream to be deposited on agricultural lands, leading to long-term declines in soil fertility. The disruption of downstream ecosystems that evolved to take advantage of a river's sporadic flood pattern typically reduces the diversity of aquatic species.

Solar Energy

Each year, the Earth intercepts solar energy that is equivalent to many thousands of times the energy people currently use. Inexhaustible and nonpolluting, **solar energy** is the ultimate origin of most forms of utilized energy: fossil fuels and plant life, water power, and wind power. It is, however, the direct capture of solar energy that is seen by many as the best hope of satisfying a proportion of future energy needs with minimal environmental damage and maximum conservation of Earth resources. The chief drawback to solar power is its diffuse and intermittent nature. Being diffuse, it must be collected over a large area to make it practical to use, and because it is intermittent, it requires some means of storage.

The technology for using solar energy for domestic uses such as hot-water heating and space heating is well known. In the United States, both passive and active solar-heating technologies have secured a permanent foothold in the marketplace. More than one-half million homes use solar radiation for water and space heating. Japan is reported to have some 4 million solar hot-water systems, and in China and Israel, more than one-half of all homes heat water with solar energy. The use of solar panels for individual homes is best in climates that are warm and sunny, with not too much cloud cover and not too many hours of darkness in winter.

A second type of solar energy use involves converting concentrated solar energy into thermal energy to generate electricity. Research efforts are focusing on a variety of thermal-electric systems, including power towers, parabolic troughs, and solar ponds. Most involve the concentration of the sun's rays onto a collecting system. In a parabolic trough system, long troughs of curved mirrors guided by computers follow the sun, focusing solar energy onto a steel tube filled with synthetic oil (**Figure 12.20**). Heated to 390°C (735°F), the oil in turn heats water to produce steam to power generators. Several plants of this type are operating in Spain and the Mojave Desert in California where sunshine is abundant.

The plants just described generate electricity indirectly by first converting light to heat, but electricity can also be generated directly from solar rays by **photovoltaic (PV) cells** (also called *solar cells*), semiconductor devices made of silicon. In North America, such solar-electric cells were first used for a variety of specialized purposes where cost is not a constraint, such as powering spacecraft, mountaintop communications relay stations, navigation buoys, and foghorns. As the price of the PV cells has declined, and their efficiency has increased, they have found a market in powering highway signs, cellphone towers, and small electronics such as calculators and radios. In developing countries, they power irrigation pumps, run refrigerators for remote health clinics, and charge batteries. The introduction of small PV cells in rural areas of developing countries has been revolutionary. This technology allows families to have electric light for the first time and to charge cell phones. The latest advance in solar technology is a photovoltaic roofing material that is integrated with solar cells, in effect making the roof the power plant for the building. In Japan, Germany, and France, PV roof systems are being installed in many new houses. Several states and cities in the United States offer tax rebates, subsidies, or other incentives to homeowners and businesses (e.g., factories, warehouses, shopping malls) to install solar panels on their roofs.

As of 2014, Germany had the greatest installed solar power capacity, followed by China, Japan, Italy, and the United States. Although improvements in manufacturing technology have led to a significant drop in the cost of PV solar powezr systems, considerable research and development will be necessary before they make a significant contribution to a country's supply of electricity.

Other Renewable Energy Resources

In addition to biomass, hydropower, and solar power, a number of other renewable sources of energy can be exploited. Two of these are geothermal energy and the wind. Although neither appears able to make a major contribution to the world's energy needs, each has limited, often localized, potential.

Geothermal Energy

People have always been fascinated by volcanoes, geysers, and hot springs, all of which are manifestations of **geothermal** (literally, "Earth-heat") **energy**. There are several methods of deriving energy from the Earth's heat as it is captured in hot water and steam trapped a mile or more beneath the Earth's surface. Conventional methods of tapping geothermal energy depend on the fortuitous availability of hot-water reservoirs beneath the Earth's surface. Deep wells drilled into these reservoirs use the heat energy either for generation of electricity or for direct heat applications, such as heating houses or drying crops.

Geothermal fields are usually associated with areas where magmas are relatively near the Earth's surface—that is, in areas of recent volcanic activity above the subduction zones and volcanic hotspots. Thus, Iceland, Mexico, the United States, the Philippines, Japan, and New

Figure 12.20 **Parabolic trough reflectors at a solar thermal energy plant in the Mojave Desert near Daggett, California.** The facility uses sunlight to produce steam to generate electricity. Guided by computers, the parabolic reflectors follow the sun, focusing solar energy onto a steel tube filled with heat-transfer fluid. Together, nine solar energy plants in the Mojave Desert generate enough electricity for 232,500 households. © Doug Sherman/Geofile RF.

Dammed Trouble

Dams are built for four primary purposes: to generate power, to improve navigation, to prevent floods, and to provide a reliable source of water for agriculture and municipalities. They offer many advantages, including recreation on the reservoirs they create. They exploit a free and endlessly renewable source of energy, flowing water. After the initial capital investment, the costs of operation are relatively low, and hydroelectric dams are often the least-expensive means of generating electricity. Dams permit large numbers of people to inhabit and cultivate arid regions—to make the desert bloom. Some open up shipping passages on formerly wild rivers. Dams on the Columbia and Lower Snake rivers, for example, allow barge traffic to penetrate almost 800 kilometers (500 mi) into the interior, making Lewiston, Idaho, the West Coast's most inland seaport.

The United States built its largest dams in the 1930s and 1940s, when the Tennessee, Colorado, Columbia, and other rivers were dammed. It was a time when environmentalists held little sway, and the negative consequences of dam construction went largely unrecognized. The benefits, however, were evident.

The best sites for water development in the United States have already been exploited, but this is not the case in much of the rest of the world. Large dam construction projects are underway on the Narmada River in India and the Tigris and Euphrates rivers in Turkey, and others have been proposed in Brazil and Laos. In 2010, China completed the world's largest hydroelectric project, the Three Gorges Dam. The enormous structure blocks the Chang Jiang (Yangtze) River in China's scenic Three Gorges area, honored by painters and poets, where the river flows through a series of sheer chasms. Although the dam itself is the world's biggest, its hydroelectric output will be more than 18,000 megawatts (the equivalent of 18 large coal or nuclear power plants), by far the largest in the world.

The Three Gorges Dam has created a wide range of unintended negative consequences that illustrate the disadvantages of large dams.

- When valleys are flooded for water storage, people and wildlife are displaced and forests are destroyed. About the length of Lake Superior, the huge reservoir—600 kilometers (375 mi) long—behind the Three Gorges Dam submerged the cultural artifacts, homes, fertile river bottom farmland, and villages housing an estimated 1.2 million people. Critics of the project argue that the successful resettlement of so many people is impossible. Moving them up on the surrounding hillsides is unrealistic; the slopes cannot support new farming. Resettling people far from their homes hasn't worked in other regions. Many drift back to their native homes; others live as impoverished refugees.
- Although its proponents embrace hydropower as a source of clean energy, the flooding that accompanies big dams often submerges large forests. As the vegetation decays, it releases methane, a potent greenhouse gas, thus contributing to global warming.
- As we learned in Chapter 3, running water is a powerful erosional agent. Upstream of the dam, the water will slow down and drop an enormous amount of silt on the bottom of the reservoir; as it fills with sediment, upstream flooding increases.
- Within a few decades, silting-up can reduce the original generating capacity of a dam significantly; in the long run, reservoirs will fill with silt, ending their useful life.
- The higher water levels in the reservoir have destabilized the adjacent riverbanks and even increased earthquakes. This leads to landslides, as the riverbank's walls collapse in places, and heavy silting of the riverbed.

- There will be a loss of nutrients to river life because the sediment trapped behind the dam and turbines contains organic matter that nourishes downriver food webs.
- Many places have experienced an increase in the incidence of waterborne diseases in reservoirs, irrigation canals, and rivers following dam construction. In Egypt, the Aswan High Dam expanded the breeding areas for snails that transmit schistosomiasis, a debilitating disease that is almost impossible to cure. Malaria typically increases following dam construction in tropical regions because hundreds of square kilometers of ideal mosquito breeding area are exposed during periods of reservoir drawdown.
- The reduced water flow during the dry season will disrupt the Chang Jiang's estuary, causing seawater to move upstream. Because sediment is no longer deposited by floodwaters, the shoreline may erode. As estuaries, beaches, and wetlands shrink, wildlife habitat will be diminished.
- Large dams affect the temperature and oxygen content of downstream waters, altering the mix of aquatic species. Rivers become less livable for some species, better for others. Some analysts predict a dramatic reduction in the number of fish in the Chang Jiang because of the lower water temperaturer.
- Large dams block fish migration, although fish ladders can help some fish make it past dams. The dams of the U.S. Pacific Northwest have proved lethal to the region's wild salmon, which need to migrate from the rivers out to the Pacific Ocean and then back again to spawn where they were born. The Army Corps of Engineers has struggled for years with little success to find ways for salmon and the dams to coexist.
- Finally, there is a possibility of a catastrophic dam failure, the result of age, poor design and construction, or an earthquake.

Options in the construction of huge dams do exist. The efficiency of existing dams can be increased. Replacing old, inefficient generators with newer, highly efficient ones can triple a dam's electrical output. Small dams can deliver many of the benefits of hydropower at a fraction of the cost and without displacing people. Many developing countries are building small dams and installing small generators to supply power to areas far from electrical utility lines.

Considering the Issues

1. The advantages of hydropower are manifold, yet some observers charge that water development has amounted to a Faustian bargain between people and the natural world. What do they mean?

(continued)

353

2. If you were the Minister of Water Resources in China, would you have recommended that the Three Gorges Dam be built? What questions would you want answered in order to assess the likely benefits and costs of constructing a large dam?
3. In the United States, there is an emerging belief that old dams that have outlived their usefulness should be dismantled, returning rivers to their free-flowing nature. More than 1000 dams had been decommissioned and removed, mostly older, smaller dams in the Northeast and Midwest. About one-quarter of the dams in the United States are at least 50 years old, and many are in dire need of repair. Would you favor repairing or dismantling them? Why or why not?
4. Dams were often built for economic development purposes, particularly hydropower, navigation, and irrigation. The negative consequences of dams often are environmental such as blocking fish migration or changes to river sediment loads, water temperatures, or oxygen levels. Yet, as we have seen, hydropower is the largest source of renewable energy. How can environmental and economic development objectives be balanced in the management of rivers?
5. About 100,000 dams regulate American rivers, yet according to the U.S. Geological Survey, flooding continues to be the most destructive and costly type of natural disaster in the country, there has been no reduction in the average number of flood deaths each year, and even when adjusted for inflation, flood damage to property has almost tripled since 1951. Can you think of some reasons why this is so?

Zealand are among the 21 countries that produce geothermal energy (**Figure 12.21**). In Iceland, one-half of the geothermal energy is used to produce electricity; one-half is used for heating. Almost all buildings in Reykjavík, the capital of Iceland, are heated by geothermal steam. Outdoor public geothermal-heated pools and hot tubs are also part of Icelandic culture.

Although relatively few places have geothermal steam that can be exploited to generate electricity, geothermal energy can also be used directly for heating and cooling. *Geothermal heat pumps* (also known as ground-source heat pumps) take advantage of the constant temperature found in soil below the frost line to heat or cool air pumped through a building. Loops of piping are buried in the ground; an electric compressor circulates refrigerant through them and then cools or heats the air, which is distributed throughout a building. Energy-efficient and environmentally clean, geothermal heating systems have grown in popularity in the United States in recent years, particularly for new construction.

Wind Power

Although wind power was used for centuries to pump water, grind grain, and drive machinery, its contribution to the energy supply in the United States virtually disappeared more than a century ago, when windmills were replaced first by steam and later by the fossil fuels. Windmills offer many advantages as sources of electrical power. They can turn turbines directly, do not use any fuel, and can be built and erected rather quickly. They need only strong, steady winds to operate, and these exist at many sites. Furthermore, wind turbine generators do not pollute the air or water and do not deplete scarce natural resources. Technological advances in design have lowered the cost of using wind turbines to generate electricity, so they are becoming increasingly competitive with conventional power plants. Wind power now costs from 3 to 6 cents per kilowatt-hour, about the same as power plants fired by fossil fuels.

Development of larger, more-efficient turbines and renewable energy mandates in several states encouraged utility companies to invest in **wind farms,** clusters of wind-powered turbines producing commercial electricity (**Figure 12.22a**).

The open prairies of the central United States proved to be ideal for wind energy production. Iowa now leads the country with the greatest wind energy capacity on a land area basis while Texas has the greatest total installed capacity (**Figure 12.22b**). In the 1990s, a commitment to reducing dependence on fossil fuels and developing renewable resources stimulated the growth of wind-power installations in several European countries, particularly Germany, Spain, and Denmark. In terms of megawatts of installed capacity, Germany and the United States are the world's leading producers of wind-powered energy, followed by Spain. Denmark, however, has the highest per capita output of wind energy in the world. With the exception of India, Asian countries have been somewhat slower to build wind farms, but sizable projects are under construction in China and Japan.

Offshore wind turbine installations play a growing role in the supply of renewable energy, particularly in northern Europe. The world's biggest offshore windmill park is at the entrance to Copenhagen's harbor. It and other installations provide about 20% of the country's electricity. The Netherlands and Sweden also have offshore windmill parks, and planned or under construction are parks off the coasts of Britain, Ireland, Belgium, Germany, and Spain.

The chief disadvantage of wind power is that it is unreliable and intermittent; because its energy cannot be easily stored, it requires a backup system. Additionally, in some countries wind power is most plentiful in remote areas, far from the existing power grid, and costly new transmission lines would have to be installed to bring it to consumers. Detractors point out that it takes thousands of wind turbines to produce the same amount of electricity as a single nuclear power plant. Environmental concerns include the aesthetic impact of wind farms (they are very visible, often covering entire hillsides and dominating the landscape) and the hazard they might pose to migrating birds.

12.5 Nonfuel Mineral Resources

The mineral resources already discussed provide the energy that enables people to do their work. Equally important to our economic well-being are the *nonfuel* minerals, for they can be processed into steel, aluminum, and other metals and into glass,

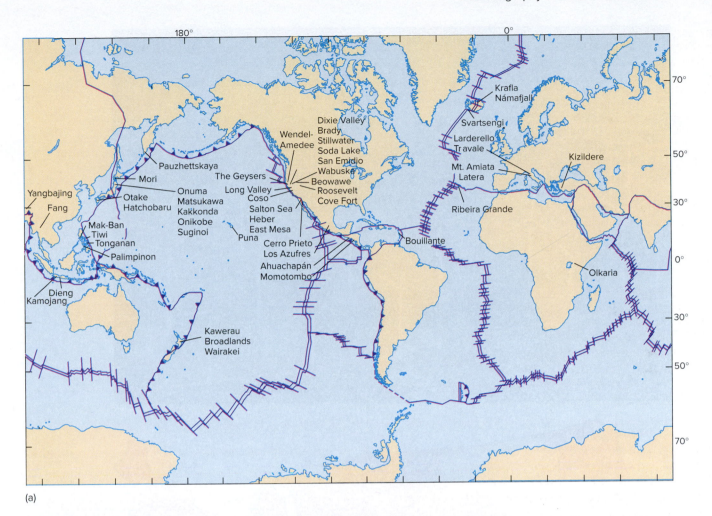

(a)

(b)

Figure 12.21   Geothermal energy. (a) Geothermal power plants worldwide. Most areas where geothermal energy is tapped are along or near plate boundaries. Because the number of suitable sites is limited, and most are far from large cities where the demand for power is great, geothermal power is likely to remain a minor contributor to world energy production. **(b)** The Blue Lagoon in Iceland is a popular geothermal spa. It uses the leftover hot water from a geothermal energy plant. Like Hawaii, Iceland is located above a volcanic hot spot. This creates an ideal setting for capturing geothermal energy. Superheated water is vented from lava flows and used to run turbines that generate electricity or to heat homes, swimming pools, and greenhouses. Geothermal energy heats 90% of the houses in Iceland and supplies 25% of the country's electricity. © *robas/Getty Images RF*.

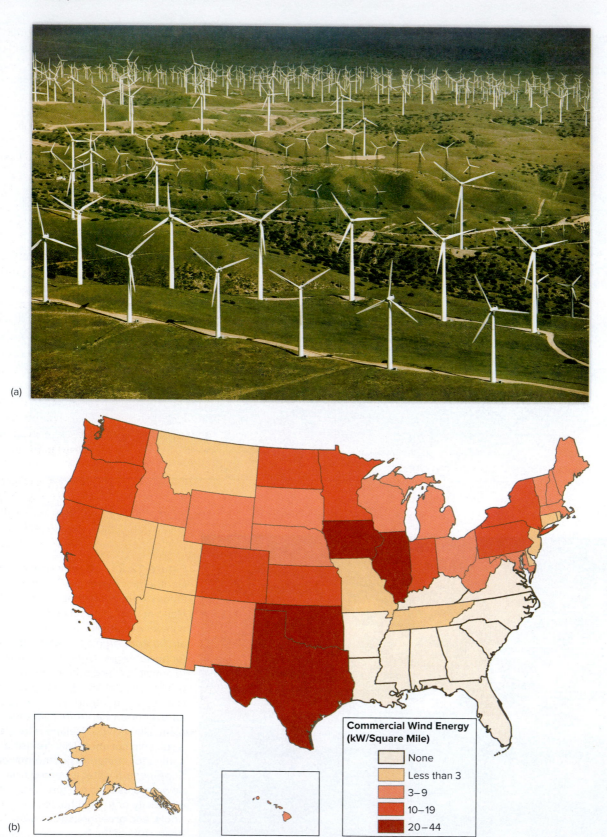

Figure 12.22 **(a) A wind farm in California.** The wind turbine generators harness wind power to produce electricity. **(b) Wind-generating capacity by state, in kilowatt capacity per square kilometer, 2016.** The uneven spatial distribution of wind-power generation stems from both physical factors and political factors. Wind speeds are typically higher in the Great Plains, along shorelines, and near mountain passes. Some governments have supported wind power through tax credits and mandates for renewable energy, and others have restricted it. Although only a small portion of the country's wind potential has been tapped, new wind-power projects are underway in many states. Wind turbine technology advanced dramatically during the 1990s; modern turbines are significantly more efficient and cost effective than earlier ones. *(a) Source: Photo by M. Smith/USGS. (b) Source of data: American Wind Energy Association.*

cement, and other products. Our buildings, tools, and weapons are chiefly mineral in origin.

Virtually all the resources we deem essential, including metals, nonmetallic minerals, rocks, and the fuels, are contained in the Earth's crust, the thin outer skin of the planet. Of the 92 natural elements, just 8 account for more than 98% of the mass of the Earth's crust (**Figure 12.23**). They can be thought of as geologically abundant, and all others as geologically scarce. In most places, minerals exist in concentrations too low to make their exploitation profitable. If the concentration is high enough to make mining feasible, the mineral deposit is called an **ore**. Thus what is or is not an ore depends on demand, price, and technology and changes over time.

Exploitation of a mineral resource typically involves six steps:

1. *exploration* (finding concentrated deposits of the material)
2. *extraction* (removing it from the Earth)
3. *concentration* (separating the desired material from the ore)
4. *smelting* and/or *refining* (breaking down the mineral to the desired pure material)
5. *transporting* it to where it will be used
6. *manufacturing* the ore into a finished product

Each step requires inputs of energy and materials.

Five factors help determine the practicality and profitability of mining a specific deposit of a mineral: its value, the quantity available, the richness of the ore in a particular deposit, the distance to market, and land acquisition and royalty costs. Even if these conditions are favorable, mines may not be developed or even remain operating if supplies from competing sources are available more cheaply. In the 1980s, more than 25 million tons of iron ore–producing capacity was permanently shut down in the United States and Canada. Similar declines occurred in North American copper, nickel, zinc, lead, and molybdenum mining as market prices fell below domestic production costs. Beginning in the early 1990s, as a result of both resource depletion and low-cost imports, the United States became a net importer of nonfuel minerals for the first time. Although increases in mineral prices may lead to the opening or reopening of mines that have been deemed unprofitable, the developed industrial countries with market economies find themselves at a competitive disadvantage against developing country producers with lower-cost labor and state-owned mines with abundant reserves.

Natural processes produce minerals so slowly that they fall into the category of nonrenewable resources, existing in finite deposits. The supplies of some, however, are so abundant that a ready supply will exist far into the future. These include coal, sand and gravel, and potash. The supply of others, such as tin and mercury, is small and getting smaller as industrial societies place ever-greater demands on them. **Table 12.1** gives one estimate of "years remaining" for some important metals. It should be taken as suggestive rather than definitive, because mineral reserves are difficult to estimate. As we noted in the case of fossil fuels, such estimates are based on economic and technological conditions, and we cannot predict either future prices for minerals or improvements in technology. The depletion of the currently identified, usable reserves of a valuable mineral will drive up the price of the mineral, which will make it profitable to mine some of what are now classified as subeconomic deposits (see Figure 12.4). Those deposits would then be reclassified as proved reserves. The discovery of new deposits and/or improvements in mineral-processing technology would also increase the reserve figure and thus its projected lifetime.

Although human societies began to use metals as early as 3500 B.C., world demand remained small until the Industrial

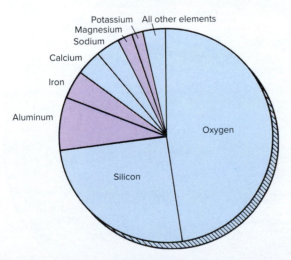

Figure 12.23   The relative abundance, by weight, of elements in the Earth's crust. Only four of the economically important elements—those shown in purple—are geologically abundant, accounting for more than 1% of the total weight of the Earth's crust. Fortunately, these and other commercially valuable minerals have been concentrated in specific areas within the crust. Were they uniformly disseminated throughout the crust, their exploitation would not be feasible.

Table 12.1	Projected Lifetimes of Reserves of Selected Minerals	
	Years Remaining of Proved Reserves	
Mineral	**World**	**United States**
Bauxite	100	0
Copper	39	26
Lead	19	13
Manganese	34	31
Nickel	31	6
Phosphate	310	40
Potash	95	160
Silver	21	23
Tin	16	0
Zinc	15	13

Source: Mineral Commodity Summaries 2016, U.S. Geological Survey.
Note: These figures reflect the approximate number of years the proved reserves of selected minerals will last, based on current rates of production and consumption. World values are global reserves divided by annual global mine production. United States values are U.S. reserves divided by U.S. mine production. Such figures are only suggestive, because reserve totals and consumption rates fluctuate over time. A decline in the costs of exploitation, increase in value of the material, and/or new discoveries of deposits will extend the lifetimes shown here.

Revolution. It was not until after World War II that increasing shortages and rising prices (and in the United States, increasing dependence on foreign sources) began to impress themselves on the general consciousness. Worldwide technological development has established ways of life in which minerals are the essential constituents. That industrialization has proceeded so rapidly and so cheaply is the direct result of the earlier ready availability of rich and accessible deposits of the requisite materials. Economies grew fat by skimming the cream. The question, yet unanswered, is whether the remaining supplies of scarce minerals will limit the expansion of industrialized and developing economies or whether, and how, people will find a way to cope with shortages.

The Distribution of Nonfuel Minerals

Because the distribution of mineral resources is the result of long-term geologic processes that concentrated certain elements into commercially exploitable deposits, it follows that, the larger the country, the more likely it is to contain such deposits. And in fact, Russia, China, Canada, the United States, Brazil, and Australia possess abundant and diverse mineral resources. As **Figure 12.24** indicates, these are the leading mining countries. They contain roughly one-half of the nonfuel mineral resources and produce the bulk of the metals (e.g., iron, manganese, and nickel) and nonmetals (e.g., potash and sulfur).

Many types of nonfuel minerals are concentrated in a small number of countries, and some scarce elements occur in just a few regions of the world. Thus, extensive deposits of cobalt and diamonds are largely confined to Russia and central-southern Africa. South Africa has nearly one-half of the world's gold ore and more than three-quarters of the chromium and platinum-group metals.

China produces 85% of the world's supply of rare earth elements. Rare earth elements are important components in many high technology products including the wind turbines used for renewable energy production. Some countries contain only one or two exploitable minerals—Morocco has phosphates, for example, and New Caledonia, nickel. Several countries with large populations are at a disadvantage with respect to mineral reserves. They include industrialized countries such as France and Japan, which are able to import the resources, as well as developing countries such as Nigeria and Bangladesh, which are less able to afford imports.

It is important to note that no country contains all of the economically important mineral resources. Some, such as the United States, which were bountifully supplied by nature, have spent much of their assets and now depend on foreign sources. Although the United States was virtually self-sufficient in mineral supplies in the 1940s and 1950s, it is not today. Because of its history of use of domestic reserves and its continually expanding economy, the United States now depends on other countries for more than 50% of its supply of a number of essential minerals, some of which are shown in **Table 12.2**.

The increasing costs and declining availability of metals encourage the search for substitutes. The fact that industrial chemists and metallurgists have been so successful in the search for new materials that substitute for the traditional resources has tended to allay fears of possible resource depletion. But it must be understood that no adequate replacements have been found for some minerals, such as cobalt and chromium. Many other substitutes are synthetics, often employing increasingly scarce and costly hydrocarbons in their production. Many, in their use or disposal, constitute environmental hazards, and all have their own high and increasing price tags.

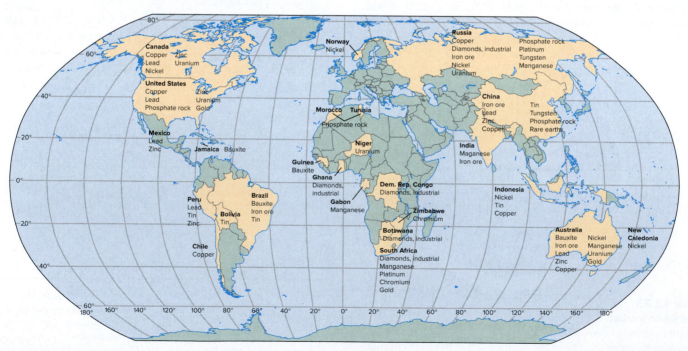

Figure 12.24 **Leading producers of selected minerals.** The countries shown are not necessarily those with the largest deposits. India, for example, contains reserves of bauxite, Brazil of uranium, and South Africa of phosphates, but none is yet a major producer of those materials. *Data from World Resources Institute and United States Geological Survey.*

Table 12.2	U.S. Reliance on Foreign Supplies of Selected Nonfuel Minerals, 2016		
Mineral	**% Imported**	**Major Sources**	**Major Uses**
Bauxite and alumina	100%	Jamaica, Brazil, Guinea	Aluminum production
Manganese	100	Gabon, Australia, S. Africa	Steelmaking, batteries, agricultural chemicals
Quartz crystal	100	China, Japan, Russia	Electronics, optical applications
Strontium	100	Mexico, Germany	Ferrite magnets, fireworks
Yttrium	100	China	Light emitting diodes (LEDs) in TV sets, lights
Platinum	90	Germany, S. Africa, United Kingdom	Catalysts in the automotive and chemical industries
Potash	84	Canada, Russia	Fertilizers, chemicals
Zinc	82	Peru, Canada, Mexico	Galvanizing, brass and bronze
Rare Earths	76	China	Computers, cell phones, solar panels
Cobalt	75	China, Norway, Russia	Superalloys for jet engines, cutting tools, magnets, chemicals
Silver	72	Mexico, Canada, Peru	Electrical products, coins, antibacterial agents
Chromium	66	S. Africa	Stainless steel

Source: U.S. Geological Survey, Mineral Commodities Survey 2016.

Copper: A Case Study

Table 12.1 indicates that the world reserves of copper will last only another generation or so, based on current rates of production and consumption and assuming that no new extractable reserves appear. Copper is a relatively scarce mineral, and its importance to industrialized societies is evidenced by the fact that more copper is mined annually than any other nonferrous metal except aluminum.

Three properties make copper desirable: it conducts both heat and electricity extremely well, it is malleable and thus can be hammered or drawn into thin films or wires, and it resists corrosion. Copper is a major industrial metal with many applications. It is used in building construction, industrial and farm machinery, power transmission and generation, telecommunications, electrical wire and equipment, electronic products, transportation, plumbing fittings and pipes, coinage, and consumer and general products (e.g., cookware, musical instruments, and statues). A significant proportion of copper is also used as alloys in bronze, brass, and other metals.

Like most minerals, copper is unevenly distributed in the Earth's crust. The largest copper deposits are found at convergent tectonic margins in western North America, western South America, and Australia. Copper deposits in sedimentary basins include those extending across northern Europe, from England to Poland, and the copper belt of central Africa (Zambia and Democratic Republic of Congo). Chile leads the world in copper production (about 36% of the total), followed by the United States, Indonesia, and Peru. Because the United States consumes more copper than it produces, it imports significant amounts to meet its demand for the metal.

Globally, the production of copper increased fairly steadily from 9.5 million metric tons in 1984 to 19 million metric tons in 2015. The demand for copper is strong due both to its increased use in motors and electronic equipment and to rising consumption of the metal in Russia, India, and China.

The scarcity of copper supplies has had several effects that suggest how societies will cope with shortages of other raw materials. First, in the United States, the grade of mined ores has decreased steadily. Those with the highest percentage of copper (2% and above) were mined early (**Figure 12.25**). Now, ore

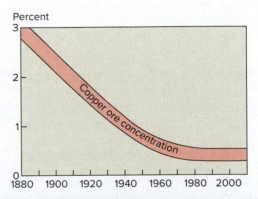

Figure 12.25   Concentration of copper needed in order to be mined economically. In 1880, 3% copper ore rock was necessary, but today, rock with 0.4% or less copper is mined. As the supply of a metal decreases and its price increases, the concentration needed for economic recovery goes down. *Source: Data from the U.S. Bureau of Mines, U.S. Department of the Interior.*

of 0.4% grade is the average. Thus, 1000 tons of rock must be mined and processed to yield 4 tons of copper—or, in more practical terms, 3 tons of rock are necessary to equip one automobile with the copper used in its radiator and its various electrical components. The remaining 2.985 tons is waste, generated at the mine, the concentrator, and the smelter.

Second, the recovery of copper by recycling has increased. Virtually all products made from copper can be recycled. Unlike some minerals, copper lends itself to recycling because much of it is used in pure form in sizable pieces. It is less expensive to remanufacture recycled, or "secondary," copper into new products than to produce it by mining and refining new ore. Recycling contributes significantly to the copper supply in the United States.

Increasing demand for copper has led to the search for additional deposits. Two companies are actively exploring the feasibility of extracting copper, gold, silver, and other metals from underneath the South Pacific Ocean, where the minerals are concentrated in mounds at midoceanic ridges and the crests of undersea volcanoes. Nearly 2 kilometers (about 1 mi) below sea level, the high-grade sulfide deposits are thought to contain 12% copper. Critics of deep seabed mining argue that it has the potential to be environmentally destructive, with adverse effects on marine life, while its advocates contend that it is less disruptive than land-based mining.

Finally, price rises have spurred the search for substitutes. In many of its applications, copper is being replaced by other, less expensive materials. Aluminum is replacing copper in some electrical applications and in heat exchangers. Plastics are supplanting copper in plumbing pipes and building materials. Glass fibers are employed in many telephone transmission lines, and steel can be used in shell casings and coinage. Thus, as with many other mineral resources, prices have undergone wide fluctuations.

12.6 Land Resources

The mineral resources we have just discussed are nonrenewable. We turn our attention now to land resources that are potentially renewable, examining the distribution and status of three of those resources: soils, wetlands, and forests. Because they support or are living things, they are sometimes called *biological resources*.

Soils

By design or by accident, people have brought about many changes in the physical, chemical, and biochemical nature of the soil and altered its structure, fertility, and drainage characteristics. The exact nature of the changes in any area depends on past practices as well as on the original nature of the land.

Over much of the Earth's surface, the thin layer of topsoil upon which life depends is only a few inches deep, usually less than 30 centimeters (1 ft). Below it, the lithosphere is a complex mixture of rock particles, inorganic mineral matter, organic material, living organisms, air, and water. Under natural conditions, soil is constantly being formed by the physical and chemical decomposition of rock material and by the decay of organic matter. It is simultaneously being eroded, for **soil erosion**—the removal of soil particles, usually by wind or running water—is as natural a process as soil

formation, and it occurs even when land is totally covered by forests or grass. Under most natural conditions, however, the rate of soil formation equals or exceeds the rate of soil erosion, so that soil depth and fertility tend to increase with time.

When land is cleared and planted to crops, or when the vegetative cover is broken by overgrazing or other disturbances, the process of erosion accelerates. When its rate exceeds that of soil formation, the topsoil becomes thinner and eventually disappears, leaving behind only sterile subsoil or barren rock. At that point, the renewable soil resource has been converted through human impact into a nonrenewable and dissipated asset. Carried to the extreme of bare rock hillsides or wind-denuded plains, erosion spells the total end of agricultural use of the land.

Such massive destruction of the soil resource could endanger the survival of the civilization it has supported. For the most part, however, farmers devise ingenious ways to preserve and even improve the soil resource upon which their lives and livelihoods depend. Farming skills have not declined in recent years, but pressures upon farmlands have increased with population growth. Farming has been forced higher up onto steeper slopes, more forest land has been converted to cultivation, grazing and crops have been pushed farther and more intensively into semiarid areas, and existing fields have had to be worked more intensively and less carefully. Many traditional agricultural systems and areas that were ecologically stable and secure as recently as 1950, when world population stood at 2.5 billion people, are disintegrating under the pressures of more than 7 billion people.

The pressure of growing population numbers is having an especially destructive effect on tropical rain forests. Expanded demand for fuel and commercial wood and a midlatitude market for beef that can be satisfied profitably by replacing tropical forest with cleared grazing land are responsible for some of the loss, but the major cause of *deforestation* is the clearing of the land for crops. Extending across parts of Asia, Africa, and Latin America, the tropical rain forests are the most biologically diverse places on Earth, but vast expanses are being destroyed every year. More than one-half of their original expanse has already been cleared or degraded. Deforestation is discussed in greater detail on pages 365–371, but it is important to note here that accelerated soil erosion quickly removes tropical forest soils from deforested areas. Lands cleared for agriculture soon become unsuitable for that use, partially because of soil loss (**Figure 12.26**).

The tropical rain forests can succumb to deliberate, massive human assaults and be irretrievably lost. With much less effort, and with no intent to destroy or alter the environment, humans are similarly affecting the arid and semiarid regions of the world. The process is called **desertification**, the expansion or intensification of areas of degraded or destroyed soil and vegetation cover; it usually occurs in arid and semiarid environments. Climatic change—unpredictable cycles of rainfall and drought—is often a contributing cause, but desertification accelerates because of human activity, mainly overgrazing, deforestation for fuelwood, clearing of original vegetation for cultivation, and burning. Desertification implies a continuum of ecological alteration from slight to extreme (**Figure 12.27**).

Whatever its degree of development, when the process results from human rather than climatic change, it begins in the same

Figure 12.26 **Soil degradation following deforestation.** The tropical rain forest was cleared on this tract in Amazonia, Brazil, to make room for a tin-mining operation. Exposed soils quickly deteriorate in structure and fertility and are easily eroded. © *Photodisc/Getty Images RF.*

seeping in, carrying soil particles with it. When the water is lost through surface flow rather than seepage downward, the water table is lowered. Eventually, even deep-rooted bushes are unable to reach groundwater, and all natural vegetation is lost. The process is accentuated when too many grazing animals pack the Earth down with their hooves, blocking the passage of air and water through the soil. When both plant cover and soil moisture are lost, desertification has occurred.

It happens with increasing frequency in many areas of the Earth as pressures upon the land continue. Since semiarid regions on the margins of deserts are at greatest risk, Africa is most vulnerable to desertification. The United Nations has estimated that 40% of that continent's nondesert land is in danger of human-induced desertification. But parts of Asia and Latin America are similarly endangered. In North African and Middle East countries where desertification is particularly extensive and severe (Algeria, Ethiopia, Iraq, Jordan, Lebanon, Mali, and Niger), per capita food production has declined significantly. The resulting threat of starvation spurs populations of the affected areas to increase their farming and livestock pressures on the denuded land, further contributing to the cycle of desertification.

Desertification is but one example of the global problem of soil degradation (**Figure 12.29**). Evidence of human-caused soil degradation is found in all parts of the world. In Guatemala, for example, some 40% of the productive capacity of the land has been lost through erosion, and several areas of the country have been abandoned because agriculture has become economically

fashion: the disruption or removal of the native cover of grasses and shrubs through farming or overgrazing (**Figure 12.28**). If the disruption is severe enough, the original vegetation cannot reestablish itself, and the exposed soil is made susceptible to erosion during the brief, heavy rains that dominate precipitation patterns in semiarid regions. Water runs off the land surface instead of

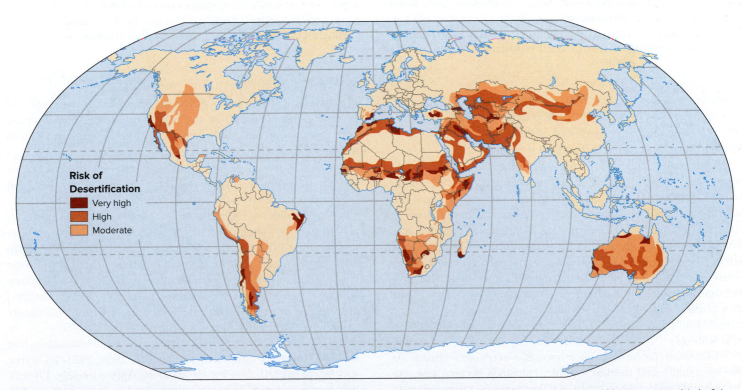

Risk of Desertification
- Very high
- High
- Moderate

Figure 12.27 **Desertification** affects about 1 billion people in more than 100 countries. According to the United Nations, one-third of the world's land surface is at risk of becoming desert wasteland. Most at risk are dry regions on the edges of deserts. Many factors contribute to the desertification of dry areas: climate shifts, population growth, overexploitation of water supplies, diversion of rivers for irrigation, destructive farming practices that expose topsoil to wind and water erosion, overgrazing of grasslands by livestock, and cutting of trees and shrubs for firewood. *Source: John Allen,* Student Atlas of World Geography, *5/e, Map 81, p. 97.*

Figure 12.28   Desertification in Africa. Windblown dust is engulfing the scrub forest in this drought-stricken area of Mali, near Timbuktu. The district is part of the Sahel region of Africa, where desertification has been accelerated by both the climate and human pressures on the land. The cultivation of marginal land, overgrazing by livestock, and recurring droughts have led to the destruction of native vegetation, erosion, and land degradation. © *Lissa Harrison RF.*

impracticable. The figure is 50% in El Salvador, and Haiti has little high-value soil left at all. In Turkey, about one-half of the land is severely or very severely eroded. A full one-quarter of India's total land area has been significantly eroded.

In recent decades, soil erosion in the United States has decreased and yet remains unacceptably high (see "Maintaining Soil Productivity," p. 364"). Wind and water are blowing and washing soil off croplands in Iowa and Missouri, pasturelands in the Great Plains, and ranches in Texas. America's croplands lose almost 1.7 billion tons of soil per year to erosion. Of the roughly 167 million hectares (357 million acres) of land that are intensively cropped in the United States, more than one-quarter are losing topsoil faster than it can be replaced naturally (it can take up to a century to replace an inch of topsoil). In parts of Iowa and Illinois where the topsoil was once a foot deep, less than one-half of it remains.

Like most processes, soil erosion has secondary effects. As the soil quality and quantity decline, croplands become less productive and yields drop. Streams and reservoirs experience accelerated siltation. In countries where the topsoil is heavily laden with agricultural chemicals, erosion-borne silt pollutes water supplies. The danger of floods increases as bottomlands fill with silt, and the costs of maintaining navigation channels grow.

Accelerated erosion is a primary cause of agricultural soil deterioration, but in arid and semiarid areas, salt accumulation can be a contributing factor. **Salinization** is the concentration of salts in the topsoil as a result of the evaporation of surface water. It occurs in poorly drained soils in dry climates, where evaporation exceeds precipitation. As water evaporates, some of the salts are left behind to form a white crust on the surface of the soil (**Figure 12.30**).

Like erosion, salinization is a natural process that has been accelerated by human activities. Poorly drained irrigation systems are the primary culprit, because irrigation water tends to move slowly and thus to evaporate rapidly. All irrigation water contains dissolved salts, which are left behind on the surface when water evaporates. Mild or moderate salinity makes soil less productive and lowers crop yields; extreme salinity ultimately can render the land unsuitable for agriculture.

Thousands of once fertile acres have been abandoned in Iran and Iraq; over 25% of the irrigated areas of India, Pakistan, Syria, and Egypt are affected by salinization. Approximately 1.6 million hectares (4 million acres) of cultivated soils in the Canadian provinces of Saskatchewan and Alberta are classified as overly saline. Areas of serious salinization also appear in the U.S. Southwest, particularly in the Colorado River drainage basin and in the Central Valley of California. Ironically, the irrigation water that

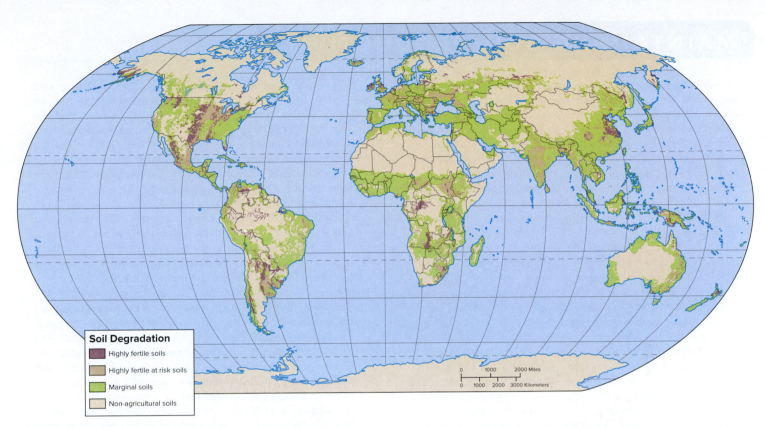

Figure 12.29   **Soil Degradation.** About 11% of the earth's land area is suitable for agriculture. Of this total, the largest category consists of marginal soils that are less fertile and high susceptible to wind and water erosion. *Source: Christopher Sutton, Student Atlas of World Geography, 8/e, Map 98, p. 121.*

Figure 12.30   **Salinization lowers crop yields and can destroy soils.** Salinization has left a white crust of salt on the surface of this soil in Colorado. The problem is most severe in hot, dry areas where irrigation is practiced. *Source: USDA Natural Resources Conservation Service. Photo by Tim McCabe.*

transformed that arid, 430-kilometer-long (270-mi-long) valley into one of the country's most productive farming regions now threatens to make portions of it worthless again.

Wetlands

Vegetated land surfaces that are periodically or permanently covered by or saturated with standing water are called **wetlands**. Transitional zones between land and water, wetlands take a variety of forms, including grassy marshes, wooded swamps, tidal flats, and estuaries, and they are found on all continents. Some are permanently wet; others have standing water for only part of the year. In North America, wetlands range in size from the small prairie potholes of the Midwest and Alberta, Canada, to areas as large as Florida's Everglades. Among the best known wetlands of the United States are the Okefenokee Swamp in Georgia and the bayous of Louisiana and Mississippi. Alaska has large expanses of wetlands, mostly peatlands.

The two major categories of wetlands are *inland* and *coastal*. In the United States, most wetlands are freshwater inland wetlands. They include bogs, marshes, swamps, and floodplains adjacent to rivers. Either fresh or salt water covers coastal wetlands, which are crucial to marine ecosystems. The part of the sea lying above the continental shelf is the most productive of all ocean waters, supporting the major commercial marine fisheries. Because it is not very deep, this *neritic zone* is penetrated and warmed by sunlight. It also receives the nutrients flowing into oceans from streams

MAINTAINING SOIL PRODUCTIVITY

In much of the world, increasing population numbers are largely responsible for accelerated soil erosion. In the United States, economic conditions rather than population pressures have often contributed to excessive rates of erosion. Federal tax laws and the high farmland values of the 1970s encouraged farmers to plow virgin grasslands and to tear down windbreaks to increase their cultivable land and yields. The secretary of agriculture exhorted farmers to plant all of their land, "from fencerow to fencerow," to produce more grain for export. Land was converted from cattle grazing to corn and soybean production as livestock prices declined.

When prices of both land and agricultural products declined in the 1980s, farmers felt impelled to produce as much as they could in order to meet their debts and make any profit at all. To maintain or increase their productivity, many farmers neglected conservation practices, plowing under marginal lands and using fields for the same crops every year.

Conservation techniques were not forgotten, of course. They are practiced by many and persistently advocated by farm organizations and soil conservation groups. Techniques to reduce erosion by holding the soils in place are well known. They include contour plowing, terracing, strip-cropping and crop rotation, erecting windbreaks and constructing water diversion channels, and practicing no-till or reduced-tillage farming (the practice of sowing seeds without turning over the soil, thus allowing plants to grow amid the stubble of the previous year's crop).

Another method of reducing erosion and reclaiming land that is badly damaged is to pay farmers to stop producing crops on highly erodible farmland. After the mid-1980s, federal farm programs attempted to reverse some of the damage resulting from past economic pressures and farming practices. One objective has been to retire for conservation purposes some 18.6 million hectares (46 million acres) of croplands that were eroding faster than three times the natural rate of soil formation. The Conservation Reserve Program of 1985 pays farmers to take highly erodible land out of production; in return, farmers agree to plant the land in trees or a grass or legume cover and then retire it from further use for 10 years.

In 1996, however, Congress passed the Federal Agriculture Improvement and Reform Act, which contained as its cornerstone a program known as "Freedom to Farm." The premise of the program is that the market and not the federal government should determine which crops are planted. Almost immediately, farmers began expanding their acreage, planting crops with a high market value on land that had been set aside. Subsidies for ethanol production and high corn prices have been shown to result in reduced acreage of conservation lands. It remains to be seen how much land that was previously protected by conservation measures will be cultivated using more destructive farming methods.

Only by reducing the economic pressures that lead to abuse of farmlands and by continuing to practice known soil conservation techniques can the country maintain the long-term productivity of soil, the resource base upon which all depend.

Contour plowing and strip-cropping. © Jerry Irwin/Science Source.

No-till farming. Source: Gene Alexander/Soil Conservation Service/USDA.

and rivers, so that vegetation and a great variety of aquatic life can flourish. However, the neritic zone depends to a considerable extent on the continued functioning of the **estuarine zone**, the relatively narrow area of wetlands along coastlines where salt water and fresh water meet and mix (**Figure 12.31**).

Extremely valuable ecological systems, wetlands perform a number of vital functions. Trapping and filtering the silt, pollutants, and nutrients that rivers bring downstream, wetlands improve water quality and help replenish underground aquifers. Among the most diverse and productive of all ecosystems, wetlands provide habitat and food for a wide variety of plants and animals. Indeed,

wetlands are essential to the survival of many species of fish and shellfish by serving as their spawning grounds. Wetlands are also major breeding, feeding, nesting, and wintering grounds for many types of birds (**Figure 12.32**). Not only are these areas extraordinarily productive themselves, but also they contribute to the productivity of the neritic zone, where fish feed on the life that flows from wetlands into the sea.

Importantly, wetlands absorb floodwaters and help stabilize shorelines by providing barriers to coastal erosion. One reason Hurricane Katrina did so much damage to New Orleans was that much of the wetlands that used to buffer the city from storms were

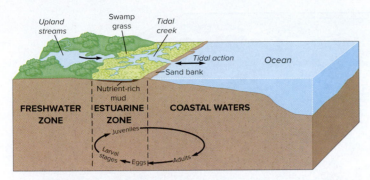

Figure 12.31 **The estuarine zone.** The outflow of fresh water from streams and the action of tides and wind mix deep ocean waters with surface waters in estuaries, contributing to their biological productivity. The saline content of estuaries is lower than that of the open sea. Many fish and shellfish require water of low salinity at some point in their life cycles.

gone—due to the draining of swamps and marshes to create new land for houses and businesses and to U.S. Army Corps of Engineers (USCOE) projects that reshaped the Mississippi River and built canals and locks to make shipping easier.

People have not always recognized or appreciated the value of wetlands, tending to view them as swampy, smelly areas that provide breeding grounds for mosquitoes and impeded settlement—wastelands that should be reclaimed for productive uses such as agriculture and commercial development. Indeed, in the United States, Congress in the mid-1800s passed the Swamp Land Acts, which made it national policy to drain and fill wetlands. Scientists estimate that one-half of the wetlands of the world have been destroyed. Australia and New Zealand are thought to have lost about 90% of their original wetlands, and Europe at least 60%. Since the 1780s, the contiguous United States has lost more than one-half of its wetlands, going from some 87 million hectares (215 million acres) to 42 million hectares (105 million acres).

Wetlands have been drained, dredged, filled and built upon, converted to cropland, and used as garbage dumps. They are polluted by chemicals, excess nutrients, and other waterborne wastes. Natural shorelines have been bulldozed, and artificial levees and breakwaters interfere with the flooding that nurtures wetlands with fresh infusions of sediment and water. Habitat alteration or destruction inevitably disrupts the intricate ecosystems of the wetlands.

Growing awareness of the importance of wetlands and of how much wetland acreage has been damaged or destroyed has led, in the United States and elsewhere, to efforts to preserve and protect them. In the United States, the Clean Water Act of 1972 and subsequent amendments gave wetlands a measure of federal protection. The act prohibits the filling of wetlands without a permit issued by the U.S. Army Corps of Engineers, a significant change in mission because the Corps of Engineers has long emphasized dredging, stream straightening, and dike building. Officially, the

government since 1989 has had a "no net loss" policy. If a development project destroys wetlands, that loss must be offset by restoring or creating a comparable amount of wetlands elsewhere.

Federal protection does not mean a wetland cannot be developed, however, and the Corps of Engineers has issued thousands of permits letting homeowners and developers fill in hundreds of thousands of acres of wetlands. Prosecutions for degrading or filling wetlands, or for failing to offset their loss, have been extremely rare. As a result, although the rate of wetland loss has slowed in recent years, many of the wetlands that remain are in danger of degradation or loss.

Forest Resources

Wetlands are only one of the renewable resources in danger of irreparable damage by human action. In many parts of the world, forests are similarly endangered.

After the retreat of continental glaciers some 12,000 years ago, and before the rise of agriculture, the world's forests and woodlands probably covered some 45% of the Earth's land area exclusive of Antarctica. They were a sheltered and productive environment for earlier societies that subsisted on gathered fruits, nuts, berries, leaves, roots, and fibers collected from trees and woody plants. Few such cultures remain, though the gathering of forest products is still an important supplemental activity, particularly among subsistence agricultural societies.

Even after millennia of land clearance for agriculture and, more recently, commercial lumbering, cattle ranching, and fuelwood gathering, forests still cover about one-third of the world's land area. As an industrial raw material source, however, forests are more restricted in area. Although forests of some type reach discontinuously from the equator northward to beyond the Arctic Circle and

Figure 12.32 **A salt marsh in Louisiana.** Tidal marshlands have been subjected to dredging and filling for residential and industrial development. The loss of such wetland areas reduces the essential habitat of waterfowl, fish, crustaceans, and mollusks. Many waterfowl breed and feed in coastal marshes and use them for rest during long migrations. © *Franke Keating/Science Source.*

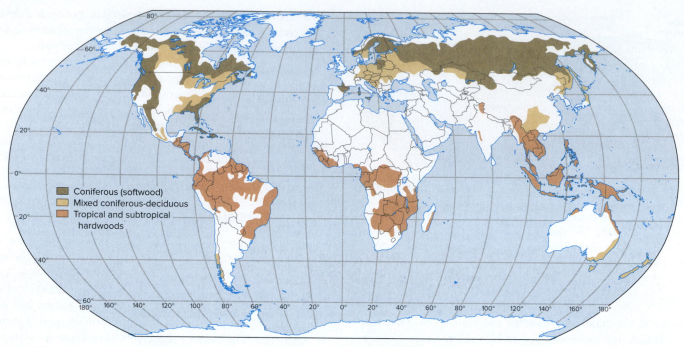

Figure 12.33   **Major commercial forest regions.** Much of the original forest, particularly in midlatitude regions, has been cut over. Many treed landscapes that remain do not contain commercial stands. Significant portions of the northern forests are not readily accessible and at current prices cannot be considered commercial.

southward to the tips of the southern continents, *commercial forests* are restricted to two very large global belts (**Figure 12.33**). One, nearly continuous, occupies the upper-middle latitudes of the Northern Hemisphere. The second straddles the equatorial zones of South and Central America, Central Africa, and Southeast Asia. These forest belts differ in the types of trees they contain and in the type of market or use they serve.

The *northern coniferous,* or softwood, forest is the largest and most continuous stand, circling the globe below the polar regions. Its pine, spruce, fir, and other types of conifers are used for construction lumber and to produce pulp for paper, rayon, and other cellulose products. To its south are the *temperate hardwood* forests, containing deciduous species such as oak, hickory, maple, and birch. These and the trees of the mixed forest lying between the hardwood and softwood belts have been greatly reduced in areal extent by centuries of agricultural and urban settlement and development, though they still are commercially important for hardwood applications: furniture, veneers, railroad ties, and so on.

The *tropical lowland hardwood* forests are exploited primarily for fuelwood and charcoal, on which the populations of developing countries are heavily dependent. About 90% of world fuelwood production comes from the forests of Africa, Asia, Oceania, and Latin America. An increasing quantity of special-quality woods from the tropical forests is cut for export as lumber, however. Southeast Asian countries such as Myanmar and Indonesia now account for much of the world's hardwood log exports (**Figure 12.34**).

The adage about not being able to see the forest for the trees is applicable to those who view forests only for the commercial value of the trees they contain. Forests are more than trees, and timbering is only one purpose that forests serve. Chief among the other purposes are soil and watershed conservation, the provision of a

Figure 12.34   **Logging trucks in Indonesia.** © *Corbis RF.*

habitat for wildlife, and recreation. Forests also play a vital role in the global recycling of water, carbon, and oxygen.

Because forests serve a variety of purposes, the kinds of management techniques employed in any one area depend on the particular use(s) to be emphasized. Thus, if the goal is to maintain a diversity of native plant species in order to provide a maximum number of ecological niches for wildlife, the forest will be managed differently than if it is designed for public recreation or the protection of watersheds. Even if the use to be emphasized is timber production, different management approaches may be taken. Logging techniques for the production of plywood or wood chips, for example, differ from those used for the production of high-quality lumber.

Figure 12.35 **Clear cutting versus selective cutting. (a)** Clear cut logging in Snoqualmie National Forest and on private land contrasts with forests inside Mount Rainier National Park where logging is prohibited. Cutting every tree, regardless of species or size, drives out wildlife, damages watersheds, disrupts natural regeneration, and removes protective ground cover, exposing slopes to erosion. **(b)** Selective cutting in Humboldt County, California. Older, mature specimens are removed at first cutting. Younger trees are left for later harvesting. *(a) © James P. Blair/National Geographic/Getty Images; (b) © Manuel Willequet/Alamy Stock Photo.*

Commercial forests can be considered a renewable resource only if sustained-yield techniques are practiced—that is, if harvesting is balanced by new growth (see the term *maximum sustainable yield* in Chapter 9). Timber companies employ a number of methods of tree harvesting and regeneration. Two quite different practices, clear cutting and selective cutting, illustrate the diversity of such approaches (**Figure 12.35**).

Clear cutting is one of the most controversial logging practices. As the name implies, all the trees are removed from a given area at one time. The site is then left to regenerate naturally or is replanted, often with fast-growing seedlings of a single species. Excessive clear cutting, particularly on steep slopes, destroys wildlife habitats, accelerates soil erosion and water pollution, replaces a mixed forest with a wood plantation of no great genetic diversity, and reduces or destroys the recreational value of the area.

Selective cutting is practiced in mixed-forest stands containing trees of varying ages, sizes, and species. Medium and large trees are cut either singly or in small groups, encouraging the growth of younger trees that will be harvested later. Over time, the forest will regenerate itself. From the point of view of the harvester, selective cutting is not as efficient or economical as clear cutting. Moreover, the practice is often followed very loosely, and the construction of logging roads can still cause extensive damage to the forest.

U.S. National Forests

Roughly one-third of the United States is forested, the same proportion for the world as a whole. Only some 40% of those forests provide the annual harvest of commercial timber. The remaining forests do not contain economically valuable species; are in small, fragmented holdings; are inaccessible; or are in protected areas. Of that 40% of commercial forest land, almost one-half is in 155 national forests owned by the public and managed by the U.S. Forest Service (**Figure 12.36**). Logging by private companies is permitted; timber companies pay for the right to cut designated amounts of timber. Currently, the Forest Service is at the center of debates over how the forests should be managed. Among the issues are methods of harvesting, the cutting of very old tree stands, road building, and rates of reforestation.

By law—the 1960 Multiple Use Sustained Yield Act—the national forests are to be managed for four purposes: recreation, timber production, watershed protection, and wildlife habitat preservation. Although no use is to be particularly favored over others, conservationists charge that the Forest Service increasingly supports commercial logging and that the forests are being cut at an unprecedented rate.

In recent years, billions of board feet of timber have been taken from the national forests. Environmentalists are especially concerned that nearly one-half of this has come from national forests in Oregon and Washington, most of it irreplaceable "old growth." These virgin forests contain trees that are among the tallest and oldest in the world, indeed that were alive when Pilgrims set foot on Plymouth Rock. Although companies plant new seedlings to replace those they cut, the character of an old-growth forest is lost forever. Further, traditional logging practices, including clear cutting, road building, and harvesting after decades, not centuries, of regeneration, prevent the development of a true old-growth forest ecosystem.

Old-growth forests include trees of every age and size, both living and dead. Some ancient trees are immense, capable of

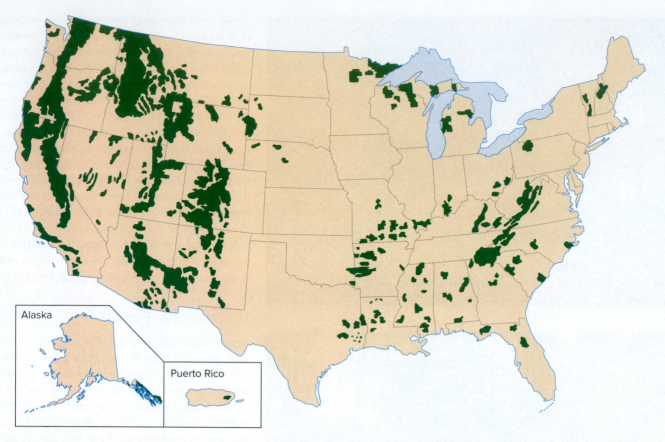

Figure 12.36 **National forests of the United States.** Every year, more than 180,000 hectares (450,000 acres) of trees are cut down within these forests—about 4 square kilometers (2 sq mi) of deforestation every day. To accommodate the cutting, 547,000 kilometers (340,000 mi) of logging roads (10 times the length of the U.S. interstate highway system) had been built by 2001 in formerly pristine areas.

growing 90 meters (300 ft) high, and they may live for more than 1000 years. They include the Douglas fir, Western red cedar, sequoia, and redwood. Tons of dead and decaying logs carpet the forest floor, where, sodden with moisture, they help control erosion and protect the forest from fire. As they decay, the logs release nutrients back into the soil. Such forests provide a habitat for hundreds of types of insects and animals, some of them threatened or endangered species.

The only large expanses of old-growth forest remaining in the United States are in the Pacific Northwest, most of them owned by the federal government. These ancient forests once covered about 60% of the forested areas between the Cascade Mountains and the Pacific Ocean, stretching 3200 kilometers (2000 mi) from California to Alaska. Today, only 6% of the old-growth forests of the Pacific Northwest remain. Controversy broke out in the 1990s when a U.S. District Court ruling halted timber sales on state and federal lands in Washington, Oregon, and northern California to protect the habitat of the northern spotted owl that had recently been added to the endangered species list. A compromise brokered by President Clinton resulted in the 1994 Northwest Forest Plan. The plan set aside old-growth reserves and conservation areas and established limits on annual timber harvests.

It is ironic that many Americans condemn the burning of the tropical rain forests while the U.S. government not only permits

the destruction of forests just as ecologically precious but, in fact, subsidizes that destruction. The federal government usually loses substantial amounts of money on timber sales, because building and maintaining the logging roads costs far more than the timber companies pay for the wood. A study of the U.S. Forest Service's timber program in the 1990s showed annual losses of $333 million per year.

Perhaps the worst abuse has occurred in the Tongass National Forest, North America's largest temperate rain forest, which stretches along 800 kilometers (500 mi) of Alaska's southeastern coastline (**Figure 12.37**). It is thought to be one of the last remaining places in the United States containing every plant and animal species that existed in it before European colonization. A storehouse of biodiversity, the Tongass is home to such threatened species as grizzly bears and bald eagles; its waters teem with aquatic life, from salmon to whales. Clear cutting and road building, however, are endangering the viability of those wildlife habitats. Loggers have selectively focused on the largest, old-growth trees which are the most profitable. The federal government has spent hundreds of millions of dollars to build some 7400 kilometers (4600 mi) of access roads and to promote commercial timbering and has received a pittance in return. Five-hundred-year-old trees 3 meters (10 ft) in diameter and more than 32 meters (105 ft) tall have been sold for $3 each and ground into pulp that is sent overseas and converted into products such as rayon and cellophane.

Figure 12.37   **The Tongass National Forest,** the largest national forest in North America, covers approximately 7 million hectares (17 million acres). The innumerable islands, inlets, and fiords of the Tongass are set against a backdrop of coastal mountains. Since the late 1950s, the federal government has subsidized the timber industry by promising companies a long-term supply of cheap wood from the Tongass. Conservationists contend that old-growth trees are not needed for forest products and that the Tongass should be managed for wildlife, fishing, and tourism, as well as logging. © *Blaine Harrington/age fotostock/Superstock.*

Tropical Rain Forests

It is not only in the United States that government economic policies accelerate the rate of forest destruction. Much of the deforestation occurring in tropical areas also has government sanction. Brazil, Indonesia, and the Philippines are among the countries where governments subsidize projects aimed at converting forests to other uses, such as farming, cattle ranching, and mining. Their economic policies are driven by the pressure of growing population numbers, the need for more agricultural land, expanded demand for fuel and commercial wood, an overseas market for beef that can be satisfied profitably by replacing tropical forest with cleared grazing land, and an increasing demand in China for soybeans, soy oil, and soy meal by-products.

The tropical forests extend across parts of Asia, Africa, and Latin America (**Figure 12.38**). Millions of acres are being completely cleared every year, and almost one-half of their original expanse has already been either cleared or degraded. In Central America and the Caribbean, 70% of the rain forests have disappeared. The Amazon River Basin, a vast region of more than 4 million square kilometers (1.5 million sq mi) extending across part of Brazil and adjacent parts of eight other countries, is thought to contain about one-half of the world's remaining tropical rain forests. Most of the rain forests of Africa now exist mainly in central Africa because those of west Africa, from Sierra Leone eastward to Cameroon, have been largely destroyed. The remaining tropical rain forests are found in the Asia/Pacific region. Here again, the picture is bleak. India, Malaysia, and the Philippines have already lost much of their forests, and rates of deforestation recently have risen sharply in Myanmar, Kampuchea (Cambodia), Thailand, Vietnam, and Indonesia. Overall, it is estimated that nearly one-half of Asia's natural forest is gone.

Although no one doubts that the tropical rain forests are being cleared, there is considerable uncertainty about the rate at which that is happening. The United Nations estimates that on average about 40,000 square kilometers (15,000 sq mi) have been cleared annually in recent years, or slightly less than 1% per year of the remaining forests. Satellite imagery of the Brazilian forests in the Amazon yields a similar figure. In Brazil, this means that, on average, an area roughly the size of Connecticut is deforested each year.

Deforestation in Brazil has become the focus of international attention because it has both the largest area of tropical rain forests *and* one of the highest rates of clearing. No country better illustrates the fact that deforestation is a complicated issue that pits the governments of developing countries against those of industrialized countries, international development organizations against environmental groups, and multinational corporations against advocates of human rights. To critics of its policy of deliberately developing the Amazon Basin, Brazilians respond that it is justified because it relieves high population densities in its crowded northeast; gives farmers and ranchers land on which to build a livelihood; permits the country to tap unexploited natural resources; and helps Brazil repay its huge foreign debt. Pointing out that most of the forests in Western Europe and the United States were cleared decades if not

Figure 12.38 **The world's tropical rainforests.** Tropical rain forests exist in tropical latitude regions with high temperatures and high levels of precipitation year-round. The Amazon River Basin has the world's largest continuous area of rain forests. Three countries contain more than one-half of the tropical rain forests: Brazil, the Democratic Republic of Congo, and Indonesia. Large tracts are being cleared to make way for farming, cattle ranching, commercial logging, and development projects.

| 1975 | 1989 | 2001 |

Figure 12.39 **Deforestation in the Amazon Basin between 1975 and 2001.** These satellite images of a portion of the state of Rondônia, Brazil, near the city of Ariquemes reveal the effect of the systematic destruction of the tropical rain forests. In 1975, the area was virtually untouched. Typically, clearing of the forest vegetation began along roads and then fanned out to create the "fishbone" pattern evident in the 1989 image. By 2001, the extent of deforestation had more than doubled. The fastest and cheapest way to clear areas for farms and ranches is to burn them. About the size of Oregon, Rondônia is the Brazilian state with the largest percentage of its forest cover destroyed by fire, although the states of Mato Grosso and Pará also show significant deforestation. Likened by some to an environmental holocaust, the fires generate hundreds of millions of tons of gases that contribute to global warming and depletion of the Earth's protective ozone layer. *Source: United Nations Enviroment Programme.*

centuries ago, and that subsequent resource exploitation led to wealth and prosperity, Brazilians ask why they shouldn't be able to use their natural resources in what they deem to be their own best interest.

There are good reasons, however, that North Americans should care what happens to the tropical rain forests. Their destruction raises three principal global concerns and a host of local ones. First, all forests play a major role in maintaining the oxygen and carbon balance of the Earth. People and their industries consume oxygen; vegetation both extracts the carbon from atmospheric carbon dioxide and releases oxygen back into the atmosphere. Indeed, the forests of the Amazon have been called the "lungs of the world" for their major contribution to the oxygen breathed by humankind. When the tropical forest is cleared, its role both as a carbon "sink" and as an oxygen replenisher is lost.

A second global concern is the contribution of forest clearing to air pollution and climate change. Deforestation by burning

releases vast quantities of carbon dioxide into the atmosphere. Brazilian scientists estimate that the thousands of fires that are set to clear the Amazon forest account for one-tenth of the global production of carbon dioxide, contributing to the warming of the atmosphere (**Figure 12.39**). In addition, the fires generate gases (nitrogen oxides and methane) that create acid rain and contribute to the depletion of the ozone layer, topics discussed in Chapter 13.

Finally, the eradication of tropical forests is already leading to the loss of a major part of the biological diversity of the planet. The forests are one component in an intricate ecosystem that has developed over millions of years. The trees, vines, flowering plants, animals, and insects depend on one another for survival. The destruction of the habitat by clearance annually causes the extinction of thousands of plant and animal species that exist nowhere else. Although the tropical rain forests now occupy less than 10% of

TROPICAL FORESTS AND MEDICAL RESOURCES

Tropical forests are biological cornucopias, possessing a stunning array of plant and animal life. Costa Rica, about the size of South Carolina, contains as many bird species as all of North America, more species of insects, and nearly one-half the number of plant species. Scientists have identified 1300 butterfly species in Peru's rain forest. One stand of rain forest in Kalimantan (formerly Borneo) contains more than 700 species of tree, as many as exist in North America. Seventy-two species of ant inhabit a single tree species in Peru, dependent on it for food and shelter and providing, in return, protection from other insects.

The tropical forests yield an abundance of chemical products used to manufacture alkaloids, steroids, anesthetics, and other medicinal agents. Indeed, one-half of all modern drugs, including strychnine, quinine, curare, and ipecac, come from the tropical forests. A single flower, the Madagascar periwinkle, produces two drugs used to treat leukemia and Hodgkin's disease.

As significant as these and other modern drugs derived from tropical plants are, scientists believe that the medical potential of the tropical forests remains virtually untapped. They fear that deforestation will eradicate medicinal plants and traditional formulas before their uses become known, depriving humans of untold potential benefits that may never be realized. Tribal peoples make free use of plants of the rain forest for such purposes as treating stings and snakebites, relieving burns and skin fungi, reducing fevers, and curing earaches; yet botanists have only recently begun to identify tropical plants and study traditional herbal medicines to discover which of them might contain medically important compounds.

the Earth's land surface, they are thought to contain anywhere from 50% to 70% of all the species of plants, animals, and microorganisms in the world. Many of the plants have become important world staple food crops, among them rice, corn, cassava, squash, banana, pineapple, and sugarcane. Unknown additional potential food species remain as yet unexploited. In addition, the tropical forests yield an abundance of industrial products (oils, gums, latexes, and turpentines) and are the world's main storehouse of medicinal plants (see "Tropical Forests and Medical Resources").

Deforestation also incurs heavy environmental, economic, and social costs on a more local basis. All forests anchor topsoil and absorb excess moisture. In a vicious cycle, forest clearance accelerates soil erosion and siltation of streams and irrigation channels, leaving the area vulnerable to flooding and drought, leading in turn to future shortages of food and fuelwood. Within a matter of years, land that has been cleared for agriculture can become unsuitable for that use. In the Himalayan watershed, in the Ethiopian highlands, and in numerous other places, deforestation, erosion, and rainfall runoff have aggravated floods that have killed tens of thousands of people and left millions of others homeless.

12.7 Resource Management

The destruction of the rain forests is a tragedy that yields no long-term benefits. The world is approaching the end of a period in which resources were cheap, readily available, and lavishly used. For the first two centuries of industrialization, the Earth was viewed as an almost inexhaustible storehouse of resources for humans to exploit and, simultaneously, as a vast repository for the waste products of society. Now there is a growing realization that resources can be depleted, even renewable ones, such as forests; and that the air and water—also resources—cannot absorb massive amounts of pollutants yet retain their life-supporting abilities.

That realization was reflected in June 1992, at the Earth Summit in Rio de Janeiro, when the world's governments agreed to form the UN Commission on Sustainable Development. Since that time, more than 70 countries, including the United States, have launched efforts to chart a path toward **sustainable development**, which is generally defined as development that satisfies current needs without jeopardizing the ability of future generations to meet their needs. The principles of sustainability are straightforward. The sustainable use of renewable resources means using them at rates within their capacity for regeneration. Over the long term,

- soil erosion cannot exceed soil formation
- forest destruction cannot exceed forest regeneration
- species extinction cannot exceed species evolution
- fish catches cannot exceed the regenerative capacity of fisheries
- pollutants cannot exceed the capacity of the system to absorb them

A society can violate the principles of sustainability in the short run but not in the long run and still endure.

Defining *sustainable development* is easy, but achieving it will be difficult. The implementation of sustainable development policies will require, among other things, educating the public about the need for such policies; forging agreement on policy initiatives among government leaders, the business community, and environmentalists; providing financial incentives to resource users to practice sustainable development; and ensuring consistency in government implementation of agreements. Developed countries are in the best position to promote sustainable development, because they are wealthy enough to invest in the necessary research and technology. At the same time, they will have to recognize that developing countries (1) see the consumption (some would say overconsumption) of material resources by developed countries as in large measure responsible for the dwindling supplies of many of those resources and (2) do not want to be told they cannot follow the path toward economic development and prosperity that other countries already have taken.

Moving toward the wise management of resources is not impossible, however. It entails three strategies: conservation, reuse, and substitution. By **conservation** we mean the careful use of resources so that future generations can obtain as many benefits from them as we now enjoy. It includes decreasing our consumption of resources, avoiding their wasteful use, and preserving their quality. Thus, soils can be conserved and their fertility maintained by contour plowing, crop rotation, and a variety of other practices. Properly managed, forests can be preserved even as their resources are tapped.

Opportunities to reduce the consumption of energy resources are many and varied. Nearly everything can be made more energy-efficient. Motor vehicles use a significant portion of the world's oil output. Doubling their fuel efficiency by reducing vehicle weight and using more-efficient engines and tires would save at least 20% of the world's total annual oil output. Industries have an enormous potential for saving energy by using more-efficient equipment and processes. The Japanese steel industry, for example, uses one-third less energy to produce a ton of steel than does that industry in most other countries. Energy used for heating, cooling, and lighting homes and office buildings could be reduced by half if they were properly constructed and furnished. Switching to compact fluorescent, halogen, or light emitting diode (LED) light-bulbs reduces energy use by 25% to 80% compared to traditional incandescent bulbs.

The *reuse* of materials also reduces the consumption of resources. Instead of being buried in landfills, waste can be burned or decomposed and fermented to provide energy (**Figure 12.40**). Recycling of steel, aluminum, copper, glass, and other materials can be greatly increased, not only to recover the materials themselves but also to recoup the energy invested in their production. It takes only 5% as much electricity to make aluminum from scrap as from raw materials. In other words, manufacturers can make 20 cans out of recycled material with the same energy it takes to make a single can out of new material. See "Source Reduction and Recycling" in Chapter 13 (pp. 408 for a further discussion of this topic.

The *substitution* of other materials for nonfuel minerals in short supply can extent the lifetimes of their reserves. Energy supplies are a more pressing challenge. The high standard of living enjoyed by the developed countries has been built upon fossil fuels which are nonrenewable and finite in quantity. Even our food supply is highly dependent upon fossil fuels, primarily oil and gas. If the technology to exploit them economically can be developed, coal and oil shale could supply energy needs well into the future. However, consuming all of the earth's fossil fuels could spell disastrous global warming. In addition, the renewable energy resources, such as biomass, solar, and geothermal power, are virtually infinite in their amount and variety. Although no single renewable source is likely to be as important as oil or gas, collectively they could make a significant contribution to energy needs.

Summary of Key Concepts

- Natural resources can be classified as renewable, those that can be regenerated in nature as fast as or faster than societies exploit them, and nonrenewable, those that are generated so slowly that they are thought of as existing in finite amounts. The proved reserves of a resource are the amounts that have been identified and that can be extracted profitably.

- Advanced industrial countries depend heavily on resources derived from nonrenewable minerals, chiefly the fossil fuels crude oil, natural gas, and coal, all of which are unevenly distributed. Some countries receive more than one-half of their electricity from nuclear power plants, while others have none.

- Renewable natural resources are more widely and evenly distributed than the nonrenewable ones. Wood and other forms of biomass are the primary source of energy for more than one-half of the world's people. Hydropower is the leading source of renewable electricity. The availability of renewable resources, including hydropower, solar power, geothermal energy, and wind power, depends upon factors of physical geography such as river flow rates, topography, sunshine, plate boundaries, and wind speeds.

- The nonfuel mineral resources from which people fashion metals, glass, stone, and other products are nonrenewable. Some exist in vast amounts, others in relatively small quantities; some are widely distributed, others concentrated in just a few locations.

- Human activities have had, and continue to have, a severe impact on wetlands and forests, both of which play vital ecological roles.

- The growing demand for resources, induced by population growth and economic development, strains the Earth's supply of raw materials. The wise and careful management of natural resources of all types involves conservation, reuse, and substitution.

Figure 12.40 **Cardboard ready for recycling.** Cardboard can be recycled into a variety of new cardboard or paper products. Recycling reduces the volume dumped in landfills and conserves energy and forest resources. © *Eric Raptosh Photography/Blend Images/Getty Images RF.*

Key Words

biomass fuel 349	nonrenewable resource 336
conservation 371	nuclear fission 347
desertification 360	nuclear fusion 348
energy 338	oil sands 346
estuarine zone 364	oil shale 346
geothermal energy 352	ore 357
hydraulic fracturing 341	perpetual resource 336
hydropower 350	photovoltaic (PV) cell 352
liquefied natural gas (LNG) 344	potentially renewable resource 336

Thinking Geographically

1. What is the basic distinction between a *renewable* and a *nonrenewable* resource? Why do estimates of *proved reserves* vary over time?

2. Why are energy resources considered the most essential of all natural resources? What is the relationship between energy consumption and industrial production? Briefly describe historical energy consumption patterns in the United States.

3. Why has oil become the dominant form of commercial energy? Which countries are the main producers of crude oil? Why is it difficult to predict how long proved reserves of oil are likely to last?

4. What is hydraulic fracturing and how has it changed the geographic patterns of energy production? What are its advantages and disadvantages?

5. Since 2000, the proportion of U.S. energy supplied by coal has decreased while the proportion supplied by natural gas has increased. What is causing this change? What are the benefits of the switch from coal to natural gas?

6. What are the different methods of generating nuclear energy? Why is there public opposition to nuclear power?

7. Which are the most widely employed ways of using renewable resources to generate energy? What are the advantages and disadvantages of using such resources? What sorts of physical environments and locations are best suited to each of the renewable energy sources?

8. What, in general, are the leading mining countries? What role do developing countries play in the production of critical raw materials? How have producing countries reacted to the threatened scarcity of copper?

9. Since soil erosion is a natural process, why is it of concern? What are some commonly used methods of reducing erosion? Under what conditions and in what types of areas do desertification and salinization occur?

10. What are some types of wetlands? Why are they important, and why have so many disappeared?

11. What vital ecological functions do forests perform? Where are the tropical rain forests located, and what concerns are raised by their destruction?

12. Discuss three ways of reducing demands on resources.

CHAPTER 13

Human Impact on the Environment

Bulldozers compact municipal solid waste in a California landfill. Residents of the United States threw out 230 million metric tons (254 million tons) of household solid waste in 2013, or about 2.0 kilograms (4.4 lbs) of solid waste per person per day. Fortunately, a growing proportion of that waste is recycled. © *Al Franklin/Corbis Premium RF/Alamy Stock Photo RF.*

CHAPTER OUTLINE

LEARNING OBJECTIVES

After studying this chapter you should be able to:

13.1 Use the IPAT equation to explain the different factors that influence a society's impact on the environment.

13.2 Describe how human impacts on the environment change with rising wealth.

13.3 List the three components of the biosphere and describe how are they interrelated.

13.4 Given that the supply of water on Earth is constant, explain how human activities can result in depleted water supplies.

13.5 Summarize the major sources of air and water pollution.

13.6 Summarize the different threats to plant and animal biodiversity.

13.7 Describe the different types of wastes generated by advanced societies and how each type is best managed.

Late in the summer of 2014 nearly half a million residents of Toledo, Ohio area were told not to drink their tap water or even use it for cooking. A toxin produced by an algal bloom had entered the city's water intakes in Lake Erie. The toxins were produced by *Microcystis aeruginosa*, a species of cyanobacteria (also known as blue-green algae). Cyanobacteria are photosynthetic aquatic bacteria that grow in dense colonies in warm, nutrient-rich waters. During a bloom, ponds and lakes resemble pea soup or a layer of green paint floating atop the water (**Figure 13.1**). Ingesting or

contacting water containing the microsystin toxin can cause skin rashes, vomiting, nausea, dizziness, diarrhea, and numbness. In rare cases it can cause liver, cardiac, or respiratory failure.

Lake Erie is the shallowest of the Great Lakes, making it more vulnerable to human impacts since it warms more quickly and has less volume to dilute pollutants. In recent years, harmful algal blooms have become more frequent in Lake Erie. There is a sad, double irony to Lake Erie's troubles. The Great Lakes are among the world's greatest freshwater resources and are considered well positioned to handle the challenges of a warming climate. Furthermore, Lake Erie was considered one of the greatest environmental success stories. It was notorious for its pollution in the 1960s and 1970s when it was declared "dead." But, the Clean Water Act, international agreements between the United States and Canada, bans on phosphorus in detergents, and the construction of municipal wastewater treatment systems led to marked water quality improvements. However, while cities and industries cleaned up their discharges, the agricultural runoff that caused recent algal blooms went unaddressed.

The harmful algal blooms in Lake Erie are a harbinger of what the future holds in a global environment that has been highly altered by human activities. China, Brazil, and many other countries experience harmful algal blooms. On the Atlantic, Pacific, and Gulf coasts of North America, red tide algal blooms have sickened humans and killed fish, sea birds, dolphins, manatees, sea otters, and other marine mammals. During a red tide event, smelly piles of biomass accumulate on beaches and humans are advised to avoid to stay indoors. Commercial and recreational fishing and shellfish harvesting are shut down to avoid the risk of paralytic or diarrheic poisoning. Closures of shellfish beds, fisheries, and beaches have enormous economic impacts; a single red tide event on the Gulf of Mexico coast of Florida is estimated to result in a loss of $20 million in economic activity.

Globally, the frequency, duration, and geographic extent of harmful algal blooms are all on the increase. While toxic algal species are naturally occurring and blooms depend upon particular

(a)   (b)

Figure 13.1   **The toxic algal blooms** shown here on satellite imagery are increasingly common in Lake Erie. The bright green areas on the image are caused by a green algal scum on the water surface. Harmful algal blooms like these make the water unsafe for drinking or recreation and can cause fish kills. The left image was acquired by the National Aeronautics and Space Administration (NASA) on October 11, 2011, the right close-up image on August 4, 2015. *(a) Source: NASA image courtesy Jeff Schmaltz, MODIS Rapid Response Team at NASA GSFC; (b) Source: NASA Earth Observatory images by Joshua Stevens, using Landsat data from the U.S. Geological Survey.*

wind conditions and water currents, the human factor is considerable. Essentially, harmful algal blooms are the result of an over-fertilized world. The primary culprit is phosphorus or nitrogen in runoff from agricultural fields, livestock manure, leaking septic systems, and urban wastewaters. In North America, it is the commercial agriculture described in Chapter 9 that is responsible for most of the excess nutrients causing the blooms. Exacerbating the problem, climate warming increases the growth rate of harmful blooms and expands the geographic range of toxic species. For example, the massive toxic bloom of *Pseudo-nitzschia* in 2015 that lead to shellfish bed closures from Alaska to California was blamed on unusually warm waters. The growing number of harmful algal blooms is just one of many illustrations of how human actions affect the quality of the environment on which all life depends.

Terrestrial features and ocean basins, elements of weather and characteristics of climate, flora, and fauna constitute the building blocks of that complex mosaic called the *environment,* or the totality of things that in any way may affect an organism. Plants and animals, landforms, soils and nutrients, weather, and climate all constitute an organism's environment. The study of how organisms interact with one another and with their physical environment is called **ecology**. It is critically important in understanding environmental problems, which usually arise from a disturbance of the natural systems that make up our world.

Humans exist within a natural environment that they have modified by their individual and collective actions. Forests have been cleared, grasslands plowed, dams built, and cities constructed. On the natural environment, then, has been erected a cultural environment, modifying, altering, or destroying the natural communities and processes that existed before human impact was expressed. This chapter is concerned with the interrelation between humans and the natural environment that they have so greatly altered.

Since the beginning of agriculture, humans have changed the face of the Earth, have distorted delicate interrelationships within nature, and, in the process, have both enhanced and endangered the societies and the economies that they have erected. The essentials of the natural balance and the ways in which humans have altered it are not only our topics here but also matters of social concern that rank among the principal domestic and international issues of our time. As we shall see, the fuels we consume, the raw materials we use, the products we create, and the wastes we discard all contribute to the harmful alteration of the **biosphere**, the thin film of air, water, and Earth within which we live.

13.1 Ecosystems

The biosphere is composed of three interrelated parts:

1. the *troposphere,* the lowest layer of the Earth's atmosphere, extending about 9.5–11.25 kilometers (6–7 mi) above the ground

2. the *hydrosphere,* including surface and subsurface waters in oceans, streams, lakes, glaciers, or groundwater—much of it locked in ice or Earth and not immediately available for use
3. the *lithosphere,* containing the soils that support plant life, the minerals that plants and animals require to exist, and the fossil fuels and ores that humans exploit.

The biosphere is an intricately interlocked system, containing all that is needed for life, all that is available for living things to use, and, presumably, all that ever will be available. The ingredients of the biosphere must be, and are, constantly recycled and renewed in nature. Plants purify the air, the air helps purify the water, and the water and the minerals are used by plants and animals and are returned for reuse.

The biosphere, therefore, consists of two intertwined components: (1) a nonliving outside energy source—the sun—and requisite chemicals and (2) a living world of plants and animals. In turn, the biosphere may be subdivided into specific **ecosystems**, self-sustaining units that consist of all the organisms (plants and animals) and physical features (air, water, soil, and chemicals) existing together in a particular area. The most important principle concerning all ecosystems is that everything is interconnected. Any intrusion or interruption inevitably results in cascading effects elsewhere in the system. Each organism occupies a specific *niche,* or place, within an ecosystem. In the energy exchange system, each organism plays a definite role; individual organisms survive because of other organisms that also live in that environment. The problem lies not in recognizing the niches but in anticipating the chain of causation and the readjustments of the system consequent on disturbing the occupants of a particular niche.

Life depends on the energy and nutrients flowing through an ecosystem. The transfer of energy and materials from one organism to another is one link in a **food chain**, a sequence of organisms, such as green plants, herbivores, and carnivores, through which energy and materials move within an ecosystem (**Figure 13.2**). Most food chains have three or four links, although some have only two—for example, when human beings eat rice. Because the ecosystem in

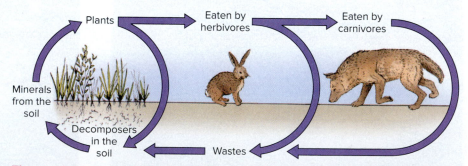

Figure 13.2 **The supply of food in an ecosystem is a hierarchy of "who eats what"—a hierarchy that creates a food chain.** In this simplified example, green plants are the *producers* (autotrophs), using nutrients and energy from the sun to make their own food. Herbivorous rabbits (*primary consumers*) feed directly on the plants, and carnivorous foxes (*secondary consumers*) feed on the rabbits. A food chain is one thread in a complex *food web,* all the feeding relationships that exist in an ecosystem. For example, a mouse might feed on the plants shown here and then be eaten by a hawk, another food chain in this food web.

nature is in a continuous cycle of integrated operation, there is no start or end to a food chain. There are, simply, nutritional transfer stages in which each lower level in the food chain transfers part of its contained energy to the next higher-level consumer.

The *decomposers* pictured in Figure 13.2 are essential in maintaining food chains and the cycle of life. They cause the disintegration of organic matter—animal carcasses and droppings, dead vegetation, waste paper, and so on. In the process of decomposition, the chemical nature of the material is changed, and the nutrients contained within it become available for reuse by plants or animals. *Nutrients,* the minerals and other elements that organisms need for growth, are never destroyed; they keep moving from living to nonliving things and back again. Our bodies contain nutrients that were once part of other organisms, perhaps a hare, a hawk, or an oak tree.

Ecosystems change constantly whether people are present or not, but humans have affected them more than has any other species. The impact of humans on ecosystems was small at first, with low population size, energy consumption, and technological levels. It has increased so rapidly and pervasively as to present us with widely recognized and varied ecological crises. The **IPAT equation** is a helpful way of summarizing the different factors influencing the degree of human impact on the environment. The formula is written as

$$I = PAT$$

where:

   I = Impact on the environment
   P = Population
   A = Affluence or standard of living
   T = Technology

The IPAT equation shows that growing populations and rising standards of living both contribute to greater strain on the environment. Technologies, however, can increase or decrease that impact. For example, with rising standards of living often come preferences for a cleaner environment; thus, pollution controls can be added to reduce environmental impacts. Unfortunately, some changes aimed at improving the quality of the local environment, such as switching to taller smokestacks, come at the expense of the environment elsewhere (**Figure 13.3**). As awareness of environmental pollution has increased in developed countries, there has also been a tendency to shift polluting industries and waste disposal to poorer regions. Persons in poorer regions often have less control over the environment where they live and work and thus may be disproportionately exposed to the negative consequences of environmental pollution. The environmental justice movement (see "Environmental Justice", pp. 407) has arisen to demand a clean, healthy environment for everyone. Each of the elements in the IPAT equation relates to topics in geography such as population, economic, and cultural geography. Thus, understanding human impacts on the environment requires knowledge of both human and physical geography. Some of the negative effects of humans on the natural environment are the topic of the remainder of this chapter.

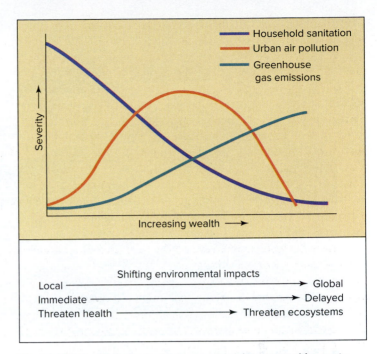

Figure 13.3 **Standard of living and environmental impacts.** For the world's poor, the most significant environmental issues are clean water, sanitation, and smoke from cooking fires. As standards of living increase, communities can afford water and sewage treatment systems, but also increase their consumption of raw materials, synthetic chemicals, and fossil fuel energy. This shifts environmental problems from local, immediate threats to human health to longer-term, delayed, global-scale impacts such as destruction of the ozone layer, acid precipitation, and global climate change.

13.2 Impacts on Water

The world's supply of water is constant, and most of it is found in the oceans, where it is not available for direct human consumption. The system by which it continuously circulates through the biosphere is called the **hydrologic cycle** (**Figure 13.4**). In that cycle, water changes form and is purified or distilled through evaporation and condensation so that it is available with appropriate properties to the ecosystems of the Earth. *Evaporation, transpiration* (the emission of water vapor from living things), and *precipitation* are the mechanisms by which water is redistributed. Water vapor collects in clouds, condenses, and then falls again to the Earth. There it is reevaporated and retranspired, only to fall once more as precipitation.

People's dependence on water has long led to efforts to control its supply. Such manipulation has altered the quantity and quality of water in rivers and streams.

Availability of Water

Globally, water is abundant. Enough rain and snow fall on the continents each year to cover the Earth's total land area with 83 centimeters (33 in.) of water. It is usually reckoned that the

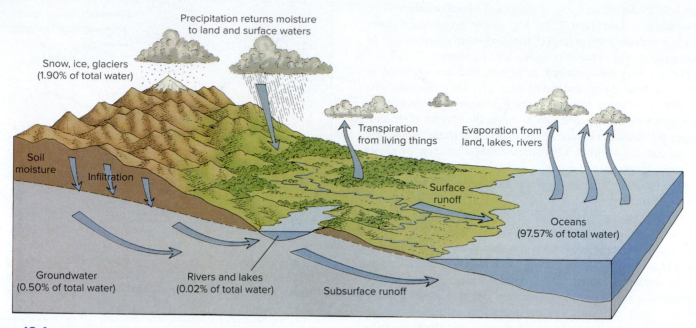

Precipitation returns moisture
to land and surface waters

Snow, ice, glaciers
(1.90% of total water)

Transpiration
from living things

Evaporation from
land, lakes, rivers

Soil
moisture

Infiltration

Surface
runoff

Oceans
(97.57% of total water)

Groundwater
(0.50% of total water)

Rivers and lakes
(0.02% of total water)

Subsurface runoff

Figure 13.4   **The hydrologic cycle.** The sun provides energy for the evaporation of fresh and ocean water. The water is held as vapor until the air becomes supersaturated. Atmospheric moisture is returned to the Earth's surface as solid or liquid precipitation to complete the cycle. Because precipitation is not uniformly distributed, moisture is not necessarily returned to areas in the same quantity as it has evaporated from them. The continents receive more water than they lose. The excess returns to the seas as surface water or groundwater. A global water balance, however, is always maintained.

volume of fresh water annually renewed by the hydrologic cycle could meet the needs of a world population 5 to 10 times its present size.

In many parts of the world, however, water supplies are inadequate and dwindling. The problem is not with the global amount of water but with its geographic distribution, reliability (the variability of precipitation from year to year), and quality. Regional water sufficiency is also a function of the size of the population using the water and the demands it places on the resource. For the world as a whole, agriculture accounts for about 80% of freshwater use (**Figure 13.5**). Industry uses about 20%, and household and municipal use (water for drinking, bathing, watering lawns, and so on) accounts for the remainder. Irrigation expands the amount of arable land and increases yields on existing farmland. Irrigated agriculture produces about 40% of the world's harvests on about 15% of its cropland. Since 1950, the acreage of land under irrigation has tripled, as have withdrawals of fresh water from streams, lakes, aquifers, and other sources. (An *aquifer* is a zone of water-saturated sands and gravels beneath the Earth's surface; the water it contains is called *groundwater,* in contrast to surface waters, such as rivers and lakes.)

Scarcity is the word increasingly used to describe water supplies in parts of both the developed and developing world (**Figure 13.6**). About 2 billion people live in regions facing a scarcity of water today, and an estimated 3.5 billion will live in water-scarce regions by 2025. Insufficient water for irrigation periodically endangers crops and threatens famine; permanent streams have become intermittent in flow; lakes are shrinking; and from throughout the world come reports of rapidly falling water tables and wells that have gone dry. According to the World Bank, chronic water shortages that

Figure 13.5   **Irrigation sprinklers in the Central Valley of California.** Expanded irrigation could increase crop yields to feed a growing world population. Unfortunately, this would stress scarce water supplies. During spray irrigation, as shown in the photo, much water is lost to evaporation. *Drip irrigation,* which delivers water directly to plant roots through small perforated tubes laid across the field, is one method of reducing water consumption. © *Corbis RF.*

threaten to limit food production, economic development, sanitation, and environmental protection already plague 80 countries. Ten countries in North Africa and the Middle East actually run a *water deficit:* they consume more than their annual renewable supply, usually by pumping groundwater faster than it is replenished by rainfall.

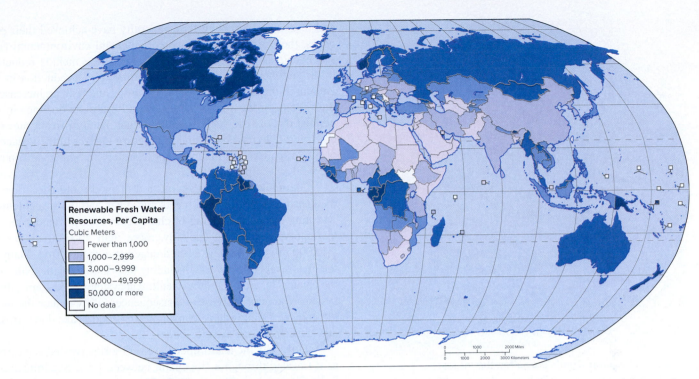

Figure 13.6   Availability of renewable water per capita. Renewable water resources are those available from streams, lakes, and aquifers. Although the United States has tremendous quantities of fresh water, it also uses tremendous quantities. Water consumption rises with population growth, increases in the standard of living, expansion of irrigation farming, and the enlarged demands of industry and municipalities that come with development. In a growing number of world areas, water is scarce, constraining sustainable development and requiring hard choices in the allocation of water among competing users. *Source: Sutton,* Student Atlas of World Geograpy, *8e, Map 79, p. 101.*

National data can mask water scarcity problems at the local level. A number of countries have major crop-producing regions where groundwater overpumping and aquifer depletion have led to serious water shortages and restricted supplies.

- In northern China, for example, irrigated farmland and urban and industrial growth are so depleting water supplies that for much of the year the Huang He (Yellow River) runs dry in its lower reaches before arriving at the Yellow Sea. More than 100 of China's large cities, most of them in the north, already have serious water shortages. Across much of northern China, water tables are dropping an average of 1–2 meters (3–6 ft) each year.
- In India, the world's second most populous country, the overall rate of groundwater withdrawal is twice the rate of recharge. In several northern states, including Haryana and Punjab, the country's most important agricultural regions, excessive water use is causing water tables to fall and wells to dry up.
- Lake Chad has been severely reduced in size by irrigation withdrawals for agriculture, poor management, and drought. Once Africa's second-largest lake, with a surface area of 25,000 square kilometers (9650 sq mi) in 1965, Lake Chad has shrunk by more than 90%.
- Lake Chapala, the largest body of fresh water in Mexico, has lost about 80% of its intake water since the mid-1970s. The lake is fed primarily by the Rio Lerma, but almost all of the river's flow is now diverted for irrigation and industry. Mexico City

and its suburbs extract water from their aquifers more than twice as fast as nature can replenish the water. As a result of removing so much groundwater, the city sank 9 meters (30 ft) during the 20th century.

Scientists using sophisticated satellites have discovered that groundwater supplies are being depleted across much of the western United States as pumping exceeds replenishment. Demands upon the only significant source of surface water in the southwestern United States, the Colorado River, are so great that the stream often reaches the ocean (at the Gulf of California) as little more than a briny trickle. Aqueducts and irrigation canals siphon off its water for use in seven western states and northern Mexico. A number of western cities, including Denver and Santa Fe, have imposed mandatory restrictions on water use.

The country's largest underground water reserve, the Ogallala aquifer, which supplies about 25% of all U.S. irrigated land, is being depleted three times faster than it is being replenished (**Figure 13.7**). Stretching from South Dakota to west Texas, the aquifer supports nearly one-half of the country's cattle industry, one-fourth of its cotton crop, and a great deal of its corn and wheat. More than 150,000 wells now puncture the aquifer, pumping water for irrigation, industry, and domestic use.

Shortages of water pit developers against environmentalists, and users—agriculture, industry, and municipalities—against one another. Tensions are exacerbated in **transboundary river basins**—areas where two or more countries share a river system.

Figure 13.7 The giant Ogallala aquifer, the largest underground water supply in the United States, provides about one-fourth of all groundwater used for irrigation in the country. Depletion of the aquifer is most severe in its southern portion, where the Ogallala gets little replenishment from rainfall. Water tables have declined by 32 meters (100 ft) or more in many parts of north Texas. The capacity of the aquifer has decreased by about one-third since 1950.

The world's 263 transboundary river basins cover an estimated 40% of the Earth's land area and are home to 60% of the world's population. Rivers where water conflict is a distinct possibility include the Indus, Jordan, Nile, and Tigris-Euphrates. Mexico, for example, is angered by U.S. depletion of the Colorado River before it reaches the international border; and by building dams and irrigating fields, Turkey and Syria have taken much of the water of the Tigris and Euphrates Rivers, reducing the amount of water available in downstream Syria and Iraq. India's diversion of a large portion of the Ganges River has left Bangladesh with water shortages.

Because the oceans contain a never-ending supply of water, some have looked to *desalination* (cleansing seawater of its salt and mineral content) to provide a technological solution to water short-fall. Desalination, however, requires large quantities of energy and is cost-prohibitive. Therefore, it is unlikely to supply more than a small portion of humanity's needs. If in the future the costs fall far enough, desalination could help augment domestic water supplies; it is doubtful that it could ever become cheap enough to provide water for agriculture.

Modification of Streams

To prevent flooding, to regulate the water supply for agriculture and urban settlements, or to generate power, for thousands of years people have manipulated rivers by constructing dams, canals,

and reservoirs. Although they generally have achieved their purposes, these structures can have unintended environmental consequences, as discussed in Chapter 12. These include reduction in the sediment load downstream, followed by a reduction in the amount of the nutrients available for crops and fish; an increase in the salinity of the soil; subsidence; and dramatic declines in the populations of fish, such as salmon, that hatch in freshwater rivers, migrate to the ocean, and then return upriver to their place of origin to spawn (see "Blueprint for Disaster: Stream Diversion and the Aral Sea," pp. 382–383).

Channelization, another method of modifying river flow, is the construction of embankments and dikes and the straightening, widening, and/or deepening of channels to control floodwaters or to improve navigation. Globally, more than 500,000 kilometers (310,000 mi) of river have been dredged and channelized. Many of the great rivers of the world, including the Mississippi, the Nile, and the Huang He, are lined by embankment systems. Like dams, these systems can have unforeseen consequences. They reduce the natural storage of floodwaters, can aggravate flood peaks downstream, and can cause excessive erosion.

Until 1960, the Kissimmee River in Florida twisted and turned as it traveled through a floodplain between Lake Kissimmee and Lake Okeechobee (**Figure 13.8**). The habitat supported hundreds of species of birds, reptiles, mammals, and fish. At the request of developers, ranchers, and farmers who had moved onto the wetlands and were disturbed by the tendency of the river to flood, the Army Corps of Engineers dredged the stream, turning 166 kilometers (103 mi) of meandering river into a straight, dirt-lined canal only 90 kilometers (56 mi) long. After the completion of the canal in 1971, the wetlands disappeared, alien plant species moved in, fish populations declined drastically, and 90% of the waterfowl, including several endangered species, vanished.

Channelization and dam construction are deliberate attempts to modify river regimes, but other types of human action also affect river flow. Urbanization, for example, has significant hydrologic impacts, including a lowering of the water table, pollution, and increased flood runoff. Likewise, the removal of forest cover increases runoff, promotes flash floods, lowers the water table, and hastens erosion.

Nevertheless, the primary adverse human impact on water is felt in the area of water quality. People withdraw water from lakes, rivers, or underground deposits to use for drinking, bathing, agriculture, industry, and many other purposes. Although the water that is withdrawn returns to the water cycle, it is not always returned in the same condition as it was at the time of withdrawal. Water, like other segments of the ecosystem, is subject to serious problems of pollution.

Water Quality

As a general definition, **environmental pollution** by humans means the introduction into the biosphere of wastes that, because of their volume, their composition, or both, cannot be readily disposed of by natural recycling processes. In the case of water, the central idea is that pollution exists when water composition has been so modified by the presence of one or more substances

(a) Before 1960

(b) After 1972

(c)

Figure 13.8 **The Kissimmee River** in Florida (**a**) before and (**b**) after the Army Corps of Engineers turned it into a straight canal 9 meters (30 ft) deep. The channelization largely eliminated wetlands, degraded bird and fish populations, and contributed to the deterioration of Lake Okeechobee and the Everglades. The federal government and the state of Florida have embarked on a long-term project to reestablish a natural flow of water to portions of the river by returning it to a meandering path. *(a) and (b) South Florida Water Management District.*

that either it cannot be used for a specific purpose or it is less suitable for that use than it was in its natural state. Pollution is brought about by the discharge into water of substances that cause unfavorable changes in its chemical or physical nature or in the quantity and quality of the organisms living in the water. *Pollution* is a relative term. Water that is not suitable for drinking may be completely satisfactory for cleaning streets. Water that is too polluted for fish may provide an acceptable environment for certain water plants.

Human activity is not the only cause of water pollution. Leaves that fall from trees and decay, animal wastes, oil seepages, and other natural phenomena may affect water quality. There are natural processes, however, to take care of such pollution. Organisms in water are able to degrade, assimilate, and disperse such substances in the amounts in which they naturally occur. Only in rare instances do natural pollutants overwhelm the cleansing abilities of the recipient waters. What is happening now is that the quantities of wastes discharged by humans often exceed the ability of a given body of water to purify itself. In addition, humans are introducing pollutants, such as metals or synthetic organic substances, that take a very long time to break down or cannot be broken down at all by natural mechanisms.

As long as there are people on Earth, there will be pollution. Thus, the problem is one not of eliminating pollution but of controlling it. Such control can be a matter of life or death. Waterborne pathogens and pollution kill millions of people (mostly children) each year. The people die from diarrhea and water-related diseases, such as cholera and typhoid. One column in Appendix 3 at the back of this book shows, by country, the percentage of the population with access to safe water.

The four major contributors to water pollution are agriculture, industry, mining, and municipalities and residences. It is helpful to distinguish between "point" and "nonpoint" sources of pollution. As the name implies, *point sources* enter the environment at specific sites, such as a sewage treatment facility or an industrial discharge pipe. *Nonpoint sources* are more diffuse and, therefore, more difficult to control; examples include runoff from agricultural fields and road salts.

Agricultural Sources of Water Pollution

On a worldwide basis, agriculture probably contributes more to water pollution than does any other single activity. In the United States, agriculture is estimated to be responsible for about two-thirds of stream pollution. Agricultural runoff carries three main types of pollutants: fertilizers, biocides, and animal wastes.

Fertilizers

Agriculture is a chief contributor of *excess nutrients* to water bodies. Pollution occurs when nitrogen or phosphorus in fertilizers or present in animal manure drain into streams and rivers, eventually accumulating in ponds, lakes, and estuaries. The nutrients hasten the process of *eutrophication*, or the enrichment of waters by nutrients. Eutrophication occurs naturally when nutrients in the

BLUEPRINT FOR DISASTER: STREAM DIVERSION AND THE ARAL SEA

In 1960 it was the fourth-largest inland sea in the world, covering an area of 70,000 square kilometers (27,000 sq mi)—larger than the state of West Virginia. Now the Aral Sea has less than one-half of its original surface area, and the volume of the water is only one-fourth of what it used to be. The level of the sea has dropped more than 18 meters (59 ft), and the sea has split into two: the Little Aral Sea in the north and the Large Aral Sea in the south.

One of the world's greatest human-induced environmental catastrophes, the shrinkage of the lake is just one result of diverting nearly all the water from its primary sources, the Amu Darya and the Syr Darya, to irrigate the agricultural fields of central Asia. These are some of the other consequences:

- As the water receded, it left behind thousands of square kilometers of polluted waste on the seabed. The rivers that feed it are contaminated by agricultural runoff and untreated industrial and municipal wastewater containing high levels of fertilizers, pesticides, heavy metals, and other toxic compounds. The Large Aral Sea is now considered a dead sea. Unable to survive in the salty, toxic water, all 20 known fish species have died out, and the commercial fishing industry has collapsed. Once a fishing port and popular resort, Mo'ynoq is now more than 100 kilometers (63 mi) from the sea. Aralsk was once the main northern port for a sea that is now 80 kilometers (50 mi) away.
- Fierce windstorms whip up salty grit and toxic chemicals from the dried-up seabed and the land around it and deposit them on cropland hundreds of kilometers away, reducing the soil's fertility. Ironically, the agricultural crops for which the Aral Sea was sacrificed—mainly cotton and rice—are themselves at risk.
- Forests and wetlands upon which the region's animal life depend have been decimated. It has been estimated that three-quarters of the animal species in the basin have vanished.
- The health of the human population is threatened by the contaminated water, air, and soil. The incidence of such serious diseases as cholera, typhus, and cancers of the stomach and esophagus has increased significantly, as have respiratory system illnesses (asthma, bronchitis). High rates of miscarriages and birth defects prevail, and child mortality rates are among the highest in the world.

In an effort to save the Little Aral Sea, the World Bank and the Kazakhstan government financed the building of a dam and a series of dikes on the Syr Darya. Since the completion of the dam in 2005, the level of the sea has risen, the surface area has expanded by some 30%, and fish stocks are being replenished. Although the Large Aral Sea is still dying, efforts to save the Little Aral Sea show that it is possible to mitigate the harm people do to the environment.

---

surrounding area are washed into the water; but when the sources of enrichment are human activities, as is true of commercial fertilizers, the body of water may become overloaded with nutrients. When human activities accelerate the process, it is called **cultural eutrophication**. Algae and other plants are stimulated to grow abundantly, blocking the sunlight that other organisms need. When they die and decompose, the level of dissolved oxygen in the water decreases. Fish and plants that cannot tolerate the poorly oxygenated water suffocate.

It has been estimated that, worldwide, cultural eutrophication affects one-half of all lakes and reservoirs in North America, Europe, and Southeast Asia, and a smaller but growing proportion in South America and Africa. Symptoms of a eutrophic lake are prolific weed growth, large masses of algae, fish kills, rapid accumulation of sediments on the lake bottom, and water that has a foul taste and odor.

High levels of nutrients from agricultural runoff are a primary cause of about 200 "dead zones" around the world, although sewage plants, storm water runoff, and air pollution also contribute the nitrogen and phosphorus that spur the explosive growth of phytoplankton and zooplankton, which die and sink to the bottom, and then are eaten by bacteria that use up the oxygen in the water. A dead zone is an area of oxygen depletion in which fish, crabs, and other aquatic creatures cannot survive. The majority of dead zones are found in bays and along coastlines near the large population centers of developed countries. Examples include the Baltic Sea in northern Europe; the East China Sea between China, South Korea, and Japan; Chesapeake Bay; and the Gulf of Mexico off the coast of Louisiana (**Figure 13.9**). The zones, whose sizes vary from year to year, reach their peak during the summer months, as the water grows warmer and solar radiation increases, causing algal populations to bloom.

1975, *Source: EROS Data Center, USGS*

1987, *Source: NASA*

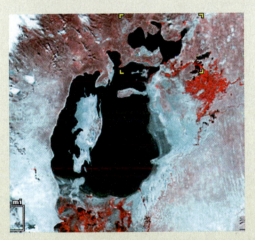

1997, *Source: EROS Data Center, USGS*

2010, *Source: NASA*

Herbicides and pesticides

The herbicides and pesticides used in agriculture are another source of the chemical pollution of water bodies. Runoff from farms where such *chemicals* have been applied contaminates both groundwaters and surface waters. One of the problems connected with the use of biocides is that the long-term effects of such usage are not always immediately known. Dichlorodiphenyl-trichloroethane (DDT), for example, was used for many years before people discovered its effect on birds, fish, and water plant life. Another problem is that thousands of these products, containing more than 600 active ingredients, are now in wide use, yet very few have been reviewed for safety by the Environmental Protection Agency (EPA). Finally, herbicides and pesticides that leach into aquifers can remain there long after the chemicals are no longer used. Thus, DDT is still found in some U.S. waters even though its use was banned in the late 1960s.

Animal Wastes

A final agricultural source of chemical pollution is animal wastes, especially in countries where animals are raised intensively. This is a problem both in feedlots, where animals are crowded together at maximum densities to be fattened before slaughter, and on the factory-like farms where beef, hog, and poultry production is increasingly concentrated (**Figure 13.10**). American farms and large feedlots produce vast quantities of manure—nearly 1.4 billion kilograms (3 billion lbs) a day—but usually lack sewage treatment facilities. The main method for disposing of the waste is to put it into open lagoons (containment ponds) and then spray it onto surrounding fields as a fertilizer, from which it can leach into the water table and rivers.

The water pollution that occurs from spreading manure on land is suspected by many to be responsible for recent outbreaks

Figure 13.9 **The dead zone in the Gulf of Mexico off the Louisiana-Texas coast forms during the summer months.** In 2015, the dead zone covered roughly 16,800 square kilometers (6500 sq mi), a little less than the size of New Jersey. More than 40% of the United States is drained by the Mississippi River and its tributaries, and each year the Mississippi washes about 1.5 million metric tons (1.65 million tons) of excess nutrients—nitrogen—into the Gulf of Mexico. The largest sources of the excess nitrogen are fertilizer runoff and animal waste, largely from the Corn Belt. The nutrients stimulate the growth of algae, which eventually die, sink to the seafloor, and decompose, creating low oxygen (hypoxia) in the lower waters. When oxygen falls below 2 parts per million, the water cannot sustain sea life.

of the microorganism *Pfiesteria piscicida* in North Carolina and the Chesapeake Bay and its tributaries. Dubbed the "cells from hell," these single-celled algae proliferate and become toxic when exposed to high levels of nitrogen and phosphorus, by-products of animal waste. The manure comes from the large poultry and hog farms of Maryland, Virginia, and North Carolina. *Pfiesteria* has killed millions of fish, and there is some evidence that the organism also sickens humans.

Other Sources of Water Pollution

As noted earlier, agriculture is only one of the human activities that contribute to water pollution. Other sources are industry, mining, cities, and residences.

Industry

Many industries generate liquid wastes containing acids, highly toxic metals, such as mercury or arsenic, or, in the case of petroleum refineries, toxic organic chemicals. When such wastes are dumped into water bodies without treatment, the resulting pollution

can kill aquatic organisms, the water may become unsuitable for domestic use or irrigation, or the waste chemicals may enter the food chain, with deleterious effects on humans. One of the most notorious pollution cases, which focused international attention on the dangers of industrial pollution, occurred in the village of Minamata in southwest Japan. From 1932 to 1968, a chemical plant that used mercury in its manufacturing process discharged tons of waste mercury into Minamata Bay, where it settled into the mud. Fish and shellfish that fed on organisms in the mud absorbed the mercury and concentrated it; the fish were in turn eaten by humans. More than 700 people died, and at least 9000 others suffered deformity or other permanent disability.

Most developed countries have strict regulations on the discharge of industrial waste to water bodies. Testing and treatment requirements for industrial wastes have resulted in significant water quality improvements Such pollution control laws are usually not in place in developing countries. As discussed in Chapter 10, many advanced countries are undergoing deindustrialization while manufacturing jobs are shifted to developing countries. The result is a shifting geography of pollution with the health of waters improving in advanced economies and deteriorating in newly industrializing countries.

Mercury pollution comes not just from industrial wastewater but also from coal-fired power plants, waste incinerators, and chemical plants that pump mercury into the air. Much of the mercury returns to Earth, contaminating lakes and streams, where it accumulates in fish and shellfish. Canada, China, and Brazil are among the countries with areas of significant mercury pollution. In the United States, the EPA reports that virtually all of the country's lakes and rivers are contaminated with mercury, and 45 states have issued advisories on eating fish.

Among the pollutants that have been discharged into the water supply in the United States are **polychlorinated biphenyls (PCBs)**, a family of related chemicals used as lubricants in pipelines and in a wide variety of electrical devices, paints, and plastics. During the manufacturing process, companies discharged PCBs into rivers, from which they entered the food chain. Several states have banned commercial fishing in lakes and rivers where fish have higher levels of PCBs than are considered safe. Although not all of the effects of PCBs on human health are known, they have been linked to birth defects, damage to the immune system, liver disease, and cancer. In 1977, the Environmental Protection Agency banned the direct discharge of PCBs into U.S. waters, but because they do not decompose readily, immense quantities of the chemicals remain in water bodies.

The petroleum industry is a significant contributor to water pollution. Oceans and inlets where offshore oil drilling takes place or where oil tankers travel are at risk of oil spills. The world's largest oil spill was in 2010 when British Petroleum's *Deepwater Horizon* offshore oil rig exploded and gushed oil for 3 months, leaking 4.9 million barrels (210 million U.S. gal) of crude oil into the Gulf of Mexico. The BP spill contaminated 790 kilometers (490 mi) of shoreline, devastated marine species, and put toxic organic chemicals into the food chain. While large spills command media attention, smaller spills routinely dump millions of gallons of oil into American waters each year. Over one-half of the oil normally comes from oil tankers and barges, usually because of ruptures in accidents. Much of the rest comes from refineries, the discharge of tank flushings and oil-drenched ballast water from tanker holds, and seepage from offshore drilling platforms. The Gulf of Mexico, the site of extensive offshore drilling, is among the most seriously polluted major bodies of water in the world.

Many industrial processes, as well as electric power production, require the use of water as a coolant. **Thermal pollution** occurs when water that has been heated is returned to the environment and has adverse effects on the plants and animals in the water body. If the heated wastewaters are significantly warmer than the waters into which they are discharged, they can disrupt the growth, reproduction, and migration of fish populations. Many plants and fish cannot survive changes of even a few degrees in water temperature. They either die or migrate. The species that depend on them for food must also either die or migrate. Thus, the food chain has been disrupted. In addition, the higher the temperature of the water, the less oxygen it contains, which means that only lower-order plants and animals can survive.

Mining

Surface mining for coal, copper, gold, uranium, and other substances contributes to contamination of the water supply through the wastes it generates. Rainwater reacts with the wastes, and dissolved minerals seep into nearby water bodies. The exact chemical changes produced depend on the composition of the coal or ore slag heaps and the reaction of the minerals with sediments or river water.

Heap-leach gold mining, for example, involves pouring large amounts of cyanide onto piles of low-grade ore in order to extract the gold. Major mining companies practice this technique in many countries, among them are the United States, Peru, Romania, Tanzania, and Indonesia. The companies dump massive amounts of the cyanide-laced waste into local rivers, contaminating the water. In addition to altering the quality of the water, the contaminants have secondary effects on soils, vegetation, and animals. Each year, for example, thousands of animals and migratory birds die in such western states as Arizona, Nevada, and California after drinking the cyanide-polluted water that has settled into ponds and lakes at gold mines.

A similar contamination of the water supply is occurring in the Amazon River and its tributaries, but with mercury rather than cyanide. Because mercury attaches itself to gold, an estimated one-half million small-scale miners (called *garimpeiros*) in Brazil, Venezuela, and neighboring countries use the toxic liquid to separate gold from Earth and rock. They pour mercury over the crushed ore they have dredged from riverbeds, press out the mercury with their hands, then burn the mixture to evaporate the rest of the heavy metal. The gold mining is estimated to send about 91 metric tons (100 tons) of mercury into the Amazon each year, poisoning the water and the fish in it, and another 91 metric tons (100 tons) into the atmosphere. Because it can take decades for concentrations to reach toxic levels, mercury pollution of streams is like a delayed-action time bomb, and thus the full effects of mercury poisoning in the Amazon Basin may not be known for many years. What is known, however, is that through

(a)

(b)

Figure 13.10 **An industrial hog-production operation in Georgia.** (**a**) On completely automated factory-style hog farms such as the one pictured here, hundreds or thousands of animals are fed and raised indoors in large, rectangular barns for the 4 or 5 months it takes them to grow to 113 kilograms (250 lbs). (**b**) Feces and urine are washed through slots in the floor into pipes that carry the waste outdoors into manure lagoons. If released improperly, the waste can trigger fish kills and pollute nearby waterways with disease-causing pathogens and high concentrations of nitrogen and phosphorus. *(a) Source: Jeff Vanuga/Natural Resources Conservation Service/U.S. Department of Agriculture; (b) Source: Jeff Vanuga, USDA Natural Resources Conservation Service.*

absorption or inhalation a high percentage of miners have extremely high levels of mercury in their bodies and that they and others are exposed to mercury by eating fish.

Cities and Residences

A host of pollutants derives from the activities associated with urbanization. The use of detergents has increased the phosphorus content of rivers, and salt (used for deicing roads) increases the chloride content of runoff. Water runoff from urban areas contains contaminants from lawn chemicals, pet waste, litter, motor oil, and the like. Because the sources of pollution are so varied, the water supply in any single area is often affected by diverse contaminants. This diversity complicates the problem of controlling water quality.

Contaminated drinking-water wells have been found in many locations across the United States. Chemicals have reached groundwater by seeping into aquifers from landfills, leaking underground gasoline and fuel-oil storage tanks, septic tanks, and fields sprayed with pesticides and herbicides. The pollution of aquifers is particularly troublesome because, unlike surface waters, groundwater has a low capacity for purifying itself; it can remain contaminated for centuries.

Sewage can also be a major water pollutant, depending on how well it is treated before being discharged. This is not simply an environmental concern; it directly affects human health. Raw, untreated human waste contains viruses responsible for dysentery, polio, hepatitis, spinal meningitis, and other diseases.

Although municipal wastewater treatment is the norm in the most-developed countries, more than 90% of sewage in the developing world is discharged directly into waterways without treatment of any kind.

- A survey of 200 major Russian rivers shows that 80% of them are polluted by raw sewage and have dangerously high levels of bacterial and viral agents.
- Mexico City treats less than 10% of its wastewater, sending its sewage into rivers that irrigate farmland.
- The majority of surface waters in India are polluted, in large part because only about 200 of its more than 3000 cities have full or partial sewage collection and treatment facilities. In Delhi, India with 25 million residents only 40% of wastewater is treated. Instead, open sewers—ideal breeding grounds for mosquitoes that carry malaria and dengue fever—line the narrow lanes of slums, and although the water is unsafe for drinking, women use it to launder clothes, wash vegetables, and bathe their children.

Although sewage-treatment plants are found in nearly all U.S. communities, antiquated combined stormwater and sanitary sewage systems in many older cities present a lingering problem. When heavy rains occur, sewers and sewage-treatment plants are overwhelmed and raw sewage must be diverted into rivers, lakes, and oceans. Beaches near these cities are sometimes subject to contamination—and closure—when untreated sewage overflows from aged combined sewer systems. New York City alone has more than 500 storm water outlets that overflow in heavy rains, pouring untreated sewage into the East River, Hudson River, and Long Island Sound each year. Communities with combined sewer overflows are working to separate storm sewers from sanitary sewers and to provide additional treatment capacity to prevent the discharge of untreated sewage.

Controlling Water Pollution

As the IPAT equation reminds us, technologies can help reduce the environmental impacts of growing populations and rising standards of living. Water pollution can be mitigated, although doing so requires the concerted effort of a variety of actors. Public education programs and environmental organizations foster concern over polluted water. In many communities, periodic manual cleanup efforts remove trash and debris from waterways, riverbanks, and beaches. Some companies have adopted "green" plans to eliminate or reduce their production of pollutants. Many countries have enacted water quality legislation, and many have entered into international agreements that have already begun to bear fruit.

The most useful strategy for controlling water pollution is to avoid producing or releasing it into the environment in the first place. Industries can recycle or reuse materials that otherwise would be discarded into the waste stream. Communities can employ secondary or tertiary treatment of their sewage and recycle the sludge. They can also clean up leaking landfills and abandoned dumps. Legislation can ban the use of DDT, PCBs, phosphate detergents, and other toxins; it can also require the preservation and/or restoration of wetlands, which filter out sediment and contaminants. The use of double- instead of single-hulled oil tankers, mandated by international convention, already has led to a significant decline in oil spills.

Once referred to as the sewer of Europe, the Rhine River is now much cleaner than it was in the 1970s and 1980s because of the International Commission for the Protection of the Rhine. The countries along the river have worked together to cut the quantities of pollutants such as nitrogen, phosphorus, lead, and ammonium entering the river and, eventually, the North Sea. They have also spent billions of dollars on wastewater treatment plants. Atlantic salmon have returned to the river after one-half century when it was too poisonous for them to survive, and the populations of other fish have increased significantly.

The Mediterranean Sea is also on its way to gradual recovery. In 1976, when the 18 countries that border the sea signed the Convention for the Protection of the Mediterranean Sea against Pollution, all coastal cities dumped their untreated sewage into the sea; tankers spewed oily wastes into it; and tons upon tons of phosphorus, detergents, lead, and other substances contaminated the waters. Now many cities have built or are building sewage-treatment plants, ships are prohibited from indiscriminate dumping, and national governments are beginning to enforce control of pollution from land-based sources.

In the United States, the federal government in 1972 took the lead in regulating water pollution with the enactment of the Clean Water Act. Its objective was "to restore and maintain the chemical, physical, and biological integrity of the nation's waters." Congress established uniform nationwide controls for each category of major polluting industry and directed the government to pay most of the cost of new sewage-treatment plants. Since 1972 such plants have been built to serve more than 80 million Americans, and industries have spent billions of dollars to comply with the Clean Water Act by reducing organic waste discharges.

The gains have been impressive. Many rivers and lakes that were ecologically dead or dying are now thriving. Once dumping grounds for all kinds of human and industrial waste, the Hudson, Potomac, Cuyahoga, and Trinity are among the rivers that are cleaner, more inviting, and more productive than before, and they now support fishing, swimming, and recreational boating. Similarly, Seattle's Lake Washington and the Great Lakes have made dramatic comebacks since the 1960s. Efforts are underway to clean up the waters of Chesapeake Bay, the country's largest estuary, and to undo much of the damage that has been inflicted on Florida's Everglades by improving the water quality of the Kissimmee River and Lake Okeechobee.

Nonpoint runoff is the main culprit in the harmful algal blooms in Lake Erie and remains a difficult challenge as we restore rivers and lakes. Because they are so diffuse, nonpoint sources of water pollution, such as agricultural and urban runoff, are more difficult to monitor and control than those from point sources. Again, the most useful tactic is to reduce the volume of pollutants before they reach waterways. Farmers and homeowners can apply less fertilizer, pesticides, and irrigation water to fields and lawns. Cities can restore wetlands and build rain gardens to absorb stormwater. Although pollution control is expensive, the long-term costs of pollution are even higher.

13.3 Impacts on Air

The **troposphere**, the thin layer of air just above the Earth's surface, contains all the air that we breathe. Every day thousands of tons of pollutants are discharged into the air by cars and incinerators, factories, and airplanes. Air is polluted when it contains substances in sufficient concentrations to have a harmful effect on living things.

Air Pollutants

Truly clean air has probably never existed. Just as there are natural sources of water pollution, so are there substances that pollute the air without the aid of humans. Ash from volcanic eruptions, marsh gases, smoke from forest fires, and windblown dust are natural sources of air pollution.

Normally, these pollutants are of low volume and are widely dispersed throughout the atmosphere. On occasion, a major volcanic eruption may produce so much dust that the atmosphere is temporarily altered. In general, however, the natural sources of air pollution do not have a significant, long-term effect on air, which, like water, is able to cleanse itself.

Far more important than naturally occurring pollutants are the substances that people discharge into the air. These pollutants result primarily from burning fossil fuels (coal, gas, and oil) and other materials. Fossil fuels are burned in power plants that generate electricity, in many industrial plants, in home furnaces, and in cars, trucks, buses, and airplanes. Scientists estimate that about three-quarters of all air pollutants come from burning fossil fuels. The remaining pollutants largely result from industrial processes other than fuel burning, the incineration of solid wastes, forest and agricultural fires, and the evaporation of solvents. **Table 13.1** summarizes the major sources of six *primary* pollutants emitted in large quantities. Once they are in the atmosphere, these may

Table 13.1	Primary Air Pollutants		
Type of Pollutant	**Symbol**	**Major Effect(s)**	**Major Sources**
Carbon dioxide	CO_2	Greenhouse gas	Combustion of fossil fuels
Carbon monoxide	CO	Reduces oxygen-carrying capacity of blood, can be fatal at high levels	Incomplete combustion of fossil fuels, nearly all from vehicles
Lead	Pb	Neurological damage; cardiovascular damage; kidney damage; IQ loss	Leaded gasoline combustion; smelters
Nitrogen oxides	NO_x	Respiratory effects; ozone formation; acid precipitation	Vehicle exhaust; power plants
Particulates	PM	Cardiovascular damage, lung damage, visual haze	Dust, vehicle exhaust; coal-fired power plants; oil-refining; farming and construction operations
Sulfur dioxide	SO_2	Respiratory damage; acid precipitation	Combustion of sulfur-containing fuels, especially from coal-fired power plants

react with other primary pollutants or with normal atmospheric constituents, such as water vapor, to form *secondary* pollutants.

Air pollution is a major environmental risk to human health and a global problem. A recent study by the World Health Organization (WHO) concluded that 80% of the people who live in urban areas are exposed to unhealthful air. Particularly at risk of breathing bad air are residents of low-income cities and such megacities as Mexico City, Cairo, Delhi, Seoul, Beijing, and Jakarta. In cities with high air pollution levels, residents have higher rates of respiratory and cardiovascular problems. Like polluted water, dirty air kills people. According to the WHO, each year 7 million people die prematurely from illnesses caused by air pollution. In the United States, California, and the Megalopolis region stand out for their nonattainment of multiple air quality standards (**Figure 13.11**).

Factors Affecting Air Pollution

Many geographic factors affect the type and the degree of air pollution found at a given place. Those over which people have relatively little control are physical geographic factors of climate, weather, wind patterns, and topography. These determine whether pollutants will be blown away, will be diluted, or will accumulate. For example, a city on a plain is less likely to experience air pollution buildup than a city surrounded by mountains.

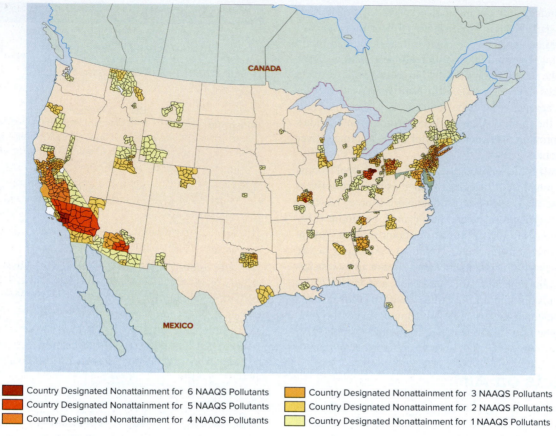

Country Designated Nonattainment for 6 NAAQS Pollutants	Country Designated Nonattainment for 3 NAAQS Pollutants
Country Designated Nonattainment for 5 NAAQS Pollutants	Country Designated Nonattainment for 2 NAAQS Pollutants
Country Designated Nonattainment for 4 NAAQS Pollutants	Country Designated Nonattainment for 1 NAAQS Pollutants

Figure 13.11   **U.S. counties violating air quality standards.** The map shows the counties that do not attain federal air quality standards for one or more criteria pollutants. The criteria pollutants are ground-level ozone, carbon monoxide, nitrogen dioxide, sulfur dioxide, particulate matter, and lead. Despite significantly reducing harmful emissions in recent years, the Los Angeles basin leads the country in the number of criteria pollutants violating standards. The most extreme ozone violations are all in California. *Source of data: U.S. Environmental Protection Agency.*

Unusual weather can alter the normal patterns of pollutant dispersal. A *temperature inversion* magnifies the effects of air pollution. Under normal circumstances, the higher you go above the Earth's surface, the cooler the air becomes. However, when the surface air gets very cold, usually in winter, the temperature can actually increase with height—an inversion. The warm layer of air on top of the colder surface air acts as a lid, preventing the normal rising and mixing of air. As described in Chapter 4 (see Figure 4.12 and "The Donora Tragedy," p. 82), the air becomes stagnant during an inversion. Pollutants accumulate in the lowest layer, so the air becomes more and more contaminated. Normally, inversions last for only a few hours, although certain areas experience them much of the time. Temperature inversions occur often in Los Angeles in the fall and Denver in the winter. If an inversion lingers long enough, over several days, it can contribute to the accumulation of air pollutants to levels that seriously affect human health.

The air pollutants generated in one place may have their most serious effect in areas hundreds of kilometers away, as atmosphere circulaton moves pollutants freely without regard to political boundaries. The prevailing wind patterns shown in Figure 4.16 strongly influence the movement of air pollutants. Thus, the worst effects of the air pollution that originates in New York City are felt in Connecticut and parts of Massachusetts. The chemical reaction that produces smog takes a few hours, and by that time air currents have carried the pollutants away from New York. In a similar fashion, New York is the recipient of pollutants produced in other places. Much of the acid rain that affects New England and eastern Canada originates in the coal-fired power plants along the lower Great Lakes and in the Ohio Valley that use extremely high smokestacks to disperse sulfurous emissions. Air pollution from China travels across the Pacific Ocean and ends up on the West Coast of the United States. And the coal-based industries in Russia and Europe produce sulfate, carbon, and other pollutants that are transported by air currents to the land north of the Arctic Circle, where they result in a contamination known as Arctic haze.

Other factors that affect the type and the degree of air pollution at a given place are the levels of urbanization, industrialization, and adoption of pollution control technologies. Population densities, traffic densities, the type and density of industries, and home-heating practices all help determine the kinds of substances discharged into the air at a single point. In general, the more urbanized and industrialized a place is, the more responsible it is for pollution. Where burning for farmland improvement or expansion combines with rapidly expanding urban and industrial development—increasingly common in large areas of the developing world—widespread atmospheric contaminations results. For example, full-color satellite cameras regularly reveal a nearly continuous blanket of soot, organic compounds, dust, ash, and other air debris almost 3.2 kilometers (2 mi) thick that stretches across much of India, Bangladesh, and Southeast Asia, reaching northward to the industrial heart of China.

The sources of pollution are so many and varied that we cannot begin to discuss them all in this chapter. Instead, we will examine three types of air pollution and their associated effects.

Acid Rain

Although acid *precipitation* is a more precise description, **acid rain** is the term generally used for pollutants, chiefly oxides of sulfur and nitrogen, that are created by burning fossil fuels and that change chemically as they are transported through the atmosphere and fall back to Earth as acidic rain, snow, fog, or dust. The main sources of these pollutants are vehicles, industries, power plants, and ore-smelting facilities. When sulfur dioxide reacts with water vapor in the atmosphere, it becomes sulfuric acid, which is highly corrosive. Sulfur dioxide contributes about two-thirds of the acids in the rain. About one-third comes from nitrogen oxides, transformed into nitric acid in the atmosphere.

Once the pollutants are airborne, winds can carry them hundreds of kilometers, depositing them far from their source. In North America, most of the prevailing winds are westerlies, which means that much of the acid rain that falls on the eastern seaboard and eastern Canada originates in the Midwest (**Figure 13.12**). Similarly, airborne pollutants from Great Britain, France, and Germany cause acidification problems in Scandinavia.

Acid rain has three kinds of effects: *terrestrial, aquatic,* and *material.* The acids change the *pH factor* (the measure of acidity/alkalinity on a scale of 1 to 14) of both soil and water, setting off a chain of chemical and biological reactions (**Figure 13.13**). It is important to note that the pH scale is logarithmic, which means every step on the scale represents a factor of 10. Thus, 4.0 is 10 times more acidic than 5.0, and it is 100 times more acidic than 6.0. The average pH of normal rainfall is 5.6, categorized as slightly acidic, but acid rainfalls with a pH of 1.5 (far more acidic than vinegar or lemon juice) have been recorded.

Acid deposition does not kill plants or trees directly. Rather, acid deposition harms vegetation by dissolving and washing away needed soil nutrients and minerals, and by leaching toxic levels of aluminum out of rocks and soils. It also kills microorganisms in the soil that break down organic matter and recycle nutrients through the ecosystem. Significant forest damage has occurred in the Appalachian Mountains in the eastern United States, northern and western Europe, Russia, and China.

The aquatic effects of acid rain are manifold. The acidity of a lake or stream need not increase much before it begins to interfere with the early reproductive stages of fish. Also, the food chain is disrupted as acidification kills the plants and insects upon which fish feed. Acid rains have been linked to the disappearance of fish in thousands of lakes and streams in the United States, Canada, and Scandinavia and to a decline of fish populations elsewhere.

The material effects of atmospheric acid are evident in damage to buildings and monuments. The acid etches and corrodes many building materials, including marble, limestone, steel, and bronze (**Figure 13.14**). Worldwide, tens of thousands of structures are slowly being dissolved by acid precipitation.

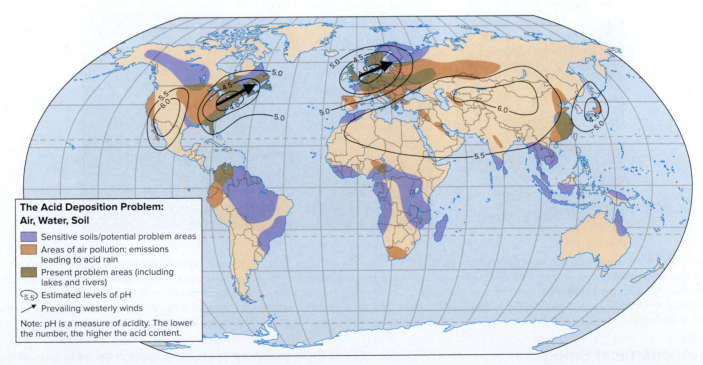

The Acid Deposition Problem: Air, Water, Soil

- Sensitive soils/potential problem areas
- Areas of air pollution: emissions leading to acid rain
- Present problem areas (including lakes and rivers)
- 5.5 Estimated levels of pH
- Prevailing westerly winds

Note: pH is a measure of acidity. The lower the number, the higher the acid content.

Figure 13.12   Acid precipitation: points of origin and current problem areas. Prevailing winds can deposit acid precipitation across international boundaries and far from its area of origin. Acid precipitation harms or destroys vegetation, aquatic life, and buildings. Occurring primarily in industrialized countries, acid precipitation has become a serious problem in many parts of North America, Europe, and East Asia. *Source: Data from* Student Atlas of World Geography, *5th ed., John Allen, Map 65, p. 81. McGraw-Hill/Dushkin, 2008.*

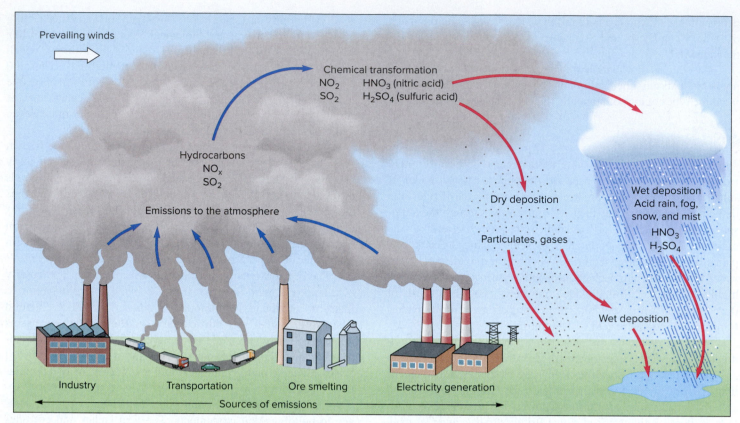

Figure 13.13   The formation of acid precipitation. Sulfur dioxide and nitrogen oxides produced by the combustion of fossil fuels are transformed into sulfate and nitrate particles; when the particles react with water vapor, they form sulfuric and nitric acids, which then fall to Earth. *Source: Data from* Biosphere 2000, *by Donald G. Kaufman and Cecilia M. Franz (NY: HarperCollins College Publishers, 1993, Fig. 14.5, p. 259).*

(a)   (b)

Figure 13.14   Damage due to acid deposition. (a) Acidic fumes in the air, especially sulfuric and nitric acids, exact a heavy toll on buildings, monuments, and statues. The destructive effect of acid rain is evident on this limestone statue at the cathedral in Reims, France. Other structures slowly being destroyed by acid rain are the Lincoln Memorial and the Washington Monument in Washington, D.C., the Parthenon in Athens, Greece, and the Taj Mahal in Agra, India. **(b)** Forest death here at Mount Mitchell in North Carolina is thought to be the result of pollution carried eastward from the Ohio and Tennnessee valleys. *(a) © TopFoto/The Image Works; (b) © Susan Reisenweaver.*

Photochemical Smog

While sulfur oxides are the chief cause of acid rain, oxides of nitrogen are responsible for the formation of **photochemical smog**. This type of air pollution is created when nitrogen oxides react with the oxygen present in water vapor in the air to form nitrogen dioxide. In the presence of sunlight, nitrogen dioxide reacts with hydrocarbons from automobile exhausts, industry, and natural sources to form new compounds, such as **ozone**. The primary component of photochemical smog, ozone is a molecule consisting of three oxygen atoms rather than the two of normal oxygen. Warm, dry weather and poor air circulation promote ozone formation. The hotter and sunnier the weather, the more ozone and smog are created. In general, therefore, more ozone is produced during the summer months than during the rest of the year.

Because the primary sources of the nitrogen oxides and hydrocarbons are motor vehicles and industries, photochemical smog tends to be an urban problem. The severity of the problem in any single area depends on climate, landforms, and traffic. Smog occurs around the world, affecting cities such as Ankara, Turkey; New Delhi, India; Mexico City; and Santiago, Chile. According to the World Bank, 16 of Asia's cities with the worst smog are in China.

Some 120 million Americans live in counties that violate the federal standards for ground-level ozone. The warm, sunny climate and topography of California are particularly conducive to ozone pollution (**Figure 13.15**). California's valleys are encircled by mountains, which help hold air pollutants in the basins. When temperature inversions occur, the pollutants are effectively trapped, unable to escape to the stratosphere. And like their U.S. counterparts, more than one-half of all Canadians live in areas where ozone pollution can reach unacceptable levels. The region from Windsor to Quebec has the worst air quality. Approximately one-half of the region's ozone is generated locally; the other half comes from the Ohio Valley and the Cleveland and Detroit areas.

Photochemical smog damages both human health and vegetation. Chronic exposure to smog causes permanent damage to lungs, aging them prematurely, and is believed to increase the incidence of such respiratory ailments as asthma, bronchitis, pneumonia, and emphysema. Because children have smaller breathing passages and less-developed immune systems than adults, they are especially susceptible to damage from the polluted air.

In addition to its effects on humans, ozone harms vegetation. Exposure over several days to ozone concentrations as low as 0.1 part per million damages trees, plants, and crops. Although smog originates in urban industrial centers, it can affect areas downwind from them. Damage associated with photochemical smog has been documented in forests downwind from Tokyo and Osaka in Japan; Beijing, China; Karachi, Pakistan; and Los Angeles, California, among other places.

Depletion of the Ozone Layer

Ozone, the same chemical that is a noxious pollutant near the ground, is essential in the stratosphere. There, approximately 10 to 24 kilometers (6–15 mi) above the ground, ozone forms a protective blanket called the **ozone layer**, which shields all forms of life on Earth from overexposure to lethal ultraviolet (UV) radiation from the sun. Mounting evidence indicates that emissions from a variety of chemicals are destroying the ozone layer. Most important is a family of synthetic chemicals developed in 1931 and known as **chlorofluorocarbons (CFCs)**, including halons, carbon tetrachloride, and methyl chloroform. CFCs are found in hundreds of products. They are used as coolants for refrigerators and air conditioners; as aerosol spray propellants; as a component in foam packaging, home insulation, and upholstery; as fire retardants; and as cleaning agents. In liquefied form, they are used to sterilize surgical equipment and to clean computer chips and other microelectronic equipment. Also implicated in the depletion of the ozone layer is methyl bromide, a pesticide used to sterilize soils and grain silos and to fumigate shipments of perishable goods.

After the gases are released into the air, they rise through the lower atmosphere and, after a period of 7 to 15 years, reach the stratosphere (**Figure 13.16**). There, UV radiation breaks the molecules apart, producing free chlorine and bromine atoms.

(a)

(b)

Figure 13.15 **Los Angeles (a) on a clear day and (b) when it is cloaked by photochemical smog.** When air over the city remains stagnant, it can accumulate increasing amounts of automobile and industrial exhaust, reducing afternoon sunlight to a dull haze and sharply lifting ozone levels. Mandates of the Clean Air Act and more-stringent restrictions on automobile emissions have resulted in reducing peak levels of ozone to one-quarter of their 1955 levels. *(a) © Purestock/SuperStock RF; (b) © Robert Landau/Getty Images.*

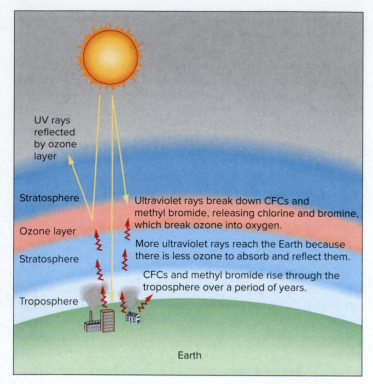

UV rays reflected by ozone layer

Stratosphere

Ultraviolet rays break down CFCs and methyl bromide, releasing chlorine and bromine, which break ozone into oxygen.

Ozone layer

Stratosphere

More ultraviolet rays reach the Earth because there is less ozone to absorb and reflect them.

CFCs and methyl bromide rise through the troposphere over a period of years.

Troposphere

Earth

Figure 13.16   How ozone is lost. CFCs and methyl bromide released into the air rise through the troposphere without breaking down (as most pollutants do) and eventually enter the stratosphere. Once they reach the ozone layer, UV rays break them down, releasing chlorine (from CFCs) and bromine. These elements, in turn, disrupt ozone molecules, breaking them up into molecular oxygen, and thus deplete the ozone layer.

Over time, a single one of these atoms can destroy tens of thousands, if not a potentially infinite number, of ozone molecules.

Each year, beginning in July, the atmosphere over the Antarctic starts to lose more and more ozone. In most parts of the world, horizontal winds tend to keep chemicals in the air well mixed. But circulation patterns are such that the freezing whirlpool of air over Antarctica is not penetrated by air currents from warmer Earth regions. In the absence of sunlight and atmospheric mixing, the CFCs and other gases work to destroy the ozone.

In 1985, researchers discovered what is popularly termed a *hole* (actually, a broad area of low ozone concentrations) as big as the continental United States in the ozone layer over Antarctica, extending northward as far as populated areas of South America (**Figure 13.17**). The ozone depletion intensifies during August and September before tapering off in October as temperatures rise, the winds change, and the ozone-deficient air mixes with the surrounding atmosphere. A less dramatic but still serious depletion of the ozone shield occurs over the North Pole, and the ozone shield over the midlatitudes has dropped significantly since 1978.

A depleted ozone layer allows more UV radiation to reach the Earth's surface. Increased exposure to UV radiation raises the incidence of skin cancer and, by suppressing bodily defense mechanisms, increases risk from a variety of infectious diseases. Because UV radiation also causes cell and tissue damage in plants, it is likely to reduce agricultural production. The most serious damage

may occur in oceans. Increased amounts of UV radiation affect the photosynthesis and metabolism of the microscopic plants called phytoplankton that flourish just below the surface of the Antarctic Ocean. Phytoplankton form the base of the oceanic food chain and play a central role in the Earth's CO_2 cycle.

The production of CFCs and other ozone-depleting substances is being phased out under the *Montreal Protocol on Substances That Deplete the Ozone Layer* of 1987, an international treaty signed by 196 countries that required developed countries to stop production and consumption of ozone-depleting substances after 1995 and to cease in developing countries in 2010. The Montreal Protocol spurred a rapid decline in CFC output. CFC use decreased 96% between 1986 and 2005, and the use of methyl bromide by 66%.

Even if no additional ozone-depleting chemicals were emitted to the air, however, past emissions will continue to cause ozone degradation for years to come. The two most widely used forms of CFCs stay in the stratosphere, breaking down ozone molecules, for up to 120 years. Thus, while recovery of the ozone layer has begun, full recovery cannot be expected until the middle of the 21st century.

Controlling Air Pollution

A number of developments in recent years have given rise to the hope that people can reverse the decline in air quality. The total amount of lead added to gasoline has dropped worldwide by 75% since 1970. Several countries, both industrialized and developing, eliminated leaded gasoline from their markets. Many others reduced the lead content and/or introduced unleaded gasoline. The development is significant because exposure to the microscopic particles of lead that are emitted into the atmosphere when leaded gasoline is burned contributes to mental retardation, high blood pressure, and an increased risk of heart attacks and strokes.

As discussed earlier, the Montreal Protocol of 1987 called for global efforts to reduce the release of ozone-depleting substances in order to protect the ozone layer. The treaty has made a significant difference. The protocol was signed by 196 governments, a remarkable international achievement on behalf of the environment. Challenges remain, however, for some of the chemicals used as transitional replacements for CFCs turned out to be potent greenhouse gases that could accelerate climate change. In response, the protocol was amended to speed the phase-out of this second group of chemicals.

Another successful international accord is the 1979 Convention on Long-Range Transboundary Air Pollution, signed by 33 countries in Europe and North America and intended to reduce the emissions of nitrous oxides and sulfur dioxides. During the 1980s, air pollution in Europe was reduced as, for example, Austria, West Germany, Sweden, and Norway cut their SO_2 emissions by more than 50%. Emissions of nitrous oxides have proved more difficult to control.

Air pollution remains a serious problem in many developing countries, however, particularly those where major metropolitan areas are growing at explosive rates. In Mexico City and Santiago, Chile, for example, pollution exceeds WHO health standards 300 or more days per year. As noted earlier, China has 16 of the world's 20 most polluted cities, a situation that is not likely to improve in

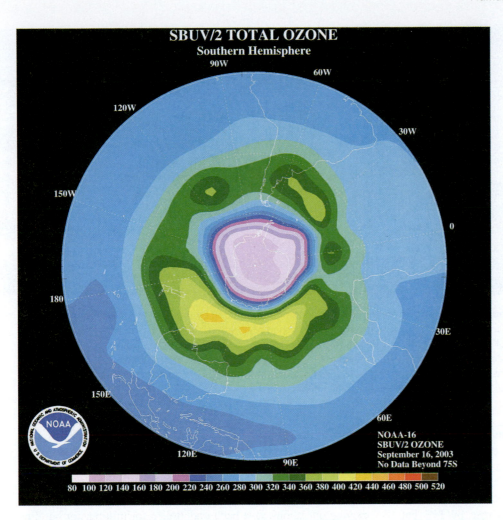

SBUV/2 TOTAL OZONE
Southern Hemisphere

80 100 120 140 160 180 200 220 240 260 280 300 320 340 360 380 400 420 440 460 480 500 520

NOAA-16
SBUV/2 OZONE
September 16, 2003
No Data Beyond 75S

Figure 13.17 **Ozone depletion over the Southern Hemisphere on September 16, 2003.** The color scale below the image shows the total ozone levels. The pinks and purples indicate the areas of greatest ozone depletion. The area of significant depletion varies in size from year to year, depending on weather conditions. In 2003, the hole grew to near record size, 29 million square kilometers (11.1 million sq mi). Although depletion of the ozone layer has been particularly severe over Antarctica, a decline in stratospheric ozone has also been observed over other parts of the world. *National Centers for Environmental Prediction, NOAA.*

the near future as long as the country's economy continues to grow. Around 2010, China surpassed the United States as the largest emitter of carbon dioxide. It is building an average of one new coal-fired power plant per week and, in another 20 years, will have millions more cars on the road.

Nonetheless, the United States has made significant progress in cleaning up its air. A series of Clean Air Acts (1963, 1965, 1970, 1977) and amendments identified major pollutants and established national air quality standards. After many years of debate, Congress in 1990 passed a Clean Air Act that represented the most sweeping legislation to date. It set forth goals to protect public health and the environment by reducing the amount of air pollutants that can be released, and it established a timetable for reaching those goals. Major provisions call for

- reducing urban smog by setting limits on allowable levels of particulates and the concentration of ozone in the air
- using cleaner-burning fuels in the most-polluted cities
- lowering nitrogen oxide and hydrocarbon emissions from motor vehicles
- requiring utilities to reduce their emissions of NO_x and SO_2

The country's air is cleaner now than it was when the first several Clean Air Acts were enacted, despite the fact that the population, economy, and number of vehicles have grown. Between 1980 and 2010, lead levels in the air at U.S. monitoring stations dropped an average of 89%, sulfur dioxide by 83%, and carbon monoxide by

82%. Nevertheless, the air in many parts of the country still violates the public health standards.

Reaching the goals of the Clean Air Act will require reducing the type and volume of air pollutants from both stationary and non-stationary sources. A number of strategies can be employed to clean up stationary sources. Technological options include switching to cleaner-burning fuels; coal washing, which removes much of the sulfur in coal before it is burned; and removing pollutants from the smokestack by using scrubbers, precipitators, and filters. Switching to renewable energy sources such as wind would reduce the need for coal-fired electrical generation. Another approach is to reduce energy consumption by using more-efficient appliances, installing weatherstripping and insulation, and strengthening energy performance standards in the building codes for new buildings.

Reducing emissions from nonstationary sources—mainly motor vehicles of all types—can be accomplished in a number of ways. These include conforming to tighter tailpipe emission standards by retiring older automobiles, driving fuel-efficient automobiles, phasing out leaded gas, and implementing rigorous vehicle inspection programs. Catalytic converters have sharply reduced smog from vehicles. Better yet, through better improved urban planning and telecommuting, the need to travel can be reduced. Cities that have added higher-density housing close to jobs, sidewalks, bike lanes, carpool lanes, and public transit systems have made progress in reducing emissions and cleaning the urban air.

Figure 13.18   **Technologies for reducing air pollutant emissions** are part of the solution as suggested by the IPAT equation. Many European countries maintain high standards of living with per capita carbon emissions one-half or one-third those of U.S. residents. Denmark is a world leader in wind energy, generating about 20% of its electricity with renewable wind power, mostly in offshore turbine installations. The Beddington Zero Energy Development (BedZed) is one of many examples of an ecocommunity. At BedZed in South London, passive solar heating, solar panels to generate electricity, a car-sharing club, and rooftop gardens allow residents to reduce significantly or even eliminate their emissions of greenhouse gases. © Mark Bjelland; © BioRegional.

European countries have been leaders in adopting renewable energy sources and other techniques to reduce the human impact on the air and climate (**Figure 13.18**). Sweden demonstrates the progress that can be made with strong personal and political commitments to environmental sustainability. Between 1980 and 2009, carbon dioxide emissions in the United States rose by 14% but decreased by 38% in Sweden. While enjoying a standard of living comparable to residents of the United States, Swedes emit about one-third as much carbon dioxide per person.

13.4 Impacts on Landforms

People have affected the Earth wherever they have lived. Whatever we do, or have done in the past, to satisfy our basic needs has had an impact on the landscape (**Figure 13.19**). To provide food, clothing, shelter, transportation, and defense, we have cleared the land and replanted it, rechanneled waterways, and built roads, fortresses, and cities. We have mined the Earth's resources, logged entire forests, terraced mountainsides, even reclaimed land from the sea. The nature of the changes made in any single area depends on what was there to begin with and how people have used the land.

Landforms Produced by Excavation

Although we tend to think of landforms as "givens," created by natural processes over millions of years, people have played and continue to play a significant role in shaping local physical landscapes. Some features are created deliberately, others unknowingly or

indirectly. Pits, ponds, ridges and trenches, subsidence depressions, canals, and reservoirs are the chief landform features resulting from excavations. Some date back to neolithic times, when people dug into chalk pits to obtain flint for toolmaking. Excavation has had its greatest impact within the last two centuries, however, as earthmoving operations have been undertaken for mining; for building construction and agriculture; and for the construction of transport facilities such as railways, ship canals, and highways.

Surface mining, which involves the removal of vegetation, topsoil, and rocks from the Earth's surface in order to get at the resources underneath, has had perhaps the greatest environmental impact. Open-pit mining and strip mining are the most commonly used methods of surface mining. *Open-pit mining* is used primarily to obtain iron and copper, sand, gravel, and stone. As **Figure 13.20a** indicates, an enormous pit remains after the mining has been completed, because most of the material has been removed for processing.

Strip mining is increasingly employed in the United States to obtain coal; more coal per year now comes from strip mines than from underground mines. Phosphate is also mined in this way. A trench is dug, the material is excavated, and another trench is dug, the soil and waste rock being deposited in the first trench, and so on. Unless reclamation is practiced, the result is a ridged landscape (**Figure 13.20b**). In a variation of strip mining called *mountaintop removal*, the tops of mountains are blasted off with dynamite to expose coal seams. Bulldozers push the mountaintops and slag waste into the valleys and streams below (see "Mountaintop Removal: Good or Bad?" pp. 398).

Damage to the aesthetic value of an area is not the only problem with surface mining. If the area is large, wildlife habitats are

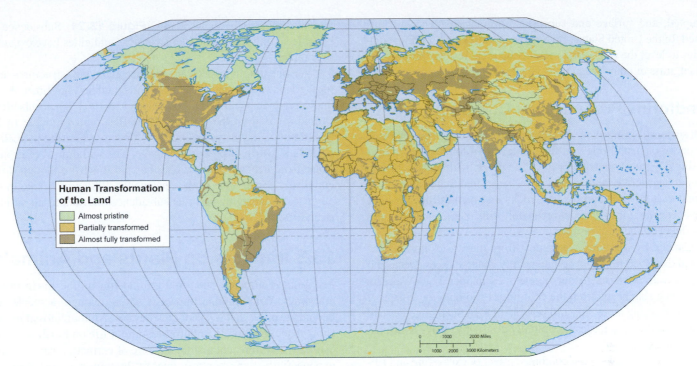

Figure 13.19   Human transformation of the land. Humans have altered much of the Earth's surface in some way. The "almost pristine" areas, covered with original vegetation, tend to be too high, dry, cold, or otherwise unsuitable for human habitation in large numbers. They generally have very low population densities. "Partially transformed" describes areas of secondary vegetation, grown after removal of original vegetation. Most are used for agriculture or livestock grazing. "Almost fully transformed" areas are those of permanent and intensive agriculture and urban settlement; little or no original vegetation remains. *Sutton,* Student Atlas of World Geography, *Map 98.*

(a)   (b)

Figure 13.20   (a) An aerial view of the Bingham Canyon open-pit copper mine in Utah, said to be the largest human-made excavation in the world. Since mining began at the site, approximately 15 billion tons of material have been removed, creating a pit more than 800 meters (2600 ft) deep and 4 kilometers (2.5 mi) wide at the top. The mine has yielded vast quantities of copper, gold, silver, and molybdenum. **(b) About 400 square kilometers (154 sq mi) of land surface in the United States are lost each year to the strip mining of coal and other resources;** far more is disrupted worldwide. On flat or rolling terrain, strip mining leaves a landscape of parallel ridges and trenches, the result of stripping away the unwanted surface material, or overburden. The material taken from one trench to reach the underlying mineral is placed in an adjacent one, leaving the wavelike terrain shown here. Besides altering the topography, strip mining interrupts surface and subsurface drainage patterns, destroys vegetation, and places sterile and often highly acidic subsoil and rock on top of the new ground surface. *(a) © Dr. Parvinder Sethi RF; (b) © hsvrs/Getty Images RF.*

disrupted, and surface and subsurface drainage patterns are disturbed. In the United States, current law requires strip-mining companies to level the ridges and regrade the area, restore the soil, and replant grass or other vegetation. The law is not always obeyed.

Landforms Produced by Dumping

Both surface mining and subsurface mining produce tons of waste and enormous spoil piles. In fact, in terms of tonnage, mining is the single greatest contributor to solid wastes, with about 2 billion tons per year left to be disposed of in the United States alone. The normal custom is to dump waste rocks and mill tailings in huge heaps near the mine sites. Unfortunately, this practice has secondary effects on the environment.

Carried by wind and water, dust from the wastes pollutes the air, and dissolved minerals pollute nearby water sources. Occasionally, the wastes cause greater damage, as happened in Wales in 1966, when slag heaps from the coal mines slid onto the village of Aberfan, burying more than 140 schoolchildren. Such tragedies call attention to the need for less potentially destructive ways of disposing of mine wastes.

Another example of the combined effect of excavating and filling on the landscape is the agricultural terrace characteristic of parts of Asia. In order to retain water and increase the amount of arable land, terraces are cut into the slopes of hills and mountains. Low walls protect the patches of level land.

Human impact on land has been particularly strong in areas where land and water meet. Dredging and filling operations create landscape features such as embankments and dikes. In many places, the actual shape of the shoreline has been altered, as builders in need of additional land have dumped solid wastes into landfills. In the Netherlands, millions of acres of land have been reclaimed from the sea by the building of dikes to enclose polders and canals to drain them. Farming practices in river valleys have had significant effects on deltas. For example, increased sedimentation has often extended the area of land into the sea.

Subsidence

The extraction of material from beneath the ground can lead to **subsidence**, the settling or sinking of a portion of the land surface. Many of the world's great cities are sinking because of the removal of *fluids* (groundwater, oil, and gas) from beneath them. Cities threatened by such subsidence are located on unconsolidated sediments (New Orleans and Bangkok), coastal marshes (Venice and Tokyo), or lake beds (Mexico City). When the fluids are removed, the sediments compact and the land surface sinks. Because many of the cities are on coasts or estuaries and are often only a few feet above sea level, subsidence makes them more vulnerable to flooding from the sea and rising sea levels associated with climate change.

The removal of *solids* (such as coal, salt, and gold) by underground mining may result in the collapse of land over the mine. *Sinkholes,* or *pits* (circular, steep-walled depressions), and *sags* (larger and shallower depressions) are two types of landscape features produced by such collapse (**Figure 13.21**). Subsidence has become a more serious problem as towns and cities have expanded over mined-out areas.

As one might expect, subsidence damages structures built on the land, including buildings, roads, and sewage lines. A dramatic example occurred in Los Angeles in 1963, when subsidence caused the dam at the Baldwin Hills Reservoir to crack. In less than 2 hours, the water emptied into the city, resulting in millions of dollars' worth of property damage. The withdrawal of groundwater from beneath Mexico City has led to severe though differential subsidence. One of the reasons the 1985 earthquake in that city was so damaging was that subsidence had weakened building structures.

13.5 Impacts on Plants and Animals

People have affected plant and animal life on the Earth in several ways. When human impact is severe enough, a species can become *extinct;* that is, it no longer exists. Although fossil records show that extinction is a natural feature of life on Earth, scientists estimate that in recent history the rate of extinction has increased exponentially—an increase due to human activity. Scientists describe the current period as the sixth great extinction. The last great extinction saw the demise of the dinosaurs about 65 million years ago. The likelihood of future extinction is assessed based on trends in population, the number of remaining members, the geographical range of the species, and the suitability of remaining habitat. Three levels of *threatened* species are recognized: *critically endangered, endangered,* and *vulnerable. Critically endangered* species are those at extremely high risk of extinction in the wild, while endangered species are at high risk of extinction. *Vulnerable* species have decreasing populations and are likely to become endangered within the foreseeable future. Other species maybe classified as *near threatened* or if trends are positive, *least concern.*

Certain kinds of species have a greater probability than others of becoming extinct. They usually exhibit one or more of the following characteristics. They exist in small populations of dispersed individuals; have low reproductive rates, especially as compared with the species that prey upon them; live in a small geographic area; and/or are specialized organisms that rely for their survival on a few key factors in their environment.

The World Conservation Union, whose members include 81 governments and more than 850 nongovernment groups, reported in 2008 on assessments of 45,000 different species of animals and plants. The data show that more than one-third of all species are threatened and more than 800 are extinct. Species at risk of disappearing include about 25% of all known mammal species and 12% of all bird species. Between 1980 and 2008, an average of 52 species per year moved one category closer to extinction. Although endangered mammals and birds have attracted public attention, many species of plants are also in jeopardy.

Figure 13.22 shows the location of **biodiversity hot spots** — areas with an exceptionally high number of endemic species (those that occur nowhere else) *and* that is at high risk of disruption by

four major ways human activities modify animal and plant life and threaten to reduce biodiversity .

Habitat Loss or Alteration

The main cause of extinction has been the loss or alteration of wildlife habitat. About three-fourths of all threatened species are affected by habitat degradation or loss. Agricultural activities (crop and livestock farming), extractive activities (logging and mining), and various forms of development (e.g., draining wetlands, clearing forests and grasslands for urban growth) all modify or destroy the habitats in which plants and animals live. The destruction of the world's rain forests, the most biologically diverse places on Earth, by some estimates is causing the extinction of hundreds of plant and animal species every year.

As countries in Africa, Asia, and South America become more industrialized and more urbanized and expand their areas under cultivation, there will be an increasingly negative impact on wildlife. In Africa, wild animals are vanishing fast, in part the victims of habitat destruction. In Botswana, for example, 250,000 antelope and zebra died in a decade, disoriented by fences erected to protect cattle.

The tropical rain forests of Indonesia, home to such animals as the silvery gibbon and the orangutan, are being destroyed by logging, both legal and illegal; by the conversion of forests to plantations to grow trees for palm oil and paper production; and by surface mining for gold and zircon. The silvery gibbon lives only

Figure 13.21   **A sinkhole in Guatemala City, Guatemala.** Sinkholes form when the ground above an underground cavern suddenly collapses. They can be triggered by the lowering of the water table or by heavy rains that enlarge a crack in the cavern roof. Areas underlain by limestone, which is soluble in water, are particularly susceptible to the formation of sinkholes, as the many such depressions in Florida, Kentucky, Indiana, and Missouri attest. © *Moises Castillo/AP Images.*

human activities. Although the hot spots occupy only a small percentage of the world's land area, they house 60% of the total world's terrestrial plant and animal species. In this section, we examine

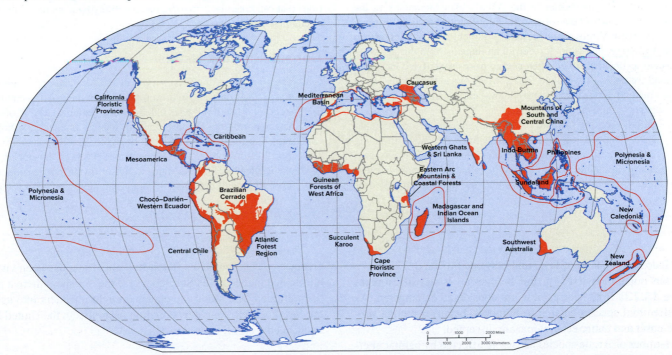

Figure 13.22   **Biodiversity hot spots, Earth's biologically richest and most threatened ecosystems.** These biodiversity hotspots provide critical habitats for plants and vertebrate animals. Each has a high number of species found nowhere else and has already lost at least 70% of its original vegetation. Some areas with high biodiversity, such as Amazonia and the Congo Basin, are not included on the map because most of their land area is relatively undisturbed. Notice that the hot spots are in areas that have been relatively isolated from other areas for a long time, either because they are islands or because they have special climatic conditions (e.g., Mediterranean or tropical). *Sutton,* Atlas of World Geography 8e, *Map 95, p. 118.*

Mountaintop Removal: Good or Bad?

As its name suggests, mountaintop removal (MTR)—also called mountaintop mining, or MTM—is a type of coal mining that requires the removal of a mountaintop to get at the coal seams lying beneath. The first step is to clear-cut upper-elevation forests and strip off the topsoil. The lumber is burned or sold, and the topsoil is supposed to be removed and set aside for later reclamation. Once the area is cleared, miners use explosives to blast away the overburden (rock and subsoil) to expose the coal seams beneath. Enormous machines scrape away the rubble and dump it down the sides of the mountain into valleys and streambeds. As much as 240 meters (800 ft) of mountaintop may be removed, leading critics of the practice to call it strip mining on steroids.

Federal law requires that, after active mining has been completed and the mountaintop has been regraded, most sites must be reclaimed to the land's premining contour and use; but waivers can be granted so that reclamation creates a level plateau or a gently rolling contour. Topsoil or a substitute is placed on top, and grass seed is spread in a mixture of seed, fertilizer, and mulch. Trees may be planted, or the land can be reclaimed to be used for other purposes if a greater use than natural restoration can be found for it. Most reclamation efforts have focused on stabilizing rock formations and controlling for erosion, not on reforestation of the affected area. Fast-growing, nonnative grasses planted to quickly provide vegetation compete with tree seedlings, and trees have difficulty establishing root systems in compacted landfill.

Mountaintop mining began in the Appalachian Mountains in the 1970s as an extension of conventional strip-mining techniques. It occurs primarily in West Virginia and eastern Kentucky; other sites are in Virginia, Tennessee, and Ohio. It received an impetus from the shortages of petroleum in 1973 and 1979, when energy prices soared and increased the demand for coal in the United States. MTR expanded further in the 1990s as a method of retrieving relatively low-sulfur coal, made desirable by amendments to the U.S. Clean Air Act.

Mountaintop mining has two distinct advantages for coal companies. Highly mechanized, it requires relatively few workers, significantly reducing the labor costs of a company. Second, many of the coal seams mined in the MTR method are too thin to be mined using more conventional techniques. Because miners can repeat the blasting process over and over, MTR makes it economically feasible to mine the seams, which often exist in multiple stratified layers on a single mountain. This allows for almost complete recovery of coal from a single mountain.

Critics of MTR contend that it has such substantial environmental, water, and health impacts and is so destructive that the government should stop giving out new permits for it. "The science is so overwhelming that the only conclusion one can reach is that mountaintop mining needs to be stopped," says a leading environmental scientist. According to studies by the Environmental Protection Agency, mining operations have permanently altered stream flows and wildlife habitats in the watersheds. More than 3200 kilometers (2000 mi) of streams across central Appalachia have been buried by valley fills. Removal of the forests, which used to support several endangered species, leads to a loss of biodiversity.

MTR also causes water pollution. When rainwater falls on the rubble dumped down the mountainside, it picks up pollutants from the rocks that came from deep underground. The water that emerges is polluted with metals and chemicals called sulfates, which can be toxic to the insects and fish in Appalachian streams.

Communities near mining sites have a high risk of health problems. Exposed to airborne dust and debris, to contaminated water wells, and to fish tainted with toxins, people show increased rates of mortality, lung cancer, and chronic heart, lung, and kidney diseases. In addition, residents are forced to contend with continual blasting from mining operations that can take place up to 92 meters (100 yds) from their houses and operate 24 hours a day.

Considering the Issues

1. The president of the West Virginia Coal Association contends that MTR is only an "interim disruption to the existing environment" and that it provides much needed flatter areas that offer new opportunities for development, such as an airport, a shopping mall, a residential development, or a golf course. In actuality, less than 20% of land set aside for such uses has been developed. As one resident said, "Who wants to live or invest in a development where little will grow and the area is an ugly scar?" Do you think

on the Indonesian island of Java, and between 80% and 90% of the orangutan population lives on the Indonesian island of Sumatra (**Figure 13.23**). Both species now are considered endangered. As selected animal species decline in numbers, balances among species are upset and entire ecosystems are disrupted.

A number of Arctic species are being affected by habitat alteration during the current warming trend. Springtime arrives in the Arctic a month earlier than it did a decade ago. As the amount of ice has declined, so have the populations of polar bears, walruses, and other ice-dependent marine mammals.

Habitat disruption characterizes developed as well as developing countries. Tidal marshes in the United States, for example, have been dredged and filled for residential and industrial development. The loss of such areas reduces the essential habitat of fish, crustaceans, and mollusks. The whooping crane has been virtually eliminated in the United States because the marshes where it nested were drained, and roads and canals brought intruders into its habitat. Its comeback, sought by breeding programs in the United States and Canada, is still uncertain.

Hunting and Commercial Exploitation

Another way in which people have affected plants and animals is through their deliberate destruction. We have overhunted and overfished, for food, fur, hides, jewelry, and trophies. In the past, unregulated hunting harmed wildlife all over the world and was responsible for the destruction of many populations and species. Beavers, sea otters, alligators, and buffaloes are among the species

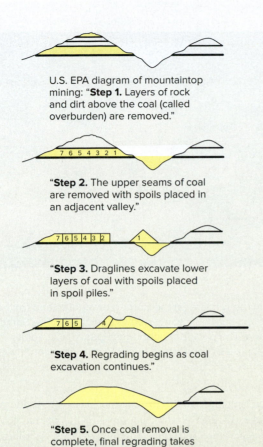

U.S. EPA diagram of mountaintop mining: "**Step 1.** Layers of rock and dirt above the coal (called overburden) are removed."

"**Step 2.** The upper seams of coal are removed with spoils placed in an adjacent valley."

"**Step 3.** Draglines excavate lower layers of coal with spoils placed in spoil piles."

"**Step 4.** Regrading begins as coal excavation continues."

"**Step 5.** Once coal removal is complete, final regrading takes place and the area is revegetated."

© Photo courtesy Vivian Stockman/www.ohvec.org. Flyover courtesy SouthWings.org.

the government should stop granting waivers for restoration of the mountaintop to its premining contour and use?

2. A spokesman for the Coal Association maintains that after a mountaintop mine is finished, the damage to nearby streams "is usually very short-term," lasting no more than 18 months. Not so, say environmental scientists. "It obliterates stream ecosystems," and the damage could last hundreds or even thousands of years. How could one determine which statement is accurate?

3. If you were a spokesperson for a mining company, how would you respond to this statement by a Kentucky resident? "Five years ago the area around my home was very beautiful. Now the trees are cut and the animals are gone. The water on my land is contaminated with high levels of arsenic. I am forced to bathe my 3-year-old in it. There is no other water; I have no choice. I am continuously threatened by mud slides; my house could be washed away at anytime. Now they want to build sludge ponds behind my house. My dreams have been ruined."

4. Conversely, if you lived near an MTR site, how would you react to this statement by a representative of a mining company? "Mountaintop mining provides a resource on which the country depends. Only the topmost portion of the mountain is mined and leveled for the maximum recovery of coal. Once reclaimed, it's hard to tell that mining had ever occurred there. What's left is flatter, more useful land on the top of the mountain. The coal industry does an excellent job of reclamation. The amount of actual stream loss from fills is minimal. Mountaintop mining is simply the right thing to do—both for the environment and for the local economy. A true win-win."

5. Is MTR a temporary disruption of the environment or a destructive practice that benefits a small number of corporations at the expense of local communities and the environment? Should the government stop issuing permits for MTR? Defend your answers.

For more information about the practice and to see photographs and videos, enter "mountaintop removal mining" or "mountaintop mining" in your search engine.

brought to the edge of extinction in the United States by thoughtless exploitation. Under protective legislation, their populations are now increasing, but hunting in developing countries still poses a threat to a number of species. In Central and West Africa, hunting rather than habitat destruction constitutes the main threat to some species.

Four African animals whose existence is threatened by hunting, most of it illegal, are the elephant, the rhinoceros, the mountain gorilla, and the hippopotamus.

- Prized for its ivory tusks, the African elephant has been ruthlessly slaughtered. Between 5 and 10 million of the elephants are estimated to have been alive in the 1930s, a number that declined to 1.3 million by 1979. Now, only 300,000 to 500,000 remain.

- The black rhinoceros, killed for its horn, is now an endangered species. The population has declined from nearly 1 million in sub-Saharan Africa a century ago to about 3500 today.

- All three subspecies of gorillas are endangered; the rarest of all is the mountain gorilla. About 700 mountain gorillas are believed to exist, most of them in national parks in the Democratic Republic of Congo, Uganda, and Rwanda.

- Since 1994, hunting has caused a precipitous decline in the population of the common hippopotamus in the Democratic Republic of the Congo, and the species is now ranked as vulnerable. The pygmy hippopotamus of West Africa was classified as vulnerable in 2000; by 2006 hunting for meat had reduced the population to between 2000 and 3000, and the species had become endangered, at risk of extinction.

(a)

(b)

Figure 13.23   Two victims of habitat loss. Among the animals facing extinction in the wild are **(a)** the Borneo orangutan and **(b)** the polar bear. The number of orangutans has declined due to loss of habitat as tropical rainforest has been converted to agriculture. The Sumatran orangutan is listed as critically endangered while the Borneo orangutan is endangered. The polar bear depends upon sea ice for its livelihood and was the first species protected under the Endangered Species Act because of habitat loss due to global warming. *(a) © Kjorgen/iStock/Getty Images RF; (b) © Vadim Balakin/Moment/Getty Images RF.*

The decline in the productivity of marine fisheries, discussed in Chapter 12, is due in large part to the widespread use of modern fishing technology, which has made hunting easier and more efficient. The technology includes the use of sonar, radar, helicopters, and Global Positioning Systems to locate schools of fish; more-efficient nets and tackle; and factory trawlers to follow fishing fleets to prepare and freeze the catch.

According to the UN Food and Agriculture Organization, three-fourths of all ocean fish stocks are now being fished at or above sustainable levels. The plundering of U.S. coastal waters has imperiled a number of the most desirable fish species, including haddock, yellowtail flounder, and cod in New England waters; Spanish mackerel, grouper, and red snapper off the Gulf of Mexico; halibut and striped bass off California; and salmon and steelhead in the Pacific Northwest.

Commercial fishermen from Japan, South Korea, Taiwan, and other countries use drift nets to capture squid, tuna, and salmon. The nets, which can stretch up to 65 kilometers (40 mi), have been called "curtains of death" because they are devastatingly effective, essentially scooping up all life in their path—not just the target species but also millions of nontarget fish, sea birds, turtles, and marine mammals. The World Wildlife Fund estimates that at least 60,000 dolphins, porpoises, and whales drown each year after becoming entangled in fishing nets and other equipment. These accidental captures by crews fishing for other species are known as *bycatch;* they have significantly depleted the populations of many of the 80 species of the cetacean (fishlike) groups of sea mammals, critically endangering some of them.

Introduction of Exotic Species

A plant, an animal, or another organism (such as a microbe) that has been released into an ecosystem in which it did not evolve is nonindigenous (nonnative)—an **exotic species**. An exotic species

that causes economic or environmental harm is considered *invasive* (**Figure 13.24**). Human actions are the primary way that invasive species are introduced to new places. With increased human interaction between distant places on Earth, we can expect more and more exotic species, often carried inadvertently on container ships.

The deliberate or inadvertent introduction of a species into an area where it did not previously exist can have damaging and unforeseen consequences. Introduced species have often left behind their natural enemies—predators and diseases—giving them an advantage over native species that are held in check by biological controls. The rabbit, for example, was purposely introduced into Australia in 1859. The original dozen pairs multiplied to a population in the thousands in only a few years and, despite programs of control, to an estimated 1 billion by 1950. Inasmuch as five rabbits eat about as much as one sheep, a national problem had been created. Rabbits became an economic burden and environmental menace, competing with sheep for grazing lands and stimulating soil erosion.

Invasions of exotic (also called *alien*) species have multiplied as the speed and range of world trade and travel have increased. In the United States alone, hundreds of harmful invaders have been discovered in recent years—harmful because they consume or outcompete native species. They include

- the Asian tiger mosquito, which was discovered in 1985 in a Japanese container of tires headed for a recapping plant in Texas and has since spread to 40 states; it carries a number of tropical viruses, including the zika virus, yellow fever, dengue fever, and various forms of encephalitis
- zebra mussels, first detected in Lake St. Clair, near Detroit, in 1985, which were released from ballast water from Eastern European ships and have spread rapidly to northeastern Canada, all the Great Lakes, hundreds of inland lakes, the Mississippi River and its tributaries; prolific reproducers, the

Purple loosestrife   Asian longhorn beetle   Round goby   Kudzu vine   Multiflora rose   Eurasian milfoil

Zebra mussel   Grass carp   Leafy spurge   Gypsy moth   Mongoose   Asian tiger mosquito

Scotch broom   Glossy buckthorn   Cheat grass   Sea lamprey   Water hyacinth

Boll weevil   Nutria   *Caulerpa taxifolia* alga   European green crab   Canadian thistle   Russian thistle

Figure 13.24   A few of the approximately 50,000 nonnative species that have become established in North America over the past 300 years. At least 750 of these species cause significant environmental and/or economic damage. You can check on the Web to see if any of these exist where you live.

mussels clog the underwater intake pipes of water treatment plants, power plants, and industrial facilities, smother native mollusks, and compete with algae-eating fish for food and oxygen

- Asian carp escaped fish farms in the southern United States during floods and spread into the Mississippi River and beyond. The Asian carp outcompetes native fish and can injure boaters when they jump out of the water and threatens to disrupt the Great Lakes' ecosystem

- the Asian gypsy moth, a voracious eater of trees and some crops, which arrived in Oregon, Washington, and British Columbia in 1991

- Africanized honeybees, more aggressive and venomous than European bees; they reached the southwestern United States in the 1990s after escaping from an experimental station in Brazil in 1957

- the Asian long-horned beetle, which probably arrived in Brooklyn, New York, in wood pallets from China in 1996 and has since appeared in several other states; it threatens species such as maple, elm, willow, and birch

- the emerald ash borer, an Asian beetle that arrived in wood packing material carried in cargo ships or airplanes; it was detected in Michigan in 2002 and quickly spread to other states in the Midwest and Northeast as well as Ontario, Canada, where it has killed millions of ash trees

In the United States, exotic species are second only to habitat loss as a cause of endangerment. About one-half of the species threatened with extinction in the country are in trouble at least partly because of them.

Plants and animals found on islands are particularly prone to extinction. Native plants and animals evolved in isolation, with few diseases or predators. Furthermore, because many island species occur on just one or a few islands, the loss of only a few individuals can be devastating to small populations. In the United States, Hawaii has more endangered species than any other state. One-half of the islands' 140 native breeding bird species are now extinct, and of the 71 remaining, 32 are considered threatened. Bird declines were due both to habitat alteration—deforestation—and to such alien predators as tree-climbing rats, mongooses, and feral cats, dogs, and pigs. Plants introduced to Hawaii include eucalyptus trees, ginger, and gorse. Miconia, a large-leaved plant, was brought in as a tropical ornamental. Outside the confines of a flowerpot, it grows to 15 meters (50 ft) or more, and its mammoth leaves (1 meter, or 3 ft, across) cast dense shade, killing native vegetation beneath it, promoting water runoff and erosion. Small colonies of these plants are scattered throughout the islands, and there is a massive eradication program underway in an attempt to curb its spread. The fear is that miconia might do to Hawaii what it has done in Tahiti, where it has replaced 70% of the native rain forest and is threatening 25% of the island's native wildlife species.

As this discussion indicates, introduced plants, as well as animals, can alter vegetative patterns. Some 300 species of invasive plants now threaten native ecosystems in the mainland United States and Canada. At least one-half were deliberately imported, including purple loosestrife, the melaleuca tree, buckthorn, and water hyacinth. These and other imports have arrived without their natural enemies and have spread uncontrolled, driving out native species.

An aquatic vine, hydrilla, imported into Florida from Sri Lanka for use in aquariums, was dumped into a canal in Tampa in 1951. Also known as water thyme, it has overgrown more than 40% of Florida's rivers and lakes and continues to spread rapidly. Transported on motorboat propellers and boat trailers, hydrilla has spread through the lakes and rivers of the South, as far west as California and as far north as Maine and the state of Washington. The plant creates dense canopies at the surface, interfering with swimming, fishing, and boating. The mats clog boat propellers and water intake pipes and reduce the amount of sunlight reaching the water bottom. By monopolizing the dissolved oxygen that fish and aquatic plants require to thrive, hydrilla reduces native diversity.

In an attempt at *biological control,* fighting an invader by importing a predator known to keep it in check elsewhere, Florida officials imported tilapia, a weed-eating fish, to solve the problems created by hydrilla. While the fish have done little to clear Florida waterways, they have driven out many types of native fish, especially large-mouth bass.

Disease-carrying microorganisms can also become invasive and do great harm. The Asiatic chestnut blight destroyed most of the native American chestnut trees in the United States, trees with significant commercial as well as aesthetic value. The cause was the imported chestnut trees from China that carried a fungus fatal to the American chestnut tree but not to the Asiatic variety. Similarly, Dutch elm disease devastated the American elms that once graced streets throughout the country.

These are just a few of the many examples that illustrate an often ignored ecological truth: plant life and animal life are so interrelated that when people introduce a new species to a region, whether by choice or by chance, there may be unforeseen and far-reaching consequences.

It is important to note, however, that not all introduced species are harmful. They are of little concern if they assimilate well and coexist with the native stock. Further, an import may become a problem in one area but not be highly invasive elsewhere.

Poisoning and Contamination

Humans have also affected plant and animal life by poisoning and contamination. In the last 50 years, we have become increasingly conscious of the effect of insecticides, rodenticides, and herbicides, known collectively as *biocides.*

Some of the side effects of these biocides have been well enough documented for scientists to question their indiscriminate use. Once used, a biocide travels through the environment. Depending on its chemical properties, it may break down or persist. The chemical may remain in the soil, dissolve in water, be carried on the wind, or be stored in the tissues of living organisms. If an organism is unable to excrete the biocide, its concentration within the animal continues to increase, a process called *bioaccumulation.* Toxins present in

very small amounts in the environment can reach dangerous levels inside cells and tissues.

Food webs magnify the effect of toxins in the environment. When a predator eats a large number of plants or animals at a lower trophic level, the predator collects and concentrates the toxins from its prey. **Biological magnification** (also called **biomagnification**) is the accumulation of a chemical in the fatty tissue of an organism and its concentration at progressively higher levels in the food chain. Zooplankton and small fish, for example, collect and retain the toxins from water, sediments, and organic debris. They are eaten in turn by small fish, shrimp, and clams, which build up higher concentrations of the toxins, as shown in **Figure 13.25**. The higher

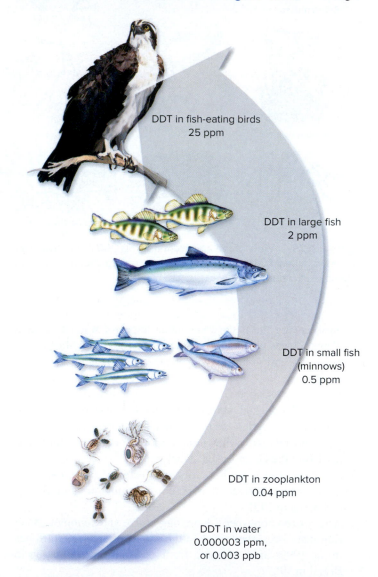

DDT in fish-eating birds
25 ppm

DDT in large fish
2 ppm

DDT in small fish
(minnows)
0.5 ppm

DDT in zooplankton
0.04 ppm

DDT in water
0.000003 ppm,
or 0.003 ppb

Figure 13.25   The bioaccumulation and biomagnification of DDT. The numbers are parts per million (ppm). Although the level of DDT in the water may be low, notice how the amount of DDT in the bodies of fish and birds increases as we go up the food chain. In this simplified example, birds at the top of the chain have concentrations of residues 50 times greater than those of small fish. Mercury, aldrin, chlordane, and other chlorinated hydrocarbons, such as PCBs, undergo magnification in the food chain. *Source: Cunningham, Cunningham, and Saigo,* Environmental Science, *7th ed. Boston: McGraw-Hill, 2003.*

the level of a plant or an animal in the food chain, the greater the concentration of the poisons. The carnivores at the top of the food chain, large fish, fish-eating birds, and humans can accumulate such high levels of the biocide that it adversely affects their health and reproduction.

One of the oldest and most dangerous pesticides is dichloro-diphenyl-trichloroethane (DDT). Hailed as a chemical miracle, DDT was first used during World War II as a delousing agent and to clear malaria-bearing mosquitoes from the paths of Allied troops. After the war, one of the first major undertakings of the World Health Organization was the Global Malaria Eradication Program, launched in 1955; its principal tool was DDT. Soon, tons of the pesticide were being used, both to combat disease and to increase agricultural yields, but within a few years two significant problems emerged: (1) biomagnification, or the increasing concentration of the chemical at higher levels of the food chain, and (2) insecticide resistance—the reproduction within a few years of mosquitoes that were immune to the poison.

The publication of Rachel Carson's *Silent Spring* in 1962 alerted the public to the devastating effect of the pesticide on birds and other wildlife. The chemical caused a decrease in the thickness of the eggshells of some of the larger birds, so that a greater number of eggs broke than normally would. Peregrine falcons, bald eagles, and brown pelicans were among the birds nearly made extinct by this disruption of the reproductive process. DDT was banned in the United States in 1972, and many bird species have made impressive comebacks (**Figure 13.26**). However, a number of countries, most of them in Africa, still use it. Because malaria is the biggest killer of children under age 5, and because DDT is one of the most effective insecticides against mosquitoes, the indoor spraying of DDT is supported by the World Health Organization and other agencies.

Meanwhile, other chlorinated hydrocarbon compounds have been developed and are in wide use. They include chlordane, aldrin, and dieldrin. Because they do not break down easily, they accumulate in the bodies of animals in the food chain, thus affecting many nontarget organisms. The deaths of thousands of seals, Pacific sea lions, and Mediterranean striped dolphins are thought to be caused by accumulation of the chemicals.

The use of pesticides has increased steadily, largely in commercial agriculture. According to the U.S. EPA, more than 2.4 million metric tons (5.2 billion lbs) containing hundreds of active ingredients are applied each year. The pesticides pollute water supplies, often contaminate the crops they are meant to protect, can cause allergic reactions, and sometimes sicken the farmworkers who apply them. In addition, too often the pesticides are only temporarily effective.

Biocides may, in fact, exacerbate the problem their use is designed to eradicate. By altering the natural processes that determine which pests (insects, rodents, weeds) in a population will survive, biocides spur the development of resistant species. If all but 5% of the mosquito population in an area are killed by an insecticide, the ones that survive are the most-resistant individuals, and they are the ones that will produce the succeeding generations.

There are now insects whose total resistance to certain pesticides has led some scientists to conclude that the entire process of insecticide development may be self-defeating. Despite the enormous growth in the use of biocides, crop loss to insect and weed pests has actually grown. According to Department of Agriculture figures, 32% of crops were lost to pests in 1945; 40 years later, such losses had increased to 37%.

Biocides have also been known to increase problems by destroying the natural enemies of the intended target, leaving it to breed unchallenged. Examples include the tobacco bud-worm and the brown planthopper, which were relatively minor pests before intensive crop spraying destroyed rival pests.

Preserving Biodiversity

As people have become more aware of how their activities affect plant and animal life in often deleterious ways, public concern with the preservation of biodiversity has mounted. Ideally, the goal is to maintain diversity in all biomes, not just in the biodiversity hot spots identified in Figure 13.22. In the broadest sense, efforts are aimed at controlling or reducing the impact of the four main causes of endangerment: habitat loss or alteration, hunting and commercial exploitation, the introduction of exotic species, and the overuse of biocides. More narrowly, specific attempts tend to focus either on preventing the extinction of individual species or on preserving habitats that support maximum biological diversity. Governments, nongovernmental organizations, and other agencies and groups use a variety of methods to further their goals.

Legal Protection

Three of the many legal measures for preserving biodiversity are the following.

- *The UN Convention on Biological Diversity* (1992): an international treaty that by 2007 had been signed by 189 countries, its objective is for countries to develop national strategies for the conservation and sustainable use of biodiversity.
- *Convention on International Trade in Endangered Species* (1973): by 2007 signed by more than 160 countries, it regulates trade in living specimens of wild flora and fauna and the products made from some 700 threatened species. Due to its ban

Figure 13.26   Bald eagle recovery. The bald eagle is a North American species whose range extends from Mexico to Alaska. It was nearly extinct by the late 1960s and early 1970s. However, the ban on DDT and the eagle's placement on the U.S. Endangered Species list helped it make a remarkable recovery. © *Fuse/Getty Images RF.*

on international trade in elephant products—the most important being ivory—two African countries (Botswana and South Africa) have experienced a rise in their elephant populations.

- *The Endangered Species Act* (1973) is the most significant piece of biodiversity legislation in the United States. Among other things, it prohibits hunting or commercial harvesting of threatened or endangered species; prohibits importing them into or exporting them out of the United States; and can mandate protection of a habitat that is critical to the survival of a species. The act has saved several species—including the bald eagle, the peregrine falcon, and the American alligator—from possible extinction.

In addition, most countries have established national parks or other government-protected areas (e.g., refuges, nature preserves, marine reserves). Many, particularly in developing countries, attract significant numbers of ecotourists, who shoot with a camera rather than a gun.

Nongovernmental Organizations (NGOs)

Private organizations pursue a number of activities to preserve and protect species and habitats and slow the accelerating rate of species extinction. Among these NGOs are The Nature Conservancy, the World Wildlife Fund, The Wildlife Conservation Union, and Conservation International. The Nature Conservancy has joined with governments, private corporations, and indigenous people to purchase critical habitats. With branches in Africa, Asia, the Caribbean, and the Americas, it has undertaken projects in more than 30 countries and all 50 United States to preserve natural landscapes and the plants and animals they house. Greenpeace, established in 1971, has been actively involved in convincing world leaders to, among other things, ban commercial whaling, stop the destruction of ancient forests and the deterioration of oceans, and designate marine reserves in the Bering Sea.

A number of zoos, aquariums, and botanical gardens have breeding programs to save severely threatened species. Two species saved by captive breeding programs are the condor and the nene. Hunting and fragmentation of their habitat had reduced the wild population of California condors to less than 20 birds by 1987. All were captured and taken to the San Diego and Los Angeles Zoos to serve as breeding populations. Today, there are about 400 condors, more than one-half of which were living in the wild. Similarly, the number of the Hawaiian nene goose had fallen from the tens of thousands in the 1700s to less than 30 by 1950. They were bred in captivity, and the wild population now numbers more than 2500.

13.6 Waste Management

People have always been faced with the problem of disposing of materials they no longer want. Prehistoric dwelling sites are located and analyzed by their *middens,* the refuse piles containing the kitchen wastes, broken tools, and other debris of human settlement. We have learned much about Roman and medieval European urban life by examining the refuse mounds that grew as hills in their vicinities. Modern societies differ from their predecessors in the volume and character of their wastes. As suggested

by the IPAT equation, the greater a society's population and material wealth, the greater the amount and variety of its garbage. Disposing of the waste is a problem that each individual and each municipality must deal with.

Municipal Waste

The wastes that communities must somehow dispose of include newspapers and beer cans, toothpaste tubes and old television sets, broken refrigerators, cars, and tires. American communities face two major hurdles in disposing of these wastes: the sheer volume of trash and the toxic nature of much of it. Americans throw away more trash per person than any other country in the world, about 2 kilograms (4.5 lbs) per person per day—a slight decrease from 2000. Solid-waste disposal costs are now the second-largest expenditure of most local governments. Americans generate more than twice as much waste per person as do Japanese and Europeans, four times as much as Pakistanis or Indonesians.

This volume of trash is the result of three factors—affluence, packaging, and population density. Craving convenience, Americans rely on disposable goods that they throw away after very limited use. Thus, although readily available substitutes are more economical, Americans annually throw out billions of baby diapers, razors, pens, and paper plates, cups, towels, and napkins. People in less affluent countries repair and recycle a far greater proportion of domestic products. In addition, nearly all consumer goods are encased in some sort of wrapping, whether it be paper, cardboard, plastic, or foam. One-third of the yearly volume of municipal trash consists of these packaging materials. Finally, with its relatively low population density, the United States has traditionally had ample space in which to dump unwanted materials. Countries with higher densities ran short of such space decades ago and have made greater progress in reducing the volume of waste.

Although ordinary household trash does not meet the government designation of **hazardous waste**—defined as discarded material that may pose a substantial threat to human health or to the environment when improperly stored, transported, or disposed of—much of it is hazardous nonetheless. Products containing toxic chemicals include paints, paint thinners, old TV sets and computers, bleaches, oven and drain cleaners, used motor oil, and garden weed killers and pesticides (see "E-Waste," p. 406).

Countries use various methods of disposing of solid wastes; each has its own impact on the environment. Loading wastes onto barges and dumping them in the sea, long a practice for coastal communities, inevitably pollutes oceans. Open dumps on land are a menace to public health, for they harbor disease-carrying rats and insects. Burning the combustibles discharges chemicals and particulates into the air. In the United States, three methods of solid-waste disposal are employed: landfills, incineration, and recycling (**Figure 13.27**).

Landfills

Most U.S. municipal solid waste is deposited in *sanitary landfills,* where each day's waste is deposited in a natural depression or excavated trench, compacted, and then covered by a layer of soil (**Figure 13.28**). *Sanitary* is a relative word. Until recently, there were no federal standards to which local landfills had to adhere, and while some communities and states regulated the environmental impact of dumps, many did not. Even if no commercial or industrial waste has been dumped at the site, most landfills eventually produce *leachate* (chemically contaminated drainage), liquids that pollute the groundwater when they leak from the landfill. Leachate forms when precipitation entering the landfill percolates through the decomposing materials. Thus, heavy metals are leached from batteries and old electrical parts, while organic chemicals drain out of leftover paint and other household products. Typical leachate contains a multitude of organic and inorganic chemicals, many of them poisonous. Environmental

regulations now require that clay or plastic liners be used to protect groundwater supplies and that systems to extract water and methane be installed.

New York City's largest landfill, Fresh Kills on Staten Island, illustrates the problem. For more than one-half century, trucks and barges daily carried approximately 9980 metric tons (11,000 tons or 22 million lbs) of residential waste to the site. Opened in 1948 as a temporary, 200-hectare (500-acre) facility, Fresh Kills was not constructed to hold its contents securely. Located in an ecologically sensitive wetland area and adjacent to residential communities, it expanded to become a malodorous 1200 hectares (3000 acres) of decomposing garbage. Its four large mounds are as much as 70 meters (225 ft) high, taller than the Statue of Liberty. Every day thousands of gallons of leachate seeped into the groundwater beneath the landfill. Fresh Kills finally closed as a depository for municipal solid waste in 2001, although it temporarily reopened the same year to accept debris from the towers of the World Trade Center. New York City is turning the site into the largest park in the city, with sports fields, concert arenas, wildlife sanctuaries, and the like. The conversion is technically complicated and will take years. Construction cannot be completed until the garbage decomposes, the mountains of trash settle, and the polluted water dissipates.

The closing of Fresh Kills made disposal of New York City's trash more complex and expensive. Sanitation trucks, tractor trailers, and railroads now carry most of the waste out of state to landfills and incinerators in Pennsylvania, Virginia, South Carolina, Georgia, and Ohio. Many communities face similar problems in disposing of their solid waste. Finding acceptable locations for landfills is a major problem because **Not in My Backyard (NIMBY)** protesters rise up in opposition nearly everywhere the locations

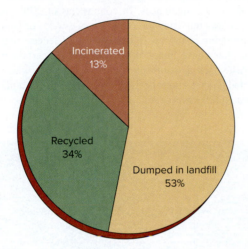

Figure 13.27   **Methods of solid-waste disposal in the United States, 2010.**

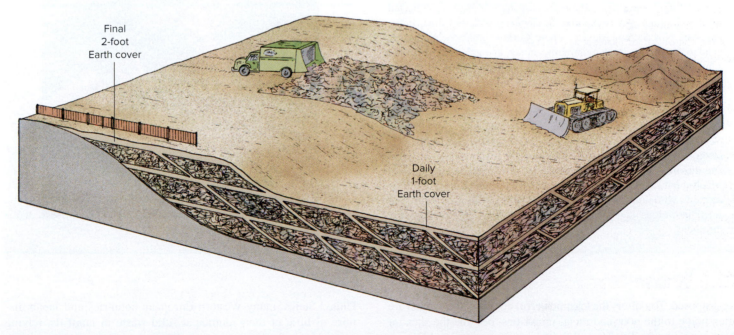

Figure 13.28   **A sanitary landfill.** Each day's deposit of refuse is compacted and isolated in a separate cell by a covering layer of soil or clay. Although far more desirable than open dumps, sanitary landfills pose environmental problems of their own, including potential groundwater contamination and seepage of methane and hydrogen sulfide, gaseous products of decomposition. By federal law, modern landfills must be lined with clay and plastic, equipped with *leachate* (chemically contaminated drainage from the landfill) collection systems to protect the groundwater, and monitored regularly for underground leaks—requirements that have increased significantly the cost of constructing and operating landfills.

E-WASTE

How many cell phones, gaming systems, and computers have you owned in your lifetime? Where are those old devices now? What impact do those obsolete devices have on the environment?

Technological innovation gives consumers a constant stream of new electronic devices to choose from, each with a dizzying array of new features. Unfortunately, this stream of innovation also produces a steady stream of discarded, obsolete devices. Electronic waste, commonly called **e-waste**, consists of discarded consumer electronic products used for data processing, telecommunications, and entertainment in private households and businesses. Those products include computers, computer monitors and printers, copiers, cellular phones, televisions and the like. Because it is often cheaper and easier to buy a new device than to repair or upgrade an old one, millions are replaced each year. In the United States, the average cell phone is replaced after just 18 months. The United Nations estimates that 42 million metric tons of e-waste is generated each year.

Some of the components in electronics can be removed and recycled—glass, steel, aluminum, copper, and gold, for example. But the products also contain toxic metals more difficult to separate, including lead, arsenic, barium, selenium, cadmium, and mercury. If the products are not recycled properly, the chemicals inevitably will make their way from landfills and incinerators into the soil, air, and water.

Fax machines, VCRs, camcorders, BlackBerries, CRT televisions and computer monitors, and iPods may have disappeared from use, but they are not completely gone. Because people are uncertain how to dispose of their electronic goods, particularly if the devices still work, an estimated 75% of the discards are stored in our closets, garages, offices, and warehouses. The U.S. EPA estimates that 5 million tons of electronic waste is in storage. About 25% are taken to recycling centers. By some estimates, 80% of e-waste collected for recycling actually is exported to countries such as China, India, Pakistan, Nigeria, and Mexico, where workers break them apart by hand to recover the steel, aluminum, copper, and gold. The child and adult laborers probably don't realize they are handling toxins that can damage their health and the environment.

In Guiyu, in the Guandong province of China, for example, thousands of scavengers who make their living this way are exposed to poisonous waste that creeps into their skin and lungs. The leftover plastics they burn or discard into rivers, irrigation canals, or fields along with other processing residues release toxics such as chlorine and dioxins into the air and create piles of contaminated ash, from which chemicals seep into the land and water. Guiyu's drinking water contains 2400 times the level of lead the World Health Authority considers hazardous, and its soil contains more than 200 times the lead threshold.

E-waste piles are accumulating around the world. Governments are beginning to draft legislation for the environmentally friendly disposal of this waste. In 2005, the European Union issued several directives addressing e-waste recycling and toxic-content problems. Among other things, the directives ban more than trace amounts of six toxic substances, including lead, cadmium, mercury, and chromium-6, from most new electronic products; forbid disposing of untreated e-waste in landfills; and ban the export of toxic wastes. Additionally, they make the collection, recycling, and disposal of e-waste the responsibility of the manufacturers, to whom individuals may return electronic appliances. In contrast, as of 2016, there are no U.S. federal laws requiring the recycling of e-waste or banning its export to developing countries. In the absence of federal regulation, over half the states have imposed landfill bans on certain types of electronic waste such as computer monitors. In states with e-waste rules, regulated products must be taken to a recycler or to a municipal hazardous waste collection center.

© Don Hammond/DesignPics RF

---

are proposed. Too often, the least powerful communities, which are often home to the poor and racial minorities, become the sites for these unwanted facilities (see "Environmental Justice," p. 407).

The number of municipal landfills in the United States shrank significantly during the 1990s, falling from about 6500 in 1990 to fewer than 1800 today. They closed either because they were full or because they could not comply with environmental regulations and were replaced by huge, regional landfills. In contrast to the United States, many Western European countries and Japan dispose of little of their municipal solid waste in landfills, relying instead on incineration and recycling.

Incineration

The quickest way to reduce the volume of trash is to burn it, a practice that was common at open dumps in the United States until it

Environmental Justice

In Houston, a city of about 2 million people, about a quarter of the population is African American. Yet, when researchers examined the location of the city's garbage facilities, they found that 11 of the 13 solid-waste facilities were in mostly African American areas and all five of the city's garbage incinerators were in African American or Hispanic neighborhoods. Thus, when the city proposed building a new dump in a primarily African American neighborhood in 1979 close to houses and a high school, local residents protested and took the waste management company to court, charging it with racial discrimination in the selection of the landfill site. The court decided in favor of the company and the landfill was built.

In 1982, in Warren County, North Carolina, the rural, mostly African American residents were shocked to learn that the state was proposing their county as the site of a hazardous waste landfill for disposing of polychlorinated biphenyls (PCBs). Their protests resulted in more than 500 arrests, but the effort to block the landfill failed. The Warren County activists were the first to use the term *environmental racism*.

Environmental racism refers to any policy or practice that differentially affects or harms individuals, groups, or communities because of their race or color. The harm may be intentional or unintentional. **Environmental justice** is the fair treatment of all people regardless of race, color, national origin, or income with respect to environmental laws, regulations, and policies. No group should bear a disproportionate share of negative environmental burdens and every group should exercise meaningful participation in decision making regarding the environments where they live, work, and play. In many cases, environmental injustices result from long-established and unexamined structural inequalities that shape access to political power and to housing in desirable locations. In many cases, locally unwanted land uses such as waste disposal facilities have been located in low-income and minority communities because there they faced the least political opposition.

The problem of environmental injustice is global. In a controversial 1991 internal memorandum, World Bank chief economist Lawrence Summers pointed out the economic logic of encouraging polluting industries to move from developed countries to low-income countries. Within many developing countries, residents of communities most burdened with polluting industries, waste disposal facilities, or toxic air, soils, and water tend to be the poor and ethnic minorities. In South Africa, poor black African residents of the Kagiso Township near Johannesburg live beside gold mine tailings piles. Dust high in silica blows from the tailings piles into the community, leading to an irreversible respiratory disease called silicosis which is marked by shortness of breath, fatigue, and vulnerability to lung cancer and tuberculosis. In Cairo, Egypt, the persecuted Coptic Christian minority live in the garbage dumps and make their living collecting and recycling the city's waste.

The United States has taken some steps to rectify environmental injustices. During the Clinton administration, an Executive Order made environmental justice a priority and the U.S. Environmental Protection Agency established an Office of Environmental Justice. Still, researchers using spatial analyses of environmental conditions and human populations continue to find disturbing inequities. The American Lung Association found that African American neighborhoods had air pollution levels 1.5 times higher than other neighborhoods and that pollution levels rose as incomes decreased. The Environmental Justice and Health Alliance for Chemical Policy Reform studied the populations living just outside the fenceline at industrial facilities using or storing hazardous chemicals. They found that the residents vulnerable to a chemical disaster were

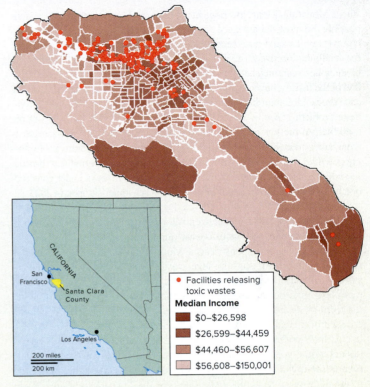

Poverty and environmental risk. The distribution of household income and facilities that release toxic materials in Santa Clara County, California. (*Source: Data from M. R. Meusar and A. Szosz. "Environmental Inequality in Silicon Valley." www.mapcruzin.com/EI/index.hml.*)

disproportionately African American or Hispanic, low income, and less educated. Or, consider Louisiana's Chemical Corridor, or Cancer Alley, a 130-kilometer (80-mi) stretch of the Mississippi River that is home to a chemical plant every half mile and a predominantly poor, African American population. The environmental justice movement represents a broadening of the environmental movement by combining the concerns of the environmental movement with those of the civil rights movement. While the environmental justice movement has had some victories, we will still have environmental injustice as long as there are imbalances in political power and in access to housing opportunities.

Considering the Issues

1. Environmentally hazardous facilities—landfills, incinerators— have to be placed somewhere. How should communities and governments make decisions about where to locate such facilities? Should the activism of local residents affect government decisions?

2. How would you or your university react if a company wanted to build a new toxic waste landfill nearby?

3. Where are the landfills, incinerators, wastewater treatment plants, and other less-desirable land uses within your community? Are low-income residents and minorities exposed to greater than average amounts of pollution than the rest of the population?

Yucca Mountain

Yucca Mountain, a long, low ridge in the basin-and-range region of Nevada is possibly the most studied piece of ground on earth. If the U.S. Congress had had its way it would have become America's first permanent repository for the deadly radioactive waste that nuclear power plants generate. Despite years of study and planning, and the expenditure of billions of dollars on construction of the facility, however, opponents of the project appear to have prevailed, and Yucca Mountain has become a symbol of society's inability to solve a basic problem posed by nuclear power production: where to dispose of the used fuel. At present, no permanent disposal site for radioactive waste exists anywhere in the world. Canada, Japan, and the United Kingdom are exploring potential disposal sites. Only Finland has approved a permanent disposal site and begun construction. A deep underground bunker is under construction in an island off the coast of Finland and is scheduled to open in 2023.

In 1982, Congress ordered the Department of Energy (DOE) to construct by 1998 a permanent repository for the spent fuel of civilian nuclear power plants as well as vast quantities of waste from the production of nuclear weapons. Yucca Mountain in southern Nevada was selected as the site for this high-level waste facility, which was intended to store wastes safely for 10,000 years, until radioactive decay had rendered them less hazardous than they are today. Most of the waste would be in the form of radioactive fuel pellets sealed in metal rods; these would be encased in extremely strong glass and placed in steel canisters entombed in chambers 300 meters (1000 ft) below the Nevada desert. The steel containers would corrode in one or two centuries, after which the volcanic rock of the mountain would be responsible for containing the radioactivity. In addition to its geological properties, the Yucca Mountain site was selected because of the arid climate and remote location.

Three areas of concern about the site emerged. First, the area is vulnerable to both volcanic and earthquake activity, which could cause groundwater to well up suddenly and flood the repository. Yucca Mountain itself was formed from volcanic eruptions that occurred about 12 to 15 million years ago; some geologists contended that a new volcano could erupt within the mountain. Seven small cinder cones in the immediate area have erupted in recent times, the latest just 10,000 years ago. In addition, a number of seismic faults lie close to Yucca Mountain. One, the Ghost Dance fault, runs right through the depository site. The epicenters of the 1992 and 2002 earthquakes at Little Skull Mountain were only 19 kilometers (12 mi) from the proposed dump site.

Second, rainwater percolating down through the mountain could penetrate the vaults holding the waste. Over centuries, water could dissolve the waste itself, and the resulting toxic brew could be carried into the groundwater under the mountain and then beyond the boundaries of the repository.

Finally, opponents of the project pointed out that transporting high-level nuclear waste across the country is dangerous and irresponsible. Because each truck, barge, or railcar container would carry more radioactive material than was released by the atomic bombs used in World War II, breaching even one of them in an accident or a terrorist attack would be catastrophic. As one of Nevada's congressional representatives stated, "This is an open invitation to terrorists around the world."

Despite these concerns, the Energy Department in 2002 recommended that Yucca Mountain be designated as the site for a national nuclear waste depository, a plan President Bush and Congress later approved. The project still faced substantial technical, legal, and

was halted by the Clean Air Act of 1970. Concern over air pollution also forced the closure of old, inefficient *incinerators* (facilities designed to burn waste), providing an impetus to design a new generation of the plants. Municipal incinerators in the United States, mostly in the more densely populated Northeast, burn about 12% of the national total of trash; Canada incinerates about 8%. Most municipal incinerators are of the waste-to-energy type, which use extra-high (980°C; 1800°F) temperatures to reduce trash to ash and simultaneously generate electricity or steam, which is sold to help pay operating costs.

A decade ago, incinerators were hailed as the ideal solution to overflowing landfills, but it has become apparent that they pose environmental problems of their own by generating toxic pollutants in both air emissions and ash. Air emissions from incinerator stacks have been found to contain an alphabet soup of highly toxic elements, ranging from *a* (arsenic) to *z* (zinc), and including, among other toxins, cadmium, dioxins, lead, and mercury, as well as significant amounts of such gases as carbon monoxide, sulfur dioxide, and nitrogen oxides. Emissions can be kept to acceptably low limits by installing electrostatic precipitators, filters, and scrubbers to capture pollutants before they are released into the outside air, although the devices add significantly to the cost of the plant.

A greater problem is created by the concentration of toxins, particularly lead and cadmium, in the ash residue after burning.

Incinerators typically reduce trash by 90%; one-tenth remains as ash, which then must be buried in a landfill. In 1994, the U.S. Supreme Court ruled that the ash must be tested for toxicity and handled as hazardous waste if it exceeds federal safety standards. This means it must be disposed of in licensed hazardous waste landfills, which have double plastic linings, moisture collection systems, and tighter operating procedures than do ordinary municipal landfills.

Although the likelihood of pollution from incinerator by-products has sparked strong protest to their construction in the United States, they have become more accepted abroad. The seriousness of the air pollution and toxic ash problems, however, has aroused concern everywhere. In Japan, where more than three-fifths of municipal waste is incinerated, high atmospheric dioxin levels led the Ministry of Health to strengthen earlier emission guidelines. Some European countries called at least temporary halts to incinerator construction while their safety was reconsidered, and increasingly landfills are refusing to take their residue.

Source Reduction and Recycling

The problems and costs associated with landfills and incinerators have spurred interest in two alternative waste-management strategies: source reduction and recycling. By *source reduction,* we mean producing less waste in the first place so as to shrink the

political challenges, however, and the election of President Obama in 2008 brought construction of the depository to an end. Obama's budget eliminated almost all of the funding for the facility, the DOE withdrew its application for a license to build the depository, and a commission was appointed to devise a new strategy for permanent storage of nuclear waste.

Considering the Issues

1. What are the ethics of permitting nuclear power plants to operate when no system exists for disposing of their hazardous wastes?
2. Comment on the paradox that nuclear waste will remain dangerously radioactive for longer than any civilization has existed. Or, consider that plutonium remains dangerously radioactive for 240,000 years, yet it might have been stored in a repository that was intended to safely hold wastes for only 10,000 years.
3. React to the fact that, even if Yucca Mountain opened, it would have been too small. It was designed to accept 70,000 metric tons (77,000 tons) of civilian and military wastes, but the radioactive waste that has accumulated already exceeds that amount.
4. Considering the uncertainties that would attend the irreversible underground entombment of high-level waste, do you think the government should instead pursue aboveground storage in a form that would allow for the continuous monitoring and retrieval of the wastes? Why or why not?

Figure 13.29 **Recycling bins in Barcelona, Spain.** With its high population densities, Europe has high rates of solid-waste recycling. In Barcelona, waste is separated into general waste (gray), plastics (yellow), glass (green), and paper and cardboard (blue). Tactile markers allow the blind to recycle their wastes. © *Aaron Roeth Photography RF.*

volume of the waste stream and lower the monetary and environmental costs associated with landfills and incinerators. Manufacturers can reduce the amount of paper, plastic, glass, and metal they use to package food and consumer products. Since 1977, for example, the weight of plastic soft-drink bottles and of aluminum beverage cans has been reduced by 20% to 30%. Detergents produced in concentrated form and packaged in smaller containers require less water and less plastic to make, and because they are smaller, more of the plastic bottles can be shipped at the same time, reducing fuel use and greenhouse gas emissions.

Another way to reduce the amount of waste needing disposal is by **recycling**, the recovery and reprocessing or reuse of previously used material into new products for the same or another purpose. Aluminum beverage cans usually are recast into new cans, for example, and glass bottles are crushed, melted, and made into new bottles (**Figure 13.29**). Old tires can be shredded and turned into rubberized road surfacing or used as mulch around playground equipment. Many communities collect leaves and other yard waste, which accounts for almost 15% of the typical municipal waste stream, and turn it into compost. Recycling plastics is complicated by the many types of plastic containers in common use, which must be separated before being recycled. Recycled plastic can be converted into fiber for carpeting, playground equipment, insulation for clothing, and other products.

The widespread adoption of municipal recycling programs in the United States now diverts an estimated 34% of trash away from landfills and incinerators. The country currently recycles some 63% of the paper, 27% of the glass bottles, 20% of the aluminum cans, but only 8% of the plastics entering the waste stream. Compared with other methods of solid-waste disposal, recycling has a relatively benign impact on the environment, despite the fact that it takes water, energy, and other resources to recover, process, and convert the materials into new products. Recycling saves natural resources by making it possible to cut fewer trees, burn less oil, and mine less coal. Because it typically takes less energy to make things out of recycled material than out of virgin materials, recycling saves energy. It also reduces the pollution of air, water, and land that stems from the manufacture of new materials and from other methods of waste disposal, and it saves landfill space for materials that cannot be recycled.

With all of these advantages, why isn't recycling more widely practiced in the United States? Japan and a number of Western European countries recycle a far greater proportion of their waste stream. Analysts have offered a number of explanations: (1) the cost of collecting the goods to be recycled, (2) fluctuations in market prices for commodities, and (3) the lack of a ready market for products manufactured from recycled material. Perhaps the most important factor, however, is that in the United States the price of energy historically has been low and the supply abundant, and the true monetary and environmental costs of making things from raw materials rather than recycled ones are hidden.

In less-affluent countries, where manufactured products are expensive but labor is cheap, recycling plays an important role in reducing the amount of solid waste. In many cities in developing countries, such as Manila (the Philippines), Phnom Penh (Cambodia), Cairo (Egypt), and Mexico City, thousands of poor people make a living sifting through the city's trash, looking for recyclable goods—tin cans, copper, wood, electronics, clothing, and the like—that they can sell to middlemen for businesses and industries (**Figure 13.30**). The wastepickers, or scavengers, play a vital role in reducing the amount of trash that must be compacted and covered with a fresh layer of dirt each day. It is estimated that in Indonesian cities, for example, wastepickers reduce total urban refuse by one-third.

Figure 13.30 **Wastepickers.** Some of the estimated 80,000 wastepickers who make a living by scavenging the Promised Land dump in Manila, the Philippines, looking for items to resell. Metal, glass, plastic, paper, cloth, broken toys, and bits and pieces of machinery are all candidates. Wastepicking is an important source of income for many poor families, but the pickers, many of them women and children, usually work long hours in unhealthful conditions. Hundreds of people died here in July 2000, when the mountain of garbage, loosened by a week of monsoon rains, collapsed on their shantytown at the edge of the dump. © Digital Vision/PunchStock RF.

Unfortunately, the environmental conditions under which wastepickers work are not good, and their life expectancy is considerably less than that of the general population. The health risks include injuries from accidents, infections from disease-carrying organisms, and exposure to gases that seep out of the layers of fermenting trash and to hazardous wastes such as dioxin and heavy metals.

Hazardous and Radioactive Wastes

The problems of municipal and household solid-waste management are daunting; those of treatment and disposal of hazardous and radioactive wastes seem overwhelming.

Hazardous Waste

As we have seen, the waste stream often contains highly toxic and hazardous materials that can harm human health and/or the environment. The terms *toxic* and *hazardous* frequently are used interchangeably, as we shall do here. More strictly defined, *toxic waste* is a relatively limited concept, referring to substances that are poisonous and can cause death or serious injury to humans and other organisms. *Hazardous waste* is a broader term referring to all wastes that pose an immediate or long-term human health risk or that endanger the environment. The discarded material, liquid or solid, contains substances that have one or more of the following characteristics: (1) ignitability (e.g., gasoline), (2) corrosiveness (e.g., strong acids), (3) explosiveness (e.g., nitroglycerine), and (4) toxicity (e.g., PCBs).

The major producers of hazardous wastes are the chemical and petrochemical industries, mining, and electric generation. The EPA has classified more than 400 substances as hazardous, and currently about 10% of industrial waste materials are so categorized.

Such wastes contaminate the environment in different ways and by different routes. Because most hazardous debris is disposed of by dumping or burial on land, groundwater is at greatest risk of contamination. In all industrial countries, at least some drinking water contamination from highly toxic solvents, hydrocarbons, pesticides, trace metals, and PCBs has been detected. Finally, careless or deliberate distribution of hazardous materials outside of confinement areas can cause unexpected, but deadly, hazards. Although methods of disposal other than containment techniques have been developed—including incineration, infrared heating, and bacterial decomposition—none is fully satisfactory.

Radioactive Waste

Every facility that either uses or produces radioactive materials generates *low-level waste*, material whose radioactivity will decay to safe levels in 100 years or less. Nuclear power plants produce about one-half of the total low-level waste in the form of material from decommissioned reactors, used resins, filter sludges, lubricating oils, and detergent wastes. Industries that manufacture radio-pharmaceuticals, smoke alarms, and other consumer goods produce such wastes in the form of machinery parts, plastics, and organic solvents. Research establishments, universities, and hospitals also produce radioactive waste materials.

Because low-level waste is generated by so many sources, its disposal is particularly difficult to control. Evidence indicates that much of it has been placed in landfills, often the local municipal dump, where the waste chemicals may leach through the soil and into the groundwater. By EPA estimates, the United States contains at least 25,000 legal and illegal dumps with hazardous waste. As many as 2000 are deemed potential ecological disasters.

Waste can remain radioactive for 10,000 years and more; plutonium stays dangerously radioactive for 240,000 years. High-level waste consists primarily of spent fuel assemblies of nuclear power reactors—termed *civilian waste*—and waste generated as a by-product of the manufacture of nuclear weapons, or *military waste*. The volume of civilian waste alone is not only great but increasing rapidly, because approximately one-third of a reactor's rods need to be disposed of every year.

By 2016, nearly 76,000 metric tons (84,000 tons) of spent-fuel assemblies were being stored either indoors in the containment pools of America's commercial nuclear power reactors or outdoors in sealed containers made of steel and concrete, awaiting more permanent disposition (**Figure 13.31**). Several thousand tons more are added annually. *Spent fuel* is a misleading term: the assemblies are removed from commercial reactors not because their radiation is spent but because they have become too radioactive for further use. The assemblies will remain radioactively "hot" for thousands of years.

Unfortunately, no country except possibly Finland has yet solved the problem of how to safely dispose of the radioactive waste it already has, not to mention the added waste an expansion of nuclear power would produce. Until 1970, the United States, Britain, France, and Japan sealed wastes in protective tanks and dumped them at sea, a practice that now has been banned worldwide. Cardboard boxes containing wastes contaminated with plutonium have been rototilled into the soil, on the assumption that the Earth would dilute and absorb the radioactivity. The Netherlands is reported to have incinerated some radioactive wastes at sea.

In the United States, much low-level radioactive waste has been placed in tanks and buried in the Earth at 13 sites operated by the U.S. Department of Energy and 3 sites run by private firms. Millions of cubic feet of high-level military waste are temporarily stored in underground tanks at 4 sites: Hanford, Washington; Savannah River, South Carolina; Idaho Falls, Idaho; and West Valley, New York. Several of these storage areas have experienced leakages, with seepage of waste into the surrounding soil and groundwater. The most problematic may be the 177 tanks buried underground at the Hanford site. At least 66 of the giant tanks—some with a capacity of 4.2 million liters (1.1 million gals)—have leaked, and about 4.2 million liters (1 million gals) have seeped into the soil, raising the fear that the highly radioactive waste had already reached underground water supplies and was flowing toward the Columbia River.

Many scientists believe that deep geologic burial is the safest way to permanently dispose of long-lived radioactive waste. This would involve packaging high-level solid and liquid wastes in containers that are buried deep underground. The United States has

(a)

(b)

Figure 13.31   Radioactive waste. (a) Highly radioactive spent-fuel rods from nuclear power plants are stored in water-filled pools for an extended period of time until they have cooled sufficiently. Water must be recirculated through heat exchangers to prevent the water from boiling away and the fuel from overheating. The 2011 Fukushima nuclear accident in Japan was triggered when a power failure prevented the recirculation pumps from cooling the spent-fuel ponds. **(b)** Typically after about 5 years in a cooling water pond, spent nuclear fuel has cooled enough that it can be transferred to dry, aboveground casks. *(a) © Steve Allen/DigitalVision/Getty Images RF; (b) Source: Office of Civilian Waste Management. Department of Energy.*

halted construction on its first permanent high-level waste repository in Nevada (see "Yucca Mountain"). Finland has selected a site for its repository—the island of Olkiluoto—and Sweden is expected to do so soon. Winning local support for a burial site is difficult, however, and attempts to do so in Britain, France, Switzerland, and Japan have not yet been successful.

Solid waste will never cease to be a problem, but its impact on the environment can be lessened by reducing the volume of waste that is generated, eliminating or reducing the production of toxic residues, halting irresponsible dumping, and finding ways to reuse the resources that waste contains. Until then, current methods of waste disposal will continue to pollute soil, air, and water.

Summary of Key Concepts

- We began this chapter with a description of the harmful algal blooms in Lake Erie. This was a reminder that human activities can despoil the environment upon which all life depends. We saw that what humans do on the land affects the health of the waters.

- Humans are part of the natural environment and depend, literally, for their lives on the water, air, and other resources contained in the biosphere. But people have subjected the intricately interconnected systems of that biosphere—the troposphere, the hydrosphere, and the lithosphere—to profound and frequently unwittingly destructive alteration.

- All human activities have effects on the environment, effects that are complex and never isolated. An external action that impinges on any part of the web of nature inevitably triggers chain reactions, the ultimate consequences of which appear never to be fully anticipated.

- Efforts to control the supply of water alter both the quantity and the quality of water in streams, and structures such as dams and reservoirs often have unintended side effects. In many parts of the world, increased demand for fresh water has led to a lack of adequate supplies.

- Pollutants associated with agriculture, industry, and other activities have degraded the quality of freshwater supplies, although regulatory efforts have brought about major improvements in some areas in recent years.

- Combustion of fossil fuels has contributed to serious problems of air pollution. Some manifestations of that pollution, such as acid rain and depletion of the ozone layer, are matters of global concern.

- Activities such as agriculture and mining have long helped shape local landscapes, producing a variety of landforms. In the 20th century, world population growth and economic expansion accelerated the degradation of air, water, and soil over much of the world.

- People affect other living things—plants and animals—by importing them to areas where they did not previously exist, and by disrupting their habitats, hunting them, and using biocides to eradicate them. The greatest human impact is occurring in the tropical rain forests, where agriculture, industrialization, and urbanization are contributing to the extinction of hundreds of species a year.

- All methods of waste disposal have potential effects on air quality or water quality. Often the poor and ethnic minorities are disproportionately affected by polluted environments.

Key Words

Thinking Geographically

1. Sketch and label a diagram of the *biosphere*. Briefly indicate the content of its component parts. Is that content permanent and unchanging? Explain.

2. How are the concepts of *ecosystem, niche,* and *food chain* related? How does each add to our understanding of the "web of nature"?

3. Draw a diagram of or briefly describe the *hydrologic cycle*. How do population growth, urbanization, and industrialization affect that cycle?

4. Is all environmental pollution the result of human action? Given that many pollutants are naturally occurring substances, when can we say that pollution of a part of the biosphere has occurred?

5. Describe the chief sources of water pollution. What steps have the United States and other countries taken to control water pollution?

6. What factors affect the type and degree of air pollution found at a place? What is *acid rain* and where is it a problem? Describe the relationship of *ozone* to *photo-chemical smog*. Why has the ozone layer been depleted?

7. What kinds of landforms has excavation produced? Dumping? What are the chief causes and effects of subsidence?

8. Briefly describe the chief ways that humans affect plant and animal life. What is meant by *biological magnification*? Why may the use of biocides be self-defeating?

9. What methods do communities use to dispose of solid waste? What ecological problems does solid-waste disposal present? How does the government define *hazardous waste,* and how is it disposed of?

10. What activities result in the production of high-level radioactive waste, why is it difficult to manage and dispose of?

11. Using the IPAT equation, discuss how the environmental impacts of a society vary from country to country.

A map projection is a system for displaying the curved surface of the earth on a flat sheet of paper. No matter how one tries to "flatten" the world, it can never be done in such a way as to show all earth details in their correct relative sizes, shapes, distances, or directions. Something is always wrong, and the cartographer's task is to select and preserve those earth relationships important for the purpose at hand and to minimize or accept those distortions that are inevitable but unimportant.

If we look at a globe directly, only the front is visible; the back is hidden. To make a world map, we must decide on a way to flatten the globe's curved surface on the hemisphere we can see. Then we have to cut the globe map down the middle of its hidden hemisphere and place the two back quarters on their respective sides of the already visible front half. In simple terms, we have to "peel" the map from the globe and flatten it in the same way we might try to peel an orange and flatten the skin (**Figure A.1**). Inevitably, the peeling and flattening process will produce a map that either shows tears or breaks in the surface or is subject to uneven stretching or shrinking to make it lie flat.

Of course, mapmakers do not physically engage in cutting, peeling, flattening, or stretching operations. Their task, rather, is to construct or project on a flat surface the network of parallels and meridians of the globe grid, or *graticule*. This can be done in a number of ways. Before we discuss them, it is important to note that two types of circles appear on a globe's spherical grid. A **great circle** is formed on the surface of a sphere by a plane that passes through the center of the sphere. Thus, the equator is a great circle, and each meridian is one-half of a great circle. Every great circle bisects the globe, dividing it equally into hemispheres. An arc segment of the great circle joining them is the shortest distance between any two points on the earth's surface. A **small circle** is the line created by the intersection of a spherical surface with a plane that does *not* pass through its center. Except for the equator, all parallels of latitude are small circles. Different projections will represent great and small circles in different ways.

Geometric Projections

Although all projections can be described mathematically, some can be thought of as being constructed by geometric techniques rather than by mathematical formulas. In geometric projections, the grid system is, in theory, transferred from the globe to a geometric figure, such as a cylinder or a cone, which, in turn, can be cut and then spread out flat (or *developed*) without any stretching

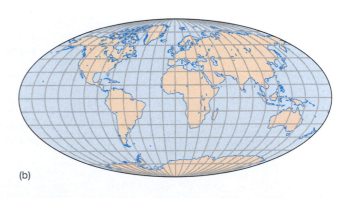

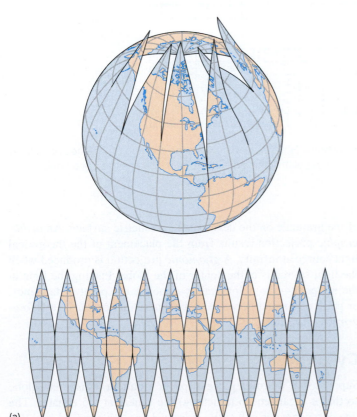

Figure A.1 **(a)** A careful "peeling" of the map from the globe yields a set of tapered gores, which, although individually not showing much stretching or shrinking, do not collectively result in a very useful world map. **(b)** It is usually considered desirable to avoid or reduce the number of interruptions by depicting the entire global surface as a single flat circular, oval, or rectangular shape. That continuity of area, however, can be achieved only at the cost of considerable alteration of true shapes, distances, directions, and/or areas. Although the Mollweide projection shown here depicts the size of areas correctly, it distorts shapes.

(a)

(b)

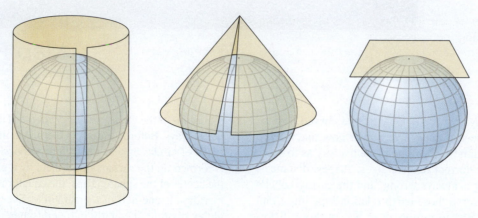

Figure A.2   Geometric projections. The three common geometric forms used in such projections are the cylinder, the cone, and the plane. One can think of making a map projection by imagining a transparent globe with a light source inside it and a sheet of paper touching the globe in one of the ways shown here. The globe grid and the outline of the continents would be silhouetted on the paper to form the map.

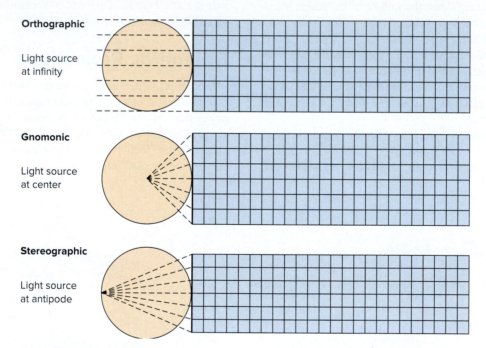

Figure A.3   The effect of the light source location on *planar* surface projections. Note the variations in the spacing of the lines of latitude that occur when the light source is moved. Completely different map grids would result from using a cylinder or cone as the developable surface.

or tearing (**Figure A.2**). The surfaces of cylinders, cones, and planes are said to be **developable surfaces**—cylinders and cones can be cut and laid flat without distortion, and planes are flat to begin with. In actuality, geometric projections are constructed not by tracing shadows but by applying geometry and using lines, circles, arcs, and angles drawn on paper.

Imagine a transparent globe with a light source located either inside or outside the globe. Lines of latitude and longitude (or of coastlines or any other features) drawn on that globe will cast shadows on any nearby surface. A tracing of that shadow globe grid would represent a geometric map projection. As **Figure A.3** shows, the location of the light source in relation to the globe surface causes significant variation in the projection

of the graticule on the developable geometric surface. An *orthographic* projection results from the placement of the theoretical light source at infinity. A *gnomonic* projection is produced when the light source is at the center of the globe. Placing the light at the antipode—the point exactly opposite the point of tangency, or point of contact between globe and map—produces a *stereographic* projection.

Cylindrical Projections

Suppose that we roll a piece of paper around a transparent globe so that it is tangent to (touching) the sphere at the equator. The line of tangency is called the *standard line* (or *standard parallel,* if it is a parallel of latitude); along it, the map has no distortion. Instead of the paper being the same height as the globe, however, it extends far beyond the poles. For a gnomonic projection, we would place a light source at the center of the globe, and the light would project a shadow map upon the cylinder of paper. The result is one of many **cylindrical projections**, all of which are developed geometrically or mathematically from a cylinder wrapped around the globe.

Note the variance between the grid we have just projected and the true properties of the globe grid. The grid lines of latitude and longitude intersect one another at right angles, as they do on the globe, and they are all straight north–south or east–west lines. But the meridians do not converge at the poles, as they do on a globe. Instead, they are equally spaced, parallel, vertical lines. Because the meridians are everywhere equally far apart, the parallels of latitude have all become the same length. Although there is no scale distortion along the line of tangency, the equator, distortion increases with increasing distance away from it. The polar regions are stretched north and south as well as east and west, and their sizes are enormously exaggerated. The poles themselves can never be shown on a cylindrical projection tangent at the equator with a light source at the center of the globe.

The mathematically derived **Mercator projection** was inspired by the idea of a cylinder tangent at the equator. It is one of the most commonly used (and misused) cylindrical projections. Invented in 1569 by Gerardus Mercator, the Mercator projection was created to serve as a navigational chart during a time when European exploration of other parts of the world was at its height. The Mercator is the standard projection used by navigators because

of a peculiarly useful property: a straight line drawn anywhere on the map is a line of constant compass bearing. If such a line, called a **rhumb line**, is followed, a ship's or a plane's compass will show that the course is always at a constant angle with respect to geographic north (**Figure A.4**). On no other projection is a rhumb line both straight and true as a direction.

Although it is an excellent navigational aid, the Mercator projection frequently has been misused in book maps or on wall maps as a general-purpose world map—misused because it gives grossly exaggerated impressions of the size of land areas away from the tropics. Notice on Figure A.4 that Greenland appears many times bigger than Mexico, when in fact it is only slightly larger, and that Alaska and Brazil appear to be about the same size; in actuality, Brazil is more than five times as large.

A number of cylindrical projections that are neither equal area nor conformal, such as the *Miller cylindrical projection* shown in **Figure A.5**, are often used as bases for world maps. The spacing between parallels of latitude on the Miller projection does not increase as rapidly toward the poles as it does on the Mercator, so there is less distortion of the size of areas in the higher latitudes. Although it retains no globe qualities, the Miller cylindrical projection is used in atlases and wall maps.

Conic Projections

Of the three developable geometric forms—cylinder, cone, and plane—the cone is the closest in form to one-half of a globe. **Conic projections**, therefore, often are used to depict hemispheres or smaller parts of the earth.

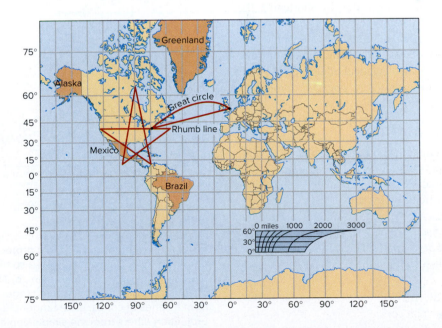

Figure A.4   Distortion on the Mercator projection. A perfect five-pointed star was drawn on a globe, and the latitude and longitude of the points of the star were transferred to the Mercator map shown here. The manner in which the star is distorted reflects the way the projection distorts land areas. The enlargement of areas with increasing latitude is so great that a Mercator map should not be published without a diagram showing the varying scale of distance at different latitudes. The most significant property of the Mercator projection is that it is the only one on which any straight line is a line of constant compass bearing, or *rhumb line*. Although a rhumb line usually is not the shortest distance between two points, navigators can draw a series of straight lines between the starting point and the destination that will approximate the great circle route.

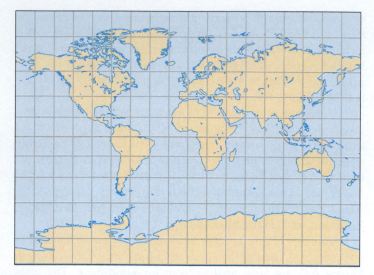

Figure A.5   **The Miller cylindrical projection** is mathematically derived.

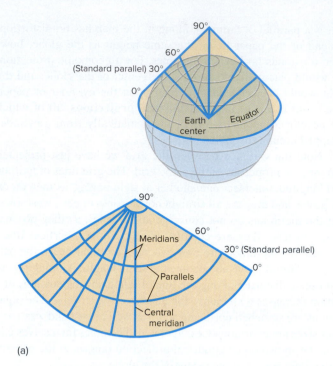

(a)

A useful projection in this category, and the easiest to visualize, is the *simple conic* projection. Imagine that a cone is laid over one-half of the globe, as in **Figure A.6a**, tangent to the globe at the 30th parallel. Distances are true only along this standard parallel. When the cone is developed, of course, the standard parallel becomes an arc of a circle, and all other parallels become arcs of concentric circles. With a central light source, the parallels become increasingly far apart as they approach the pole, and distortion is accordingly exaggerated.

One can lessen the amount of distortion by shortening the length of the central meridian, spacing the parallels of latitude at equal distances on that meridian, and making the 90th parallel (the pole) an arc rather than a point. Most of the conic projections that are in general use employ such mathematical adjustments. When more than one standard parallel is used, a *polyconic* projection results (**Figure A.6b**).

Conic projections are widely used because they can be adjusted to minimize distortions and become either equal-area or conformal. By their nature, however, they can never show the whole globe. In fact, they are most useful for and generally restricted to maps of midlatitude regions of greater east–west than north–south extent. Many official map series use types of conic projections. The U.S. Geological Survey, for example, selected the *Albers equal-area conic projection,* shown in **Figure A.7**, for its *National Atlas of the United States of America.* It is an equivalent projection with very little distortion of shape even in an area as large as the United States.

Planar Projections

Planar (or **azimuthal**) **projections** are constructed by placing a plane surface tangent to the globe at a single point. Although the plane may touch the globe anywhere the cartographer wishes, the polar case with the plane centered on either the North Pole or the South Pole is easiest to visualize (**Figure A.8a**).

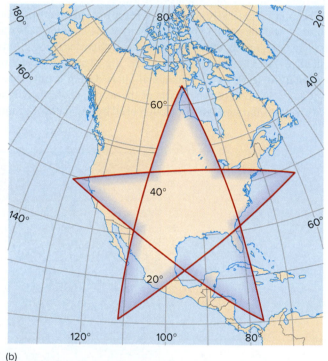

(b)

Figure A.6   (a) **A simple conic projection with one standard parallel.** Most conics are adjusted so that the parallels are spaced evenly along the central meridian. (b) **The polyconic projection.** The map is produced by bringing together east–west strips from a series of cones, each tangent at a different parallel. This projection differs from the simple conic in that the parallels of latitude are *not* arcs of concentric circles and the meridians are curved rather than straight lines. Although neither equivalent nor conformal, the projection portrays shape well. Note how closely the star resembles a perfect five-pointed star, unlike the star shown in Figure A.4.

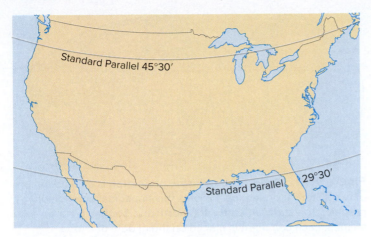

Figure A.7 **The Albers equal-area conic projection**, used for many official U.S. maps, has two standard parallels. All parallels of latitude are concentric arcs of circles, the meridians are straight lines, and parallels and meridians intersect at right angles. The projection is best suited for regions of considerably greater east–west than north–south extent.

This *equidistant* projection is useful because it can be centered anywhere, facilitating the correct measurement of distances from that point to all others. For this reason, it is often used to show air navigation routes that originate from a single place. When the plane is centered on places other than the poles, the meridians and the parallels become curiously curved, as is evident in **Figure A.8b**.

Because they are particularly well suited for showing the arrangement of polar landmasses, planar maps are commonly used in atlases. Depending on the particular projection used, true shape, equivalency, or a compromise between them can be depicted. In addition, one of the planar projections is widely used for navigation and telecommunications. The *gnomonic planar projection,* shown in **Figure A.9**, is the only one on which all great circles (or parts thereof) appear as straight lines. Because great circles are the shortest distances between two points, navigators need only connect the points with a straight line to find the shortest route.

Mathematical Projections

The geometric projections we have just discussed can all be thought of as developed from the projection of the globe grid onto a cylinder, cone, or plane. Many projections, however, cannot be classified in terms of simple geometric shapes. They are derived from mathematical formulas and usually have been developed to display the world or a portion thereof in a fashion that is visually acceptable. Ovals are most common, but hearts, trapezoids, stars, butterflies, and other—sometimes bizarre—forms have been devised for special purposes.

One such projection is *Goode's Homolosine,* developed by the geographer J. Paul Goode as an equal-area projection for statistical mapping. Usually shown in its interrupted form, as in **Figure A.10**, it is actually a product of fitting together the least distorted portions of two different projections (the sinusoidal projection and the Mollweide, or homolographic, projection) and centering the split

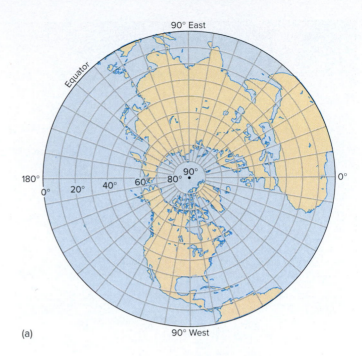

(a)

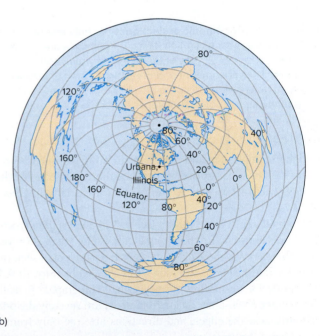

(b)

Figure A.8 **(a) The planar equidistant projection.** Parallels of latitude are circles equally spaced on the meridians, which are straight lines. This projection is particularly useful because distances from the center to any other point are true. If the grid is extended to show the Southern Hemisphere, the South Pole is represented as a circle instead of a point. **(b) A planar equidistant projection centered on Urbana, Illinois.** The scale of miles applies only to distances from Urbana or on a line through it. The scale on the rim of the map, representing the antipode of Urbana, is infinitely stretched. *(b) Source: Copyright 1977, Brooks and Roberts.*

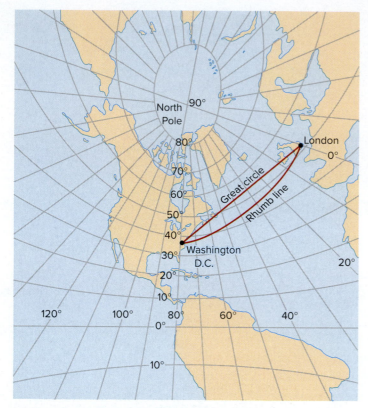

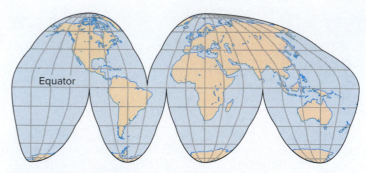

Figure A.9 The **gnomonic projection** is the only one on which all great circles appear as straight lines. Rhumb lines are curved. In this sense, it is the opposite of the Mercator projection, on which rhumb lines are straight and great circles are curved (see Figure A.4). Note that the distortion of shapes and areas increases away from the center point. The map is not conformal, equal-area, or equidistant.

Figure A.10 **Goode's homolosine projection** is a combination of two different projections. It joins the sinusoidal and homolographic projections at about 40° North and South. To improve shapes, each continent is placed on the middle of a lobe approximately centered on its own central meridian. This projection can also interrupt the continents to display the ocean areas intact. *Source: Copyright by the Committee on Geographic Studies, University of Chicago.*

Orthographic projection
centered on Washington, D.C.

Figure A.11 The **azimuthal orthographic projection** is formed by the projection of a hemisphere onto a plane, with the light source infinitely far away. It shows one-half of the globe as it appears from deep space. An old projection, it was used by the Greek astronomer Hipparchus in the second century B.C. but probably was known earlier.

map along multiple standard meridians to minimize the distortion of either land or ocean surfaces. This equal-area projection, which also represents shapes well, is widely used, especially in *Goode's World Atlas.*

One kind of planar projection (see p. A5), the *azimuthal orthographic,* was known to Egyptians and Greeks more than 2000 years ago. It shows a hemisphere of the earth as it would look if seen from deep space, thus matching a person's view of a globe (**Figure A.11**). Used mainly for illustration purposes, the orthographic is neither conformal nor equal-area. Shapes and areas are severely distorted, particularly near the edges, and directions are true only from the center point of the projection.

The azimuthal orthographic and Goode's projections show how projections can be manipulated or adjusted to achieve desired objectives. Since most projections are based on a mathematically consistent rendering of the actual globe grid, the possibilities for such manipulation are nearly unlimited. The map properties to be

retained, the size and shape of areas to be displayed, and the overall map design influence the cartographer's choices in reproducing the globe grid on the flat map.

Some very effective projections are non-Euclidean in origin, transforming space in unconventional ways. Distances may be measured in nonlinear fashion (in terms of time, cost, number of people, or even perception), and maps that show relative space may be constructed from these data. One example of such a transformation is shown in **Figure A.12**.

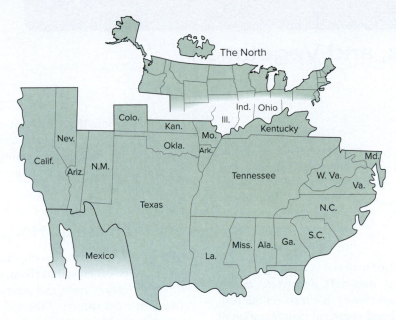

Figure A.12   **Places mentioned in country music lyrics.** In this map transformation, the United States is shown as it might look if the largest states were those most often cited in country music lyrics.

Summary of Key Concepts

It is impossible to transform the globe into a flat map without distortion. Cartographers have devised hundreds of possible geometric and mathematical projections to display to their best advantage the variety of the world's features and relationships they wish to emphasize. Some projections are highly specialized and restricted to a single, limited purpose; others achieve a more general acceptability and utility.

Key Words

azimuthal projection A-4
conic projection A-3
cylindrical projection A-3
developable surface A-2
great circle A-1

Mercator projection A-3
planar projection A-4
rhumb line A-3
small circle A-1

Climate, Soils, and Vegetation

Soils and Climate

Soil is one of the most important components of the physical environment. Life as we know could not exist without soil. Soils play an important role by storing and purifying water and are essential to plant, and thus animal and human, life.

Soil Formation

Soil can be defined as a layer of fine material containing *organic matter* (dead plant and animal material), *inorganic matter* (weathered rock materials), air, and water that rests on the bedrock underlying it. The physical and chemical disintegration of rock, called *weathering* (see Chapter 3), begins the process of soil formation. Weathering breaks down solid rock, ultimately producing finely fragmented mineral particles. These particles, together with the decomposed organic matter that comes to rest on top of them, are transformed into soils by water, heat, and the various living agents (such as bacteria and fungi) that decompose the organic matter. Soil formation is a dynamic process. The physical, chemical, and biological activities that produce it are constantly at work.

Although broad classes of soils extend over large areas of the Earth's surface, the individual characteristics of soil in a given place may differ markedly over very short distances. Variation is due chiefly to the five major factors involved in soil formation.

1. The *geologic factor* is the parent (underlying) rock, which influences the depth, texture, drainage, and nutrient content of a soil.
2. The *climatic factor* refers to the effects of temperature and precipitation on soil. Temperature affects the length of the growing season, the speed of vegetation decay, and the rate of evaporation. Precipitation totals and intensity influence the type of vegetation that grows in an area and thus the supply of *humus,* decomposed organic matter.
3. The *topographic factor* refers to the height of the land, the *aspect* of a slope (which direction it faces), and the angle of slope. All of these affect the amounts of precipitation, cloud cover, and wind; temperature; and the surface runoff of water, drainage, and rate of soil erosion.
4. The *biological factor* refers to both living and dead plants and animals, which add organic matter to the soil and interact in the nutrient cycle. Plants take up mineral nutrients from the soil and return them to it when they die. Microorganisms, such as bacteria and fungi, assist in the decomposition of dead organic matter, whereas larger organisms, such as ants and worms, mix and aerate the soil.
5. The *chronological factor* indicates the length of time the preceding four factors have been interacting to create a particular soil. Recall that soil formation is an ongoing but gradual process. Soils that have been forming for short periods of time retain many of the characteristics of their parent materials. Soils that have been forming for long periods of time are more influenced by soil-forming factors such as climate and organisms.

Soil Profiles and Horizons

Over time, soils tend to develop into layers of various thicknesses. These layers, called **soil horizon**, differ from one another in their structure, texture, color, and other characteristics. A **soil profile** is a vertical cross section of the soil, from the earth's surface down into the underlying parent-material, showing its different horizons (**Figure B.1**).

- The surface layer, or *O-horizon* (*O* for *organic*), consists largely of fresh and decaying organic matter: leaves, twigs, animal droppings, dead insects, and so on.
- Beneath the O-horizon lies the fertile mineral-based *A-horizon,* representing topsoil. Plant nutrients abound, and biological activity and humus content are at their maximum. The humus helps give this horizon a dark color.
- Water percolating through the soil removes some of the organic and mineral matter from the bottom of the A-layer in a process called *eluviation* (outwashing), yielding the lighter-colored *E-horizon*.
- The material removed from the E-horizon is deposited in the *B-horizon,* or zone of illuviation (inwashing). Because it contains little organic matter, it is less fertile than the A-horizon. Depending on the types of minerals that have been deposited in this layer, it may be darker or more brightly colored than the E-horizon.
- The *C-horizon* is where weathering is slowly transforming bedrock into soil particles. The older the soil becomes, and the warmer and wetter the climate, the deeper and more discernible the C-horizon will be.
- The lowest layer, the unaltered bedrock, is designated the *R-horizon* (*R* for *regolith*). The designation is used only if the bedrock is within about 2 meters (6 ft) of the surface.

Soil Properties

The four major components of soil—minerals, organic matter, water, and air—interact to produce distinctive soils. **Soil properties** are the characteristics that enable us to distinguish one type of soil from another.

Soils contain both *organic* and *inorganic matter.* The latter, produced by weathering, consists of minerals, such as quartz, silicate clays, and iron and aluminum oxides. When weathering breaks down rocks into soil particles, the minerals are released to nourish plant growth.

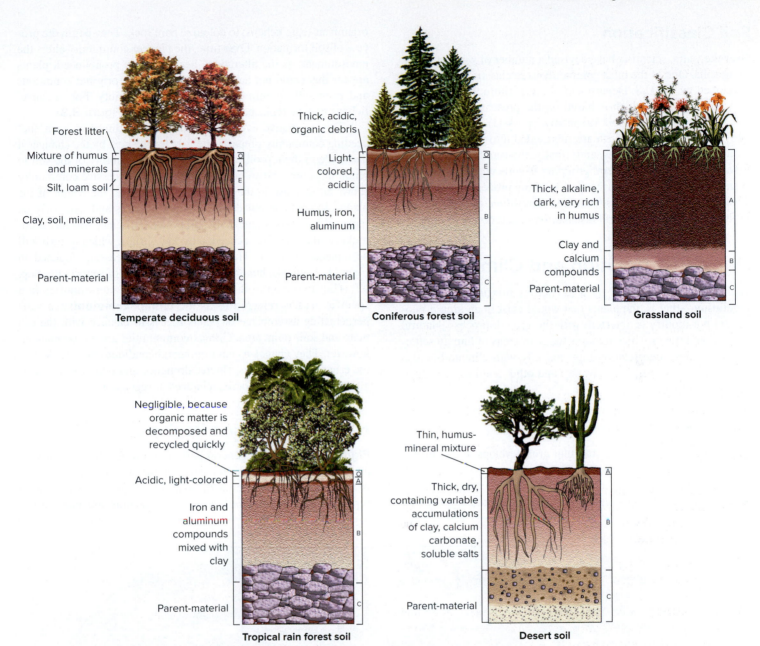

Temperate deciduous soil

Forest litter

Mixture of humus and minerals

Silt, loam soil

Clay, soil, minerals

Parent-material

Coniferous forest soil

Thick, acidic, organic debris

Light-colored, acidic

Humus, iron, aluminum

Parent-material

Grassland soil

Thick, alkaline, dark, very rich in humus

Clay and calcium compounds

Parent-material

Tropical rain forest soil

Negligible, because organic matter is decomposed and recycled quickly

Acidic, light-colored

Iron and aluminum compounds mixed with clay

Parent-material

Desert soil

Thin, humus-mineral mixture

Thick, dry, containing variable accumulations of clay, calcium carbonate, soluble salts

Parent-material

Figure B.1   Generalized soil profiles found in five major ecosystems. The number, composition, and thickness of the soil horizons vary depending on the type of soil. Not shown on these profiles is the lowest layer, the unaltered bedrock, or R-horizon. *Source: From* Biosphere 2000: Protecting Our Global Environment, *3d ed. by Donald G. Kaufman and Cecilia M. Franz (New York: HarperCollins Publishers, 2000), Fig. 16.3, p. 313.*

Texture refers to the size of the mineral matter in the soil and is determined by the proportion of sand, silt, and clay particles. Sand is the largest particle type, followed by silt and then clay. The most agriculturally productive soil texture, called *loam,* is about 40% sand, 40% silt, and 20% clay.

Texture influences *soil structure,* which is defined by the way individual particles aggregate into larger clumps. The size, shape, and alignments of clumps affect the capacity of the soil to hold water, air, and plant nutrients.

Soils vary considerably in the *nutrients* they contain. These chemical elements—such as nitrogen, phosphorus, and calcium—are essential for plant growth and to maintain the fertility of the soil. Soils deficient in nutrients can be made productive by adding fertilizer artificially.

As we have noted, organic matter, or humus, is derived mainly from dead and decaying plant and animal matter. Humus holds water and supplies nutrients to plants. The highest amounts of humus are found in the fertile prairies of North America, the Argentinian pampas, and the treeless grasslands (steppes) of Russia.

High humus content gives the soil a dark brown or black *color,* another soil property. In tropical and subtropical regions, iron compounds can give soil a yellow or reddish color. Light colors (gray or white) often indicate highly leached soils in wet areas and alkaline soils in dry areas. Leached soils are those in which groundwater has dissolved and removed the water-soluble minerals.

The pH scale, discussed in Chapter 13, measures the *acidity* or *alkalinity* of a soil. The most agriculturally productive soils tend to be balanced between being very acidic and very alkaline.

Soil Classification

Over the years, scientists have devised a number of ways of classifying soils. One of the most commonly used classifications is that developed by the U.S. Department of Agriculture, known simply as *Soil Taxonomy,* which is based on the present-day characteristics of the soil (see "Soil Taxonomy," p. A-11). It divides soils into 12 orders, which in turn are subdivided into suborders, great groups, subgroups, families, and, finally, thousands of soil series. A **soil order** is a very general grouping of soils with similar composition, horizons, weathering, and leaching processes. The *sol* at the end of each soil order is from the Latin *solum,* meaning "soil." **Figure B.2** shows the world distribution of soil orders.

Natural Vegetation and Climate

Each climate is typified by a particular mixture of **natural vegetation**—that is, the plants that would exist in an area if people did not modify or interfere with the growth process. Natural vegetation, little of which remains today in areas of human settlement, has close interrelationships not only with climate but also with soils, landforms, groundwater, and other features of habitat, including animals.

Succession

The natural vegetation of a particular area develops in a sequence of stages known as **succession** until a final stage of equilibrium has been reached with the natural environment. Succession usually begins with a relatively simple *pioneer* plant community, the first

organisms (e.g., lichens) to colonize bare rock. They begin the process of soil formation. Over time, the pioneer community alters the environment; as the alterations become more pronounced, plants appear that could not have survived under the original conditions and eventually dominate the pioneer community. For example, lichens may be replaced by mosses and ferns (**Figure B.3**).

This vegetative evolutionary process continues as each succeeding community prepares the way for the next by the changes it has made in the topsoil, the soil structure, the ability of the soil to retain moisture, and so on. In general, each successive community shows an increase in the number of species and the height of the plants. To continue our earlier example, mosses and ferns may be replaced by grasses and herbs, and once sufficient soil has accumulated, they will be succeeded by low shrubs, which in turn will be replaced by trees. The idealized plant succession depicted in Figure B.3 can take hundreds or even thousands of years to develop.

The final step in the succession of plant communities in a specific area is referred to as the **climax community**, a self-perpetuating assemblage of plants that is in balance with the climate and soils of an area. Climax communities are not permanent, however. They change as environmental conditions change. Volcanic eruptions, forest fires, floods, droughts, and other disturbances alter the environment, forcing changes in vegetation.

Natural Vegetation Regions

Figure B.4 shows the general pattern of the Earth's natural vegetation regions. In the hotter parts of the world, where rainfall is heavy and well scattered throughout the year, the vegetation type is *tropical rain forest. Forests,* in general, are made up of

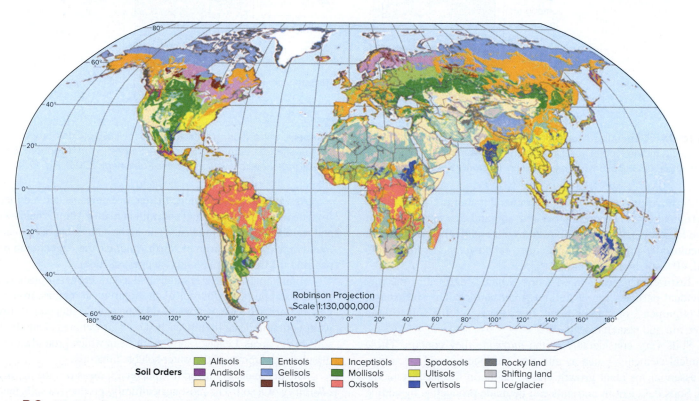

Soil Orders					
Alfisols	Entisols	Inceptisols	Spodosols	Rocky land
Andisols	Gelisols	Mollisols	Ultisols	Shifting land
Aridisols	Histosols	Oxisols	Vertisols	Ice/glacier

Figure B.2   **World soils map.** *Source: Data from U.S. Department of Agriculture, Natural Resources Conservation Service, Soil Survey Division.*

SOIL TAXONOMY

Soil Order	Brief Description
Oxisols	Red, orange, and yellow; highly weathered, leached, and acidic; low fertility; found in the humid tropics of South America and Africa
Ultisols	Red and yellow; highly weathered and leached; less acidic than oxisols; low fertility; develop in warm, wet or dry tropics and subtropics
Alfisols	Gray-brown; moderately weathered and leached; fertile, extremely rich in nutrients; found in humid midlatitudes
Spodosols	Light-colored, sandy A-horizon, red-brown B-horizon; moderately weathered, leached, and acidic; form beneath coniferous forests
Mollisols	Dark brown to black; moderately weathered and leached; extremely rich in nutrients; world's most fertile soil; form beneath grasslands in midlatitudes
Aridisols	Light in color; sandy; usually saline or alkaline; dry and low in organic matter but agriculturally productive if irrigated properly
Inceptisols	Immature, poorly developed soils; form in high-latitude cold climates, especially tundra and mountain areas; limited agricultural potential except in river valleys where seasonal floods deposit fresh layers of sediment
Vertisols	Dark in color; high in clay content; form under grasses in tropical and subtropical areas with pronounced wet and dry periods; fertile but difficult to cultivate
Entisols	Thin and sandy; immature, poorly developed; low in nutrients; found on tundra, mountain slopes, and recent floodplains
Histosols	Black; acidic; composed primarily of organic matter in various stages of decay; waterlogged all or part of the year; found in poorly drained areas (peat bogs, swamps, and meadows) of the upper midlatitudes and tundra; can be fertile if drained
Andisols	Young, immature soils developing on parent-materials of volcanic origin, such as ash and basalt deposits; high organic content; acidic
Gelisols	Formed in areas of permafrost

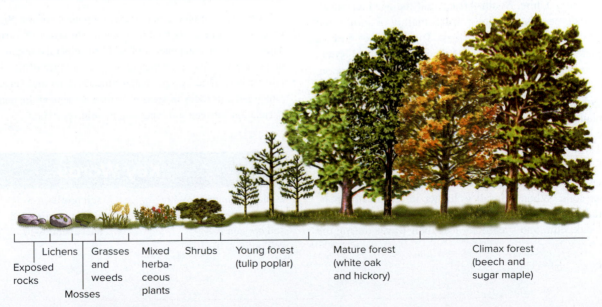

Exposed rocks — Lichens — Mosses — Grasses and weeds — Mixed herbaceous plants — Shrubs — Young forest (tulip poplar) — Mature forest (white oak and hickory) — Climax forest (beech and sugar maple)

Figure B.3 **An idealized plant succession in a temperate deciduous region.** An increasing number of species are usually found at each stage of the succession. The particular species in any one location are determined by local differences in such factors as bedrock, elevation, temperature, sunlight, and rainfall. *Source: From* Biosphere 2000: Protecting Our Global Environment, *3d ed. by Donald G. Kaufman and Cecilia M. Franz (New York: HarperCollins Publishers, 2000), Fig. 5.3, p. 86.*

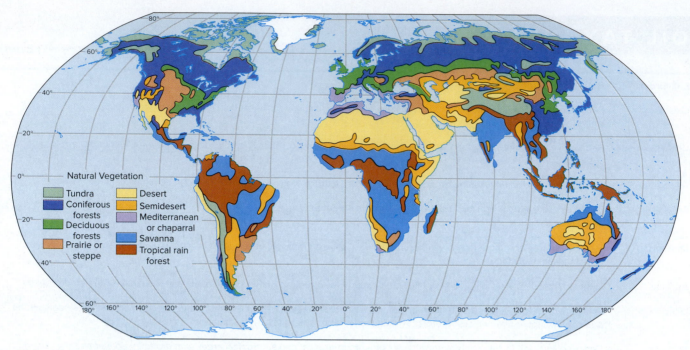

Figure B.4   **World map of natural vegetation.**

trees growing closely together, creating a continuous and overlapping leaf canopy. In the tropics, the forest consists of hundreds of tree species in any small area. Because the canopy blocks the sun's rays, only sparse undergrowth exists. When tropical rainfall is seasonal, *savanna* vegetation occurs, characterized by a low grassland with occasional patches of forests or individual trees. The high evaporation rate denies the savanna region sufficient moisture for dense vegetation.

Mediterranean, or *chaparral,* vegetation is found in the hot summer and mild, damp winter midlatitudes characterizing California, Australia, Chile, South Africa, and the Mediterranean Sea regions. This type of vegetation consists mainly of shrubs and trees of limited size, such as the live oak. Together, they form a low, dense vegetation that is green during the wet season and brown during the dry season. Most dryland areas support some vegetation. *Semidesert* and *desert* vegetation is made up of dwarf trees, shrubs, and various types of cactus, although in gravelly and sandy areas, virtually no plants exist.

In temperate parts of the world with modest year-round rainfall, such as central North America, southern South America, and south-central Asia, the most prevalent type of vegetation is *prairie,* or *steppe.* These are extensive grasslands, usually growing on high-humus-content soils. When rainfall is higher in temperate areas, natural vegetation turns to *deciduous woodlands.* These types of trees, such as oak, elm, and sycamore, lose their leaves during the cold season.

Beyond temperate zones, in northerly regions that have mild summers and very cold winters, *coniferous forests* are common. Evaporation rates are low. Usually, only a few species of trees predominate, such as pines and spruces. Still farther north, forests yield to *tundra* vegetation, a complex mix of very low-growing shrubs, mosses, lichens, and grasses.

Summary of Key Concepts

Soils and vegetation are important components of the physical environment. Over time, soils tend to develop into layers of various thickness. The four major components of soil are minerals, organic matter (vegetation), water, and air, which interact to produce distinctive soils. Each climate is typified by a particular mixture of natural vegetation, little of which remains today in areas of human settlement. In general, there is a correlation between soil types and natural vegetation.

Key Words

climax community A-10
natural vegetation A-10
soil horizon A-8
soil order A-10

soil profile A-8
soil properties A-8
succession A-10

2015 World Population Data

	Population mid–2015 (millions)	Crude Birth Rate (Births per 1000 Population)	Crude Death Rate (Deaths per 1000 Population)	Rate of Natural Increase (%)	Net Migration Rate per 1000	Projected Population (millions) mid-2030	Projected Population (millions) mid-2050	Infant Mortality Rate[a]	Total Fertility Rate[b]	Maternal Mortality Rate, 2013[c]	Percent of Population <15 Years	Percent of Population >64 Years	Life Expectancy at Birth, Both Sexes (years)	Percent Urban	GNI PPP per Capita ($US) 2014[d]	Physiological Density (Persons per Square Kilometer of Arable Land)	Crude Population Density (Persons per Square Kilometer)
WORLD	**7336**	**20**	**8**	**1.2**	**—**	**8505**	**9804**	**37**	**2.5**	**136**	**26**	**8**	**71**	**53**	**15,030**	**523**	**54**
MORE DEVELOPED	1254	11	10	0.1	2	1295	1310	5	1.7	15	16	17	79	77	39,020	238	27
LESS DEVELOPED	6082	22	7	1.5	−1	7210	8495	40	2.6	159	28	6	69	48	9870	696	73
LESS DEVELOPED (Excl. China)	4702	24	7	1.7	0	5779	7120	44	3.0	200	32	5	68	46	8740	612	64
LEAST DEVELOPED	938	34	9	2.5	−1	1300	1887	62	4.3	384	40	4	62	29	2270	521	46
AFRICA	**1171**	**36**	**10**	**2.6**	**0**	**1658**	**2473**	**59**	**4.7**	**412**	**41**	**4**	**60**	**40**	**4720**	**487**	**38**
SUB–SAHARAN AFRICA	**949**	**38**	**11**	**2.7**	**0**	**1369**	**2081**	**64**	**5.0**	**488**	**43**	**3**	**57**	**38**	**3480**	**484**	**40**
NORTHERN AFRICA	**222**	**29**	**6**	**2.3**	**−1**	**289**	**392**	**29**	**3.4**	**118**	**31**	**5**	**71**	**51**	**9740**	**500**	**26**
Algeria	39.9	26	6	2.0	−1	49.9	60.4	21	3.0	89	28	6	74	73	13,540	524	17
Egypt	89.1	31	6	2.5	0	117.9	162.4	22	3.5	45	31	4	71	43	11,020	3196	89
Libya	6.3	21	4	1.7	−11	7.5	8.4	14	2.4	15	29	5	71	78	16,190	359	4
Morocco	34.1	22	6	1.6	−2	38.7	41.9	26	2.5	120	25	6	74	60	7180	425	76
Sudan	40.9	38	9	2.9	−2	61.7	105.0	52	5.2	360	43	3	62	33	3980	193	22
Tunisia	11.0	19	6	1.3	−1	12.3	12.9	16	2.1	46	23	8	76	68	10,600	388	67
Western Sahara[e]	0.6	20	6	1.4	9	0.8	0.8	37	2.4	—	26	3	68	82	—	—	2
WESTERN AFRICA	**349**	**39**	**12**	**2.7**	**−1**	**509**	**784**	**64**	**5.4**	**539**	**43**	**3**	**55**	**45**	**4040**	**402**	**57**
Benin	10.6	37	10	2.7	0	15.1	21.5	67	4.9	340	45	3	59	45	1850	393	94
Burkina Faso	18.5	44	11	3.3	−1	28.4	46.6	69	6.0	400	45	2	56	27	1660	308	68
Cape Verde	0.5	21	6	1.5	−2	0.6	0.7	22	2.4	53	31	6	75	62	6320	1090	126
Côte d'Ivoire	23.3	37	14	2.3	0	32.0	46.3	74	4.9	720	41	3	51	50	3350	805	72

	Population mid–2015 (millions)	Crude Birth Rate (Births per 1000 Population)	Crude Death Rate (Deaths per 1000 Population)	Rate of Natural Increase (%)	Net Migration Rate per 1000	Projected Population (millions) mid-2030	Projected Population (millions) mid-2050	Infant Mortality Rate[a]	Total Fertility Rate[b]	Maternal Mortality Rate, 2013[c]	Percent of Population <15 Years	Percent of Population >64 Years	Life Expectancy at Birth, Both Sexes (years)	Percent Urban	GNI PPP per Capita ($US) 2014[d]	Physiological Density (Persons per Square Kilometer of Arable Land)	Crude Population Density (Persons per Square Kilometer)
WESTERN AFRICA (continued)																	
Gambia	2.0	42	10	3.2	−1	3.1	5.0	47	5.6	430	46	2	59	57	1580	459	180
Ghana	27.7	33	8	2.5	−2	37.7	52.6	41	4.2	380	39	5	61	51	3960	588	116
Guinea	11.0	38	12	2.6	0	16.0	24.2	67	5.1	650	42	3	60	36	1140	366	45
Guinea-Bissau	1.8	37	13	2.4	−1	2.5	3.5	92	4.9	560	43	3	54	49	1430	594	51
Liberia	4.5	36	9	2.7	−1	6.4	9.4	54	4.7	640	42	3	60	47	820	899	41
Mali	16.7	44	15	2.9	−4	26.1	43.6	56	5.9	550	47	3	53	39	1660	245	14
Mauritania	3.6	34	9	2.5	−1	5.0	7.1	72	4.2	320	40	3	63	59	3700	883	4
Niger	18.9	50	11	3.9	0	33.8	68.0	60	7.6	630	52	4	60	22	950	118	15
Nigeria	181.8	39	14	2.5	0	261.7	396.5	69	5.5	560	43	3	52	50	5680	520	197
Senegal	14.7	37	8	2.9	−1	21.5	32.3	33	5.0	320	42	4	65	45	2290	439	75
Sierra Leone	6.5	37	14	2.3	−1	8.3	10.6	92	4.9	1100	41	3	50	41	1830	375	91
Togo	7.2	38	11	2.7	0	10.5	16.3	49	4.8	450	42	3	57	38	1310	273	127
EASTERN AFRICA	**388**	**36**	**9**	**2.7**	**0**	**562**	**841**	**52**	**4.8**	**440**	**43**	**3**	**61**	**24**	**1930**	**560**	**61**
Burundi	10.7	43	10	3.3	0	17.2	30.4	65	6.2	740	46	3	59	10	790	977	383
Comoros	0.8	33	9	2.4	−3	1.0	1.3	36	4.3	350	41	3	61	28	1530	868	346
Djibouti	0.9	27	9	1.8	−3	1.1	1.2	58	3.4	230	34	4	62	77	—	38,827	40
Eritrea	5.2	37	7	3.0	−5	7.3	10.4	46	4.4	380	43	2	63	21	1180	981	44
Ethiopia	98.1	31	8	2.3	0	130.5	165.1	49	4.1	420	41	4	64	17	1500	641	89
Kenya	44.3	31	8	2.3	0	60.1	81.4	39	3.9	400	41	3	62	24	2890	794	76
Madagascar	23.0	34	7	2.7	0	34.3	52.8	38	4.4	440	41	3	65	33	1400	660	39
Malawi	17.2	37	11	2.6	0	24.7	36.6	53	5.0	510	44	3	61	16	780	458	145
Mauritius	1.3	11	8	0.3	−1	1.3	1.2	15	1.4	73	20	9	74	41	18,290	1663	633
Mayotte	0.2	31	2	2.9	−5	0.3	0.5	4	4.1	—	44	3	79	50	—	—	583
Mozambique	25.7	45	13	3.2	0	41.0	72.9	83	5.9	480	45	3	54	31	1170	455	33
Reunion	0.9	17	5	1.2	−3	1.0	1.2	8	2.4	—	24	10	80	94	—	—	341
Rwanda	11.3	31	8	2.3	−1	15.8	21.0	32	4.2	320	41	3	65	28	1530	959	430
Seychelles	0.1	17	8	0.9	6	0.1	0.1	13	2.4	—	22	8	73	54	24,630	9173	184
Somalia	11.1	44	12	3.2	−7	16.9	27.1	79	6.6	850	47	3	55	38	—	985	18
South Sudan	12.2	36	12	2.4	11	17.3	24.8	77	6.9	730	42	3	55	17	2030	—	19
Tanzania	52.3	39	9	3.0	−1	79.4	129.4	37	5.2	410	45	3	62	30	2530	360	55

	Population mid–2015 (millions)	Crude Birth Rate (Births per 1000 Population)	Crude Death Rate (Deaths per 1000 Population)	Rate of Natural Increase (%)	Net Migration Rate per 1000	Projected Population (millions) mid-2030	Projected Population (millions) mid-2050	Infant Mortality Rate[a]	Total Fertility Rate[b]	Maternal Mortality Rate, 2013[c]	Percent of Population <15 Years	Percent of Population >64 Years	Life Expectancy at Birth, Both Sexes (years)	Percent Urban	GNI PPP per Capita ($US) 2014[d]	Physiological Density (Persons per Square Kilometer of Arable Land)	Crude Population Density (Persons per Square Kilometer)
EASTERN AFRICA (continued)																	
Uganda	40.1	40	9	3.1	−1	63.4	104.1	54	5.9	360	48	2	59	18	1690	582	167
Zambia	15.5	43	13	3.0	0	23.7	42.0	75	5.6	280	46	3	53	40	3860	408	20
Zimbabwe	17.4	33	9	2.4	−3	25.2	37.5	55	4.3	470	43	3	61	33	1710	436	44
MIDDLE AFRICA	**149**	**44**	**14**	**3.0**	**0**	**229**	**378**	**96**	**6.1**	**672**	**46**	**3**	**52**	**46**	**2680**	**569**	**22**
Angola	25.0	46	14	3.2	1	39.4	65.5	95	6.1	460	47	2	52	62	7150	421	20
Cameroon	23.7	37	11	2.6	0	34.4	51.9	57	4.9	590	43	3	57	52	2940	383	50
Central African Republic	5.6	45	16	2.9	0	8.5	13.9	109	6.2	880	45	3	50	39	610	307	9
Chad	13.7	48	14	3.4	1	21.8	37.4	95	6.5	980	48	2	51	22	2130	279	10
Congo	4.8	37	10	2.7	−8	6.7	10.2	61	4.8	410	41	3	58	64	5120	870	14
Congo, Dem. Rep.	73.3	46	16	3.0	0	114.9	193.6	108	6.6	730	46	3	50	42	700	1044	31
Equatorial Guinea	0.8	37	13	2.4	5	1.2	1.8	70	5.1	290	39	3	57	39	22,480	667	30
Gabon	1.8	32	9	2.3	1	2.4	3.3	43	4.1	240	38	5	63	86	16,500	523	7
Sao Tome and Principe	0.2	36	7	2.9	−6	0.3	0.4	43	4.3	210	42	4	66	67	3030	2239	190
SOUTHERN AFRICA	**63**	**23**	**10**	**1.3**	**3**	**69**	**77**	**36**	**2.7**	**156**	**31**	**5**	**61**	**59**	**12,290**	**464**	**23**
Botswana	2.1	26	8	1.8	2	2.3	2.5	31	2.9	170	33	5	64	57	17,460	755	3
Lesotho	1.9	31	20	1.1	−5	2.3	3.0	59	3.3	490	36	5	44	27	3260	682	63
Namibia	2.5	29	7	2.2	0	3.3	4.7	39	3.6	130	35	4	64	46	9880	301	3
South Africa	55.0	22	10	1.2	3	59.8	65.2	34	2.6	140	30	6	61	62	12,700	458	45
Swaziland	1.3	30	14	1.6	−1	1.5	1.8	50	3.3	310	37	4	49	21	5940	733	76
AMERICAS	**987**	**16**	**7**	**0.9**	**1**	**1116**	**1221**	**14**	**2.0**	**61**	**24**	**10**	**76**	**80**	**29,900**	**266**	**23**
NORTHERN AMERICA	**357**	**12**	**8**	**0.4**	**3**	**401**	**445**	**6**	**1.8**	**26**	**19**	**15**	**79**	**81**	**54,620**	**178**	**16**
Canada	35.8	11	7	0.4	6	41.0	46.9	5	1.6	11	16	16	81	80	43,400	79	3
United States	321.2	13	8	0.5	3	359.4	398.3	6	1.9	28	19	15	79	81	55,860	207	34
LATIN AMERICA AND THE CARIBBEAN	**630**	**18**	**6**	**1.2**	**−1**	**716**	**776**	**17**	**2.1**	**79**	**27**	**7**	**75**	**80**	**15,260**	**371**	**31**

	Population mid–2015 (millions)	Crude Birth Rate (Births per 1000 Population)	Crude Death Rate (Deaths per 1000 Population)	Rate of Natural Increase (%)	Net Migration Rate per 1000	Projected Population (millions) mid-2030	Projected Population (millions) mid-2050	Infant Mortality Rate[a]	Total Fertility Rate[b]	Maternal Mortality Rate, 2013[c]	Percent of Population <15 Years	Percent of Population >64 Years	Life Expectancy at Birth, Both Sexes (years)	Percent Urban	GNI PPP per Capita ($US) 2014[d]	Physiological Density (Persons per Square Kilometer of Arable Land)	Crude Population Density (Persons per Square Kilometer)
CENTRAL AMERICA	**173**	**20**	**5**	**1.5**	**−2**	**205**	**231**	**14**	**2.4**	**63**	**29**	**6**	**75**	**74**	**14,420**	**585**	**70**
Belize	0.4	21	4	1.7	4	0.5	0.5	13	2.4	45	36	4	74	44	7870	475	19
Costa Rica	4.8	15	4	1.1	2	5.6	6.1	8	1.9	38	23	7	79	73	13,900	1972	94
El Salvador	6.4	18	5	1.3	−8	6.8	6.8	17	2.0	69	31	7	73	67	7720	904	303
Guatemala	16.2	25	5	2.0	−1	21.4	27.5	19	3.1	140	40	5	73	52	7260	1056	149
Honduras	8.3	24	5	1.9	−2	10.2	11.7	22	2.7	120	34	5	74	54	4120	819	74
Mexico	127.0	19	5	1.4	−2	148.1	163.8	13	2.3	49	28	7	75	79	16,710	526	65
Nicaragua	6.3	23	5	1.8	−4	7.4	8.4	16	2.4	100	32	5	75	59	4670	416	48
Panama	4.0	19	5	1.4	2	4.9	5.8	17	2.7	85	28	8	78	78	19,630	744	53
CARIBBEAN	**43**	**18**	**8**	**1.0**	**−4**	**47**	**50**	**28**	**2.3**	**169**	**26**	**9**	**73**	**68**	**12,800**	**793**	**183**
Antigua and Barbuda	0.1	14	6	0.8	0	0.1	0.1	16	1.5	—	24	8	77	30	21,120	2248	177
Bahamas	0.4	15	6	0.9	1	0.4	0.5	14	1.9	37	26	7	74	85	22,310	4708	26
Barbados	0.3	12	9	0.3	2	0.3	0.3	19	1.7	52	20	13	75	46	14,750	2525	644
Cuba	11.1	11	8	0.3	−2	11.2	10.6	4	1.7	80	17	13	78	75	18,710	348	100
Curaçao	0.2	13	8	0.5	1	0.2	0.2	9	2.1	—	19	15	78	—	—	—	444
Dominica	0.1	14	9	0.5	−5	0.1	0.1	20	2.1	—	22	10	75	68	10,300	1133	67
Dominican Republic	10.5	21	6	1.5	−3	11.3	12.2	31	2.5	100	31	6	73	72	12,450	1310	216
Grenada	0.1	17	8	0.9	−2	0.1	0.1	15	2.1	23	26	7	76	41	11,650	3710	334
Guadeloupe	0.4	13	7	0.6	−2	0.4	0.4	9	2.2	—	21	14	81	98	—	—	236
Haiti	10.9	28	9	1.9	−3	13.6	16.9	42	3.2	380	35	4	64	59	1750	1092	392
Jamaica	2.7	18	7	1.1	−5	2.9	2.7	21	2.3	80	24	9	74	52	8490	2268	247
Martinique	0.4	11	8	0.3	−10	0.4	0.4	8	1.9	—	19	17	82	89	—	—	359
Puerto Rico	3.5	10	8	0.2	−15	3.5	3.4	7	1.5	—	18	17	79	99	23,960	5806	394
St. Kitts-Nevis	0.1	14	8	0.6	1	0.1	0.1	13	1.8	—	21	8	75	32	21,990	921	104
St. Lucia	0.2	12	6	0.6	0	0.2	0.2	18	1.5	34	22	9	79	15	10,230	5855	314
St. Vincent and the Grenadines	0.1	17	8	0.9	−8	0.1	0.1	20	2.0	45	25	6	71	51	10,610	2204	278
Trinidad and Tobago	1.4	14	8	0.6	−1	1.3	1.2	13	1.7	84	21	9	75	15	26,220	5375	276

	Population mid–2015 (millions)	Crude Birth Rate (Births per 1000 Population)	Crude Death Rate (Deaths per 1000 Population)	Rate of Natural Increase (%)	Net Migration Rate per 1000	Projected Population (millions) mid-2030	Projected Population (millions) mid-2050	Infant Mortality Rate[a]	Total Fertility Rate[b]	Maternal Mortality Rate, 2013[c]	Percent of Population <15 Years	Percent of Population >64 Years	Life Expectancy at Birth, Both Sexes (years)	Percent Urban	GNI PPP per Capita ($US) 2014[d]	Physiological Density (Persons per Square Kilometer of Arable Land)	Crude Population Density (Persons per Square Kilometer)
SOUTH AMERICA	**414**	**17**	**6**	**1.1**	**0**	**464**	**496**	**18**	**2.0**	**78**	**26**	**8**	**75**	**84**	**14,850**	**309**	**23**
Argentina	42.4	18	8	1.0	0	49.4	58.4	11	2.2	69	24	11	77	93	—	108	16
Bolivia	10.5	26	7	1.9	−1	13.0	15.8	39	3.2	200	31	6	67	69	6130	242	10
Brazil	204.5	15	6	0.9	0	223.1	226.3	19	1.8	69	24	7	75	86	15,900	281	24
Chile	18.0	14	6	0.8	2	19.6	20.2	7	1.8	22	21	10	79	90	21,570	1347	24
Colombia	48.2	19	6	1.3	−1	53.2	54.9	16	1.9	83	27	7	75	76	12,600	3104	43
Ecuador	16.3	21	5	1.6	0	19.8	23.4	17	2.6	87	31	7	75	70	11,120	1425	57
French Guiana	0.3	26	3	2.3	5	0.4	0.6	9	3.5	—	34	5	80	77	—	—	5
Guyana	0.7	21	7	1.4	−7	0.8	0.7	32	2.6	250	27	6	66	29	6930	180	4
Paraguay	7.0	23	6	1.7	−1	8.5	10.1	29	2.8	110	33	5	72	64	8010	159	17
Peru	31.2	20	5	1.5	−1	35.9	40.1	17	2.5	89	29	6	75	79	11,510	761	24
Suriname	0.6	18	7	1.1	−2	0.7	0.7	17	2.3	130	28	6	71	71	15,960	923	4
Uruguay	3.6	14	10	0.4	−1	3.7	3.8	9	1.9	14	21	14	77	93	20,220	204	20
Venezuela	30.6	20	5	1.5	0	36.1	40.5	13	2.5	110	28	6	75	94	17,140	1120	34
ASIA	**4397**	**18**	**7**	**1.1**	**0**	**4939**	**5324**	**33**	**2.2**	**108**	**25**	**8**	**72**	**47**	**11,450**	**938**	**138**
ASIA (Excl. China)	**3017**	**21**	**7**	**1.4**	**0**	**3507**	**3949**	**38**	**2.4**	**145**	**28**	**6**	**70**	**44**	**10,480**	**832**	**135**
WESTERN ASIA	**257**	**22**	**5**	**1.7**	**3**	**321**	**387**	**22**	**2.9**	**54**	**30**	**5**	**74**	**71**	**25,130**	**705**	**53**
Armenia	3.0	14	9	0.5	−6	2.9	2.5	9	1.5	29	19	11	75	63	8550	675	100
Azerbaijan	9.7	18	6	1.2	0	11.0	12.1	11	2.2	26	22	6	74	53	16,910	510	112
Bahrain	1.4	15	2	1.3	5	1.7	1.9	8	2.1	22	21	2	76	100	38,140	88,490	2073
Cyprus	1.2	12	6	0.6	−12	1.3	1.4	5	1.4	10	17	12	80	67	29,800	1260	127
Georgia	3.8	14	12	0.2	−2	4.9	4.7	10	1.7	41	17	14	75	54	7510	944	55
Iraq	37.1	31	4	2.7	2	53.4	76.5	37	4.2	67	41	3	69	71	14,670	1080	85
Israel	8.4	21	5	1.6	1	10.6	13.9	3	3.3	2	28	11	82	91	32,550	2846	380
Jordan	8.1	28	6	2.2	3	9.0	11.4	17	3.5	50	37	3	74	83	11,910	3810	91
Kuwait	3.8	17	2	1.5	22	5.0	6.1	8	2.3	14	23	2	74	98	87,700	35,893	212
Lebanon	6.2	15	5	1.0	31	5.5	5.6	8	1.7	16	26	6	77	87	17,330	2993	597
Oman	4.2	21	3	1.8	45	5.2	5.7	10	2.9	11	22	3	77	75	36,240	13,574	14
Palestinian Territory	4.5	32	4	2.8	−2	6.6	9.2	18	4.1	—	40	3	73	83	5080	9925	742
Qatar	2.4	12	1	1.1	28	2.8	3.0	7	2.0	6	15	1	78	100	133,850	18,750	216

	Population mid–2015 (millions)	Crude Birth Rate (Births per 1000 Population)	Crude Death Rate (Deaths per 1000 Population)	Rate of Natural Increase (%)	Net Migration Rate per 1000	Projected Population (millions) mid-2030	Projected Population (millions) mid-2050	Infant Mortality Rate[a]	Total Fertility Rate[b]	Maternal Mortality Rate, 2013[c]	Percent of Population <15 Years	Percent of Population >64 Years	Life Expectancy at Birth, Both Sexes (years)	Percent Urban	GNI PPP per Capita ($US) 2014[d]	Physiological Density (Persons per Square Kilometer of Arable Land)	Crude Population Density (Persons per Square Kilometer)
WESTERN ASIA (continued)																	
Saudi Arabia	31.6	20	4	1.6	5	39.0	47.1	16	2.9	16	30	3	74	81	53,760	979	14
Syria	17.1	23	7	1.6	−26	26.1	31.2	16	2.8	49	33	4	70	54	—	366	93
Turkey	78.2	17	5	1.2	3	88.4	93.5	11	2.2	20	24	8	77	77	19,040	381	100
United Arab Emirates	9.6	14	1	1.3	8	12.3	15.5	6	1.8	8	16	1	77	83	63,750	19,093	115
Yemen	26.7	33	7	2.6	1	35.7	46.1	43	4.4	270	41	3	65	34	3820	2110	50
SOUTH CENTRAL ASIA	**1903**	**22**	**7**	**1.5**	**−1**	**2227**	**2526**	**45**	**2.5**	**174**	**30**	**5**	**68**	**34**	**6010**	**776**	**176**
CENTRAL ASIA	**69**	**25**	**6**	**1.9**	**−1**	**82**	**96**	**37**	**2.9**	**40**	**29**	**5**	**69**	**47**	**9930**	**219**	**17**
Kazakhstan	17.5	25	8	1.7	0	20.7	24.6	25	3.0	26	25	7	70	53	21,580	76	6
Kyrgyzstan	6.0	27	6	2.1	−1	8.2	11.6	24	4.0	75	32	4	70	36	3220	463	29
Tajikistan	8.5	33	7	2.6	−3	11.2	14.8	40	3.8	44	36	3	67	27	2630	990	59
Turkmenistan	5.4	21	8	1.3	−1	6.2	6.6	46	2.3	61	28	4	65	50	14,520	279	11
Uzbekistan	31.3	23	5	1.8	−1	36.0	38.3	44	2.4	36	28	4	68	51	5840	721	70
SOUTH ASIA	**1834**	**22**	**7**	**1.5**	**−1**	**2145**	**2430**	**45**	**2.5**	**179**	**30**	**5**	**68**	**33**	**5870**	**857**	**270**
Afghanistan	32.2	34	8	2.6	2	45.8	64.3	74	4.9	400	45	2	61	25	1980	415	49
Bangladesh	160.4	20	6	1.4	−3	185.1	201.9	38	2.3	170	33	5	71	23	3340	2089	1114
Bhutan	0.8	18	7	1.1	2	0.9	1.1	47	2.2	120	31	5	68	38	7560	764	17
India	1314.1	21	7	1.4	−1	1512.9	1660.1	42	2.3	190	29	5	68	32	5760	842	400
Iran	78.5	19	5	1.4	−1	90.2	99.3	15	1.8	23	24	5	74	71	16,080	442	48
Maldives	0.3	22	3	1.9	0	0.4	0.6	9	2.2	31	26	5	74	45	12,770	11,565	1110
Nepal	28.0	22	7	1.5	−1	32.4	36.0	33	2.4	190	33	6	67	18	2420	1322	190
Pakistan	199.0	30	7	2.3	−2	254.7	344.0	69	3.8	170	36	4	66	38	5100	939	250
Sri Lanka	20.9	18	6	1.2	−4	22.5	23.0	9	2.3	29	25	8	74	18	10,270	1672	318
SOUTHEAST ASIA	**628**	**20**	**7**	**1.3**	**0**	**737**	**839**	**28**	**2.4**	**131**	**27**	**6**	**71**	**47**	**10,720**	**906**	**139**
Brunei	0.4	17	3	1.4	1	0.5	0.5	4	1.6	27	25	5	79	77	71,020	9796	72
Cambodia	15.4	24	6	1.8	−2	18.1	21.3	28	2.7	170	31	6	64	21	3080	376	85
Indonesia	255.7	21	6	1.5	−1	307.6	366.5	31	2.6	190	29	5	71	54	10,250	1086	135
Laos	6.9	27	6	2.1	−3	8.8	10.6	68	3.1	220	37	4	68	38	4910	475	30
Malaysia	30.8	17	5	1.2	3	36.0	42.3	7	2.0	29	26	6	75	74	23,850	3231	93
Myanmar	52.1	19	9	1.0	−1	56.5	56.5	62	2.3	200	24	5	65	34	—	481	77
Philippines	103.0	23	6	1.7	−1	127.8	157.1	23	2.9	120	34	4	69	44	8300	1857	344

	Population mid–2015 (millions)	Crude Birth Rate (Births per 1000 Population)	Crude Death Rate (Deaths per 1000 Population)	Rate of Natural Increase (%)	Net Migration Rate per 1000	Projected Population (millions) mid-2030	Projected Population (millions) mid-2050	Infant Mortality Rate[a]	Total Fertility Rate[b]	Maternal Mortality Rate, 2013[c]	Percent of Population <15 Years	Percent of Population >64 Years	Life Expectancy at Birth, Both Sexes (years)	Percent Urban	GNI PPP per Capita ($US) 2014[d]	Physiological Density (Persons per Square Kilometer of Arable Land)	Crude Population Density (Persons per Square Kilometer)
SOUTHEAST ASIA (continued)																	
Singapore	5.5	10	5	0.5	14	6.5	7.0	2	1.3	6	16	11	83	100	80,270	879,543	8043
Thailand	65.1	12	8	0.4	0	69.8	66.1	11	1.6	26	18	11	75	49	13,950	393	127
Timor-Leste	1.2	36	8	2.8	−9	1.8	2.8	45	5.7	270	42	5	68	32	5680	775	83
Viet Nam	91.7	17	7	1.0	0	103.2	108.2	16	2.4	49	24	7	73	33	5350	1436	276
EAST ASIA	**1609**	**12**	**7**	**0.5**	**0**	**1654**	**1572**	**11**	**1.6**	**31**	**17**	**12**	**76**	**59**	**16,040**	**1380**	**137**
China	1372	12	7	0.5	0	1422.5	1365.7	12	1.7	32	17	10	75	55	13,130	1293	143
China, Hong Kong SAR[f]	7.3	9	6	0.3	3	8.1	8.6	2	1.2	—	11	15	84	100	56,570	231,314	6670
China, Macao SAR[f]	0.7	12	3	0.9	11	0.7	0.8	3	1.2	—	11	8	83	100	118,460	—	25,463
Japan	126.9	8	10	−0.2	1	116.6	96.9	2	1.4	6	13	26	83	93	37,920	3000	336
Korea, North	25.0	14	9	0.5	0	26.7	27.0	25	2.0	87	22	10	70	61	—	1064	207
Korea, South	50.7	9	5	0.4	3	52.2	48.1	3	1.2	27	14	13	82	82	34,620	3339	509
Mongolia	3.0	28	6	2.2	−1	3.7	4.4	21	3.1	68	27	4	69	68	11,230	487	2
Taiwan	23.5	9	7	0.2	1	23.4	20.4	4	1.2	—	14	12	80	73	—	—	652
EUROPE	**742**	**11**	**11**	**0.0**	**2**	**744**	**728**	**6**	**1.4**	**12**	**16**	**17**	**78**	**73**	**31,650**	**269**	**32**
NORTHERN EUROPE	**103**	**12**	**9**	**0.3**	**4**	**112**	**120**	**4**	**1.8**	**7**	**18**	**17**	**81**	**79**	**40,340**	**522**	**57**
Channel Islands	0.2	10	7	0.3	3	0.2	0.2	3	1.7	—	16	16	82	31	—	3819	804
Denmark	5.7	10	9	0.1	7	6.0	6.3	4	1.7	5	17	19	81	87	46,160	235	132
Estonia	1.3	10	12	−0.2	−1	1.3	1.2	3	1.5	11	16	19	77	68	25,690	212	30
Finland	5.5	10	10	0.0	3	5.8	6.1	2	1.7	4	16	20	81	85	40,000	244	16
Iceland	0.3	13	6	0.7	3	0.4	0.4	2	1.9	4	20	14	82	95	42,530	275	3
Ireland	4.6	15	6	0.9	−5	5.2	5.8	4	2.0	9	22	13	81	60	40,820	395	66
Latvia	2.0	11	14	−0.3	−4	1.6	1.4	4	1.6	13	15	19	74	68	23,150	168	32
Lithuania	2.9	11	14	−0.3	−4	2.7	2.4	4	1.7	11	15	18	74	67	25,390	129	44
Norway	5.2	12	8	0.4	7	5.9	6.7	2	1.8	4	18	16	82	80	65,970	646	14
Sweden	9.8	12	9	0.3	8	11.4	12.4	2	1.9	4	17	20	82	84	46,710	376	22
United Kingdom	65.1	12	9	0.3	4	71.0	77.0	4	1.9	8	18	17	81	80	38,370	1047	268
WESTERN EUROPE	**191**	**10**	**10**	**0.0**	**4**	**198**	**199**	**3**	**1.7**	**7**	**16**	**19**	**81**	**77**	**44,790**	**566**	**173**
Austria	8.6	10	9	0.1	6	9.2	9.5	3	1.5	4	14	18	81	67	45,040	638	102
Belgium	11.2	11	10	0.1	5	12.3	13.1	4	1.8	6	17	18	80	99	43,030	1397	367

	Population mid–2015 (millions)	Crude Birth Rate (Births per 1000 Population)	Crude Death Rate (Deaths per 1000 Population)	Rate of Natural Increase (%)	Net Migration Rate per 1000	Projected Population (millions) mid-2030	Projected Population (millions) mid-2050	Infant Mortality Rate[a]	Total Fertility Rate[b]	Maternal Mortality Rate, 2013[c]	Percent of Population <15 Years	Percent of Population >64 Years	Life Expectancy at Birth, Both Sexes (years)	Percent Urban	GNI PPP per Capita ($US) 2014[d]	Physiological Density (Persons per Square Kilometer of Arable Land)	Crude Population Density (Persons per Square Kilometer)
WESTERN EUROPE (continued)																	
France	64.3	12	8	0.4	0	68.5	72.3	4	2.0	9	19	18	82	78	39,720	352	116
Germany	81.1	8	11	−0.3	5	81.1	76.4	3	1.5	7	13	21	80	73	46,840	685	227
Liechtenstein	0.04	9	7	0.2	4	0.04	0.05	3	1.5	—	15	16	82	15	—	1249	229
Luxembourg	0.6	11	7	0.4	19	0.7	0.7	3	1.5	11	17	14	82	90	57,830	908	245
Monaco	0.04	6	7	−0.1	13	0.04	0.05	—	1.4	—	13	24	—	100	—	—	36,356
Netherlands	16.9	10	9	0.1	2	17.6	17.9	4	1.7	6	17	17	81	90	47,660	1675	408
Switzerland	8.3	10	8	0.2	11	8.7	9.0	4	1.5	6	15	18	83	74	59,600	2057	201
EASTERN EUROPE	**292**	**12**	**13**	**−0.1**	**1**	**280**	**260**	**8**	**1.6**	**19**	**16**	**14**	**73**	**69**	**21,130**	**153**	**16**
Belarus	9.3	13	13	0.0	2	9.1	8.7	4	1.7	1	16	14	73	76	17,610	173	46
Bulgaria	7.2	9	15	−0.6	0	6.6	5.8	8	1.5	5	14	20	75	73	15,850	216	65
Czech Republic	10.6	10	10	0.0	2	10.8	11.1	2	1.5	5	15	17	79	74	26,970	334	134
Hungary	9.8	9	13	−0.4	−3	9.7	9.4	5	1.4	14	15	18	76	69	23,830	224	106
Moldova	4.1	11	11	0.0	−1	3.7	2.9	10	1.3	21	16	10	72	42	5480	227	122
Poland	38.5	10	10	0.0	0	37.2	34.0	4	1.3	3	15	15	78	60	24,090	352	123
Romania	19.8	9	13	−0.4	−4	18.6	16.4	9	1.3	33	16	17	75	54	19,030	226	83
Russia[g]	144.3	13	13	0.0	2	140.4	134.2	9	1.8	24	16	13	71	74	24,710	121	8
Slovakia	5.4	10	9	0.1	0	5.4	5.0	6	1.4	7	15	14	76	54	25,970	389	110
Ukraine[g]	42.8	11	15	−0.4	1	38.2	32.3	10	1.5	23	15	15	71	69	8560	132	70
SOUTHERN EUROPE	**156**	**9**	**10**	**−0.1**	**0**	**154**	**149**	**4**	**1.4**	**6**	**15**	**19**	**81**	**68**	**29,730**	**517**	**119**
Albania	2.9	12	7	0.5	−6	3.0	2.8	8	1.8	21	19	12	78	56	10,260	467	103
Andorra	0.1	9	4	0.5	−7	0.1	0.1	3	1.3	—	15	18	—	86	—	3254	122
Bosnia-Herzegovina	3.7	7	9	−0.2	0	3.5	3.2	5	1.2	8	15	16	75	40	10,020	363	73
Croatia	4.2	9	12	−0.3	−2	4.0	3.6	4	1.5	13	15	18	77	56	20,560	468	74
Greece	11.5	9	10	−0.1	−1	11.1	9.7	4	1.3	5	15	21	81	78	26,130	454	87
Italy	62.5	8	10	−0.2	2	63.5	63.5	3	1.4	4	14	22	83	68	34,710	878	207
Kosovo[h]	1.8	13	4	0.9	−12	1.9	1.9	12	2.3	—	28	7	77	38	9410	—	164
Macedonia[i]	2.1	11	10	0.1	0	2.0	1.8	10	1.5	7	17	13	75	57	12,600	500	80
Malta	0.4	10	8	0.2	3	0.4	0.4	6	1.4	9	15	16	82	95	27,020	4799	1262
Montenegro	0.6	12	10	0.2	−1	0.7	0.8	4	1.6	7	18	14	77	64	14,510	362	45

	Population mid–2015 (millions)	Crude Birth Rate (Births per 1000 Population)	Crude Death Rate (Deaths per 1000 Population)	Rate of Natural Increase (%)	Net Migration Rate per 1000	Projected Population (millions) mid-2030	Projected Population (millions) mid-2050	Infant Mortality Rate [a]	Total Fertility Rate [b]	Maternal Mortality Rate, 2013 [c]	Percent of Population <15 Years	Percent of Population >64 Years	Life Expectancy at Birth, Both Sexes (years)	Percent Urban	GNI PPP per Capita ($US) 2014 [d]	Physiological Density (Persons per Square Kilometer of Arable Land)	Crude Population Density (Persons per Square Kilometer)
SOUTHERN EUROPE (continued)																	
Portugal	10.3	8	10	−0.2	−3	9.9	9.1	3	1.2	8	14	19	80	61	28,010	950	112
San Marino	0.03	9	8	0.1	5	0.03	0.03	2	1.5	—	15	18	87	94	—	3293	530
Serbia	7.1	9	14	−0.5	−2	6.8	6.1	6	1.6	16	14	18	75	60	12,150	216	92
Slovenia	2.1	10	9	0.1	0	2.1	2.0	2	1.6	7	15	18	81	50	28,650	1206	102
Spain	46.4	9	9	0.0	−2	45.4	43.7	3	1.3	4	15	18	83	77	32,860	373	91
OCEANIA	**40**	**18**	**7**	**1.1**	**6**	**48**	**59**	**22**	**2.5**	**54**	**24**	**12**	**77**	**70**	**31,600**	**82**	**4**
Australia	23.9	13	7	0.6	8	28.5	34.0	4	1.9	6	19	15	82	89	42,880	51	3
Federated States of Micronesia	0.1	24	5	1.9	−14	0.1	0.1	29	3.5	96	34	4	70	22	3680	5074	152
Fiji	0.9	21	8	1.3	−6	0.9	1.0	15	3.1	59	29	5	70	51	8030	527	52
French Polynesia	0.3	16	5	1.1	0	0.3	0.3	6	2.0	—	24	7	77	56	—	10,265	69
Guam	0.2	21	6	1.5	−6	0.2	0.2	13	2.9	—	26	8	79	93	—	17,953	291
Kiribati	0.1	30	9	2.1	−1	0.2	0.2	45	3.8	130	36	4	65	54	2580	5600	145
Marshall Islands	0.1	30	4	2.6	−17	0.1	0.1	26	4.1	—	41	3	72	74	4630	2753	182
Nauru	0.01	35	8	2.7	−9	0.01	0.02	33	3.9	—	37	1	66	100	—	—	485
New Caledonia	0.3	15	6	0.9	4	0.3	0.3	5	2.3	—	24	9	77	70	—	4959	14
New Zealand	4.6	13	7	0.6	11	5.2	5.7	6	1.9	8	20	15	81	86	33,760	794	17
Palau	0.02	13	11	0.2	0	0.02	0.02	13	1.7	—	20	6	72	84	14,280	1779	45
Papua New Guinea	7.7	33	10	2.3	0	10.5	14.2	47	4.3	220	39	3	62	13	2510	2443	17
Samoa	0.2	29	5	2.4	−28	0.2	0.2	16	4.7	58	39	5	74	19	5600	2451	66
Solomon Islands	0.6	30	5	2.5	0	0.9	1.4	26	4.1	130	39	3	70	20	2020	3276	19
Tonga	0.1	27	7	2.0	−19	0.1	0.1	17	3.9	120	37	6	76	23	5300	646	138
Tuvalu	0.01	25	9	1.6	0	0.01	0.02	10	3.2	—	33	5	70	59	5260	—	433
Vanuatu	0.3	33	5	2.8	0	0.4	0.5	28	4.2	86	39	4	71	24	2870	1423	21

NOTES

(—) Indicates data unavailable or inapplicable.

[a] Infant deaths per 1000 live births. Rates shown with decimals indicate reported national statistics, those without are estimates.

[b] Average number of children born to a woman during her lifetime.

[c] Number of maternal deaths due to complications in pregnancy or childbirth per 100,000 live births.

[d] Gross national income (GNI) per capita has been adjusted for purchasing power parity (PPP) to account for differences in relative purchasing power.

[e] The status of Western Sahara is disputed by Morocco.

[f] Special Administrative Region.

[g] Does not include the population of Crimea, estimated at 2.3 million.

[h] Kosovo declared independence from Serbia on Feb. 17, 2008. Serbia has not recognized Kosovo's independence.

[i] The former Yugoslav Republic.

Table modified from the Population Reference Bureau's 2015 *World Population Data Sheet*. Crude population densities calculated by authors.

GLOSSARY

A

absolute direction Direction with respect to cardinal east, west, north, and south reference points.

absolute distance The shortest-path separation between two places measured on a standard unit of length (miles or kilometers, usually); also called real distance.

absolute location (*syn:* mathematical location) The exact position of an object or a place stated in spatial coordinates of a grid system designed for locational purposes. In geography, the reference system is the global grid of parallels of latitude north or south of the equator and of meridians of longitude east or west of a prime meridian.

accessibility The relative ease with which a destination may be reached from other locations; the relative opportunity for spatial interaction. May be measured in geometric, social, or economic terms.

acculturation The cultural modification or change resulting from one culture group or individual adopting traits of a more advanced or dominant society; cultural development through "borrowing."

acid rain Precipitation that is unusually acidic; created when oxides of sulfur and nitrogen change chemically as they dissolve in water vapor in the atmosphere and return to Earth as acidic rain, snow, fog, or dry particles.

activity space The area within which people move freely on their rounds of regular activity.

adaptation A presumed modification of heritable traits through response to environmental stimuli.

agglomeration The spatial grouping of people or activities for mutual benefit.

agglomeration economies (*syn:* external economies) The savings to an individual enterprise that result from spatial association with other, similar economic activities.

agricultural density The number of rural residents per unit of agriculturally productive land. The measure excludes a region's urban population and its nonarable land from the density calculation.

agriculture Cultivating the soil, producing crops, and raising livestock; farming.

air mass A large body of air with little horizontal variation in temperature, pressure, and humidity.

air pressure The weight of the atmosphere as measured at a point on the earth's surface.

alluvial fan A fan-shaped accumulation of alluvium deposited by a stream at the base of a hill or mountain.

alluvium The sediment carried by a stream and deposited in a floodplain or delta.

amalgamation theory In human geography, the concept that multiethnic societies become a merger of the culture traits of their member groups.

anaerobic digestion The process by which organic waste is decomposed in an oxygen-free environment to produce methane gas (biogas).

anecumene *See* nonecumene.

animism A belief that natural objects may be the abode of dead people, spirits, or gods who occasionally give the objects the appearance of life.

antecedent boundary A boundary line established before the area in question is well populated.

aquaculture Producing and harvesting of fish and shellfish in freshwater ponds, lakes, and canals or in fenced-off coastal bays and estuaries; fish farming.

aquifer Underground porous and permeable rock that is capable of holding groundwater, especially rock that supplies economically significant quantities of water to wells and springs.

arable land Land that is or can be cultivated.

Arctic haze Air pollution resulting from the transport by air currents of combustion-based pollutants to the area north of the Arctic Circle.

area analysis tradition One of the four traditions of geography, that of regional geography.

area cartogram (*syn:* value-by-area map) A type of map in which the areas of the units are proportional to the data they represent.

arithmetic density *See* crude density.

arroyo A steep-sided, flat-bottomed gully, usually dry, carved out of desert land by rapidly flowing water.

artifacts The material manifestations of culture, including tools, housing, systems of land use, clothing, and the like. Elements in the technological subsystem of culture.

artificial boundary *See* geometric boundary.

assimilation The social process of merging into a composite culture, losing separate ethnic or social identity, and becoming culturally homogenized.

asthenosphere The partially molten, plastic layer above the core and lower mantle of the earth.

atmosphere The gaseous mass surrounding the earth.

atoll A near-circular low coral reef formed in shallow water enclosing a central lagoon; most common in the central and western Pacific Ocean.

azimuthal projection *See* planar projection.

B

barchan A crescent-shaped sand dune; the horns of the crescent point downwind.

basic sector Those products or services of an urban economy that are exported outside the city itself, earning income for the community.

bench mark A surveyor's mark indicating the position and elevation of a stationary object; used as a reference point in surveying and mapping.

bioaccumulation The buildup of a material in the body of an organism.

biocide A chemical used to kill plant and animal pests and disease organisms. *See also* herbicide, pesticide.

biodiversity hot spot An area with an exceptionally high number of endemic species that are at high risk of disruption by human activities.

biological magnification The accumulation of a chemical in the fatty tissue of an organism and its concentration at progressively higher levels in the food chain.

biomagnification *See* biological magnification.

biomass Living matter, plant and animal, in any form.

biomass fuel Any organic material produced by plants, animals, or microorganisms that can be used as a source of energy through either direct burning or conversion into a liquid or gas.

biome The total assemblage of living organisms in a single major ecological region.

biosphere (*syn:* ecosphere) The thin film of air, water, and earth within which we live, including the atmosphere, surrounding and subsurface waters, and the upper reaches of the earth's crust.

birth rate (*syn:* crude birth rate) The ratio of the number of live births during 1 year to the total population, usually at the midpoint of the same year, expressed as the number of births per year per 1000 population.

blizzard A heavy snowstorm accompanied by high winds.

boundary A line separating one political unit from another.

boundary definition A general agreement between two states about the allocation of territory between them.

boundary delimitation The plotting of a boundary line on maps or aerial photographs.

boundary demarcation The actual marking of a boundary line on the ground; the final stage in boundary development.

business services The portion of the tertiary (services) sector of the economy engaged in providing services to other businesses.

Examples include accounting, advertising, commercial real estate, consulting engineering, and legal work performed for corporations.

butte   A small, flat-topped, isolated hill with steep sides, common in dry climate regions.

C

carcinogen   A substance that produces or incites cancerous growth.

carrying capacity   The numbers of any population that can be adequately supported by the available resources on which that population subsists; for humans, the numbers supportable by the known and utilized resources—usually agricultural—of an area.

cartogram   A map that has been simplified to present data in a diagrammatic way; the base normally is not true to scale.

cartography   The art, science, and technology of making maps.

caste   One of the hereditary social classes in Hinduism that determines one's occupation and position in society.

central business district (CBD)   The center, or "downtown," of an urban unit, where retail stores, offices, and cultural activities are concentrated and where land values are high.

central city   That part of the metropolitan area contained within the boundaries of the main city around which suburbs have developed.

central place   A nodal point for the distribution of goods and services to a surrounding hinterland population.

central place theory   A deductive theory formulated by Walter Christaller (1893–1969) to explain the size and distribution of settlements through reference to a competitive supply of goods and services to dispersed rural populations.

centrifugal force   In political geography, a force that disrupts and destabilizes a state, threatening its unity.

centripetal force   In political geography, a force that promotes unity and national identity.

CFCs   *See* chlorofluorocarbons.

chain migration   The process by which migration movements from a common home area to a specific destination are sustained by links of friendship or kinship between first movers and later followers.

channelization   The modification of a stream channel; specifically, the straightening of meanders or dredging of the stream channel to deepen it.

channelized migration   The tendency for migration to flow between areas that are socially and economically allied by past migration patterns, by economic trade considerations, or by some other affinity.

chemical weathering   The decomposition of earth materials because of chemical reactions that include oxidation, hydration, and carbonation.

chlorofluorocarbons (CFCs)   A family of synthetic chemicals that have significant commercial applications but whose emissions are contributing to the depletion of the ozone layer.

choropleth map   A map that depicts quantities for areal units by varying pattern and/or color.

circumpolar vortex   High-altitude winds circling the poles from west to east.

city   A multifunctional nucleated settlement with a central business district and both residential and nonresidential land uses.

climagraph   A bar and line graph used to depict average monthly temperatures and precipitation.

climate   The long-term average weather conditions in a place or region.

climax community   An association of grasses, shrubs, and/or trees that is in equilibrium with the climate and soil of the site; the last stage of an ecological succession.

cogeneration   The simultaneous use of a single fuel for the generation of electricity and low-grade central heat.

cognition   The process by which an individual gives mental meaning to information.

cohort   A population group unified by a specific common characteristic, such as age, who are treated as a statistical unit during their lifetimes.

commercial economy   (*syn:* market economy) The production of goods and services for exchange in competitive markets where price and availability are determined by supply and demand forces.

commercial energy   Commercially traded fuels such as coal, oil, or natural gas and excluding wood, vegetable or animal wastes, or other biomass.

commodity chain   The set of activities involved in the production of a single good. A commodity chain encompasses the relationships between buyers and suppliers and the flows of materials, finance, and knowledge.

Common Market   *See* European Union.

compact state   A state whose territory is nearly circular.

comparative advantage   A region's profit potential for a productive activity compared with alternative areas of production of the same good or with alternative uses of the region's resources.

concentric zone model   A model describing urban land uses as a series of circular belts or rings around a core central business district, each ring housing a distinct type of land use.

conformal projection   A map projection on which the shapes of small areas are accurately portrayed.

conic projection   A map projection based on the projection of the grid system onto a cone as the presumed developable surface.

connectivity   The directness of routes linking pairs of places; all of the tangible and intangible means of connection and communication between places.

consequent boundary   (*syn:* ethnographic boundary) A boundary line that coincides with some cultural divide, such as religion or language.

conservation   The wise use or preservation of natural resources so as to maintain supplies and qualities at levels sufficient to meet present and future needs.

consumer services   The portion of the tertiary (services) sector of the economy engaged in providing services to individuals and households. Examples include hair salons and retailers.

contagious diffusion   The spread of a concept, a practice, or an article from one area to others through contact and/or the exchange of information.

continental drift   The hypothesis that an original single landmass (Pangaea) broke apart and that the continents have moved very slowly over the asthenosphere to their present locations.

contour interval   The vertical distance separating two adjacent contour lines.

contour line   A map line along which all points are of equal elevation above or below a datum plane, usually mean sea level.

conurbation   An extended urban area formed by the coalescence of two or more formerly separate cities.

convection   The circulatory movement of rising warm air and descending cool air.

convectional precipitation   Rain produced when heated, moisture-laden air rises and then cools below the dew point.

Convention on the Law of the Sea   *See* Law of the Sea Convention.

coral reef   A rocklike landform in shallow tropical water composed chiefly of compacted coral and other organic material.

core   The nucleus of a region or country, the main center of its industry, commerce, population, political, and intellectual life; in urban geography, that part of the central business district characterized by intensive land development.

core area   The nucleus of a state, containing its most developed area, greatest wealth, densest populations, and clearest national identity.

Coriolis effect   A fictitious force used to describe motion relative to a rotating Earth; specifically, the force that tends to deflect a moving object or fluid to the right (clockwise) in the Northern Hemisphere and to the left (counterclockwise) in the Southern Hemisphere.

countermigration   *See* return migration.

country   *See* state.

creole   A language developed from a pidgin to become the native tongue of a society.

critical distance   The distance beyond which cost, effort, and/or means play an overriding role in the willingness of people to travel.

crude birth rate (CBR)   *See* birth rate.

crude death rate (CDR)   *See* death rate.

crude density   (*syn:* arithmetic density, population density) The number of people per unit area of land.

crude oil   A mixture of hydrocarbons that exists in a liquid state in underground reservoirs; petroleum as it occurs naturally, as it comes from an oil well, or after extraneous substances have been removed.

cultural convergence   The tendency for cultures to become more alike as they increasingly share technology and organization structures in a modern world united by improved transportation and communication.

cultural divergence The likelihood or tendency for isolated cultures to become increasingly dissimilar with the passage of time.

cultural ecology The study of the interactions between societies and the natural environments they occupy.

cultural eutrophication The overnourishment of a water body with nutrients stemming from human activities, such as agriculture, industry, and urbanization.

cultural integration The interconnectedness of all aspects of a culture; no part can be altered without impact upon other culture traits.

cultural lag The retention of established culture traits despite changing circumstances rendering them inappropriate.

cultural landscape The natural landscape as modified by human activities and bearing the imprint of a culture group or society; the built environment.

culture A society's collective beliefs, symbols, values, forms of behavior, and social organizations, together with its tools, structures, and artifacts; transmitted as a heritage to succeeding generations and undergoing adoptions, modifications, and changes in the process.

culture complex An integrated assemblage of culture traits descriptive of one aspect of a society's behavior or activity.

culture-environment tradition One of the four traditions of geography; in this text, identified with population, cultural, political, and behavioral geography.

culture hearth A nuclear area within which an advanced and distinctive set of culture traits develops and from which there is diffusion of distinctive technologies and ways of life.

culture realm A collective of culture regions sharing related culture systems; a major world area having sufficient distinctiveness to be perceived as set apart from other realms in its cultural characteristics and complexes.

culture region A formal or functional region within which common cultural characteristics prevail. It may be based on single culture traits; on culture complexes; or on political, social, or economic integration.

culture system A generalization suggesting shared, identifying traits uniting two or more culture complexes.

culture trait A single distinguishing feature of regular occurrence within a culture, such as the use of chopsticks or the observance of a particular caste system; a single element of learned behavior.

cyclone A type of atmospheric disturbance in which masses of air circulate rapidly about a region of low atmospheric pressure.

cyclonic (frontal) precipitation The rain or snow that is produced when moist air of one air mass is forced to rise over the edge of another air mass.

cylindrical projection A map projection based on the projection of the globe grid onto a cylinder as the presumed developable surface.

D

database *See* geographic database.

DDT A chlorinated hydrocarbon that is among the most persistent of the biocides in general use.

death rate (mortality rate) A mortality index usually calculated as the number of deaths per year per 1000 population.

decomposers Microorganisms and bacteria that feed on dead organisms, causing their chemical disintegration.

deforestation The clearing of land through total removal of forest cover.

delta A triangular deposit of mud, silt, or gravel created by a stream where it flows into a body of standing water.

demographic equation A mathematical expression that summarizes the contribution of different demographic processes to the population change of a given area during a specified time period: $P_2 = P_1 + B_{1-2} - D_{1-2} + IM_{1-2} - OM_{1-2}$, where P_2 is population at time 2; P_1 is population at beginning date; B_{1-2} is the number of births between times 1 and 2; D_{1-2} is the number of deaths during that period; IM_{1-2} is the number of in-migrants and OM_{1-2} the number of out-migrants between times 1 and 2.

demographic momentum *See* population momentum.

demographic transition A model of the effect of economic development on population growth. The first stage involves both high birth and high death rates; the second stage displays high birth rates and falling mortality rates and population increases. The third stage shows reduction in population growth as birth rates decline to the level of death rates. The final and fourth stage implies again a population stable in size but larger in numbers than at the start of the transition cycle.

demography The scientific study of population, with particular emphasis on quantitative aspects.

density of population *See* population density.

dependency ratio The number of dependents, old or young, that each 100 persons in the productive years must, on average, support.

deposition The process by which silt, sand, and rock particles accumulate and create landforms, such as stream deltas and talus slopes.

desertification The conversion of arid and semiarid lands into deserts as a result of climatic change or human activities such as overgrazing or deforestation.

developable surface A geometric form, such as a cylinder or cone, that may be flattened without distortion.

devolution The transfer of certain powers from the state central government to separate political subdivisions within the state's territory; decentralization of political control.

dew point The temperature at which condensation forms, if the air is cooled sufficiently.

dialect A regional or socioeconomic variation of a more widely spoken language.

diastrophism The earth force that folds, faults, twists, and compresses rock.

dibble Any small hand tool or stick used to make a hole for planting.

diffusion *See* spatial diffusion.

distance The amount of separation between two objects, areas, or points; an extent of areal or linear measure.

distance decay The exponential decline of an activity or a function with increasing distance from its point of origin.

domestication The successful transformation of plant or animal species from a wild state to a condition of dependency on human management, usually with distinct physical change from wild forebears.

doubling time The time period required for any beginning total, experiencing a compounding growth, to double in size.

dune A wavelike desert landform created by wind-blown sand.

E

earthquake The movement of the earth along a geologic fault or at some other point of weakness at or near the earth's surface.

earth science tradition One of the four traditions of geography, identified with physical geography in general.

ecology The scientific study of how living creatures affect one another and what determines their distribution and abundance.

economic base The manufacturing and service activities performed by the basic sector of the labor force of a city to satisfy demands both inside and outside the city and earn income to support the urban population.

economic geography The study of how people earn a living, how livelihood systems vary by area, and how economic activities are spatially interrelated and linked.

ecosphere *See* biosphere.

ecosystem A population of organisms existing together in a particular area, together with the energy, air, water, soil, and chemicals upon which it depends.

ecotourism Tourism to relatively pristine natural environments that attempts to conserve the quality of the environment and improve the local quality of life. In theory, ecotourism destinations generate resources for conservation, educate visitors, and hire local workers and local companies.

ecumene The permanently inhabited areas of the earth. *See also* nonecumene.

edge city A distinct modal concentration of retail and office space situated on the outer fringes of a metropolitan area.

electoral geography The study of the delineation of voting districts and the spatial patterns of election results.

electromagnetic spectrum The entire range of radiation, including the shortest as well as the longest wavelengths.

El Niño The periodic (every 3 to 7 or 8 years) buildup of warm water along the west coast of

South America; replacing the cold Humboldt current off the Peruvian coast, El Niño is associated with both a fall in plankton levels (and decreased fish supply) and short-term, widespread weather modification.

elongated state A state whose territory is long and narrow.

enclave A territory that is surrounded by, but is not part of, a state.

endangered species Species that are present in such small numbers that they are in imminent danger of extinction.

energy The ability to do work. *See also* kinetic energy, potential energy.

energy efficiency The ratio of the output of useful energy from a conversion process to the total energy inputs.

environment Surroundings; the totality of things that in any way may affect an organism, including both physical and cultural conditions; a region characterized by a certain set of physical conditions.

environmental determinism The view that the physical environment, particularly climate, molds human behavior and conditions cultural development.

environmental justice The goal of providing healthy surroundings for all people, regardless of their skin color, level of poverty, or place of residence.

environmental perception The way people observe and interpret, as well as the ideas they have about near or distant places.

environmental pollution *See* pollution.

environmental racism Any policy or practice that differentially affects or harms individuals, groups, or communities because of their race or color; the harm may be intentional or unintentional.

epidemiologic transition Long-term shifts in health and disease patterns as mortality moves from high to low levels.

equal-area projection *See* equivalent projection.

equator An imaginary line that encircles the globe halfway between the North and South Poles.

equidistant projection A map projection on which true distances in all directions can be measured from one or two central points.

equivalent projection A map projection on which the areas of regions are represented in correct or constant proportions to earth reality; also called equal-area projection.

erosion The result of processes that loosen, dissolve, wear away, and remove earth and rock material. Those processes include weathering, solution, abrasion, and transportation.

erosional agents The forces of wind, moving water, glaciers, waves, and ocean currents that carve, wear away, and remove rock and soil particles.

estuarine zone The relatively narrow area of wetlands along coastlines where salt water and fresh water mix.

estuary The lower course or mouth of a river where tides cause fresh water and salt water from the sea to mix.

ethanol Organic matter that has been fermented and distilled into alcohol; an alternative automotive fuel that, when blended with gasoline, forms gasohol.

ethnic cleansing The killing or forcible relocation of one traditional or ethnic group by a more powerful one.

ethnicity The social status afforded to, usually, a minority group within a national population. Recognition is based primarily on culture traits, such as religion, distinctive customs, or native or ancestral national origin.

ethnic religion A religion identified with a particular ethnic group and largely exclusive to it.

ethnoburb A suburban ethnic enclave. Ethnoburbs differ from earlier urban ethnic enclaves both in their suburban location and middle-class status.

ethnocentrism The belief that one's own ethnic group is superior to all others.

ethnographic boundary *See* consequent boundary.

European Union (EU) An economic association established in 1957 of a number of Western European states that promotes free trade among member countries; often called the Common Market.

eutrophication The enrichment of a water body by the addition of nutrients received through erosion and runoff from the watershed. *See also* accelerated eutrophication.

evapotranspiration The return of water from the land to the atmosphere through evaporation from the soil surface and transpiration from plants.

e-waste A popular name for discarded electrical or electronic products.

exclave A portion of a state that is separated from the main territory and surrounded by another country.

exclusive economic zone (EEZ) As established in the United Nations Convention on the Law of the Sea, a zone of exploitation extending 200 nautical miles seaward from a coastal state that has exclusive mineral and fishing rights over it.

exotic species A plant, an animal, or another organism that has been deliberately or inadvertently introduced into an ecosystem in which it did not evolve; a nonindigenous species.

extensive agriculture A crop or livestock system in which land quality or extent is more important than capital or labor inputs in determining output. It may be part of either a commercial or a subsistence economy.

extensive commercial agriculture Extensive farming in a commercial economy; examples include large-scale wheat farming and livestock ranching.

extensive subsistence agriculture Extensive farming in a subsistence economy; examples include nomadic herding and shifting cultivation.

external economies *See* agglomeration economies.

extinction The elimination of all of the individuals of a particular species.

extractive industries Primary activities involving the mining and quarrying of nonrenewable metallic and nonmetallic mineral resources.

extrusive rock Rock solidified from molten material (magma) that has flowed from beneath the earth's surface onto it.

F

fault Break in rock produced by stress or the movement of lithospheric plates.

fault escarpment A steep slope formed by the vertical movement of the earth along a fault.

fiord A glacial trough whose lower end is filled with seawater.

flexible production Production systems that use computer information technology and just-in-time manufacturing to produce smaller runs of customized products. Flexible production is a post-Fordist industrial strategy in contrast to the vertical integration and mass production associated with Fordism.

floodplain A valley area bordering a stream that is subject to inundation by flooding.

flow-line map A map used to portray linear movement between places; may be qualitative or quantitative.

fold A bend or wrinkle in rock resulting from compression and formed when the rock was in a plastic state.

folk culture The body of institutions, customs, dress, artifacts, collective wisdoms, and traditions of a homogeneous, isolated, largely self-sufficient, and relatively static social group.

food chain A sequence of organisms through which energy and materials move within an ecosystem.

food security The condition where all people have access to safe and nutritious food of sufficient quantity for an active and healthy lifestyle.

footloose A descriptive term applied to manufacturing activities for which the cost of transporting material or product is not important in determining the location of production.

Fordism The manufacturing economy and system derived from assembly line mass production and the mass consumption of standardized goods. Named after Henry Ford, who innovated many of its production techniques.

foreign direct investment (FDI) The purchase or construction of foreign factories and other fixed assets by transnational corporations; also the purchase of or merging with foreign companies.

formal (uniform) region A region distinguished by uniformity of one or more characteristics that can serve as the basis for an areal generalization and of contrast with adjacent areas.

form utility A value-increasing change in the form—and therefore in the utility—of a raw material or commodity.

forward-thrust capital A capital city deliberately sited in a state's frontier zone.

fossil fuel Any of the fuels derived from decayed organic material converted by earth processes; especially coal, petroleum, and natural gas, but also including oil sands and oil shales.

fragmented state A state whose territory contains isolated parts, separated and discontinuous.

frictional effect In climatology, the slowing of wind movement due to the frictional drag of the earth's surface.

friction of distance A measure of the retarding effect of distance on spatial interaction. Generally, the greater the distance, the greater the "friction" and the less the interaction or exchange, or the greater the cost of achieving the exchange.

front The line or zone of separation between two air masses of different temperatures and humidities.

frontal precipitation See cyclonic (frontal) precipitation.

frontier That portion of a country adjacent to its boundaries and fronting another political unit.

frontier zone A belt lying between two states or between settled and uninhabited or sparsely settled areas.

Fujita scale A scale for categorizing tornado intensity.

functional (nodal) region A region differentiated by what occurs within it rather than by a homogeneity of physical or cultural phenomena; an earth area recognized as an operational unit based on defined organizational criteria.

G

gated community A restricted-access subdivision or neighborhood, often surrounded by a barrier, with entry permitted only for residents and their guests; usually totally planned in land use and design.

gathering industries Primary activities involving the harvesting of renewable natural resources of land or water; commercial gathering usually implies forestry and fishing industries.

gender The socially created, not biologically based, distinctions between femininity and masculinity.

gender empowerment measure (GEM) A statistic summarizing the extent of economic, political, and professional participation of women in the society of which they are members; a measure of relative gender equality.

gene flow The passage of genes characteristic of one breeding population into the gene pool of another by interbreeding.

genetic drift A chance modification of gene composition occurring in an isolated population and becoming accentuated through inbreeding.

gentrification The process by which middle- and high-income groups refurbish and rehabilitate housing in deteriorated inner-city areas, thereby displacing low-income populations.

geocaching The pursuit of a cache using a GPS unit.

geodetic control data The information specifying the horizontal and vertical positions of a place.

geographic database In cartography, a digital record of geographic information.

geographic grid The set of imaginary lines of latitude and longitude that intersect at right angles to form a system of reference for locating points on the surface of the earth.

geographic information system (GIS) A configuration of computer hardware and software for assembling, storing, manipulating, analyzing, and displaying geographically referenced information.

geometric boundary (syn: artificial boundary) A boundary without obvious physical geographic basis; often a section of a parallel of latitude or a meridian of longitude.

geomorphology The scientific study of landform origins, characteristics, and evolutions and their processes.

geothermal energy Energy that is generated by harnessing the naturally occurring steam and hot water produced by contact with heated rocks in the earth's crust.

gerrymandering Dividing an area into voting districts in such a way as to give one political party an unfair advantage in elections, to fragment voting blocks, or to achieve other nondemocratic objectives.

GIS See geographic information system.

glacial till The deposits of rocks, silt, and sand left by a glacier after it has receded.

glacial trough A deep, U-shaped valley or trench formed by glacial erosion.

glacier A huge mass of slowly moving land ice.

globalization The increasing interconnection of all parts of the world as the full range of social, cultural, political, economic, and environmental processes and patterns of change becomes international in scale and effect.

Global Positioning System (GPS) A method of using satellite observations for the determination of extremely accurate locational information.

global warming A rise in surface temperatures on Earth, a process believed by some to be caused by human activities that increase the concentration of greenhouse gases in the atmosphere, magnifying the greenhouse effect.

globe grid See geographic grid.

globe properties The characteristics of the grid system of longitude and latitude on a globe.

GPS See Global Positioning System.

gradational processes The processes of weathering, gravity transfer, and erosion that are responsible for the reduction of the land surface.

grade (of coal) A classification of coals based on their content of waste materials.

graphic scale A graduated line included in a map legend by means of which distances on the map may be measured in terms of ground distances.

great circle A circle formed by the intersection of the surface of a globe with a plane passing through the center of the globe. The equator is a great circle; meridians are one-half of a great circle.

greenhouse effect The heating of the earth's surface as shortwave solar energy passes through the atmosphere, which is transparent to it but opaque to reradiated longwave terrestrial energy. Also refers to increasing the opacity of the atmosphere through the addition of increased amounts of carbon dioxide, nitrous oxides, methane, and chlorofluorocarbons.

greenhouse gases Heat-trapping gases added to the atmosphere by human activities; carbon dioxide, chlorofluorocarbons, methane gas, and nitrous oxide.

Green Revolution The great increases in food production, primarily in subtropical areas, accomplished through the introduction of very high-yielding grain crops, particularly wheat and rice.

Greenwich mean time (GMT) Local time at the prime meridian (zero degrees longitude), which passes through the observatory at Greenwich, England.

gross national income (GNI) See gross national product (GNP).

gross national product (GNP) The total value of all goods and services produced by a country per year; also called gross national income (GNI).

groundwater Underground water that accumulates in aquifers below the water table in the pores and cracks of rock and soil.

gyre In oceanography, a large spiral oceanic surface current.

H

half-life The time required for one-half of the atomic nuclei of an isotope to decay.

hazardous waste Discarded solid, liquid, or gaseous material that may pose a substantial threat to human health or the environment when it is improperly disposed of, stored, or transported.

herbicide A chemical that kills plants, especially weeds. See also biocide, pesticide.

hierarchical diffusion The process by which contacts between people and the resulting diffusion of things or ideas occurs first among those at the same level of a hierarchy and then among elements at a lower level of the hierarchy (e.g., small-town residents acquire ideas or articles after they are common in large cities).

hierarchical migration The tendency for individuals to move from small places to larger ones.

hierarchy of central places The steplike series of urban units in classes differentiated by both size and function.

high-level waste Nuclear waste that can remain radioactive for thousands of years, produced principally by the generation of nuclear power and the manufacture of nuclear weapons.

hinterland An outlying region that furnishes raw materials or agricultural products to the heartland; the market area or region served by a town or city.

homeostatic plateau The equilibrium level of population that can be supported adequately by available resources; equivalent to carrying capacity.

human interaction The communication and interdependencies between people. Sometimes the term *spatial interaction* is used to more specifically identify the locations that are interacting.

humid continental climate A climate of east coast and continental interiors of midlatitudes, displaying large annual temperature ranges resulting from cold winters and hot summers; precipitation at all seasons.

humid subtropical climate A climate of the east coast of continents in lower-middle latitudes, characterized by hot summers with convectional precipitation and cool winters with cyclonic precipitation.

humus Dark brown or black decomposed organic matter in soils.

hunting-gathering An economic and social system based primarily or exclusively on the hunting of wild animals and the gathering of food, fiber, and other materials from uncultivated plants.

hurricane A severe tropical cyclone with winds exceeding 120 kilometers per hour (75 mph) originating in the tropical region of the Atlantic Ocean, Caribbean Sea, or Gulf of Mexico.

hydrologic cycle The system by which water is continuously circulated through the biosphere by evaporation, condensation, and precipitation.

hydropower The kinetic energy of moving water converted into electrical power by a power plant whose turbines are driven by flowing water.

hydrosphere All water at or near the earth's surface that is not chemically bound in rocks; includes the oceans, surface water, groundwater, and water held in the atmosphere.

I

iconography In political geography, the study of symbols that unite a country.

ideological subsystem The complex of ideas, beliefs, knowledge, and means of their communication that characterizes a culture.

igneous rock Rock formed from cool, solidified magma; may solidify beneath or at the earth's surface.

incinerator A facility designed to burn waste.

inclination The tilt of the earth's axis about 23½° away from the perpendicular.

Industrial Revolution The rapid economic and social changes in agriculture and manufacturing that followed the introduction of the factory system to the textile industry of England in the last quarter of the 18th century.

infant mortality rate A refinement of the death rate to specify the ratio of deaths of infants age 1 year or less per 1000 live births.

informal economy Economic activity that operates without official recognition, government regulation, or tax collection. It is not recorded in official statistics and includes barter, unlicensed vendors, and "under the table" employment.

infrared Electromagnetic radiation having wavelengths greater than those of visible light.

infrastructure The basic structure of services, installations, and facilities needed to support industrial, agricultural, and other economic development.

innovation Introduction into an area of new ideas, practices, or objects; an alteration of custom or culture that originates within the social group itself.

insolation The solar radiation received at the earth's surface.

intensive agriculture The application of large amounts of capital and/or labor per unit of cultivated land to increase output; may be part of either a commercial or a subsistence economy.

intensive commercial agriculture Intensive farming in a commercial economy; crops have high yields and high market value.

intensive subsistence agriculture Intensive farming in a subsistence economy; the cultivation of small landholdings through the expenditure of great amounts of labor.

International Date Line By international agreement, the designated line where each new day begins; generally following the 180th meridian.

intrusive rock Igneous rock from magma that has hardened beneath the earth's surface and that has penetrated or been forced into or between preexisting rocks.

IPAT equation An equation relating the environmental impact of a society to the key issues of population, affluence, and technology.

irredentism The desire of a state to gain or regain territory inhabited by people who have historic or cultural links to the country but who now live in a neighboring state.

isochrone A line connecting points that are equidistant in travel time from a common origin.

isoline A map line connecting points of equal value, such as a contour line or an isobar.

isotropic plain A hypothetical portion of the earth's surface assumed to be an unbounded, uniformly flat plain with uniform distribution of population, purchasing power, transport costs, accessibility, and the like.

J

J-curve A curve shaped like the letter J, depicting exponential, or geometric, growth (1, 2, 4, 8, 16, . . .).

jet stream A meandering belt of strong winds in the upper atmosphere; significant because it guides the movement of weather systems.

just-in-time (JIT) manufacturing Systems of manufacturing that rely on fast response and quick delivery of raw materials and component parts just as they needed for final assembly, rather than maintaining large stockpiles and warehouse inventories.

K

karst topography A limestone region marked by sinkholes, caverns, and underground streams.

kerogen A waxy, organic material occurring in oil shales that can be converted into crude oil by distillation.

kinetic energy The energy that results from the motion of a particle or body.

L

land breeze Airflow from the land toward the sea, resulting from a nighttime pressure gradient that moves winds from the cooler land surface to the warmer sea surface.

landform region A large section of the earth's surface characterized by a great deal of homogeneity among types of landforms.

landlocked state A state that lacks a seacoast.

Landsat satellite One of a series of continuously orbiting satellites that carry scanning instruments to measure reflected light in both the visible and near infrared portions of the spectrum.

landscape The appearance of an area and the items comprising that appearance. A distinction is often made between "physical landscapes" confined to landforms, natural vegetation, soils, etc., and "cultural landscapes" (q.v.).

language An organized system of speech by which people communicate with one another with mutual comprehension.

language family A group of languages thought to have descended from a single, common ancestral tongue.

lapse rate The rate of change of temperature with altitude in the troposphere; the average lapse rate is about 6.4°C per 1000 meters (3.5°F per 1000 ft).

large-scale map A representation of a small land area, usually with a representative fraction of 1:75,000 or less.

latitude The angular distance north or south of the equator, measured in degrees ranging from 0° (the equator) to 90° (the North and South Poles).

lava Molten material that has emerged onto the earth's surface.

Law of the Sea Convention A code of sea law approved by the United Nations in 1982 that authorizes, among other provisions, territorial waters extending 12 nautical miles from shore and 200-nautical-mile-wide exclusive economic zones; generally referred to as UNCLOS.

leachate The contaminated liquid discharged from a sanitary landfill to either surface or subsurface land or water.

leaching The downward movement of water through the soil layer, resulting in the removal of soluble minerals from the upper soil horizons.

least-cost theory (*syn:* Weberian analysis) The view that the optimum location of a manufacturing establishment is at the place where the costs of transport and labor and the advantages of agglomeration or dispersion are most favorable.

levee In agriculture, a continuous embankment surrounding areas to be flooded. *See also* natural levee.

lingua franca Any of the various auxiliary languages used as common tongues among people of an area where several languages are spoken.

liquefied natural gas (LNG) Methane gas that has been liquefied by refrigeration for storage or transportation.

lithosphere The outermost layer of the earth, composed of the crust and upper mantle.

loam Agriculturally productive soil containing roughly equal parts of sand, silt, and clay.

locational tradition One of the four traditions of geography; in this text, identified with economic, urban, and environmental geography.

loess A deposit of windblown silt.

longitude The angular distance east or west of the prime (zero) meridian, measured in degrees ranging from 0° to 180°.

long lot A farm or other property consisting of a long, narrow strip of land extending back from a river or road.

longshore current A current that moves roughly parallel to the shore and transports the sand that forms beaches and sand spits.

low-level waste Hazardous material whose radioactivity will decay to safe levels in 100 years or less, produced principally by industries and nuclear power plants.

M

magma Underground molten material.

malnutrition Food intake insufficient in quantity or deficient in quality to sustain life at optimal conditions of health.

Malthus Thomas R. Malthus (1766–1834), English economist, demographer, and cleric, suggested that, unless checked by self-control, war, or natural disaster, population will inevitably increase faster than will the food supplies needed to sustain it.

mantle The layer of earth between the crust and the core.

map projection A method of transferring the grid system from the earth's curved surface to the flat surface of a map.

map scale *See* scale.

marine west coast climate A regional climate found on the west coast of continents in upper midlatitudes, rainy all seasons with relatively cool summers and relatively mild winters.

market economy *See* commercial economy.

mass movement The downslope movement of earth materials due to gravity.

mass wasting *See* mass movement.

material culture The tangible, physical items produced and used by members of a specific culture group and reflective of their traditions, lifestyles, and technologies.

mathematical location *See* absolute location.

maximum sustainable yield The maximum rate at which a renewable resource can be exploited without impairing its ability to be renewed or replenished.

mechanical weathering The physical disintegration of earth materials, commonly by frost action, root action, or the development of salt crystals.

Mediterranean climate A climate of lower midlatitudes characterized by mild, wet winters and hot, dry, sunny summers.

megalopolis An extensive, heavily populated urban complex with contained open, nonurban land, created through the spread and merging of separate metropolitan areas; (*cap.*) the name applied to the continuous functionally urban area of the northeastern seaboard of the United States from Maine to Virginia.

megawatt A unit of power equal to 1 million watts (1000 kilowatts) of electricity.

mental map The maplike image of the world, country, region, or other area that a person carries in his or her mind; includes knowledge of actual locations and spatial relationships and is colored by personal perceptions and preferences related to place.

mentifacts The central, enduring elements of a culture that express its values and beliefs, including language, religion, folklore, artistic traditions, and the like. Elements in the ideological subsystem of culture.

Mercator projection A true conformal cylindrical projection first published in 1569, useful for navigation.

meridian A north–south line of longitude; on the globe, all meridians are of equal length and converge at the poles.

mesa An extensive, flat-topped elevated tableland with horizontal strata, a resistant cap rock, and one or more steep sides; a large butte.

metamorphic rock Rock transformed from igneous and sedimentary rocks into a new type of rock by earth forces that generate heat, pressure, or chemical reaction.

metes-and-bounds survey A system of property description that uses natural features (trees, boulders, streams, etc.) to describe and define the boundaries of individual properties.

metropolitan area A large, functional entity, perhaps containing several urbanized areas, discontinuously built up but operating as a coherent economic whole.

migration The permanent (or relatively permanent) relocation of an individual or a group to a new, usually distant place of residence.

migration field An area that sends major migration flows to or receives major flows from a given place.

mineral A natural inorganic substance that has a definite chemical composition and characteristic crystal structure, hardness, and density.

ministate An imprecise term for a state or territory small in both population and area. An informal definition accepted by the United Nations suggests a maximum of 1 million people combined with a territory of less than 700 square kilometers (270 sq mi).

monoculture An agricultural system dominated by a single crop.

monotheism The belief that there is only one God.

monsoon A wind system that reverses direction seasonally, producing wet and dry seasons; especially describes the wind system of South, Southeast, and East Asia.

moraine Any of several types of landforms composed of debris transported and deposited by a glacier.

mortality rate *See* death rate.

mountain breeze The downward flow of heavy, cool air at night from mountainsides to lower valley locations.

multiple-nuclei model The idea that large cities develop by peripheral spread, not from one central business district but from several nodes of growth, each of specialized use.

multiplier effect The expected addition of nonbasic workers and dependents to a city's total employment and population that accompanies new basic sector employment.

N

nation A culturally distinctive group of people occupying a particular region and bound together by a sense of unity arising from shared ethnicity, beliefs, and customs.

nationalism A sense of unity binding the people of a state together; devotion to the interests of a particular nation; an identification with the state and an acceptance of national goals.

nation-state A state whose territory is identical to that occupied by a particular nation.

natural boundary (*syn:* physical boundary) A boundary line based on recognizable physiographic features, such as mountains, rivers, or deserts.

natural gas A mixture of hydrocarbons and small quantities of nonhydrocarbons existing in a gaseous state or in solution with crude oil in natural reservoirs.

natural increase The growth of a population through excess of births over deaths, excluding the effects of immigration or emigration.

natural landscape The physical *environment* unaffected by human activities. The duration and near totality of human occupation of the earth's surface assure that little or no "natural landscape" so defined remains intact. Opposed to *cultural landscape*.

natural levee   An embankment on the sides of a meandering river formed by deposition of silt during floods.

natural resource   A physically occurring item that a population perceives to be necessary and useful to its maintenance and well-being.

natural selection   The process of survival and reproductive success of individuals or groups best adjusted to their environment, leading to the perpetuation of their genetic qualities.

natural vegetation   The plant life that would exist in an area if humans did not interfere with its development.

neo-Malthusianism   The advocacy of population control programs to preserve and improve general national prosperity and well-being.

neritic zone   The relatively shallow part of the sea that lies above the continental shelf.

net migration   The difference between in-migration and out-migration of an area.

new international division of labor (NIDL)   A spatial arrangement of production in which developing countries with lower wages capture more of the world's manufacturing activity while developed countries shift to services. The new international division of labor is linked to the rise of transnational corporations and made possible by improved global transport and communication technologies.

niche   The place an organism or a species occupies in an ecosystem.

nodal region   *See* functional region.

nomadic herding   The migratory but controlled movement of livestock solely dependent upon natural forage.

nonbasic sector   The economic activities of an urban unit that supply the resident population with goods and services.

nonecumene   (*syn:* anecumene) The portion of the earth's surface that is uninhabited or only temporarily or intermittently inhabited. *See also* ecumene.

nonfuel mineral resource   A mineral used for purposes other than providing a source of energy.

nongovernmental organization (NGO)   A group of people acting outside government and major commercial agencies and advocating or lobbying for particular causes.

nonmaterial culture   The oral traditions, songs, and stories of a culture group along with its beliefs and customary behaviors.

nonpoint source of pollution   Pollution from a broad area, such as one of fertilizer or pesticide application, rather than from a discrete source.

nonrenewable resource   A natural resource that is not replenished or replaced by natural processes or is used at a rate that exceeds its replacement rate.

North and South Poles   The end points of the axis about which the earth spins.

North Atlantic drift   The massive movement of warm water in the Atlantic Ocean from the Caribbean Sea and Gulf of Mexico in a north-easterly direction to the British Isles and the Scandinavian peninsula.

not in my backyard (NIMBY)   Protests from local residents aimed at stopping the siting of an unwanted land use such as a waste incinerator. NIMBY protests arise from fear that the proposed new facility will damage the quality of the local environment or lower property values.

nuclear fission   The controlled splitting of an atom to release energy.

nuclear fusion   The combining of two atoms of deuterium into a single atom of helium in order to release energy.

nuclear power   Electricity generated by a power plant whose turbines are driven by steam produced by the fissioning of nuclear fuel in a reactor.

nutrient   A mineral or another element an organism requires for normal growth and development.

O

offshoring   The relocation of business processes and services to a lower-cost foreign location; the offshore outsourcing of, particularly, white-collar technical, professional, and clerical services.

oil sands   Sand and sandstone impregnated with heavy oil.

oil shale   Sedimentary rock containing solid organic material (kerogen) that can be extracted and converted into a crude oil by distillation.

ore   A mineral deposit that can be extracted at a profit.

organic   Derived from living organisms; plant or animal life.

Organization of Petroleum Exporting Countries (OPEC)   An international cartel composed of 11 countries that aims at pursuing common oil marketing and pricing policies.

orographic precipitation   The rain or snow caused when warm, moisture-laden air is forced to rise over hills or mountains in its path and is thereby cooled.

orthophotomap   A multicolored, distortion-free aerial photographic image to which certain supplementary information has been added.

outsourcing   (1) Producing abroad parts or products for domestic use or sale; (2) subcontracting production or services rather than performing those activities "in house."

outwash plain   A gently sloping area in front of a glacier composed of neatly stratified glacial till carried out of the glacier by meltwater streams.

overburden   Soil and rock of little or no value that overlie a deposit of economic value, such as coal.

overpopulation   A value judgment that the resources of an area are insufficient to sustain adequately its present population numbers.

oxbow lake   A crescent-shaped lake contained in an abandoned meander of a river.

ozone   A gas molecule consisting of three atoms of oxygen (O_3) formed when diatomic oxygen (O_2) is exposed to ultraviolet radiation. In the lower atmosphere, it constitutes a damaging component of photochemical smog; in the upper atmosphere, it forms a normally continuous, thin layer that blocks ultraviolet light.

ozone layer   A layer of ozone in the high atmosphere that protects life on Earth by absorbing ultraviolet radiation from the sun.

P

parallel of latitude   An east–west line indicating the distance north or south of the equator.

PCBs   Polychlorinated biphenyls; compounds containing chlorine that can be biologically magnified in the food chain.

peak value intersection   The most accessible and costly parcel of land in a central business district and, therefore, in the entire urbanized area.

perceptual (vernacular, popular) region   A region perceived to exist by its inhabitants or the general populace. It has reality as an element of popular culture or folk culture represented in the mental maps of average people.

perforated state   A state whose territory is interrupted ("perforated") by a separate, independent state totally contained within its borders.

peripheral model   A model describing metropolitan area land uses in the circumferential belt around a city; nodes on the belt are centers for employment or services, and residents of the developments on the belt lead their lives largely on the periphery of the city.

permafrost   Permanently frozen subsoil.

perpetual resource   A resource that comes from an inexhaustible source, such as the sun, wind, and tides.

pesticide   A chemical that kills insects, rodents, fungi, weeds, and other pests. *See also* biocide, herbicide.

Peters projection   An equal-area cylindrical projection developed by Arno Peters that purports to show developing countries in proper proportion to one another.

petroleum   Oil and oil products in all forms, such as crude oil and unfinished oils.

pH factor   The measure of the acidity/alkalinity of soil or water, on a scale of 0 to 14, rising with increasing alkalinity.

photochemical smog   A form of air pollution produced by the interaction of hydrocarbons and oxides of nitrogen in the presence of sunlight.

photovoltaic (PV) cell   A device that converts solar energy directly into electrical energy. *See also* solar power.

physical boundary   *See* natural boundary.

physiological density   The number of persons per unit area of agricultural land. *See also* population density.

pidgin   An auxiliary language derived, with reduction of vocabulary and simplification of structure, from other languages. Not a native tongue, it is employed to provide a mutually intelligible vehicle for limited transactions of trade or administration.

pixel An extremely small sensed unit of a digital image.

place utility (1) The perceived attractiveness of a place in its social, economic, or environmental attributes; (2) the value imparted to goods or services by tertiary activities that provide things needed in specific markets.

planar (azimuthal) projection A map projection based on the projection of the globe grid onto a plane as the presumed developable surface.

planned economy A system of the production of goods and services, usually consumed or distributed by a government agency, in quantities and at prices determined by government programs.

plantation A large agricultural holding, frequently foreign-owned, devoted to the production of a single export crop.

plate tectonics The theory that the earth's crust consists of lithospheric plates that carry the continents and the ocean floor and float slowly on the plastic upper mantle, colliding with and scraping against one another.

playa A temporary lake or lake bed found in a desert environment.

Pleistocene The geologic epoch dating from 2 million to about 10,000 years ago during which four stages of continental glaciation occurred.

point source of pollution Pollution originating from a discrete source, such as a smokestack or the outflow from a pipe.

political geography The branch of human geography concerned with the spatial analysis of political phenomena.

pollution The presence in the biosphere of substances that, because of their quantity, chemical nature, or temperature, have a negative impact on the ecosystem or that cannot be readily disposed of by natural recycling processes.

polychlorinated biphenyls See PCBs.

polytheism The belief in or worship of many gods.

popular culture The constantly changing mix of material and nonmaterial elements available through mass production and the mass media to an urbanized, heterogeneous, nontraditional society.

popular region See perceptual region.

population density A measurement of the numbers of persons per unit area of land within predetermined limits, usually political or census boundaries. See also physiological density.

population geography The branch of human geography dealing with the number, composition, and distribution of humans in relation to variations in earth-space conditions.

population momentum (syn: demographic momentum) The tendency for population growth to continue despite stringent family planning programs because of a relatively high concentration of people in the childbearing years.

population projection A report of future size, age, and sex composition of a population based on assumptions applied to current data.

population pyramid A graphic depiction of the age and sex composition of a (usually national) population.

possibilism The philosophical viewpoint that the physical environment offers human beings a set of opportunities from which (within limits) people may choose according to their cultural needs and technological awareness.

potential energy The energy stored in a particle or body.

potentially renewable resource A resource that can last indefinitely if its natural replacement rate is not exceeded; examples include forests, groundwater, and soil.

precipitation All moisture, solid and liquid, that falls to the earth's surface from the atmosphere.

pressure gradient force Differences in air pressure between areas that induce air to flow from areas of high pressure to areas of low pressure.

primary activity The part of the economy involved in making natural resources available for use or further processing; includes mining, agriculture, forestry, fishing or hunting, and grazing.

primary air pollutant Material, such as particulates or sulfur dioxide, that is emitted directly into the atmosphere in sufficient quantities to adversely affect human health or the environment.

primate city A country's leading city, much larger and functionally more complex than any other; usually the capital city and a center of wealth and power.

prime meridian An imaginary line passing through the Royal Observatory at Greenwich, England, serving by agreement as the zero degree line of longitude.

projection An estimate of future conditions based on current trends. See also map projection.

prorupt state A state of basically compact form that has one or more narrow extensions of territory.

proto-language An assumed, reconstructed, or recorded language ancestral to one or more contemporary languages or dialects.

proved (usable) reserves The portion of a natural resource that has been identified and can be extracted profitably with current technology.

psychological distance The way an individual perceives distance.

pull factor A characteristic of a region that acts as an attractive force, drawing migrants from other regions.

purchasing power parity (PPP) A monetary measurement that takes account of what money actually buys in each country.

push factor A characteristic of a region that contributes to the dissatisfaction of residents and impels their migration.

Q

quaternary activity Employment concerned with research, with the gathering or disseminating of information, and with administration, including administration of the other economic activity levels.

quinary activity A sometimes separately recognized subsection of tertiary activity management functions involving highest-level decision making in all types of large organizations. Also deemed the most advanced form of the quaternary subsector.

R

race A subset of human population whose members share certain distinctive, inherited biological characteristics.

radioactive waste Solid, liquid, or gaseous waste containing radioactive isotopes, whose half-lives can range from less than 1 second to millions of years; usually classified as low level or high level according to the amount and types of radioactivity in the substance.

rank (of coal) A classification of coals based on their age and energy content; those of higher rank are more mature and richer in energy.

rank-size rule An observed regularity in the city-size distribution of some countries. In a rank-size hierarchy, the population of any given town will be inversely proportional to its rank in the hierarchy; that is, the nth-ranked city will be $1/n$ the size of the largest city.

rate The frequency of occurrence of an event during a specified time period.

rate of natural increase The birth rate minus the death rate, suggesting the annual rate of population growth without considering net migration.

recycling The reuse of disposed materials after they have passed through some form of treatment (e.g., melting down glass bottles to produce new bottles).

redistricting The drawing of new electoral district boundary lines in response to changing patterns of population or changing legal requirements.

reflection The process of returning to outer space some of the earth's received insolation.

region In geography, an area of the earth that displays a distinctive grouping of physical or cultural phenomena or is functionally united as a single organizational unit.

regional autonomy A measure of self-governance for a subdivision of a country.

regional concept The view that physical and cultural phenomena on the surface of the earth are rationally arranged by complex but comprehensible spatial processes.

regionalism In political geography, minority group identification with a particular region of a state rather than with the state as a whole.

relative direction (*syn:* relational direction) A culturally based locational reference, such as the Far West, the Old South, or the Middle East.

relative distance A transformation of *absolute distance* into such relative measures as time or monetary costs. Such measures yield different explanations of human spatial behavior than do linear distances alone. Distances between places are constant by absolute terms, but relative distances may vary with improvements in transportation or communication technology or with different psychological perceptions of space.

relative humidity A measure of the moisture content of the air, expressed as the amount of water vapor present relative to the maximum that can exist at the current temperature.

relative location The position of a place or an activity in relation to other places or activities.

relic boundary A former boundary line that is still discernible and marked by a cultural landscape feature.

religion A value system that involves formal or informal worship and faith in the sacred and divine.

relocation diffusion The transfer of ideas, behaviors, or articles from one place to another through the migration of those possessing the feature transported; also, spatial relocation in which a phenomenon leaves an area of origin as it is transported to a new location.

remote sensing Any of several techniques of obtaining images of an area without having the sensor in direct physical contact with it, as by air photography or satellite sensors.

renewable resource A naturally occurring material that is potentially inexhaustible, because either it flows continuously (such as solar radiation or wind) or it is renewed within a short period of time (such as biomass). *See also* sustained yield.

replacement level The number of children per family just sufficient to keep the total population constant. Depending on mortality conditions, replacement level is usually calculated to be between 2.1 and 2.5 children.

representative fraction (RF) The scale of a map expressed as a ratio of a unit of distance on the map to distance measured in the same unit on the ground (e.g., 1:250,000).

reradiation A process by which the earth returns solar energy to space; some of the shortwave solar energy that is absorbed into the land and water is returned to the atmosphere in the form of longwave terrestrial radiation.

resource *See* natural resource.

return migration (countermigration) The return of migrants to the region from which they had earlier emigrated.

rhumb line A line of constant compass bearing; it cuts all meridians at the same angle.

Richter scale A logarithmic scale used to express the magnitude of an earthquake.

S

Sahel The semiarid zone between the Sahara Desert and the savanna area to the south in West Africa; district of recurring drought, famine, and environmental degradation.

salinization The concentration of salts in the topsoil as a result of the evaporation of surface water; occurs in poorly drained soils in dry climates, often as a result of improper irrigation.

sandbar An offshore shoal of sand created by the backwash of waves.

sanitary landfill The disposal of solid wastes by spreading them in layers covered with enough soil or ashes to control odors, rats, and flies.

savanna A tropical grassland characterized by widely dispersed trees and experiencing pronounced yearly wet and dry seasons.

scale In cartography, the ratio between length or size of an area on a map and the actual length or size of that area on the earth's surface; map scale may be represented verbally, graphically, or as a fraction. In more general terms, *scale* refers to the size of the area studied, from local to global.

S-curve The horizontal bending, or leveling, of an exponential J-curve.

sea breeze Airflow from the sea toward the land, resulting from a daytime pressure gradient that moves winds from the cooler sea surface onto the warmer land surface.

secondary activity The part of the economy involved in the processing of raw materials derived from primary activities; includes manufacturing, construction, power generation.

sector model A description of urban land uses as wedge-shaped sectors radiating outward from the central business district along transportation corridors. The radial access routes attract particular uses to certain sectors.

secularism An indifference to or rejection of religion and religious belief.

sedimentary rock Rock formed by the accumulation of particles of gravel, sand, silt, and clay that were eroded from already existing rocks and laid down in layers.

seismic waves Vibrations within the earth set off by earthquakes.

self-determination The concept that nationalities have the right to govern themselves in their own state or territory, a right to self-rule.

shaded relief A method of representing the three-dimensional quality of an area by use of continuous graded tone to simulate the appearance of sunlight and shadows.

shale oil The crude oil resulting from the distillation of kerogen in oil shales.

shamanism A form of tribal religion based on belief in a hidden world of gods, ancestral spirits, and demons responsive only to a shaman, or interceding priest.

shifting cultivation (*syn:* slash-and-burn agriculture, swidden agriculture) Crop production of forest clearings kept in cultivation until their quickly declining fertility is lost. Cleared plots are then abandoned and new sites are prepared.

sinkhole A deep surface depression formed when ground collapses into a subterranean cavern.

site The place where something is located; the immediate surroundings and their attributes.

situation The location of something in relation to the physical and human characteristics of a larger region.

slash-and-burn agriculture *See* shifting cultivation.

small circle The line created by the intersection of a spherical surface with a plane that does not pass through its center.

small-scale map A representation of a large land area on which small features (e.g., highways, buildings) cannot be shown true to scale.

sociofacts The institutions and links between individuals and groups that unite a culture, including family structure and political, educational, and religious institutions; components of the sociological subsystem of culture.

sociological subsystem The totality of expected and accepted patterns of interpersonal relations common to a culture or subculture.

soil The complex mixture of loose material, including minerals, organic and inorganic compounds, living organisms, air, and water, found at the earth's surface and capable of supporting plant life.

soil depletion The loss of some or all of the vital nutrients from soil.

soil erosion The wearing away and removal of soil particles from exposed surfaces by agents such as moving water, wind, or ice.

soil horizon A layer of soil distinguished from other soil zones by color, texture, and other characteristics resulting from soil-forming processes.

soil order A general grouping of soils with broadly similar composition, horizons, and weathering and leaching processes.

soil profile A vertical cross section of soil horizons.

soil properties The characteristics that distinguish types of soil from one another, including organic and inorganic matter, texture, structure, and nutrients.

solar energy Radiation from the sun, which is transformed into heat primarily at the earth's surface and secondarily in the atmosphere.

solar power The radiant energy generated by the sun; sun's energy captured and directly converted for human use. *See also* photovoltaic cell.

solid waste The unwanted materials generated in production or consumption processes that are solid rather than liquid or gaseous in form.

source region In climatology, a large area of uniform surface and relatively consistent temperatures where an air mass forms.

southern oscillation The atmospheric conditions occurring periodically near Australia that create the El Niño condition off the coast of South America.

spatial diffusion The outward spread of a substance, a concept, a practice, or a population from its point of origin to other areas.

spatial distribution The arrangement of things on the earth's surface.

spatial interaction The movement (e.g., of people, goods, information) between different places; an indication of interdependence among areas.

spatial margin of profitability The set of points delimiting the area within which a firm's profitable operation is possible.

special-purpose map *See* thematic map.

spring wheat Wheat sown in spring for ripening during the summer or autumn.

stage in life Membership in a specific age group.

standard language A language substantially uniform with respect to spelling, grammar, pronunciation, and vocabulary and representing the approved community norm of the tongue.

standard parallel The tangent circle, usually a parallel of latitude, in a conic projection; along the standard line, the scale is as stated on the map.

state (*syn:* country) An independent political unit occupying a defined, permanently populated territory and having full sovereign control over its internal and foreign affairs.

step migration A migration in which an eventual long-distance relocation is undertaken in stages, as from farm to village to small town to city.

steppe Treeless midlatitude grassland.

stratosphere The layer of the atmosphere that lies above the troposphere and extends outward to about 56 kilometers (35 mi).

stream load The eroded material carried by a stream in one of three ways, depending on the size and composition of the particles: (1) in dissolved form, (2) suspended by water, or (3) rolled along the streambed.

subduction The process by which one lithospheric plate is forced down beneath another into the asthenosphere as a result of a collision with that plate.

subnationalism The feeling that one owes primary allegiance to a traditional group or nation rather than to the state.

subsequent boundary A boundary line that is established after the area in question has been settled and that considers the cultural characteristics of the bounded area.

subsidence The settling or sinking of a portion of the land surface, sometimes as a result of the extraction of fluids, such as oil or water, from underground deposits.

subsistence agriculture Any of several farm economies in which most crops are grown for food, nearly exclusively for local consumption.

subsistence economy A system in which goods and services are created for the use of producers or their immediate families. Market exchanges are limited and of minor importance.

substitution principle In industry, the tendency to substitute one factor of production for another in order to achieve optimum plant location and profitability.

suburb A functionally specialized segment of a large urban complex located outside the boundaries of the central city.

succession A natural process in which an orderly sequence of plant species will occupy a newly established landform or a recently altered landscape.

superimposed boundary A boundary line placed over, and ignoring, an existing cultural pattern.

supranationalism The acceptance of the interests of more than one state, expressed as associations of states created for mutual benefit and to achieve shared objectives.

surface water Water that is on the earth's surface, such as in rivers, streams, reservoirs, lakes, and ponds.

sustainable development Economic development and resource use that satisfy current needs without jeopardizing the ability of future generations to meet their own needs.

sustained yield The practice of balancing harvesting with the growth of new stocks in order to avoid depletion of the resource and ensure a perpetual supply.

swidden agriculture *See* shifting cultivation.

syncretism The development of a new form of, for example, religion or music, through the fusion of distinctive parental elements.

syntax The way words are put together in phrases and sentences.

systems analysis An approach to the study of large systems through (1) segregation of the entire system into its component parts, (2) investigation of the interactions between system elements, and (3) study of inputs, outputs, flows, interactions, and boundaries within the system.

T

talus slope A landform composed of rock particles that have accumulated at the base of a cliff, hill, or mountain.

technological subsystem The complex of material objects together with the techniques of their use by means of which people carry out their productive activities.

technology An integrated system of knowledge and skills developed within a culture to carry out successfully purposeful and productive tasks.

tectonic forces The processes that shape and reshape the earth's crust, the two main types being diastrophic and volcanic.

temperature inversion The condition caused by rapid reradiation in which air at lower altitudes is cooler than air aloft.

territoriality The persistent attachment of most animals to a specific area; the behavior associated with the defense of the home territory.

territorial production complex In the economic planning of the former Soviet Union, a design for large regional industrial, mining, and agricultural development leading to regional self-sufficiency and the creation of specialized production for a larger national market.

terrorism The calculated use of violence against civilians and other symbolic targets in order to publicize a cause or to diminish people's support for a leader, a government, a policy, or a way of life that the perpetrators of violence find objectionable.

tertiary activity The part of the economy that fulfills the exchange function and that provides market availability of commodities; includes wholesale and retail trade and associated transportation, government, and information services.

thematic map (*syn:* special-purpose map) A map that shows a specific spatial distribution or category of data.

thermal pollution The introduction of heated water into the environment, with consequent adverse effects on aquatic life.

Third World Originally (in the 1950s), designating countries uncommitted to either the "First World" Western capitalist bloc or the Eastern "Second World" communist bloc; subsequently, countries considered not yet fully developed or in a state of underdevelopment in economic and social terms.

threatened species A species that has declined significantly in total numbers and may be on the verge of extinction; an endangered or vulnerable species.

threshold In economic geography, the minimum market needed to support the supply of a product or service.

topographic map A map that portrays the shape and elevation of the terrain, often in great detail.

toponym A place-name.

toponymy The place-names of a region or, especially, the study of place-names.

tornado A small, violent storm characterized by a funnel-shaped cloud of whirling winds that can form beneath a cumulonimbus cloud in proximity to a cold front and that moves at speeds as high as 480 kilometers per hour (300 mph).

total fertility rate (TFR) The average number of children that would be born to each woman if, during her childbearing years, she bore children at the current year's rate for women that age.

town A nucleated settlement that contains a central business district but that is smaller and less functionally complex than a city.

township and range system A rectangular survey system whose basic units of areas are the township and section; it identifies townships as being north or south of a particular baseline and east or west of a particular principal meridian. Townships are subdivided into square sections 1 mile on a side.

traditional religion *See* tribal religion.

tragedy of the commons The observation that, in the absence of collective control over the use of a resource available to all, it is to the advantage of all users to maximize their separate shares even though their collective pressures may diminish total yield or destroy the resource altogether.

transboundary river basin   The area drained by a river that flows through or collects water from two or more countries. Such a basin requires international cooperation over the management of its water resources.

transform fault   A break in rocks that occurs when one lithospheric plate slips past another in a horizontal motion.

transition economies   Economies in formerly communist countries that are in the process of shifting away from central planning toward free market exchange.

transnational corporation (TNC)   A large business organization operating in at least two separate national economies.

tribal religion   (*syn:* traditional religion) An ethnic religion specific to a small, localized, preindustrial culture group.

tropical rain forest   The tree cover composed of tall, high-crowned evergreen deciduous species, associated with the continuously wet tropical lowlands.

tropical rain forest climate   The continuously warm, frost-free climate of tropical and equatorial lowlands, with abundant moisture year-round.

troposphere   The atmospheric layer closest to the earth, extending outward about 11 to 13 kilometers (7 to 8 mi) at the poles to about 26 kilometers (16 mi) at the equator.

truck farming   The intensive production of fruits and vegetables for market rather than for processing or canning.

tsunami   Sea waves generated when an earthquake, eruption, or underwater landslide abruptly moves the seabed, jolting the waters above.

tundra   The treeless area lying between the tree line of Arctic regions and the permanently ice-covered zone.

typhoon   A hurricane occurring in the western Pacific Ocean region.

U

ubiquitous industry   A market-oriented industry whose establishments are distributed in direct proportion to the distribution of population (market).

underpopulation   A value statement reflecting the view that an area has too few people in relation to its resources and population-supporting capacity.

unitary state   A state in which the central government dictates the degree of local or regional autonomy and the nature of local government units; a country with few cultural conflicts and a strong sense of national identity.

United Nations Convention on the Law of the Sea (UNCLOS)   *See* Law of the Sea Convention.

universalizing religion   A religion that claims global truth and applicability and seeks the conversion of all humankind.

urban hierarchy   The steplike series of urban units (e.g., hamlets, villages, towns, cities, metropolises) in classes differentiated by size and function.

urban influence zone   An area outside of a city that is nevertheless affected by the city.

urbanization   The transformation of a population from rural to urban status; the process of city formation and expansion.

urbanized area   A continuously built-up urban landscape defined by building and population densities with no reference to the political boundaries of the city; it may contain a central city and many contiguous towns, cities, suburbs, and unincorporated areas.

usable reserves   *See* proved reserves.

V

valley breeze   The flow of air up mountain slopes during the day.

value-by-area map   *See* area cartogram.

variable costs   In economic geography, the costs of production inputs that change as the level of production changes. They differ from the costs incurred by agricultural or industrial firms that are fixed and do not change as the amount of production changes.

verbal scale   A statement of the relationship between units of measure on a map and distance on the ground, such as "1 inch represents 1 mile."

vernacular   (1) The nonstandard indigenous language or dialect of a locality; (2) of or related to indigenous arts and architecture, such as a vernacular house; (3) of or related to the perceptions and understandings of the general population, such as a vernacular region.

vernacular region   *See* perceptual region.

volcanism   The earth force that transports subsurface materials (often heated, sometimes molten) to or toward the surface of the earth.

von Thünen model   The model developed by Johann H. von Thünen (1783–1850) to explain the forces that control the prices of agricultural commodities and how those variable prices affect patterns of agricultural land utilization.

von Thünen rings   The concentric zonal pattern of agricultural land use around a single market center proposed in the von Thünen model.

vulnerable species   Species whose numbers have been so reduced that they could become threatened or endangered.

W

warping   The bowing of a large region of the earth's surface due to the movement of continents or the melting of continental glaciers.

wash   A dry, braided channel in the desert that remains after the rush of rainfall runoff water.

water table   The upper limit of the saturated zone and therefore of groundwater; the top of the water within an aquifer.

weather   The state of the atmosphere at a given time and place.

weathering   The mechanical and chemical processes that fragment and decompose rock materials.

Weberian analysis   *See* least-cost theory.

Weber model   The analytical model devised by Alfred Weber (1868–1958) to explain the principles governing the optimum location of industrial establishments.

wetland   A vegetated inland or coastal area that is either occasionally or permanently covered by standing water or saturated with moisture.

wind farm   A cluster of wind-powered turbines producing commercial electricity.

wind power   The kinetic energy of wind converted into mechanical energy by wind turbines that drive generators to produce electricity.

winter wheat   Wheat planted in autumn for early summer harvesting.

world city   One of a small number of interconnected, internationally dominant centers (e.g., New York, London, Tokyo) that together control the global systems of finance and commerce.

Z

zero population growth (ZPG)   A situation in which a population is not changing in size from year to year, as a result of the combination of births, deaths, and migration.

zoning   Designating by ordinance areas in a municipality for particular types of land use.

Note: Page numbers followed by *f* or *t* indicate material in figures or tables, respectively.